T0186378

FRACTURE MECHANICS OF CONCRETE AND CONCRETE STRUCTURES

BALKEMA – Proceedings and Monographs
in Engineering, Water and Earth Sciences

PROCEEDINGS OF THE 6TH INTERNATIONAL CONFERENCE ON FRACTURE MECHANICS OF CONCRETE AND CONCRETE STRUCTURES, CATANIA, ITALY, 17–22 JUNE 2007

Fracture Mechanics of Concrete and Concrete Structures

VOLUME 2: Design, Assessment and Retrofitting of RC Structures

Editors

Alberto Carpinteri
Politecnico di Torino, Department of Structural Engineering and Geotechnics, Italy

Pietro G. Gambarova
Politecnico di Milano, Department of Structural Engineering, Italy

Giuseppe Ferro
Politecnico di Torino, Department of Structural Engineering and Geotechnics – Italian Group of Fracture, Italy

Giovanni A. Plizzari
University of Brescia, Department of Civil Engineering, Architecture, Land and Environment, Italy

Taylor & Francis
Taylor & Francis Group

LONDON / LEIDEN / NEW YORK / PHILADELPHIA / SINGAPORE

Cover illustrations
Courtesy of A. Carpinteri, G. Lacidogna & A. Manuello

Taylor & Francis is an imprint of the Taylor & Francis Group, an informa business

© 2007 Taylor & Francis Group, London, UK

Typeset by Charon Tec Ltd (A Macmillan Company), Chennai, India
Printed and bound in Great Britain by Bath Press Ltd (a CPI-group company), Bath

Published by: Taylor & Francis/Balkema
              P.O. Box 447, 2300 AK Leiden, The Netherlands
              e-mail: Pub.NL@tandf.co.uk
              www.balkema.nl, www.taylorandfrancis.co.uk, www.crcpress.com

ISBN 13 Set (3 volumes): 978-0-415-44066-0
ISBN 13 Vol-1:   978-0-415-44065-3
ISBN 13 Vol-2:   978-0-415-44616-7
ISBN 13 Vol-3:   978-0-415-44617-4

Fracture Mechanics of Concrete and Concrete Structures – Design, Assessment and Retrofitting of RC Structures – Carpinteri, et al. (eds)
© 2007 Taylor & Francis Group, London, ISBN 978-0-415-44616-7

Table of contents

VOLUME 2 – Design, Assessment and Retrofitting of RC Structures

VOLUME 3 – High-Performance Concrete, Brick-Masonry and Environmental Aspects

XV

XVII

Fracture Mechanics of Concrete and Concrete Structures – Design, Assessment and Retrofitting of RC Structures – Carpinteri, et al. (eds)
© 2007 Taylor & Francis Group, London, ISBN 978-0-415-44616-7

Preface

The present volume is the second part of the Proceedings of the 6th International Conference on Fracture Mechanics of Concrete and Concrete Structures (FraMCoS-6), co-organized by the Politecnico di Torino, the Politecnico di Milano, the University of Brescia, and the Italian Group of Fracture (IGF), under the auspices of the International Association of Fracture Mechanics for Concrete and Concrete Structures (IA-FraMCoS). FraMCoS-6 was also scientifically supported by FIB, ICF, RILEM, ACI, ESIS, and JCI, as well as by the National Research Council of Italy (CNR). The venue was the Conference Centre of Catania Sheraton Hotel, in Catania (Italy), a very attractive place on the island of Sicily, not far from the well-known archaeological sites of Syracuse and Taormina.

Since the IA-FraMCoS' establishment by Professor Zdenek Bazant, Founding and Honorary President of our Scientific Society, the FraMCoS series have taken place in different parts of the World on a triennial basis. The previous conferences were held in:

- Breckenridge (Colorado, USA) – FraMCoS-1, organised in 1992 by Z.P. Bazant;
- Zurich (Switzerland) – FraMCoS-2, organised in 1995 by F.H. Wittmann;
- Gifu (Japan) – FraMCoS-3, organised in 1998 by H. Mihashi and K. Rokugo;
- Cachan (France) – FraMCoS-4, organised in 2001 by R. de Borst, J. Mazars, G. Pijaudier-Cabot, and J.G.M. van Mier;
- Vail (Colorado, USA) – FraMCoS-5, organised in 2004 by V.C. Li, C.K.Y. Leung, K.J. Willam, and S.L. Billington.

It should be reminded that, before the formal foundation of IA-FraMCoS in 1992, significant and pioneering conferences on the same subject had been held since 1984, in Evanston, Lausanne, Houston, Vienna, Cardiff, Torino, and Noordwijk, organised by the same Scientific Community, and, in particular, by S.P. Shah, F.H. Wittmann, S. Swartz, H.P. Rossmanith, B.I.G. Barr, A. Carpinteri, and J.G.M. van Mier.

As always, the organisers were responsible for pointing out the main themes and the open problems on which our Scientific Community should work and debate. These themes and research directions usually change in time, and represent the genetic mutations that are necessary for the natural selection of ideas and for the scientific evolution of the IA-FraMCoS. This Community, on the other hand, has produced important ideas, that have been appreciated by other Communities also interested in Material Strength and Structural Integrity. Among the many possible examples, the Fictitious Crack Model by Arne Hillerborg, proposed in the Seventies, may be cited, since it is nowadays utilised even for materials other than concrete, under the more general denomination of "Cohesive Crack Model".

The present volume, titled "Design, Assessment and Retrofitting of RC Structures", is divided into four Parts: (1) Theoretical and Experimental Investigation on the Mechanical Behaviour of RC Structures; (2) Practical Problems in RC Structural Applications; (3) Monitoring and Assessment of RC Structures ; (4) Maintenance and Retrofitting of RC Structures.

Fracture Mechanics is used to interpret different problems: anchor fastening, plastic rotation capacity in RC beams, minimum reinforcement and ductility, size effect, flexural-shear-crushing failure mode transition, cohesive crack modelling, rebar corrosion. Practical and traditional problems in RC structures are also reconsidered and reinterpreted: crack width evaluation, dynamic and impact loading, fire and thermal degradation, fatigue strength assessment, punching and spalling. Monitoring and assessment issues in RC structures are also debated, like those based on Acoustic Emission and on Ultra Sounds, whose many advantages are well known, although other techniques are available. Maintenance and retrofitting techniques are presented as well, since the application of fibre-reinforced polymer sheets to cracked structures (like in beams and columns wrapped by jackets) is becoming increasingly popular.

Finally, as at the end of the first volume, the Editors would like to express their sincere and warm thanks to all the Sponsors of the FraMCoS-6, whose generous financial support was instrumental in making this Event feasible, as well as to all the actors of the Event: the Board of Directors, the Advisory

Board, the International Scientific Committee, the Local Organising Committee, the Chairmen of the various Sessions, the Plenary Lecturers, the Participants and, in particular, the Authors, for their excellent contributions.

The Editors of the Volume
Alberto Carpinteri, Pietro G. Gambarova,
Giuseppe Ferro, and Giovanni A. Plizzari

Fracture Mechanics of Concrete and Concrete Structures – Design, Assessment and Retrofitting of RC Structures – Carpinteri, et al. (eds)
© *2007 Taylor & Francis Group, London, ISBN 978-0-415-44616-7*

Sponsors

Organized by

 Politecnico di Torino

 University of Brescia

 Politecnico di Milano

 Italian Group of Fracture (IGF)

Under the auspices of

 International Association of Fracture Mechanics for Concrete and Concrete Structures (IA-FraMCoS)

With the scientific support of

 Fédération Internationale du Béton (fib)

 International Congress on Fracture (ICF)

 International Union of Laboratories and Experts in Construction Materials, Systems and Structures (RILEM)

XXI

American Concrete Institute (ACI)

 European Structural Integrity Society (ESIS)

 Japan Concrete Institute (JCI)

 National Research Council of Italy (CNR)

 Italian National Agency for New Technologies, Energy and the Environment (ENEA)

 University of Applied Science of Southern Switzerland, Lugano, Switzerland

 University of Bologna

aicap Italian Association for Reinforced and Prestressed Concrete (AICAP)

 Italian Society of Building Experts (CTE)

With the support of

 City of Catania

 Provincia Regionale di Catania

 CONSULTA ORDINI INGEGNERI SICILIA

Association of Engineers
of Sicily

Association of Engineers
of Catania

This Volume was edited in the framework of the ILTOF Project (EU Leonardo da Vinci Programme): Innovative Learning and Training On Fracture – www.iltof.org.

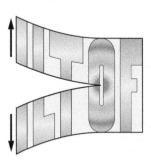

Main sponsors

ANAS S.p.A.   Italian Agency of National Roads (ANAS)

   MAPEI, Milan (Italy)

   Saint-Gobain Vetrotex Espana, Madrid (Spain)

   Technochem, Barzana (BG, Italy)

Part V
Theoretical and experimental investigation on the mechanical behavior of RC structures

Fracture Mechanics of Concrete and Concrete Structures – Design, Assessment and Retrofitting of RC Structures – Carpinteri, et al. (eds)
© *2007 Taylor & Francis Group, London, ISBN 978-0-415-44616-7*

Behavior and design of fastenings with bonded anchors: Numerical analysis and experimental verification

R. Eligehausen & J. Appl
Institute for Construction Materials, University of Stuttgart, Stuttgart, Germany

ABSTRACT: This paper presents the results of extensive numerical and experimental work performed to establish a behavioral model that provides the basis for developing design provisions for anchorages to concrete using adhesive bonded anchors. These types of anchorage systems are used extensively. The behavioral model is compared to a worldwide data base containing 415 tests on adhesive anchor groups, 133 tests on adhesive anchors located near a free edge.

1 INTRODUCTION

A wide spectrum of bonded anchor systems are currently available. A distinction can be made between so called capsule systems and injection systems. For both systems the bonding materials may consist of polymer resins, cementitious materials, or a combination of the two.

Capsule anchor systems employ a threaded rod equipped with a 45° chisel- or roof shaped tip and a hexagonal nut and washer. The glass capsule is filled with the constituent bonding materials. It contains polymer resin, hardener and quartz aggregate in a defined mix ratio.

For injection systems resin and hardener are contained in separate chambers.

The embedment depth of capsule systems can be 8 to 10 times of the diameter of the rod; for injection systems the embedment depth can be user defined, but it should not exceed the limit of 20 times the rod diameter. The capsule is placed in a hole from which all drilling dust has been removed. When driving the threaded rod into the hole, the glass capsule is broken and fragmented, the resin, hardener and aggregates and capsule fragments are mixed and the annular gap around the threaded rod is filled with the polymeric matrix.

For injection systems the injection of the components into the drilled hole is accomplished with the aid of a mechanical or pneumatic dispenser. Conventional bonded anchors are not ideal for resisting tension loads in concrete that is subjected to cracking. Therefore special bonded anchors for uncracked concrete have been developed, so called bonded undercut or bonded expansion anchors. More details for the different anchoring systems are included in Eligehausen et al. (2006).

2 BACKROUND

2.1 *Headed cast – in place and post-installed mechanical anchors*

Fuchs et al. (1995) proposed a behavioral model for concrete breakout failure. This model was created to predict the failure loads of cast-in-place headed anchors and post-installed mechanical anchors loaded in tension or in shear that exhibit concrete breakout failure. According to Fuchs et al. (1995), the mean concrete breakout capacity for single cast-in-place anchors and post-installed mechanical anchors in uncracked concrete unaffected by edge influences or overlapping cones of neighboring anchors loaded in tension is given by the following equations:

Cast-in-place anchors:

$$N_{u,c}^0 = 15.5 \cdot f_{cc}^{0.5} \cdot h_{ef}^{1.5} \quad \text{(N)} \qquad \text{(1a)}$$

Post-installed mechanical anchors:

$$N_{u,c}^0 = 13.5 \cdot f_{cc}^{0.5} \cdot h_{ef}^{1.5} \quad \text{(N)} \qquad \text{(1b)}$$

where f_{cc} = concrete compressive strength measured on cubes with a side length of 200 mm; and h_{ef} = embedment depth.

If fastenings are located so close to an edge that there is not enough space for a complete concrete cone to develop, the load-bearing capacity of the anchorage is also reduced. For anchor groups with a spacing

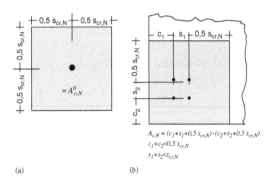

$$A_{c,N} = (c_1 + s_1 + 0.5\, s_{cr,N}) \cdot (c_2 + s_2 + 0.5\, s_{cr,N})$$
$$c_1 + c_2 < 0.5\, s_{cr,N}$$
$$s_1 + s_2 < s_{cr,N}$$

(a)   (b)

Figure 1. Calculation of effective areas: (a) single anchors away from edges and anchors; and (b) groups of 4 closely spaced anchors located near a corner.

smaller than a critical value it is found, that the failure load is reduced with decreasing spacing, which is called group effect. The critical value is called characteristic spacing $s_{cr,N}$. The concrete breakout capacity of anchor groups $N_{u,c}$ and anchors located near free edges with a tension load applied concentrically to the anchors is given by Equation 2 where $N_{u,c}^0$ is taken from Equation 1:

$$N_{u,c} = N_{u,c}^0 \cdot \frac{A_{c,N}}{A_{c,N}^0} \cdot \psi_{s,N} \qquad \text{(N)} \qquad (2)$$

with

$$\psi_{s,N} = 0.7 + 0.3 \cdot \frac{c}{c_{cr,N}} \leq 1.0 \quad (-) \qquad (2b)$$

where c = edge distance; $c_{cr,N}$ = characteristic edge distance; $\psi_{s,N}$ = factor to consider disturbance of radial symmetric stress distribution caused by an edge; $A_{c,N}^0$ = projected area of one anchor at the concrete surface unlimited by edge influences or neighboring anchors, idealizing the failure cone as a pyramid with the base length $s_{cr,N}$; and $A_{c,N}$ = actual projected area at the concrete surface, assuming the failure surface of the individual anchors as a pyramid with a base length $s_{cr,N}$. Figure 1 provides information on how projected areas $A_{c,N}^0$ and $A_{c,N}$ are determined. According to Fuchs et al. (1995), for cast-in-place and post-installed mechanical anchors, the characteristic spacing, $s_{cr,N}$, is $3.0 h_{ef}$ and the characteristic edge distance, $c_{cr,N}$, is $1.5 h_{ef}$.

One of the principal advantages of this model is that calculation of the changes in capacity due to factors such as edge distance, spacing, geometric arrangement of groupings, and similar variations can be readily determined though use of relatively simple geometrical relationships based on rectangular prisms.

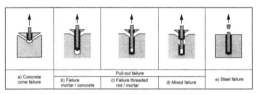

Figure 2. Possible embedment-related failure modes of bonded anchors (Cook et al. 1998).

2.2 Single bonded anchors

Bonded anchors resist tension load by adhesion and micro-keying of the resin and by mechanical interlock to the anchors rod to the sides of the drilled hole and by mechanical interlock to the anchor rod. The tension forces are transferred to the surrounding concrete by radially symmetric compression struts that spread out from the anchor. This in turn generates tension stresses perpendicular to the compression struts. Due to the load-transfer mechanism and available bond strength different failure modes can be observed. Figure 2 shows typical failure modes of single bonded anchors. If bond strength is high enough to utilize the tension strength of the concrete, concrete failure will occur which is characterized by cone-shaped concrete breakout originating at the base of the anchor (Fig. 2a). The slope of the cone envelope with the respect to the surface of the concrete member is approximately 25° to 35°. Normally, this failure mode can be observed at small embedment depth ($h_{ef} \sim 3d\text{-}5d$). For greater embedment depth the failure mode shifts from a concrete cone to a mixed mode type of failure. A concrete cone with a depth of approximately $2d$ to $3d$ forms at the top end of the anchor and bond failure occurs along the remaining length of the anchor. Bond failure occurs either at the boundary between threaded rod and mortar (Interface 1, Fig. 2c) or between the mortar and the sides of the drilled hole (Interface 2, Fig. 2b). Often a mixed interface failure can be observed (Fig. 2d). For large embedment depths steel failure can occur.

Experimental studies discussed in Eligehausen et al. (2004), indicate that the actual bond stress distribution along the embedment length at peak load is non-linear with lower bond stresses at the concrete surface and higher bond stresses at the embedded end of the anchor. However, in Cook et al. (1998) a comparison of suggested behavioral models with a worldwide data base for single adhesive anchors indicates that their failure load is best described by a uniform bond stress model incorporating the nominal anchor diameter (d) with the mean bond stress ($\tau_{u,m}$) associated with each product. This is confirmed by experimental and numerical studies of Meszaros (1999) and McVay et al. (1996). The uniform bond stress model for adhesive anchors is given by Equation 3. This equation is valid for $4 \leq h_{ef}/d \leq 20$.

$$N^0_{u,m,p} = \pi \cdot d \cdot h_{ef} \cdot \tau_{u,m} \quad \text{(N)} \qquad (3)$$

where $N^0_{u,m,p}$ = average failure load of single bonded anchor failing by pullout (Fig. 2); d = diameter of the anchor rod; h_{ef} = embedment depth; and $\tau_{u,m}$ = average bond strength.

Assuming that the maximum failure load of bonded anchors is limited to the concrete breakout failure load of post-installed mechanical anchors as given by Equation 1 the upper limit of the bond strength to be used for single anchors can be determined by equating Equation 1 with Equation 3:

$$\tau_{u,max} = 4.2 \cdot \frac{f_{cc}^{0,5} \cdot h_{ef}^{0,5}}{d} \quad \text{(N/mm}^2) \qquad (4)$$

3 NUMERICAL ANALYSIS AND EXPERIMENTAL TESTS OF ADHESIVE ANCHORS

3.1 Microplane model

In the following numerical study, the finite element code MASA which is based on the microplane model, was used. MASA was developed for two-and three-dimensional analysis of quasi-brittle materials. In the model the material is characterized by a uniaxial relation between the stress and strain components of planes of various orientations. At each integration point these planes may be imagined to represent the damage planes or weak planes of the microstructure (Fig. 3). Tensorial invariance restrictions need to be directly enforced. Superimposing the response from all microplanes in a suitable manner automatically satisfied them. The model allows for a realistic prediction of the material behavior in case of three-dimensional stress-strain states. A smeared crack approach is employed. To ensure mesh independent results the crack band approach is used. More details about the model can be found in Ozbolt (1998).

3.2 Modelling

3.2.1 General

To understand the behavior of adhesive anchors under tension loading, three-dimensional non-linear finite element analyses were performed. The mortar behavior was simulated using the microplane model with a proper calibration of the model parameters to represent the measured macroscopic mortar properties. The load transfer between threaded rod and mortar results from mechanical interlock between thread and mortar consoles. In this case the maximum load that can be transferred at the interface is mainly influenced by the macroscopic properties of the mortar. On the contrary, the maximum load which can be transferred

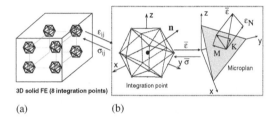

(a)

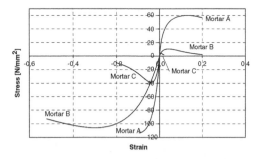

(b)

Figure 3. Concept of microplane model: (a) Integration points; and (b) unit volume sphere and strain components (Ozbolt 1998).

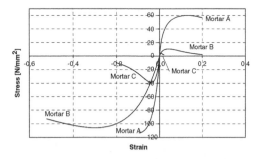

Figure 4. Macroscopic stress-strain relationship of mortar elements.

between mortar and concrete is mainly influenced by the microscopic properties of the mortar and the concrete. However, the material's behavior of the interface mortar/concrete can hardly be described by means of theoretical model. At the moment there is no all-purpose adhesion theory. Main influencing parameters on the adhesion are bonding forces, ratio of adhesion and cohesion, polarity, porosity and roughness of the adjacent materials. Consequently the complicated process of penetration and curing of the mortar in the concrete must be taken into account by special contact-elements which are located between mortar and concrete elements. Figure 4 shows the stress-strain relationships of macroscopic mortar elements for three different mortar types which depend on the polymeric resins and the fillers used.

Simulated were single adhesive anchors near to and far away from an edge, as well as quadruple anchor groups with four and six bonded anchors. Parameters varied for single anchors were anchor diameter, embedment depth, bond strength of the mortar, concrete compressive strength and edge distance. For anchor groups the spacing of the anchors was also varied. In all numerical simulations, the member thickness was large enough to avoid splitting failures. The loading process was displacement-controlled, by applying incremental displacements to the anchor at the concrete surface.

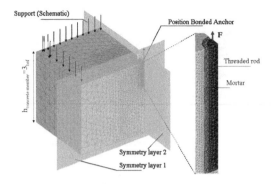

Figure 5. Typical FE meshes of concrete block and adhesive anchor.

3.2.2 Single bonded anchors

The geometry of the specimen and the finite element mesh are shown in Figure 5. A concrete slab with a width of $6h_{ef}$ and a height of $3h_{ef}$ is analysed. The materials properties of the concrete are equal to the properties measured in experiments. The existing symmetry planes were used to reduce the analysis time. The mesh was refined within the area of the bonded anchor. The modeled test specimen is restrained in vertical direction at a distance of $3 h_{ef}$ from the anchor. The numerical analysis considers the same boundary conditions as in experiments with unconfined tests. The anchor was discretized by three-dimensional linear elastic finite elements.

Figure 6 shows the modeled threaded rod and the mortar layers. The geometry of the thread was simplified compared to a real threaded rod. The modeled thread runs perpendicular to the shaft.

3.2.3 Groups with bonded anchors

The geometry of the specimen for anchor groups and the finite element mesh are shown in Figure 7. The geometry is similar to that for single anchors. Only a slab with a width of $6h_{ef} + s$ and a thickness of $3h_{ef}$ is analysed. The parameters under investigation included anchor diameter, embedment depth, concrete strength, bond strength and anchor spacing.

3.3 Pullout failure

3.3.1 Single bonded anchors

Figure 8 show the numerical results concerning the principal strains in the concrete after the peak load, for a single bonded anchor wit a bond strength of $\tau_{u,m} = 9.3$ N/mm^2 and diameter of d = 24 mm, as well as a photograph of the anchor after tension test. Dark areas in Figure 8a characterize crack formation. With a embedment depth of $10d$ just to the peak load a shallow cone is formed at the concrete surface and bond failure occurs along the remaining length of the anchor.

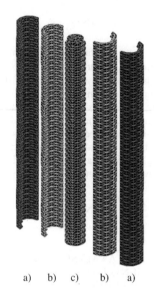

a) b) c) b) a)

Figure 6. Typical FE meshes of an adhesive anchor: (a) Mortar elements (microscopic properties); (b) Mortar elements (macroscopic properties); and (c) Steel elements.

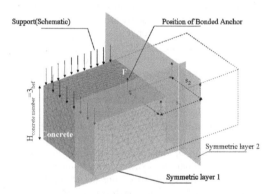

Figure 7. Typical FE mesh of concrete block for a group of adhesive anchors.

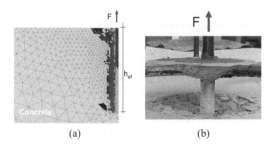

Figure 8. Failure mode of bonded anchor: (a) Numerical result; and (b) Test result.

(a)

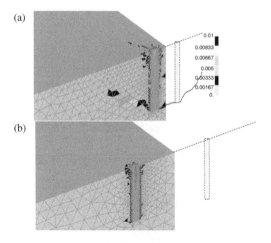

(b)

Figure 9. Failure modes of quadruple anchor group with bonded anchors ($d = 12$ mm, $h_{ef} = 10d$, $\tau_{u,m} = 9.3$ N/mm^2): (a) $s = 4d$; and (b) $s = 16\,d$.

3.3.2 Groups with bonded anchors

Figure 9 shows the principal tensile strains for a group of four bonded anchors with $d = 12$ mm, $h_{ef} = 10d$, and $\tau_{u,m} = 9.3$ N/mm^2 after passing peak load. With a small spacing of $s = 4d$, a common concrete cone breakout cone starting at the base of the anchors is formed (Fig. 9a). With a larger spacing ($s = 16d$), the individual anchors of the group fail in the same way as single anchors with a pullout failure similar to that shown in Figure 8. The failure load of anchorages with adhesive anchors can be calculated by means of Equation 2 that needs to be modified with reference to bond strength, anchor spacing and anchor distance from a nearby edge.

As indicated by Equation 3, the average bond strength ($\tau_{u,m}$), the anchor diameter (d), and the anchor embedment length (h_{ef}) represent the parameters that influence the characteristic spacing and the characteristic edge distance.

As a result of the numerical studies by Li et al. (2002), it was determined that the characteristic spacing is not significantly influenced by the embedment depth (h_{ef}) of the anchors. This is shown in Figure 10 where the ratios between the numerically obtained failure loads of groups with bonded anchors to the failure load of single anchors with the same embedment depth are plotted as a function of the anchor spacing, for different values of embedment depth h_{ef}.

If the characteristic spacing were influenced by the embedment depth, groups with a smaller embedment depth would reach the capacity of four single anchors at a smaller spacing than those with a larger embedment depth. However, for a given spacing, the related failure load is hardly dependent on the embedment depth. This behavior is explained by Figure 11

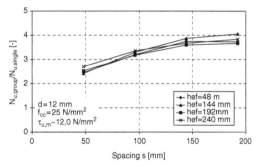

Figure 10. Numerically-obtained failure load of quadruple groups in case of pullout failure related to the failure load of a single anchor as a function of the spacing (Li et al. 2002).

(a)

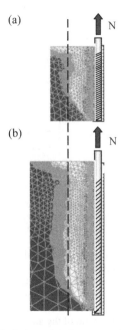

(b)

Figure 11. Principal compression stresses in concrete. Single adhesive anchors: (a) $d = 12$ mm, $h_{ef}/d = 10$; and (b) $d = 12$ mm, $h_{ef}/d = 20$.

where – in the case of pullout failure – the width of the principal compression stress field of single bonded anchors with constant bond strength and significantly different embedment lengths is nearly identical. The width of the compression stress field is directly related to the characteristic spacing.

Li et al. (2002) found that the characteristic spacing is dependent on anchor diameter (d). This can be seen in Figure 12 that shows related failure loads of groups of anchors with different diameters as a function of the related spacing (s/d).

The related group failure load is almost independent of the anchor diameter for a constant ratio of s/d and

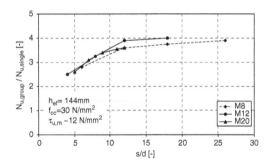

Figure 12. Numerically-obtained failure load of quadruple groups in case of pullout failure related to the failure load of a single anchor as a function of the related spacing (Li et al. 2002).

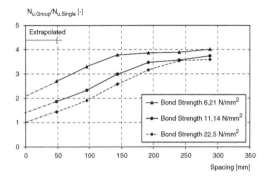

Figure 13. Numerically-obtained failure load of quadruple groups in case of pullout failure related to the failure load of a single anchor as a function of the spacing.

reaches the full capacity of four individual anchors at about the same value of s/d.

Studies by Li et al. (2002) indicated that the characteristic spacing is also influenced by the average bond strength $\tau_{u,m}$.

In Figure 13, the ratio of the anchor group strength to the single anchor strength is plotted as a function of anchor spacing. According to the analysis, the anchor diameter and embedment depth were kept constant and bond strength was varied. For anchorages with the highest bond strength, failure occurred by concrete breakout. Therefore the assumed bond strength was not fully utilized. The characteristic spacing obtained in the analysis is about 300 mm: For the minimum embedment depth $h_{ef} = 96$ mm the characteristic spacing is about $s_{cr,Np} = 3h_{ef}$. With decreasing bond strength the characteristic spacing decreases. The assumption that the characteristic spacing is influenced by the bond strength is confirmed by Figure 14 which shows that the width of the principal compression stress field of a single anchor with constant embedment depth increases with increasing bond strength.

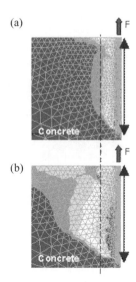

Figure 14. Principal compression stresses in concrete. Single adhesive anchors ($d = 24$, $h_{ef}/d = 20$): (a) $\tau_{u,m} = 9.9$ N/mm^2; and (b) $\tau_{u,m} = 19.6$ N/mm^2.

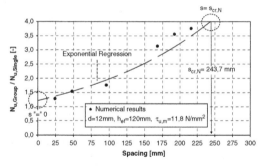

Figure 15. Numerically-obtained failure loads of quadruple groups in related to the failure load of a single anchor as a function of spacing.

To determine the characteristic spacing, ($s_{cr,Np}$) for each individual numerical test series, the ratios $N_{u,group}/N_{u,single} = f_{(scr,N)}$ were approximated by an exponential function which was found by regression analysis (Fig. 15).

The characteristic spacing was determined by extrapolating this function to the value of $N_{u,group}/N_{u,single} = 4$. Figure 16 provides a summary of results. The values of the characteristic spacing found from each test series divided by the diameter ($s_{cr,Np}/d$) are plotted as a function of bond strength. For comparison the characteristic spacing evaluated in the same way from results of tests are shown as well.

It is obvious that the characteristic spacing found by numerical analysis is smaller than that evaluated by means of experimental tests. This is due to the fact,

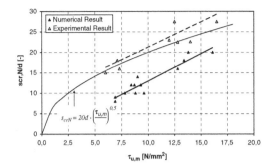

Figure 16. Related critical spacing $s_{cr,N}/d$ as a function of mean bond strength.

that the numerical analysis does not take into account the stiffness of the plate and the scattering of the load-displacement curves of the single bonded anchors of a group. Based on the test results, the characteristic spacing can be approximated by Equation 5.

$$s_{cr,Np} = 20 \cdot d \cdot \left(\frac{\tau_{u,m}}{10} \right)^{0.5} \text{ (mm)} \qquad (5)$$

The characteristic edge distance ($c_{cr,Np}$) may be taken as one half of the characteristic spacing.

Based on the above considerations the failure load of adhesive anchor groups and/or anchorages located near edges can be calculated by Equation 2 with $N_{u,c}^0$ replaced by $N_{u,m,p}^0$ from Equation 3 and using $s_{cr,Np}$ and $c_{cr,Np}$ determined from Equation 5.

3.3.3 Concrete cone and pullout failure
In the case of concrete cone failure, the failure load of a group of anchors with a theoretical spacing $s = 0$ is equal to the value valid for a single anchor (Eqn. 2). However in case of combined concrete cone and pull-out failure when extrapolating the regression lines that describe the failure loads of bonded anchor, the group failure load for a theoretical spacing of $s = 0$ is larger than that of a single anchor (Figs. 10,12,13,15). This increase is denoted by the factor $\psi_{g,N}^0$ in Figure 18. It is explained in Figure 17. If the bond strength is low the failure of two adjacent anchors is caused by bond failure resulting in anchor pullout. The bond failure area of the two adjacent anchors is approximately equal to $\sqrt{2}$ times the effective bond area of a single anchor and for a group of n anchors approximately $\sqrt{n}$ times the bond area of a single anchor. Therefore, the failure load of the group is $\sqrt{n}$ times the failure load of a single anchor ($\psi_{g,N}^0 = \sqrt{n}$). On the contrary, the failure load of a group of adjacent anchors is not increased over that of a single anchor when failure is controlled by concrete breakout ($\psi_{g,N}^0 = 1$). The value of $\psi_{g,N}^0$ should be related to the bond strength. If the bond strength is

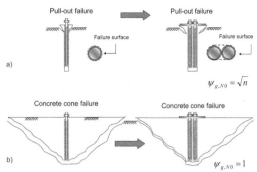

Figure 17. Increase of failure area and failure load: (a) pull-out failure; and (b) concrete cone failure.

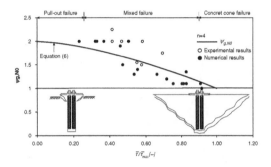

Figure 18. Factor $\psi_{g,N}^0$ as a function of mean bond strength $\tau_{u,m}$ related to the maximum value $\tau_{u,max}$ according to Equation 4.

equal to $\tau_{u,max}$ according to Equation 4 then a single anchor will fail by concrete breakout and $\psi_{g,N}^0 = 1.0$. If the bond strength is very small (e.g. $\tau < 0.3\ \tau_{u,max}$) then failure of the group will be caused by anchor pull-out resulting in $\psi_{g,N}^0 \approx \sqrt{n}$. Values for $\psi_{g,N}^0$ between these limiting cases were determined from the results of the individual numerical test series of quadruple anchor groups. They are plotted in Figure 18 as a function of the ratio $\tau_u/\tau_{u,max}$ and can be approximated by Equation 6.

$$\psi_{g,N}^0 = \sqrt{n} \cdot \left(\sqrt{n} - 1 \right) \left(\frac{\tau_{u,m}}{\tau_{u,max}} \right)^{1.5} \geq 1.0 \qquad (6)$$

Increasing the spacing among the anchors diminishes the favorable effects that a larger bonded area has on the failure load. This effect is taken into account by the factor $\psi_{g,N}$. It is assumed that this factor potentially decreases between $s = 0$ where $\psi_{g,N} = \psi_{g,N}^0$ and $s = s_{cr,Np}$ where $\psi_{g,N} = 1.0$. This leads to Equation 7.

$$\psi_{g,N} = \psi_{g,N}^0 - \left(\frac{s}{s_{cr,Np}} \right)^{0.5} \left(\psi_{g,N}^0 - 1 \right) \qquad (7)$$

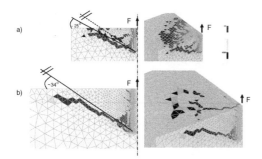

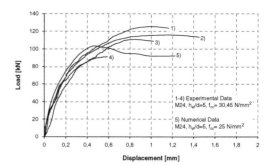

Figure 19. Concrete cone failure indicated by numerical modeling: (a) $d = 6$ mm, $h_{ef}/d = 5$; and (b) $d = 24$ mm, $h_{ef}/d = 5$.

Figure 21. Numerically-obtained load-displacement curve in comparison to test result.

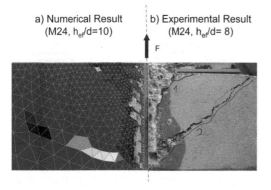

a) Numerical Result
(M24, h_{ef}/d=10)

b) Experimental Result
(M24, h_{ef}/d= 8)

Figure 20. Concrete cone failure predicted by numerical modeling and comparison with test result.

On the basis of the afore-mentioned considerations, the mean failure load of anchorages using adhesive anchors in case of pullout or combined concrete pullout failure may be calculated as follows:

$$N_{u,m,p} = N^0_{u,m,p} \cdot \frac{A_{p,N}}{A^0_{p,N}} \cdot \psi_{s,N} \cdot \psi_{g,N} \qquad (8)$$

In Equation 8, $A_{p,N}$ and $A^0_{p,N}$ are determined according to Figure 1, $\psi_{s,N}$ is given by Equation 2b, $\psi_{g,N}$ is given by Equation 7, and $N^0_{u,m,p}$ is determined from Equation 3. The characteristic spacing $s_{cr,Np}$ and characteristic edge distance $c_{cr,Np}$ provided by Equation 5 should be used when calculating $A_{p,N}$ and $A^0_{p,N}$, $\psi_{s,N}$ and $\psi_{g,N}$.

3.4   Concrete cone failure

3.4.1   Single bonded anchors

Figure 19 and Figure 20 show the numerically obtained principal strains in the concrete after reaching the peak load for single anchors with a diameter $d = 6$ mm, $d = 24$ mm and a bond strength which is high enough to utilize the concrete cone strength. Figures 19a and 19b are valid for an anchor with $h_{ef}/d = 5$, while Figure 20 is valid for $h_{ef}/d = 10$. Dark areas in these figures characterize crack formation.

Similar to the experiments with an embedment length of $5d$ a crack forms at the base which grows with increasing imposed displacement resulting in a concrete breakout failure. The average angle of crack propagation in respect to the surface increases with increasing embedment depth from about 25° ($h_{ef} = 30$ mm) to about 35° ($h_{ef} = 120$ mm). This increase of the angle can be explained by fracture mechanics. Based on fracture mechanics, tension loaded anchors can be classified as a mixed-mode problem. With increasing crack length the ratio of the stress intensity factor for Mode 1 and for Mode 2 is changes. As a result, the mean value of the angle changes as well.

For a related embedment depth of $h_{ef}/d = 10$ similar as in experiment two internal cracks are formed. Failure is caused by the crack starting at the bottom of the anchor.

Figure 21 shows a comparison between load displacement curves measured in experiments and obtained numerically. The agreement is acceptable.

In Figure 22 the numerically obtained failure loads of bonded anchors, related to the average concrete cone failure load of headed anchors (Eqn. 1) with the same embedment depth are plotted as a function of the ratio embedment depth to bar diameter. In all cases failure was caused by concrete cone breakout. The figure shows, that the ratio of numerically-obtained failure loads in case of concrete cone failure ($N^0_{u,c,BondedAnchor}$) to the calculated average failure load of headed anchors ($N^0_{u,c,HeadedAnchor}$) decreases with increasing related embedment depth h_{ef}/d. This can be explained by the fact, that with a small ratio h_{ef}/d only one crack at the end of the anchor is formed (Fig. 19), while with a long embedment depth two or more internal cracks are formed (Fig. 20), which reduce the tensile capacity of the concrete in the region where failure occurs.

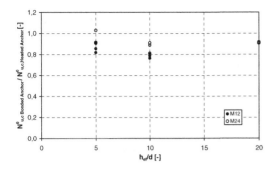

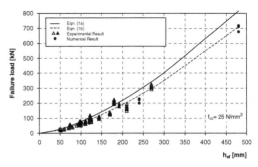

Figure 22. Ratio of numerically-obtained failure loads in case of concrete cone failure to the calculated average failure load of headed anchors as a function of the related embedment dept (h_{ef}/d).

Figure 24. Numerically and experimentally obtained failure loads of bonded anchors in case of concrete cone failure as function of the embedment depth predicted according to Equation 1b.

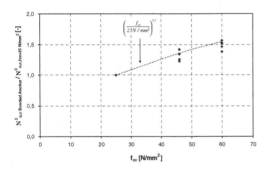

Figure 23. Ratio $N^0_{u,c,Bonded\ Anchor}/N^0_{u,c,fcc=25\,N/mm^2}$ as a function of the concrete compression strength.

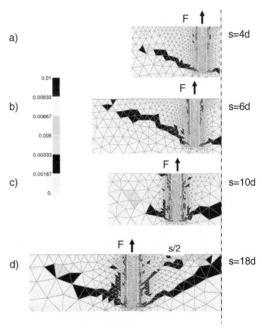

Figure 25. Concrete cone failure of quadruple anchor group predicted by numerical modeling for different spacing.

In Figure 23 the normalized failure loads yield by the analysis are plotted as a function of concrete compressive strength, the normalizing parameter being the failure load for $f_{cc} = 25\,\text{N/mm}^2$. As with both cast-in-place headed anchors and post-installed mechanical anchors, increasing concrete compressive strength increases bonded anchors failure load, but not linearly, as shown by Equation 9:

$$\alpha_c = \left(\frac{f_{cc}}{25\,N/mm^2}\right)^{0,5}\ (-) \tag{9}$$

This indicates that failure of bonded anchors is caused by concrete cone breakout.

Figure 24 shows the measured and numerically obtained failure loads of single bonded anchors normalized to $f_{cc} = 25\,\text{N/mm}^2$ plotted as a function of the embedment depth. In the tests and the numerical analysis, failure was classified as concrete cone breakout. The failure loads increases in proportion to $h_{ef}^{1,5}$. On average they agree with concrete cone failure load of post-installed mechanical anchors. Note that in case of pullout failure the failure load increases linearly with increasing embedment depth.

3.4.2 Groups with bonded anchors

Figure 25 shows principle tensile strains for a group of four adhesive anchors with $d = 12\,\text{mm}$, $h_{ef} = 5d$, and $\tau_u \geq \tau_{u,max}$ beyond the peak load. A single anchor with the same embedment depth would fail by a concrete cone failure. With a small spacing of $s = 4d$, the usual concrete breakout cone starting at the base of the anchors is formed (Fig. 25a). With a larger spacing ($s = 10d$) the angle of the crack between the neighboring bonded anchors increase (Fig. 25c). For a large spacing ($s = 18d$), the individual anchors of the group fail in the same way as a single anchors.

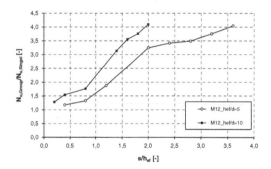

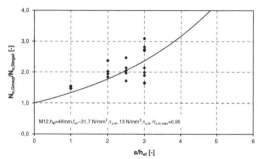

Figure 26. Numerically-obtained failure loads of quadruple groups related to the numerically failure load of a single anchor. Single anchors and anchor groups fail by a concrete cone failure.

Figure 28. Measured failure loads of quadruple groups related to the measured failure load of a single anchor ($d = 12$, $h_{ef}/d = 4$, $\tau_u \sim \tau_{u,max}$).

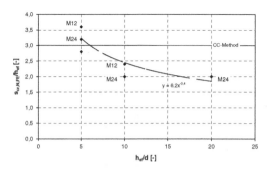

Figure 27. Numerically-obtained related characteristic spacing $s_{cr,N}/h_{ef}$ as a function of the related embedment depth (h_{ef}/d) in case of concrete cone failure.

Figure 26 shows numerically-obtained failure loads of quadruple anchor groups with adhesive anchors for different values of the normalized embedment depth (s/h_{ef}).

As with cast-in-place headed and postinstalled mechanical anchors the failure load of adhesive anchor groups increases with increasing spacing until it reaches a limit of 4 times the single anchor strength at a critical spacing $s_{cr,N}$. Similarly, the failure load of anchorages with adhesive anchors located near edges decreases when the edge distance is smaller than a critical value $c_{cr,N}$.

For each individual numerical test series, the critical spacing was evaluated from the results of the numerical analysis. The characteristic spacing related to the embedment depth is given as a function of the embedment depth related to d in Figure 27, where the characteristic spacing given in the CC–Method ($s_{cr,N} = 3h_{ef}$) is plotted as well. For a small embedment depth the characteristic spacing is larger than $s_{cr,N} = 3h_{ef}$ while for larger embedment depth it approaches the value of $s_{cr,N} = 2h_{ef}$. This is due to the fact that the average slope of the cone for single bonded anchors with

respect to the surface is increasing with increasing embedment depth (Fig. 19). This is generally confirmed by tests with groups of bonded anchors with $h_{ef} = 40$ mm which failed by concrete cone breakout. The average failure load of the group at a spacing of $s = 3h_{ef}$ is only 2.5 times the failure load of a single anchor (Fig. 28). The results are more accurately predicted, when the characteristic spacing is assumed as $s_{cr,N} = 5h_{ef}$.

In all numerical simulations, the calculated failure load of anchorages with adhesive anchors was smaller than the numerically obtained failure load of the same anchorages with headed anchors. Therefore, in Equation 2 and 8 the mean bond failure load, $N_{u,m,p}$ and $N_{u,m,p}$ is limited to the mean concrete breakout failure load, $N_{u,c}^0$ and $N_{u,c}$, given by Equation 1 and 2 for post-installed mechanical anchors.

4 BEHAVIORAL MODEL

The afore-mentioned numerical and experimental results made it possible to develop a behavioral model, that can accurately predict the failure loads of anchorages with adhesive anchors where the effects of anchor groups and/or edges needs to be accounted for. The proposed model can describe the two possible failure modes, either by concrete-cone breakout or by pullout failure. The behavioral model is represented by Equation 8 however, the characteristic spacing $s_{cr,Np}$ and characteristic edge distance $c_{cr,Np}$ provided by Equation 5 should be used when calculating $A_{p,N}$ and $A_{p,N}^0$ according to Figure 1, $\psi_{s,N}$ according to Equation 2b and $\psi_{g,N}$ according to Equation 7.

For design purposes, appropriate capacity-reduction factors and nominal strengths must be introduced in developing code provisions to implement the findings of this research. It is suggested that the 5% fractile of the bond strength be used for the design of bonded anchors which should be adjusted to consider several influencing factors on anchor performance such as

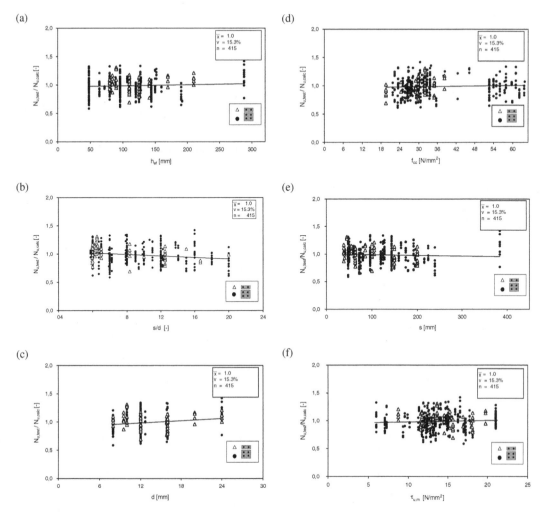

Figure 29. Proposed Model for anchor groups: fitting of tests results concerning: (a) effective embedment depth h_{ef}; (b) normalized anchors spacing s/d; (c) anchor diameter d; (d) concrete compressive strength f_{cc}; (e) anchor spacing s; and (f) average bond strength $\tau_{u,m}$.

sensitivity to hole cleaning procedures and increased temperature as well as long term behavior.

4.1 Proposed Model versus available experimental results

In Figures 29 and 30, the ratios of the measured failure loads divided by the strengths predicted by the proposed model ($N_{u,test}/N_{u,calc}$) are plotted as a function of several parameters investigated in the tests. Figures 29 and 30 also show the "best fit" trend lines. If these lines are horizontal and are located at ($N_{u,test}/N_{u,calc}$) = 1.0 then the influence of the varied parameter on the failure load is well taken into account by the behavioral model. As indicated by these Figures the behavioral model provides an excellent fitting of the experimental

results with groups. For the 415 tests the mean value of ($N_{u,test}/N_{u,calc}$) is 0.99 with a coefficient of variation of 15.4%. An equally good prediction is obtained when the tests are divided into two groups, depending on whether anchor failure is due to concrete-cone failure or pullout failure (Table 1). Summing up, the proposed behavior model for groups with adhesive anchors is as accurate as the behavior model for headed anchors. However, as shown in Figure 30 the predicted failure loads are conservative for anchorages located very close to a free edge.

The proposed model for adhesive anchors (Eqn. 8) is very similar to the behavioral model for headed anchors (Eqn. 2) except for the $\psi_{g,N}$-factor in Equation 7.

(a)

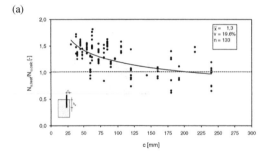

(b)

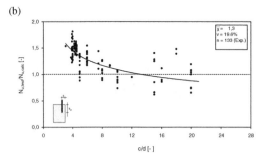

Figure 30. Failure loads of single anchors placed near a free edge: fitting of the tests results with the proposal mode, as a function of the edge distance c (a), and (b) of the normalized distance c/d.

Table 1. Comparison of measured failure loads with predicted values.

Type		Pullout and concrete breakout failure		
		n [−]	$\bar{x}$ [−]	v [%]
Group tests	$N_{u,test}/N_{u,calc}$	415	0.99	15.4
Edge tests	$N_{u,test}/N_{u,calc}$	133	1.3	19.6
		Pullout failure		
		n [−]	$\bar{x}$ [−]	v [%]
Group tests	$N_{u,test}/N_{u,calc}$	377	0.98	15.1
Edge tests	$N_{u,test}/N_{u,calc}$	133	1.3	19.6
		Concrete breakout failure		
		n [−]	$\bar{x}$ [−]	v [%]
Group tests	$N_{u,test}/N_{u,calc}$	38	1.05	16.3
Edge tests	$N_{u,test}/N_{u,calc}$	−	−	−

5 SUMMARY AND CONCLUSIONS

Based on the results of both numerical and experimental investigations, a behavioral model to predict the average failure load of fastenings using adhesive bonded anchors is proposed. The model is similar to the behavioral model that predicts the concrete breakout failure load of cast-in-place and post-installed mechanical anchors but with the following modifications.

The basic strength of a single adhesive anchor predicts the pullout capacity and not the concrete breakout capacity. It is based on the uniform bond stress model as given by Equation 3. The characteristic spacing and characteristic edge distance of adhesive anchorages depend on the anchor diameter and the bond strength and not on the anchor embedment depth. Furthermore, an additional factor, $\psi_{g,N}$ is used, that takes into account the larger bond area of closely spaced adhesive anchors in comparison to a single anchor. The failure load of anchorages with adhesive anchors is limited to the concrete cone failure load of post-installed mechanical anchors.

The proposed behavioral model agrees very well with the results of 415 group tests contained in a worldwide data base. Based on a comparison to 133 tests with single anchors very near to an edge, the behavioral model is conservative for anchorages located very close to an edge.

REFERENCES

Appl, J. & Eligehausen, R. 2003. Gruppenbefestigungen mit Verbunddübeln − Bemessungskonzept- Groups of Bonded Anchors -Design Concept-), *Report* No. 03/27-2/55, Institute of Construction Materials, University of Stuttgart. Stuttgart

Cook, R.A., Kunz, J., Fuchs, W. & Konz, R.C. 1998. Behavior and Design of single Adhesive Anchors under Tensile Load in Uncracked concrete, *ACI Structural Journal*, V.95: pp. 9–62

Eligehausen R., Mallée, R. & Silva J.F. 2006. *Anchorages in Concrete Construction*. Berlin: Ernst & Sohn

Eligehausen R., Appl, J., Lehr, B., Meszaros, J. & Fuchs, W. 2004. Tragverhalten und Bemessung von Verbunddübeln unter Zugbeanspruchung, Teil 1: Einzeldübel mit großem Achs- und Randabstand (Load bearing behavior of bonded anchors under tension loading, Part 1: Single bonded anchors far away from edges), *Beton- und Stahlbetonbau 99*, No. 7: pp. 561–571, Berlin: Ernst & Sohn

Fuchs, W., Eligehausen, R. & Breen, J.E. 1995. Concrete Capacity Design (CCD) Approach for Fastenings to Concrete, *ACI Structural Journal*, V.92, pp. 365–379

Li, Y.-J., Eligehausen, R., Ozbolt, J. & Lehr, B. 2002. Numerical Analysis of Quadruple Fastenings with bonded anchors, *ACI Structural Journal*, V.99, pp. 149-156

McVay, M., Cook, R.A. & Krishnamurthy, K. 1996. Pullout Simulation of Post-Installed Chemically Bonded Anchors, *Journal of Structural Engineering, ASCE*, V.122, pp. 1016–1024

Meszaros, J. 1999. Tragverhalten von Verbunddübeln in ungerissenem und gerissenem Beton (Load-Bearing behavior of bonded anchors in uncracked and cracked concrete). *Doctoral Thesis*, University of Stuttgart (in German)

Ozbolt, J. 1998. MASA-Finite Element Program for Nonlinear Analysis of Concrete and Reinforced Concrete Structures, *Research Report*, Institute of Construction Materials, University of Stuttgart

Fracture Mechanics of Concrete and Concrete Structures – Design, Assessment and Retrofitting of RC Structures – Carpinteri, et al. (eds)
© 2007 Taylor & Francis Group, London, ISBN 978-0-415-44616-7

Cohesive versus overlapping crack model for a size effect analysis of RC elements in bending

A. Carpinteri, M. Corrado, M. Paggi & G. Mancini
Department of Structural and Geotechnical Engineering, Politecnico di Torino, Italy

ABSTRACT: The well-known *Cohesive Crack Model* describes strain localization with a softening stress variation in concrete members subjected to tension. An analogous behaviour is also observed in compression, when strain localization takes place in a damaged zone and the stress reaches the compressive strength with surface energy dissipation. In the present paper, we propose the new concept of *Overlapping Crack Model,* which is analogous to the cohesive one and permits to simulate material compenetration.

The two aforementioned elementary models are merged into a more complex algorithm able to describe both cracking and crushing growths during loading processes in RC members. A numerical procedure based on elastic coefficients is developed, taking into account the proposed constitutive laws in tension and compression. With this algorithm, it is possible to effectively capture the flexural behaviour of RC beams by varying the reinforcement percentage and/or the beam depth.

1 INTRODUCTION

The description of the behaviour of reinforced concrete members is complicated by the contemporaneous presence of different nonlinear contributions: crack opening in tension, concrete crushing in compression and steel yielding or slippage.

In most practical applications, the concrete contribution in tension is totally neglected or just considered with a linear-elastic stress-strain law until the ultimate tensile strength is reached. The non-linear concrete behaviour in compression is certainly not negligible. Many different laws may be used to model the concrete behaviour in compression: elastic-perfectly plastic, parabolic-perfectly plastic, Sargin's parabola, etc. The most utilised constitutive laws for steel are the elastic-perfectly plastic or the elastic-hardening stress-strain relationships. Using these constitutive laws with a FEM program, it is possible to fully describe the behaviour of a reinforced concrete member, though it is difficult to catch the size-scale effects. The reason is that the above-mentioned constitutive laws consider only an energy dissipation over the volume in the nonlinear regime.

On the other hand, the application of Fracture Mechanics concepts has been proved to be very effective for the analysis of size-scale effects. In particular, Carpinteri (1986) proposed a cohesive formulation to explain the ductile-brittle transition in unreinforced concrete beams under three-point bending test. This

algorithm can also be used to describe the nonlinear behaviour of concrete in RC members.

With regard to the behaviour of concrete in compression, Hillerborg (1990) firstly introduced a model based on the concept of strain localization. According to his approach, when the ultimate compressive strength is achieved, a strain localization takes place within a characteristic length proportional to the width of the compressed zone. This model permits to study the problem of size effects, although the definition of the length over which the strain localization occurs is not clear.

Furthermore, many experimental tests, see e.g. Van Vliet and Van Mier (1996), Carpinteri et al. (2005), Suzuki et al. (2006) put into evidence that a significant scale effect on dissipated energy density takes place. This parameter can be assumed as a material constant only if it is defined as a crushing surface energy. Hence, it emerges that the process of concrete crushing can be analysed with an approach similar to the cohesive model, which is valid for the tensile behaviour of concrete. In particular, we can define a linear-elastic stress-strain law, before achieving the compressive strength. Afterwards, a descending stress-displacement law can be introduced, for the analysis of the nonlinear behaviour in compression.

Finally, a special mention has to be given to the steel-concrete interaction. The most appropriate law for the reinforcing bar bridging a tensile crack propagating in concrete is a nonlinear relationship between force and crack opening. For this reason, it is appropriate to consider a bond-slip law characterizing the interaction between steel and concrete.

In the next sections we propose a method able to describe step-by-step the behaviour of a reinforced concrete member during both fracturing and crushing. Firstly, we introduce separately the elementary models that we use to describe the whole concrete behaviour: the *Cohesive Crack Model* for concrete in tension, the *Overlapping Crack Model* for concrete in compression, and the stress-displacement relationship for steel in tension. A closed-form solution is also proposed to demonstrate that the strain localization in compression causes ductile-brittle transition by varying the structural dimension. Finally, the proposed numerical algorithm and some applications are reported to investigate on the influence of the various model's parameters.

2 MATHEMATICAL FORMULATION

Let us consider the reinforced concrete member shown in Fig. 1, subjected to a bending moment M. We assume that the midspan cross-section is fully representative of the mechanical behaviour of the whole element. The stress distribution is linear-elastic until the tensile stress at the intrados achieves the concrete tensile strength. When this threshold is overcome, a cohesive crack propagates from the bottom side toward the upper side. Correspondingly, the applied moment increases. Outside the crack, the material is assumed to behave elastically (Fig. 2). The stresses in the cohesive zone depend on the crack opening displacement and become equal to zero when the crack opening reaches a critical value and beyond.

On the other hand, concrete crushing takes place when the maximum stress in compression reaches the concrete compressive strength. Damage is described as a compenetration between the two half-beams representing the cause of the localization of the dissipated energy (Fig. 3). Larger the compenetration, also called *overlapping*, lower the transferred forces along the damaged zone.

2.1 Cohesive crack model

Linear Elastic Fracture Mechanics has been proven to be a useful tool for solving fracture problems, provided that a crack-like notch or flaw exists in the body and that the nonlinear zone ahead of the tip is negligible. These conditions are not always fulfilled and, both for metallic and cementitious materials, the size of the nonlinear zone due to plasticity or microcracking may be not negligible with respect to other dimensions of the cracked geometry. The localized damaged material

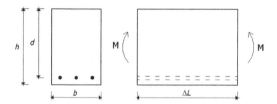

Figure 1. Scheme of a reinforced concrete element.

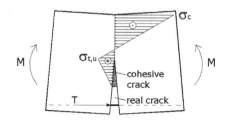

Figure 2. Cohesive stress distribution in tension with linear-elastic distribution in compression.

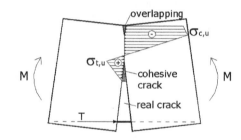

Figure 3. Nonlinear behaviour in tension and compression.

may be modelled as a pair of restrained fracture surfaces. This idea has been extensively applied to materials which are commonly classified as quasi-brittle such as concrete, glass, polymers, rocks, etc. In particular, the most suitable model for concrete was firstly proposed by Hillerborg et al. (1976) with the name of *Fictitious Crack Model*. Carpinteri (Carpinteri et al. (1985) and Carpinteri (1989a)), introduced the terminology *Cohesive Crack Model* and applied an updated algorithm to the study of ductile-brittle transition and snap-back instability in concrete.

The hypotheses of the model can be summarized as follows:

(1) The constitutive law used for the non-damaged zone is the σ-ε linear-elastic law shown in Fig. 4a.
(2) The process zone develops when the maximum stress reaches the ultimate tensile strength.
(3) (3) The process zone is perpendicular to the main tensile stress.
(4) In the process zone, the damaged material is still able to transfer a tensile stress across the crack surfaces. The cohesive stresses are considered as

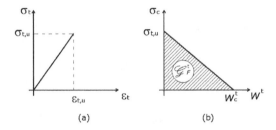

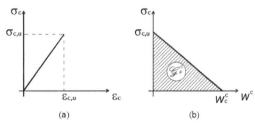

Figure 4. Concrete constitutive laws in tension: linear-elastic (a); post-peak softening (b).

Figure 5. Concrete constitutive laws in compression: linear-elastic (a); post-peak softening (b).

decreasing functions of the crack opening w^t, (see Fig. 4b):

$$\sigma_t = \sigma_{t,u}\left(1-\frac{w^t}{w^t_c}\right), \tag{1}$$

where w^t is the crack opening, w^t_c is the critical value of the crack opening and $\sigma_{t,u}$ is the ultimate tensile strength of concrete.

The shaded area under the stress vs. displacement curve in Fig. 4b represents the fracture energy, $\mathscr{G}^t_F$.

2.2 Overlapping Crack Model for concrete crushing

The most frequently adopted constitutive laws for concrete in compression describe the material behaviour in terms of stress and strain. This approach imply that the energy is dissipated over a volume, whereas experimental results reveal that the energy is mainly dissipated over a surface. Hillerborg (1990) firstly proposed to model the crushing phenomenon as a strain localization over a length proportional to the depth of the compressed zone. However, the evaluation of this characteristic length is rather complicated by the fact that the depth of the compressed zone varies during the loading process. As a result, it is difficult to formulate a material constitutive law describing the mechanical response of concrete in compression.

In our formulation, we introduce a stress-displacement relationship between the compressive stress and the compenetration, in close analogy with the cohesive model. The main hypotheses are the following:

(1) The constitutive law used for the undamaged material is a linear-elastic stress-strain relationship, see Fig. 5a.
(2) The crushing zone develops when the maximum compressive stress achieves the material strength for concrete.
(3) The process zone is perpendicular to the main compressive stress.

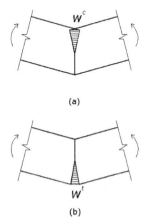

Figure 6. Compression crushing with overlapping (a); tensile fracture with cohesive zone (b).

(4) The damaged material in the process zone is assumed to be able to transfer a stress between the overlapping surfaces. Concerning the crushing stresses, they are assumed to be a decreasing function of the compenetration w^c (Fig. 5b):

$$\sigma_c = \sigma_{c,u}\left(1-\frac{w^c}{w^c_c}\right), \tag{2}$$

where w^c is the compenetration, w^c_c is the critical value of overlapping and $\sigma_{c,u}$ is the ultimate compressive strength. This zone is represented by a fictitious overlapping, that is analogous to the fictitious crack in tension, as shown in Fig. 6.

In analogy with the cohesive crack model, we can define the area under the stress-displacement curve as the crushing energy, $\mathscr{G}^c_F$.

A more sophisticated stress-displacement law considering the phenomenon of compacting, was recently proposed by Suzuki et al. (2006). In this case, the crushing energy is computed according to the following empirical equation, which considers the confined concrete compressive strength, by means of the stirrup yield strength and the stirrup volumetric content:

$$\frac{\mathscr{G}_F^c}{\sigma_{c,0}} = \frac{\mathscr{G}_{F,0}^c}{\sigma_{c,0}} + 10000\frac{k_a^2 p_e}{\sigma_{c,0}^2}. \tag{3}$$

Parameter $\mathscr{G}_{F,0}^c$ is the crushing energy for unconfined concrete, $\sigma_{c,0}$ is the average compressive strength, k_a is the parameter depending on the stirrup strength and ratio and p_e is the effective lateral pressure.

A comparison between the crushing energy and the fracture energy for different compressive strengths is proposed in Tab. 1. The crushing energy is calculated according to Eq. (3) for concrete without stirrups, while the fracture energy is calculated according to the CEB-FIP Model Code 90 in case of maximum aggregate dimension of 16 mm. It is worth noting that $\mathscr{G}_F^c$ is between 2 and 3 orders of magnitude higher than $\mathscr{G}_F^t$.

Finally, we remark that the critical values for crushing compenetration and crack opening are respectively $w_c^c \approx 1$ mm and $w_c^t \approx 0.1$ mm.

2.3 Steel-concrete interaction

In order to model the steel contribution to the load carrying capacity of the beam, it is necessary to introduce a suitable bond-slip law for the characterization of steel-concrete interaction. Typical bond-slip relationships are defined in terms of a tangential stress along the steel-concrete interface as a function of the relative tangential displacement between the two materials (see Jenq and Shah (1989), Model Code 1990, Carpinteri (1999)). The integration of the differential slip over the transfer length, l_{tr}, is equal to half the opening crack at the reinforcement level, as shown in Fig. 7a. On the other hand, the integration of the bond stresses gives the reinforcement reaction. In order to simplify the calculation, the stress-displacement law is assumed to be linear until the yield stress (or until the critical opening crack for steel w_c^y) is achieved. After that, the reinforcement reaction is considered as constant (Fig. 7b).

3 PURE CRUSHING COLLAPSE

Let us consider a reinforced concrete beam under three-point bending. We assume a rigid-perfectly plastic constitutive law for steel, whereas for concrete we use the constitutive law shown in Fig. 5. The linear elastic behaviour, with the stress distribution shown in Fig. 8, may be represented by the following nondimensional equation:

$$\widetilde{P} = \frac{16}{\lambda^3}\widetilde{\delta}, \tag{4}$$

where the nondimensional load and the nondimensional mid-span deflection are respectively given by

$$\widetilde{P} = \frac{Pl}{\sigma_{c,u}td^2}, \tag{5a}$$

$$\widetilde{\delta} = \frac{\delta l}{\varepsilon_{c,u}d^2}. \tag{5b}$$

The parameter l is the beam span, d is the beam effective depth, t is the beam thickness and λ is the beam slenderness.

Once the ultimate compressive stress, $\sigma_{c,u}$, is reached at the extrados, crushing develops in the middle cross-section. Such a nonlinear phenomenon

Table 1. Comparison between crushing energy and fracture energy for different concrete compressive strengths.

$\sigma_{c,0}$ N/mm^2	$\mathscr{G}_F^c$ N/mm	$\mathscr{G}_F^t$ N/mm
30	30	0,065
50	40	0,090
70	51	0,117
90	58	0,140

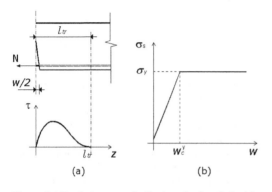

(a)   (b)

Figure 7. Bond stresses τ in the transfer length l_{tr} (a); stress-displacement law for steel (b).

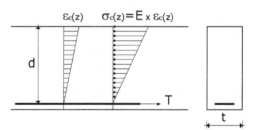

Figure 8. Linear elastic strain and stress distribution.

admits the limit configuration shown in Fig. 9. In this situation, two rigid half-beams are connected by a hinge placed at the reinforcement level. The rotational equilibrium is ensured by the applied load, the reaction forces and the linear distribution of the crushing forces. Clearly, this distribution depends on the amount of overlapping: the higher the penetration w^c, the lower the crushing forces until they vanish for $w^c = w_c^c$.

The rotational equilibrium around point A (see Fig. 9), provides the following nondimensional equation:

$$\widetilde{P} = \frac{1}{6}\left(\frac{s_E^c \lambda^2}{\varepsilon_{c,u}\widetilde{\delta}}\right)^2 , \tag{6}$$

where

$$s_E^c = \frac{w_c^c}{2d} = \frac{\mathscr{G}_F^c}{\sigma_{c,u}d} . \tag{7}$$

While the linear Eq. (4) describes the mechanical response of the beam in the elastic regime, the hyperbolic Eq. (6) represents the asymptotic behaviour of the beam when it is completely failed in compression.

Figure 10 shows two possible situations. When the domains are separated, the two P-δ branches, the linear and the hyperbolic one, may be connected by a

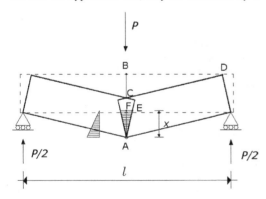

$$P$$

$$B \qquad D$$

$$C$$

$$F \backslash E$$

$$x$$

$$P/2 \qquad\qquad P/2$$

$$A$$

$$l$$

Figure 9. Limit situation of complete overlapping with linear distribution of crushing stresses.

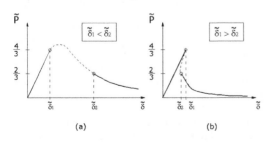

(a)   (b)

Figure 10. Load – Deflection diagrams: ductile condition (a); brittle condition (b) ($\delta_1 = \lambda^3/12$; $\delta_2 = s_E^c\lambda^2/2\varepsilon_{c,u}$).

regular curve (Fig 10a). On the other hand, when the two domains are partially overlapped, a snap-back instability may occur (Fig. 10b).

In conclusion, unstable behaviours and catastrophic events are expected when $\delta_2 \leq \delta_1$, i.e., for:

$$\frac{s_E^c \lambda^2}{2\varepsilon_{c,u}} \leq \frac{\lambda^3}{12} , \tag{8}$$

which gives the following brittleness condition:

$$\frac{s_E^c}{\varepsilon_{c,u}\lambda} \leq \frac{1}{6} . \tag{9}$$

The mechanical system is expected to be brittle for a low brittleness number, S_E^c, a high ultimate strain, $\varepsilon_{c,u}$, and large slenderness, λ.

It is worth noting that the brittleness condition in Eq. (9), as compared with that obtained for plain concrete beams, where $S_E^f/\varepsilon_{c,u}\lambda \leq 1/3$, (Carpinteri (1989b)) suggests that the crushing phenomenon is more ductile than pure tensile flexural failure.

4 NUMERICAL ALGORITHM

A discrete form of the elastic equations governing the mechanical response of the two half-beams is introduced. To this aim the finite element method is used. At the middle cross-section the nodes of the finite element mesh are distributed along the potential fracture line (nodes from 1 to l) and the potential crushing line (nodes from $l+1$ to n) (Fig. 11). The position of the ligament between nodes l and $l+1$ is arbitrary; it may depend on the ultimate configuration, i.e. on the material parameters and the reinforcement percentage. In this scheme, cohesive and overlapping stresses are replaced by equivalent nodal forces. They depend on the nodal opening or closing displacements according to the cohesive or overlapping softening laws respectively shown in Fig. 4b and in Fig. 5b.

The horizontal nodal displacements, w, along the middle cross-section can be computed as follows:

$$\{w\} = [H]\{F\} + \{C\}M \tag{10}$$

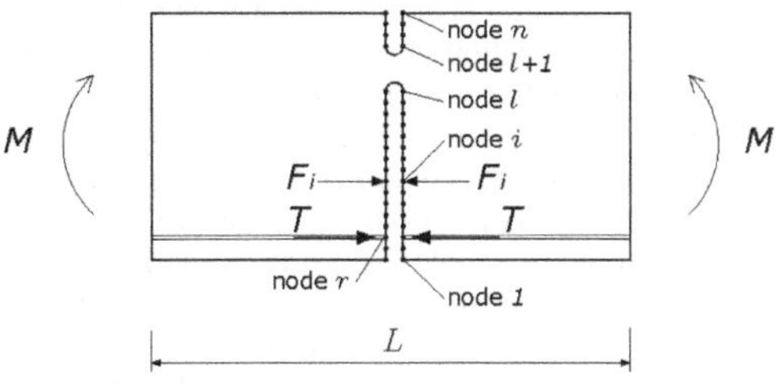

$$M \left(\begin{array}{c} \text{node } n \\ \text{node } l+1 \\ \text{node } l \\ \text{node } i \\ F_i \rightarrow \quad \leftarrow F_i \\ T \qquad T \\ \text{node } r \\ \text{node } 1 \end{array} \right) M$$

$$L$$

Figure 11. Finite element nodes along potential fracture (1 to l) and crushing ($l+1$ to n) lines.

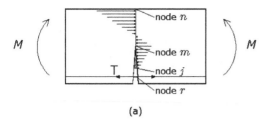

(a)

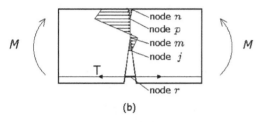

(b)

Figure 12. Force distribution with: cohesive crack in tension and linear elastic behaviour in compression (a); cohesive crack in tension and crushing in compression (b).

where $\{w\}$ is the vector of nodal displacements, $[H]$ is the matrix of the coefficients of influence for the nodal forces, $\{F\}$ is the vector of nodal forces, $\{C\}$ is the vector of the coefficients of influence for the applied moment and M is the applied moment.

The coefficients of influence representing node opening or overlapping displacements are computed by a finite element analysis in which the fictitious structure shown in Fig. 11 is subjected to $n + 1$ different loading conditions.

In Eq. (10) the reinforcement contribution is included in the nodal force corresponding to the r-th node.

In the generic situation shown in Fig. 12a, we can consider the following equations:

$$F_i = 0 \qquad \text{for } i = 1, 2, ..., (j-1); \quad i \neq r \qquad (11a)$$

$$F_i = F_{t,u}\left(1 - \frac{w_i^t}{w_c^t}\right) \qquad \text{for } i = j, ..., (m-1) \qquad (11b)$$

$$w_i^t = 0 \qquad \text{for } i = m, ..., n \qquad (11c)$$

Equations (10) and (11) constitute a linear algebraic system of $(2n)$ equations and $(2n + 1)$ unknowns, i.e., the elements of the vectors $\{w\}$ and $\{F\}$ and the applied moment, M. The additional equation required to solve the problem is obtained by setting the value of the force at the fictitious crack tip, m, equal to the ultimate tensile force. The driving parameter of the process is the position of the fictitious crack tip, defined by the position of the node m in Fig. 12a, that is increased by one step at each iteration. The position of the real crack tip, j, turns out to be a function of the crack opening.

When crushing takes place, (see Fig. 12b), Eqs. 11 are replaced by:

$$F_i = 0 \qquad \text{for } i = 1, 2, ..., (j-1); \quad i \neq r \qquad (12a)$$

$$F_i = F_{t,u}\left(1 - \frac{w_i^t}{w_c^t}\right) \qquad \text{for } i = j, ..., (m-1) \qquad (12b)$$

$$w_i^t = 0 \qquad \text{for } i = m, ..., p \qquad (12c)$$

$$F_i = F_{c,u}\left(1 - \frac{1}{2}\frac{w_i^c}{w_c^c}\right) \qquad \text{for } i = (p+1), ..., n \qquad (12d)$$

Again, Equations (10) and (12) constitute a linear algebraic system of $(2n)$ equations and $(2n + 1)$ unknowns. In this case, there are two possible additional equations: either the force in the fictitious crack tip, m, equal to the ultimate tensile force, or the force in the fictitious crushing tip, p, equal to the ultimate compressive force. In the numerical scheme, we choose the situation which is closer to one of these critical conditions. The driving parameter of the process is the tip that in the considered step has reached the limit resistance. Only this tip is moved passing to the next step.

Finally, at each step of the algorithm it is possible to calculate the beam rotation, ∂, as follows:

$$\vartheta = \{D_F\}^T\{F\} + D_m M \qquad (13)$$

where D_F is the vector of the coefficients of influence for the nodal forces and D_m is the coefficient of influence for the applied moment. The physical dimensions of the coefficients D_{Fi} and D_m are, respectively, $[F]^{-1}$ and $[F]^{-1}[L]^{-1}$.

5 NUMERICAL EXAMPLES

In this section we show the results of numerical simulations carried out to investigate on the influence of two fundamental parameters on the global mechanical behaviour, namely the beam depth and the steel percentage. The midspan cross-section was subdivided into 160 intervals and it was constrained in four nodes. This last region represents the elastic core of the beam at the ultimate condition. In all the following examples the slenderness, λ, and the thickness, λ, are kept constant.

The diagrams depicted in Fig. 13 show the mechanical response in terms of bending moment vs. localised beam rotation, by varying the beam depth for a steel percentage equal to 2%.

In the case of $h = 200$ mm, we have a ductile mechanical response, where the moment-rotation diagram has a first part with positive slope then followed by a plastic flow. Increasing the depth from 200 mm

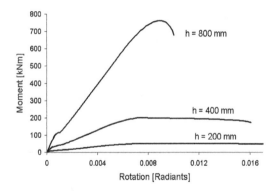

Figure 13. Moment-Rotation diagrams for a constant steel percentage, $\rho = 2\%$, and different beam depths, h.

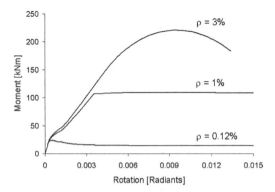

Figure 14. Moment-Rotation diagrams for a constant beam depth, $h = 400$ mm, and different steel percentages, ρ.

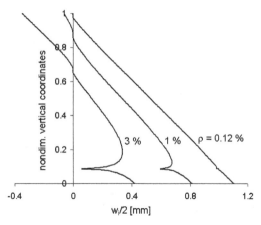

Figure 15. Horizontal nodal displacements for a constant beam depth, $h = 400$ mm, and different steel percentages, ρ, at the ultimate condition.

$\rho = 0.12\%$ to $\rho = 1\%$ we have a transition from brittle to ductile and then, from $\rho = 1\%$ to $\rho = 3\%$, the mechanical behaviour becomes brittle again. For the highest reinforcement ratio, a softening branch occurs due to concrete crushing. In Figs. 15 and 16 we see that for a very low steel percentage ($\rho = 0.12\%$) the ultimate resistant moment is provided by a tensile plastic flow in the reinforcement and a compressive zone concentrated in the extrados. In this case the ductile behaviour is due to the reinforcement yielding. With $\rho = 1\%$, the reinforcement is yielded at the ultimate condition, but a lower value of crack opening is observed. On the other hand, the crushing contribution becomes more relevant.

For a very high steel percentage ($\rho = 3\%$) the reinforcement is not yielded and so the only contribution to the ductility results from crushing. In this situation the ultimate resistant moment is not proportional to the reinforcement ratio, because it is limited by the maximum compressive force.

Finally, it is worth noting that, in the case of low steel percentages, crushing does not take place and a single nondimensional parameter, N_P (Carpinteri (1981) and (1984)) can be used to describe the transition from ductile to brittle behaviours:

$$N_P = \frac{\rho \sigma_y h^{0.5}}{\sqrt{\mathscr{G}_F^t E_c}} \tag{14}$$

A brittle to ductile transition clearly emerges in the diagrams of Fig. 17, when the brittleness number increases. The diagrams reveal an unstable response, represented by a strain-softening relationship, for the lower brittleness number, and a stable response, represented by a strain-hardening relationship, for the higher brittleness number.

to 400 mm, the plastic range decreases and a softening branch characterized by a low negative slope appears. An analogous behaviour, although more emphasised, is obtained when the depth is equal to 800 mm. Summarizing, we observe that the higher the beam depth, for a given reinforcement ratio, the higher the global stiffness. Obviously, the ultimate resistant moment is an increasing function of the beam depth, whereas the plastic range is progressively diminished with the appearance of steeper and steeper softening branches. The algorithm puts into evidence that the crushing zone increases with the beam depth, while the crack opening at the intrados decreases.

Another set of simulations was carried out by considering a constant beam depth equal to 400 mm and by varying the reinforcement ratio from 0.12% to 3%. The resulting moment-rotation diagrams are shown in Fig. 14. According to the numerical and experimental results concerning the minimum reinforcement in RC beams (see Bosco et al. (1990), Bosco and Carpinteri (1992)), the global behaviour with a very low steel percentage (e.g. 0.12%) is brittle. By increasing the steel ratio, two transitions may be highlighted. From

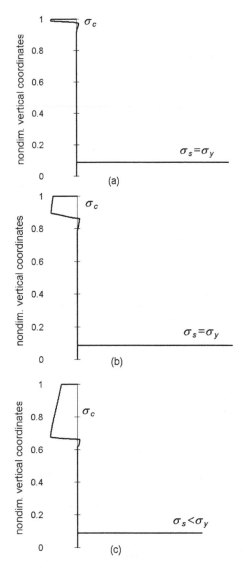

Figure 16. Stress distribution at the ultimate condition referred to: $\rho = 0.12\%$ (a), $\rho = 1\%$ (b), $\rho = 3\%$ (c) ($h = 400\,\mathrm{mm}$).

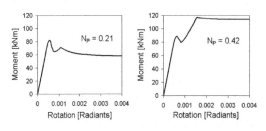

Figure 17. Moment-Rotation diagrams for a low reinforcement ratio, $\rho = 0.25\%$, and two different brittleness numbers, N_P.

6 CONCLUSIONS

In this paper we have proposed a new mathematical and numerical approach to the problem of crushing in RC beams, which is treated in analogy with the cohesive formulation. The effect of the reinforcement contribution, which is not usually included in the cohesive models, has been taken into account as a local modification of the tensile cohesive law.

This very general approach has permitted to investigate on the competition between compressive crushing and tensile cracking in RC beams, with special focus on size-scale effects. Future developments are expected as far as the analysis of the size effects on the rotational capacity is concerned. In this framework, a comparison with the experimental results in Bosco, Carpinteri and Debernardi (1992) will be proposed.

ACKNOWLEDGEMENT

The financial support provided by the Ministry of University and Research (MIUR) is gratefully acknowledged.

REFERENCES

Bosco, C., & Carpinteri, A. 1992. Softening and snap-through behavior of reinforced elements. *Journal of Engineering Mechanics (ASCE)*, Vol. 118, 1564–1577.

Bosco, C., Carpinteri, A. & Debernardi, P.G. 1990. Minimum reinforcement in high strength concrete. *Journal of Structural Engineering (ASCE)*, Vol. 116, 427–437.

Bosco, C., Carpinteri, A. & Debernardi, P.G. 1992. Scale effect on plastic rotational capacity of R.C. beams. In *Fracture Mechanics of Concrete Structures* (Proceedings of the 1st FraMCoS Conference, Breckenridge, USA, 1992), Ed. Z.P. Bazant, Elsevier Applied Science, London, 735–740.

Carpinteri, A. 1981. A fracture mechanics model for reinforced concrete collapse. In *Advanced Mechanics of Reinforced Concrete* (Proceedings of a IABSE Colloquium, Delft, the Netherlands, 1981), Delft University Press, Delft, 17–30.

Carpinteri, A. 1984. Stability of fracturing process in R.C. beam. *Journal of Structural Engineering* (ASCE), Vol. 110, 544–558.

Carpinteri, A. 1986. Limit analysis for elastic-softening structures: scale and slenderness influence on the global brittleness. In *Structure and Crack Propagation in Brittle Matrix Composite Materials* (Proceedings of the Euromech Colloquium 204, Warsaw, Poland, 1985), Eds. A. M. Brandt, I. H. Marshall, Elsevier Applied Science, London, 497–508.

Carpinteri, A. 1989a. Cusp catastrophe interpretation of fracture instability. *Journal of the Mechanics and Physics of Solids*, Vol. 37, 567–582.

Carpinteri, A. 1989b. Size effects on strength, toughness, and ductility. *Journal of Engineering Mechanics (ASCE)*, Vol. 115, No. 7, July, 1375–1392.

Carpinteri, A., Di Tommaso, A. & Fanelli, M. 1985. Influence of material parameters and geometry on cohesive crack propagation. In *Fracture Toughness and Fracture Energy of Concrete,* (Proceedings of an International Conference on Fracture Mechanics of Concrete, Lausanne, Switzerland, 1985), Ed. F.H. Wittmann, Elsevier, Amsterdam, The Netherlands, 1986, 117–135.

Carpinteri, A. (ed.) 1999. *Minimum Reinforcement in Concrete Members,* Elsevier Science Ltd, VIII + 203, Oxford, UK.

Carpinteri, A., Ferro, G. & Ventura, G. 2005. Scale-independent constitutive law for concrete in compression. *Proceeding of the 11th International Conference on Fracture (ICF),* Torino, Italy, 2005, CD-ROM, Paper N. 5556.

Comité Euro-International du Beton. 1993. CEB-FIP Model Code 1990.

Hillerborg, A. 1990. Fracture mechanics concepts applied to moment capacity and rotational capacity of reinforced concrete beams. *Engineering Fracture Mechanics,* Vol. 35, 233–240.

Hillerborg, A., Modeer, M. & Petersson, P. E. 1976. Analysis of crack formation and crack growth in concrete by means of fracture mechanics and finite elements. *Cement and Concrete Research,* Vol. 6, 773–782.

Jenq, Y. S. & Shah, S. P. 1989. Shear resistance of reinforced concrete beams – a fracture mechanics approach. In *Fracture Mechanics: Application to Concrete* (Special Report ACI SP-118), eds. Li, V. C. and Bazant, Z. P., American Concrete Institute, Detroit, 327–358.

Suzuki, M., Akiyama, M., Matsuzaki, H. & Dang, T.H. 2006. Concentric loading test of RC columns with normal- and high-strength materials and averaged stress-strain model for confined concrete considering compressive fracture energy. *In FIB, Proceedings of the 2nd International Conference,* June 5–8, 2006 – Naples, Italy.

Van Vliet, M. & Van Mier, J. 1996. Experimental investigation of concrete fracture under uniaxial compression. *Mechanics of Cohesive-Frictional Materials,* Vol. 1, 112–127.

Fracture Mechanics of Concrete and Concrete Structures – Design, Assessment and Retrofitting of RC Structures – Carpinteri, et al. (eds)
© 2007 Taylor & Francis Group, London, ISBN 978-0-415-44616-7

Numerical evaluation of plastic rotation capacity in RC beams

A.L. Gamino & T.N. Bittencourt
Structural and Geotechnical Engineering Department, Polytechnical School of São Paulo University, São Paulo, Brazil

ABSTRACT: The objective of the present paper is the analysis of the parameter "θ_{pl}" in reinforced-concrete beams by means of smeared crack approaches. The effects of the compressive strength and the size dependence in bending are re-investigated. The non-linear analyses were carried out by means of the Finite Element Method. The programs DIANA, CASTEM2000, and QUEBRA2D/FEMOOP were used in the numerical simulations. Appropriate constitutive models for concrete, rebars and bond-slip interfaces were implemented to represent more realistically the behavior of the structural system. Interface and reinforcement (discrete and embedded approaches) finite elements were used to accomplish an explicit representation of the concrete-reinforcement bond.

1 INTRODUCTION

Since the publication of Kani in 1967, it is known that the resistant capacity of beams is highly dependent on size effects. In 1966 Corley (1966) advocated that size effect did not significantly affect the plastic rotation capacity "θ_{pl}" in R/C beams.

However, in 1988 Hillerborg (1988) concluded, using fracture mechanics concepts, that "θ_{pl}" in R/C sections is inversely proportional to the height of the structural element.

Over the last decades research on ductility of beams has evolved and demonstrated the influence of the reinforcement ratio, concrete grade, stirrup spacing and finally size effects.

The importance of the size effect in beams of reinforced-concrete is in fact that this can cause reductions in the capacities of inelastic deformation and plastic rotation as well as transition of types of brittle/ductile ruptures.

These effects can appear in two forms:

Modifications in the "a/d" relations ("a" is the shear span; "d" is the effective beam depth) that over all affect the load capacity of the beams to the shear Walraven & Lehwalter (1994);

Variations in the slenderness inside or out of pure bending regions;

The first research on ductility of beams had involved to portray the influence of the reinforced ratio Leslie, Rajagopalan & Everard (1976), strength of the concrete Tognon et al. (1980), Ashour (2000), span

between stirrups Shin et al. (1989) and finally size effect Alca et al. (1997).

This work is concentrated on the evaluation of the ductility in 2D beams using smeared crack models in the concrete. The numerical 2D analysis is based on the finite element method implemented in CASTEM 2000 developed by the Département de Mécanique et de Technologie (DMT) du Commissariat Français à l'Energie Atomique (CEA). This program uses the constitutive elastoplastic perfect model for the steel, the Drucker-Prager two parameter model for the concrete and the Newton-Raphson for the solution of non-linear systems.

In the QUEBRA2D/FEMOOP system appropriate constitutive models for concrete, rebars and bond-slip interfaces have been implemented to represent more realistically the behavior of the structural system. Interface and reinforcement (discrete and embedded approaches) finite elements have been used to accomplish an explicit representation of the concrete-reinforcement bond.

In the numerical 2D simulations, the established computational program in the Finite Element Method DIANA was developed by the TNO Building and Constructions Research of the Department of Computational Mechanics (Netherlands). First, the influence of each smeared crack model on the numerical results was presented. Later, a parametric study is carried out. An analytical model is presented in order to calculate the plastic rotation capacity "θ_{pl}" of the analyzed beams. Later the experimental results are compared with the results from the numerical simulations.

2 NUMERICAL EVALUATION OF PLASTIC ROTATION CAPACITY

2.1 Concrete constitutive model in CASTEM 2000

The concrete constitutive model used was the Drucker-Prager two parameters model (derived from the Ottosen four parameters model). This model was formulated in 1952 and can be seen as a simple modification of the criterion of Von Mises, including the influence of the hydrostatic pressure, according to Equation (1).

$$f(\sigma) = \alpha I_1 + \sqrt{J_2} - k_{dp} = 0 \qquad (1)$$

where "α" and "k_{dp}" are the material constants, "I_1" and "J_2" the invariants that depend on the normal stress on a body. Figure 1 brings the graphical representation of the surface of plasticity in the plan $\sigma_1 - \sigma_2$ where "ft" is the tensile strength, "fc" is the compressive strength and "fbc" is the biaxial strength in compression.

2.2 Concrete constitutive models in DIANA

2.2.1 Hardening effect
For the concrete under compression the model described in the code (CEB-Fip, 1993) was used in some modeling represented in the Equation 2.

$$\frac{\sigma_c}{f_{cm}} = \frac{\dfrac{E_{ci}}{E_{c1}}\dfrac{\varepsilon_c}{-0,22\%} - \left(\dfrac{\varepsilon_c}{-0,22\%}\right)^2}{1 + \left(\dfrac{E_{ci}}{E_{c1}} - 2\right)\dfrac{\varepsilon_c}{-0,22\%}} \qquad (2)$$

In other modeling the non-linear hardening model of *Thorenfeldt* was used. This model presented in Figure 2 uses an equation between compressive stress and the deformations based in the adoption of diverse parameters.

2.2.2 Softening effect
For the corresponding curve to the behavior of the concrete under traction the described bilinear model for the code (CEB-Fip, 1993) was used in some analyses in agreement with to Equation 3.

$$\bar{\sigma} = -\frac{(f_{ctm} - 2,12)}{\left(0,015 - \bar{\varepsilon}\right)} \cdot 0,015 + f_{ctm} \qquad (3)$$

In other simulations the non-linear curve of Hordijk was used. This model presented in Figure 3 uses an

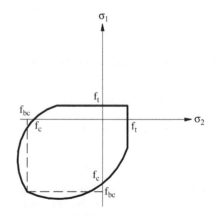

Figure 1. Drucker-Prager yield criteria.

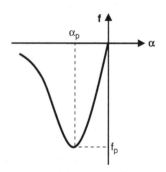

Figure 2. *Thorenfeldt* hardening curve.

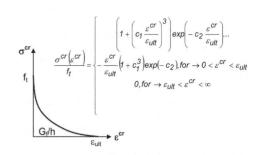

$$\frac{\sigma^{cr}(\varepsilon^{cr})}{f_t} = \begin{cases} -\frac{\varepsilon^{cr}}{\varepsilon_{ult}}\left(1 + c_1^3\right)\exp(-c_2), \text{for} \to 0 < \varepsilon^{cr} < \varepsilon_{ult} \\ 0, \text{for} \to \varepsilon_{ult} < \varepsilon^{cr} < \infty \end{cases}$$

Figure 3. Hordijk softening curve.

exponential relation between the normal tensile stress and the deformations, with "c_1" and "c_2" assuming the values respectively of 3,0 and 6,93.

Three smeared crack models were available in this program: fixed, multi-directional and rotating.

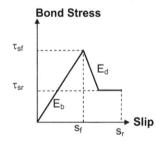

Bond Stress (on figure, vertical axis)

τ_{sf}, E_d, τ_{sr}, E_b, s_f, s_r, **Slip**

Figure 4. *Homayoun* bond-slip curve.

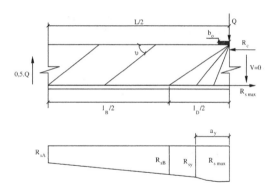

Figure 5. Equivalent beam.

2.3 Concrete constitutive model in FEMOOP

A four parameter model of Ottosen (1977) was used to represent an uncracked material structure. The rupture surface is given by:

$$F = A\frac{J_2}{f_{cm}^2} + \lambda\frac{\sqrt{J_2}}{f_{cm}} + B\frac{I_1}{f_{cm}} - 1 = 0 \tag{4}$$

A linear softening curve law was utilized to represent the cracked region with a rotating smeared-crack model.

The contact and truss finite elements were implemented (discrete and embedded approach in agreement with Elwi & Hrudey (1989) formulations) for the numerical representation of rebars and interfaces.

2.4 Steel reinforcements

In all programs the reinforcements were modeled with the Von Mises plasticity model.

2.5 Steel bonding model in FEMOOP

The Homayoun & Mitchell (1996) multilinear bond-slip model described in Figure 4 was implemented for the representation of interfaces (concrete/steel) behavior. The following parameters are given: τ_{sf} is the interface bond strength, τ_{sr} is the interface residual bond strength, s_f is the interface slip at the peak stress, s_r is the residual interface slip, E_b is the pre-peak bond modulus and E_d is the post-peak bond modulus.

3 THEORETICAL ANALYSIS OF PLASTIC ROTATING CAPACITY

Using the MC-90 the plastic rotation capacity can be defined as:

$$\theta_{pl} = \int_{a=0}^{l_{pl}} \frac{1}{d - x(a)}\left[\varepsilon_{sm}(a) - \varepsilon_{smy}\right]da \tag{5}$$

where $\varepsilon_{sm}(a)$ is the reinforced strain in the member plastification place, ε_{smy} is the reinforced deformation for crack tension equal to f_{yk}.

Using the definition and applying the model $\sigma s(\varepsilon sm)$ by Kreller (1989):

$$\theta_{pl} = \int_{a=0}^{l_{pl}} \frac{0,8}{d - x}\left(1 - \frac{\sigma_{sr1}}{f_{yk}}\right)(\varepsilon_{s2} - \varepsilon_{sy})dx \tag{6}$$

where σ_{sr1} is the reinforced stress in the crack, ε_{s2} and ε_{sy} respectively are the yield deformations in steel reinforcements and cracks.

The Figure 5 represents the process of the plastic rotating capacity using an equivalent beam.

By using the Figure 5 can be calculated the concrete and steel forces:

$$R_c = -M/z + N.z_s/z + 0,5.V.\cot\upsilon$$
$$R_s = M/z + N.(1 - z_s/z) + 0,5.V.\cot\upsilon \tag{7}$$

where υ is the inclination angle of the compression field.

The tensile forces can be derived from the following form:

$$R_s(x) = R_{s,max} - 4.(R_{s,max} - R_{sB}).(x/l_D)^2 \rightarrow x \leq l_D/2$$

$$R_s(x) = R_{sA} + \frac{R_{sB} - R_{sA}}{l_B}.(L - 2x) \rightarrow L/2 \geq x \geq l_D/2 \tag{8}$$

And from this form is found the length of the plasticized zone:

$$a_y = \frac{l_{pl}}{2} = 0,5.l_D.\sqrt{\frac{R_{s\,max} - R_{sy}}{R_{s\,max} - R_B}} \rightarrow R_{sy} > R_{sB} \tag{9}$$

$$a_y = \frac{l_{pl}}{2} = 0,5.\left(L - l_B\frac{R_{sy} - R_{sA}}{R_B - R_{sA}}\right) \rightarrow R_{sy} \leq R_{sB}$$

Solving the integral of Equation (6) the plastic rotation capacity can be calculated:

$$\theta_{pl} = \frac{2.\delta}{3(d-x)}\left(1 - \frac{\sigma_{sr1}}{f_{yk}}\right)\frac{l_D}{A_s E_s}\sqrt{\frac{(R_{s\,max} - R_{sy})^3}{R_{s\,max} - R_{sB}}}$$

$$R_{sy} \geq R_{sB}$$

(10)

$$\theta_{pl} = \frac{2.\delta}{d-x}\left(1 - \frac{\sigma_{sr1}}{f_{yk}}\right)\frac{l_D}{A_s E_s}\left[\frac{l_D}{3}(2.R_{s\,max} - 3.R_{sy} + R_{sB}) + \frac{l_B}{2}\frac{(R_{sB} - R_{sy})^2}{(R_{sB} - R_A)}\right]$$

$$R_{sy} \leq R_{sB}$$

Making the width of the support plate ($b_o = 0$) and leaving the effect of the shear forces ($\cot \upsilon = 0$) arrived the following expressions are found:

$$a_y = \frac{l_{pl}}{2} = \frac{L}{2}\left(1 - \frac{M_y}{M_u}\right)$$

(11)

$$\theta_{pl} = a_y . \frac{M_u - M_y}{(EI)_{pl}} = \frac{L}{2.(EI)_{pl}}\frac{(M_u - M_y)^2}{M_u}$$

4 ANALYSIS OF THE SMEARED CRACK APPROACH

Three reinforced-concrete beams were analyzed in the computational program DIANA, analyzed in the experimental program of Barbosa (1998) every 3,60 m of span, width of 15 cm and height of 28,3 cm, with stirrups of 6 mm a diameter spaced in 8 cm and with applied loads in 1/3 and 2/3 of span. The longitudinal bottom reinforcements consist of two bars of 16 mm a diameter and the longitudinal top reinforcements consist of two bars of 8 mm a diameter. The mechanical properties of these beams is shown in Table 1.

In the modeling the three models of smeared cracks (fixed, rotating and multi-directional) without softening (brittle model) and hardening model in agreement to proposal of Thorenfeldt for two distinct values of the shear retention factor (β). In the fixed model (Figure 6) the use of a smaller shear retention factor implied in a more extended cracking; the variations in β had not modified the maximum values of the crack deformations (ε^{nncr}). In the multi-directional model (Figure 7) the use of a smaller shear retention factor implied in a more extended cracking mainly in the region of pure bending (for $\beta = 0,2$ notices eight wider cracks in this region were noticed; for $\beta = 0,9$ six wider cracks in the same region were noticed); the variations in "β" had modified the maximum values of the crack deformations ("β" is inversely proportional to "ε^{nncr}" in this model).

In the rotating model (Figure 8) the use of a smaller shear retention factor implied in a more extended

Table 1. Material properties.

Beam	f_c(MPa)	f_y(MPa)	E_c(GPa)	E_s(GPa)
1	40	620	38,2	210
2	75	830	42	210
3	100	830	51,2	210

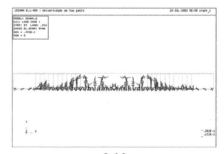

$\beta = 0,2$

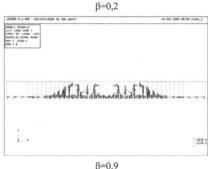

$\beta = 0,9$

Figure 6. Crack pattern for fixed-crack approach.

cracking mainly is of the region of pure bending; variations in "β" had very affected the values of the deformations in the cracks. The fixed, multi-directional and rotating models predicted respectively six, eight and four wider cracks (for $\beta = 0,2$) in the region of pure bending. The experimental crack pattern (Figure 9) shows four wider cracks similar to the ones found with the rotating smeared crack model.

5 ANALYSIS OF THE PLASTIC ROTATION CAPACITY

The plastic rotation capacity for the force-displacement curves was evaluated for experimental and numerical modeling in DIANA, CASTEM and QUE-BRA2D/FEMOOP programs for the three previous beams.

Each "θ_{pl}" was calculated using the Equation (11). The calculated values were disposed in Table 2. In

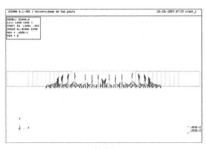

β=0,2

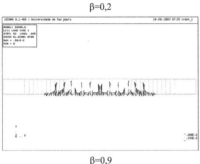

β=0,9

Figure 7.   Crack pattern for multi-directional approach.

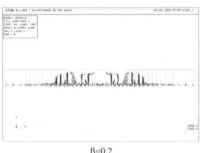

β=0,2

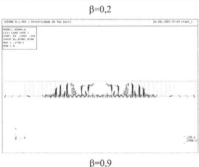

β=0,9

Figure 8.   Crack pattern for rotating model.

Figure 9.   Experimental crack pattern.

Table 2.   Plastic rotation capacity values.

	Beam	a_y(cm)	θ_{pl}(mrad)
Experiment	1	2,63	21,69
	2	13,17	25,86
	3	31,31	33,93
CASTEM	1	1,87	21,02
	2	7,48	28,32
	3	16,62	38,90
DIANA	1	4,32	22,93
	2	1,32	21,29
	3	6,14	28,30
FEMOOP	1	3,36	22,66
	2	9,66	24,29
	3	27,77	29,77

Table 3.   Ductility ratio values.

	Beam	δ_u(mm)	δ_y(mm)	$\mu_d = \delta_u/\delta_y$
Experiment	1	29,52	12,31	2,40
	2	73,53	21,87	3,36
	3	71,94	13,55	5,30
CASTEM	1	29,18	15,16	1,92
	2	70,15	22,89	3,06
	3	69,23	22,30	3,10
DIANA	1	20,5	13,30	1,54
	2	38	23,60	1,61
	3	45,7	21,9	2,09
FEMOOP	1	26,8	14,6	1,83
	2	44,2	22,6	1,96
	3	49,3	18,3	2,69

observed in the values of found for the plastic rotation capacity: bigger values for the numerical simulation in the DIANA and smaller values for CASTEM. The results found with FEMOOP were located between the DIANA and the CASTEM results (closer to the experimental results).

Table 3 illustrates the displacements at the steel yield stress (δy), displacements in the rupture (δu) and global ductility ratio ($\mu = \delta u/\delta y$) from the experimental and numerical analysis.

The results point to a magnifying of the inelastic deformation capacity of beam 2 with regard to beam 1 when it extended concomitantly with the compressive strength and reinforced yield tension.

general, the results from the DIANA were a little more rigid than the experimental results and the results in platform CASTEM were a little less rigid than the experimental ones, whose influence can be

Another magnifying of this ductility can be observed in beam 3 compared to beam 2 where the compressive strength of the concrete only extended itself.

6 CONCLUSIONS

For the fixed rotation crack model the use of a smaller shear retention factor implicates wider cracks; the variations in β don't modify the maximum values of the crack deformations (ε^{nncr}). For the multi-directional crack model the use of a smaller shear retention factor implicates also wider cracking mainly in the region of pure bending (for $\beta = 0{,}2$ eight wider cracks were noticed in this region; for $\beta = 0{,}9$ six wider cracks in the same region were noticed); the variations in "β" modify the maximum values of the crack deformations ("β" is inversely proportional to "ε^{nncr}" in this model).

For the rotating crack model the use of a smaller shear retention factor implicates more extended cracking in the region of pure bending; variations in "β"affect the deformation values in the cracks.

The fixed, multi-directional and rotating models predict respectively six, eight and four wider cracks (for $\beta = 0{,}2$) in the region of pure bending.

The experimental crack pattern (Figure 9) shows four main cracks in agreement with the rotating smeared crack model.

The results can be observed in the values of the plastic rotation capacity parameters: larger values for the numerical simulation in the DIANA and smaller values for the CASTEM. The results obtained with the FEMOOP were found to be between the ones obtained with the DIANA and the CASTEM (closer to the experimental results).

The results point to a magnification of the inelastic deformation capacity of beam 2 with regard to beam 1 when the compressive strength and steel yield stress are increased.

Another magnification of ductility can be observed in beam 3 compared to beam 2 where the compressive strength of the concrete only is increased.

REFERENCES

Alca, N., Alexander, S.D.B. & MacGregor, J.G. 1997. Effect of size on flexural behavior of high-strength concrete beams, ACI Structural Journal, 94(1), 59–67.

Ashour, S.A. 2000. Effect of Compressive Strength and Tensile Reinforcement Ratio on Flexural Behavior of High Strength Concrete Beams. Elsevier Science Ltd., Engineering Structures, pp. 413–423.

Barbosa, M.P. 1998. An Experimental and Numerical Contribution on High Reinforced-concrete Structures: Study of the Anchorage and the Behavior of Bending Beams. Engineering School, São Paulo State University, FEIS, Unesp, 174 p.

Comité Euro – International du Béton. 1993. Model Code for Concrete Structures. CEB – FIP MC 90, 437 p.

Corley, G.W. 1966. Rotational capacity of reinforced-concrete beams, ASCE, 92(5), 121–146.

Elwi, A.E. & Hrudey, T.M. 1989. Finite element model for curved embedded reinforcement. Journal of Engineering Mechanics, v. 115, n. 4, pp. 740–754.

Hillerborg, A. 1988. Rotational capacity of reinforced-concrete beams, Norwegian Concrete Research, publication n. 7, 121–134.

Homayoun, H.A. & Mitchell, D. 1996. Analysis of bond stress distributions in pullout specimens. Journal of Structural Engineering, v. 122, n. 3, pp. 255–261.

Kani, G.N.J. 1967. How safe are our large reinforced-concrete beams, ACI Journal, 64(3), 128–141.

Kreller, H. 1989. Zum nicht-linearen Trag- und Verformungsverhalten von Stahlbetonstabtragwerken unter Last- und Zwangeinwirkung. Dissertation, Universitat Stuttgart.

Leslie, K.E., Rajagopalan, K.S. & Everard, N.J. 1976. Flexural behavior of high strength concrete beams, ACI Structural Journal, 73(9), 517–521.

Ottosen, N.S. 1977. A failure criterion for concrete. Journal of the Engineering Mechanics Division, v. 103, n. 4, pp. 527–535.

Shin, S.W., Gosh, S.K. & Moreno, J. 1989. Flexural ductility of ultra high strength concrete members, ACI Structural Journal, 86(4), 394–400.

Tognon, G., Ursella, P. & Coppetti, G. 1980. Design and properties of concretes with stregth over 1500 kgf/cm². ACI Structural Journal, 77(3), 171–178.

Walraven, J. & Lehwalter, N. 1994. Size effects in short beams loaded in shear, ACI Structural Journal, 91(5), 585–593.

Fracture Mechanics of Concrete and Concrete Structures – Design, Assessment and Retrofitting of RC Structures – Carpinteri, et al. (eds)
© *2007 Taylor & Francis Group, London, ISBN 978-0-415-44616-7*

Studies on ductility of RC beams in flexure and size effect

G. Appa Rao* & I. Vijayanand
University of Stuttgart, Stuttgart, Germany
Indian Institute of Technology Madras, Chennai, India

R. Eligehausen
University of Stuttgart, Stuttgart, Germany

ABSTRACT: This paper reports on some experimental investigations on ductility of reinforced concrete beams in flexure and evaluation of size effect. The minimum flexural reinforcement has been evaluated from experimental observations on ductility of reinforced concrete (RC) beams. Beams of depth 100 mm, 200 mm, 400 mm at different flexural reinforcements namely 0.15, 0.30, 0.60 and 1.0% were tested under uniform bending moment. The beams were made of 30 MPa concrete. The cracking and ultimate flexural strength, influence of beam depth on ductility and rotation capacity have been analyzed. The size of RC members has a significant influence on flexural behaviour. The variation of cracking strength is not very conclusive in small-size beams, where as it decreases as depth of beam increases beyond 200 mm. The ultimate flexural strength has been observed to decrease as the beam size increases. As the flexural reinforcement ratio in beams increased the ductility of beams was observed to increase. Ductility of RC members decreases with increase of beam size. The optimum flexural reinforcement has been obtained from an optimum ductility number, N_p, equal to 0.20. The minimum flexural reinforcement was observed to decrease as the beam depth increased, and decreased as the yield strength of steel reinforcement increased.

1 INTRODUCTION

The expressions for minimum steel ratios, both in flexure and shear, prescribed by various codes of practice are basically empirical. Due to lack of rational approach, the code provisions for design of RC members show vast variation in the design values. A minimum area of reinforcement is generally required in flexural members to prevent cracking and excessive deflections due to various loading effects. Two parameters such as tensile strength of concrete and yield strength of reinforcement are incorporated in the expressions for predicting minimum flexural reinforcement in RC beams. However, the effect of size of RC members is not considered. In some lightly reinforced beams, the cracking moment, as a plain concrete beam, may exceed the yielding moment of the beam, as an RC beam, at first cracking. This criterion has been considered for evaluating the minimum flexural steel ratio by ACI code (ACI-318-2005). The Indian standard (IS: 456-2000) specifies minimum flexural reinforcement to avoid sudden failures of RC members based on simple assumption that the yielding moment of RC beam, M_y is greater than or equal to cracking moment, M_{cr}, of plain concrete beams.

Recently, a theoretical evaluation of minimum longitudinal and transverse reinforcement ratios in beams subjected to flexure, shear and torsion, associated with ductility and minimum strength at ultimate limit state (ULS) has been reported (Shehata et al., 2003). The behavior of RC beams is useful to postulate a rational approach to estimate minimum flexural reinforcement based on rational approach. Carpenteri (1984) defined a brittleness number to determine the minimum flexural reinforcement in RC beams using fracture mechanics principles. The brittleness number is defined as "N_p"

$$N_p = \frac{f_y h^{\frac{1}{2}}}{K_{Ic}} \left[\frac{A_s}{A} \right] \tag{1}$$

where f_y = yield strength of steel reinforcement, K_{Ic} = concrete fracture toughness, A_s = area of steel reinforcement, A = area of c/s of beam, h = beam depth.

The brittleness of structural member increases as its size increases and/or the reinforcement ratio decreases. Physically similar behaviour was revealed in some cases where the brittleness number N_p was the same. At a value of N_p equal to 0.26 using 91 MPa concrete, the yielding moment was more or less equal to the first cracking moment of the beam. The reinforcement corresponding to this condition was considered for evaluating the minimum reinforcement for flexural members. The minimum percentage reinforcement tends to be inversely proportional to the beam depth, while the values by codes are independent of the beam depth. Strain localization has been taken into account for the analysis of RC beams by Hillerborg (1990). The descending portion occurs due to crack formation within fracture process zone. The analysis of RC beams for balanced reinforcement ratio decreases with increasing beam depth.

Bosco et al. (1990) reported that the minimum flexural reinforcement corresponds to a condition at which the formation of first flexural cracking and yielding of steel reinforcement occur simultaneously. The brittleness of RC beams increases as the size of beam increases and as steel ratio decreases. An optimum value of N_p has been observed for estimating the minimum flexural reinforcement corresponding to the above condition. The minimum steel ratio is inversely proportional to depth of beam.

Bosco et al. (1990) reported that the brittleness of beams increases by increasing size-scale and/or decreasing steel area. For low N_p values in lightly reinforced beams or for small cross sections the fracture moment decreases while the crack extends. The peak or first cracking load is lower than the steel-yielding load only at high brittleness number. The size of beam seems to govern the post peak behavior, especially for low brittleness number for larger beam depth.

Baluch et al. (1992) proposed a criterion for minimum flexural reinforcement with which unstable crack propagation was avoided. This is achieved by ensuring that the moment corresponding to the maximum load, in a reinforced concrete beam is greater than its cracking moment as a plain concrete beam. The expression proposed to predict the minimum flexural reinforcement is

$$\rho_{min} = \frac{1.9134 K_{lc}^{0.82}}{f_y^{0.9922}\left(1.7 - 2.6c_s / D\right)} \qquad (2)$$

Gerstle et al. (1992) used fictitious crack model to study tensile cracking behaviour of singly reinforced concrete beams in flexure. A theoretical analysis was performed to plot normalised moment with the normalised crack length for different values of "β" (a measure of brittleness) and "α" (a measure of steel percentages). Stable crack propagation has been associated with a continuously increasing curve and that value of "α" corresponds to minimum flexural reinforcement. An expression for minimum reinforcement

Table 1. Properties of cement.

S.No.	Property	Results
1	Compressive	43.0
2	Strength	N/mm^2
3	Fineness	3.5
4	Initial setting time	205 min
5	Final setting time	335 min
6	Specific gravity	3.12

ratio has been proposed defined but it does not contain f_y as shown below.

$$\rho_{min} = \frac{E_c}{E_s}\left(\sqrt{0.0081 + 0.0148\,\frac{f_t D}{E_c w_c}} - 0.0900\right)^{1/2} \qquad (3)$$

2 RESEARCH SIGNIFICANCE

The design of members based on strength criteria does not consider fracture mechanics theory. Failures in RC members exhibit different modes due to change of beam depth and percentage reinforcement. The ductility of RC members changes with size of member and strength of concrete. However, this is not considered in the design of structural members. In other words, failure according to strength theory should not show any size dependence, nor should the size of beam have any effect on its ductility. The effect of size of member and ductility in design of RC members can be predicted by fracture mechanics. However, there exists a controversy in the evaluation of minimum flexural reinforcement in RC members. An attempt has been made to understand size effect on ductility and minimum reinforcement of lightly reinforced beams.

3 EXPERIMENTAL PROGRAMME

3.1 Materials

An Ordinary Portland Cement (OPC) was used for the present study. The properties of cement are presented in Table 1. The fine aggregate was obtained from a natural river bed. The aggregate fraction passing through sieve size 1.18 mm and retained on 600 μ size was used in concrete. The specific gravity and fineness modulus of sand are given in Table 2. The machine crushed granite aggregate was used for concreting, consisting of mixture of 10 and 20 mm size particles. The properties of aggregate are given in Table 3. Potable water was used for mixing of concrete and curing of specimens. The pH value of water was 7.8.

The flexural reinforcement was high strength steel reinforcement with 415 N/mm^2 guaranteed yield strength. The diameter of the bars varied from 3 to

Table 2. Properties of fine aggregate.

S.No	Property	Result
1	Specific gravity	2.78
2	Fineness modulus	2.82

Table 3. Properties of coarse aggregate.

S. No	Property	Results
1	Specific gravity	2.70
2	Fineness Modulus	6.84

Table 4. Beam dimensions and reinforcement details.

Beam	σ_y, MPa	A_{st} Provided (%) mm^2	No. of bars	Stirrups Spacing, mm
A1	637	(0.30) 14	2–3 mm	3 mm @150
B1	637	(0.15) 28	4–3 mm	3 mm @ 300
C1	389	(0.15) 113	4–6 mm	6 mm @ 150
A2	637	(0.3) 14	2–3 mm	3 mm @150
B2	389	(0.3) 56.5	2–6 mm	6 mm @ 140
C2	459	(0.3) 226	2–12 mm	6 mm @ 300
A3	637	(0.6) 28	4–3 mm	3 mm @150
B3	389	(0.59) 113	4–6 mm	6 mm @ 140
C3	459	(0.59) 452	4–12 mm	6 mm @ 130
A4	389	(1.19) 56.5	2–6 mm	6 mm @ 70
B4	577	(1.0) 756	6–12 mm +1–10 mm	8 mm @ 175
C4	459	(1.0) 756	6–12 mm +1–10 mm	8 mm @ 175

12 mm depending on the size of beam. The nominal shear reinforcement consists of MS bars of diameter from 3 mm to 6 mm depending on the size of beam.

3.2 Specimen details

Rectangular beam specimens of different depths were adopted. In order to maintain geometric similarity, the aggregate size was varied depending on the depth. In small beams of size 50 mm × 100 mm × 500 mm and 100 mm × 200 mm × 1000 mm, 10 mm aggregate was used, while 20 mm aggregate was used in large beams of size 200 mm × 400 mm × 2000 mm. The ratio of reinforcement cover-to-depth was 0.05 in all the beams.

The details of beam specimen and steel reinforcement are given in Table 4. The concrete mix proportions were 1: 2.75: 5.1 respectively cement: fine aggregate: coarse aggregate. The cement content was 250 kg/m^3 and the water cement ratio was 0.75. The compressive strength of concrete at 28

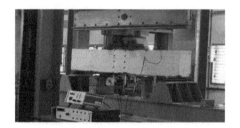

Figure 1. Experimental set-up.

days on 100 mm size cubes was 30 N/mm^2. The split tensile strength of 150 mm × 300 mm cylinders was 2.62 N/mm^2. The beam as well as companion cube and cylindrical specimens were cured in water for 28 days. The specimens were tested after 28 days. The steel reinforcement consisted of 3 mm, 6 mm, 10 mm, and 12 mm diameter bars as flexural reinforcement. The actual yield strength of reinforcement was used to calculate the flexural strength of beams. The yield strengths of 6 mm, 10 mm, and 12 mm diameter bars were 577, 483 and 459 N/mm^2 respectively. The ductility number defined by Carpenteri (1984) was used for evaluation of minimum flexural reinforcement. Fracture energy, G_F of plain concrete was determined on three-point bend specimens (depth, $d = 100$ mm, width, $b = 100$ mm, and span, $l = 500$ mm). The notch-to-depth ratio was 0.5 and the notch width was 3 mm. The mean value of fracture energy was 150 N/m. The critical stress intensity factor was evaluated as $K_{1C} = \sqrt{G_F} \sqrt{E}$ and was equal to 64.09 N/mm$^{3/2}$. The modulus of elasticity of concrete was 27.40 GPa.

A total of 20 RC beams with depth equal to 100, 200 mm and 400 mm, maintaining the depth-to-width ratio 2.0 were tested. For a particular parameter, two beams in class A and B and only one beam in class C were tested. The span between supports was equal to five times beam depth, d. Therefore, the spans measured 500, 1000 and 2000 mm respectively in beams (for class A: d = 100 mm, b = 50 mm; l = 500; class B: d = 200 mm, b = 100 mm; l = 1000 mm; and class C: d = 400 mm, b = 200 mm; l = 2000 mm). The reinforcement was estimated from ductility numbers selected i.e. 0.091, 0.183, 0.366 and 0.732. Steel moulds were used for casting the beams of required dimensions.

3.3 Experimental set-up and testing

Four-point loading set-up was used for testing of RC beams as shown in Figure 1. Statically determinate system was ensured by adopting hinge and roller supports at two ends. The load was applied through a hydraulic jack at constant load increments. The load was applied symmetrically at one third points. LVDT was used to measure the deflection at the center of beams.

The load at first cracking was visualized by means of magnifying glass in the uniform bending moment region between two central loading points where the first flexural crack was formed. The strains along the depth of the beam were measured using demountable mechanical gage. The crack propagation was monitored on the beam surface and the crack width was measured by a microscope. The ultimate moments of the beam were estimated both theoretically and experimentally.

4 RESULTS AND DISCUSSION

4.1 Load-deflections curves

It was noticed that the deflection of beams at failure increased as the percentage steel reinforcement increased from 0.15% to 1.0%. It was observed that the ductility of beams increased as the flexural reinforcement increased. The beam with 0.30% reinforcement exhibited large deflection at failure showing increased rotation capacity and ductility. Similar trend was observed in all the beams i.e. deflections increase with percentage flexural reinforcement keeping other parameters constant. At 0.30% reinforcement the beams exhibited improved ductility. Interestingly, the nature of failure changed from ductile to brittle as the depth of beam increased. The large size beams exhibited relatively small deflection at failure. As the size of beam increased, keeping the percentage flexural reinforcement constant, the deflections at failure decreased. This shows that as the beam size increases the failure of beams turned from ductile to brittle. The ductility of beams was found to decrease with increase of depth. Further it was observed that as the percentage reinforcement increased at a given beam depth the ductility increased.

4.2 Flexural strength

The flexural strength is defined as the flexural capacity of the beam at the ultimate load. In this case, nominal strength of beam at the ultimate load was represented by its flexural strength. The nominal strength was calculated by dividing the ultimate load by square of depth of beams in three dimensional similarities.

Figure 2 shows the variation of flexural strength calculated as $(M_u/f_{ck}\ bd^2)$ with depth of beam at different flexural reinforcements i.e. 0.15, 0.30, 0.60 and 1.0%. The flexural strength has been observed to decrease with depth at 0.15 and 0.30% reinforcements. However, the nondimensional flexural strength has been observed to decrease as the beam depth increased at 0.60 and 1.0% reinforcements. This shows that a general size effect law has been possible in R.C. beams in flexure at small percentage reinforcements. However, at heavy flexural reinforcements, the effect of size needs further studies. At this juncture it would not

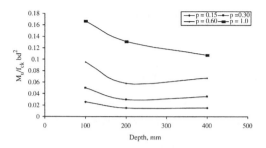

Figure 2. Moment ratio vs. depth at different steel ratio.

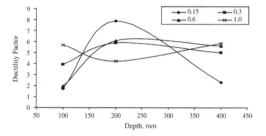

Figure 3. Ductility factor vs. beam depth.

be possible to conclude the exact size effect on flexural strength of reinforced concrete beams with heavy reinforcement.

4.3 Ductility factor

Ductility factor may be defined as the ratio of deflection at failure to the deflection at yield or at the first crack. As there is no information on the effect of size on ductility of reinforced concrete beams, the present study was undertaken. In codes of practice, the design strength of RC members in flexure is considered to be constant. When the concepts of fracture mechanics are used, there could be an improved safety margin against failure and the prediction of failure could be possible with reasonable reliability. Figure 3 shows variation of ductility factor with size of structure at 0.15, 0.30, 0.60 and 1.0% reinforcement. It demonstrates that the ductility factor increases as the beam size increases from 100 mm to 200 mm. Thereafter, the ductility factor decreases with size.

At small flexural reinforcement ratios, the ductility factor has been observed to be the highest at 200 mm beam depth. At higher percentages of reinforcement, the trend seems to be increasing with size of structure. At 100 mm depth, the ductility at all percentages of reinforcement was found to be the lowest. However, as the depth of the beam increases beyond 200 mm, the ductility factor has been showing size dependence. Further at small percentage of reinforcement, the ductility factor increases with increasing percentage reinforcement.

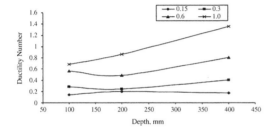

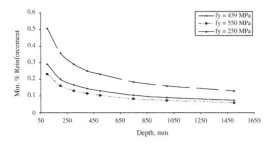

Figure 4. Ductility number vs. beam depth.

Figure 5. Min reinforcement vs. beam depth.

4.4 *Ductility number*

The ductility number is a measure of brittleness of R.C. beams. The study was designed to investigate the effect of beam size and percentage reinforcement on ductility of R.C. beams. In order to evaluate the effect of beam size on ductility three different beam sizes i.e. 100 mm, 200 mm and 400 mm were adopted by varying the flexural reinforcement at 0.15, 0.30, 0.60 and 1.0%. Figure 4 shows the ductility number with beam depth at all percentages of reinforcement. It demonstrates that the ductility number increases with size of beam at a given percentage flexural reinforcement. The ductility number has been found to increase as the beam depth increased. Similar trend has been observed in the case of beams reinforced with 0.6 and 1.0% flexural reinforcement. It was observed from the load-deflection curves that the beams turned brittle with increase in depth. The increase of the ductility number at 0.6 and 1.0% was significantly higher at a given size of beams. The ductility number for 200 mm deep beams was 0.198, beyond which it was found to decrease with increase in beam depth. This value of 0.198 is the optimum value for achieving minimum required ductility for evaluation of the minimum reinforcement.

Figure 5 shows the minimum flexural reinforcement with beam depth at the optimum ductility, which corresponds to the value of N_p equal to $0.198 \approx 0.20$. As the strength of concrete increases the brittleness of beam increases due to material brittleness. Similar observations have been made in high strength concrete beams by Bosco et al. (1990),where the ductility number for

achieving the minimum flexural reinforcement was 0.26. Therefore, in order to maintain the minimum ductility in RC members, the percentage reinforcement should be a function of strength of concrete, depth of beam and yield strength of steel reinforcement.

5 CONCLUSIONS

The following conclusions were drawn from the studies on lightly reinforced concrete beams.

1. The effect of size and percentage reinforcement on the ultimate strength of RC beams has been found to be significant. The ultimate strength is inversely proportional to the beam depth.
2. As the percentage flexural reinforcement increases, the ultimate load and the corresponding the beam deflections increase. As the depth of beam increases the ductility factor decreases.
3. The ductility number of RC beams increases with increasing beam depth and with decreasing percentage reinforcement. The optimum ductility number is 0.20 in 30 MPa concrete.
4. The minimum percentage reinforcement is inversely proportional to beam depth. It indicates that the formula for minimum steel reinforcement provided by the codes needs to be modified.

REFERENCES

ACI 318-2005. Building code requirements for structural concrete and commentary, *ACI* 1995 Farmington Hills, MI:

Baluch, M. H., Azad, A. K. and Ashmawi, W. 1992. Fracture mechanics application to reinforced concrete members in flexure. *Proc. Int Workshop on application of fracture mechanics to reinforced concrete*, Italy, 413–436.

Bosco, C., Carpinteri, A., and Debernardi, P.G. 1990. Minimum reinforcement in high strength concrete. *Journal of Structural Engg* 116(2): 427–437.

Bosco, C., Carpinteri, A. and Debernardi, P.G. 1990. Fracture of reinforced concrete; scale effects and snap – back instability. *Engg. Frac Mech* 35(4/5): 665–677.

Carpinteri, A. 1984. Stability of fracturing process in RC beams, ASCE Jl of Str Engg, 110 (3), 544–558

Gerstle, W. H., Dey, P. P., Prasad, N. N. V., Rahulkumar, P. and Xie, M. 1992. Crack growth in flexural members-A fracture mechanics approach. *ACI Str Jl*, 89(6): 617–625.

Hillerborg, A. 1990. Fracture mechanics concepts applied to moment capacity and rotational capacity of reinforced concrete beams. *Eng. Fra. Mech* 35(1/2/3): 233–240.

IS 456-2000. Code of practice for design of plain and reinforced concrete structures, *BIS,* New Delhi.

Shehata, I.A.E.M., Shehata, L.C.D. and Garcia, S.L.G. 2003. Minimum steel ratio in RC beams made of concrete with different strength–theoretical approach. *Mat Str* 36: 03–11.

Fracture Mechanics of Concrete and Concrete Structures – Design, Assessment and Retrofitting of RC Structures – Carpinteri, et al. (eds)
© 2007 Taylor & Francis Group, London, ISBN 978-0-415-44616-7

Flexural to shear and crushing failure transitions in RC beams by the bridged crack model

A. Carpinteri & G. Ventura
Politecnico di Torino, Torino, Italy

J.R. Carmona
Universidad de Castilla-La Mancha, Ciudad Real, Spain

ABSTRACT: The *bridged crack model* has been developed for modelling the flexural behaviour of reinforced concrete beams and related size effects explaining brittle-ductile-brittle failure mode transitions. In the present paper the model is extended to analyze shear cracks and concrete crushing, introducing a given shape for the hypothetical crack trajectory and determining the initial crack position and the load versus crack length curve for three point bending problems. The proposed formulation reproduces the pure Mode I flexural behaviour as a particular case, so that the flexural and the diagonal tension (shear) failures modes can be immediately compared to detect which one dominates and determine the relevant failure load. A concrete crushing criterion completes the model. All the mutual transitions between the different collapse mechanisms can be predicted. In the paper these transitions are shown by varying the governing nondimensional parameters.

1 INTRODUCTION

For a long time, the transitions between flexural and diagonal tension failures in reinforced concrete elements – inside a consistent theoretical framework – have represented an unsolved problem. The main issue for the present analysis is to get a consistent modelling of shear cracks behavior and diagonal tension failure as well as concrete crushing mechanisms. These problems, despite numerous extensive studies over the past 50 years, still remain unsolved for a completely satisfying framework, unifying all the failure modes, so that a direct relation between failure mode transition could be drawn.

Shear crack propagation and diagonal tension failure have been addressed in the literature by several authors with different approaches. In the field of Fracture Mechanics and using a cohesive model to describe concrete behaviour, some analyses have been performed by Gustafson and Hillerborg (Gustafsson and Hillerborg 1983) and Niwa (Niwa 97) among others. In the framework of Linear Elastic Fracture Mechanics and in order to avoid finite element computations, some models are especially remarkable in the long list of literature contributions. In particular, Jenq and Shah (Jenq and Shah 1989) analysed the diagonal shear fracture superposing the contribution of concrete and steel bars, with a technique that is somehow conceptually

close to the *bridged crack model* (Carpinteri 1984). Some further development of this work with other original contributions were made by So and Karihaloo (So and Karihaloo 1993).

The *bridged crack model* has been originally proposed by Carpinteri (Carpinteri 1981; 1984) for the study of reinforced concrete beams by Fracture Mechanics. The problem of the size effect and the brittle-ductile transition were analyzed with reference to the problem of minimum reinforcement (Carpinteri et al. 1999; Bosco and Carpinteri 1992). Subsequently, the action of cohesive stresses has been introduced in addition to that of the reinforcing bars (Carpinteri et al. 2003). More recently, the model has been further extended analysing concrete crushing by Fracture Mechanics concepts (Carpinteri et al. 2004) and leading to analyse in a consistent way the interaction between flexural (yielding) and crushing failures. Moreover, while limit state analysis yields only the ultimate load, the *bridged crack model* reveals in addition scale effects, instability phenomena and brittle-ductile failure transition of the structural member.

In the present work, the behaviour of reinforced concrete beams without stirrups is analyzed, using the *bridged crack model*. To extend the model to account for the shear cracks behavior and to evaluate diagonal tension failure load, some additional hypotheses about

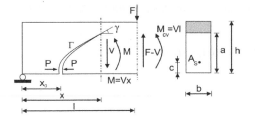

Figure 1. Cracked element.

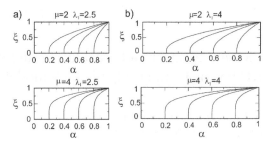

Figure 2. Crack trajectories, (a) $\lambda_l = 2.5$; (b) $\lambda_l = 4$.

the crack trajectory and for the evaluation of the stress-intensity factors are assumed. In this way the different collapse modes are joined together into a unified general model, so that the simulation of the transitional phenomena is naturally accomplished. The model is analysed showing the influence of the variation in the nondimensional parameters on the mechanical response of the reinforced concrete element and the related failure mode transitions.

2 MODELLING OF FLEXURAL AND SHEAR CRACKS

The *bridged crack model* can be applied for studying the propagation of a crack in a reinforced concrete beam assuming as monotonically increasing control parameter the length of the crack path. Linear Elastic Fracture Mechanics is assumed for concrete with a crack propagation condition ruled by the comparison of the stress intensity factors K_I to the concrete toughness K_{IC}. Neither closed form solution nor nonlinear regressions of numerical data are available for evaluating the stress-intensity factors of the geometry given in Figure 1. Therefore, some assumptions have been made to derive suitable approximations.

The adopted model scheme is reported in Figure 1 along with the used symbols. The geometric dimensions are converted into nondimensional quantities, after dividing by the height h in the case of vertical distances and by the shear span l in the case of horizontal distances. Thus, the following nondimensional parameters are defined: let $\alpha = \frac{x}{l}$ be the nondimensional horizontal distance from the support to the crack tip, $\xi = \frac{a}{h}$ the crack depth, $\alpha_0 = \frac{x0}{l}$ the initial crack mouth position and $\zeta = \frac{c}{h}$ the reinforcement cover. All these nondimensional parameters range from 0 to 1. Additionally, let λ_l be the shear span slenderness ratio ($\lambda_l = \frac{l}{h}$).

The crack trajectory Γ is considered as formed by a first vertical segment Γ_1 from the bottom to the reinforcement layer. A second part Γ_2 is assumed being a power law with some given exponent, going from the end of the first part to the load point, Figure 2.

The crack trajectory $\Gamma = \Gamma_1 \cup \Gamma_2$ is defined by the nondimensional function:

$$\alpha(\zeta, \xi) = \begin{cases} \alpha_0 & 0 \leq \xi \leq \zeta \\ \alpha_0 + \left(\dfrac{\xi - \zeta}{1 - \zeta}\right)^{\mu}(1 - \alpha_0) & \zeta \leq \xi \leq 1 \end{cases} \quad (1)$$

The constitutive relation for the reinforcement bars is assumed rigid-perfectly plastic with no upper limit to the maximum deformation. The maximum for the bridging reinforcement reaction is defined by $P_P = A_s \sigma_y$, where A_s is the reinforcement area and σ_y the minimum between yielding and sliding stress for the bars.

With reference to Figure 1, let K_I be the stress intensity factor at the crack tip. By superposition, it is given by the sum of the stress intensity factor K_{IV}, due to the bending moment associated to the shear force V, and K_{IP_γ}, due to the closing force at the reinforcement position:

$$K_I = K_{IV} - K_{IP_\gamma} \quad (2)$$

To evaluate the stress intensity factor due to the external load K_{IV}, Jenq and Shah (Jenq and Shah 1989) assumed that it can be approximated by the stress-intensity factor of a bent beam with a symmetric edge notch of depth a subjected to the bending moment corresponding to the cross section at the mouth of the crack. Here a similar approach is followed, although the variation of the bending moment at each section due to the crack path will be accounted for. Therefore:

$$K_{IV} = \frac{V l \alpha(\zeta, \xi)}{h^{\frac{3}{2}} b} Y_M(\xi) = \frac{V}{h^{\frac{1}{2}} b} Y_V(\zeta, \xi) \lambda_l \quad (3)$$

The stress-intensity factor produced at the crack tip by the applied forces P acting at the level of the reinforcement is obtained from the case of vertical crack, K_{IP}. Several numerical analyses by boundary elements (Portela and Aliabadi 1992) have been made to get an approximation to the stress-intensity factor for different positions of the crack tip. It is observed that the stress intensity factor is a function of the angle γ, see

Figure 1. Consequently, a function $\beta(\gamma)$ to approximate the variation Y_P with γ has been defined. Finally, the stress intensity factor due to the reinforcement reaction P is given by

$$K_{IP_\gamma} = \frac{P}{h^{\frac{1}{2}}b} Y_P(\zeta,\xi)\beta(\gamma) = \frac{P}{h^{\frac{1}{2}}b} Y_{P_\gamma}(\zeta,\xi) \qquad (4)$$

with $\beta(\gamma) = (\frac{\gamma}{90})^{0.2}$ and γ expressed in degrees.

The functions $Y_M(\xi)$ and $Y_P(\zeta,\xi)$ are given in the stress intensity factor handbook (Okamura et al. 1975).

Let ρ be the bar reinforcement percentage defined as $\rho = \frac{A_s}{bh}$ and w the crack opening at reinforcement level. The following nondimensional parameters can be defined

$$N_P = \frac{\sigma_y h^{\frac{1}{2}}}{K_{IC}}\rho; \qquad \tilde{w} = \frac{wE}{K_{IC}h^{\frac{1}{2}}} \qquad (5)$$

where N_P is the *brittleness number* defined by Carpinteri (Carpinteri 1981; 1984) and E is the Young's modulus of the material.

Substituting Eqs. (3) and (4) into Eq. (2), the following nondimensional equilibrium equation is obtained

$$\tilde{V}_F = \frac{1}{\lambda_l Y_V(\xi)} \left[1 + N_P \tilde{P} Y_{P_\gamma}(\zeta,\xi) \right] \qquad (6)$$

where $\tilde{V}_F = V_F/(K_{IC}h^{\frac{1}{2}}b)$ and $\tilde{P} = P/P_P$.

The crack opening, $\tilde{w}$ at the nondimensional coordinate ζ can be determined by adding the two contributions of shear $\tilde{V}$ and bar reaction $\tilde{P}$. The nondimensional opening evaluated at the crack propagation shear $V = V_F$, presents the following expression:

$$\tilde{w} = 2\lambda_l \tilde{V}_F \int_\zeta^\xi Y_V(z)Y_{P_\gamma}(\zeta,z)g(\zeta,z)\,\mathrm{d}z -$$

$$2N_P \tilde{P} \int_\zeta^\xi Y_{P_\gamma}^2(\zeta,z)g(\zeta,z)\,\mathrm{d}z \qquad (7)$$

where $g(\zeta,\xi)$ is the Jacobian mapping the curvilinear integral along the crack trajectory onto the interval $[0,\xi]$ (Carpinteri et al. 2006).

If the relative displacement in the cracked cross-section at the level of reinforcement is assumed to be equal to zero, up to the yielding or slippage of the reinforcement ($\tilde{w} = 0$), we obtain the displacement compatibility condition that allows us to obtain the unknown force $\tilde{P}$ as a function of the applied shear $\tilde{V}$. In fact, from Eq. (7), we may define:

$$r''(\zeta,\xi) = \frac{\lambda_l \tilde{V}_F}{N_P \tilde{P}} = \frac{\int_\zeta^\xi Y_{P_\gamma}^2(\zeta,z)g(\zeta,z)\,\mathrm{d}z}{\int_\zeta^\xi Y_V(z)Y_{P_\gamma}(\zeta,z)g(\zeta,z)\,\mathrm{d}z} \qquad (8)$$

If the force transmitted by the reinforcement is equal to $P_P = \sigma_y A_s$, in other words, if the reinforcement traction limit has been reached ($\tilde{V}_F = \tilde{V}_P$), from (6) we obtain:

$$\tilde{V}_P = \frac{1}{\lambda_l Y_V(\xi)} \left[1 + N_P Y_{P_\gamma}(\zeta,\xi) \right] \qquad (9)$$

On the other hand, if $\tilde{V}_F < \tilde{V}_P$, the following relation holds from Eqs. (6) and (8):

$$\tilde{V}_F = \frac{1}{\lambda_l \left[Y_V(\xi) - \dfrac{Y_{P_\gamma}(\zeta,\xi)}{r''(\zeta,\xi)} \right]} \qquad (10)$$

Therefore, according to the model, when $\tilde{V}_F < \tilde{V}_P$ the shear of crack propagation $\tilde{V}_F$ depends only on the relative crack depth ξ, and is not affected by the brittleness number N_P.

3 MODELLING CONCRETE CRUSHING

The problem of concrete crushing in the upper part of the beam is analyzed evaluating the compressive stress in the cracked element. Concrete crushing will be detected by comparing the stress σ_c to the crushing strength σ_{cu}.

The compressive stress at the upper edge of the cracked section is the sum of the contributions due to shear and reinforcement reaction:

$$\sigma_c = \sigma_c^V + \sigma_c^P \qquad (11)$$

Introducing two suitable shape functions $Y_\sigma^M(\xi)$ and $Y_\sigma^P(\zeta,\xi)$ (Carpinteri et al. 2003) and letting $Y_\sigma^V(\xi) = \alpha(\zeta,\xi)Y_\sigma^M(\xi)$, the following expression is derived

$$\sigma_c = \lambda_l \frac{V}{bh} Y_\sigma^V(\xi) - \frac{P}{bh} Y_\sigma^P(\zeta,\xi) \qquad (12)$$

Let $V = V_C$ be the concrete crushing load, attained when $\sigma_c = \sigma_{cu}$. In nondimensional form we may write:

$$\frac{\sigma_{cu} h^{\frac{1}{2}}}{K_{IC}} = \tilde{V}\lambda_l Y_\sigma^{Vc}(\xi) - N_P \tilde{P} Y_\sigma^P(\zeta,\xi) \qquad (13)$$

Consequently, in the same way as for the steel yielding mechanism, a *brittleness number* for the crushing failure can be naturally defined:

$$N_C = \frac{\sigma_{cu} h^{\frac{1}{2}}}{K_{IC}} \qquad (14)$$

so that the nondimensional shear for compression failure is given by

$$\tilde{V}_C = \frac{1}{\lambda_l Y_\sigma^V(\xi)} \left[N_C + N_P \tilde{P} Y_\sigma^P(\zeta,\xi) \right] \qquad (15)$$

To eliminate the dependence on $\tilde{P}$ in (15), we may observe that, at steel yielding, it is $\tilde{P} = 1$ and therefore, from Eq. (15):

$$\tilde{V}_C = \frac{1}{\lambda_l Y_\sigma^V(\xi)}\left[N_C + N_P Y_\sigma^P(\zeta,\xi)\right] \quad (16)$$

In the same way, when $\tilde{P} < 1$, from (8) it is:

$$\tilde{V}_C = \frac{1}{\lambda_l Y_\sigma^V(\xi)}\left[N_C + \frac{Y_\sigma^P(\zeta,\xi)}{Y_V(\xi)r''(\zeta,\xi) - Y_{P_\gamma}(\zeta,\xi)}\right] \quad (17)$$

The nondimensional shear of Eqs. (16) and (17) produces, for a given crack depth ξ, the crushing stress $\sigma_c = \sigma_{cu}$ in the uppermost part of the beam. On the other hand, for equilibrium and compatibility being satisfied, only the non-dimensional shear of crack propagation, Eqs. (9) and (10), is compatible with a given crack depth ξ, so that crushing failure occurs when the crack propagation non-dimensional shear, Eqs. (9) and (10), is equal to the non-dimensional crushing shear, Eqs. (16) and (17) respectively.

For $\tilde{V}_C = \tilde{V}_F \geq \tilde{V}_P$, we have:

$$\frac{N_C + N_P Y_\sigma^P(\zeta,\xi)}{Y_\sigma^V(\xi)} = \frac{1 + N_P Y_P(\zeta,\xi)}{Y_V(\xi)} \quad (18)$$

as well as, for $\tilde{V}_C = \tilde{V}_F < \tilde{V}_P$, it is:

$$\frac{N_C + \frac{Y_\sigma^P(\zeta,\xi)}{[Y_V(\xi)r''(\zeta,\xi) - Y_{P_\gamma}(\zeta,\xi)]}}{\lambda_l Y_\sigma^V(\xi)} = \frac{1}{Y_V(\xi) - \frac{Y_{P_\gamma}(\zeta,\xi)}{r''(\zeta,\xi)}} \quad (19)$$

Equations (18) and (19) determine the points of crushing failure in in the crack depth vs. non-dimensional shear diagram.

4 FLEXURAL AND SHEAR CRACK PROPAGATION

In this section it will be shown how the value of the initial crack position α_0 affects the mechanical response of the beam and implies the stability/instability of the cracking process.

For the sake of clarity, reference is made to a real example, based on experimental results. A more detailed explanation can be found in (Carpinteri et al. 2006). The experimental test has been performed by Bosco and Carpinteri (Bosco and Carpinteri 19992), and it was labeled as B100-06. The material properties and beam geometry of this test are shown in Figure 3a.

The following nondimensional parameters characterize the simulation case: $\lambda_l = 2.5$, $\zeta = 0.1$, $N_P = 1.41$ and a 4th order crack trajectory curve ($\mu = 4$) is

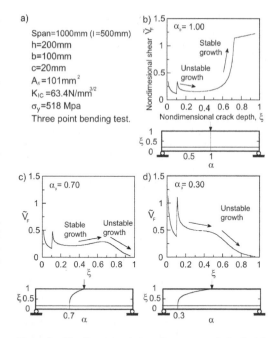

Figure 3. Nondimensional shear force vs. crack depth: (a) material properties and beam geometry; (b) initial crack position $\alpha_0 = 1.0$; (c) initial crack position $\alpha_0 = 0.70$; (d) initial crack position $\alpha_0 = 0.30$.

assumed. Figures 3b-d show the nondimensional shear force vs. crack depth curves for the initial crack position α_0 in the interval [0.3, 1.0] and a sketch of the crack trajectories.

It is well-known that a crack growth may present stable or unstable behaviour. When the crack growth is stable, an increase in the crack depth requires a load increase to fulfil the model equations. On the contrary, unstable crack growth implies a load decrease. Both kinds of behavior can occur at different load levels during crack propagation, see Figures 3b, c and d.

When the initial crack position is at midspan ($\alpha_0 = 1.0$), Figure 3b, the model converges to the original *bridged crack* for beams in flexure (no shear). Immediately after the crack crosses the reinforcement, an unstable branch begins. This turns stable for a crack depth $\xi \simeq 0.3$ Then the nondimensional shear force grows until the yielding of steel takes place ($\xi \simeq 0.7$). Physically the reinforcement reaction stabilizes the initial unstable crack propagation and finally produces the steel yielding.

The second plot, Figure c, computed for an initial crack position $\alpha_0 = 0.70$, shows the same characteristic behaviour for low values of the crack depth, an unstable branch follows the stable branch for a crack depth value of 0.65. From this point on the crack growth is unstable leading the beam to failure. In this case, as for the flexural crack $\alpha_0 = 1.00$, the

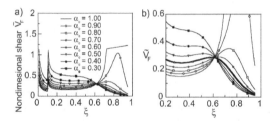

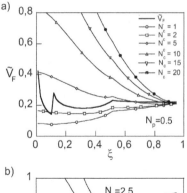

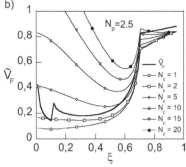

Figure 4. (a) Nondimensional shear force, $\tilde{V}_F$ vs. crack depth, ξ; (b) detail, the thick line is the curve of the minimum critical shear load.

reinforcement reaction stabilizes the crack propagation for low crack depth but there is a point where the propagation becomes unstable. The change in the nature of the propagation provokes the relative maximum that is observed in Figure 3c. This change for shear crack in reinforced concrete beams without stirrups has been reported experimentally by Carmona (Carmona et al. 2006).

Figure 4a shows a superposition of the plots for different initial crack positions. Observe that, neglecting the singularity at the reinforcement position, each curve presents a relative maximum with the exception of cracks near the support which are unstable during all the propagation process. The thick line in Figure 4b represents the crack having the property that its relative maximum is minimum among the maxima. This relative maximum is assumed as the shear failure load and the curve where it is located allows to determine the initial crack position as well.

The minimum in the shear force when the crack initiation position is changed along the shear span has been reported also by Niwa (Niwa 1997) in a Finite Element numerical study. Niwa fixed the shear span to depth ratio λ_l to 2.4 and assumed a linear crack path from the initiation point to the load point. The position for the crack initiation reported for the minimum shear resistance in his study was 0.62 and the nondimensional shear was 0.33. These results compare fairly well with the results of the present model.

5 CONCRETE CRUSHING FAILURE

The equations (16) and (17) reported in Section 3, give the shear producing crushing failure. But, to satisfy equilibrium and compatibility, the crushing shear must be equal to the shear of crack propagation, as expressed by (18) and (19). Therefore, the crushing points expressed by (18) and (19) can be found by intersecting the crushing curve (16) and (17) with the crack propagation curve given by (9) and (10). For a better clarity, this is done in the hypothesis that failure by crushing occurs at the central crack ($\alpha_0 = 1.0$), but the model is not restricted to this situation.

Figure 5. Influence of N_C in nondimensional shear as function of the crack depth for $\lambda_1 = 2$ and $\zeta = 0.1$; (a) $N_P = 0.5$; (b) $N_P = 2.5$.

In Figure 5a the curve showing a discontinuity at the reinforcement position ($\xi = 0.1$) represents the nondimensional shear of crack propagation, $\tilde{V}_F$, for a vertical crack at the midspan ($\alpha_0 = 1.0$). The other family of curves represent the shear of crushing failure, $\tilde{V}_C$, for different values of N_C, Eqs. (16) and (17). As appears from Figure 5a, if N_C is less than 2 (see the two lowest curves), the RC beam exhibits an unstable behaviour, and the fracture process cannot occur because the shear necessary for crushing is less than the shear necessary to the cracking process. For higher values of N_C the beam exhibits first yielding ($\xi \simeq 0.5$), and then crushing when the curves with varying N_C intersect the thick curve of the load vs. crack depth diagram. In this example it is therefore useless to increase the concrete strength above say $N_C \simeq 5$ as yielding will always precede crushing and the failure mode is flexural.

In Fig. 5b the beam brittleness number N_P is varied from 0.5 to 2.5 with respect to Figure 5a. For values of N_C smaller that 2 the beam presents the same unstable behaviour of the previous example. In contrast, for N_C higher than 4, the crushing failure occurs before yielding. As N_C is increased, the crushing collapse progressively approaches the yielding point. Only for $N_C = 20$ the yielding precedes crushing failure, as the curve for $N_C = 20$ intersects the thick curve of

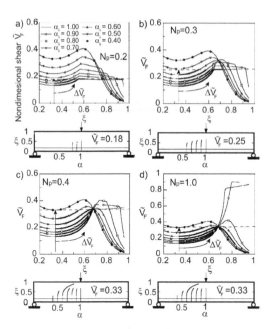

Figure 6. Transition from flexural to diagonal tension failure in RC beams as function of the governing nondimensional parameter N_P; (a) $N_P = 0.2$; (b) $N_P = 0.3$; (c) $N_P = 0.4$; (d) $N_P = 1.0$.

the cracking process only after yielding. Therefore, a variation in the brittleness number N_C can change the collapse mechanism from yielding to concrete crushing and viceversa.

6 TRANSITION BETWEEN FAILURES MODES

The proposed model covers the three fundamental failure mechanisms of RC beams: steel yielding (flexural), diagonal tension (shear) and concrete crushing. As shown in the following, the transition between the aforementioned mechanisms is ruled by the nondimensional model parameters N_P, N_C and λ_l. For the sake of clarity, first the transition from flexural to shear failure is analysed, then the transition from shear to crushing failure.

Figure 6 shows four $\tilde{V}_F$–ξ curves obtained by increasing the brittleness number N_P from 0.2 to 1.0 and keeping constant all the remaining parameters ($\lambda_l = 2.5$, $\zeta = 0.1$, $\mu = 6$). A sketch illustrating the crack trajectories at failure is reported for each beam model.

In Figure 6a the model response for a brittleness number $N_P = 0.2$ is shown. When the nondimensional shear force reaches a value of 0.14 the flexural crack ($\alpha_0 = 1.00$) begins its stable growth. As the load is increased, some other neighboring cracks develop with a stable growth. The more marked lines in the plot

represent the growing cracks. When the nondimensional shear force is equal to 0.18 the steel yields at the flexural crack. We assume that this value of the nondimensional shear force represents the flexural failure load. Thus the beam modeled in Figure 6a shows a flexural failure due to the yielding of the steel at the midspan crack. In the same way, when the brittleness number is increased to 0.3, the beam collapses by flexural failure at the midspan crack, although an increment in nondimensional shear force from 0.18 to 0.25 is observed. Comparing the crack pattern sketches at failure, we observe that the increment in the nondimensional shear also allows for new neighboring cracks to develop through the reinforced concrete element.

If the brittleness number is increased to 0.4, see Figure 6c, initially the cracking process is similar to the previous cases. Nevertheless, when the nondimensional shear force reaches the value 0.33, flexural and diagonal tension failure occur at the same time. In fact, as pointed out in Section 4, diagonal tension (shear) failure occurs when a shear crack develops an instability process after a stable crack growth. For higher values of the brittleness number flexural failure needs a higher nondimensional shear than diagonal tension failure, as illustrated in Figure 6d, where the brittleness number is set to 1.0: the load required to provoke flexural collapse for the crack situated at midspan is 0.8 while the load to provoke diagonal tension failure is 0.33 for the crack in $\alpha_0 = 0.6$.

Therefore, for low values of N_P, cracks at midspan (flexural cracks) need lower nondimensional shear force to provoke flexural failure than shear cracks situated along the span to develop diagonal shear failure. As N_P is increased, the opposite case occurs: cracks along the span need lower nondimensional shear force to provoke beam collapse than the crack at midspan. Thus there is a point where the transition between these types of failure takes place.

Figure 7 shows a conceptual sketch of all the failure mode transitions in reinforced concrete elements without stirrups predicted by the model by varying the nondimensional parameters.

Based on the definition of the brittleness number N_P, an increase in N_P can be read as:

- an increase in the reinforcement area, transition from (d) to (e);
- a decrease in the scale with constant reinforcement area, transition from (a) to (e);
- an increase in the scale with a constant reinforcement percentage, transition from (g) to (e).

When a crushing failure is considered, the behavior of the RC element is controlled by the three nondimensional parameters N_P, N_C and λ_l. As defined in Section 3, crushing failure occurs when the crushing vs. shear plot intersects the crack propagation vs. shear plot.

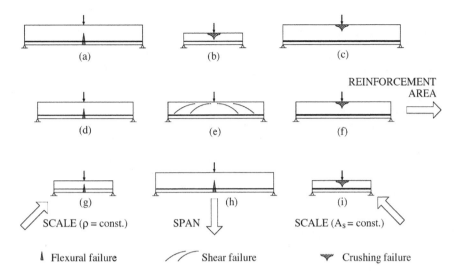

Figure 7. The global conceptual scheme illustrating failure mode transitions.

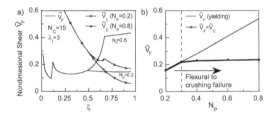

Figure 8. Transition from flexural to crushing failure in RC beams as function of governing nondimensional parameters, increment of N_P; (a) $\tilde{V}_F$-ξ curves; (b) $\tilde{V}_F$-N_P curve.

To simplify the explanation of the transition and for reasons of space in the present paper, a direct transition from flexural to crushing failure is illustrated, although intermediate shear failure transition can be demonstrated to exist as reported in the general scheme of Figure 7.

In Figure 8a the transition process is shown when N_P is varied and the rest of parameters remains constant. The nondimensional shear $\tilde{V}_F$ at yielding increases as N_P increases. At the same time, the shear for crushing failure increases, although in a smoother way. Thus the transition from flexural to crushing failure appears clearly as shown in Fig.8b, where we can read the brittleness number in the abscissas against the nondimensional shear at failure. Two different areas are delimitated. For low values of the brittleness number failure is due to steel yielding (flexure). For $N_P \simeq 0.3$, the transition takes place and then shear for crushing failure needs a lower value compared to shear for flexural failure, i.e. crushing precedes yielding. Physically this transition appears when the reinforcement ratio ρ is increased and the rest of parameters remains

constant: we have the transition from (d) to (f) in the scheme of Figure 7.

According to the scheme, another transition can be demonstrated when the beam is scaled keeping constant the reinforcement ratio ρ. Looking at the definitions of N_P and N_C, the condition of an increment in the scale can be expressed keeping constant the ratio $\frac{N_C}{N_P}$.

Finally, the transition by size effect can be demonstrated in the hypothesis that the reinforcement area A_s is constant. This condition can be expressed by considering the ratio $\frac{N_C}{N_P}$ as a linear function of the scaled beam depth $\frac{h}{h_0}$, where h_0 is a reference depth.

The conceptual scheme in Figure 7 summarizes the failure transitions predicted by the model, from flexural crushing failure.

The transitions to crushing can take place as:

- an increase in the reinforcement area, transition from (d) to (e) and finally (f);
- an increase in the scale with constant reinforcement percentage, transition from (g) to (e) and finally (c);
- a decrease in the scale with constant reinforcement area, transition from (a) to (e) and finally (i).

In some cases, depending on material and geometrical properties, the intermediate transition through (e) may be skipped and a direct transition from yielding to crushing can be observed.

7 CONCLUSIONS

This paper presents an extension of the *bridged crack model* to analyse flexural-shear-crushing failure modes in R.C. beams. The failure mode transitions have been illustrated by varying the controlling

683

nondimensional parameters: the brittleness numbers N_P and N_C and the slenderness λ_l. The study demonstrates that the diagonal tension failure is a consequence of unstable crack propagation. The shear failure initiation point and collapse load are determined analytically by the present model without using empirical parameters.

The model gives rational explanation to the transitions between all the failure modes and size effects in failure transitions are shown by varying the brittleness numbers, N_P and N_C.

ACKNOWLEDGEMENTS

Jacinto R. Carmona gratefully acknowledges the financial support for this research provided by the *Ministerio de Educación y Ciencia*, Spain, under grant MAT2003-00843, and by the *Ministerio de Fomento*, Spain, under grant BOE305/2003.

REFERENCES

Bosco, B. and A. Carpinteri (1992). Fracture mechanics evaluation of minimum reinforcement in concrete structures. In *Applications of Fracture Mechanics to reinforced concrete*, London, pp. 347–377. A. Carpinteri, ed., Elsevier Applied Science.

Carmona, J. R., G. Ruiz, and del Viso. J. R. (submitted 2006). Mixed-mode crack propagation through reinforced concrete. *Engineering Fracture Mechanics*.

Carpinteri, A. (1984). Stability of fracturing process in RC beams. *Journal of Structural Engineering-ASCE 110*, 544–558.

Carpinteri, A., J. R. Carmona, and G. Ventura (submitted 2006). Propagation of flexural and shear cracks through reinforced concrete beams by the bridged crack model. *Magazine of Concrete Research*.

Carpinteri, A., G. Ferro, C. Bosco, and M. Elkatieb (1999). Scale effects and transitional failure phenomena of reinforced concrete beams in flexure. In A. Carpinteri (Ed.), *Minimum Reinforcement in Concrete Members*, Volume 24 of *ESIS Publications*, pp. 1–30. Elsevier Science Ltd.

Carpinteri, A., G. Ferro, and G. Ventura (2003). Size effects on flexural response of reinforced concrete elements with a nonlinear matrix. *Engineering Fracture Mechanics 70*, 995–1013.

Carpinteri, A., G. Ferro, and G. Ventura (2004). A fracture mechanics approach to over-reinforced concrete beams. In V. Li, C. Leung, K. Willam, and S. Billington (Eds.), *Proceedings of the* 5th Fracture Mechanics of Concrete and Concrete Structures (FraMCoS-5), pp. 903–910.

Gustafsson, P. and A. Hillerborg (1983). Sensitivity in shear strength of longitudinally reinforced concree beams to fracture energy of concrete. *ACI Structural Journal 85*(3), 286–294.

Jenq, Y. S. and S. P. Shah (1989). Shear resistance of reinforced concrete beams – a fracture mechanics approach. In *Fracture Mechanics: Applications to Concrete*, Detroit, pp. 237–258. V. Li and Bažant, Z.P., eds., American Concrete Institute.

Niwa, J. (1997). Size effect in shear of concrete beams predicted by fracture mechanics. In *CEB Bulletin d'Information n 237 – Concrete Tension and Size Effects*, Lausanne, Switzerland, pp. 147–158. Comite Euro-International du Béton (CEB).

Okamura, H., K. Watanabe, and T. Takano (1975). Deformation and strength of cracked member under bending moment and axial force. *Engineering Fracture Mechanics 7*, 531–539.

Portela, A. and M. H. Aliabadi (1992). *Crack Growth Analysis Using Boundary Elements*. Southampton: Computational Mechanics Publications.

So, K. O. and B. Karihaloo (1993). Shear capacity of longitudinally reinforced beams – A fracture mechanics approach. *ACI Structural Journal 90*, 591–600.

Fracture Mechanics of Concrete and Concrete Structures – Design, Assessment and Retrofitting of RC Structures – Carpinteri, et al. (eds)
© 2007 Taylor & Francis Group, London, ISBN 978-0-415-44616-7

Experimental study of mixed-mode crack propagation in RC beams without stirrups

J.R. Carmona, G. Ruiz & J.R. del Viso
E.T.S. de Ingenieros de Caminos, Canales y Puertos, Universidad de Castilla-La Mancha, Spain

ABSTRACT: This paper presents the results of a very recent experimental research program aimed at investigating mixed-mode fracture of longitudinally reinforced concrete beams. The tests were designed so that only one single mixed-mode crack generates and propagates through the specimen, as opposed to the usual dense crack pattern found in most of the tests in scientific literature. The specimens were three-point-bend beams of three different sizes. They were notched asymmetrically and reinforced with various ratios of longitudinal reinforcement. These experiments may help to understand the mechanisms of mixed-mode crack propagation in longitudinally reinforced concrete elements. Finally an analytical model based on the experimental results model is presented to analyze size effect and hyper-strength in this kind of elements.

1 INTRODUCTION

This paper presents some very recent results of an experimental program aimed at disclosing some aspects of the propagation of mixed-mode cracks through longitudinally reinforced concrete elements and its consequences. Specifically, the program was designed to investigate the influence of the size of the specimen and of reinforcement detailing on mixed-mode crack propagation. This research is an extension of previous works on the nucleation and propagation of mode I cracks in reinforced concrete (Ruiz et al., 1998; Ruiz and Carmona, 2006). By focusing on mixed-mode cracks we aim at completing the study of the generation and development of the different types of cracks that may appear in longitudinally reinforced concrete beams.

In reinforced concrete, mixed-mode crack propagation is mainly addressed from a technological standpoint. The dense crack pattern that results from the usual reinforcement detailing and element geometry may somehow make it difficult to induce direct relations between causes and effects. That is why we focus on the propagation of one single mixed-mode crack. Of course, there are some other excellent studies with common points with our methodology. They addressed problems related to the shear resistance of reinforced elements, like the study on failure by diagonal tension performed by Bažant and Kazemi (Bažant and Kazemi, 1991), or the work by Kim and White (Kim and White, 1999) on the generation of shear-damaged in reinforced concrete.

The article is organized as follows. An outline of the experimental program is given in Section 2. In Section 3 we describe the characterization tests performed on the materials used to make the reinforced beams. Section 4 deals with the experimental set-up for the mixed-mode tests. The experimental results are presented and discussed in Section 5. Section 6 include a simple analysis of size effect in reinforced notched concrete beams. Finally, in Section 7 some conclusions are extracted.

2 OVERVIEW OF THE EXPERIMENTAL PROGRAM

The program was designed to study the propagation of mixed-mode cracks through reinforced concrete. Specifically, we wanted to disclose the influence of the amount of reinforcement and specimen size on crack propagation. We also intended to analyze the variations in the crack pattern and in the mechanical behavior due to the size of the specimens. In addition, the program had to provide an exhaustive material characterization to allow a complete interpretation of the test results.

With these intentions in mind, we chose the beam sketched in Figure 1 as a convenient specimen for this research. Our choice revisits the geometry tested by Jenq and Shah to study mixed-mode crack propagation in plain concrete (Jenq and Shah, 1988). It is a notched beam that exhibits a single mixed-mode crack when subjected to bending at three points. In

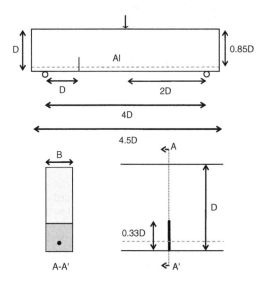

Figure 1. Beam geometry.

their work, Jenq and Shah provide plenty of insights on the generation and propagation of the crack which are of use here. We reinforce the beams with several ratios of longitudinal ($\rho_l = \frac{A_l}{BD}$, being A_l the area of the cross section of the longitudinal reinforcing bars, B the beam width and D the beam depth). The reinforcement provokes changes in the orientation of the main crack and in the global mechanical response, but the presence of a notch avoids a dense crack pattern that would blur our perception of such changes. At most, some reinforcement configurations generate a secondary flexural crack at midspan that competes with the one that starts from the notch tip.

Regarding the size of the beams, we wanted even the largest beams to be reasonably easy to handle and test. At the same time, the behavior of the laboratory beams should be representative of the behavior of beams of a normal size made of ordinary concrete. In order to fulfill both requirements, Hillerborg's brittleness number β_H (Bažant and Planas, 1998) was used as the comparison parameter. It is defined as the ratio between the size of the beams – represented by their depth D – and the characteristic length of the concrete, ℓ_{ch} (Petersson, 1981), i.e.:

$$\beta_H = \frac{D}{\ell_{ch}}, \text{ where } \ell_{ch} = \frac{E_c G_F}{f_t^2}, \quad (1)$$

E_c is the elastic modulus, G_F the fracture energy and f_t the tensile strength. As a first approximation, two geometrically similar structures display a similar fracture behavior if their brittleness numbers are equal. According to this, a relatively brittle micro-concrete was selected with a characteristic length of approximately $\ell_{ch} = 90$ mm (the details of the micro-concrete

are given in the next section), while the beams were made to be 75, 150 and 300 mm in depth. Since ℓ_{ch} of ordinary concrete is 300 mm on average, our 150 mm depth laboratory beams are expected to simulate the behavior of ordinary concrete beams 500 mm in depth, which is considered as a reasonable size for the study.

The dimensions were scaled to the beam depth D, please see Figure 1. We made small (S, $D = 75$ mm), medium (M, 150 mm) and large (L, 300 mm) specimens reinforced with several ratios of longitudinal reinforcement. The beam width is in all cases equal to 50 mm. Each specimen was named by a letter indicating the size (S, M or L) and one figure indicating the number of bars used for the reinforcement. For example, L2 names a large beam with two longitudinal bars. We performed at least two tests for each type of beam.

3 MATERIALS CHARACTERIZATION

3.1 Micro-concrete

A single micro-concrete mix was used throughout the experimentation, made with a lime aggregate of 5 mm maximum size and ASTM type II/A cement. The mix proportions by weight were 3.2 : 0.45 : 1 (aggregate : water : cement). We made characterization specimens out of all batches.

Compression tests were carried out according to ASTM C 39 and C 469 on 75×150 mm cylinders (diameter × height). Brazilian tests were also carried out on these kind of cylinders following the procedures recommended by ASTM C 496. Stable three-point bend tests on $75 \times 50 \times 337.5$ mm notched beams were carried out to obtain the fracture properties of concrete. We followed the procedures devised by Elices, Guinea and Planas (that are minutely explained in (Bažant and Planas, 1998)). Particularly, during the tests the beams rested on anti-torsion devices. They consist of two rigid-steel semi-cylinders laid on two supports permitting rotation out of the plane of the beam and rolling along the longitudinal axis of the beam with negligible friction. Table 1 shows the mechanical parameters of the micro-concrete determined in the various characterization and control tests.

3.2 Steel

For the beam dimensions selected and the desired steel ratios, the diameter of the steel bars had to be smaller than that of standard rebars, so commercial ribbed wires with a nominal diameter of 2.5 mm were used as reinforcing bars. Table 2 shows the mechanical properties of the ribbed wires. The elastic modulus E_s, the ultimate strength f_u, the 0.2% offset yield strength $f_{y,0.2}$, and the ultimate strain ε_u. The nominal value of the diameter was used to calculate the stress-related parameters in Table 2.

Table 1. Micro-concrete mechanical properties.

	$f_c^{(a)}$ MPa	$f_{ts}^{(b)}$ MPa	E_c GPa	G_F N/m	ℓ_{ch} mm
Mean	36.3	3.8	28.3	43.4	86.8
Std. dev.	1.9	0.3	2.7	5.8	–

(a) Cylinder, compression tests.
(b) Cylinder, splitting tests.

Table 2. Steel mechanical properties.

E_s GPa	$f_{y,0.2}$ MPa	f_u MPa	ε_u %
174	563	632	4.6

The ultimate strain in ribbed bars is considerably lower than in mild steel bars, due to the defects in the material resulting from the ribbing process.

3.3 Steel-concrete interface

Pullout tests were carried out by pulling the wire at a constant displacement rate while keeping the concrete surface compressed against a steel plate. Figure 2a sketches the pull out specimen, a prism of $50 \times 50 \times 75$ mm with a wire embedded along its longitudinal axis. The bonded length was 25 mm ($=10 \times$ nominal diameter of the bars) to allow a constant shear stress at the interface of the reinforcement wires (Losberg and Olsson, 1979; RILEM/CEB/FIP, 1970). The relative slip between the wire and the concrete surfaces was measured at the bottom end. The tests were carried out at a constant displacement rate of 2 μm/s. Figure 2b, shows the upper and lower limit of the bond-slip curves. The bond strength τ_c deduced from these tests was 6.4 ± 1.8 MPa. The scatter is over 30%, typical for this kind of test.

4 MIXED-MODE TESTS

As we already described in Section 2, the specimens for the mixed-mode tests were notched beams reinforced with longitudinal bars. Figure 1 sketches the geometry and reinforcement detailing of the beams.

All the beams were supported and tested in three-point bending tests, as illustrated in Figure 1. During the tests beams rested on anti-torsion supports like the ones used to measure G_F (see Section 3.1). For loading, a hydraulic servo-controlled test system was employed. The test were performed in position-control. We ensured that the evolution of the cracking process was very slow. The maximum load was achieved for each size within about 60–80-min. Each complete test had a duration of 120–140 min.

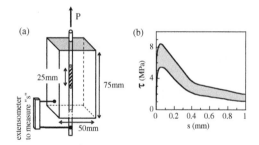

Figure 2. (a) Pull-out specimen to obtain steel-concrete interface properties; (b) upper and lower limit of the τ–s curves.

The load, P, and the displacement under the load point, δ, were continually monitored and recorded. We also used a resistive extensometer centered on the tensioned face of the beam at the mouth of the notch to measure the crack opening displacement, CMOD, in all the tests. In order to complete the experimental information, we also drew the crack pattern resulting from each test copying it directly from both sides of beam.

5 DISCUSSION

Figure 3 shows experimental P-δ and P-CMOD curves. Plain beams, as a limit case of this category of beams, are also considered in the figures. In this kind of beams, the reinforcing bars are far from the tip of the notch. The crack starting from the notch should behave like a crack that has already crossed the flexural reinforcement layer and goes on progressing under mixed-mode conditions. To facilitate the comparison between P-δ curves corresponding to the same kind of beams, the initial slope of the curves is corrected to the theoretical value obtained from finite element calculations. Different experimental initial slopes in a P-δ curve are usual even for the same kind of beams and the same set-up. This is due to the sensitivity of the elastic flexibility of the beam to the boundary conditions in the application of a concentrated load (Planas et al., 1992).

The crack propagation process can be understood with the help of Figures 4 and 5. They show the evolution of the crack related to the P-δ and the P-CMOD curves for a L4 specimen (Fig. 4) and L8 beam (Fig. 5). Please remember that L4 is a large beam – $D = 300$ mm – the reinforcement ratio is 0.13%. Figure 4a shows a picture of the specimen after being tested. The crack trajectory was digitalized and sketched in Figure 4b. The marks and figures on the sketch refer to the corresponding points in the P-δ and P-CMOD curves (Figs. 4c and d respectively) and to the load in kN that the beam was standing when the crack tip reached that

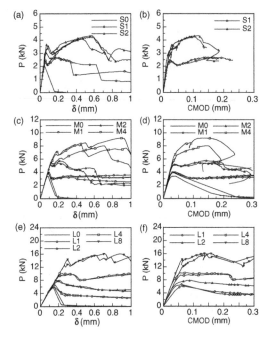

Figure 3. Experimental results given by beams with various ratios of longitudinal bars: (a) P-δ curves of the small beams; (b) P-CMOD curves of the small beams; (c) P-δ curves of the medium beams; (d) P-CMOD curves of the medium beams; (e) P-δ curves of the large beams; (f) P-CMOD curves of the large beams.

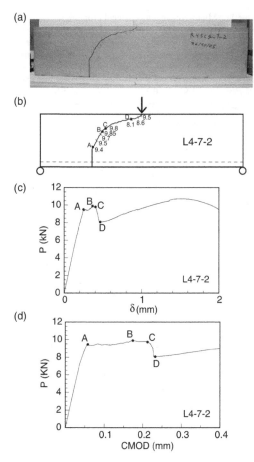

Figure 4. Crack propagation in the specimen L4-7-2: (a) photograph; (b) crack trajectory (the marks denote the extent of cracking at the given loads in kN); (c) P-δ curve; (d) P-CMOD curve.

position. Thus, Figure 4 contains all the experimental information recorded during the test of beam L4-7-2 (the last two numbers indicate that the beam was the second of its kind taken from batch number 7).

The behavior of the beam is almost linear up to the cracking load, P_c, which is assigned the label A in Figures 4b–d. From then on the crack propagates in a slow and stable manner until its tip reaches the point labelled as C. Please note that the propagation between A and C implies almost no increase in the external load. The displacement δ in C is twice the elastic δ that corresponds to A, whereas the crack opening in C is four times longer that the one in A. From C the crack goes on propagating stably towards D, but the curves show that the type of propagation has changed. Indeed, the crack length between C and D equals the growth between A and C but the loads drop from 9.8 to 8.1 kN and, strikingly, the crack growth is not associated to any δ neither CMOD increase. The change in the nature of the propagation can also be noticed by a deviation in the crack trajectory (Fig. 4b). In reinforced concrete technology the behavior just described is referred to as failure due to diagonal tension. It implies a redistribution of the way of resisting shear within the beam. Part of the load carried by the concrete ligament is

transferred to the steel bars and that is why the beam recovers some strength at D. From them on the crack goes on propagating slowly towards the loading point. Most of the shear is withstood by the bars that sew the crack. Depending on the ratio and cover of longitudinal reinforcement and on the geometry of the beam, the concrete around the bars may not be strong enough to resist the shear transferred by the reinforcement. In such cases the bars provoke the generation of a longitudinal crack at the level of the reinforcement, which implies a sudden drop in the load capacity.

Figure 5 provides additional insights on the propagation process, since a L8 beam is reinforced with 8 longitudinal bars, thus doubling the reinforcement ratio. Interestingly, the increase in the reinforcement ratio provokes the generation of flexural cracks that initially grow faster that the mixed-mode crack, as the stretch AB in Figure 5b shows. From B to D flexural

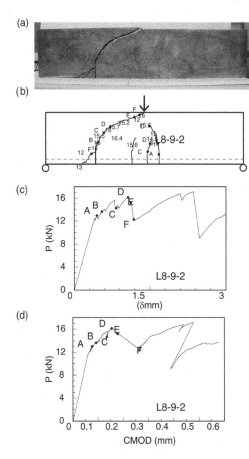

(a)

(b)

8-9-2

(c)

L8-9-2

(d)

CMOD (mm)

Figure 5. Crack propagation in the specimen L8-9-2: (a) photograph; (b) crack trajectory (the marks denote the extent of cracking at the given loads in kN); (c) P-δ curve; (d) P-CMOD curve.

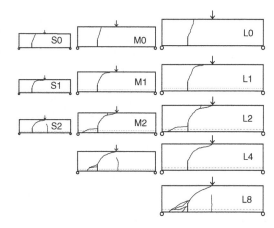

Figure 6. Crack pattern observed in beams with various ratios of longitudinal bars and no inclined bars.

cracks compete with the mixed-mode crack generated at the notch tip, the propagation being slow and stable. At this point, like in the previous case, the nature of the propagation changes. Figure 5b–d show that the mixed-mode crack grows rapidly in a stable way (stretch DE) whereas the flexural cracks arrest; likewise, points D and E are very close both in the P-δ and P-CMOD curves. There has been a redistribution of the shear carrying capacity from the concrete ligament to the steel bars. The concrete that surrounds the reinforcement is not able to stand the load transmitted by the bars and then there starts a longitudinal crack at the reinforcement level. The big jump between E and F in the CMOD indicates the opening of this new crack.

The trajectory of the crack is sensitive to the presence and amount of flexural reinforcement, as Figure 6 clearly shows. The sketches to represent the crack pattern of the beams of different size do not keep

the proportionality between the actual beams, which is 1 : 2 : 4. To facilitate the comparison we represent the sketches scaled following the ratio 1 : 1.5 : 2. Although two tests have been done per each beam type, we have selected only one of the resulting patterns to represent the beam type, since crack patterns for the same beam type are quite similar in all cases. The angle at which the crack starts propagating is almost independent from the number of bars, but as the reinforcement ratio increases, the crack gets inclined so as to reach the loading point. In this case, crack trajectories for different beam sizes are alike. All the beams broke due to the propagation of the mixed-mode crack.

The influence of the reinforcement ratio in the cracking load is analyzed in Figures 7a and b. They represent the cracking load in a nondimensional way versus the reinforcement ratio. The geometry and reinforcement arrangement in these beams facilitate that the bars work as soon as the beam starts to be loaded, which provokes a hyper-strength associated to the ratio of reinforcement. For the ratios considered, a linear relation between reinforcement ratio and cracking load fits very well the test results (please, note that the Pearson's correlation coefficient, R in Figures 7a and b is very close to 1).

Figure 7c shows the cracking load against the size of the beam in a non-dimensional form. Plain beam results and the Bažant's law (Bažant, 1984) fitted to these results are also plotted to facilitate the comparison. As reinforcement ratio increases, size effect is less noticeable.

Figure 7d plots the maximum load (at diagonal tension failure), P_{max}, versus size in a nondimensional form. The strength decreases with size in a smoother way than for Bažant's law, that is, than for plain beams.

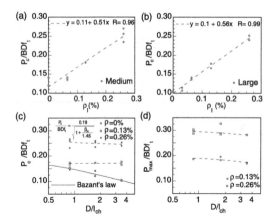

Figure 7. (a) Cracking strength versus reinforcement ratio for the medium beams; (b) cracking strength versus reinforcement ratio for the large beams; (c) cracking strength versus size; (d) maximum strength versus size.

6 SIMPLE MODEL TO EXPLAIN SIZE EFFECT AND HYPER STRENGTH

We observe in Figures 3 and 7 a hyper-strength effect in cracking and maximum load due to the reinforcement action. According to (Ruiz, 2001) the cracking load, P_c, is a function, apart from beam geometry and boundary conditions, of concrete properties, element size and steel location and properties, including the bond-slip behavior of the steel-concrete interface. If reinforcement has reached its yield strength f_y, we can write:

$$P_c = f\left(\frac{D}{l_{ch}}, \frac{c}{l_{ch}}, \rho, \eta, \frac{f_y}{f_t}\right) \quad (2)$$

Where D is the depth of the beam, c is the length of the concrete cover and η is a nondimensional parameter that represents the strength of the interface (Ruiz, 2001). In this investigation only the reinforcement ratio and the size of the element are varied. So, we are going to consider as constants the rest of the parameters and thus Eq. 2 can be rewritten as:

$$P_c = f\left(\frac{D}{l_{ch}}, \rho, \frac{f_y}{f_t}\right) \quad (3)$$

To determine a simple expression to evaluate P_0, we can decompose the load capacity of the beam in two terms. The first one represents the load stood by the plain concrete. The second one includes the hyper-strength attributable to the presence of longitudinal reinforcement. Thus we can write:

$$P_c = P_0 + \Delta P \quad (4)$$

where P_c is the part on the load due to plain concrete and ΔP is the hyper-strength.

For the sake of simplicity, concrete carrying capacity is represented according to linear elastic fracture mechanics, the simplest fracture hypothesis. We can write:

$$\sigma_0 = \frac{P_0}{BD} = K_0 \beta_H^{-\frac{1}{2}} f_t \quad (5)$$

where K_0 is a dimensionless constant for scaled plain beams, and β_H is the Hillerborg's brittleness number as defined in Eq. 1. The exponent $-\frac{1}{2}$ represents the strongest possible size effect. Applicability of such size effect to shear fracture was first analyzed in a pioneering study by Reinhardt (Reinhardt, 1981). It must be emphasized that Eq. 5 only wants to catch a trend of the actual behavior. Concrete response could be modelled with other expressions like Bažant's law (Bažant, 1984) or using another exponent like $-\frac{1}{4}$ (Hillerborg and Gustafsson, 1988).

ΔP in Eq. 4 can be considered as a function of the beam geometry, the position and mechanical properties of the steel rebars and of the bond-slip behavior of the steel-concrete interface. In our experimental program we have used a very low reinforcement ratio and the steel was most of the times yielded at the cracking load. We derived a linear relation between the hyper-strength and the longitudinal reinforcement ratio, based in the results showed in Figures 7a and b.

$$\sigma_\Delta = \frac{\Delta P}{BD} = K_\Delta \rho f_y \quad (6)$$

where K_Δ is another dimensionless constant provided the beams keep the same proportions and reinforcement ratio. The cracking strength can be rewritten as:

$$\frac{P_c}{BD} = \sigma_c = \sigma_0 + \sigma_\Delta = K_0 \beta_H^{-\frac{1}{2}} f_t + K_\Delta \rho f_y, \quad (7)$$

which can be expressed in a non-dimensional fashion as:

$$\frac{P_c}{BDf_t} = \frac{\sigma_c}{f_t} = K_0 \beta_H^{-\frac{1}{2}} + K_\Delta \rho \frac{f_y}{f_t} \quad (8)$$

Applicability of Eqs. 7 and 8 requires not only that the beams are scaled to each other, but also that the shape of the cracks be similar. In our case the crack patterns are very similar for the cracking load, as we showed in section 5.

Figures 8a and c show the linear regression made with some of test results to get the constants K_0 and K_Δ. Figures 8b and d show tests results compared to the obtained size effect law. It may be pointed out that in notched reinforced concrete beams size effect tends to

690

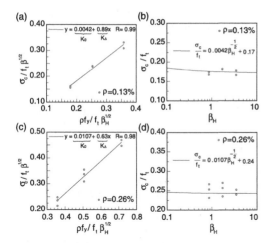

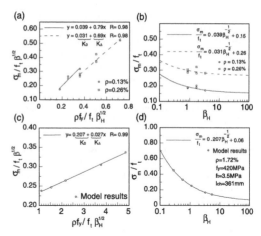

Figure 8. Size effect plots for cracking load: (a) results for the regression to calibrate K_0 and K_Δ coefficients for $\rho = 0.13\%$; (b) size effect law for $\rho = 0.13\%$; (c) results for the regression to calibrate K_0 and K_Δ coefficients for $\rho = 0.26\%$; (d) size effect law for $\rho = 0.26\%$.

Figure 9. Size effect plots for maximum load: (a) results for the regression to calibrate K_0 and K_Δ coefficients, $\rho = 0.13\%$ and $\rho = 0.26\%$; (b) size effect law for $\rho = 0.13\%$ and $\rho = 0.26\%$; (c) results for the regression to calibrate K_0 and K_Δ coefficients for Ožbolt and Eligehausen results; (d) size effect law for Ožbolt and Eligehausen results.

disappear when $D \rightarrow \infty$. The nominal shear strength converges to a value different from zero, which is a function on steel properties. Further analysis would be necessary to evaluate the influence of the mechanical behavior of the steel-concrete interface and of the geometry of the beam.

In reinforced concrete elements, the largest crack at maximum load has the same effect as the notches in fracture specimens. If we consider that diagonal tension failure is caused by fracture propagation and maximum load is attained only after a large fracture growth (and not at fracture initiation), then Eq. 7 will be susceptible to represent the ultimate strength at failure by diagonal tension. Test results available in the literature show that for diagonal tension failure, the LEFM asymptote of slope $-1/2$ fits results better than a horizontal asymptote. This means that shear failure of beams is predominantly brittle (Bažant and Yu, 2005a; Bažant and Yu, 2005b) and so the hypothesis made in Eq. 5 can be considered accurate enough for our proposal.

To apply Eq. 7 to analyze diagonal tension failure we have to make some additional hypothesis. The first one is that the main crack in similar beams of various sizes has to be geometrically similar. This is a reasonable assumption having in mind our results, as it is observed in Figure 6 for $\rho = 0.13\%$ and $\rho = 0.26\%$. The second hypothesis is that steel stress in similar structures of various sizes at failure have to be similar. In beams without any notch diagonal tension failure occurs when the steel is still in the elastic range. In Eq. 6 we consider that the steel yielded because this is what happens in our tests due to the low reinforcement

ratio. In un-notched beams f_y has to be changed by σ_s, steel tension at failure.

Figure 9a shows the linear regression made to get K_0 and K_Δ constants for the beams that failed by diagonal tension. Figure 9b shows the tests results against the obtained law. We have followed the same procedure with results obtained by Ožbolt and Eligehausen (Ožbolt and Eligehausen, 1997) (Figs. 9c and d). We selected these results due to the wide size range that they cover (0.1–2.0 m) and the accuracy obtained from their model. The size effect model that we propose fits their results quite well, which proves that the model catches the trends of the response. Summarizing, tests results indicate that the strength tends to converge to a constant value different from zero. The proposed model follows this tendency, based on experimental observations. It can be of use to develop recommendations on shear reinforcement requirements.

7 CONCLUSIONS

This article presents very recent experimental results on the propagation of mixed-mode cracks through reinforced concrete. The tests were designed so that only one single mixed mode crack generates and propagates through the specimen, as opposed to the usual dense crack pattern found in most of the tests in scientific literature. The specimens were three-point-bend beams with an asymmetrical notch of three different sizes reinforced with various ratios of longitudinal (flexural) reinforcement.

The cracking load of beams was very sensitive to the amount of reinforcement and the crack propagated towards the point where the load was applied. Another observation is that after a large crack progress the final stretch of the crack propagation induced a sudden drop in the carrying capacity of the beam, similar to the so-called diagonal tension failure. Also the effect of the size of the beams is noticeable in our tests. On the one hand, large beams resisted less load in terms of stress. On the other hand the larger the beam, the more leaned towards the load point the crack trajectory was. These experimental results can be used profitably for modeling the behavior of mixed mode crack propagation on reinforced concrete beams.

Finally, the size effect in both cracking and maximum load (at the failure by diagonal tension) is accurately described by a simple model. It discloses the influence of the ratio of longitudinal reinforcement on the hyper-strength and subsequently, can enlighten code developers on updating recommendations for shear reinforcement provisions.

ACKNOWLEDGEMENTS

The authors gratefully acknowledge financial support for this research provided by the *Ministerio de Fomento*, Spain, under grant BOE305/2003, and the *Junta de Comunidades de Castilla -La Mancha*, Spain, under grant PAI05-028.

REFERENCES

Bažant, Z. P. (1984). Size effect in blunt fracture: Concrete, rock, metal. *Journal of Engineering Mechanics-ASCE*, 110:518–535.

Bažant, Z. P. and Kazemi, M. P. (1991). Size effect in diagonal shear failure. *ACI Structural Journal*, 88(3):268–276.

Bažant, Z. P. and Planas, J. (1998). *Fracture Size Effect in Concrete and Other Quasibrittle Materials*. CRC Press, Boca Raton.

Bažant, Z. P. and Yu, Q. (2005a). Designing against size effect on shear strength of reinforced concrete beams without stirrups: I. Formulation. *Journal of Structural Engineering-ASCE*, 131(12):1877–1885.

Bažant, Z. P. and Yu, Q. (2005b). Designing against size effect on shear strength of reinforced concrete beams without stirrups: II. Verification and calibration. *Journal of Structural Engineering-ASCE*, 131(12):1886–1897.

Hillerborg, A. and Gustafsson, P. J. (1988). Sensitivity in shear strength of longitudinally reinforced concrete beams to fracture energy of concrete. *ACI Structural Journal*, 85(3):286–294.

Jenq, Y. S. and Shah, S. P. (1988). Mixed mode fracture of concrete. *International Journal of Fracture*, 38:123–142.

Kim, W. and White, R. N. (1999). Shear-critical cracking in slender reinforced concrete beams. *ACI Structural Journal*, 96(5):757–765.

Losberg, A. and Olsson, P. A. (1979). Bond failure of deformed reinforcing bars based on the longitudinal splitting effect of the bars. *ACI Journal*, 76(1):5–17.

Ožbolt, J. and Eligehausen, R. (1997). Size effects in concrete and RC structures – Diagonal shear and bending. In *CEB Bulletin d'Information n 137 – Concrete Tension and Size Effects*, pages 103–145, Lausanne, Switzerland. Comite Euro-International du Béton (CEB).

Petersson, P. E. (1981). *Crack Growth and Development of Fracture Zones in Plain Concrete and Similar Materials*. Report No. TVBM-1006, Division of Building Materials, Lund Institute of Technology, Lund, Sweden.

Planas, J., Guinea, G. V., and Elices, M. (1992). Stiffness associated with quasi-concentrate loads. *Materials and Structures*, 27:311–318.

Reinhardt, H. W. (1981). Similitude of brittle fracture of structural concrete. In *Advanced Mechanics of Reinforced Concrete*, pages 175–184, Delf. IASBE Colloquium.

RILEM/CEB/FIP (1970). Test and specifications of reinforcement for reinforced and prestressed concrete: Four recomendations of the RILEM/CEB/FIB,2: Pullout test. *Materials and Structures*, 3(15):175–178.

Ruiz, G. (2001). Propagation of a cohesive crack crossing a reinforcement layer. *International Journal of Fracture*, 111(3):265–282.

Ruiz, G. and Carmona, J. R. (2006). Experimental study on the influence of the shape of the cross-section and of the rebar arrangement on the fracture of lightly reinforced beams. *Materials and Structures*, 39:343–352.

Ruiz, G., Elices, M., and Planas, J. (1998). Experimental study of fracture of lightly reinforced concrete beams. *Materials and Structures*, 31:683–691.

Fracture Mechanics of Concrete and Concrete Structures – Design, Assessment and Retrofitting of RC Structures – Carpinteri, et al. (eds)
© 2007 Taylor & Francis Group, London, ISBN 978-0-415-44616-7

Shear strength of RC deep beams

G. Appa Rao* & K. Kunal
University of Stuttgart, Stuttgart, Germany
Indian Institute of Technology Madras, Chennai, India

R. Eligehausen
University of Stuttgart, Stuttgart, Germany

ABSTRACT: This paper reports on some experimental investigations on the shear behaviour of reinforced concrete (RC) deep beams without and with shear (web) reinforcement. Twelve large scale deep beams made of 60 MPa concrete were tested. Three different beams of depth 250 mm, 500 mm and 750 mm were tested to understand size effect. The behaviour of deep beams including load-deflection curves, web strains and crack width, shear ductility and reserve strength has been investigated. The beams tested under three-point loading failed in shear and failure modes were influenced by the beam depth and amount of shear reinforcement. The shear strength was found to decrease with increase of beam size and large size beams exhibited brittle failure, which was attributed to size effect. Sufficient shear reinforcement in beams turned brittle failure in to ductile. The load-deflection curves are regular in small size beams with heavy shear reinforcement. The web strains and the width of shear cracks increase at failure with web reinforcement. With increased quantity of shear reinforcement, more confinement is offered to sustain greater web strains and crack widths. Shear ductility (=capability of withstanding severe cracking and deformation) decreases in deep beams and increases in highly shear-reinforced deep beams. Significant reserve strength beyond diagonal cracking was observed in deep beams. As a matter of fact, this reserve strength was to two times larger in small-size beams, compared to large-size beams.

1 INTRODUCTION

The shear behaviour of deep beams is very complex and there is still no agreement on the role of size effect in shear due to lack of information. Deep beams are classified as nonflexural members, in which plane sections do not remain plane in bending. Therefore, the principles of stress analysis developed for slender beams are neither applicable nor adequate to determine the strength of deep beams. An important characteristic of deep beams is their high shear strength. The greater shear strength of deep beams is due to internal arch action, which transfers the load directly to a support through concrete struts. The reinforcement acts as a tie and, hence RC beams are analogous to steel trusses. Deep beams are also classified as disturbed regions, which are characterized by nonlinear strain distribution. Elastic solutions of deep beams provide good description of their behaviour before cracking. However, after cracking major redistribution of strains and stresses takes place and the beam strength must be predicted by nonlinear analysis. For a simple deep beam with concentrated load on top, the top load and bottom reactions create large compressive stresses at

a right angle to beam axis. These stresses interact with shear stresses to form complicated stress field in the web. Because of short horizontal distance between top and bottom load points i.e. small a/d ratios, the effect of such stresses result in arch action unique in deep beams. Because of these complexities, study of deep beams has become a special interest. Over the years various models have been proposed by many researchers and extensive test campaigns have been carried out.

2 REVIEW OF LITERATURE

Several research efforts have been made to understand the shear strength of deep beams and size effect. Due to complex behaviour of deep beams limited information has been reported over the years and further evidence is needed on the role of the many parameters involved, as demonstrated by some recent studies. The study on deep beams has been an interesting topic by varying the parameters. However, some studies have been reported on the investigations on the behaviour of deep beams in shear recently. As for the definition of deep beams, ACI 318 defines deep beams as those loaded on one face and supported on other face and the shear span-to-depth ratio is less than or equal to two. Due to their geometric proportions deep beams fail in shear. A disturbance in internal stresses is caused by shear action with compression in one direction and tension in the perpendicular direction. This leads to an abrupt shear failure of beam as the beam depth increases (Yang and Chung, 2003). The development of crack pattern is much faster than small size deep beams and then leading to sudden failure (Bakir and Boduroglu, 2004).

Several modifications have been incorporated in the shear design of deep beams in the codes of practice. ACI 318-2005 and IS 456-2000 consider the contribution of concrete, percentage longitudinal and transverse reinforcement, shear span-to-depth ratio for estimating the shear strength of deep beams, while BS 8110 does not specify any guidelines for design of deep beams. However, it explicitly says that for design of deep beams specialist literature should be referred. Unlike in ACI 318 and IS 456, BS 8110 considers size effect in shear design of RC beam. However, the maximum depth is limited to 400 mm. Therefore, in order to understand the shear design of deep beams and to evaluate size effect serious research efforts are needed.

Failure in deep beams is generally due to crushing of concrete in either reduced region of compression zone at the tip of inclined cracks or by fracture of concrete along the crack. In deep beams with shear span-to-depth ratio 2.5, there seems to be some reserve strength in the post-cracking region, resulting in relatively less brittle in nature (Khaldoun, 2000, Lin and Lee, 2003). Therefore, to estimate the reserve strength and ductility of deep beams in shear, the influence of various parameters need to be investigated. This paper presents some experimental observations on behaviour and size effect in RC beams with different shear reinforcements.

Ashour and Morley (1996) carried out an upper bound mechanism analysis on continuous reinforced concrete deep beams. The effect of horizontal and vertical web reinforcement on the load carrying capacity is mainly influenced by the shear span-to-effective depth ratio. In deep beams, the horizontal shear reinforcement is effective than the vertical shear reinforcement. Ashour (2000) reported analysis of shear mechanism in simply supported RC deep beams. Concrete and steel reinforcement are modeled as rigid perfectly plastic materials. The failure modes were idealized as assemblage of rigid blocks separated by failure zones of displacement discontinuity. The shear strength of deep beams is derived as a function of location of the instantaneous center of relative rotation of moving blocks.

Tang and Tan (2004) proposed an approach to account for the effect of transverse stresses to the load carrying capacity of concrete in the diagonal strut based on strut-and tie concept. This involves an interaction between two modes of failure; diagonal tensile splitting and diagonal crushing of concrete due to compression. Russo et al. (2005) proposed an explicit expression that considers the shear strength based on strut and tie mechanism due to diagonal concrete strut and longitudinal reinforcement as well as vertical stirrups and horizontal web reinforcement. Bakir et al. (2004) recommended the strut and tie model for the design of short and deep beams. The model consists of three separate mechanisms; direct strut mechanism, truss mechanism which takes in to account the horizontal shear reinforcement and truss mechanism, which takes in to account of the stirrups.

Several fracture mechanics models have been proposed in order to characterize the failure of concrete (Hillerborg and co-workers (1976), Bazant and Oh (1984), Jenq and Shah (1989). Each one of these models introduces some material fracture properties regardless of the structural geometry and size. Concrete structures exhibit size effect, which has been explained as a consequence of the randomness of material strength. In large structures it is more likely to encounter a material point of smaller strength. Bazant proposed that whenever the failure does not occur at the initiation of cracking, size effect should properly be explained by energy release caused by macro-crack growth, and that the randomness of strength plays a meager role. Nevertheless, size effect in concrete structures ought to be explained by a non-linear form of fracture mechanics that takes in to account the localization of damage in to a fracture process zone (FPZ) of a non negligible size. Bazant's size effect law (Bazant and Oh, 1984) is based on the ductile-brittle transition of the failure mode of geometrically similar fracture specimens. For most practical cases, Bazant's size effect law can be described by the following equation

$$\sigma_N = \frac{B f_0}{\sqrt{1 + \frac{d}{d_0}}} = B f_0 (1 + \beta)^{-\frac{1}{2}} \quad (1)$$

where $\beta = d/d_0$, B and d_0 are empirical constants to be obtained by fitting equation to the experimental values from different sizes of specimens.

Smith and Vantsiotis (1982) tested 52 RC deep beams under two point loading to study the effect of shear span-to-depth (a/d) ratio and vertical and horizontal web reinforcement on ultimate shear strength and crack width. The web reinforcement produces no effect on formation of inclined cracks but affects the ultimate shear strength. The addition of vertical web reinforcement improves the ultimate shear strength, but addition of horizontal web reinforcement has negligible influence on ultimate shear strength. Iguro et al. (1984) carried out some experimental studies on uniformly loaded reinforced concrete beams of depth varying between 100 to 3000 mm without shear reinforcement, in order to study size effect on shear strength of beams. As the effective depth increases the shear strength gradually decreases. Collins and Kuchma (1999) reported that for large, lightly reinforced concrete beams, reduction in shear stress at failure was related more directly to the maximum spacing between the layers of longitudinal bars rather than overall depth of the member. It has been observed that high strength concrete (HSC) beams exhibit strong size effect in shear. Accordingly, some modifications to ACI shear design provisions are recommended. Karim (1999) proposed an alternative shear strength prediction equation, at both ultimate and cracking stage, for an RC member without web reinforcement. From 350 beam test results collected from the existing literature of RC beams in shear covering a wide range of beam properties and test methods, a technique of dimensional analysis, interpolation function, and multiple regression analysis was carried out, for both normal strength concrete (NSC) and HSC members. An interpolation function was used to account for the difference in behaviour between arch action of short beam and beam action of long beams.

Raghu et al. (2000) conducted a comprehensive experimental and technical investigation to asses the concrete component of shear resistance in beams made of HSC. The experimental program consists of testing of 24 beams, with and without shear reinforcement, to determine the contribution of concrete to shear strength. The data from the experimental observations and literature were compared with shear provisions in codes of practice. When extrapolated the current provisions for shear resistance of HSC, the safety margins for structural designs are reduced. Angelakos et al. (2001) reported on tests of 21 large RC beams in shear. It has been revealed that concrete strength is the most important parameter influencing shear stress at failure and the longitudinal reinforcement has only negligible effect. The shear stress at failure decreases substantially as member size increases and as the longitudinal reinforcement ratio decreases. Aguilar et al. (2002) studied RC deep beams. The experimental results have been compared with the shear design procedures laid down in ACI 318-99. Yang et al. (2003) tested twenty one beam specimens to investigate the shear characteristics with various variables such as concrete strength, shear span-to-depth ratio, and beam depth. It has been found that decrease in shear span-to-depth ratio and increase in beam depth at a shear span-to-depth ratio resulting in more brittle failure with wide diagonal cracks and high energy release rate related to size effect. Also, HSC deep beams exhibited more remarkable size effects with brittle behavior.

Zararis (2003) reported that the shear failure of RC deep beams is due to crushing of concrete in compression zone with restricted depth above the tip of the critical diagonal crack. This theory has been applied to evaluate the shear strength of RC slender beams subject to shear and flexure. According to this, the reason for shear failure is the loss of shear force of the main tension reinforcement, which occurs due to horizontal splitting of concrete cover along the main reinforcement. Lubell et al. (2004) used the specific situation of Bahen Center beams (University of Toronto) to investigate the possibility of shear failure of large size thick deep beams. The conclusions by earlier researchers that the shear strength of wide beams is directly proportional to the width of the beam are found to be correct. Accordingly, modifications have been suggested to ACI code for shear design of large wide beams. Khaldoun et al. (2004) reported the experimental results on shear behaviour of 11 beams made of 65 MPa concrete, reinforced with transverse and longitudinal reinforcement. Performance of specimens based on cracking pattern, crack widths at estimated service load, and on post cracking reserve strength have been evaluated. A significant reduction in crack width was observed with increase in amount of longitudinal reinforcement. The quantity of longitudinal reinforcement provided in the beams can demonstrate what should be limit of minimum transverse reinforcement. The shear strength equations in current ACI, CSA, and AASHTO LRFD specifications are conservative. Russo et al. (2005) proposed an explicit formula that considers the shear strength provided by the strut-and-tie mechanism due to diagonal concrete and the longitudinal main reinforcement as well as the vertical stirrups and horizontal web reinforcement.

The objective of the study is to understand size effect in RC deep beams in shear with and without web reinforcement and also to evaluate the shear ductility of RC deep beams failing in shear. The scope of this study is limited to RC deep beams with shear span-to-depth ratio 1.5, concrete compressive strength of 60 MPa, with longitudinal reinforcement of 2.0% and comparison of existing code values with the experimental values.

3 RESEARCH SIGNIFICANCE

The design of deep beams is rather complex, since the very behaviour of these structural members is complex and is still not totally clarified. Due to geometric proportions, the behaviour of RC deep beams is governed mainly by shear strength. The shear strength of deep beams seems to be significantly greater than that of the slender beams due to redistribution of internal stresses. Several parameters affect the strength of RC beams in shear, which include shear span-to-depth ratio, concrete strength, anchorage of reinforcement into the supports, size effect, amount and arrangement of tensile and web reinforcement. The disturbance of internal stresses due to heavy concentrated loads causes reduction of load carrying capacity of deep beams and fosters an abrupt shear failure as the depth increases. Thus, it is necessary to investigate the shear behavior of deep beams with different sizes. The design codes are developed from experimental test results using low strength concrete and on RC beams without shear reinforcement and with depth less than 350 mm.

4 EXPERIMENTAL PROGRAMME

4.1 Materials

Concrete used for this program was designed to achieve compressive strength of 60MPa for all the beams. Mix proportions of the concrete used for achieving the required strength were 1: 1.5:2.9 using Portland Pozzolana Cement (PPC). Table 1 shows the constituent materials used for the concrete. The water cement ratio used was 0.32. Along with each set of RC deep beams, six companion plain concrete cubes of size $150 \times 150 \times 150$ mm were cast and tested to find the characteristic compressive strength of concrete. The coarse aggregate was 20 mm maximum size aggregate with specific gravity 2.70 and fineness modulus of 6.93. Sand was naturally obtained with specific gravity of 2.73 and fineness modulus of 2.84. Potable water was used for mixing of concrete and curing purpose. The steel reinforcement consists of high strength deformed bars for longitudinal flexural reinforcement in all the beams. The steel ratio of the flexural reinforcement was 2.0% in all beams. The properties of reinforcement are shown in Table 2.

4.2 Casting of test beams

Well seasoned wooden beam moulds were fabricated for casting beams of 250, 500 and 750 mm depth and 150 mm width. Superplasticizer was used to produce flowable concrete in order to pour the concrete in to the beam moulds to avoid sand pockets. Needle vibrators were used to compact the concrete in beam specimens. After 24 hours, the beams were removed from the moulds and cured for 28 days. The curing was done using gunny bags covered around the beams and water was sprinkled in every 3 hrs intervals to avoid evaporation of moisture from the beam surfaces. After curing all the beams were white washed and square grids were drawn on the beam surface in order to visualize the crack pattern and to make crack-width measurements easier.

Table 1. Constituent materials used for concrete.

Mix	Cement kg/m^3	Sand kg/m^3	Aggregate kg/m^3	w/c Ratio
M60	474	710	1373	0.32

Table 2. Mechanical properties of reinforcement.

S No.	ϕ (mm)	f_y (MPa)	ε_Y ($\times10^{-3}$)	E (10^3, MPa)	σ_{ut} MPa
1	4	400	2.0	200	480
2	5	479	2.4	200	521
3	6	425	2.1	200	600
4	16	607	2.8	217	657
5	20	543	3.2	199	663

4.3 Reinforcement and beam dimensions

Two variables are considered in this study; beam depth and web or shear reinforcement. All the beams were rectangular in cross section with a width of 150 mm. The shear span-to-depth (a/d) ratio was 1.5. The beams are grouped in to four series. These series are designated as HSCB-0, HSCB-0.4, HSCB-0.6 and HSCB-0.8. "HSCB" indicates "High Strength Concrete Beam" and the number following HSCB indicates the shear reinforcement index (SRI) which is the measure of amount of shear reinforcement provided in the beam. Each series consists of three beams of depth 250, 500 and 750 mm designated by S, M and L respectively to indicate small, medium and large size beams. The flexural reinforcement has been adopted after evaluating the flexural strength of beams and comparing with the shear strength so that the failure could be initiated by shear failure only. Sufficient reinforcement was provided near the support including for shear and anchorage length. All the flexural reinforcement bars were bent up vertically at the supports to achieve adequate end anchorage. The clear cover of the flexural reinforcement was kept as 25 mm in all the beams. 6mm diameter mild steel bars were used as top corner steel for hanging the shear reinforcement. The stirrups were made from mild steel bars of 6, 5 and 4 mm diameter depending on the beam size according to the code provisions for minimum shear reinforcement and minimum spacing of shear reinforcement in beams.

The reinforcement arrangement in typical RC beams is shown in Figure 1. For the first series of beams designated as HSCB-0, no web reinforcement was provided. However, in order to maintain the longitudinal bars in their position stirrups are provided one each at the ends and at the center of the beam. This series of beams was tested in order to understand the shear behaviour of deep beams without shear

Figure 1. Typical reinforcement in RC Beam.

Figure 2. Beam specimen and experimental set up.

Table 3. Details of test specimens.

Beam designation	D	b	l	P_t %	f_{yv}	SRI
HSCB-S0.0	250	150	930	2		0.0
HSCB-M0.0	500	150	1680	2		0.0
HSCB-L0.0	750	150	2430	2		0.0
HSCB-S0.4	250	150	930	2	400	0.40
HSCB-M0.4	500	150	1680	2	479	0.40
HSCB-L0.4	750	150	2430	2	479	0.40
HSCB-S0.6	250	150	930	2	479	0.60
HSCB-M0.6	500	150	1680	2	425	0.60
HSCB-L0.6	750	150	2430	2	425	0.60
HSCB-S0.8	250	150	930	2	479	0.80
HSCB-M0.8	500	150	1680	2	425	0.80
HSCB-L0.8	750	150	2430	2	425	0.80

reinforcement for comparison with those with shear reinforcement. The subsequent series of beams were designated as HSCB-0.4, HSCB-0.6 and HSCB-0.8 with different stirrup spacing to achieve the required SRI. The spacing of shear reinforcement was varied in the beam specimens in order to achieve the required shear reinforcement index. All the beams were reinforced with the same steel ratio for the flexural reinforcement. The stirrups were provided with an end hook of 135°. Details of all the shear reinforcements are given in Table 3. The yield strength of longitudinal reinforcement is 521 MPa.

A shear reinforcement index (SRI) is defined to represent the shear reinforcement, which is given by

$$SRI = R. f_{yv} \qquad (2)$$

where $R = A_{sv}/b.s_v$

4.4 Experimental setup and testing of beams

Twelve simply supported RC deep beams were tested up to failure under three-point loading. Each beam was loaded with a central concentrated load and supported on two simply supported ends as shown in Figure 2.

Ends of all beams were extended by 150 mm from the line of action of support reaction. Bearing plates of dimensions $100 \times 150 \times 20$ mm were provided at the supports and below the point loading. All the beams were tested using 1000kN capacity displacement controlled actuators. LVDT was attached at the mid span to measure the deflection of beams under the point loading. At each displacement increment, the load applied on the beam, mid span deflection, maximum crack width and diagonal strain in concrete were measured.

5 DISCUSSION OF RESULTS

5.1 Modes of failure

All the beams tested under three-point loading failed in perfect diagonal shear. Typical crack pattern and modes of failure are shown in Figure 3. In all the beams, cracks started as flexural cracks, but no cracks were observed up to 20% of the ultimate load. The first vertical flexural crack was formed in the region of maximum bending moment within a load range of 20–30% of ultimate load. In the range of load between 40–70% of the ultimate load a major diagonal tension crack formed at the middle of shear span. With further increase in the applied load, new inclined cracks appeared within the shear span, their orientation being the same as that of the previously-formed major inclined cracks. Eventually, beam failure occurred due to crushing of concrete in either reduced region of compression zone at the tip of inclined crack or by the fracture of the concrete along the inclined crack.

The modes of beam failure were influenced by the depth of beam and the amount of shear reinforcement. It has been observed that for all smaller size beams i.e. HSCB-S0.0, HSCB-S0.4, HSCB-S0.6 and HSCB-S0.8 and also in medium size beams HSCB-M0.0 and HSCB-M0.4 having relatively smaller amount of shear reinforcement failed by fracture of concrete along the tension diagonal. However, in few medium size beams such as HSCB-M0.6 and HSCB-M0.8, and also in all large size beams HSCB-L0.0, HSCB-L0.4, HSCB-L0.6 and HSCB-L0.8, the failure was shear-compression type of failure. The failure due to crushing of concrete resulted in brittle failure. In all the beam failures, the inclined cracking pattern reveals

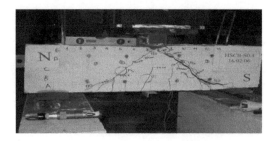

Figure 3. Crack pattern for beams with SRI 0.4.

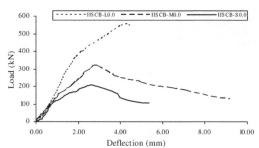

Figure 4. Load-deflection curves with SRI 0.0.

a tied-arch action, with tension reinforcement acting as a tie rod and portion of beam between the inclined cracks as struts. The cracking pattern was found to be more uniform as the amount of shear reinforcement increases and also as the beam depth increases, keeping shear span-to-depth (a/d) ratio constant. The deterioration of concrete and cracking were symmetric just before failure. However, at the stage of failure cracking propagated rapidly at only one end of the beam due to diagonal cracking. As the depth of the beam increases, the failure mode changes from diagonal tension to diagonal tension-compression type. The deeper the beam, the steeper the inclination of the diagonal crack. In all the large size beams, crushing of concrete in compression at the tip of the diagonal crack has been observed. As the shear reinforcement increases, more inclined cracks formed with small spacing in between the cracks. At failure only a major crack was widened.

5.2 Diagonal cracking and ultimate shear strength

The diagonal-cracking strength is defined as the strength at which the first fully developed major diagonal tension crack appears in the shear span. The diagonal tension cracking strength was observed to be considerably less than the ultimate strength. Many mechanisms may be responsible for such behaviour. However, the major phenomenon is attributed to the arch action. Deep RC beams exhibited significantly enhanced shear resistance after first diagonal cracking as a result of strong strut action of concrete in compression. The difference between the ultimate shear strength and diagonal cracking strength can be considered as reserve strength. The reserve strength was analyzed from the experimental observations in beams of varying sizes. Defining V_u and V_{cr} as the ultimate and diagonal cracking strength of RC beams, a ratio of V_u/V_{cr} has been evaluated to represent the reserve strength in terms of measured cracking strength. The ratio V_u/V_{cr} in all deep beams lies in the range between 2.0 to 1.08. The highest value has been observed in small size beams. As the beam depth increases beams exhibited brittle failure.

5.3 Load-deflection curves and diagonal strains

Figure 4 shows the load-deflection curves of beams with SRI 0.0. Similarly, the load-deflection curves in beams with SRI 0.0 to 0.8 respectively with different beam sizes have been drawn. It has been observed that beams of 750 mm depth have higher deflections at ultimate load and their failure is relatively brittle than that of beams of depth 500 mm and 250 mm. In different sizes of beams, the maximum deflections have been observed with SRI 0.8 followed by the beams with SRI 0.6, 0.4 and 0.0. The post-peak response seems to be more gradual showing increase of ductility of the beams with increase in percentage shear reinforcement or SRI. The failure of small size beams seems to be gradual indicating sufficient ductility before failure. This showed that being a shear failure, deep beams exhibit reasonable ductility represented by relatively larger deflections at failure and post peak response, with increase in shear reinforcement.

At a given loading, the strain in large size beams seems to be more. However, the large size beams fail at lower stains. Also, it has been observed that as the amount of web reinforcement increases the strain in concrete also increases. This was mainly because, with increase in amount of web reinforcement, the load is shared by the shear reinforcement, allowing concrete to sustain more cracking strain. At any given load, the diagonal tensile strains in large-size beams are larger than in small-size beams, but the strains at the onset of failure are smaller in the former case. Furthermore, the larger the shear reinforcement, the larger the diagonal strains, mainly because of the increasing share of the shear that is resisted by the reinforcement. As a result, concrete can absorb more distributed cracking.

5.4 Shear ductility

Though deep-beam failure is considered brittle in design provisions, under certain circumstances deep beams exhibit a reasonable ductility. To understand ductility of beams failing in shear, shear ductility is defined as the ratio of A_c/A_u, where A_u is the area under the load deflection curve up to ultimate load and A_c is the area under the load deflection curve for a beam up to its complete collapse. With certain limitations shear ductility can measure the ductility of RC beams failing in shear. It has been observed that shear ductility increases linearly as the SRI increases. However, this increase is prominent after SRI of 0.4. Further, as SRI increases beyond 0.6, the shear ductility has been found to increase significantly. Also, it has been observed that large size beams exhibited brittle failure than those of small and medium size beams, in which the failure seems to be ductile.

5.5 Comparison of code provisions

The ratio of experimental shear strength and those evaluated using ACI 318, IS 456 and BS 8110, V_u/V_{ACI} and V_u/V_{BS} are compared in all the tested beams. It has been noticed that the shear strength provisions are more conservative for deep beams according to IS 456 and ACI 318 than those of BS8110 for small and medium size beams, while BS8110 code provisions are more conservative for large size beams of depth greater than 750 mm. However, code provisions by IS 456 give the most conservative ultimate shear strength of deep RC beams. Further, it is worth mentioning that only BS 8110 considers the size effect in shear strength of RC beams. The design of RC deep beams considering size effect given by BS 8110 seems to be appropriate for shear design of RC deep beams.

6 CONCLUSIONS

1. The modes of failure in reinforced concrete deep beam are influenced by the beam size and the percentage of shear reinforcement. However, as the depth of beam and amount of web reinforcement increase the failure seems to be due to shear-compression failure.
2. Deep beams exhibit significant reserve strength in shear measured as the ratio of V_u/V_{cr}. After a fully developed diagonal crack, small beams exhibit high reserve strength than large beams.
3. Increase in shear reinforcement increases the ultimate shear strength of RC beams. However, in larger size beams, at a given shear reinforcement large size beams exhibit less strength and fail in a brittle manner.
4. As the depth of beam increases, the crack width also increases. However, with increase in amount of shear reinforcement, the crack width decreases.
5. The shear ductility of RC deep beams increases as the shear reinforcement increases. The increase is significant in beams with shear reinforcement index greater than 0.6.
6. ACI 318 shear strength provisions on deep beams are conservative and it does not consider size effect,

while BS8110 code provisions are appropriate for deep-beam design.

REFERENCES

Aguilar, G., Matamoros, A., Ramirz, J. and Wight, J. 2001. Experimental evaluation of design procedures for shear strength of deep RC beams. *ACI StrJl* 99(4): 539–548.

Anngelakos, D., Bentz, D. and Collins, M. 2001. Effect of concrete strength and minimum stirrups on shear strength of large members. *ACI StrJl* 98(3): 290–300.

Ashour, A. and Morley, C. 1996. Effectiveness factor of concrete in continuous deep beams. *Jl St Eng* 122(2): 169–178.

Ashour, A. 2000. Shear capacity of reinforced concrete deep beams. *Jl of StrEngg* 126(9): 1045–1052.

Bakir P. and Boduroglu, H. 2005. Mechanical behaviour and non-linear analysis of short beams using softened truss and direct strut and tie models. *Eng Str* 27: 639–651.

Bazant, Z.P. and Oh, B.H. 1983. Crack band theory for fracture of concrete. *Mat and Strs*, 16(93):155–177.

Collins, M. and Kuchma, D. 1999. How safe are our large, lightly reinforced concrete beams, slabs, and footings?. *ACI Str Jl*, 96(4): 482–490.

Hillerborg, A., Modeer, M. and Petersson, P.E. 1976. Analysis of crack formation and crack growth in concrete by means of fracture mechanics and finite elements. *Cem Con Res* 6(6): 773–782.

Iguro, M., Shioya, T., Nojir, I.Y. and Akiyama, H. 1984. Experimental studies on shear strength of large reinforced concrete beams under uniformly distributed loads. *JSCE Proc*, 1(345): 137–154.

Jenq, Y. and Shah, S.P. 1985. Two parameter fracture model for concrete. *Jl. of Eng Mech* 111(10): 1227–1240.

Karim, R. 1999. Shear strength prediction for concrete mebers. *Jl of Str Eng* 125(3): 301–308.

Khaldoun, R. and Khaled, A. 2004. Minimum transverse reinforcement in 65 MPa concrete beams. *ACI StrJl* 101(6): 872–878.

Kotsovos, M. 1990. Strength and behaviour of deep beams, in 'Reinforced Concrete Deep Beams' by F. K. Kong, *Van Nostrand Reinhold Publications, New York.*

Lubell, A., Sherwood, T., Bentz, E. and Collins, M. 2004. Safe shear design of large, wide beams. *Conc Intl*, Jan, 67–77.

Mau, S.T. and Hsu, Thomas, T. Shear strength prediction for deep beams with web reinforcement. *ACI Jl* 96: 513–523.

Raghu, P. and Priyan, M. 2000. Experimental study on shear strength of high strength concrete beams. *ACI StrJl*, 97(4): 564–571.

Russo, G., Venir, R. and Pauletta, M. 2005. Reinforced concrete deep beams – shear strength model and design formula. *ACI Str Jl* 102(3): 429–437.

Smith, K. and Vantsiotis, A. 1982. Shear strength of deep beams. *ACI StrJl*, 79(3): 290–300.

Tang, C. and Tan, K. 2004. Interactive Mechanical model for shear strength of deep beams. *Jl of Str Engg* 130(10): 1534–1544.

Yang, K., Chung, H., Lee, E., and Eun, H. 2003. Shear characteristics of high-strength concrete deep beams without shear reinforcements. *Engg Strs* 25: 1343–1352.

Zararis, P. 2003a. Shear compression failure in reinforced concrete deep beams. *Jl of Str Engg* 129(4): 544–553.

Fracture Mechanics of Concrete and Concrete Structures – Design, Assessment and Retrofitting of RC Structures – Carpinteri, et al. (eds)
© 2007 Taylor & Francis Group, London, ISBN 978-0-415-44616-7

Experimental study on the shear capacity of randomly-cracked longitudinally-reinforced FRC beams

T. Miki, I.O. Toma & J. Niwa
Department of Civil Engineering, Tokyo Institute of Technology, Tokyo, Japan

ABSTRACT: This study is aimed at investigating the influence of steel fibers on the shear behavior of RC beams with random cracks. Since obtaining randomly-cracked concrete elements takes a lot of time and special conditions, the use of an expansion agent came as a solution to have extensive cracking. The idea of using short fibers came from the well-known fact that they behave very well as crack arrestors. Strain gages were attached to the longitudinal steel bars to monitor the strain variation during the curing process. At the same time, concrete surface strains were monitored and recorded by a data logger. Four-point loading tests were carried out on the RC beams. Steel fibers proved to be effective in reducing the overall expansion of the beams. An increase in the shear carrying capacity and deformability of the beams with steel fibers was observed.

1 INTRODUCTION

Many studies have been carried out to analyze and understand the shear failure of RC beams (Bresler & Scordelis 1963, Kreffeld & Thurston 1966). This type of failure is due to the combined action of shear and flexure and may happen in a brittle way, without any warning signs. The shear carrying capacity of RC beams can greatly decrease when the concrete members are already cracked. Cracks can appear in concrete members due to a multitude of factors either external (severe environment, loads) or internal (chemical reactions within concrete). Engineers have looked for ways to improve concrete properties and behavior and for ways to protect it from these attacks but their efforts were not always successful.

One of the undesired phenomena occurring in concrete structures is random cracking. There are many causes leading to random cracking and one of them is Alkali Silica Reaction (ASR). Many studies have been performed in order to better understand ASR (Smaoui et al. 2005). The basics of ASR consist in the formation of ASR gel which swells, provided that enough moisture is present, creating tensile stresses in the surrounding cement matrix. When the induced tensile stresses become larger than the tensile strength of the matrix, cracking occurs. As a result of the improved understanding of ASR, methods have been developed and proposed to either slow down or completely avoid it. However, some of these methods gave unexpected results on the long run (Diamond 1997). For example, it was found out that silica fume can induce ASR rather than mitigating it (Petersen 1992).

2 RESEARCH MOTIVATION AND OBJECTIVES

While most of today's research in the field of concrete focuses on the properties and behavior of undamaged concrete elements, the present research aims at investigating the shear resistance of RC beams affected by random cracking. Moreover, the possible use of short steel fibers as a tool to improve the behavior of randomly-cracked concrete beams is also studied.

Thus, the objectives are: to create random cracks in RC beams, to investigate the role of steel fibers in RC beams with random cracks and to understand the shear behavior of steel fiber-reinforced concrete beams with random cracks.

Because obtaining random cracking in a concrete specimen requires a long time and special conditions, the use of an expansion agent came as a solution to have extensive cracking. Expansion agents, also known as shrinkage reducing admixtures, are special products developed to increase the volume of concrete by means of specific chemical reactions. According to the literature (Collepardi et al. 2005) there are two main types of expansion agents: those based on ettringite formation and those based on the formation of calcium hydroxide. In this study, an expansion agent of the second type was used.

Random cracks can hardly be controlled by using conventional reinforcement. On the other hand, the randomly-distributed steel fibers ensure a better control of cracking in any direction inside the concrete mass.

Table 1. Mix proportions for each of the concrete batches.

Concrete type	W^{*1} [kg/m³]	C^{*2} [kg/m³]	W/C [%]	S^{*3} [kg/m³]	G^{*4} [kg/m³]	EA^{*5} [kg/m³]	F^{*6} [kg/m³]	AE^{*7} [kg/m³]	SP^{*8} [kg/m³]
C	175	350	50	788	963	–	–	2.8	1.75
0F95EA	175	350	50	708	963	95	–	2.8	1.75
05F102EA	175	350	50	702	963	102	40	2.8	2.6
05F110EA	175	350	50	695	963	110	40	2.8	2.6
10F110EA	175	350	50	692	963	110	80	2.8	2.6
10F130EA	175	350	50	679	963	130	80	2.8	2.6

Table 2. Properties of concrete.

Concrete type	$f_c'^{*1}$ [N/mm²]	f_t^{*2} [N/mm²]
C	34.8	2.4
0F95EA	2.7	0.2
05F102EA	4.9	0.5
05F110EA	2.5	0.3
10F110EA	8.9	1.1
10F130EA	2.2	0.3

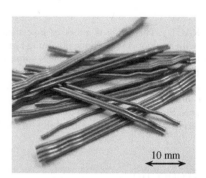

Figure 1. Steel fibers layout.

3 MATERIALS

3.1 Concrete

In this study, a concrete with a designed compressive strength of 30 N/mm², obtained from uniaxial compression tests at 7 days, was considered. In view of the fact that early-strength Portland cement was used in the concrete mixes and taking into account the standard specifications for concrete structures in Japan (JSCE 2002) the uniaxial compressive tests could be run at 7 days and not at 28 days. The early strength Portland cement allows for a rapid development of concrete strength within the first 7 days. After that, the rate of increase in concrete strength is lower than that for normal Portland cement concrete. Six different mix proportions were considered and they are presented in Table 1. The compressive and tensile strengths of each of the concrete mixes were measured at the day of testing and they are summarized in Table 2.

3.2 Reinforcement

The characteristics of the conventional reinforcement used in this study are as follows: bar size D25 (nominal diameter: $d = 25.4$ mm) and steel grade SD345 (yield strength: $f_y = 345$ N/mm²). The specifications are according to JIS G 3112.

3.3 Steel fibers

The steel fibers have crimped ends as it can be seen from Figure 1. The length is $L_f = 30$ mm and the diameter is $d_f = 0.6$ mm. The material properties are: the tensile strength $f_u = 1000$ N/mm² and the Young's modulus $E = 2.1 \times 10^5$ N/mm².

3.4 Expansion agent

The amount of expansion agent was chosen in order to replace a part of the fine aggregate mass and not of the cement mass. This was based on the fact that with the quality control commonly achieved in modern cement factories, the flaws are more likely to come from the aggregates that are used in the mixing than from the cement itself. Moreover, in some damaging processes like ASR, the silica in the aggregate reacts with the alkali in the cement to create the ASR gel. Thus, the expansion agent stands for the reactive fine aggregate.

4 TEST PROGRAM

The test program consisted in a total number of six specimens. For each concrete mix in Table 1 a beam with the dimensions of 1200 × 200 × 150 mm (L × H × B) was cast. The beams were designed to fail in shear and had a longitudinal reinforcement ratio: $p_w = 4.3\%$ and a shear span to effective depth ratio: $a/d = 2.71$. The high value for p_w was chosen to ensure shear failure of the beam. More details of the beam geometry and reinforcement are presented in Figure 2.

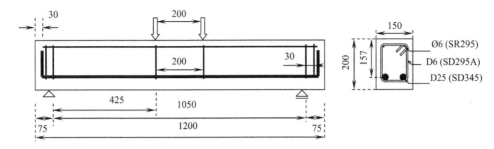

Figure 2. Beem geometry and reinforcement layout (all dimensions are in mm).

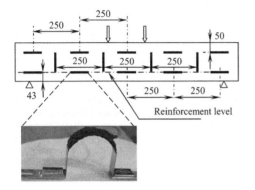

Figure 3. Distribution of PI-5-100 gages on the beam surface.

Table 3. Compressive strength given by Schmidt hammer test $f'_{c,Sch}$.

Concrete type	$f'_{c,Sch}$ [N/mm^2]
0F80EA	21.1
0F95EA	18.9
05F102EA	19.4
05F110EA	15.9
10F110EA	19.9
10F130EA	15.8

Moreover, as it can be seen from Figure 2, there are no stirrups in the shear span. The reason for such a reinforcement layout is that we wanted to evaluate the shear carrying capacity of the concrete itself without the help of shear reinforcement.

Because expansion agent was never used before to generate random cracks in an RC beam, the decision to measure the strain, both in longitudinal bars and on the concrete surface, was taken. The procedure for strain monitoring is similar to the one described in Toma et al. 2006, with the notable difference that in this case the concrete surface strain was measured by means of displacement transducers of PI-5-100 type connected to a data logger that recorded the values every 10 minutes. In this way, the evolution of the surface strains in the concrete was monitored very regularly. The location and the layout of the gages are explained Figure 3.

After a curing period of time of 7 days, the beams were subjected to a four-point loading test. The obtained results are presented in the subsequent chapter.

For a better and easier understanding of the discussion on the results in the subsequent chapters, some explanations are necessary regarding the notation of the specimens: in Table 1 and Table 2 "C" stands for the control case, that is the concrete mix containing neither expansion agent nor steel fibers. This is also the reference specimen to which all the other results are compared. All the other specimens' designations have the ffollowing meaning: the first number represents the fiber percentages contained in each specimen and they are 0, for 0% (no fiber), 05 for 0.5% and 10 for 1.0% respectively. They are followed by the letter "F" which stands for fiber. Further on, the second number represents the amount of expansion agent, in kg/m^3, that partly replaces the fine aggregates. The expansion agent amount number is followed by the letters "EA" that mean "expansion agent". Thus, 0F95EA means that the specified mix proportion contains no fiber, 0F, and 95 kg/m^3 of expansion agent were used in the mix, 95EA.

5 RESULTS AND DISCUSSION

5.1 Concrete strength

Looking at Table 2, it can be seen that except for the control case C, all the other values for the compressive and tensile strengths are quite uncommon. For this reason, a non-destructive Schmidt hammer test was conducted on the RC beams in order to evaluate the concrete strength in the beams and the obtained values are presented in Table 3.

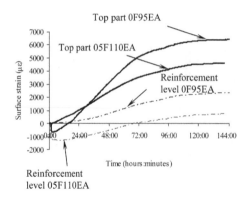

Figure 4. Concrete surface strain history.

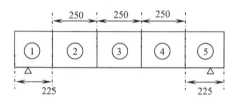

Figure 5. Subdivision of the lateral surface area to evalute the crack density.

The big differences between the values given by the uniaxial compression tests on the cylinders and the concrete strength in the beams given by the Schmidt hammer test on the concrete could be explained by the confining effect the reinforcement has on the concrete. This effect cannot be simulated in the cylinders. However, the confining effect of the steel bars cannot justify by itself the large difference between the measured and the expected values. Further studies should be carried out to fully understand the mechanisms behind such high differences.

5.2 Strain history

The steel strain was measured by means of strain gages glued to the reinforcement at the mid-span. The concrete surface strain was measured by means of displacement type strain gages attached to the beam at different locations (Fig. 3). Some of the gages were located on a horizontal line at the reinforcement level whereas others were positioned on a parallel line 50 mm from the top part of the beam. This type of arrangement gives the possibility of better monitoring the deformation of the RC beam during curing.

From the recorded data the strain history during the curing period of time, both in the longitudinal steel and of the concrete surface, is obtained for each specimen. Such an example is shown in Figure 4 for the concrete surface strain.

By looking at Figure 4, it is clear that by adding steel fibers the values for the surface strain both at the top part and at the reinforcement level are getting smaller, even though there is an increase in the amount of expansion agent. Furthermore we can conclude that the beam is bent upward during the curing and before being tested. The deformed shape of the RC beams could also be confirmed by visual inspection of the specimens. It is as if an axial force has been applied to the beam, at the reinforcement level, creating some sort of prestressing. Since no axial force was applied to the beam and the deformed shaped was the result of the chemical reactions that took place inside the concrete, this phenomenon may be called chemical prestressing.

The compressive concrete surface strain at the reinforcement level for the 05F110EA could be explained by the combined passive resistance against expansion of both the longitudinal steel bars and the steel fibers. According to Figure 2 there is only a small amount of reinforcement at the upper part of the beam. Consequently, the concrete can expand almost freely. The fiber volume in this case is quite small, 0.5%, to have a significant restraining effect on the concrete. Compared with the 0F95EA case, the occurrence of tensile strains in concrete at the reinforcement level for the beam containing steel fibers happens much later.

5.3 Crack density

Before testing, the specimens were checked for the presence of random cracking, since this was one of the purposes of this research and for which the expansion agent was used. High resolution digital pictures of the beams were taken and later processed by means of special software. After manually tracing all the visible cracks, the total crack length for each area was given by the software.

At the same time, the crack width was measured using a mobile laser crack-width reader with the reading range from 0.01 mm to 2.5 mm, with an increment of 0.05 mm. In order to increase the accuracy of the measured data, the surface of the beam was divided into five areas, Figure 5. The boundary between each individual area was chosen to be the line along which the vertical displacement type strain gages were placed. For each area readings of the crack width at different locations were taken and the average crack width for that specific area was computed. An example of the values for the crack widths obtained from the measurements is graphically presented in Figure 6 for two beams. The values of the mean crack width of each area are presented in Table 4.

For the purpose of this research, the crack density is defined as the total area of cracks divided by the area of concrete. For each area, the mean crack width

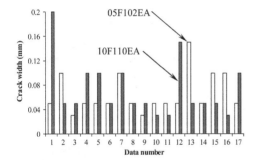

Figure 6. Crack width values for two specimens for area A_2.

Table 4. Mean crack width (mm).

Specimen	Area 1	Area 2	Area 3	Area 4	Area 5
0F95EA	0.15	0.15	0.11	0.15	0.15
05F102EA	0.067	0.065	0.089	0.08	0.082
05F110EA	0.19	0.13	0.08	0.08	0.082
10F110EA	0.08	0.07	0.07	0.09	0.09
10F130EA	0.16	0.19	0.12	0.15	0.14

is considered. This makes the method more attractive and easier to apply since there is no need to deal with each crack separately; avoiding the debate upon how one defines a crack in the case of random cracking, but with the average value.

Finally, the crack density is computed using the following equation:

$$\Omega_i = \frac{\overline{w}_i \cdot \sum_{j=1}^{m} L_j}{A_i} \quad (1)$$

in which $\overline{w}_i$ is the average crack width for the area A_i in mm, $\sum_{j=1}^{m} L_j$ is the total crack length of area A_i given by the software after manual tracing of the cracks, in mm and A_i is the area of concrete, in mm². The index i refers to a specific area the surface of the beam is divided into. The values of the total crack length for each area are summarized in Table 5.

The average crack density factor for the entire specimen can be calculated using Equation 2:

$$\overline{\Omega} = \frac{\sum_{i=1}^{n} \Omega_i}{n} \quad (2)$$

in which Ω_i is the crack density for ith area and n is the number of areas the concrete surface is divided into. Following the previously mentioned procedure,

Table 5. Total crack length (mm).

Specimen	Area 1	Area 2	Area 3	Area 4	Area 5
0F95EA	2900	1560	1110	1650	2940
05F102EA	2000	1450	850	2330	2590
05F110EA	1860	1940	1620	2810	2500
10F110EA	2430	1770	1090	2510	2710
10F130EA	2170	2040	1420	2850	3000

the crack densities for the specimens that showed random cracks before loading are computed and are presented in Figure 7 as a percentage of the area of concrete together with the random crack pattern for the respective beams.

5.4 Shear carrying capacity

Using expansion agent led to the formation of random cracking and the introduction of steel fibers proved to be effective in limiting the deformation of the RC beams. The next step was to test the specimens for their shear resisting capacity. For this, a four-point loading test was conducted.

Before loading, the ultimate load in shear for each of the specimens was computed by adding the contributions of concrete, steel fibers and chemical prestressing. It all starts from the equation of shear carrying capacity of a reinforced concrete beam, Equation 3, proposed by Niwa et al. 1986:

$$V_c = 0.2 \cdot f'^{1/3}_c \cdot p^{1/3}_w \cdot \left(\frac{d}{1000}\right)^{-1/4} \cdot b_w \cdot d \cdot \left(0.75 + \frac{1.4d}{a}\right) \quad (3)$$

where f'_c is the compressive strength of concrete in N/mm² (determined on the concrete from the RC beams by means of the Schmidt hammer test, see Table 3), p_w is the longitudinal reinforcement ratio equal to 4.3%, d is the effective depth, in mm, and b_w is the web thickness, in mm.

As mentioned previously, the use of expansion agent created some sort of chemical prestressing. Its effect was taken into account, according to the standard specifications for concrete structures (JSCE 2002), by the term β_n in the following equation:

$$V_{cp} = V_c \cdot \beta_n \quad (4)$$

where V_c is the shear carrying capacity computed using Equation 3 and β_n is computed by means of the next relationship:

$$\beta_n = 1 + \frac{2 \cdot M_0}{M_u} \quad (5)$$

705

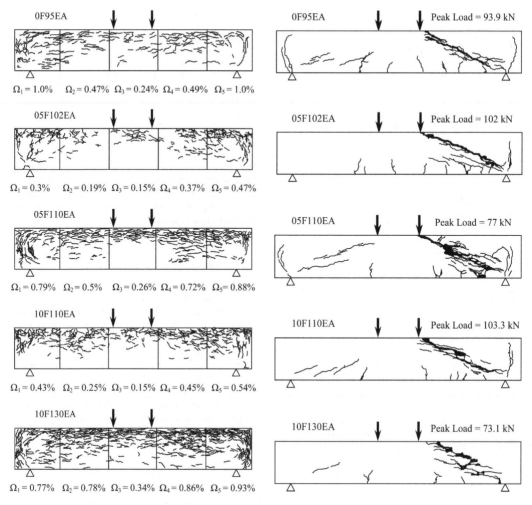

Figure 7. Crack pattern before testing and crack densities for each area as a percentage of the initial are of concrete.

Figure 8. Crack pattern at the onset of shear failure.

in which M_u is the ultimate resisting moment of the beam and M_0 is the decompression moment (the bending moment that when applied creates a zero stress in the extreme lower fibers). The procedure for computing the decompression moment is explained in detail in Toma et al. 2006.

Because fibers were also used in this research, their contribution to the shear carrying capacity must be also taken into account. This was done by means of the following empirical equation (Swamy et al. 1993):

$$v = 0.37 \cdot \tau \cdot V_f \cdot \frac{L_f}{d_f} \qquad (6)$$

where V_f is the volume of fibers, in %, L_f is the fiber length, in mm, d_f is the fiber diameter, in mm, and

τ is the bond strength equal to 4.15 N/mm² for steel fibers with crimped ends in the absence of any pull-out tests. Equation 6 was shown to provide good accuracy with respect to the experimental data (Cucchiara 2004).

Finally, the shear carrying capacity is computed as:

$$V_{cal} = V_{cp} + v \cdot b_w \cdot d \qquad (7)$$

The above-mentioned evaluation procedure was used in previous research works on RC beams, without random pre-cracking, containing both expansion agent and steel fibers and lead to good estimation of the experimental results (Toma et al. 2006).

The predicted peak loads are summarized in Table 6 where they are compared with the values obtained from the loading tests.

Table 6. Calculated results versus experimental results.

Concrete type	V_{cal} [kN]	V_{exp} [kN]	V_{cal}/V_{exp}
C	69.2	68.5	1.01
0F95EA	111.3	93.9	1.19
05F102EA	93.7	102.0	0.92
05F110EA	134.1	77.0	1.74
10F110EA	115.0	103.3	1.11
10F130EA	130.8	73.1	1.79

By looking at the values of the ratio V_{cal}/V_{exp} in Table 6 it can be seen that the evaluation of the shear carrying capacity of RC beams with random cracks using existing equations leads to an overestimation of the experimental results in most of the cases. This is due to the fact that the existing cracks before loading are not taken into account when evaluating the shear carrying capacity. Summing up, by introducing the unfavorable effect of pre-cracking, the approximation in the evaluation of the ultimate shear capacity would be better.

The crack patterns for each of the tested specimens are presented in Figure 8. A closer look at both Figures 7 and 8 and Table 6 reveals an interesting fact in terms of the location of the shear crack and in terms of the shear carrying capacity. First of all, according to the division of the RC beam surface, areas 2 and 4 are entirely located in the shear span and thus anything occurring in these areas will affect the shear carrying capacity. A comparison between the crack densities Ω_2 and Ω_4 for each of the specimens in Figure 7 shows that in every case $\Omega_2 < \Omega_4$. This means that the shear crack is more likely to happen in the shear span containing area 4 than in the shear span area 2 is located in. According to Figure 8 the failure crack is located in the shear span corresponding to area 4, whose crack density (Ω_4) is larger than that of area 2 (Ω_2).

Moreover, the values of the crack density Ω_4 for each specimen in Figure 7 are somehow related to the shear carrying capacity of each beam. For example, the values of Ω_4 for the specimens 05F110EA and 10F130EA are larger than the value of Ω_4 for 0F95EA and thus the shear carrying capacity of those beams will be smaller than for 0F95EA. This can indeed be seen in Table 6. It is clear that the peak load for 05F102EA will be higher than for 0F95EA because $\Omega_4^{05F102EA} < \Omega_4^{0F95EA}$. However, even though $\Omega_4^{10F110EA} < \Omega_4^{0F95EA}$, see Figure 7, their values are relatively close to one another and so should be their shear carrying capacities. Indeed, the experimental results, presented in Table 6, show an increase in the shear carrying capacity from 93.9 kN for 0F95EA to 103.3 kN for 10F110EA.

6 CONCLUSIONS

Using expansion agent proves to be effective in creating random cracking similar to the cases observed in actual RC members. By changing the amount of expansion agent the random crack pattern before loading can be controlled, in terms of visual observation.

Adding short steel fibers to the concrete mixes has several favorable effects, since steel fibers reduce the surface strains by reducing the mean width of randomly-distributed cracks, increase the deformability of RC members and enhance their shear carrying capacity. However, there are cases when the steel fibers could only increase the deformability of the severely damaged RC beams, with little effect on the peak load.

The notion of crack density is introduced in this research to help for a quantitative representation of the effect of random cracking. The method is relatively easy to use because it relies on the average crack width and on the total crack length from a specified area of concrete. The high values of the average crack density for the entire specimen lead to lower peak loads for the respective specimens.

The procedure for evaluating the ultimate load in shear for RC beams with random cracks leads to an excessive over-evaluation of the experimental results for the beams that exhibit random cracking before loading. While the addition of steel fibers and the chemical prestress phenomenon that occurred due to the use of expansion agent lead to an increase in the shear carrying capacity, the occurrence of random cracking before loading has a diminishing effect on the peak load. The effect of the latter should be also taken into account. Further tests are necessary (a) to improve the procedure that takes into account the crack density factor; and (b) to quantify in a better way the interaction between the steel fibers and the randomly distributed pre-cracks.

REFERENCES

Bresler, B. & Scordelis, A.C. 1963. Shear strength of reinforced concrete beams. ACI Journal. 60 (1): 51–74.
Collepardi, M., Borsoi, A., Collepardi, S., Olagot, J.J.O. & Troli, R. 2005. Effects of shrinkage reducing admixtures in shrinkage compensating concrete under non-wet curing conditions. Cement & Concrete Composites, 27: 704–708.
Cucchiara, C., La Mendola, L & Papia, M. 2004. Effectiveness of stirrups and steel fibers as shear reinforcement. Cement & Concrete Composites. 26: 777–786.
Diamond, S. 1997. Alkali silica reactions – some paradoxes. Cement & Concrete Composites. 19: 391–401.
JSCE. 2002. Standard specifications for concrete structures – structural performance verification, JSCE Guideline for Concrete 3: 25.
Kreffeld, W. J. & Thurston C. W. 1966. Study on the shear and diagonal tension strength of simply supported reinforced concrete beams. ACI Journal. 63 (4): 451–476.

Niwa, J., Yamada, K., Yokozawa, K. & Okamura, H.1983. Reevaluation of the equation for shear strength of reinforced concrete beams without web reinforcement. *Journal of Materials, Concrete Structures and Pavements*, JSCE. 372 (5): 167–176.

Petersen, K. 1992. Effects of silica fume on alkali-silica expansion in mortar specimens. *Cement & Concrete Research*, 22: 15–22.

Smaoui, N., Berube, M. A., Fournier, B., Bissonnette, B. & Durand, B. 2005. Effects of alkali addition on the mechanical properties and durability of concrete. *Cement & Concrete Research*, 35: 203–212.

Swamy, R. N., Jones, R. & Chiam, A.T.P. 1993. Influence of steel fibers on the shear resistance of lightweight concrete T-beams. *ACI Structural Journal*, 90 (1): 103–114.

Toma, I.O, Miki, T. & Niwa, J. 2006. Influence of steel fibers on the behavior of RC beams with random cracks. *Proceedings of the 10th East Asia-Pacific Conference on Structural Engineering and Construction*, Bangkok, Thailand, Materials, Experimentation, Maintenance and Rehabilitation: 413–418.

Fracture Mechanics of Concrete and Concrete Structures – Design, Assessment and Retrofitting of RC Structures – Carpinteri, et al. (eds)
© 2007 Taylor & Francis Group, London, ISBN 978-0-415-44616-7

Finite element analysis of diagonal tension failure in RC beams

T. Hasegawa
Institute of Technology, Shimizu Corporation, Tokyo, Japan

ABSTRACT: Finite element analysis of diagonal tension failure in a reinforced concrete beam is performed by using different meshes and different concrete crack models. It is found that inserting specific finite element bands in the mesh to model diagonal cracking improves crack localization and propagation. Multi-directional fixed crack and rotating crack models exhibit convergence problems, and lead to flexural or shear compression failure rather than diagonal tension failure. It is shown that the Multi Equivalent Series Phase Model clearly describes the complex mixed mode fracture that is typical of diagonal tension failure.

1 INTRODUCTION

Numerical analysis is important and effective for studying complicated mechanisms of diagonal tension failure of reinforced concrete beams without shear reinforcement, since numerical analysis can take factors influencing the failure and the causes of failure into account individually and systematically, whereas experiments cannot easily do so. In the previous study (Hasegawa 2004b) finite element analysis of diagonal tension failure in a reinforced concrete beam was performed using the Multi Equivalent Series Phase Model (MESP model; Hasegawa 1998), and the failure mechanisms were discussed by analyzing the numerical results. The first series of analysis showed that in order for diagonal tension failure of the beam to be complete, the longitudinal splitting crack should propagate unstably, leading to widening and propagation of the diagonal crack. In addition the second series of analysis with the branch-switching method was performed to simulate diagonal tension failure, assuming that the failure results from a bifurcation starting at a singular point (bifurcation or limit point) on the equilibrium path. Both series of analysis were able to simulate localization and initial propagation of diagonal cracks, but not unstable propagation of the cracks, and the formation of final shear collapse mechanism of beam could not be simulated.

In the present study (Hasegawa 2004a, 2005, 2006), based on the results of the previous analysis, another series of finite element failure analysis of a reinforced concrete beam is performed using different finite element meshes and alternative concrete crack models as factors to influence the diagonal tension failure.

2 ANALYSIS MODEL

2.1 Analysis cases D

As in the previous analysis, the diagonal tension failure of a reinforced concrete slender beam specimen, BN50, having an effective depth of 450 mm, tested at the University of Toronto (Podgorniak-Stanik 1998) is simulated in this study. The experimental cracking pattern after failure is shown in Figure 1. Figures 2 and 3 are cracking pattern results for the previous analysis cases A1 and A5. In each analysis a regular cross-diagonal (CD) mesh (finite element mesh

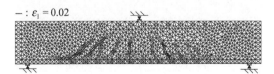

Figure 1. Experimental cracking pattern after failure.

$-: \varepsilon_1 = 0.02$

Figure 2. Crack strain at V_u in analysis case A1.

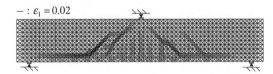

$-: \varepsilon_1 = 0.02$

Figure 3. Crack strain at V_u in analysis case A5.

type e-1: Fig. 4) or a random Delaunay triangulation mesh (finite element mesh type e-2: Fig. 5) was utilized. The plotted line in the figures indicates the maximum principal strain $\varepsilon_1 \geq 5\varepsilon_{t0}$ with the thickness proportional to its value. This represents crack strain and crack direction at maximum shear load V_u in the analysis cases, and is a good measure of crack width (ε_{t0} = the tensile strain corresponding to the tensile strength). These crack strain figures are compared with the experimental cracking pattern after failure in Figure 1. For accurate and reasonable simulation of diagonal tension failure it is necessary to reproduce a realistic shape of curved diagonal cracks and longitudinal cracks in experiments. To accelerate propagation of the curved diagonal cracks in the previous analysis cases A1 and A5, modified finite element mesh types e-1A, e-1B, and e-2A are considered by inserting CD mesh bands for the potential crack paths (Figs 6–8) in analysis cases D. The performed analysis cases are shown in Table 1. The Multi Equivalent Series Phase Model is assumed in these analysis cases as the concrete constitutive model.

2.2 Analysis cases E and F

To examine the effect of concrete crack model on diagonal tension failure, the multi-directional fixed crack (MDFC) model and the rotating crack (RC) model are assumed in the second series of analysis cases E1 and F1, utilizing the Delaunay mesh (finite element mesh type e-2).

2.3 Crack models for concrete

The Multi Equivalent Series Phase Model is a versatile nonlocal constitutive model, and is capable of describing cracking behavior under tension as well as shear and compression with good accuracy. The model is used in analysis cases A and D.

The multi-directional fixed crack model and the rotating crack model, adopted in this study, are standard ones available in the general purpose finite element system DIANA (Witte & Feenstra 1998).

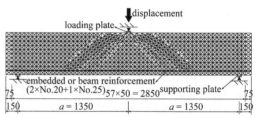

Figure 7. Finite element mesh type e-1B.

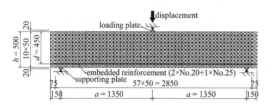

Figure 4. Finite element mesh type e-1.

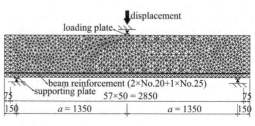

Figure 8. Finite element mesh type e-2A.

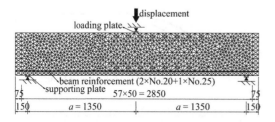

Figure 5. Finite element mesh type e-2.

Table 1. Analysis case.

Analysis case	Finite element mesh type	Reinforcement	Concrete crack model
A1	e-1	Embedded	MESP model
A5	e-2	Beam	MESP model
D1	e-1A	Embedded	MESP model
D2	e-1B	Embedded	MESP model
D3	e-1B	Beam	MESP model
D4	e-2A	Beam	MESP model
E1	e-2	Beam	MDFC model
F1	e-2	Beam	RC model

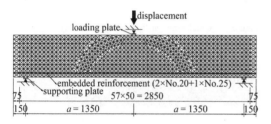

Figure 6. Finite element mesh type e-1A.

In analysis case E1 with the multi-directional fixed crack model, a strain hardening-softening type of elastoplastic model with the Drucker-Prager criterion under compression is combined. The hardening-softening parameter is determined to fit the uniaxial compression behavior to the relationship given in the CEB-FIP Model Code 1990 (Euro-International Committee for Concrete 1993). The crack band model is used for tension, and a linear elastic-bilinear softening stress-strain relationship as well as an appropriate fracture energy value are assumed together with a threshold angle of 60 degrees. Crack shear behavior is modeled with a shear retention factor of 0.01 (nearly equal to zero).

In analysis case F1 with the rotating crack model, uniaxial tension and compression stress-strain relationships are assumed to be identical to the ones for the multi-directional fixed crack model. A shear model after cracking is unnecessary since coaxiality between principal stress and strain determines incremental shear stiffness in the constitutive relation.

3 MESH DEPENDENCY

Figure 9 compares the calculated shear response in analysis cases A1, D1, D2, and D3, using CD meshes, with the experiment. Figures 11 and 13 are crack strain at maximum shear load V_u in analysis cases D1 and D2. Figures 12 and 14 show incremental deformation at V_u.

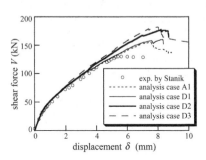

Figure 9.   Shear response in analysis cases D.

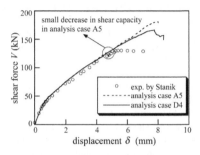

Figure 10.   Shear response in analysis case D4.

in analysis cases D1 and D2. Those crack strain figures are to be compared with the experimental cracking pattern after failure, shown in Figure 1. In analysis case D1 a main diagonal crack does not propagate in the inserted CD mesh band, but results in a very similar final cracking pattern to analysis case A1 and diagonal tension failure due to aligned elements with an inclination of about 45 degrees. Because of the similar diagonal crack shape the maximum shear loads are almost equal in analysis cases A1 and D1, which are overestimates of the experimental result. On the other hand in analysis case D2, diagonal cracks do not propagate to upper side of beam both inside and outside the CD mesh band, and a flexural failure with tensile reinforcement yielding occurs. To increase dowel action of tensile reinforcement, beam elements are used in analysis case D3 instead of embedded reinforcement, however, an improvement is not achieved.

In Figure 10 the shear response obtained in analysis cases A5 and D4, using the Delaunay meshes, is shown. Figures 15 and 16 are crack strain at step 400, which corresponds to the experimental maximum shear load,

$- : \varepsilon_1 = 0.02$

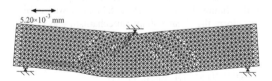

Figure 11.   Crack strain at V_u in analysis case D1.

5.20×10^{-3} mm

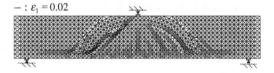

Figure 12.   Incremental deformation at V_u in analysis case D1.

$- : \varepsilon_1 = 0.02$

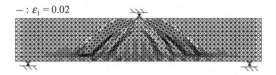

Figure 13.   Crack strain at V_u in analysis case D2.

2.27×10^{-2} mm

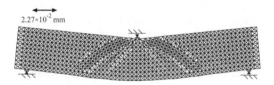

Figure 14.   Incremental deformation at V_u in analysis case D2.

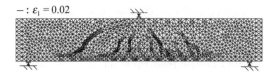

$-: \varepsilon_1 = 0.02$

Figure 15. Crack strain at step 400 in analysis case D4.

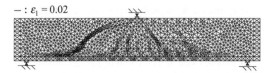

$-: \varepsilon_1 = 0.02$

Figure 16. Crack strain at V_u in analysis case D4.

and at V_u in analysis case D4. In Figure 17 incremental deformation at V_u is shown. In the previous analysis case A5 using ordinal displacement control as well as branch-switching analysis, initial propagation of the main diagonal cracks was well simulated, but subsequent further unstable propagation of those cracks could not be obtained due to mesh dependency such as stress-locking and crack diffusion. In analysis case D4, the curved CD mesh band is inserted in the Delaunay mesh for the purpose of accelerating such unstable propagation of the diagonal crack. As shown in Figures 15–17, analysis case D4 succeeds in simulating the diagonal crack localizing into the inserted CD mesh band, and the further propagation to upper side of the beam. However, the further propagation of the diagonal crack is not unstable, but still stable. Therefore, the shear collapse mechanism can not be simulated, and finally flexural failure occurs.

4 MIXED MODE FRACTURE

It is believed that Mode I tensile cracking is dominant within the fracture process zone in diagonal tension failure, and the effect of mixed mode fracture of tension and shear is regarded as less important. In this section stress-strain responses in the finite elements corresponding to the diagonal crack path are examined to study the effect of mixed mode fracture in diagonal tension failure.

Finite elements a–f (Figs 18, 25) are selected in analysis cases A1 and D4 to examine the stress-strain responses, which are shown in Figures 19–24, 26–31. Figures 32–35 are the angles θ_σ, θ_ε of the principal stress and strain s axes for elements c and d in each analysis case. Mode I fracture, observed as tensile softening responses $\sigma_1 - \varepsilon_1$, $\sigma_{xx} - \varepsilon_{xx}$, and $\sigma_{yy} - \varepsilon_{yy}$, is accompanied by mode II fracture, observed as a shear softening response $\tau_{xy} - \gamma_{xy}$, and which forms complicated mixed mode fracture. In elements c and d for both analysis cases A1 and D4 tensile cracking, as

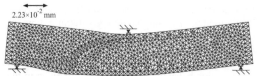

2.23×10^{-2} mm

Figure 17. Incremental deformation at V_u in analysis case D4.

observed in softening responses of relation $\sigma_1 - \varepsilon_1$, is followed by shear softening response recognized in the relation $\tau_{xy} - \gamma_{xy}$ at the earlier stage of loading. And then the shear softening turns into shear hardening because of shear friction on the crack due to shear strain increase as well as suppressed crack dilatancy.

In Figures 34 and 35 it is shown that relatively large rotation of the principal axes occurs just after the tensile softening starts, i.e. at the initial stage of the fracture process zone in elements c and d, and that coaxiality between principal stress and strain is approximately preserved. As observed above, mixed mode fracture and the effect of rotation of the principal axes start at a very early stage of the fracture process zone, but not at the last stage with wide open cracks. Therefore, crack and constitutive models that neglect or underestimate this effect should be applied with caution. Undoubtedly the observed mixed mode phenomena in analysis have not been verified by experiments. However, the consistent, rational and reasonable results shown for overall structural behavior might give the argument validity. It is worth noticing that the above-mentioned mixed mode fracture in the process zone as well as rotation of the principal axes could not be simulated with accuracy by using aggregate interlock models derived from perfectly open cracks in concrete, rotating crack models with coaxiality between principal stress and strain, or multi-directional fixed crack models either with simple crack shear models or with shear retention neglected.

Although elements b have large shear strains due to large crack mouth sliding displacement (CMSD) at the tension extreme fiber as well as dowel action of the tensile reinforcement bar, shear softening in elements b is monotonic, and does not turn into shear hardening as in elements c and d because of large tensile strain due to large crack mouth opening displacement (CMOD) at the tension extreme fiber (Figs 20, 27). In analysis case A1 where the maximum shear load and diagonal tension failure are predicted relatively well, mixed mode response similar to elements c and d is observed at element e. At the maximum shear load in analysis case A1 shear softening response at element f (Fig. 24) completes the diagonal tension failure, indicating that the shear crack reaches to underneath the loading plate. On the other hand, in analysis case D4

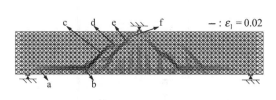 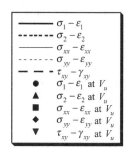

$-:\varepsilon_1 = 0.02$

Legend:
- $\sigma_1 - \varepsilon_1$
- $\sigma_2 - \varepsilon_2$
- $\sigma_{xx} - \varepsilon_{xx}$
- $\sigma_{yy} - \varepsilon_{yy}$
- $\tau_{xy} - \gamma_{xy}$
- ● $\sigma_1 - \varepsilon_1$ at V_u
- ▲ $\sigma_2 - \varepsilon_2$ at V_u
- ■ $\sigma_{xx} - \varepsilon_{xx}$ at V_u
- ◆ $\sigma_{yy} - \varepsilon_{yy}$ at V_u
- ▼ $\tau_{xy} - \gamma_{xy}$ at V_u

Figure 18.   Selected elements in analysis case A1.

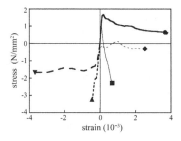

Figure 19.   Stress-strain responses of element a in analysis case A1.

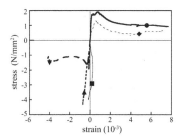

Figure 22.   Stress-strain responses of element d in analysis case A1.

Figure 20.   Stress-strain responses of element b in analysis case A1.

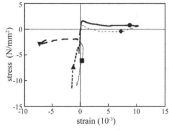

Figure 23.   Stress-strain responses of element e in analysis case A1.

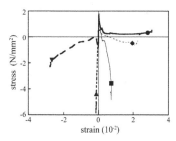

Figure 21.   Stress-strain responses of element c in analysis case A1.

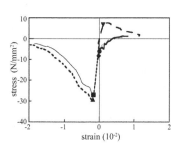

Figure 24.   Stress-strain responses of element f in analysis case A1.

where the maximum shear load and diagonal tension failure are not predicted well, neither tensile softening fracture nor shear softening fracture reaches to element e at the tip of the diagonal crack as well as element f beneath the loading plate (Figs 30, 31).

5   EFFECT OF CRACK MODELS

Figure 36 shows calculated shear response in analysis cases A5, E1, and F1, using the Multi Equivalent Series Phase Model, the multi-directional fixed crack model, and the rotating crack model. In the previous analysis case A5, the diagonal tension failure mode with unstable propagation of diagonal and longitudinal cracks became dominant when the small decrease in shear capacity occurred at step 375. However, after the

decrease a bifurcation from diagonal tension failure mode to bending mode took place. Therefore, step 375 is regarded as corresponding to the maximum shear load, and shear hardening behavior after the small decrease in shear capacity is neglected in the following discussion. Figure 37 shows incremental deformation at step 375 in analysis case A5. Figures 38 and 39 are incremental deformation at maximum shear load V_u in analysis cases E1 and F1. As pointed out in the previous study, shear response up to the small decrease of shear capacity captures the experimental results very well in analysis case A5. On the other hand analysis case F1 using the rotating crack model results in underestimation of the experimental maximum shear load and stiffness, while analysis case E1 using the multidirectional fixed crack model can achieve relatively good prediction of the experiment.

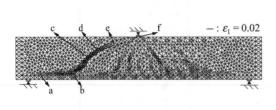 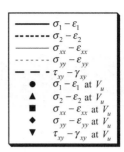

Figure 25. Selected elements in analysis case D4.

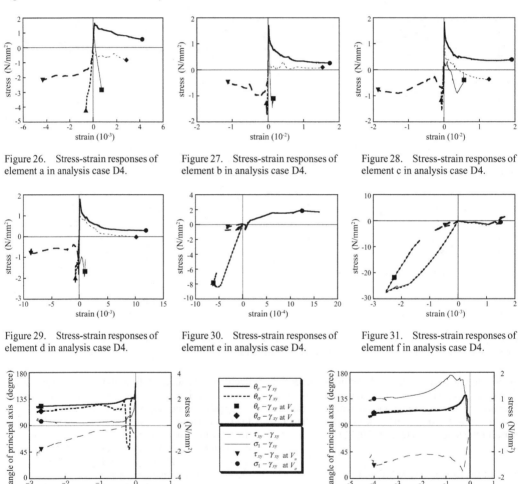

Figure 26. Stress-strain responses of element a in analysis case D4.

Figure 27. Stress-strain responses of element b in analysis case D4.

Figure 28. Stress-strain responses of element c in analysis case D4.

Figure 29. Stress-strain responses of element d in analysis case D4.

Figure 30. Stress-strain responses of element e in analysis case D4.

Figure 31. Stress-strain responses of element f in analysis case D4.

Figure 32. Responses of element c in analysis case A1.

Figure 33. Responses of element d in analysis case A1.

In analysis cases E1 and F1, using the multi-directional fixed crack model and the rotating crack model it is relatively hard to obtain convergence in iterative calculation for equilibrium. When convergence criteria (one percent of relative out-of-balance force)

can not be satisfied after a final iteration (fifty iterations) at a step the calculation is continued to the next step, bringing out-of-balance forces in the final iteration to the next step although strictly speaking the calculation should be terminated. Surprisingly at

714

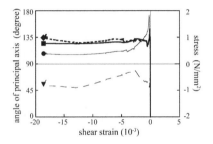

Figure 34. Responses of element c in analysis case D4.

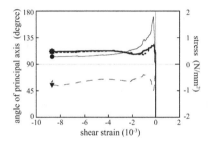

Figure 35. Responses of element d in analysis case D4.

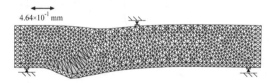

Figure 36. Shear response in analysis cases E1 and F1.

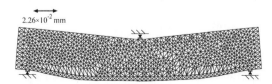

Figure 39. Incremental deformation at V_u in analysis case F1.

$-: \varepsilon_1 = 0.02$

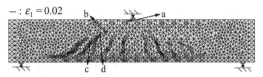

Figure 40. Selected elements in analysis case A5.

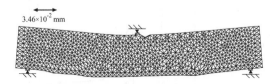

Figure 37. Incremental deformation at V_u in analysis case A5.

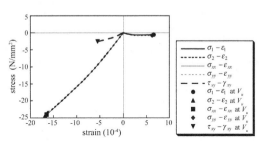

Figure 41. Stress-strain responses of element a in analysis case A5.

Figure 38. Incremental deformation at V_u in analysis case E1.

the steps corresponding to eleven and nine percent of all the steps up to maximum shear load, the convergence criteria are not satisfied in analysis cases E1 and F1. Both the multi-directional fixed crack model and the rotating crack model are considered to have serious problems in convergence and lack robustness as numerical crack models. On the other hand, in analysis case A5 using the Multi Equivalent Series Phase Model convergence criteria are satisfied at all the steps, which confirms that the Multi Equivalent Series Phase Model possesses excellent robustness as a numerical crack model.

In Figures 40, 45, 46 crack strain is shown for analysis cases A5, E1, and F1, and finite elements a-d are selected to examine the stress-strain responses in the diagonal cracks. Figures 41–44, 47–49, 50–52 are the stress-strain responses in the analysis cases. Minimum principal stress-strain responses $\sigma_2 - \varepsilon_2$ and compressive responses $\sigma_{xx} - \varepsilon_{xx}$ at elements a indicate that compressive failure occurs in the area beneath the loading plate and at the tip of the diagonal crack (Figs 47, 50). We have here a very difficult problem in judging whether this compressive failure at the upper side of the beam is compressive failure due to bending or completion of propagation of the diagonal crack. To help judge the final failure mode important points to be

715

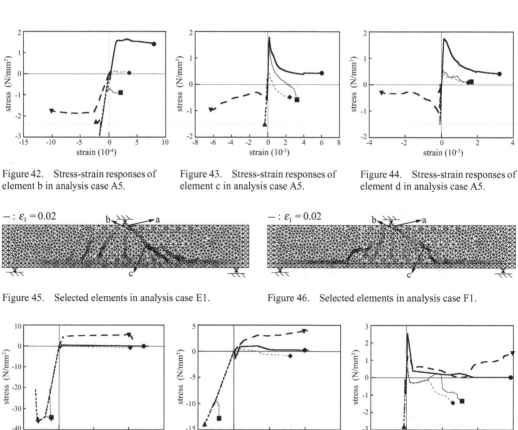

Figure 42.   Stress-strain responses of element b in analysis case A5.

Figure 43.   Stress-strain responses of element c in analysis case A5.

Figure 44.   Stress-strain responses of element d in analysis case A5.

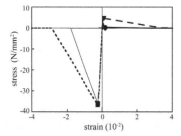

Figure 45.   Selected elements in analysis case E1.

Figure 46.   Selected elements in analysis case F1.

Figure 47.   Stress-strain responses of element a in analysis case E1.

Figure 48.   Stress-strain responses of element b in analysis case E1.

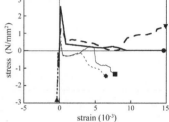

Figure 49.   Stress-strain responses of element c in analysis case E1.

Figure 50.   Stress-strain responses of element a in analysis case F1.

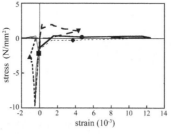

Figure 51.   Stress-strain responses of element b in analysis case F1.

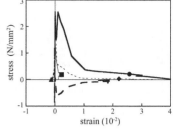

Figure 52.   Stress-strain responses of element c in analysis case F1.

discussed are incremental deformation, which represents the dominant failure mode, and crack orientation. Incremental deformation at maximum shear load V_u, shown in Figures 38 and 39, indicates that the dominant fracture deformation mode at V_u is considered to be flexural rather than diagonal tension failure. Furthermore, most of the crack strain in the area beneath the loading plate and at the tip of the diagonal crack represents splitting cracks under compression, having its direction parallel to the flexural compression fiber. Based on the above discussion it is considered that flexural failure or shear compression failure is dominant rather than the diagonal tension failure mode in analysis cases E1 and F1, using the multi-directional fixed crack model and the rotating crack model. On the other hand at the maximum shear load in analysis case A5 using the Multi Equivalent Series Phase Model a typical diagonal tension failure mode prevails, in which propagation of diagonal and longitudinal cracks is dominant (Fig. 37).

The Multi Equivalent Series Phase Model and the rotating crack model predict shear softening in elements b and d, which correspond to the tip of the diagonal crack at the maximum shear load (Figs 42, 44, 51). But a shear hardening response is observed in a similar element in the case of the multi-directional fixed crack model (Fig. 48), and which results in stiff structural behavior of the reinforced concrete beam in analysis case E1. At elements c in the middle of the beam depth, where shear slip on diagonal crack prevails, transition from shear softening to hardening is recognized in analysis cases A5 and E1 (Figs 43, 49). However, unreasonable shear response with $\tau_{xy} < 0$ for $\gamma_{xy} < 0$ is obtained from the rotating crack model since apparent negative shear stiffness emerges only due to assumption of coaxiality between principal stress and strain in the model, but without physical meaning (Fig. 52).

6   CONCLUSIONS

Finite element analysis of diagonal tension failure in a reinforced concrete beam is performed using different finite element meshes and several crack models for concrete. To accelerate propagation of curved diagonal cracks cross-diagonal mesh bands are inserted in the original meshes. This can improve localization and propagation of the cracks in some analysis cases, but it does not result in a rational shear collapse mechanism. The multi-directional fixed crack model can predict shear capacity and stiffness of the beam with relatively good accuracy, but the rotating crack model cannot. In analysis using both models, the flexural or shear compression failure mode is dominant rather than diagonal tension failure mode. Both models have serious problems in convergence and lack robustness as numerical crack models. However, the Multi Equivalent Series Phase Model results in good convergence and has excellent robustness. Stress-strain responses in finite elements corresponding to diagonal cracks give clear explanations of complicated mixed mode fracture relating to the mechanism of diagonal tension failure. It is found that mixed mode fracture and the effect of rotation of the principal axes start at a very early stage of the fracture process zone. In the case of the multi-directional fixed crack model and the rotating crack model, some of the stress-strain responses are unreasonable and are responsible for the inability to capture the diagonal tension failure.

REFERENCES

Euro-International Committee for Concrete (CEB). 1993. CEB-FIP Model Code 1990. London: Thomas Telford.

Hasegawa, T. 1998. Multi equivalent series phase model for nonlocal constitutive relations of concrete. In H. Mihashi & K. Rokugo (eds), *Fracture Mechanics of Concrete Structures:* 1043–1054. Freiburg: AEDIFICATIO Publishers.

Hasegawa, T. 2004a. Mesh dependency in finite element analysis for diagonal tension failure of reinforced concrete beam. *Proc. 59th annual conference, JSCE* 5: 771–772.

Hasegawa, T. 2004b. Numerical study of mechanism of diagonal tension failure in reinforced concrete beams. In V. C. Li, C. K. Y. Leung, K. J. Willam, & S. L. Billington, (eds), *Fracture Mechanics of Concrete Structures*: 391–398. Ia-FraMCos, USA.

Hasegawa, T. 2005. Mixed mode fracture in mechanism of diagonal tension failure of reinforced concrete beam. *Proc. 60th annual conference.* JSCE 5: 1053–1054.

Hasegawa, T. 2006. Concrete constitutive models in analysis for diagonal tension failure of reinforced concrete beam. *Proc. 61st annual conference.* JSCE 5: 857–858.

Podgorniak-Stanik, B. A. 1998. The influence of concrete strength, distribution of longitudinal reinforcement, amount of transverse reinforcement and member size on shear strength of reinforced concrete members. *M.A.S. thesis*: University of Toronto.

Witte, F. & Feenstra, P. 1998. *DIANA – Finite element analysis. Release 7.* Delft: Building and Construction Research. Netherlands Organization for Applied Scientific Research.

Fracture Mechanics of Concrete and Concrete Structures – Design, Assessment and Retrofitting of RC Structures – Carpinteri, et al. (eds)
© 2007 Taylor & Francis Group, London, ISBN 978-0-415-44616-7

Explicit cohesive crack modeling of dynamic propagation in RC beams

R.C. Yu, X.X. Zhang & G. Ruiz
ETSI de Caminos, C.y P., Universidad de Castilla-La Mancha, Ciudad Real, Spain

ABSTRACT: In this work we simulate explicitly the dynamic fracture propagation in reinforced concrete beams. In particular, adopting cohesive theories of fracture with the direct simulation of fracture and fragmentation, we represent the concrete matrix, the steel re-bars and the interface between the two materials explicitly. Therefore the crack nucleation within the concrete matrix, through and along the re-bars, the deterioration of the concrete-steel interface are modeled explicitly. The numerical simulations are validated against experiments of three-point-bend beams loaded dynamically under various strain rates. By extracting the crack-tip positions and the crack mouth opening displacement history, a two-stage crack propagation, marked by the attainment of the peak load, is observed. The first stage corresponds to the stable crack advance, the second one, the unstable collapse of the beam. The main crack propagation patterns are also illustrated for several loading velocities.

1 INTRODUCTION

Many phenomenological models have been developed to study the problem of complex fracture processes in reinforced concrete specimens, in particular, when the propagation is static (Bosco and Carpinteri 1992; Ruiz et al. 1999; Ruiz 2001; Ruiz et al. 2006). Those models focus on a crack that propagates through a reinforcement layer and that, at opening, causes the steel bars to be pulled-out. The main difficulty consists in modeling the propagation of the crack through the reinforcement layer, as such a layer—in 2D models—constitutes a discontinuity that stops the crack advance. Modeling the steel-concrete interaction is not simple either, since the pull-out process produces damage at the interface and it may also generate secondary fracture processes within the concrete bulk. Some of the aforementioned models solve both problems by substituting the rebars for closing forces applied at the crack lips, whose intensity relates to the strength of the reinforcement and of the interface (Hededal and Kroon 1991, Brincker et al. 1999). Some authors model the steel bars explicitly, however they only account for the extreme cases of perfect or null adherence (Hawkins and Hjorsetet 1992). Ruiz et al. (2006) developed a simplified model that enables propagation of the crack through the reinforcement as well as interface deterioration. They adopted a computational strategy that consisted of overlapping both materials in the same spatial position. Concrete was modeled as a continuum, whereas steel was overlapped and connected by interface elements to nodes occupying the same

initial position (Hawkins and Hjorsetet 2004). Concrete continuity allowed crack propagation through the reinforcement and interface elements transmitted shear stresses depending on the relative displacement between the two materials. However, their numerical model was not able to reproduce micro-cracking and/or damage around the main crack (Ruiz 2001) and along the steel-concrete interface (Ben Romdhane and Ulm 2002).

In dynamic regime, the fracture processes in reinforced concrete beams are more complicated than in the static case, as stress waves can be partly reflected from the steel-concrete interface and partly transmitted to the rebar. The interplay between the interface, steel re-bar and concrete matrix challenges the above numerical models, in which either the rebar is not physically represented (Hededal and Kroon 1991; Brincker et al. 1999; Ruiz et al. 2006), or the progressing deterioration of the interface is not reflected (Hawkins and Hjorsetet 1992). Indeed, such an interaction calls for an explicit modeling of those phenomena.

In this work, following previous efforts in explicitly modeling in static regime (Yu and Ruiz 2006), we attempt to simulate explicitly the dynamic fracture propagation in reinforced concrete beams. In particular, adopting cohesive theories of fracture with the direct simulation of fracture and fragmentation (Ortiz and Pandolfi 1999; Pandolfi and Ortiz 2002), we represent the concrete matrix, the steel re-bars and the interface between the two materials explicitly. Therefore the crack nucleation within the concrete matrix, through and along the re-bars, the deterioration of the

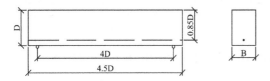

Figure 1. A reinforced concrete beam subjected to three point bending.

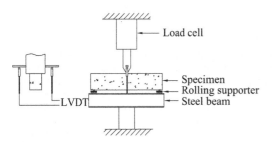

Figure 2. Experimental set-up for three-point bending tests on beams.

Table 1. Micro-concrete mechanical properties.

	f_t MPa	f_c MPa	E_c GPa	G_F N/m	ρ kg/m^3
Mean	3.74	28.43	29.64	49.48	2113
Std. Dev	0.22	0.32	6.3	4.5	–

Table 2. Mechanical properties of the steel re-bar and the interface bond strength.

E_s GPa	$\sigma_{0.2}$ MPa	σ_u MPa	ϵ_u –	τ_c MPa
133.2	434.1	465.1	0.9%	5.34 ± 0.80

concrete-steel interface are modeled explicitly. The organization of this paper is as follows. In the next section, the experimental setup is illustrated followed by the description of material characterization. A brief description of the cohesive model is provided in Section 4. In Section 5, the simulation results are discussed and finally, in Section 6, the work is summarized and some conclusions are drawn.

2 EXPERIMENTAL SETUP

Zhang et al. (2006) conducted a series of tests in order to study the combined size and strain rate effect in lightly reinforced concrete beams. The loading rate ranged from 10^{-5} to 10^{-2}/s for beams of three sizes. The tests were carried out in a three-point-bend configuration as shown in Figure 1, where the width B is 50 mm, the depth D is 75 mm, 150 mm and 300 mm for small, medium and large specimens respectively. The specimens were casted to give a constant steel-reinforcement ratio of 0.15%. Figure 2 shows the experimental set-up for three-point bending tests. The specimen rests on two rigid-steel cylinders laid on two supports permitting rotation out of the plane of the beam and rolling along the beam longitudinal axis with negligible friction. These supports roll on the upper surface of a very stiff beam fastened to the machine actuator. The load-point displacement is measured in relation to points over the supports on the upper surface of the beam. Two LVDT (linear variable differential transducer) fixed on the steel beam are directly used to measure the displacement between the loading rod and the steel beam. For loading, a hydraulic servo-controlled test system was employed. The tests

were performed in position-control. In numerical simulation, we take the experimental results for small specimens at a loading velocity of 7.5 mm/s as comparison. The strain rate at the bottom surface is related to this loading velocity through

$$V = \dot{\delta} = \dot{\epsilon} S^2 / 6D \tag{1}$$

where S and D are the span and depth of the beam, respectively, δ is the loading point displacement.

3 MATERIAL CHARACTERIZATION

A single micro-concrete was used throughout the experiment made with lime aggregates of 5 mm aggregate size and ASTM type II/A cement. The mix weight proportions were 3.2:0.5:1 for aggregate:water:cement conforming to ASTM C33. Table 1 shows the characteristic mechanical parameters of the micro-concrete determined in the various characterization and control tests, where f_c, f_t, E_c, G_F, ρ, are the compressive strength, tensile strength, elastic modulus, fracture energy, and mass density, respectively. The Hillerborg's characteristic length, calculated as $l_{ch} = E_c G_F / f_t^2$, is around 105 mm.

The mechanical properties of the steel, the elastic moduli E_s, the standard yield strength $\sigma_{0.2}$, the ultimate strength σ_u and strain ϵ_u as well as the bond strength τ_c between the concrete and the rebar are all shown in Table 2.

4 NUMERICAL MODEL

We briefly summarize here the cohesive model adopted for the concrete matrix, which is shown in Figure 3(a), further information can be found elsewhere and the references within (Ortiz and Pandolfi 1999, Ruiz et al. 2000; Ruiz et al. 2001). A variety of mixed-mode cohesive laws accounting for tension-shear coupling (Camacho and Ortiz 1996; Ortiz and Pandolfi 1999; De Andrés et al. 1999), are established by the introduction of an effective opening displacement δ,

$$\delta = \sqrt{\beta^2 \delta_s^2 + \delta_n^2}, \tag{2}$$

which assigns different weights to the normal δ_n and sliding δ_s opening displacements. Supposing that the cohesive free-energy density depends on the opening displacements only through the effective opening displacement δ, a reduced cohesive law, which relates δ to an effective cohesive traction

$$t = \sqrt{\beta^{-2} t_s^2 + t_n^2}, \tag{3}$$

where t_s and t_n are the shear and the normal tractions respectively, can be obtained (Camacho and Ortiz 1996; Ortiz and Pandolfi 1999). The weighting coefficient β is considered a material parameter that measures the relation between the shear and tensile resistance of the material. The existence of a loading envelope defining a connection between t and δ under the conditions of monotonic loading, and irreversible unloading is assumed. A linear decreasing cohesive law described in Figure 3a is adopted in the calculations.

A damage variable d, is defined as the fraction of the expended fracture energy over the total fracture energy per unit surface. Thus, a damage value of zero denotes an uncracked surface, whereas a damage value of one is indicative of a fully cracked or free surface.

The cohesive model introduces an intrinsic time described by

$$t_c = \frac{\rho c \delta_c}{2 f_t} \tag{4}$$

where ρ represents the mass density, c, the longitudinal wave velocity, δ_c, the critical opening displacement, as shown in Figure 3a. As shown in (Ruiz et al. 2001, Yu et al. 2004), due to this intrinsic time, the cohesive model is able to respond differently for a slow and fast loading process. For the material properties given in Table 1, this value is calculated as 28 μs.

In treating the steel-concrete interface, we follow the methodology illustrated in Yu and Ruiz (2006). A cohesive-like interface element is implemented. This interface element is inserted only when the slip strength is attained. The sliding between two surfaces is governed by a perfect-plastic bond-slip law, as described in Figure 3b. The steel re-bar is modeled as an elastic-perfectly plastic material.

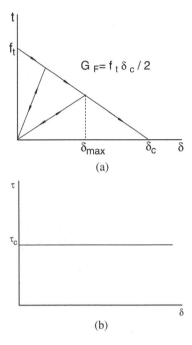

Figure 3. (a) The cohesive law for concrete bulk; (b) the bond-slip relationship for concrete-steel interface.

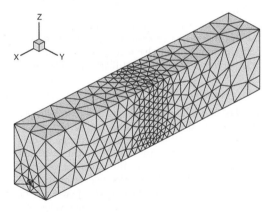

Figure 4. The initial mesh.

5 NUMERICAL RESULTS

We chose to model the small beam, 75 mm in depth to compare with the experimental results. The initial mesh consists of 6764 quadratic tetrahedrons and 10142 nodes, see Figure 4. The mesh size near the middle surface, 7 mm, is chosen to be comparable to the maximum aggregate size, 5 mm, based on the convergence analysis done by Ruiz et al. (2000, 2001) in modeling plain concrete in a dynamic regime. The average mesh size that resolves the interface—around

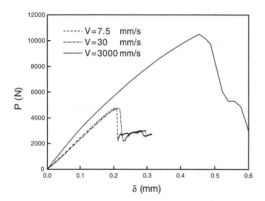

Figure 5. The load-displacement curves for loading velocity at 7.5, 30 and 3000 mm/s respectively, when static fracture properties (specific fracture energy and tensile strength) are adopted.

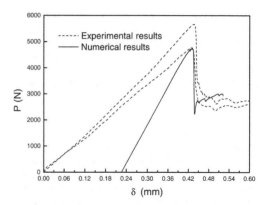

Figure 6. The numerical and experimental load-displacement comparison, when dynamic material properties (fracture energy and tensile strength) are adopted, note that the numerical curve has been translated to 0.22 mm for easier visualization of the peak load.

10 mm—is designed to resolve the effective slip length L_s, which can be described as

$$L_s = A_s f_y / \tau_c p \qquad (5)$$

according to Ruiz (2001), where f_y is the yield strength, A_s is the steel cross-section area, p the rebar perimeter, τ_c the interface bond strength. L_s is computed as 54.3 mm.

We load the beam through an elastic spring to reflect the fact that in the experiments the load transfer is realized through a mechanical system (loading rod, reaction frame, etc.), whose stiffness is 38 kN/mm.

5.1 Peak load

The first step of the simulations was to check the predicted peak load. As we mentioned before, we expected the cohesive model to respond the dynamic loading. As the strain rate increases, the predicted peak loads should increase accordingly. But as can be seen from Figure 5, when the loading velocity is increased from 7.5 mm/s to 30 mm/s, the peak load is practically the same, while when we prescribed a loading velocity of 3000 mm/s (corresponds to a loading rate of 15/s), the peak load increased from 4 kN to 10 kN. Even though we do not have the experimental data for loading velocity of 3000 mm/s for comparison, the qualitative increase of the peak load has been captured. This shows that when the crack propagation time (7000 to 3000 μs) is not comparable to the intrinsic time of the cohesive model (28 μs), the strength increase cannot be captured. This also confirms the observations of Rossi and Toulemonde (1996), who pointed out that: (a) at strain rates smaller than approximately 1/s, the moisture content is believed to play an important role in the strength increase of concrete, the free water in the micro-pores exhibits the so-called Stefan effect

causing a strengthening in concrete with increasing loading rate. The main physical mechanism counters both a micro-cracking localization, leading to increase of concrete tensile strength, and the macro propagation that leads to failure of the specimen; (b) at strain rates higher than or equal to 10/s, the micro-inertia effects in the fracture process zone might cause the rate dependence of concrete, whereas the moisture content is assumed not to be dominant. The main mechanism counters micro-cracking localization and in particular macro crack propagation. In order to capture the strength increase due to moisture content, which is not covered by the bulk constitutive law, we fit for the experimental data of Harsh et al. (1990) to obtain a 30% increase for the fracture energy and the tensile strength. This gives us a peak load of 4.8 kN, comparable to the experimental data, which is shown in Figure 6. Note that we didn't tailor the material parameter to make the numerical simulation to fit the experiments, but followed an independent source that gives the rate-dependent material properties.

5.2 Initial stiffness

We observe in Figure 6 that there is a difference in the initial stiffness between the two experimental curves, this has been explained by Guinea, Planas and Elices, who pointed out that even in the case of identical beams made of the same material, the initial slopes may show variations. This is due to the sensitivity of the elastic flexibility of the beam to the boundary conditions in the application of the concentrated load (Planas et al. 1992). Having in mind that, on the one hand, we would not know the exact experimental boundary condition but only the average, on the other hand the numerical discretization also introduces a pre-defined stiffness, this would all contribute to the deviation of

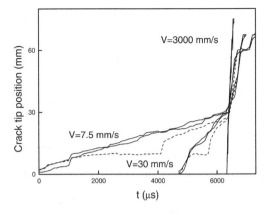

Figure 7. The crack tip position versus time for loading velocity at 7.5, 30 and 3000 mm/s respectively, where the solid lines corresponding to values measured at specimen surface, dotted lines represent values measured at specimen center.

the initial stiffness of the P-δ curve. But we want to emphasize that the post-peak behavior, for both the two experimental data and the numerical simulation, shows similarity. This gives us confidence for the succeeding analysis in extracting the crack tip velocities and the CMOD history.

5.3 Crack tip velocities

The crack tip positions for three different loading velocities, 7.5 mm/s, 30 mm/s and 3000 mm/s are plotted against time in Figure 7. The crack tip positions at the central cross section of the beam were measured separately at the two vertical edges (solid line) and at the center (dashed line) from the bottom of the specimen. The crack tip position is measured with respect to the damage variable value of 0.05 in all the cases. In addition, the axes of the graphs in the proceeding figures have been adjusted to make the comparison easier. For loading velocities 7.5 mm/s and 30 mm/s, the crack velocities at the edges are approximately the same, showing a symmetric crack propagation with respect to the central line. The crack tip position at the center exhibits a plateau due to the hindrance of the steel rebar. The initial slopes of the curves show a linear dependence on the loading velocity. For example, as shown in Table 3, the initial slopes for 7.5 and 30 mm/s are 3.5 m/s and 14.8 m/s respectively. A change in slope is observed at the peak load after which the new slope shows a marked similarity in both the cases. For loading velocity 3000 mm/s, the slope of the curve is steep since the loading velocity is faster and the run completes in a shorter time span. In short, it is evident that crack tip velocity is close after the peak load is reached and it does not depend very much on the loading velocity. The second slope for loading velocity of 7.5 mm/s is about 20 times larger than its first slope,

Table 3. Crack tip velocity (m/s)

	Loading 7.5	Velocity 30	(mm/s) 3000
	At	Surfaces	
Pre-peak	3.9	15.6	290
Post-peak	70.5	95.4	280.5
	At	Center	
Pre-peak	3.5	14.8	296
Post-peak	75.7	112.4	289

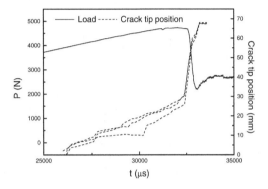

Figure 8. The zoomed view of the load history plotted against the crack tip position history for loading velocity of 7.5 mm/s.

while for 30 mm/s is about 10 times larger. We can say that the crack went through a stable-to-unstable transition when the loading velocity is relatively low. However, at higher loading rate, the whole process is unstable.

In addition, in Figure 8 we plot the load history around the peak load, against the crack tip position history for loading velocity of 7.5 mm/s. It shows that the main crack initiated around 80% (3800 N) of the maximum load (4800 N), and the attainment of the peak load marks the turning point of the two-stage crack propagation.

5.4 CMOD history

The crack mouth opening displacements for loading velocities, v = 7.5, 30, 3000 mm/s were obtained numerically at the central line of the bottom surface of the beam. The CMOD curves for aforementioned velocities are plotted in Figure 9. The curves exhibit a similar trend for all the loading velocities. They consist of two sections with different and approximately linear slopes; let us call them as the first and the second slope, the turning point is in correspondence with the time of reaching peak load. The detailed information is shown in Table 4. For loading velocity v = 7.5 and 30 mm/s, the first slope changes by approximately the

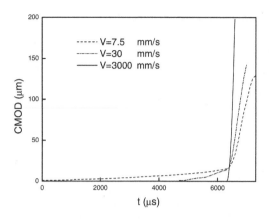

Figure 9. The CMOD history at 7.5, 30 and 3000 mm/s respectively.

Table 4. CMOD rate (m/s).

	Loading	Velocity	(mm/s)
	7.5	30	3000
Pre-peak	0.002	0.0094	0.17
Post-peak	0.13	0.242	1.26

same factor by which the velocities are scaled; after the peak load, the CMOD rate is suddenly increased by around two orders of magnitude compared with that before peak load. This also means that the crack propagation shows a transition from stable to unstable state. While for loading velocity v = 3000 mm/s, the CMOD rate before the peak load and after the peak load is similar, the whole crack propagation is unstable. As seen earlier, this trend is similar as compared to the crack tip position curves (Figure 7).

5.5 Crack patterns

We paid close attention to the crack patterns of the reinforced beams at three characteristic point of the loading process: the intermediate time between the time of crack initiation and the time of reaching the peak load, the time of reaching the peak load and the time of breaking the beam. These patterns are illustrated from the loading plane (middle surface), as shown in Figure 10. It shows that the crack patterns are similar for the two smaller loading velocities, whereas for 3000 mm/s, the crack is noticeably more advanced.

6 CONCLUSIONS

We have compared the numerical load-displacement curves to the experimental counterparts for three-point-bend beams lightly reinforced with steel re-bars. We have extracted numerically the crack tip velocity

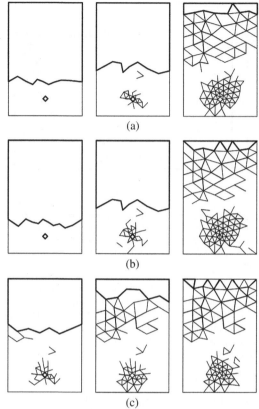

Figure 10. The main crack front advancing (the thick line within the frame) and micro crack formation (the thin lines which coincide with the element boundaries) for loading velocity of 7.5, 30, and 3000 mm/s.

and CMOD history. The results suggests a two-stage crack propagation. The first stage corresponds to the stable crack advance, the second one, the unstable collapse of the beam. Both the crack tip velocities and the CMOD rate have a linear dependence depending upon the factor by which the loading velocity is scaled for the first slopes. The second slope does not depend on the loading velocity and hence it exhibits similar slopes for all loading velocities. The crack patterns advancing at different stages are also analyzed.

ACKNOWLEDGMENTS

We would like to acknowledge the financial support from the *Dirección General de Investigación*, Spain, through Grants MAT2003-00843 and MAT2006-09105, and from the *Secretaría de Estado de Infraestructuras*, Spain, through Grant BOE305/2003. Rena C. Yu thanks the *Ministerio de Educación y Ciencia*, Spain, for the funding she receives for her research activity through the *Programa Ramón y Cajal*.

REFERENCES

Ben Romdhane, M. R. and Ulm, D. R. (2002). Computational mechanics of the steel-concrete interface. *International Journal for Numerical and Analytical Methods in Geomechanics 26*, 99–120.

Bosco, C. and Carpinteri, A. (1992). Fracture behaviour of beam cracked across reinforcement. *Theoretical and Applied Fracture Mechanics 17*, 61–68.

Brincker, R., Henriksen, M. S., Christensen, F. A. and Heshe, G. (1999) *Size effects on the bending behaviour of reinforced concrete beams*, pp. 127–180. Elsevier, London.

Camacho, G. T. and M. Ortiz, M. (1996). Computational modelling of impact damage in brittle materials. *International Journal of Solids and Structures 33 (20–22)*, 2899–2938.

De Andrés, A., J. L. Pérez, and Ortiz, M. (1999). Elastoplastic finite-element analysis of three-dimensional fatigue crack growth in aluminum shafts subjected to axial loading. *International Journal of Solids and Structures 36*(15), 2231–2258.

Harsh, S., Shen, Z. J. and Darwin, D. (1990, Sep–Oct). Strain-rate sensitive behavior of cement paste and mortar in compression. *ACI Materials Journal 87*(6), 508–516.

Hawkins N. M. and Hjorsetet, K. (1992). *Minimum reinforcement requirement for concrete flexural members.*, pp. 379–412. Elsevier, London.

Hawkins, N. M. and Hjorsetet, K. (2004). *Bond of RC members using nonlinear 3D FE analysis*, pp. 861–868. IA-FraMCoS.

Hededal, O. and Kroon, I. B. (1991). *Lightly reinforced high-strength concrete, M. Sc. Thesis*. Denmark: University of Åalborg.

Ortiz, M. and Pandolfi, A. (1999). Finite-deformation irreversible cohesive elements for three-dimensional crack-propagation analysis. *International Journal for Numerical Methods in Engineering 44*, 1267–1282.

Pandolfi, A. and Ortiz, M. (2002). An Efficient Adaptive Procedure for Three-Dimensional Fragmentation Simulations. *Engineering with Computers 18*(2), 148–159.

Planas, J., Guinea, G. and Elices, M. (1992). Stiffness associated with quasi-concentrated load. *Materials and Structures 27*, 311–318.

Rossi, P. and Toulemonde, E. (1996, Mar). Effect of loading rate on the tensile behavior of concrete:description of the physical mechanisms. *Materials and Structures 29*(186), 116–118.

Ruiz, G. (2001). Propagation of a cohesive crack crossing a reinforcement layer. *International Journal of Fracture 111*, 265–282.

Ruiz, G., Carmona, J. R. and Cedón, D. A. (2006). Local fracture and steel-concrete decohesion phenomena studied by means of cohesive interface elements. *Computer Methods in Applied Mechanics and Engineering 195*(51), 7237–7248.

Ruiz, G., Elices, and Planas, J. (1999). *Size Effect and Bond-Slip Dependence of Lightly Reinforced Concrete Beams*, pp. 67–98. Kidlington, Oxford, UK: Elsevier, ESIS Publication 24.

Ruiz, G., Ortiz, M. and Pandolfi, A. (2000). Three-dimensional finite-element simulation of the dynamic Brazilian tests on concrete cylinders. *International Journal for Numerical Methods in Engineering 48*, 963–994.

Ruiz, G., Pandolfi, A. and Ortiz, M. (2001). Three-dimensional cohesive modeling of dynamic mixed-mode fracture. *International Journal for Numerical Methods in Engineering 52*, 97–120.

Yu, R. C. and Ruiz, G. (2006). Explicit Finite Element Modeling of Static Crack Propagation in Reinforc ed Concrete. *International Journal of Fracture, 141*(3–4), 357–372.

Yu, R. C., Ruiz, G. and Pandolfi, A. (2004). Numerical investigation on the dynamic behavior of advanced ceramics. *Engineering Fracture Mechanics 71*(3–4), 897–911.

Zhang, X. X., Ruiz, G. and, Yu R. C. (2006). Combined strain rate and size effect in reinforced concrete, an experimental study. *Engineering Fracture Mechanics*, submitted.

Fracture Mechanics of Concrete and Concrete Structures – New Trends in Fracture Mechanics of Concrete – Carpinteri, et al. (eds)
© 2007 Taylor & Francis Group, London, ISBN 978-0-415-44065-3

Numerical simulation of fracture and damage in RC structures due to fire

J. Cervenka
Cervenka Consulting, Prague, Czech Republic

J. Surovec & P. Kabele
Dept. of Structural Mechanics, Czech Technical University, Prague, Czech Republic

T. Zimmerman, A. Strauss & K. Bergmeister
Dept. of Structural Engineering and Natural Hazards, Institute for Structural Engineering, University of Natural Resources and Applied Life Sciences, Vienna, Austria

ABSTRACT: Paper describes results achieved during a 4 year European research project UPTUN GRD1-2001-40739, which finished in 2006. The objective of the project was to develop new methods and procedures for upgrading European tunnels and increasing their safety against fire accidents. Within the framework of this project authors contributed by developing and extending existing fracture-plastic material models for numerical modelling of structural behaviour subjected to fire. The behaviour of the model is demonstrated on simple uni-axial tests as well as on a three-dimensional analysis of a tunnel fire.

1 INTRODUCTION

Concrete behaviour at high temperature represents a complex phenomenon that in general case requires a coupled hygro-thermo-chemo-mechanical model. An example of an advanced hygro-thermo-mechanical model is presented in Gawin, D. et al. 2003. This paper presents a modification of the combined fracture-plastic model developed by the authors and initially presented in Cervenka. J. et al. 1998. The extended model is applied for the analysis of structures subjected to fire, where the material properties of concrete as well as reinforcement are strongly dependent on temperature. The objective is to develop a model that can be applied to large-scale analyses of engineering problems at a reasonable computational cost. The presented work is an extension of a previously published thermally dependent model by Cervenka J. et al. 2006. The paper presents new results and an application to a real tunnel scenario.

The temperature distribution inside a heated structure is calculated by a separate non-linear transient thermal analysis. The obtained temperature fields are then applied in a mechanical analysis, which takes into account the thermally induced strains as well as the material degradation induced by high temperatures.

The model was developed during the European research project UPTUN. The model behaviour is tested on uni-axial tensile and compression tests as well as on experimental data of a real scale fire experiment of a tunnel suspended ceiling. At the end of the paper the model is applied to investigate the effects of fire resistant shotcrete protection during a Virgolo tunnel fire test.

The temperature dependent mechanical model is implemented in program ATENA (see ATENA 2005). This program was also used to analyse the examples presented in this paper.

2 THERMAL ANALYSIS

Temperature fields for the mechanical analysis are determined by a separate thermal analysis.

$$C_T \frac{\partial T}{\partial t} = -div(J_T), \text{ where } J_T = -K_T grad(T) \quad (1)$$

In the above standard diffusion formula, T denotes temperature and t represents time. C_T and K_T are thermal capacity and conductivity respectively. Both capacity and conductivity are functions of current temperature. A modified Crank-Nicholson integration scheme (Wood, W.L. 1990, Jendele, L. 2001) is used to integrate the non-linear set of equations. The resulting iterative correction of unknown temperatures $\Delta\psi$ at time $t + \Delta t$ is calculated by the following formula:

$$\Delta\psi = \left(\tilde{\mathbf{K}}\right)^{-1}\tilde{J} \qquad (2)$$

where:

$$\tilde{\mathbf{K}} = \left(\mathbf{K}\theta + \frac{1}{\Delta t}\mathbf{C}\right)$$

$$\tilde{J} = \overline{J} - \mathbf{K}\left(\theta^{t+\Delta t}\psi + (1-\theta)^t\psi\right) - \mathbf{C}\frac{1}{\Delta t}\left(^{t+\Delta t}\psi - {}^t\psi\right)$$

C and **K** are capacity and conductivity matrices respectively after spatial discretization by finite element method. θ is the integration parameter. For $\theta = 0.5$ the Crank-Nicholson formulation is recovered. $\theta = 0$ corresponds to the Euler explicit scheme, and the Euler implicit formulation is obtained when $\theta = 1$. The known oscillatory behaviour of the Crank-Nicholson formulation is addressed by introducing the following iterative damping (Jendele, L. 2001).

$$^{t+\Delta t}\underline{\psi} = {}^{t+\Delta t(i)}\underline{\psi} = {}^{t+\Delta t(i-1)}\underline{\psi} + {}^{t+\Delta t(i)}\eta\Delta\underline{\psi} \qquad (3)$$

The recommended value of the damping parameter η is in the interval of $\langle 0.3; 1\rangle$.

3 MATERIAL MODEL FOR CONCRETE

The concrete mechanical model follows the original theory in Cervenka at al. 1998. The material model formulation is based on the strain decomposition into elastic ε_{ij}^e, plastic ε_{ij}^p and fracturing ε_{ij}^f components (de Borst, R. 1986).

$$\varepsilon_{ij} = \varepsilon_{ij}^e + \varepsilon_{ij}^p + \varepsilon_{ij}^f \qquad (1)$$

The new stress state is then computed by the formula:

$$\sigma_{ij}^n = \sigma_{ij}^{n-1} + E_{ijkl}(\Delta\varepsilon_{kl} - \Delta\varepsilon_{kl}^p - \Delta\varepsilon_{kl}^f) \qquad (2)$$

Tensile behaviour of concrete is modelled by nonlinear fracture mechanics with a simple Rankine based criterion.

$$F_i^f = \sigma_i'' - f_{ti}'(w_i') \leq 0 \qquad (3)$$

In this method the smeared crack concept is accepted. Its parameters are: tensile strength f_t, shape of the stress-crack opening curve $f_t(w)$ and fracture energy G_F. It is assumed that strains and stresses are converted into the material directions, which in case of

rotated crack model correspond to the principal stress directions, and in case of fixed crack model, are given by the principal directions at the onset of cracking. Therefore, σ_i'' identifies the trial stress and f_{ti}' tensile strength in the material direction i. The prime symbol denotes quantities in the material directions. This approach is combined with the crack band method of Bažant, Z.P. & Oh, B.H. 1983. In this formulation, the cracking strain is related to the element size. Consequently, the softening law in terms of strains for the smeared model is calculated for each element individually, while the crack-opening law is preserved. The model uses an exponential softening law of Hordijk 1991.

Compressive behaviour is modelled using a plasticity-based model with failure surface defined by Menetrey, P. & Willam, K.J. 1995 three-parameter criterion

$$F_{3P}^p = \left[\sqrt{1.5}\frac{\rho}{f_c}\right]^2 + m\left[\frac{\rho}{\sqrt{6}f_c}r(\theta,e) + \frac{\xi}{\sqrt{3}f_c}\right] - c = 0$$

$$(4)$$

where

$$m = 3\frac{f_c^2 - f_t^2}{f_c f_t}\frac{e}{e+1},$$

$$r(\theta,e) = \frac{4(1-e^2)\cos^2\theta + (2e-1)^2}{2(1-e^2)\cos\theta + (2e-1)[4(1-e^2)\cos^2\theta + 5e^2 - 4e]^{\frac{1}{2}}}$$

In the above equations, (ξ, ρ, θ) are Heigh-Vestergaard coordinates, and f_c and f_t are compressive strength and tensile strength respectively. Parameter $e \in \langle 0.5, 1.0\rangle$ defines the roundness of the failure surface.

The surface evolves during the yielding/crushing process by the hardening/softening laws based on equivalent plastic strain defined as:

$$\Delta\varepsilon_{eq}^p = \min(\Delta\varepsilon_i^p) \qquad (5)$$

Hardening $\varepsilon_{eq}^p \in \langle -\varepsilon_c^p; 0\rangle$:

$$f_c(\varepsilon_{eq}^p) = f_{co} + (f_c - f_{co})\sqrt{1 - \left(\frac{\varepsilon_c^p - \varepsilon_{eq}^p}{\varepsilon_c^p}\right)^2} \qquad (6)$$

Softening $\varepsilon_{eq}^p \in \langle -\infty; -\varepsilon_c^p\rangle$:

$$c = \left(1 - \frac{w_c}{w_d}\right)^2, \qquad w_c \in \langle -w_d; 0\rangle$$

$$c = 0, \qquad w_c \in (-\infty; w_d) \qquad (7)$$

728

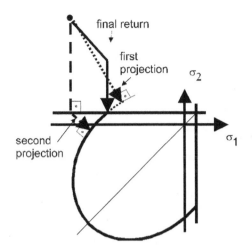

Figure 1. Iterative algorithm for solving interaction of fracture-plastic surfaces (for simplicity shown in 2D).

$$w_c = (\varepsilon_{eq}^p - \varepsilon_c^p) L_c \qquad (8)$$

When crushing enters into the softening regime, an analogous approach to the crack band model is used also for the localization in compression within the crushing band L_c. A direct return-mapping algorithm is used to solve the predictor-corrector equation of the plasticity model.

$$F^p(\sigma_{ij}^t - \sigma_{ij}^p) = F^p(\sigma_{ij}^t - \Delta\lambda l_{ij}) = 0 \qquad (9)$$

The plastic stress σ_{ij}^p is a product of plastic multiplier $\Delta\lambda$ and the return direction l_{ij}, which is defined as

$$l_{ij} = E_{ijkl}\frac{\partial G^p(\sigma_{kl}^t)}{\partial\sigma_{kl}} \text{ then } \Delta\varepsilon_{ij}^p = \Delta\lambda\frac{\partial G^p(\sigma_{ij}^t)}{\partial\sigma_{ij}}$$
$$\qquad (10)$$

Plastic potential G^p is given by:

$$G^p(\sigma_{ij}) = \beta\frac{1}{\sqrt{3}}I_1 + \sqrt{2J_2} \qquad (11)$$

where β determines the return direction. If $\beta < 0$, material is being compacted during crushing, if $\beta = 0$, the material volume is preserved, and if $\beta > 0$, the material is dilating.

A special iterative algorithm analogous to multi-surface plasticity is used to solve the plastic and fracture models such that the final stress tensors in both models are identical. This algorithm is schematically shown in two-dimensions in Figure 1.

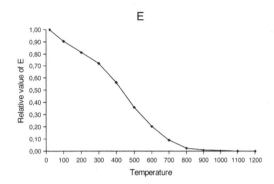

Figure 2. Temperature dependence of concrete E modulus.

4 MATERIAL MODEL FOR REINFORCEMENT

Finite element modelling of reinforced concrete structures requires special tools for modelling of all types of reinforcement (Jendele & Cervenka 2006).

Concrete is modelled by solid elements. Interface material models can model frictional contacts between structural parts. The mesh type of reinforcement can be represented as smeared reinforcement. In this approach, the individual bars are not considered, while reinforcement is modelled as a component of a general composite material.

Individual bars can be modelled by truss elements embedded in concrete elements with axial stiffness only. In this technique, the mesh is generated first for concrete. Then the bar elements are embedded into each solid element. The bar elements are then connected to the solid model by special constrain conditions. These conditions represent a kinematic dependence of reinforcement displacements on those of concrete nodes. Thus the reinforcing does not affect the mesh generation. The described family of finite elements makes it possible to cover most practical cases of reinforced concrete structures.

5 TEMPERATURE DEPENDENT MATERIAL PROPERTIES

The material models for concrete as well as reinforcement are formulated in a purely incremental manner, and the selected material parameters are temperature dependent. The temperature dependent evolution laws of these parameters are shown in subsequent figures. They have been derived based on Eurocode 2 and by matching experimental results of Castillo, C. & Durani, A.J. 1990. An analogous dependence is used also for the stress-strain law for reinforcement. The stress-strain diagram is scaled based on the highest reached temperature at each reinforcement element according to the Eurocode 2 formulas.

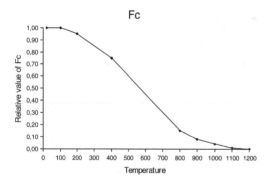

Figure 3. Temperature dependence of concrete compressive strength.

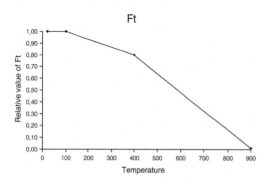

Figure 4. Temperature dependence of concrete tensile strength.

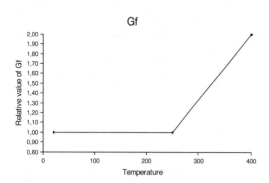

Figure 5. Temperature dependence of fracture energy.

6 EXAMPLES AND VALIDATION

Several examples are presented to demonstrate the behaviour of the proposed model on simple uniaxial tests as well as on a more complicated example of a tunnel suspended ceiling subjected to fire.

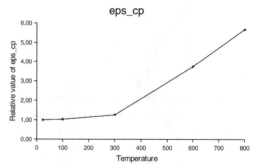

Figure 6. Temperature dependence of plastic strain at compressive strength (see Eq. (6)).

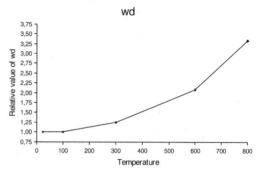

Figure 7. Temperature dependence of critical crushing displacement (see Eq. (7)).

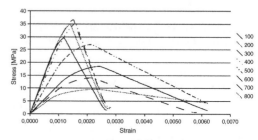

Figure 8. Uniaxial compression tests.

6.1 Uniaxial tests

Standard compression and tensile tests were simulated, in which the specimen was loaded to collapse under uniform temperature field distribution. The temperatures chosen ranged from 100°C to 800°C with a step of 100°C and they correspond to the test data published in Castillo, C. & Durani, A.J. 1990. The results show the behaviour of the material model at various temperature levels and are depicted in Figure 8 and Figure 9.

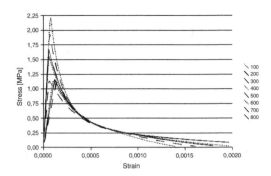

Figure 9. Uniaxial tension tests.

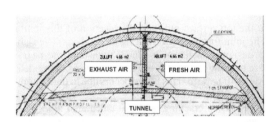

Figure 10. Typical tunnel cross-section with a suspended reinforced concrete ceiling.

6.2 Suspended ceiling simulation

The structure consisted of a reinforced concrete suspended ceiling slab with a total span of 8.0 m, suspended in the mid-span by a pair of hangers. The geometry corresponds to a specimen, which was tested during a European research project UPTUN. The thickness of the slab is 200 mm and the width is 2.0 m. The geometry of the specimen represents a suspended ceiling in a typical concrete tunnel such as is shown in Figure 10.

The symmetry of the structure is taken advantage of, so that the model in Figure 11 consists of one quarter of the structure. The deformed mesh is shown in Figure 13, and it contains 8827 nodes and 6595 linear iso-parametric brick elements.

The structure was exposed to a pool fire along the bottom part of the ceiling structure. The maximum temperature 1100°C was reached along the bottom surface during the experiment. In the analysis, however, uniform temperature was considered along the bottom face of the slab by averaging the values measured in the experiment (see Figure 12). No thermal shielding was considered. The structure was initially loaded with assumed self weight of 5 kN/m² and a live load of 3.5 kN/m², and then the fire actions were applied.

First transient thermal analysis was performed with temperature dependent material properties. Then the obtained thermal fields were used in subsequent incremental stress analysis

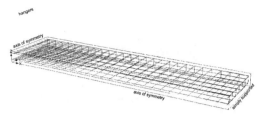

Figure 11. Three-dimensional model of a quarter of the suspended ceiling.

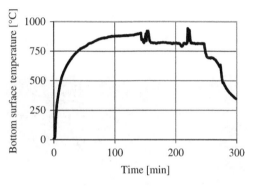

Figure 12. Average bottom surface temperature in the experiment.

Figure 13. Deformed shape of the 3D finite element model.

The subsequent figures show the main results obtained in this study. Figure 14 compares the mid-span deflections. The figure shows that a good prediction was obtained for the maximum deflection, however the deflections for times around 60 minutes were overestimated by almost 50%. The analyses showed that the main parameter affecting the deflection results is the thermal expansion. Good agreement was obtained in the case of 3D analysis with the adjusted law for free thermal strain.

In this work, no experimental data for free thermal strain was available for a given concrete type, therefore

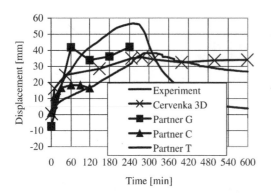

Figure 14. Midspan deflection during the experiment and comparison with numerical predictions.

Figure 15. Photo from the experiment showing an extensive damage at the ceiling corners.

the Eurocode 2 formula was used. The free thermal strain had to be reduced by 30% to obtain a better agreement with the measured data. Figure 14 includes also results from other models and other partners on the UPTUN project. These results were typically obtained by a two-dimensional thermal analysis of the slab cross-section, which was later used in an incremental plastic analysis with 2D beam elements.

The final crack pattern is shown in Figure 13, which indicates the formation of a plastic joint on the right side near the vertical hanger. The extensive cracking

Figure 16. Specimens for laboratory tests.

at the top surface indicates the location of this joint. The joint forms early on during the analysis. The slab is designed such that it cannot withstand the negative bending moment at the vertical hanger, if the slab edges lift upward. This lifting occurs due to the heating of the bottom surface.

Extensive cracking was also calculated along the bottom surface and at the front corner. The results from the 3D analysis show an additional bending in the cross-wise direction, causing the lifting of the slab corners. This lifting caused an excessive damage to the concrete due to escaping flames and smoke.

The structure suffered extensive damage, but full collapse was not reached in the investigated time period neither in the experiment nor in the numerical study.

7 APPLICATION TO VIRGOLO TUNNEL TEST

A real scale fire test was carried out on the 17th February 2005 in Bolzano/South Tyrol – Italy in a tunnel of the Brenner motorway-AG (Virgolo tunnel).

The experimental program was rather extensive. It tested human response, fire detection and suppression systems as well as material and structural behaviour. Six fire resistance shotcrete mixes were tested in order to evaluate their effectiveness as fire protection barriers to the structural concrete. The large-scale test was supplemented by laboratory experiments of the shotcrete specimens. A numerical simulation using the presented material models supported the experimental program. The results from the laboratory tests were used to calibrate the thermal properties of the shotcrete types. The obtained material laws were then used to simulate the behaviour of a tunnel wall protected by the tested materials. The main goal was to qualitatively evaluate the damage to tunnel lining and compare it to the damage of an unprotected wall.

Figure 16 shows the geometry of the laboratory specimens. The specimens were cubes with side of

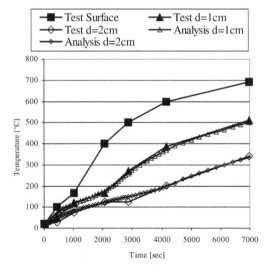

Figure 17. Temperature evolution in shotcrete material for different depths d during the laboratory test.

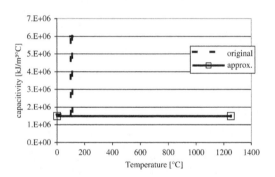

Figure 18. Capacity for the shotcrete material determined by the inverse analysis of laboratory test data.

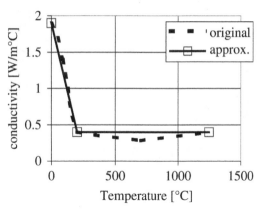

Figure 19. Conductivity for the shotcrete material determined by the inverse analysis of laboratory test data.

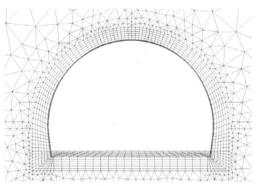

Figure 20. Numerical model for Virgolo tunnel numerical analyses.

20 cm. One side of these specimens was subjected to a fire load of up to 12 kW using gas burners. The temperature was measured at various depths, as is shown in Figure 17 for shotcrete material Meyco? Fix Fireshield 1350 from DEGUSSA. Numerical thermal analysis with ATENA was then used to determine optimal temperature dependent laws for capacitivity and conductivity in order to match the experimental temperature evolution. Probabilistic methods were used (Strauss et al. 2004, 2006) to find the optimal fit with experimental data (see Figure 17). The final evolution law for capacitivity and conductivity is shown in Figure 18 and Figure 19 respectively.

Subsequently these parameters were used to model the shotcrete performance in a real tunnel scenario.

The model shown in Figure 20 was subjected to a modified hydrocarbon fire of up to 1300°C. In the interior of the tunnel a 50 mm layer of shotcrete material was applied. The shotcrete layer was modeled by 4 finite elements while for the tunnel lining 7 elements were used through its thickness. The tunnel wall was made of concrete class C25/30 using Eurocode 2 classification.

Figure 22 describes the evolution of concrete degradation at various depths. The degradation is described as a relative value of concrete strength. The value of 1 indicates original concrete strength, i.e. no damage. The value of zero corresponds to fully damaged material with zero strength.

Table 1 summarizes the main results for the given shotcrete type (i.e. Meyco© Fix Fireshield 135). The recommended design criteria are such that the temperature at the interface between the fire protection layer and original concrete wall should not exceed 300°C, and the temperature at the depth of reinforcement location should remain below 250°C.

Figure 21. Arrangement of fire resistant concrete layer during the Virgolo tunnel test.

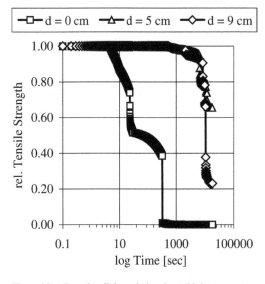

Figure 22. Tunnel wall degradation due to high temperature at various depths.

Table 1. Temperature criteria and optimal shotcrete thickness.

Location	Criteria	Value
50 mm (concrete-shotcrete interface)	300°C	214°C
90 mm (reinforcement depth)	250°C	141°C
Optimal shotcrete layer thickness	—	30 mm

Table 1 shows that both criteria are satisfied for the given material, and optimal thickness of the protective layer was determined by a parametric study to be 30 mm.

8 DISCUSSION, CONCLUSIONS AND ACKNOWLEDGEMENTS

The paper presents an extension of the combined fracture-plastic model developed previously by the authors (Cervenka, J. et al. 1998, 2006). The selected material parameters are made temperature dependent. This allows the model to be applied for the simulation of reinforced concrete structures subjected to fire. The model behavior is demonstrated on simple uniaxial tests that show the model response under various uniform temperature distributions.

The results of a large-scale fire test of a tunnel suspended ceiling were used to verify the model applicability for solving complex structures subjected to fire.

The last section contains an application to real life scenario, where a structural damage due to fire is evaluated using an example of the Virgolo tunnel fire test.

The presented work was part of a European research project UPTUN GRID-CT-2002-00766. The development of reinforcement modeling and temperature dependent material laws was supported by the research projects 103/04/2083, and 103/07/1660 from Czech Grant Agency. The financial support from European community and Czech Grant Agency is greatly appreciated.

REFERENCES

ATENA 2005. *Program Documentation*, Cervenka Consulting, Prague, Czech Republic, www.cervenka.cz

Bažant, Z.P. & Oh, B.H. 1983. Crack band theory for fracture of concrete, *Materials and Structures*, 16, 155–177. York, 360 pp.

Castillo, C. & Durani, A.J. 1990. Effect of transient high temperature on high-strength concrete, ACI Materials Journal, v 87 n 1, p 47–53.

Gawin, D., Pesavento, F., Schrefler, B.A. 2003. Modelling of hygro-thermal behaviour of concrete at high temperature with thermo-mechanical and mechanical material degradation, CMAME, 192, 1731–1771.

Cervenka, J., Cervenka, V., Eligehausen, R. 1998. Fracture-Plastic Material Model for Concrete, Application to Analysis of Powder Actuated Anchors, *Proc. FRAMCOS 3*, 1998, pp 1107–1116.

Cervenka, J., Surovec, J., Kabele, P., 2006, Modelling of reinforced concrete structures subjected to fire, Proc. EuroC-2006, ISBN 0415397499, pp 515–522.

de Borst, R. 1986. *Non-linear analysis of frictional materials*, PhD Thesis, Delft University of Technology.

Eurocode 2. 1992-1-2. *Design of concrete structures, Part 1.2. General rules – Structural fire Design*. Draft prENV 1992-1-2.

Jendele, L. 2001. Atena Pollutant Transport Module – Theory. Prague: Edited PIT, ISBN 80-902722-4-X.

Jendele, L., Cervenka, V., 2006. A General Form of Dirichlet Boundary Conditions Used in Finite Element Analysis.

Paper 165, *Proc. Eighth International Conference on Computational Structures Technology*, ed. B.H.V. Topping, G. Montero, and R. Montenegro. 2006, Civil Comp Press: Las Palmas. 373–374.

Hordijk, D.A. 1991. *Local Approach to Fatigue of Concrete*. Ph.D. Thesis, Delft University of Technology, The Netherlands.

Menetrey, P. & Willam, K.J. 1995. Triaxial failure criterion for concrete and its generalization. *ACI, Structural Journal*, 92(3), pp 311–318.

Strauss, A., Bergmeister, K., Novak, D. & Lehky, D., 2004a, Stochastic parameter identification of structural concrete for maintenance, in german, *Beton und Stahlbetonbau*, Vol. 99, No. 12, Vienna, Austria, pp. 967–974.

Strauss, A., Bergmeister, K., Lehky, D., & Novak, D., 2006, Inverse statistical nonlinear FEM analysis of concrete structures, *Proc. Comp. Modell., Concrete Structures, EuroC-2006*, eds. Meschke, de Borst, Mang & Bicanic, ISBN 0 415 39749 9, pp. 897–904.

Wood., W.L. 1990. Practical-Time Stepping Schemes. Oxford: Clarenton Press.

Fracture Mechanics of Concrete and Concrete Structures – Design, Assessment and Retrofitting of RC Structures – Carpinteri, et al. (eds)
© 2007 Taylor & Francis Group, London, ISBN 978-0-415-44616-7

3D analysis of RC frames using effective-stiffness models

C. Dundar & I.F. Kara
Department of Civil Engineering, Cukurova University, Adana, Turkey

ABSTRACT: In the present study, a computer program has been developed using rigid diaphragm model for the three dimensional analysis of reinforced-concrete frames with cracked beam and column elements. ACI, CEB and probability-based effective stiffness models are used for the effective moment of inertia of the cracked members. In the analysis, shear deformations which can be large following crack development are taken into account and the variation of the shear stiffness due to cracking is considered by reducing the shear stiffness through appropriate models. The computer program is based on an iterative procedure which is subsequently experimentally verified by fitting the results of a test on a two-story R/C frame. A parametric study is also carried out on a four-story, three-dimensional reinforced-concrete frame. The iterative analytical procedure can provide an accurate and efficient prediction of deflections in R/C structures due to cracking under service loads. The most significant feature of the proposed procedure is that the variations in the flexural stiffness of beams and columns can be directly evaluated.

1 INTRODUCTION

When designing reinforced-concrete structures, a designer must satisfy not only the strength requirements but also the serviceability requirements. To ensure the serviceability requirements of tall reinforced concrete buildings, it is necessary to accurately assess the deflection under lateral and gravity loads. In recent years high-rise and slender structures have been constructed using high-strength steel and concrete. Therefore, the serviceability limit state for lateral drift becomes much more important design criterion and must be satisfied to prevent large second-order P-delta effects. In addition, the control of the deformation in reinforced concrete beams is also important to ensure the serviceability requirements. Due to low tensile strength of concrete, cracking, which is primarily load dependant, can occur at service loads and reduce the flexural and shear stiffness of reinforced concrete members. For accurate determination of the deflections, cracked members in the reinforced concrete structures need to be identified and their effective flexural and shear stiffnesses determined.

Cracked state in reinforced concrete beam elements can be modeled in several ways (Ngo & Scordelis 1967, Channakeshava & Sundara 1988). These methods take into account the constitutive relationships of both steel and concrete together with the bond-slip relationship. However, due to the complexities of the actual behavior R/C frames and cumbersome computations to be carried out these methods cannot easily be adopted by design engineers.

In the design of tall reinforced concrete structures, the moments of inertia of the beams and columns are usually reduced at the specified ratios to compute the lateral drift by considering the cracking effects on the stiffness of the structural frame (Stafford & Coull 1991). The gross moment of inertia of columns is generally reduced to 80% of their uncracked values while the gross moment of inertia of beams is reduced to 50%, without considering the type, history and magnitude of loading, and the reinforcement ratios in the members.

A simplified and computationally more efficient method for the analysis of two dimensional reinforced concrete frames with beam elements in cracked state was developed by Tanrikulu et. al. (2000). ACI (1966) and CEB (1985) model equations, which consider the contribution of tensile resistance of concrete to flexural stiffness by moment-curvature relationships, were used to evaluate the effective moment of inertia. Shear deformations were also taken into account in the formulation and the variation of shear stiffness due to cracking was considered by reducing the shear stiffness through appropriate models. In the analysis cracking was considered only for beam elements, hence, the linear elastic stiffness equation was used for columns. However, for accurate determination of lateral deflections of tall reinforced concrete structures, it is important to consider the effects of concrete cracking on the stiffness of the columns. Therefore, the reduction of flexural and shear stiffnesses in the columns due to cracking should also be taken into account.

Two iterative analytical procedures were developed for calculating the lateral drifts in tall reinforced concrete structures (Chan et. al. 2000). A general probability-based effective stiffness model was used to consider the effects of concrete cracking. The analytical procedures named direct effective stiffness and load incremental methods were based on the proposed effective stiffness model and iterative algorithm with the linear finite element analysis. The variation of shear stiffness due to cracking was not considered in the analysis. Whereas, after the development of cracks, shear deformations, can be large and significant. Hence, the reduction of shear stiffnesses due to cracking should also be included for improving the results of the analysis.

In practice, the analysis of reinforced concrete frames is usually carried out by linear elastic models which either neglect the cracking effect or consider it by reducing the stiffness of members arbitrarily. It is also quite possible that the design of tall reinforced concrete structures on the basis of linear elastic theory may not satisfy the serviceability requirements. Therefore, an analytical model which can include the effects of concrete cracking on the flexural and shear stiffness of the members and accurately assess the deflections would be very useful. In the present study, a computer program has been developed using rigid diaphragm model for the three dimensional analysis of reinforced concrete frames with cracked beam and column elements. In the analysis, stiffness matrix method is applied to obtain the numerical solutions, and the cracked member stiffness equation is evaluated by including the uniformly distributed and point loads on the member. In obtaining the flexibility influence coefficients a cantilever beam model is used which greatly simplifies the integral equations for the case of point load. In the program, the variation of the flexural stiffness of a cracked member is evaluated by ACI, CEB and probability-based effective stiffness models. Shear deformation effect is also taken into account and reduced shear stiffness is considered by using effective shear modulus models (Cedolin & dei Poli 1977, Al-Mahaidi 1978 and Yuzugullu & Schnobrich 1973). The results have been verified with the experimental results available in the literature. Finally a parametric study is carried out on a four-story, three-dimensional reinforced-concrete frame.

2 MODELS FOR THE EFFECTIVE FLEXURAL STIFFNESS OF A CRACKED MEMBER

ACI and CEB models which consider the effect of cracking and participation of tensile concrete between cracks have been proposed to define the effective flexural behavior of reinforced concrete cracked section.

In the ACI model the effective moment of inertia is given as

$$I_{eff} = \left(\frac{M_{cr}}{M}\right)^m I_1 + \left[1 - \left(\frac{M_{cr}}{M}\right)^m\right] I_2 \quad for\, M \geq M_{cr} \quad (1a)$$

$$I_{eff} = I_1 \qquad\qquad for\ M < M_{cr} \quad (1b)$$

where $m = 3$. This equation was first presented by Bronson (1963) with $m = 4$ when I_{eff} is required for the calculation of curvature in an individual section. In the CEB model I_{eff} is also defined in the following form:

$$I_{eff} = \left[\beta_1\beta_2\left(\frac{M_{cr}}{M}\right)^2\frac{1}{I_1} + \left(1 - \beta_1\beta_2\left(\frac{M_{cr}}{M}\right)^2\right)\frac{1}{I_2}\right]^{-1} \quad for\, M \geq M_1 \quad (2a)$$

$$I_{eff} = I_1 \qquad\qquad for\ M < M_{cr} \quad (2b)$$

in which $\beta_1 = 0.5$ for plain bars and 1 for high bond reinforcement; $\beta_2 = 1$ for the first loading and 0.5 for the loads applied in a sustained manner or in a large number of load cycles (Ghali & Favre 1986).

In Eqs. (1) and (2), I_1 and I_2 are the moments of inertia of the gross section and the cracked transformed section, respectively, M is the bending moment, M_{cr} is the moment corresponding to flexural cracking considered. The cracking moment, M_{cr} is calculated by the program using the following equation:

$$M_{cr} = \frac{(f_r + \sigma_v)I_1}{y_t} \quad (3)$$

where σ_v is the axial compressive stress, f_r is the flexural tensile strength of concrete, and y_t is the distance from centroid of gross section to extreme fiber in tension.

In addition to ACI and CEB models, probability-based effective stiffness model (Chan et. al. 2000) which accounts for the effects of concrete cracking with the stiffness reduction has been considered. In the probability-based effective stiffness model, the effective moment of inertia is obtained as the ratio of the area of moment diagram segment over which the working moment exceeds the cracking moment M_{cr} to the total area of moment diagram in the following form (Fig. 1).

$$A_{uncr} = A_1 + A_2 = \int_{M(x)<M_{cr}} M(x) \quad (4a)$$

$$A_{cr} = A_3 = \int_{M(x)\geq M_{cr}} M(x) \quad (4b)$$

$$A = A_{cr} + A_{uncr} \quad (4c)$$

$$P_{uncr}[M(x) < M_{cr}] = \frac{A_{uncr}}{A} \quad (4d)$$

738

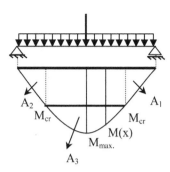

Figure 1. Cracked and uncracked regions of the simply supported beam element subjected to concentrated and a uniformly-distributed loads.

$$P_{cr}[M(x) \geq M_{cr}] = \frac{A_{cr}}{A} \qquad (4e)$$

$$I_{eff} = P_{uncr} I_1 + P_{cr} I_2 \qquad (4f)$$

where A_{cr} is the area of moment diagram segment over which the working moment exceeds the cracking moment M_{cr} and A is the total area of moment diagram. In the same equation, P_{cr} and P_{uncr} are the probability of occurrence of cracked and uncracked sections, respectively. The applicability of the proposed model was also verified with the frame test results (Chan et. al. 2000).

In the literature (Cosenza 1990, Sakai & Kakuta 1980 and Al-Shaikh & Al-Zaid 1993), the effective moment of inertia given by ACI and CEB is the best among the commonly accepted simplified methods for the estimation of instantaneous deflection. Although ACI and CEB models are usually considered for beams, in the present study these models are also used for columns, including the axial force in the determination of the cracking moment. In the analysis the axial force in the beams is also considered for the calculation of the cracking moment, no matter if it is relatively small. In the computer program developed in the present study, afore-mentioned models are used for the effective moment of inertia of the cracked section.

3 MODELS USED FOR THE EFFECTIVE SHEAR STIFFNESS OF CONCRETE

Several formulations are found in the literature for concrete effective shear modulus after cracking.

Cedolin and dei Poli (1977) observed that a value of $\overline{G}_c$ linearly decreasing with the fictitious strain normal to the crack would give better predictions for beams failing in shear, and suggested the following equation

$$\overline{G}_c = 0.24\,G_c\,(1 - 250\,\varepsilon_1) \qquad for\ \varepsilon_1 \geq \varepsilon_{cr} \qquad (5a)$$

$$\overline{G}_c = G_c \qquad for\ \varepsilon_1 < \varepsilon_{cr} \qquad (5b)$$

where G_c is the elastic shear modulus of uncracked concrete, ε_1 is the principal tensile strain normal to the crack and ε_{cr} is the cracking tensile strain.

Al-Mahaidi (1978) recommended the following hyperbolic expression for the reduced shear stiffness $\overline{G}_c$ to be employed in the constitutive relation of cracked concrete

$$\overline{G}_c = \frac{0.4\,G_c}{\varepsilon_1 / \varepsilon_{cr}} \qquad (6)$$

Yuzugullu and Schnobrich (1973) used a constant value of $\overline{G}_c$ for the effective shear modulus

$$\overline{G}_c = 0.25\,G_c \quad for\ deep\ beams \qquad (7a)$$

$$\overline{G}_c = 0.125\,G_c \quad for\ shear\ wall - frame\ systems\,. \qquad (7b)$$

The computer code developed in this research project has been implemented with the afore-mentioned formulations of the shear modulus after cracking.

In the rigid diaphragm model developed for the three dimensional analysis of reinforced-concrete frames, the basic assumption being that there are no in-plane deformations in the floor and also, the floor is assumed to be infinite rigid in its plane. Each floor plate is assumed to be translate in plan and rotate about a vertical axis as a rigid body. In this study, since rigid diaphragm model is considered, cracking occurs only in the beam element due to flexural moments in local y direction. On the other hand, the column element cracks due to the flexural moments in local y and z directions. Hence $I_{eff}, M_{cr}, M, I_1, I_2, \varepsilon_1$ and ε_2, are the values related to the flexure in local y and z directions.

4 FORMULATION OF THE PROBLEM

In this section, the flexibility coefficients of a member will first be evaluated, and then using compatibility conditions and equilibrium equations, stiffness matrix and the load vector of a member with some regions in cracked state will be obtained.

A typical member subjected to concentrated and a uniformly-distributed loads, and positive end forces with corresponding displacements are shown in Figure 2. A cantilever model is used for computing the relations between nodal actions and basic deformation parameters of a general space element (Fig. 3).

The basic deformation parameters of a general space element may be established by applying unit loads in turn in the directions of 1–3 and 7–9. Then, the

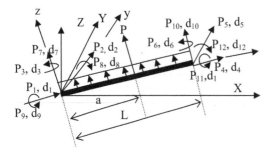

Figure 2. A typical member subjected to concentrated and a uniformly-distributed loads.

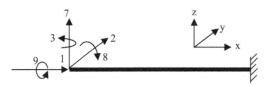

Figure 3. A cantilever model for computing the relations between the nodal actions and basic deformation parameters.

compatibility conditions give the following equation in matrix form:

$$
\begin{bmatrix}
f_{11} & 0 & 0 & 0 & 0 & 0 \\
0 & f_{22} & f_{23} & 0 & 0 & 0 \\
0 & f_{32} & f_{33} & 0 & 0 & 0 \\
0 & 0 & 0 & f_{77} & f_{78} & 0 \\
0 & 0 & 0 & f_{87} & f_{88} & 0 \\
0 & 0 & 0 & 0 & 0 & f_{99}
\end{bmatrix}
\begin{bmatrix}
P_1 \\ P_2 \\ P_3 \\ P_7 \\ P_8 \\ P_9
\end{bmatrix}
=
\begin{bmatrix}
d_1 \\ d_2 \\ d_3 \\ d_7 \\ d_8 \\ d_9
\end{bmatrix}
\quad (8)
$$

in which, f_{ij} is the displacement in i-th direction due to the application of unit loads in j-th direction, and can be evaluated by means of the principal of virtual work as follows

$$
f_{ij} = \int_0^L \left(\frac{M_{zi}M_{zj}}{E_c I_{effz}} + \frac{M_{yi}M_{yj}}{E_c I_{effy}} + \frac{V_{yi}V_{yj}}{\overline{G}_c A}s + \frac{V_{zi}V_{zj}}{\overline{G}_c A}s + \frac{M_{bi}M_{bj}}{G_c I_o} + \frac{N_i N_j}{E_c A} \right) dx \quad (9)
$$

In Eq. (9), $M_{zi}, M_{zj}, M_{yi}, M_{yj}, V_{zi}, V_{zj}, V_{yi}, V_{yj}, M_{bi}, M_{bj}, N_i$ and N_j are the bending moments, shear forces, torsional moments and axial forces due to the application of unit loads in i-th and j-th directions, respectively, E_c and I_0 denote the modulus of elasticity of concrete and torsional moment of inertia of the cross section, s and A are the shape factor and the cross sectional area, respectively.

Three dimensional member stiffness matrix is obtained by inverting the flexibility matrix in Eq. (8) and using the equilibrium conditions

The member fixed-end forces for the case of a point and a uniformly distributed loads can be evaluated by

means of compatibility and equilibrium conditions as follows

$$
P_{10} = P_{20} = P_{30} = P_{40} = P_{50} = P_{60} = P_{90} = P_{110} = 0. \quad (10a)
$$

$$
P_{70} = -(f_{88} \, f_{70} - f_{78} \, f_{80}) / (f_{77} \, f_{88} \; f_{78} \, f_{87}). \quad (10b)
$$

$$
P_{80} = -(f_{77} \, f_{80} - f_{78} \, f_{70}) / (f_{77} \, f_{88} \; f_{78} \, f_{87}). \quad (10c)
$$

$$
P_{100} = -(q \, L + P + P_{70}). \quad (10d)
$$

$$
P_{120} = -(q \, L^2 / 2 + P(L-a) + P_{70} \, L + P_{80}) \quad (10e)
$$

where f_{i0} $(i = 7,8)$ is the displacement in i-th direction due to the application of span loads which can be obtained by using the principal of virtual work in the following form

$$
f_{i0} = \int_0^L \left(\frac{M_{yi} \, M_0}{E_c \, I_{effy}} + \frac{V_{zi} \, V_0}{G_c \, A}s \right) dx \quad (11)
$$

where M_0 and V_0 are the bending moment in local y direction and shear force in local z direction due to the span loads. Finally, the member stiffness equation can be obtained as

$$
\underline{k}\,\underline{d} + \underline{P}_0 = \underline{P} \quad (12)
$$

where $\underline{k}$ (12×12) is the stiffness matrix, $\underline{d}$ (12×1) is the displacement vector, $\underline{P}_0$ (12×1) is the fixed end force vector and $\underline{P}$ (12×1) is the total end force vector of the member. Since Eq. (12) is given in the member coordinate system (x, y, z), it should be transformed to the structure coordinate system (X, Y, Z).

In the rigid diaphragm model, member equations are first obtained and then considering contributions which come from each element, the system stiffness matrix and system load vector are assembled. Finally, the system displacements and member end forces are obtained by solving the system equation. This procedure is repeated step by step in all iterations.

The flexibility influence coefficient can be evaluated by means of Eqs. (9) and (11) with the expressions of moment and shear forces obtained from the application of unit and span loads. Details of the formulation can be found in Dundar and Kara (2007).

It should be noted that, since the member has cracked and uncracked regions, integral operations in Eqs. (9) and (11) will be performed in each region. In general, the member has three cracked regions and two uncracked regions as seen in Fig. 4.

In the cracked regions where $M > M_{cr}, I_{eff}$ and $\overline{G}_c$ vary with M along the region. Hence, the integral values in these regions should be computed by a numerical integration technique. The variation of effective moment of inertia and effective shear modulus of concrete in the cracked regions necessitate the redistribution of the moments in the structure. Therefore, iterative procedure should be applied to obtain the final deflections and internal forces of the structure.

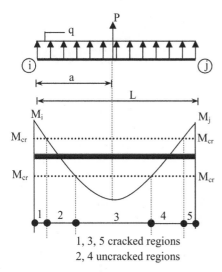

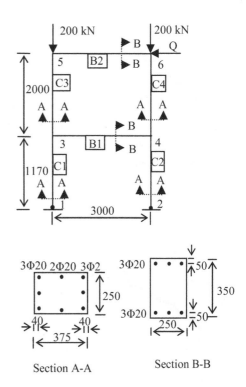

Figure 4.   Cracked and uncracked regions in a member.

5   COMPUTER PROGRAM

A general purpose computer program developed in the present study on the basis of iterative procedure is coded in FORTRAN 77 language. In the solution procedure, the member end forces used in each iteration step are taken as the mean value of the end forces of all previous iterations. This procedure accelerates the convergence of the algorithm. In the program, the following equation is used as the convergence criterion

$$\left| \frac{P_i^n - P_i^{n-1}}{P_i^n} \right| \le \varepsilon. \tag{13}$$

In Eq. (13) n is the iteration number, ε is the convergence factor and P_i^n ($i = 1,12$) is the end forces of each member of the structure for n-th iteration.

6   VERIFICATION OF ITERATIVE PROCDURE BY THE EXPERIMENTAL RESULTS AND A FOUR STORY FRAME EXAMPLE

In this part, two examples are presented. The first example is taken from the literature to verify the applicability of the analytical procedure. The second example is the application of the proposed analytical method on a three-dimensional, four-story reinforced concrete frame.

6.1   *Example 1*

In this example, the test results given by Chan et. al. (2000) for a two story reinforced concrete frame have been compared with the results of the present computer program. This reinforced concrete frame was designed

Figure 5.   Two story reinforced concrete frame tested by Chan et. al. (2000) (dimensions in mm).

with a center to center span of 3000 mm, a first story height of 1170 mm and a second story height of 2000 mm (Fig. 5). This frame is modeled by four columns of 250×375 mm and two beam elements of 250×350 mm cross sections. The reinforcing steel in the beams and columns, the span and the loads are also shown in the figure.

The test procedure involved first applying a total axial load of 200 kN to each column and maintaining this load throughout the test. The lateral load (Q) was then monotonically applied until the ultimate capacity of the frame was achieved. In the analysis, ACI and probability-based effective stiffness models are used for the effective moment of inertia and Al-Mahaidi's model is used for the shear modulus of concrete in the cracked regions. The reduction of the flexural stiffness in the beams and columns under increasing lateral loads are also obtained by using probability-based effective stiffness model. In computing flexural tensile strength and modulus of elasticity of concrete the following equations (ACI 1995) are also used.

$$E_c = 4730\sqrt{f_c} \qquad N / mm^2 \tag{14a}$$

$$f_r = 0.62\sqrt{f_c} \qquad N / mm^2 \tag{14b}$$

in which, f_c is the compressive strength of concrete.

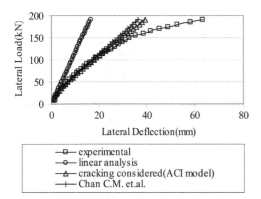

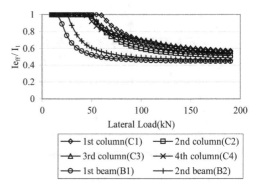

Figure 6. Comparison between experimental and analytical results concerning the drift of joint 6.

Figure 8. Flexural stiffness reduction of each member versus lateral loads.

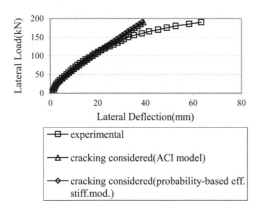

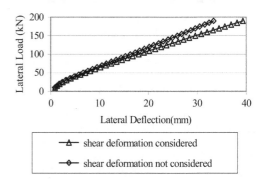

Figure 9. Effects of shear deformation on frame drift (joint 6) according to the proposed numerical procedure.

Figure 7. Plots of the drift (joint 6) according to the different models adopted for the effective moment of inertia.

In order to determine the applicability of proposed procedure, the comparison between the test and theoretical results for the lateral deflection of joint 6 obtained by the linear analysis and cracking analysis, is presented in Fig. 6.

It is seen that the deflections calculated by the developed computer program using ACI model agree well with the test results for applied lateral loads up to approximately 80% of the ultimate load (i.e., at the value of approximately 155 kN). When the lateral load is beyond this load level the difference between the experimental and theoretical results becomes significant. On the other hand, the results of the proposed analytical procedure which considers the variation of shear stiffness due to cracking gives better prediction of deflections than the other numerical study developed by Chan et. al. (2000).

Fig. 7 presents a comparison of the top deflections using the different models for the effective moment of inertia of the cracked members. As seen from the figure, although different models have been used for

the effective flexural stiffness, the results are very close to each other.

The variation of the flexural stiffness of beams and columns with respect to the lateral applied load is also shown in Fig. 8. As seen from the figure, the beams of the first and second stories crack first and then two columns on the lateral loading side, C4 and C2, start to crack followed by, in the final stage, the cracking of both columns on the opposite loading side, C3 and C1. Fig. 8 also shows that when the lateral load reaches 78% of the ultimate lateral load, the beams at the first and second stories have 45 and 47%, respectively, of the gross moment of inertia, and the two columns at the second story have 55 and 60% of their uncracked values. The other two columns at the first story have also 53 and 60% of their gross moment of inertia. The results show that considering the cracking effect on the stiffness of structural frame by assigning an 80% reduced moment of inertia to all the columns and a 50% reduced moment of inertia to all the beams does not always guarantee a conservative prediction of the lateral drift.

Fig. 9 shows the contribution of the shear deformation to the lateral drift of joint 6. This contribution

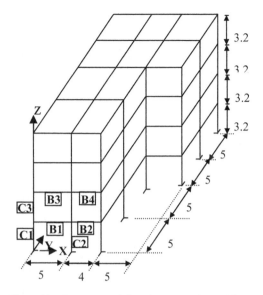

Figure 10. Four-story reinforced concrete frame (dimensions in m).

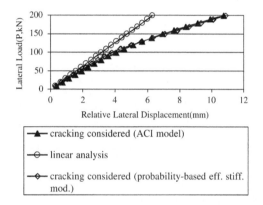

cracking considered (ACI model)

linear analysis

cracking considered (probability-based eff. stiff. mod.)

Figure 11. Plots of the drift of the second floor as a function of the lateral load.

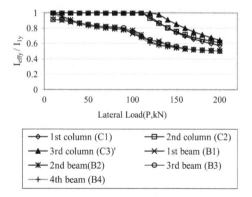

1st column (C1) 2nd column (C2)
3rd column (C3)' 1st beam (B1)
2nd beam(B2) 3rd beam (B3)
4th beam (B4)

Figure 12. Flexural stiffness reductions of beams and columns with respect to the lateral applied load.

Table 1. Dimensions of the members and loads applied to the frame.

Story	1st	2nd	3rd	4th
Dimensions of beam (mm*mm)	300*500	300*500	300*500	300*500
Dimensions of column (mm*mm)	500*500	500*500	500*500	500*500
Uniformly distributed loads (kN/m)	q	q	q	0.8 q
Lateral loads acting at master points in X direction	P	P	P	2.5 P

increases under increasing lateral loads (for instance, by 14% at 78% of the ultimate lateral load).

6.2 Example 2

In this example, the four-story reinforced concrete frame shown in Fig. 10 is analyzed by means of the proposed procedure. This three-dimensional reinforced-concrete frame is subjected to lateral loads at each story master point and uniformly distributed loads on the beams. The dimensions of the members and the loads are given in Table 1. In the analysis, the effective moment of inertia is evaluated by ACI and probability-based effective stiffness models and the effective shear modulus is predicted by Al-Mahaidi's model. In this example, the lateral loads acting at each story master point, which are expressed in terms of the values of P, are increased while the intensity of uniform loads ($q = 30$ kN/m) remain constant.

The variation of the maximum relative lateral displacement of the second floor with the lateral load, when cracking is considered and not considered for beams and columns are shown in Fig. 11. As seen from the figure, the differences in the maximum relative lateral displacement of the second floor between the two cases increase with increasing lateral loads. The difference becomes significant at higher loads such as 70% for P = 200 kN.

The flexural stiffness reductions of various members with respect to the lateral applied load are shown in Fig. 12. As seen from figure, when the stiffness of beams are reduced to 50% of their uncracked stiffness, the two first-story columns have reduced to 57–61% of their uncracked stiffness.

Fig. 13 also shows the theoretical influence of shear deformation on the maximum relative lateral displacement of the second floor. The shear deformations contribute up to approximately 10% of the total lateral displacements.

743

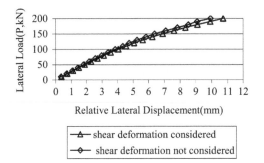

Figure 13. Theoretical influence of shear deformation on the lateral deflection of a three dimensional reinforced concrete frame.

7 CONCLUSIONS

An iterative procedure has been developed to analyze three dimensional reinforced-concrete frames with cracked beams and columns. The variation of the flexural stiffness of a cracked member has been evaluated by using ACI, CEB and probability-based effective stiffness models. Shear deformations which can be large after the development of cracks and be of practical importance in the design and behavior of the structure are also taken into account in the analysis. The role of shear stiffness in cracked regions has also been investigated taking advantage of the formulations found in the literature for the reduced shear stiffness.

The capability and the reliability of the proposed procedure have been tested by means of comparisons with the theoretical and experimental results available in the literature. The numerical results of the analytical procedure have been found to be in good agreement with the test results for applied loads up to approximately 80% of the ultimate load capacity of the frame. The analytical procedure not only predicts the deflections to a value of load equal to approximately 80% of the ultimate load with good accuracy, but it also can give an estimation of behavior at approximately 85% of the ultimate load with an acceptable degree of accuracy.

In the analysis, flexural stiffness reductions of beams and columns with respect to the lateral applied load can be obtained by the developed program using the probability-based effective stiffness model for the effective moment of inertia. This is the most significant feature of the proposed procedure and the variations in the flexural stiffness reductions of beams and columns in the reinforced concrete structure can be directly evaluated.

The theoretical deflections of reinforced concrete structure have also been evaluated using different effective flexural stiffness models. It should be noted that different models provide quite similar results.

The numerical results of the analytical procedures indicate that the proposed procedure which considers the variation of shear stiffness after the development of cracks gives better predictions of deflections than other analytical procedures. It is therefore important to consider the variation of shear stiffness in the cracked regions of members in order to obtain more accurate results.

REFERENCES

ACI Committee 435. 1966. Deflection of reinforced concrete flexural members. *ACI J.* 63: 637–674.
Al-Mahaidi, R.S.H. 1978. Nonlinear finite element analysis of reinforced concrete deep members. *Department of Struct. Engrg. Cornell University.* Report No. 79 (1): 357.
Al-Shaikh, A.H. & Al-Zaid R.Z. 1993. Effect of reinforcement ratio on the effective moment of inertia of reinforced concrete beams. *ACI Structural J.* 90: 144–149.
Branson, D.E. 1963. Instantaneous and time-dependent deflections of simple and continuous reinforced concrete beams. *Alabama Highway Deparment/US Bureau of Public Roads HPR* Report No.7(1): 78.
Cedolin, L. & dei Poli S. 1977. Finite element studies of shear critical reinforced concrete beams. *J. Engineering Mech. Div. ASCE (EM3).*
Comite Euro-International du Beton. 1985. Manual on Cracking and Deformation. *Bulletin d'Information* 158-E.
Cosenza, E. 1990. Finite element analysis of reinforced concrete elements in a cracked state. *Computers & Structures* 36(1): 71–79.
Chan, C.M., Mickleborough, N.C & Ning, F. 2000. Analysis of cracking effects on tall reinforced concrete buildings. *J. Struct. Engrg.* 126(9): 995–1003.
Channakeshava, C. & Sundara Raja Iyengar, K.T. 1988. Elasto-plastic cracking analysis of reinforced concrete. *J. Struct. Engrg. ASCE* 114: 2421–2438.
Dundar, C. & Kara, I.F. 2007. Three dimensional analysis of reinforced concrete frames with cracked beam and column elements. Accepted for publication in *Engineering Structures.*
Ghali, A. & Favre R. 1986. *Concrete Structures: Stresses and Deformations.* Chapman & Hall, N.Y.
Ngo, D. & Scordelis A.C. 1967. Finite element analysis of reinforced concrete beams. *ACI J.* 64(3): 152–163.
Sakai, K. & Kakuta, Y. 1980. Moment-curvature relationship of reinforced concrete members subjected to combined bending and axial force *ACI J.* 77: 189–194.
Stafford, S.B. & Coull, A. 1991. *Tall building structures: Analysis and design.* Wiley, New York.
Tanrikulu, A.K. Dundar, C. Cagatay, I.H. 2000. A Computer program for the analysis of reinforced concrete frames with cracked beam elements. *Structural Engineering and Mechanics* 10(5): 463–478.
Yuzugullu, O. Schnobrich, W.C. 1973. A numerical procedure for the determination of the behaviour of a shear wall frame system *ACI J.* 70(7): 474–479.

Fracture Mechanics of Concrete and Concrete Structures – Design, Assessment and Retrofitting of RC Structures – Carpinteri, et al. (eds)
© 2007 Taylor & Francis Group, London, ISBN 978-0-415-44616-7

A lattice model approach to the uniaxial behaviour of Textile Reinforced Concrete

J. Hartig, U. Häußler-Combe & K. Schicktanz
Institute of Concrete Structures, Technische Universität Dresden, Dresden, Germany

ABSTRACT: In this contribution, a lattice model for the simulation of the load-carrying and the failure behaviour of Textile Reinforced Concrete (TRC) under monotonic and cyclic tensile loading is presented. A special property of the mentioned composite TRC is that the reinforcement consists of heterogeneous multi-filament yarns of alkali-resistant glass, which are not fully penetrated with matrix of fine-grained concrete. This leads to complex bond conditions. The lattice model for the simulation consists of one-dimensional strands of bar elements with limited tensile strength, which represent the matrix as well as the reinforcement. These strands are connected with zero-thickness bond elements, which use non-linear bond laws as element characteristics. These bond laws contain a damage algorithm to include a possible degradation of the bond quality due to mechanical loading. This nonlinear problem is solved within a Finite Element Method formulation. Simulations of tensile specimens are performed and compared to experimental data for both monotonic and cyclic loading.

1 INTRODUCTION

Concerning the strengthening and retrofitting of existing steel-reinforced concrete structures built during the last century it is often desirable to apply additional thin load-carrying layers. A possible approach, which offers these properties are layers of Textile Reinforced Concrete (TRC)(Hegger 2001; Curbach 2003; Hegger et al. 2006; Brameshuber 2006). This material is a composite of textile-processed yarns of endless fibres as reinforcement and a fine-grained concrete as matrix. In this context fine-grained means a much smaller maximum aggregate size, e.g. 1 mm, than used for normal concrete. Textiles produced of yarns, made of glass filaments or carbon filaments, are typically embedded in the concrete matrix. Usually yarns of glass, which have to be alkali-resistant to sustain long-term embedded in the concrete (Schorn & Schiekel 2004), consist of up to 2000 filaments with diameters of 10–30 microns.

Unlike Fibre Reinforced Plastics (FRP), the yarns are not fully penetrated with matrix, because the empty spaces in the filament bundles are too narrow for the particles of the cement paste. Hence, there can be found two different bond zones inside a yarn. In the sleeve zone of the yarn where the filaments contact the matrix adhesional bond is dominating. Supposedly, the load transfer properties of the interface between the matrix and the filaments are load dependent, which can result in damage of the bond at higher load levels.

Almost no cement paste intrudes into the core zone. Hence, only a frictional load transfer at the contact areas of the filaments is possible. Additionally, the concrete as well as the filaments have a limited tensile strength, which leads to a crack development in the concrete and the failure of filaments or whole yarns. Altogether, even under monotonic tensile loading a complex structural behaviour is observable.

A lattice approach is developed to simulate the above-mentioned properties. Because only pure tensile loads are regarded, the model has only degrees of freedom in the load direction, which is called the longitudinal direction in the following. The whole concrete component is considered as homogeneous. Thus, it can be modelled as a serial connexion of elements. The yarn consisting of several filaments not fully penetrated with cement slurry, as mentioned above, cannot be assumed homogeneous. While the longitudinal direction is also modelled as a serial connexion of elements, the transverse direction is regarded by splitting the yarn component into a number of segments, which are also assumed to be homogeneous. These segments are regularly arranged in a lattice scheme. Among themselves, the segments are coupled with zero-thickness bond elements, which act nonlinear according to bond laws. The adhesional bond between the concrete and the filaments is subject to damage, which is applied to the bond law to ensure a proper description of the behaviour. For the core segments, which underlie only a frictional bond, a constant bond stress is assumed. Besides that, limited tensile strength is applied to the elements with the consequence of propagating cracks. This highly nonlinear problem is solved with a Finite Element Method formulation. The model is able to reproduce the results of experiments performed under both monotonic and cyclic tensile loading, as for example experiments with unidirectional reinforced tensile specimens. Thus, experimental results are used to verify the computational results and to identify possible weaknesses of the model.

2 EXPERIMENTAL OBSERVATIONS

Textile Reinforced Concrete has a tensile structural behaviour, which is in principle comparable with other continuously reinforced composites as for instance steel rebar reinforced concrete. The used multi-filament reinforcement has to bear the tensile forces after the concrete cracking. The concrete matrix has to transfer the external tensile loads to the reinforcement and has to carry compressive loads. Special characteristics regarding the load-carrying behaviour arise from the heterogeneous constitution of the multi-filament yarns. It is for example observable that the mean strength of a yarn is lower than the mean strength of a single filament (Abdkader 2004). A number of reasons are responsible for this behaviour as for instance statistical effects (Daniels 1944), unequal loading of the filaments in a yarn (Chudoba et al. 2006, Vořechovský & Chudoba 2006) or damaging of the yarns and filaments in the production and treatment processes (Abdkader 2004).

For the investigation of the structural behaviour of composites under uniaxial loading tensile specimens are often used. For the case of Textile Reinforced Concrete such investigations were made for example by (Jesse 2004; Curbach 2003) and (Molter 2005), which show similar results. Hence, only a subset of JESSE's comprehensive experimental data will be discussed here. In (Jesse 2004) results of specimens investigated under monotonic tensile loading are published. Besides these tests, JESSE made also experiments under cyclic loading, which are hitherto unpublished. The specimens under consideration are unidirectional reinforced with yarns of alkali-resistant glass with reinforcement ratios varying between 1–3 Vol. %.

In Figure 1, a typical specimen used by JESSE is shown with its dimensions. During the test, the specimen is fixed in a hydraulic testing machine with clamping devices as indicated in Figure 1 as well. The tensile loads are applied displacement-controlled with a rate of about 0.015 mm/s. On a length of 200 mm, the longitudinal displacements are measured on both sides of the specimen, see Figure 1. In Figure 2, a typical stress-strain relation is shown where the measured

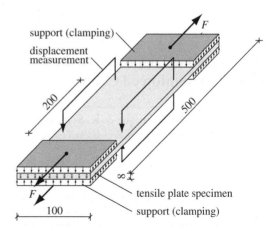

Figure 1. Typical specimen and test setup used by (Jesse 2004) for tensile tests on TRC (dimensions in mm).

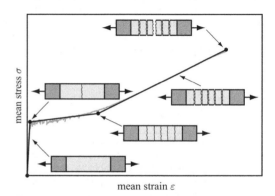

Figure 2. Typical stress-strain relation and associated crack patterns for tensile specimens under monotonic loading as shown in Figure 1.

forces F are related to the specimen's cross-sectional area leading to a mean stress σ and the measured displacements are related to the measurement length leading to a mean strain ε.

The stress-strain relation starts with a linear increase, principally according to the Young's modulus of the concrete, until the tensile strength of the concrete is reached and the matrix cracks for the first time. Upon this point, the yarn reinforcement has to bear the applied tensile load at the crack. If a sufficient amount of reinforcement with an ample bond capacity is available, further cracks will develop under increasing external tensile load. In the following part of the stress-strain relation, this is associated with a decreased slope of the curve (Figure 2). The crack development in the concrete continues until the load transferred from the reinforcement to the matrix between two cracks is too low to reach the tensile strength of the concrete again.

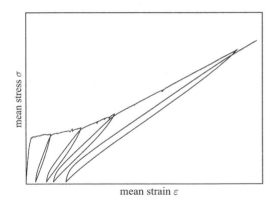

Figure 3. Typical experimental stress-strain relation under cyclic loading for tensile specimens as shown in Figure 1.

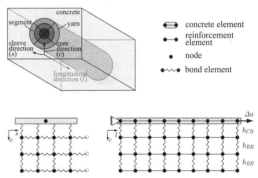

Figure 4. Geometrical model (top) and lattice discretisation in the cross section (bottom left) and the longitudinal section (bottom right).

If the final crack pattern is reached, the stress-strain relation increases again until the tensile strength of the reinforcement is reached too and the specimen finally fails, often in a brittle manner. The slope of the stress-strain relation in this state is mostly influenced by the properties of the reinforcement. Nevertheless, the concrete participates of course in the load-carrying between the cracks, which is well-known as tension stiffening.

In the case of cyclic loading, the stress-strain relation of the monotonic loading case can be seen as the envelope of the cyclic relation. The observed unloading paths of the stress-strain relation are Z-shaped. This means that the stress-strain relation decreases according to a steeper slope compared with the loading path. In the middle part of the unloading path, the slope becomes flatter and the stress-strain relation decreases almost linearly. Near the abscissa the slope of the stress-strain relation increases again. This increase of the stiffness can probably be explained with a compressional reloading of the concrete. It is also observable that during the unloading the origin is not reached again. This effect can be considered as a macroscopic plastic deformation. With reaching higher load levels the plastic deformation also increases. The increase of the plastic deformation is more pronounced in the cracking state than in post-cracking state, which can be caused by an initial stressless deformation of the reinforcement at the cracks. The mean slopes of the unloading paths are steeper at lower load levels than at higher ones. The reloading is characterised by a steep increase of the stress-strain relation, which passes into a flatter linear slope and merges into the envelope curve. The mean slopes of the reloading paths are also steeper at lower load levels than at higher load levels. According to the envelope curve, also the cyclic loaded specimen finally fails reaching the tensile strength of the reinforcement.

3 MODELLING

3.1 Mechanical model

In the following, a lattice model used for the determination of the structural behaviour of Textile Reinforced Concrete will be described, which was partly developed in previous contributions. In (Häußler-Combe & Hartig 2006a) a one-dimensional mechanical model for the determination of the load-bearing and the failure behaviour of Textile Reinforced Concrete was developed. This model was enhanced in (Häußler-Combe & Hartig 2006b) with a bond law using a damage algorithm to include the degradation of the bond quality between concrete and reinforcement.

The model is in principle a combination of strands of bar elements for the constituents of the composite, which are each assumed to be homogeneous. The concrete as well as the reinforcement elements have prescribed limited tensile strengths, which can lead to cracking while loading. The strands are connected at the nodes with zero-thickness bond elements, which act according to bond laws. The heterogeneity of the reinforcement is considered by segmentation into several strands as it is shown in principle in Figure 4. The segments are arranged in a lattice scheme, which simplifies the computational implementation. Depending on the resolution, in one segment several filaments or yarns are pooled, which are assumed to be homogeneous with effective material properties.

As mentioned before, the bond elements act according to bond laws formulated as bond stress-slip relations. Experimental investigations for the determination of the bond properties between single filaments and concrete were made for instance by (Banholzer 2004) and (Zhandarov & Mäder 2005). Such experimental results are used to estimate the magnitude of the bond force, but are not directly applicable for bond laws of a whole yarn or parts of it. The bond stress-slip relation h_{CR} used in the model for the interaction

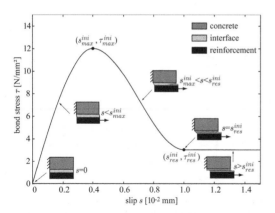

Figure 5. Bond stress-slip relation h_{CR} for the case of concrete-sleeve interaction in the initial state.

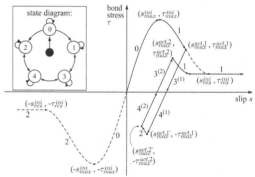

Figure 6. Degradation algorithm implemented in the bond stress-slip relation h_{CR} evolved from an arbitrarily chosen loading.

between the concrete and the sleeve segments is shown in Figure 5 in its initial state. The bond law implements a combination of damage and plasticity to take into account bond degradation due to loading. The algorithm for the degradation evolution of the bond stress is schematically shown in Figure 6 using an arbitrarily chosen loading path. Figure 6 includes also the state diagram with the starting point (●) of the algorithm and all possible state transitions. The numbers of the states in the state diagram correspond to those in the bond stress-slip relation.

The bond law h_{CR} for the concrete-sleeve interaction starts with an increase until the maximum value of the bond stress τ_{max}^{ini} corresponding to the slip s_{max}^{ini} is reached. This peak point is the maximum possible value of bond stress that can be transferred. After this peak, the bond stress-slip relation decreases until the residual bond stress τ_{res}^{ini} corresponding to the slip s_{res}^{ini} is reached. With larger slip values than s_{res}^{ini} only a frictional load transfer is assumed with a bond stress value τ_{res}. This bond stress value can be kept constant or can further decrease with increasing slip.

If a slip reduction occurs after the peak point in the softening state 1 (Figure 6), where the bond is degraded to a certain amount, the initial maximum bond stress necessarily cannot be reached again. A reduction of the slip value can occur for instance due to cracking resulting in a local stress relocation. An unloading path (state 3) different from the loading path (state 0) is used in the bond law h_{CR} to avoid a trace back to the initial peak. Therefore, the bond stress $\tau^{act,1}$ corresponding to the currently largest reached slip $s^{act,1}$ is stored and used as new maximum value $\tau_{max}^{act,1}$. The unloading path (state $3^{(1)}$) is modelled as linear decrease of the bond stress and the slip according to the slope between the origin of the coordinate system and the initial peak point $(s_{max}^{ini}, \tau_{max}^{ini})$. The value of the bond stress is limited to the negative absolute value of the currently largest

bond stress $-\tau_{max}^{act,1}$. This point is reached with the chosen load path in Figure 6 via the state $3^{(1)}$ and the state $4^{(1)}$. A further reduction of the slip will decrease the absolute maximum bond stress value, which is the case in Figure 6 where the slip decreases to $s_{max}^{act,2'}$ with the corresponding bond stress $-\tau_{max}^{act,2}$. At this point, the load in association with the slip as well as the bond stress increase again (Figure 6). Thus, the state 1 will be reached again via the states $4^{(2)}$ and $3^{(2)}$. While the slip in state 1 further increases, the bond stress decreases until the residual bond stress τ_{res}^{ini} is reached, which is equivalent with a purely frictional load transfer.

For the interaction between the sleeve and the core zone of a yarn pure friction is assumed. The corresponding bond law h_{RR} is in principle the same as h_{RR} but has no peak value τ_{max}^{ini} at s_{max}^{ini}. A constant bond stress of 3 N/mm² for both τ_{max}^{ini} and τ_{res}^{ini} is assumed.

The interpolation between the supporting points of the bond stress-slip relations can be performed according to several approaches. A multi-linear approach is used for example by (Richter & Zastrau 2006) in conjunction with an analytical modelling approach. In numerical simulations, the discontinuities of the derivatives on the transition between the intervals of the multi-linear relations lead possibly to numerical problems during computations. This can be avoided using special cubic polynomials, which show monotonicity and continuity in the first derivatives between consecutive intervals. As underlying algorithm, the Cubic Hermite Interpolating Polynomial Procedure (PCHIP) by (Fritsch & Carlson 1980), which is also published in (Kahaner et al. 1989), is used. It ensures a smooth, shape preserving interpolation of the bond stress-slip relation given by a number of data points without producing additional bumpiness or oscillations as it could be the case for example with a spline interpolation (de boor 1978). As supporting points for the bond law h_{CR} (Figure 5) the origin of the coordinate system, the peak point, the residual point and an end point are used.

3.2 Numerical model

The lattice model presented in the previous section is the basis for a numerical model formulated within the Finite Element Method (FEM). In longitudinal direction, the concrete and the yarn strand are discretised with one-dimensional bar elements of a length of 0.1 mm. This leads with the specimen length of 500 mm (Figure 1) to 5000 elements per segment strand. A finer discretisation does not affect the results significantly, whereas coarser discretisations cannot approximate the used bond law in a sufficient manner and overestimate the macroscopic stiffness of a crack bridge. The boundary conditions are given with prescribed displacements at the concrete's end nodes. The displacement is zero at $x = 0$ mm and becomes continuously increased at $x = 500$ mm for the case of monotonic loading. For the case of cyclic loading the displacement at $x = 500$ mm is increased and reduced according to a load regime as used in the experiments. As mentioned in the previous section the strands are connected with bond elements, which act according to bond laws also introduced in this section. The free value of the bond law is the slip, which is determined by the difference of the displacements between the two nodes of a bond element.

Besides the nonlinearities resulting from the bond law, additional nonlinearities arise from limited tensile strengths for both the concrete and the reinforcement. To avoid the failure of a series of elements in the case of constant or nearly constant stresses in the longitudinal direction, the failure of elements is limited to one per load step. The bar elements used to represent the concrete in the clamping zones are assumed not to crack on a length of 100 mm from the ends of the specimen to avoid failure at the concrete section's end nodes.

The resulting system of nonlinear equations is solved using the Broyden-Fletcher-Goldfarb-Shanno (BFGS) approach, which is a Quasi-Newton-Method, in combination with a line-search algorithm (Bathe 1996; Matthies & Strang 1979, Nocedal & Wright 1999). Some more details regarding the numerical implementation related to the stated problem are presented in (Häußler-Combe & Hartig 2006a).

4 COMPUTATIONS

4.1 Geometrical and material parameters

So far, the model is specified in principle. For the computations a specialised model is used, which consists of three bar element strands: one strand for the concrete and two strands for the yarn reinforcement, see Figure 7. In this model, the concrete strand is only

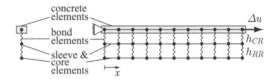

Figure 7. Specialised model with three element strands in the cross section (left) and the longitudinal section (right).

connected with the so-called sleeve strand via bond elements using the bond law h_{CR} for adhesional bond as element characteristic. The sleeve strand is additionally coupled with the core strand via bond elements acting according to the bond law h_{RR}, which represents purely frictional load transfer.

The tensile strength of the concrete is assumed with 6.5 N/mm^2 and the Young's modulus is defined with 28,500 N/mm^2 (Jesse 2004). The cross-sectional area of the concrete of 771 mm^2 is determined by the width of 99.5 mm and the thickness of 7.8 mm of the specimen. The reduction of the cross-sectional area of the concrete due to the reinforcement is neglected.

The reinforcement material are yarns of alkali-resistant glass with a unit length weight of 310 tex produced by Nippon Electric Glass. The Young's modulus was determined by (Abdkader 2004) with $E_{yarn} = 79,950$ N/mm^2. The total cross-sectional area of the reinforcement $A_{reinf} = 14.6$ mm^2 results from the number of yarns $n_{yarn} = 134$ and the cross-sectional area of a yarn $A_{yarn} = 0.11$ mm^2. Hence, the specimen was reinforced with a ratio of about 1.9 Vol. %. According to the model, the reinforcement is splitted into two parts, the sleeve strand with 25% of the total cross-sectional reinforcement area and the core strand with 75% of the total cross-sectional reinforcement area. This ratio is approximated on the base of microscopic observations on transparent cuts of yarns embedded in a cementitious matrix.

The strands of bar elements are coupled with bond elements for which the bond surface areas have to be defined. The bond surface areas S_{sleeve} and S_{core} are assumed to be the lateral surface areas of cylinders, which have the cross-sectional area of a homogeneous yarn A_{yarn}:

$$S_{sleeve} = n_{yarn} \cdot l \cdot C_{yarn}$$

$$= n_{yarn} \cdot l \cdot (2\sqrt{\pi A_{yarn}}) \quad (1)$$

$$S_{core} = n_{yarn} \cdot l \cdot C_{core}$$

$$= n_{yarn} \cdot l \cdot (2\sqrt{\pi \cdot 0.75 \cdot A_{yarn}}) \quad (2)$$

In these equations l is the element length, C_{yarn} the circumferences of a homogeneous circular yarn and C_{core} of the core fraction of a yarn respectively. In these bond

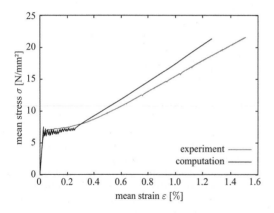

Figure 8. Computational and experimental stress-strain relation for the case of monotonic loading.

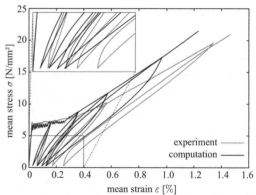

Figure 9. Computational and experimental stress-strain relation for the case of cyclic loading.

surface areas possible roughness or ellipticity of the yarn boundary is not included, because the bond forces finally result from the product of the bond surface areas and the bond stresses. The tensile strength of the yarns is defined with $1357 \, N/mm^2$ (Abdkader 2004).

4.2 Monotonic loading

In Figure 8 the computed and an experimentally obtained stress-strain relation for the case of mono-tonic tensile loading is shown. It can be seen that the uncracked state in the simulation coincides with the experimental data. Because the slope of this linear increase is mostly influenced by the Young's modu-lus of the concrete, this agreement could be expected as well as the transition point to the cracking state, which depends on the tensile strength of the concrete.

The computed mean slope of the cracking state of the concrete also agrees with the experimental data, but the drops of the stress after each concrete crack are larger than observed in the experiments. This results from the as purely brittle implemented failure of the concrete, which is not observable in reality. It is well known that concrete is able to transfer stresses over small cracks to a certain amount, which depends on the crack width. This effect is called tension softening and according to (Brockmann 2006) it also exists in the special kind of concrete used in the experiments under consideration. A further fact, which implies that tension softening would lead to more realistic com-putational results, is the calculated total number of cracks. It is lower than the number of cracks observed in the experiments.

After the crack development has finished, the com-puted stress-strain relation increases again, which agrees with the behaviour in the experiments. However, the computed slope in this state is larger than in the experiments. The reasons for this discrepancy

are currently not clear. It can be speculated that a certain number of highly bonded filaments fails pre-maturely. This has to happen simultaneous with the concrete, because if it would occur after the concrete cracking has finished the slope in the stress-strain rela-tions would have to decrease non-linearly, which is not observable in the experimental data. Another reason could be that de facto a lower number of yarns were inserted in the experiment. A reason, which is currently favoured by the authors, is that some kind of telescopic effect appears due to the heterogeneous structure of the yarns and the non-uniform bond conditions in the yarn. This could be modelled with a finer discretisa-tion of the yarn, which means that more reinforcement strands have to be connected in parallel.

The failure of the whole structure occurs if the tensile strength of firstly the sleeve strand and sec-ondly the core strand of the reinforcement are reached. While a good agreement between the simulation and the experiment regarding the ultimate stress is observ-able, the ultimate strain is according to the slope of the stress-strain relation in this state lower in the simulation as observed in the experiment.

4.3 Cyclic loading

As mentioned before, with the model it is also possi-ble to simulate the stress-strain behaviour under cyclic loading. Therefore, the model used in the previous section was loaded with four load cycles on differ-ent load levels. In Figure 9 the simulated stress-strain relation is compared with an experimental one. The stress-strain relation for the case of monotonic loading can be seen as the envelope for the cyclic stress-strain relation. Thus, the uncracked state, the cracking and the final cracking state are computed with the same quality as described for the case of monotonic loading in the previous section.

The first load cycle was executed in the state of ongoing cracking. The shape of the stress-strain relation in the computed cycle agrees in principle with the experimental observations. After a steep decrease the unloading path becomes flatter. However, the unloading path reaches a lower strain level in the simulations compared with the experiment, which means that the macroscopic observable plastic deformations are underestimated. A reason could be the participation of the concrete on the load-carrying at the cracks while unloading. It can be assumed that the cracks in the concrete do not close perfectly for example due to loosened particles and the relaxation of eigenstresses. Thus, the concrete is locally stressed compressional, which leads to a macroscopic plastic deformation. This effect cannot be reproduced with the current model. Another source of possible plastic deformations exists in the unloading path of the bond law described in Section 3.1. The slope of this unloading path in the bond law is currently arbitrarily chosen and is thus open for further improvements.

The shape of the reloading path of the experimental stress-strain relation agrees with the assumption of a moderate compressional pre-stressing of the concrete, because at the beginning of the reloading the stress-strain relation increases according to the Young's modulus of the concrete. This is not observable in the simulated stress-strain relation where the reloading starts with a flatter slope caused by the bond law and the stiffness of the reinforcement. Afterwards, in both the experimental and the simulated stress-strain relation the reloading path merges towards the monotonic stress-strain relation and follows it during further loading.

The other three load cycles are beyond the cracking state in the stress-strain relation, see Figure 9. In all three cycles, the characteristics described for the first cycle are repeated in principle, but the differences between simulation and experiment become more pronounced. Especially the compression of the concrete near the end of the unloading is clearly observable in the last cycle of the experimental data. There as well as in the previous cycles, the unloading path becomes stiffer near the abscissa. As mentioned before, this is not observable in the simulation, because the compression of the concrete after cracking is not implemented in the model. This is also the reason, why the reloading paths in the computed stress-strain relation start always flatter than in the experimentally obtained relation.

In agreement with the experimental data, the cycles on lower load levels behave stiffer than cycles on higher load levels and the hystereses become larger as well. The area in between a hysteresis is a measure of the dissipated energy. Looking on Figure 9 it can be seen that in the simulation too much energy is dissipated compared with the experiments. One reason is the simulated stiffness in the final cracking state, which is larger than in the experiment. This leads to more pronounced hystereses.

Regarding the occurring macroscopic observable plastic deformations, it must be concluded that the model underestimates these deformations. A possible reason is an initial slack, which could lead to a delayed activation of the yarn reinforcement. This could lead to stressless deformations. A fine-tuning of the parameters used in the model basing on a detailed study of several experimental will improve the agreement between the model and the experiment.

5 CONCLUSIONS

The load-carrying and failure behaviour of Textile Reinforced Concrete shows complexity even in the case of purely tensile loading. This behaviour is simulated with a lattice model reduced to the essential. The distinction between matrix and yarn, the different bond zones in the yarn and the limited tensile strength are assessed as essential properties. The spatial material distribution seems in the case of unidirectional reinforcement and loading less important. Hence, the presented model has a one-dimensional geometry but takes material-specific nonlinearities like limited tensile strengths and nonlinear bond laws with damage into account.

The presented computational results are showing a good agreement with the experimental data, although some deficiencies are still existing. A further improvement of the computational results could be reached by the implementation of tension softening for the fine-grained concrete to include the load transmission over a concrete crack and to simulate the crack patterns in the concrete more realistic. However, this will primarily ameliorate the computational results quantitatively. Regarding the cyclic loading, a modification of the unloading path of the bond law can possibly improve the computational results.

An important exercise is the estimation of the parameters used in the simulations. The material parameters for instance the tensile strengths or the Young's moduli of the concrete and the yarns are sufficiently well known but the knowledge about the interaction between matrix and reinforcement is still lacking. It is for example not experimentally confirmed how strong and durable the bond between matrix and reinforcement is. Further investigations are necessary to clarify these open questions.

ACKNOWLEDGEMENT

The authors gratefully acknowledge the financial support of this research from Deutsche Forschungsgemeinschaft DFG (German Research Foundation) within the Sonderforschungsbereich 528 (Collaborative Research Center) "Textile Reinforcement for Structural Strengthening and Retrofitting" at Technische Universität Dresden as well as their colleagues for providing the experimental data.

REFERENCES

Abdkader, A. (2004). *Charakterisierung und Modellierung der Eigenschaften von AR-Glasfilamentgarnen für die Betonbewehrung*. Ph. D. thesis, Technische Universität Dresden, Dresden.

Banholzer, B. (2004). *Bond behaviour of a Multi-Filament Yarn embedded in a cementitious Matrix*. Ph. D. thesis, RWTH Aachen, Aachen.

Bathe, K. (1996). *Finite Element Procedures*. Englewood Cliffs, New Jersey: Prentice-Hall.

Brameshuber, W. (ed.) (2006). *State-of-the Art Report of RILEM Technical Committee 201 TRC: Textile Reinforced Concrete*. RILEM Publications Report 36. Bagneux: RILEM.

Brockmann, T. (2006). *Mechanical and fracture mechanical properties of fine grained concrete for textile reinforced composites*. Ph. D. thesis, RWTH Aachen, Aachen.

Chudoba, R., Vořechovský, M. & Konrad, M. (2006). Stochastic modeling of multi-filament yarns. I. Random properties within the cross-section and size effect. *International Journal of Solids and Structures* 43: 413–434.

Curbach, M. (ed.) (2003). *Textile Reinforced Structures – Proceedings of the 2nd Colloquium in Textile Reinforced Structures (CTRS2)*, Dresden: TU Dresden.

Daniels, H. (1944). The statistical theory of the strength of bundles of threads. I. *Proceedings of the Royal Society of London* A183: 405–435.

de Boor, C. (1978). *A Practical Guide to Splines*. Number 27 in Applied Mathematical Sciences. New York: Springer-Verlag.

Fritsch, F. & Carlson, R. (1980). Monotone Piecewise Cubic Interpolation. *SIAM Journal on Numerical Analysis* 17(2): 238–246.

Häußler-Combe, U. & Hartig, J. (2006a). Structural behaviour of textile reinforced concrete. In Meschke, G.; de Borst, R.; Mang, H. & Bicanic, N. (eds), *Computational Modelling of Concrete Structures – Proceedings of the EURO-C 2006, Mayrhofen, 27–30 March 2006*: 863–872. London: Taylor& Francis.

Häußler-Combe, U. & Hartig, J. (2006b). Uniaxial structural behavior of TRC – A one-dimensional approach considering the transverse direction by segmentation. In Hegger, J.; Brameshuber, W. & Will, N. (eds), *Textile Reinforced Concrete – Proceedings of the 1st International RILEM Symposium, Aachen, 6–7 September 2006*: 203–212. Bagneux: RILEM.

Hegger, J. (ed.) (2001). *Textilbeton – 1. Fachkolloquium der Sonderforschungsbereiche 528 und 532*, Aachen: RWTH Aachen.

Hegger, J., Brameshuber, W. & Will, N. (eds) (2006). *Textile Reinforced Concrete – Proceedings of the 1st International RILEM Symposium*, Aachen, RILEM Publications PRO 50, Bagneux: RILEM.

Jesse, F. (2004). *Tragverhalten von Filamentgarnen in zementgebundener Matrix*. Ph. D. thesis, Technische Universität Dresden, Dresden.

Kahaner, D.; Moler, C. & Nash, S. (1989). *Numerical Methods and Software*. London: Prentice-Hall.

Matthies, H. & Strang, G. (1979). The solution of nonlinear finite element equations. *International Journal for Numerical Methods in Engineering* 14: 1613–1626.

Molter, M. (2005). *Zum Tragverhalten von textilbewehrtem Beton*. Ph. D. thesis, RWTH Aachen, Aachen.

Nocedal, J. & Wright, S. (1999). *Numerical Optimization*. Springer Series in Operations Research. New York: Springer-Verlag.

Richter, M. & Zastrau, B. (2006). On the nonlinear elastic properties of textile reinforced concrete under tensile loading including damage and cracking. *Materials Science and Engineering A* 422: 278–284.

Schorn, H. & Schiekel, M. (2004). Prediction of lifetime of alkali-resistant glass fibres in cementitious concretes. In di Prisco, M.; Felicetti, R. & Plizzari, G. (eds), *6th RILEM Symposium on Fibre-Reinforced Concretes (FRC) – BEFIB 2004, Volume 1, Varenna, 20–22 September 2004*: 615–624. Bagneux: RILEM.

Vořechovský, M. & Chudoba, R. (2006). Stochastic modeling of multi-filament yarns. II. Random properties over the length and size effect. *International Journal of Solids and Structures* 43: 435–458.

Zhandarov, S. & Mäder, E. (2005). Characterization of fiber/matrix interface strength: applicability of different tests, approaches and parameters. *Composites Science and Technology* 65: 149–160.

Fracture Mechanics of Concrete and Concrete Structures – Design, Assessment and Retrofitting of RC Structures – Carpinteri, et al. (eds)
© 2007 Taylor & Francis Group, London, ISBN 978-0-415-44616-7

Splitting failure mode of bonded anchors

T. Hüer & R. Eligehausen

Institute of Construction Materials, University of Stuttgart, Stuttgart, Germany

ABSTRACT: Different fastening systems are available nowadays to transfer any given load combination into a concrete member. Bonded anchors are one of the most commonly-used systems. Their failure under tensile loading occurs in two different ways, following the formation of either a conical crack or a number of splitting cracks. In this paper, the attention is focused on the latter failure mode, that may occur with fastenings located close to an edge or to a corner, particularly when the thickness of the member is small. The objective is to define a suitable design approach for bonded fastenings characterized by a splitting-type failure. To this end, numerical and experimental studies were carried out, the former being based on nonlinear finite-element modelling. The results are presented and discussed in this paper, together with the proposal of the afore-mentioned design approach. Finally, the analogy between the proposed approach and that adopted in the design of lap splices in R/C members is discussed.

1 INTRODUCTION

Fastenings are used to transfer loads, e.g. from steel constructions, in concrete members. Adhesive bonded anchors are a popular fastening system. They consist of an injection mortar or a resin capsule and a threaded rod. In case of tensile loading of bonded anchors, failure of the concrete takes place as a concrete cone breakout or, similar to reinforcing bars, by generating of splitting cracks (Eligehausen et al. 2006). A failure due to splitting cracks may occur with fastenings located near to an edge or a corner, especially in a thin member.

Splitting failure is caused by splitting forces, which are generated in the concrete member by tensile loaded fasteners (Asmus 1999). The splitting force generates cracks growing from the anchor to the edge and consequently the edge of the member spalls (Fig. 1).

Due to the load transfer of bonded anchors, compressive stresses occur in the concrete bordering the load bearing area of an anchor. This pressure generates tensile hoop stresses. These tensile stresses act like splitting forces. The splitting forces increase with the tension load.

The ultimate load at splitting failure of an anchorage is affected by the material properties of the concrete and of the mortar. As well the ultimate load depends on the size of the activated fracture surface. The size of the fracture surface is determined by geometric conditions like distances to the edges of the member, the member thickness and the anchor spacing in

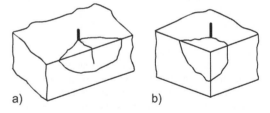

Figure 1. Splitting failure of a single anchor (a) close to an edge and (b) in the corner.

groups. Furthermore, the geometrical parameters of the anchor (diameter and embedment depth) influence its capacity.

Up to now a design concept to predict the splitting failure load is not available. With the view to develop such a design concept numerous numerical simulations and experimental tests were performed. The design approach is based on the results of the numerical investigations and validated by experimental tests.

All investigations were performed as confined tests. In confined tests concrete cone failure is eliminated by transferring the reaction force close to the anchor into the concrete. Furthermore, no bending stresses are generated. This test set-up was chosen for the investigations for two reasons: first no mixture of splitting failure and concrete cone breakout will occur and second no bending stresses are generated, which would superpose with the stresses generated by the splitting forces.

2 THE PROPOSED DESIGN CONCEPT

The proposed design concept for the splitting failure mode of bonded anchors provides to calculate initially a base value of a single anchor at the edge (Equation 1). All the geometrical parameters, which influence the base value, are shown in Figure 2. For the base value the characteristic member thickness $h_{cr,sp}$ is assumed. At this thickness the member provides the maximum capacity and a reduction of the member thickness induces a decrease of the capacity. Further on, the base value depends on a product factor k_P, the size of the load bearing area (anchor diameter d and embedment depths h_{ef}), the edge distance c_1 and the concrete compressive strength f_{cc}. The product factor has to be evaluated by tests.

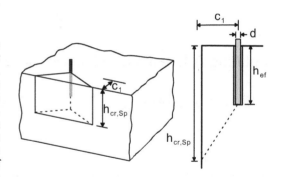

Figure 2. Geometrical parameters.

$$N^0_{u,sp} = k_P \cdot \left(\pi \cdot d \cdot h_{ef}\right)^{1/2} \cdot c_1^{3/7} \cdot h_{cr,sp}^{1/6} \cdot f_{cc}^{1/2} \quad [N] \quad (1)$$

with

$$h_{cr,sp} = 1.5 \cdot c_1 + h_{ef} \quad [mm]$$

The approach assumes that the resistance of the concrete to splitting forces is proportional to the concrete tensile strength. According to Eligehausen et al. (2006) the concrete tensile strength corresponds with sufficient accuracy throughout the whole range of concrete compressive strengths to the square root of the compression strength.

In case of the design of an actual application, e.g. an anchors group close to an edge or an anchor in the corner of the member, the corresponding fracture surface is projected onto the edge of the member. The failure load of the application is given by Equation 2.

$$N_{u,sp} = \frac{A_{c,sp}}{A^0_{c,sp}} \cdot \Psi_{h,sp} \cdot \Psi_{g,sp} \cdot N^0_{u,sp} \quad [N] \quad (2)$$

with

$$\Psi_{h,sp} = \left(\frac{h_{cr,sp}}{h}\right)^{5/6}$$

$$s_{cr,sp} = 5 \cdot c_1^{2/3} \cdot d^{1/3} \leq 32 \cdot d \quad [mm]$$

$$\Psi_{g,sp} = \sqrt{n} - \left(\sqrt{n} - 1\right) \cdot \sqrt{\frac{s}{s_{cr,sp}}} \geq 1$$

$$A^0_{c,sp} = s_{cr,sp} \cdot h_{cr,sp}$$

$A_{c,sp}/A^0_{c,sp}$ is the ratio of the projected area of the application to the projected area of the single anchor at an

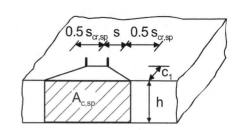

Figure 3. Projected area of a group of two anchors.

edge. $N^0_{u,sp}$ is the base value from Equation 1. The factor $\Psi_{h,sp}$ ensures that the actual member thickness h is considered with the same power as the characteristic member thickness in Equation 1. At calculating Equation 2 the actual member thickness is limited to the characteristic member thickness.

In case of anchor groups where n is the number of anchors the factor $\Psi_{g,sp}$ considers the larger load bearing area in comparison to a single anchor. The factor $\Psi_{g,sp}$ starts with a value of square root n for an anchor spacing of zero and declines to one for an anchor spacing equal to the characteristic anchor spacing. The characteristic anchor spacing $s_{cr,sp}$, where the anchors do not affect each other, depends on the edge distance and on the diameter of the threaded rod.

Figure 3 shows the example of a group of two anchors with the spacing s located close to an edge. The projected area of this application is:

$$A_{c,sp} = \left(s_{cr;sp} + s\right) \cdot h \quad (3)$$

The proposed design approach is primarily based on the results of the numerical study, that is presented in the next chapter. In addition to the results of the simulations the corresponding curves of the design approach are plotted in the figures of chapter 3.

3 NUMERICAL INVESTIGATIONS

3.1 *Finite element code*

In the following numerical study the finite-element (FE) code MASA was used. This program, developed by Ožbolt, is intended for the nonlinear two- and three-dimensional analysis of structures made of quasi-brittle materials such as concrete. It is based on the microplane model (Ožbolt et al. 2001), a macroscopic material model, and a smeared crack approach.

In the microplane model the material is characterized by an uniaxial relation between the stress and strain components on planes of various orientations. At each integration point these planes may be imagined to represent the damage planes or weak planes of the microstructure. The tensorial invariance restrictions need not be directly enforced. Superimposing the responses from all microplanes in a suitable manner automatically satisfies them.

In the analysis of materials which exhibit fracture and damage phenomena, such as concrete, one has to use a so-called localization limiter to assure mesh independent results. In the program MASA two approaches can be used: a crack band approach and a more general nonlocal approach of integral type. In the present study the crack band approach was employed. In the approach the constitutive law is related to the element size such that the specific energy consumption capacity of concrete is independent of the size of the finite element.

3.2 *The finite element model*

All constituents of the model were discretized by four-node tetrahedra elements. The mesh was refined within the area of the bonded anchor. To limit the element number only every second thread of the threaded rod was modeled and the geometry of the thread was simplified (Fig. 4b). The behavior of steel was assumed to be linear elastic with a Young's modulus $E_S = 205000$ MPa and a Poisson's ratio $v_S = 0.3$. The diameter of the threaded rod was varied.

The steel elements of the threaded rod are connected and interlocked with the elements of the mortar layer. The mortar layer was simulated using microplane parameters which are adjusted to the mechanical properties of an actual product. The Young's modulus amounts to $E_B = 5700$ MPa and the Poisson's ratio to $v_S = 0.25$. The mortar elements are coupled with the concrete elements. The material properties of the concrete are: Young's modulus $E_S = 205000$ MPa, Poisson ratio $v_S = 0.18$, tensile strength $f_t = 2.2$ MPa and uniaxial compressive strength $f_c = 28$ MPa.

To simulate a confined test set-up all nodes of the concrete elements at the upper surface within a radius of 1.5 times the embedment depth were fixed in load

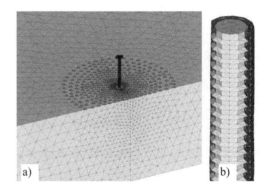

Figure 4. FE model: (a) mesh of the concrete member and (b) threaded rod with mortar layer.

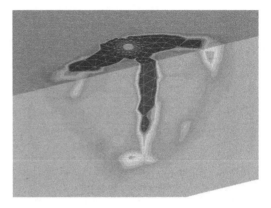

Figure 5. Crack pattern of FE simulation, single anchor near an edge, small edge distance.

direction (Fig. 4a). The tensile load was applied by incremental displacements of the threaded rod.

3.3 *FE simulations compared with design method*

Figure 5 shows the numerically obtained principal strains of a single anchor close to an edge at ultimate load. The dark regions display areas of damage or cracking. The crack pattern agrees well with the crack pattern observed in tests.

Before reaching the peak load a crack begin to form perpendicularly from the edge to the anchor. At peak load at both sides of the anchor cracks grow transversal to the edge. Figure 6 shows the crack formation of an anchor with a larger edge distance than in Figure 5. From Figure 6 can be seen that the average measured angle between edge and splitting crack is larger for the larger edge distance. That means the angle grows up with the edge distance and consequently the increase of the fracture surface is not proportional to the edge distance.

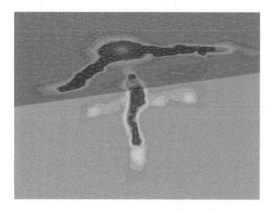

Figure 6. Crack pattern of FE simulation, single anchor at the edge, large edge distance.

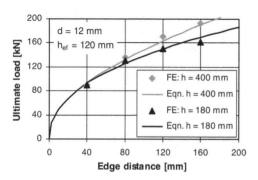

Figure 7. Influence of the edge distance.

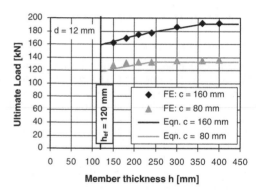

Figure 8. Influence of the member thickness.

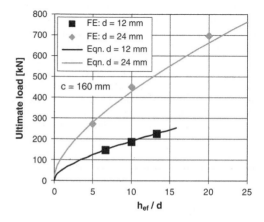

Figure 9. Influence of the embedment depth.

Initially, numerical simulations with a single anchor close to the edge were performed. The anchor diameter (12 mm) and the embedment depth (120 mm) were kept constant. The edge distance was varied. In Figure 7 the numerically obtained failure loads are plotted as a function of the edge distance for two different concrete member thicknesses. At small edge distances no influence of the member thickness on the ultimate load can be recognized. With increasing edge distance, however, the thick concrete member provides a larger increase in ultimate load than the thin member.

To investigate the influence of the concrete member thickness more precisely, numerical calculations with different member thicknesses were performed for two different edge distances. Figure 8 displays failure loads of the simulations plotted against the member thickness. For a small edge distance of 80 mm the maximum load of this edge distance is obtained at a member thickness of about 230 mm. However, for the greater edge distance (160 mm) the load increase up to

a member thickness of about 350 mm. Thus, the member thickness that provides the maximum capacity increase with the edge distance.

Figure 9 shows the numerically obtained ultimate load as a function of the embedment depth related to the anchor size. The edge distance of the simulations was kept constant (160 mm). Two anchor sizes were examined. The member thickness of the simulations with the anchor size 12 mm corresponded to the characteristic member thickness. For the anchor size 24 mm the member thickness was 120 mm larger than the embedment depth.

Figure 10 illustrates the numerically obtained splitting load of a group of two anchors near an edge (Fig. 3) as a function of the anchor spacing. While the anchor size (12 mm) and the embedment depth (120 mm) were kept constant, four different edge distances were investigated. The failure load increases with increasing spacing until it reaches a limit of n-times the capacity of a single anchor. The corresponding anchor spacing

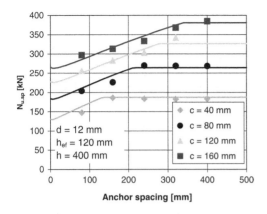

Figure 10. Anchor group at the edge, different edge distances.

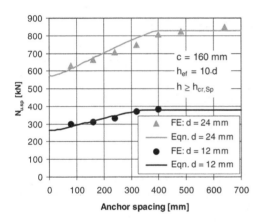

Figure 11. Anchor group at an edge, different anchor sizes.

$s_{cr,sp}$ increases with the edge distance. For the smallest edge distance the double load of the single anchor is obtained by an anchor spacing of about 150 mm, whereas for the large edge distance of 160 mm an anchor spacing of more than 300 mm is necessary.

Further on, FE simulations with an anchor group of the anchor size 24 mm were performed. The edge distance was kept constant. The results are shown in figure 11. The failure loads of the anchor size 12 mm and an edge distance of 160 mm of figure 10 are also plotted in Figure 11. For the anchor size 24 mm an increase of the spacing from 320 to 400 mm leads obviously to an increase of the failure load. However, in case of an anchor size of 12 mm the loads of these two anchor spacings show almost no difference. That means the n-times load of a single anchor is obtained at a smaller spacing for the anchor size 12 mm than for the anchor size 24 mm. It can be summarized that the characteristic anchor spacing $s_{cr,sp}$ depends on the edge distance (Fig. 10) as well as on the anchor size (Fig. 11).

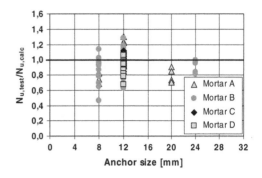

Figure 12. Ratio test to design load against anchor size.

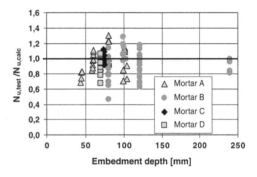

Figure 13. Ratio test to design load against embedment depth.

4 EXPERIMENTAL TESTS

The tests on single anchors close to an edge were carried out for four different adhesive anchoring systems. The systems differ in their chemical composition. That induces different mechanical properties, e.g. bond strength. The edge distance, anchor size, embedment depth and member thickness was varied. Figures 12 to 15 show the ratio of measured failure load and the load calculated in accordance with Equation 2 as a function of the anchor size (Fig. 12), the embedment depth (Fig. 13), the edge distance (Fig. 14) and the member thickness normalized by the characteristic member thickness (Fig. 15).

The respective product factor was identified from the mean value of a series with an edge distance of 40 mm, an anchor size of 12 mm and an embedment depth of 70 mm. The particular product factors are given in Table 1. The product factors of the mortars B to D have a similar value. The factor of product A is about 50% larger. System A represents an epoxy resin, which provides generally larger bond strength than the other adhesive systems.

Figure 12 illustrates the ratios test load $N_{u,test}$ to design load $N_{u,calc}$ plotted against the anchor size. The ratios $N_{u,test}/N_{u,calc}$ are located at 1.0. That means the

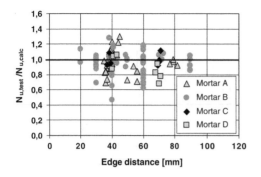

Figure 14. Ratio test to design load against edge distance.

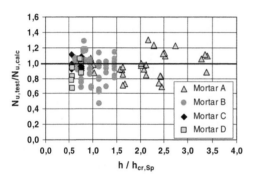

Figure 15. Ratio test to design load against member thickness.

Table 1. Product factors.

Product	Product factor
A	24.0
B	16.4
C	16.8
D	16.2

failure load is well taken into account by the proposed design approach.

The mean value of $N_{u,test}/N_{u,calc}$ of the 115 tests is 0.95 and the coefficient of variation is 16.3%. There are no noticeable tendencies in the diagrams (Fig. 12 to Fig. 15) which would indicate that one of the parameters is considered in a wrong way. Overall the design concept shows a rather good representation to the test data.

5 COMPARISON WITH REINFORCING BARS

5.1 Splitting failure of reinforcing bars

The failure of lap splices and anchorages of rebars occur typically by splitting or blasting of the concrete.

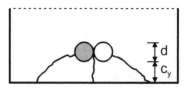

Figure 16. Failure mode C of lap spliced reinforcing bars.

Eligehausen (1979) developed a design concept to predict the ultimate steel stress of rebars at splitting failure of lap splices. He distinguishes between different failure modes.

The failure mode C (Fig. 16) occurs, if the fracture is not affected by a further edge or an adjacent lap splice. The crack pattern is similar to a single (bonded) anchor close to an edge. In the following the design concept of Eligehausen for failure mode C (Rebar) is compared with the proposed design approach for bonded anchors (BA).

According to Eligehausen the following parameters have a decisive influence on the failure load: the concrete cover c_y, the bar diameter d, the length of the lap splice h_{ef} and the concrete compressive f_{cc} strength. The member thickness has no influence, since it is for lap splices in principle larger than the characteristic member thickness.

5.2 Comparison of the design concepts

Eligehausen (1979) indicates two different equations to calculate the ultimate steel stress of rebars at splitting failure: one associated with a concrete cover smaller than 2.5 d and one for a concrete cover larger than 2.5 d. Transforming the equations from stress to load leads to the following equations:

$$N_{u,Sp} = 7.66 \cdot c_y^{1/2} \cdot h_{ef}^{2/3} \cdot d^{5/6} \cdot f_{cc}^{1/2} \cdot K \quad [N] \quad (3a)$$

for $c_y \leq 2.5 \cdot d$

$$N_{u,Sp} = 9.64 \cdot c_y^{1/4} \cdot h_{ef}^{2/3} \cdot d^{13/12} \cdot f_{cc}^{1/2} \cdot K \quad [N] \quad (3b)$$

for $c_y > 2.5 \cdot d$

with $K = 1.2 \cdot k_d$ and $0.75 \leq k_d = \left(\dfrac{10}{d}\right)^{1/2} \leq 1.10$

The comparison of the two design concepts shall disclose potential differences in the influence of the parameters on the failure load. Therefore, the curves of the equations of the two models are plotted together in

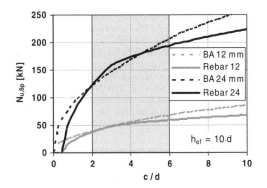

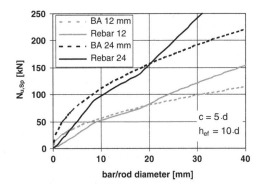

Figure 17. Influence of the edge distance.

Figure 19. Influence of the bar/rod diameter.

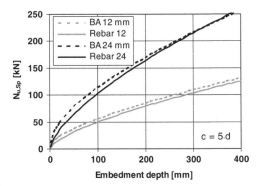

Figure 18. Influence of the embedment depth.

the Figures 17 to 19. The product factor of the proposed design model was set to 12, since that leads to similar ultimate loads of the both models.

For the comparison relative to the edge distance the diameters 12 and 24 mm were chosen. Figure 17 displays the ultimate loads of both design concepts as a function of the edge distance related to the rod/bar diameter. Since the concrete cover is a half bar diameter smaller than the edge distance, the curve of the rebar starts at an edge distance of a half bar diameter.

Up to an edge distance of 3 d the increase of the ultimate load of the lap splices is larger than of the bonded anchors. In contrast, for larger edge distances the increase is smaller. For the design of bonded anchors the range 2 d to 6 d is important: at smaller edge distances no installation is possible and at larger edge distances than 6 d a pull out failure takes place normally. In the range mentioned before the values of both design concepts are similar for the observed diameters.

Figure 18 presents that the increase of the splitting failure load with the embedment depth is very similar for both design concepts. This is valid for both

observed diameters. The underlying edge distance is the quintuple rod or bar diameter.

For a comparison of the influence of the bar and rod diameter, respectively, two different edge distances were chosen: 60 and 120 mm. In Figure 19 the splitting failure load is plotted against the diameter. For the diameters 8 to 18 the increase of both models is very similar. However, for diameters larger than 18 the design concept of Eligehausen show a greater slope than the proposed design approach.

The influence of the concrete strength is equal for both design concepts. In both concepts the concrete compressive strength is considered with a power of ½. Overall the influence of the edge distance, the embedment depth and the concrete compressive strength is considered by both design concepts in a similar way. A noteworthy difference between the compared concepts can be observed only for the influence of the diameter.

6 CONCLUSIONS

Bonded anchors subjected to tensile loads often fail by concrete splitting. To understand to what extent the ultimate capacity is affected by the many geometric and mechanical parameters controlling splitting failure, numerous FE simulations were carried out, in order to develop a new semi-empirical design approach. This approach is presented here, together with the results of the numerical simulations.

According to the proposed approach, the ultimate capacity of a single anchor close to an edge is worked out and is considered as a "base" value. The capacity of an arbitrary anchorage can be evaluated by means of projected areas. Therefore, the failure surfaces are projected on the edge of the concrete member.

As a rule, there is a satisfactory agreement between the proposed design approach (that was validated through specific tests carried out by the authors) and the numerical results.

Furthermore, comparing the proposed approach with that used in the design of lap splices shows that in both anchors and spliced bars splitting failure is governed by the same parameters.

Summing up, the proposed design approach is a useful and realistic tool to predict the ultimate capacity of bonded anchors failing because of concrete splitting.

REFERENCES

Asmus, J. 1999. *Bemessung von zugbeanspruchten Befestigungen bei der Versagensart Spalten des Betons (Design of tensile loaded anchorages at concrete splitting)*. Doctor thesis, University of Stuttgart.

Eligehausen, R. 1979. *Übergreifungsstöße zugbeanspruchter Rippenstäbe (Lapped splices of tensioned rebars)*. Berlin: Schriftenreihe DAfStB, No. 301.

Eligehausen, R., Mallée, R. & Silva, J.F. 2006. *Anchorages in Concrete Construction*. Berlin: Ernst & Sohn Verlag für Architektur und technische Wissenschaften GmbH.

Ožbolt, J., Li, Y.-J., & Kožar, I. 2001 *Microplane model for concrete with relaxed kinematic constraint*. International Journal of Solids and Structures. 38: 2683–2711.

Fracture Mechanics of Concrete and Concrete Structures – Design, Assessment and Retrofitting of RC Structures – Carpinteri, et al. (eds)
© *2007 Taylor & Francis Group, London, ISBN 978-0-415-44616-7*

3D Finite Element analysis of stud anchors with large head and embedment depth

G. Periškić, J. Ožbolt & R. Eligehausen
Institute for Construction Materials, University of Stuttgart, Stuttgart, Germany

ABSTRACT: In the present paper results of the finite element study for headed stud anchors loaded in tension (concrete cone failure) are presented and discussed. The numerical analysis was performed using a 3D FE code based on the microplane model. Considered are anchors with extremely large embedment depths (up to 1143 mm). For each embedment depth the size of the head of the anchor was varied in order to account for the influence of pressures under the anchor head. Furthermore, for the anchor group of four anchors, the influence of the head size on the characteristic spacing of anchors was investigated. The results of the finite element study are discussed and compared with the recently performed test results and with current code recommendations.

1 INTRODUCTION

In engineering practice headed anchors are often used to transfer loads into reinforced concrete. Experience and large number of experiments and numerical studies with anchors of different sizes, confirmed that fastenings are capable of transferring tension forces into a concrete member without the need for reinforcement (Eligehausen et al. 2006). Provided that the strength of anchor steel and the load bearing area of anchor head are large enough, a headed stud subjected to tensile load normally fails by a cone shaped concrete breakout.

To better understand the crack growth and to predict the concrete cone failure load of headed studs for different embedment depths, a number of experimental and theoretical studies have been carried out (Eligehausen et al. 2006). Due to the fact that the tests with large embedment depths require massive test equipment, most of the experiments were up to now performed with embedment depths in the range from $h_{ef} = 100$ to $500\,mm$. Furthermore, in the tests the size of the headed studs is usually chosen such that the compressive stress under the head at peak load is approximately 20 times larger than the uniaxial compressive strength of concrete (f_c). However, in engineering practice, especially in nuclear power plants, anchors with larger embedment depths and with larger head sizes relative to the embedment depth are frequently used. These anchors are designed according to the current design code recommendations, which are based on the experimental results obtained for fasteners with relatively small embedment depths and head sizes. Therefore, to investigate the safety of these

anchors, additional experiments are needed. Since these experiments are extremely expensive, failure capacity of large anchors can alternatively be obtained by numerical analysis.

In the last two decades significant work has been done in the development and further improvement of numerical tools. These tools can be employed in the analysis of non-standard anchorages. Unfortunately, the objectivity of the numerical simulation depends strongly on the choice of the material model. Therefore, the numerical results should be confirmed by experiments and the numerical model used should pass some basic benchmark tests. In the present paper the three-dimensional finite element analysis is carried out using the finite element (FE) code MASA. The code is based on the microplane model for concrete. On a very large number of numerical examples that have been carried out in the past, it has been demonstrated that the code is able to predict failure of concrete and reinforced concrete structures realistically (Ožbolt 2001).

2 FINITE ELEMENT CODE

The FE code employed in the present study is aimed to be used for the two- and three-dimensional non-linear analysis of structures made of quasi-brittle materials such as concrete. It is based on the microplane material model (Ožbolt et al. 2001) and smeared crack approach. To avoid mesh dependent sensitivity, either the crack band approach (Bažant & Oh 1983) or the nonlocal integral approach (Ožbolt & Bažant 1996, Pijaudier-Cabot & Bažant 1987) can be employed. The spatial discretization of concrete is performed using four or eight node solid finite elements. The reinforcement can be modelled by discrete bar elements with or without discrete bond elements or, alternatively, smeared within the concrete elements. The analysis is carried out incrementally, i.e. the load or displacement is applied in several steps. The preparation of the input data (pre-processing) and evaluation of numerical results (post-processing) are performed using the commercial program FEMAP®.

In the microplane model, material properties are characterized on planes of various orientations at a finite element integration point. On these microplanes there are only a few uniaxial stress and strain components and no tensorial invariance requirements need to be considered. The tensorial invariance restrictions are satisfied automatically since microplanes simulate the response on various weak planes in the material (interparticle contact planes, interfaces, planes of microcracks, etc.). The constitutive properties are entirely characterized by relations between normal and shear stress and strain components on each microplane. It is assumed that the strain components on microplanes are projections of the macroscopic strain tensor (kinematic constraint approach). Knowing the stress-strain relationship of all microplane components, the macroscopic stiffness and the stress tensor are calculated from the actual strains on microplanes by integrating of stress components on microplanes over all directions. The simplicity of the model lies in the fact that only uniaxial stress-strain relationships are required for each microplane component and the macroscopic response is obtained automatically by integration over all microplanes. For more details see Ožbolt et al. (2001).

Due to the loss of elipticity of the governing differential equations, the classical local smeared fracture analysis of materials, which exhibit softening (quasibrittle materials), leads in the finite element analysis to results, which are in general mesh dependent. To assure mesh independent results the total energy consumption capacity due to cracking must be independent of the element size, i.e. one has to regularize the problem by introducing the so-called localization limiter. In the present numerical study the crack band approach (Bažant & Oh 1983) is used. In this approach the constitutive law is related to the element size such that the specific energy consumption capacity of concrete (concrete fracture energy G_F) is independent of the size of the finite element.

3 NUMERICAL ANALYSIS – RESULTS AND DISCUSSION

The main purpose of the 3D FE analysis was to investigate the ultimate capacity and failure mode for single anchors with extremely large embedment depths and

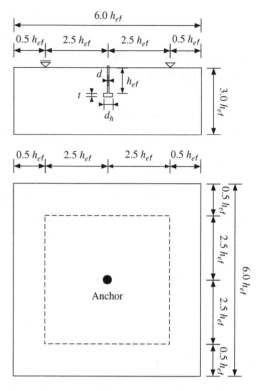

Figure 1. Geometry used in the pull-out study.

Table 1. Geometric properties.

h_{ef}	d	t	$d_{h,small}$	$d_{h,large}$
635	70	76	83	118
889	95.3	102	105	162
1143	160.8	169	171	241

with two different head sizes, which are pulled-out from a concrete block. Moreover, the influence of the head size on the critical anchor spacing for group of four anchors, obtained by Rah (2005) is also discussed.

3.1 Single anchor

The typical geometry of the concrete block and the geometry of the headed stud are shown in Figure 1. Three embedment depths were numerically investigated, namely $h_{ef} = 635$, 889 and 1143 mm. For each embedment depth two head sizes were used (small and large). The geometrical properties for all investigated cases are summarized in Table 1.

The size of the smallest head for all embedment depths is chosen such that the compressive stress under the head at peak load is 20 times larger than the uniaxial cylindrical compressive strength of concrete (f_c).

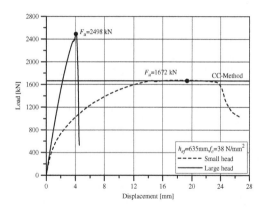

Figure 3. Typical load-displacement curves for two different head sizes (h_{ef} = 635 mm).

Table 2. Predicted peak loads.

h_{ef} [mm]	Eq. 1	Eq.2	F_u[kN] FE ($d_{h,small}$)	FE ($d_{h,large}$)	Test ($d_{h,large}$)
635	1664	1946	1675	2498	2250
889	2756	3413	2707	3806	3300
1143	4018	5193	4076	5780	5500

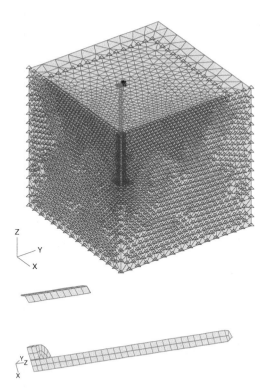

Figure 2. Typical FE meshes of concrete block and headed stud with contact elements.

The peak load is calculated based on the concrete cone capacity method (Eligehausen et al. 2006). The large sizes are chosen to be typical for engineering practice (nuclear power plants). The concrete properties are taken as: Young's modulus E_c = 28000 MPa, Poisson's ratio v_c = 0.18, tensile strength f_t = 3.0 MPa, uniaxial compressive strength f_c = 38 MPa and concrete fracture energy G_F = 0.10 N/mm. The behaviour of steel is assumed to be linear elastic with Young's modulus E_s = 200000 MPa and Poisson's ratio v_s = 0.33.

Spatial discretization is performed using four node solid finite elements. Only one quarter of the concrete block is modelled, i.e. double symmetry is utilized. Typical finite element meshes of the concrete block and headed stud are shown in Figure 2. Contact between the steel stud and concrete exists only on the top of the headed stud (compression transfer zone). To account for the confining stresses that develop in the vicinity of the head, interface elements, which can take up only compressive stresses are used (see Fig. 2). In all cases the anchor is loaded by prescribed displacements at the top of the anchor shaft. The supports were fixed in the vertical (loading) direction. The distance between the support and the anchor is taken as $2.5h_{ef}$ so that an unrestricted formation of the failure cone is possible (see Fig. 1).

Typical load-displacement curves for two different head sizes (h_{ef} = 635 mm) are shown in Figure 3. The curves show that anchors with small heads have larger displacements at peak load. According to the FE analysis this tendency is stronger if the embedment depth is larger.

The summary of the predicted peak loads (F_u) from the FE simulation as well as the available experimental results is given in Table 2. In the same table the peak loads according to Equation 1, which is the base of ETAG CC design code (Eligehausen et al. 2006):

$$F_u = 15.5\sqrt{f_{cc}}\,h_{ef}^{1.5} \tag{1}$$

and according to Equation 2, which is the base of ACI-349-01 design code (ACI Standard 349 2001):

$$F_u = \alpha\sqrt{f_c}\,h_{ef}^{\beta}$$

$$0 < h_{ef} < 279.4; \; \alpha = 16.834; \; \beta = 1.5 \tag{2a}$$

$$279.4 \le h_{ef} \le 635; \; \alpha = 6.585; \; \beta = 5/3 \tag{2b}$$

are displayed. In Equation 1 f_{cc} = $1.2 f_c$.

It should be noted that Equation 1 was calibrated using experiments in which the maximum embedment depth was 500 mm and the head size small, as defined

763

above. Moreover, the exponent 1.5 (see Eq. 1) indicates the size effect on the concrete cone failure resistance according to linear elastic fracture mechanics (LEFM), i.e. maximum possible size effect.

Numerically obtained peak loads for anchors with small head sizes show very good agreement with Equation 1 (max difference less than 2%). However, for anchors with large heads the difference between numerical results and Equation 1 is obvious (see Table 2). Furthermore, in the tests recently performed in Korea (Lee et al. 2006), where the size of the anchor head was very similar to large heads in the present FE study, measured ultimate loads are significantly higher than according to Equation 1. These ultimate loads agree better with Equation 2, i.e. the largest difference is 15% for $h_{ef} = 635$ mm and it tends to be smaller with increase of the embedment depth (6% for $h_{ef} = 1143$ mm). The test results (Lee et al. 2006) and FE results for large heads show good agreement, however, the absolute values of peak loads measured in experiments somewhat underestimate the numerical results obtained for anchors with large heads (up to 15%). There could be different reasons for this. For instance, the tests were performed on huge specimens with boundary conditions, which were possibly not the same as assumed in the analysis. Furthermore, the concrete mechanical properties adopted in the analysis were the same as the concrete properties measured on the laboratory specimen. However, the mechanical properties of test specimen (strength and fracture energy) were possibly reduced due to the effect of non-elastic deformations (shrinkage, temperature, etc.). These effects were not accounted for in the analysis. Nevertheless, the test results confirm the numerical results, which clearly show that with increase of the anchor head size the anchor resistance increases.

The typical calculated crack patterns for two different head sizes ($h_{ef} = 635$ mm) are shown in Figure 4. The cracks (dark zones) are plotted by means of maximum principal strain. A critical crack opening $w_{cr} = 0.2$ mm is assumed. This crack opening corresponds to the plotted critical principal strain of $\varepsilon_{cr} = w_{cr}/h_E$, where h_E = average element size. The crack patterns are shown for the post-peak anchor resistance. It was observed that for smaller anchor heads the crack length at peak load is shorter than the crack length obtained for the anchors with larger anchor heads. Moreover, the crack propagation angle, measured from the loading direction, increases with increase of head size. For small head sizes, the concrete cone is steeper than in the case of large head sizes (see Fig. 4). This tendency of having a flatter concrete cone in case of large head sizes was also confirmed in the tests (Lee et al. 2006), where the angle between the failure surface of the concrete cone and concrete surface varied from 20° to 30°. According to ETAG-CC method, this angle is assumed to be approximately 35°

(a)

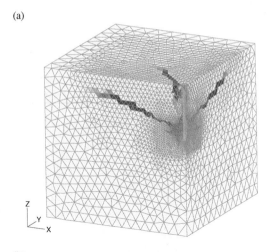

(b)

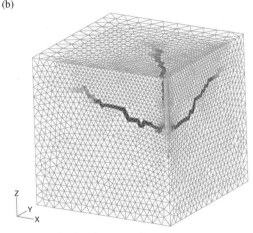

Figure 4. Typical crack patterns: (a) small head, $h_{ef} = 635$ mm and (b) large head, $h_{ef} = 635$ mm.

and it agrees well with the numerical results obtained for anchors with small heads (see Fig. 4a).

The numerical results confirm that there is a strong size effect on the concrete cone resistance. In Figure 5 the calculated relative failure resistance ($\sigma_R = \sigma_N/\sigma_{N,hef=200}$, with $\sigma_N = F_u/(h_{ef}^2\pi)$) is plotted as a function of the embedment depth. Since no FE calculation was carried out for $h_{ef} = 200$ mm, a predicted ultimate load according to Equation 1 was taken as a reference value. For comparison, the prediction according to Equation 1 is also plotted. Note that the size effect is strong if the gradient of the relative resistance with respect to the embedment depth ($\partial\sigma_R/\partial h_{ef}$) is large. As mentioned before, Equation 1 predicts the maximum size effect (LEFM). From Figure 5 it can be seen that the numerical results for anchors with small anchor heads agree well with Equation 1, i.e.

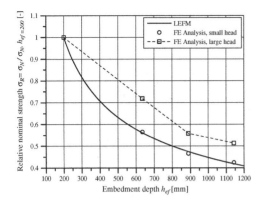

Figure 5. Relative concrete cone resistance as a function of the embedment depth.

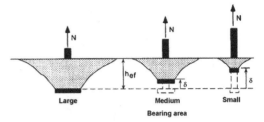

Figure 6. Concrete breakout cones of headed studs with heads of various diameters (schematic) according to Furche (1994).

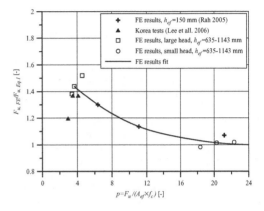

Figure 7. Relative ultimate capacity of headed studs as a function of pressures under the anchor head at peak load.

they predict strong size effect. However, with increase of the anchor head size the size effect on the relative anchor resistance decreases. The reason why the predicted size effect agrees well with the size effect prediction according to LEFM for fasteners with small heads is due to the fact that for all embedment depths the crack patterns and the crack length at peak load are geometrically similar, i.e. the crack length is relatively small and approximately proportional to the embedment depth. The main assumption of LEFM, namely the proportionality of the crack length at peak load, is fulfilled and therefore the size effect follows the prediction according to LEFM. On the contrary, for fasteners with larger heads the crack pattern at peak load is not proportional when the embedment depth increases. This is the case for both, the crack length at peak load and for the corresponding shape of the failure cone. Consequently, the size effect on the concrete cone failure load is smaller.

The fact that the concrete cone resistance increases with increase of the head size of the anchor is closely related to local pressure under the head of the stud. This has already been reported by Furche (1994) (see Fig. 6). When the head of the anchor is small, the pressure under the head is relative to fc high and the concrete under the head is significantly damaged. In the vicinity of the head, rather complex mixed-mode fracture (compression-shear) takes place and the displacement of the anchor in load direction increases due to shearing. This leads to a reduction of the effective embedment depth and causes the formation of a relatively steep concrete cone (see Fig. 6). However, in the case of anchors with large heads the concrete under the head is practically undamaged because of relatively small pressure and the crack starts to propagate in mode-I failure mode almost horizontally (see Figs. 4 and 6). Consequently, a larger cone surface forms which provide higher concrete-cone pull-out resistance.

Figure 7 shows the relation between failure load, calculated and in the experiment measured, and the failure load according to Equation 1 as a function of pressure under the anchor head at ultimate load. The pressure under the anchor head p is calculated as

$$p = \frac{F_u}{A_{ef} f_c} \qquad (3)$$

in which $A_{ef} =$ the load bearing area of the anchor. The results of the FE analysis for both head sizes and test results according to experimental tests from Korea (Lee et al. 2006) are shown. Furthermore, the results of the FE analysis obtained recently by Rah (2005) are also shown. It can be seen, that the numerically and experimentally obtained failure loads increase with decreasing pressures under the head i.e. with increasing head size.

As it can be seen from Figure 7, if the pressures under the anchor head are relatively large ($p = 20f_c$) the ultimate load agrees very well with the prediction according to Equation 1. According to the recent test results (Lee et al. 2006), as the pressure decreases the ultimate load increases and reaches about 140% ($p = 4f_c$) of failure load predicted by Equation 1.

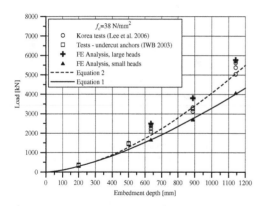

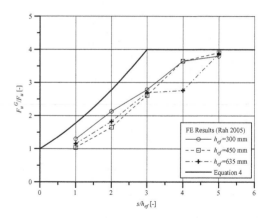

Figure 8. Ultimate load versus embedment depth comparison between calculated data, test data and design formulas.

Figure 9. Relative ultimate load of 4-anchor groups as a function of anchor spacing s divided with embedment depth h_{ef} – FE results according to Rah (2005) and ultimate load prediction according to Equation 4.

This shows that the Equation 1 is conservative in case of anchors with small pressure under the anchor head at peak load.

All results are summarized in Figure 8 in form of ultimate loads as a function of embedment depth. Note that Figure 8 also shows recent results obtained for undercut anchors with embedment depths of 200 and 500 mm (IWB 2003). The figure shows that for embedment depths up to 300 mm, test results agree well with both design codes. However, for larger embedment depths and for anchors with large head sizes the ultimate load predicted by Equation 1 is too conservative. As shown before, the anchors with large heads exhibit smaller size effect on the ultimate load than the anchors with smaller heads. Therefore, their load bearing capacity is better predicted by Equation 2, which accounts for the size effect that is weaker the prediction according to LEFM.

As already mentioned, and also discussed recently (Ozbolt et al. 2004) the pressures under the anchor head at peak load could be used to control the influence of the anchor head size on the ultimate anchor load. Principally, for relatively high pressure the Equation 1 can be used. However, for larger embedment ($h_{ef} > 300$ mm) and larger head sizes the prediction according to Equation 2 seems to be more appropriate.

3.2 Group of anchors

A possible problem in using current design codes for anchors with large embedment depths and with large heads could be the safety of anchor groups. As mentioned before, concrete cone propagates somewhat flatter in case of anchors with large heads than in the case of anchors with small heads. In recent tests the concrete cone propagation angle between concrete surface and cone surface varied from 20–30° (Lee et al. 2006). This would imply that the characteristic spacing of anchors $s_{cr,N}$ for group of anchors could be

larger than currently proposed by both design codes, i.e. $s_{cr,N} = 3h_{ef}$.

Figure 9 shows the results of FE analysis recently obtained by Rah (2005). The influence of anchor spacing s on the ultimate bearing capacity of groups of 4-anchors for large anchor heads was investigated. In the same figure the Equation 4:

$$F_u^G = F_u \cdot \frac{A_N}{A_N^o} \qquad (4)$$

$F_u^G =$ Ultimate load for group of 4 anchors

$F_u =$ Ultimate load of single anchor (Eq. 1 or Eq. 2)

$A_N^o = 9h_{ef}^2$

$A_N = (3h_{ef} + s)^2$; $s \le s_{cr,N} = 3h_{ef}$

which is the base of ETAG-CC and ACI-349-01 for predicting of ultimate load for groups of 4 anchors is plotted as well. As can be seen from Figure 9, the predicted average characteristic anchor spacing is approximately $5h_{ef}$. This means that the interaction between the anchors of the group exists for larger spacing range than according to Equation 4. Therefore, to clarify this question further numerical and experimental investigations are needed.

4 CONCLUSIONS

In the present paper the behavior of large anchor bolts embedded in concrete is numerically investigated. The numerical results are compared with available test data and current design recommendations. Based on the results, the following can be concluded: (1) The concrete cone capacities predicted by the FE analysis for anchors with small heads show for the entire investigated range of embedment depths good agreement with Equation 1, which is the based on LEFM. However, for anchors with larger heads the numerical results indicate higher resistance than predicted by Equation 1; (2) The numerical study confirms the strong size effect on the concrete cone resistance. By increasing the head size of the stud, the size effect on the failure capacity of anchors decreases; (3) The available test data for anchors with larger heads confirm the tendency observed in the numerical study, i.e. the concrete cone resistance increases with increase of the head size; (4) The influence of the size of the anchor head on the ultimate load can be suitably described using pressures under the anchor head at peak load; (5) For pressures larger than approximately $4f_c$ Equation 1 predicts realistic results. For $p < 4f_c$ Equation 1 underestimates the resistance and more realistic prediction is given by Equation 2; (6) For group of anchors with larger head sizes, the characteristic anchor spacing seem to be larger than the code prediction ($3h_{ef}$). Therefore, to improve and to extend the validity of the current design recommendations, further theoretical, experimental and numerical investigation is needed.

REFERENCES

ACI Standard 349-01/349R-01 2001. *Code Requirements for Nuclear Safety Related Concrete Structures and Commentary (ACI 349R-01)*.

Bažant, Z.P. & Oh, B.-H. 1983. Crack Band Theory for Fracture of Concrete. *Materials and Structures* 93(16): 155–177.

Eligehausen, R., Mallée, R. & Silva, J.F. 2006. *Anchorage in Concrete Construction*. Ernst & Sohn, Berlin, Germany.

Furche, J. 1994. *Load-bearing and displacement behaviour of headed anchors under axial tension loading*. Doctor thesis, Institute for construction materials, University of Stuttgart, Germany.

IWB 2003. *Ausziehversuche mit dem Hochtief Schwerlastanker HT-SHV in Beton*. Internal report of the Institute for construction materials, University of Stuttgart, Germany.

Lee, N.H., Kim, K.S., Bang, C.J. & Park, K.R. 2006. Tensile anchors with large diameter and embedment depth in concrete. *Submitted to ACI Stuctural and Materials Journals*.

Ožbolt, J. & Bažant, Z.P. 1996. Numerical smeared fracture analysis: Nonlocal microcrack interaction approach. *International Journal for Numerical Methods in Engineering* 39(4): 635–661.

Ožbolt, J., Li, Y.-J. & Kožar, I. 2001. Microplane model for concrete with relaxed kinematic constraint. *International Journal of Solids and Structures* 38: 2683–2711.

Ožbolt, J. 2001. Smeared fracture finite element analysis – Theory and examples. In Rolf Eligehausen (ed.), *International symposium on Connections between Steel and Concrete*, RILEM, SARL: 609–624.

Ožbolt, J., Periškić, G., Eligehausen, R. & Mayer, U. 2004. 3D FE Analysis of anchor bolts with large embedment depths. *Fracture mechanics of concrete structures; Proc. 5. intern. conf. on fract. mech. of conc. struct.*, Vail, Colorado, USA: 845–853.

Pijaudier-Cabot, G. & Bažant, Z.P. 1987. Nonlocal Damage Theory. *Journal of Engineering Mechanics* 113(10): 1512–1533.

Rah, K.K. 2005. *Numerical study of the pull-out behaviour of headed anchors in different materials under static and dynamic loading conditions*. Master thesis, Institute for Construction materials, University of Stuttgart, Germany.

Fracture Mechanics of Concrete and Concrete Structures – Design, Assessment and
Retrofitting of RC Structures – Carpinteri, et al. (eds)
© *2007 Taylor & Francis Group, London, ISBN 978-0-415-44616-7*

Finite element modeling of anchor subjected to static and cyclic load

H. Boussa, A. Si Chaib, H. Ung Quoc & G. Mounajed
Centre Scientifique et Technique du Bâtiment (CSTB) Pôle MOCAD, Marne la Vallée, France

C. La Borderie
Université de Pau et des Pays de l'Adour, ISA-BTP, Anglet, France

ABSTRACT: This research work deals with the study of the behavior under monotonic and alternate shear loads of a single steel bolt anchored in a concrete slab. The aim is to predict the failure modes and the failure loads on the one hand and to compare the anchor behavior under static load versus alternate load on the second hand. The study is based on a numerical resolution using the finite element method. Different types of non linearity are considered in the model: non-linear behavior laws for steel and concrete, geometrical non linearity due to the large displacements and non linearity due to the contact conditions.

1 INTRODUCTION

Metal anchor bolts are frequently used in modern construction in order to assure the connection between different building components and to allow loads transmission between different elements of a structure. Over the past 30 years, much research work has been carried out on anchors at European and International level (Klinger et al. 1982, Hawkins 1987, Mesureur et al. 1993, Eligehausen et al. 1993, ETAG 1997, Elighausen et al. 2006). The majority of the design models and methods proposed for this type of anchors are based on a statistical empirical approach. Practice and tests have shown that they are not always predictive although the values obtained are on the safe side.

The origin of this problem is that different failure modes can arise in relation to the values of the different parameters involved (the anchor characteristics and their support, the spacing between anchors, the distance to edges and the direction of the applied force). The current available models are only predictive for a restricted range of parameters.

While the qualification methods of anchors under static shear loads have been improved significantly over the past years, relatively few information exists about anchors under seismic loads (Klingner et al. 1982, Vintzeleou & Eligehausen 1991, Rodriguez 1995, David et al. 2005, Hoehler 2006). Consequently, it appeared necessary to investigate the behavior of anchors subjected to monotonic and alternate shear loads.

This research work deals with the study of the behavior under monotonic and alternate shear loads

of a single expansion anchor. The aim is to predict the failure modes, the failure load and the global load-displacement behavior on the one hand and to compare the anchor behavior under static load versus alternate load on the second hand. The study is based on a numerical simulation using the finite element method. Different types of non linearity are considered: non-linear behavior laws for steel and concrete, geometrical non linearity due to the large displacements and non linearity due to the contact conditions. A specific damage model has been adopted for the cyclic behavior of concrete.

2 CONCRETE DAMAGE MODEL

Concrete exhibits a non-linear stress strain behavior mainly because of progressive micro-cracking and void growth. The development of micro cracks results in a degradation of the material stiffness and apparition of inelastic strains. For our modeling, the cracking progress in concrete is modeled using the damage model MODEV developed at the Scientific and Technical Center for Building CSTB (Mounajed et al. 2002, Ung Quoc 2003) and implemented in the finite element software CAST3M developed by the French Atomic Energy Commission CEA. This model has been established within the framework of the damage theory (Lemaitre & Chaboche, 1988).

The MODEV damage model takes into account the specific nonlinear effects involved in the deterioration process of concrete such: unilateral effect and stiffness

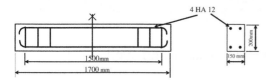

Figure 1. Schematic representation of reinforced beam geometry.

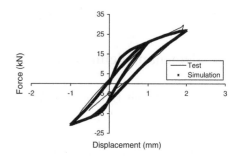

Figure 2. Global response of the reinforced beam under cyclic loading.

Figure 3. Damage profile of the reinforced beam and history of loading.

recovery due to the cracks closure, inelastic strains, and the coupling between damage and inelastic strains.

Considering the complexity of cement based materials behavior, the model was kept as simple as possible with few parameters in order to insure a an easy experimental identification. This allows the model to be used for engineering design of concrete structures.

By analogy with Mazars's model (Mazars 1984), 2 equivalent strains representing respectively the local sliding and the crack opening are introduced. They are related to deviatoric part of the strain tensors and hydrostatic one. The model considers two independent scalar damage variables, corresponding respectively to each degradation mechanisms. Each damage variable has its own evolution law.

To improve mesh objectivity, the tensile and the compressive fracture energy are introduced in the damage evolutions laws by analogy to Hillerborg's fictitious crack approach (Hillerborg et al., 1976).

The adaptation of MODEV model for cyclic loading as well as the identification of its parameters has been presented in (Si Chaib et al., 2006).

The validation of the model under cyclic loading has been conducted by using a test described in literature (La Borderie 1991). A confrontation between simulated and experimental results has been carried out. The test corresponds to a reinforced concrete beam, subjected to a flexural cyclic load. The beam has 1700 mm of length, 200 mm of cross section height and 150 mm of width. The span distance is about 1500 mm. Steel reinforcement is composed of 4 longitudinal high adherence steel, and six stirrups, three at each beam end, to overcome failing under shear load. The beam is subjected to cyclic loading at the mid span as described in figure 1.

The global force-displacement curve is plotted in figure 2. The comparison between test and simulation results shows a good agreement with both load estimation and residual strain. The damage profile is shown in figure 3.

The objectivity of the model with regard to the mesh size has been carried out.

Figure 4 shows simulation results, for the reinforced concrete beam, under cyclic loading, with 200 elements (length 20, height 10), 800 elements (length 40, height 20) and 1250 elements (length 50, height 25). It can be seen that these results both in terms of estimated load and inelastic strains are closer to

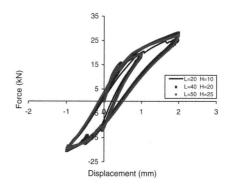

Figure 4. Global response of the reinforced beam under cyclic loading for different mesh sizes.

each other. Objectivity according to different meshes is acceptable.

3 EXPANSION ANCHOR UNDER MONOTONIC SHEAR LOAD

A mechanical expansion anchor is placed in a C20/25 concrete with an embedment depth of 80 mm located far from any edges and subjected to monotonic shear load. The anchor dimensions are given in table 1.

3.1 Mesh generation and material properties

8-nodes isoparametric hexahedral elements have been adopted for the three dimensional finite elements mesh. Additional 4-nodes isoparametric tetrahedral elements are added to the configuration. In order to

Table 1. Anchor dimensions.

External diameter (mm)	Effective embedment depth (mm)	Thickness* (mm)
18	80	25

* Thickness of the fixture.

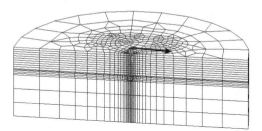

Figure 5. Global mesh.

Table 2. Concrete and steel properties.

	Concrete			Steel	
	MPa	–	J/m²	MPa	–
Young Modulus	35000	–	–	210000	–
Poisson's ratio	–	0.2	–	–	0.24
Compr. Streng	28	–	–	–	–
Tensile strength	3.5	–	–	–	–
Hardening modul.	–	–	–	2000	–
Fracture energy	–	–	80	–	–
Yield stress	–	–	–	600	–
Failure stress	–	–	–	900	–

obtain correctly the higher stress gradient around the anchor, we used a high mesh density in this area. Due to symmetry, only a half of the mesh has been modeled. Figure 5 shows the principal parts of the mesh.

The shear force is simulated by a horizontal displacement applied, on the backside of the fixture. Concrete and steel properties are given in table 2.

To solve such complex non-linear problem, the full Newton Raphson's method has been adopted. The computation is carried out in quasi static.

The contact phenomenon between steel and concrete interfaces is taken into account.

In the presence of large relative displacements between solids in contact (e.g. in the case of steel and concrete) the problem becomes highly non-linear.

Different methods can be used for numerical contact resolution. In our case, we used the double Lagrange multipliers method. Coulomb's smooth friction law has been used and the friction coefficient between steel and concrete is taken equal to 0.30.

An elasto-plastic model, with the Von Mises criterion has been adopted for the anchor behavior.

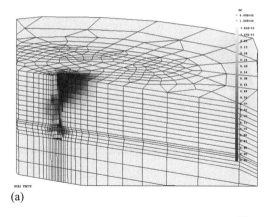

(a)

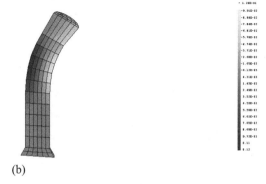

(b)

Figure 6. The failure occurs by a local concrete damage in front of the anchor followed by a steel failure: (a) deviatoric damage in the concrete structure (b) plastic strains (zz) in the anchor.

3.2 *Results*

The tests have shown that the fracture of an anchorage under shear loading is governed by the geometrical and by the mechanical properties of the anchor and its support. For a semi-infinite medium implantation, the failure may occur with concrete damage (pry-out for small embedment length) or steel failure accompanied by spalling at the concrete surface. In this study the failure occurs by the second mechanism mentioned above.

The comparison between the test results and the numerical simulation shows that the FE results are in good agreement with the tests results. In particular, the numerical simulations shows that the failure occurs by a local concrete damage in front of the anchor (Fig. 6) followed by a steel failure. The steel failure is caused by an important bending of the anchor. This bending is responsible of the excessive stresses and plastic strains in the anchor. In figure 6b and figure 7 are respectively plotted the plastic strain and the Von Mises equivalent stress in the anchor.

Figure 7. Von Mises equivalent stress in the anchor.

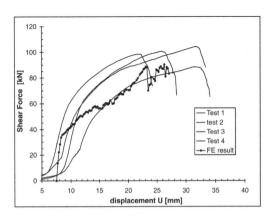

Figure 8. Load displacement curves: shear failure experimental and simulation results.

The ultimate load capacity of the anchor and the corresponding displacement comply with the experimental results as shown in figure 8.

4 ALTERNATE SHEAR LOAD

Alternate shear test has been simulated on the same anchor described above. 4 alternating displacement controlled shear cycles were performed before loading the anchor to failure. The maximum displacement corresponds to a 50% of the ultimate shear capacity of the anchor. Figure 9 shows the adopted loading pattern.

The alternate shear simulation shows (Fig. 10) that the local damage surrounding the anchor is more important compared to the damage obtained under shear load (for the same load level). This result complies with the experimental tests performed at CSTB (David et al. 2005) as shown in figure 11. It is also noticed that the local damage increases during the cycles particularly in the depth, bellow the upper surface of the slab (Fig. 12).

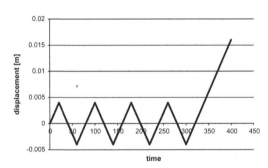

Figure 9. Alternate shear loading pattern (displacement controlled).

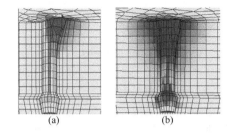

Figure 10. Local damage around the anchor: mono-tonic load (a) and alternate load (after 4 cycles) (b).

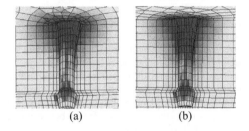

Figure 11. Local damage around the anchor with static and alternate shear tests (David et al. 2005).

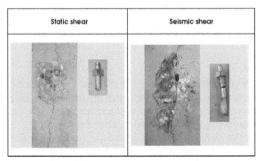

Figure 12. Local damage around the anchor: after one cycle (a) and after 4 cycles (b).

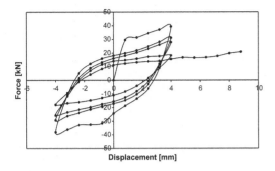

Figure 13. Load displacement curve: alternate shear simulation.

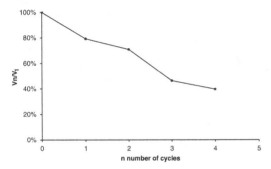

Figure 14. Force degradation ratio due to shear cycling as a function of the number of cycles.

The load displacement curve, plotted in figure 13, points up the apparition of hysteresis loops during cycling. It points up also the substantial decrease of the shear stiffness with cycling.

Figure 14 illustrates the force degradation ratio (V_n/V_1) in the nth cycle as a function of the number of cycles. This degradation is due to the local damage of the concrete surrounding the anchor. This result complies with the experimental tests performed by (Vintzeleou & Eligehausen 1991).

5 CONCLUSION

The three-dimensional modeling of an anchorage to concrete using metal anchor bolts has been achieved under static and cyclic shear loads. Different types of non linearity have been considered: non-linear behavior of steel and concrete, large displacements and contact conditions. A specific damage model, developed at CSTB and implemented in the finite element software CAST3M, has been adopted for the static and cyclic behavior of concrete.

The results of the monotonic shear simulation are summarized as follows:

– The failure occurs by a local concrete damage in front of the anchor followed by a steel failure.

– The steel failure is caused by an important bending of the anchor.
– The ultimate load capacity of the anchor and the corresponding displacement comply with the experimental results.

The key results for the alternate shear load are summarized as follows:

– For the same load level, the local damage surrounding the anchor is more important compared to the damage obtained under shear load.
– The local damage increases during the cycles.
– Apparition of hysteresis loops during cycling.
– The degradation of the shear force increased with increasing number of cycles.

REFERENCES

Cook, R., Collins, D., Klingner, E. and Polyzois. 1992. Load-deflection Behavior of cast in place and retrofit concrete anchors. *ACI Structural. Journal* Title 89-S60, pp. 639–649.

Emmanuel David, T. Guillet, B. Mesureur, P. and Rivillon, 2005. Shear seismic behavior of metal anchors for concrete: influence of loading rate, concrete strength and frequency. *FIB* Athenes, September 2005.

El Dalati, R., Mounajed, G., Mesureur, B. and et Berthaud, Y. 2000. Three dimensional modeling of anchorage subjected to shear loads, *American Concrete Institute*, 2000.

Eligehausen, R., Mallée, R. and Silva, J. F. 2006, *Anchorage in Concrete Construction,* Ernst & Sohn, Berlin.

Eligehausen, R. and Lehr, B. 1993. Shear capacity of anchors placed in non cracked concrete with large edge distance. Univ. Stuttgart, Rep. N° 10/20 E-93/11E, 1993.

EOTA, 1997. Guideline for European Technical Approval of Anchors, Metal Anchors For Use in Concrete, Part I.

ETAG, 1997. Guideline for European Technical Approval of Anchors (metal anchors) for use in concrete. Annex C: Design Methods for anchorage.

Hawkins, N., 1987. Strength in shear and tension of cast-in-place anchor bolts. Anchorage to concrete, *American Concrete Institute*, SP 103, 1987, pp. 233–257.

Fuchs, W., 1990. Tragvelhaten von Befestigungen unter Querlast in ungerissenem beton (Bearing behavior of fastenings under shear loads in uncracked concrete), dissertation Universität Stuttgart, IWB-Mitteilugen 1990/2, 1990.

Ghobarah, J. and Aziz, T.S., 2004, "Seismic qualification of ex-pansion anchors to Canadian nuclear standards", *Nuclear Engineering and Design*, 228, pp. 377–392.

Hillerborg, A., Modeer, M. and Petersson, P-E., 1976. Analysis of Crack Formation and Crack Growth in Concrete by Means of Fracture Mechanics and Finite Elements. *Cement and Concrete Research*, Vol.6, Pergamon Press.

Hoehler M. S. 2006, Behavior and Testing of Fastenings to Concrete for use in Seismic Applications, phD thesis.

Klinger, R. E. and Mendonca, J.A. 1982. Shear capacity of short anchor bolts and welded studs : A literature review. *ACI Journal.* No. 79–34, 1982, pp. 339–347.

La Borderie, C., 1991. Phénomènes Unilatéraux dans un Matériau Endommageable: Modélisation et Application à l'Analyse de Structures en Béton. PhD thesis, Paris VI University, 1991.

Lemaître, J. and Chaboche, J. L., 1988. Mécanique des matériaux solides. Dunod, 1988.

Mazars, J. 1984, "Application de la mécanique de l'endommagement au comportement non linéaire et à la rupture du béton de structure", PhD. Thesis, Paris VI University, 1984. 27.

Mesureur, B. and Guillet, T. "Fastenings loaded by a shear load and a bending moment", test report Eli/gd/1241, CSTB, 1993.

Mounajed, G., Ung Quoc, H. and Boussa, H., 2002. Development of a new concrete damage model in SYMPHONIE-CSTB. Application to metallic anchor bolts. *Second Biot International Conference on Poromechanics*, Grenoble France, August 26–28, 2002.

Ohlsson , U. and Olofsson, T. 1997. Mixed-mode fracture and anchor bolts in concrete. Analysis with inner softening bands. *J. of Eng. Mech.,* 1997, pp. 1027–1033.

Ozbolt, J. and Eligehausen, R. 1994. Bending of anchors, Final Element Studies, Universität Stuttgart, Report N° 10/23-94/6, IWB, 1994.

Rodriguez, M. 1995. Behavior of anchors in uncracked concrete under static and dynamic loading. M.S. Thesis, The Univ. of Texas at Austin.

Silva, J. F. 2001, "Test Methods for Seismic Qualification of Post Installed Anchors", *International Symposium on Connections between Steel and Concrete*, Stuttgart, 10–12 September, Vol. I, pp. 551–563.

Si Chaib A., Mounajed G., La Borderie C., Boussa H. and Ung Quoc H., 2006: adaptation of a concrete damage model for cyclic loading, *III European Conference on Computational Mechanics, Solids, Structures and Coupled Problems in Engineering,* C.A. Mota Soares et al. (eds.) Lisbon, Portugal, 5–8 June 2006.

Sang-Yun, K., Chul-Soo, Y. and Young-Soo Y., 2004, "Sleeve-type expansion anchor behavior in cracked and uncracked concrete", *Nuclear Engineering and Design*, 228, pp. 273–281.

Ung Quoc, H. 2003, "Theorie de degradation du béton et developpement d'un nouveau modèle d'endommagement en formulation incrémentale et tangente. Calcul à la rupture appliqué au cas des chevilles de fixation ancrées dans le béton", PhD. Thesis, ENPC.

Vintzeleou, E. and Elige-hausen, R. 1991. Behavior of fasteners under monotonic or cyclic displacements. *Anchors in Concrete – Design and Behavior,* Special Publication SP 130, American Concrete.

Fracture Mechanics of Concrete and Concrete Structures – Design, Assessment and
Retrofitting of RC Structures – Carpinteri, et al. (eds)
© 2007 Taylor & Francis Group, London, ISBN 978-0-415-44616-7

Studies on the pull-out strength of ribbed bars in high-strength concrete

G. Appa Rao*, K. Pandurangan & F. Sultana
*University of Stuttgart, Stuttgart, Germany
Indian Institute of Technology Madras, India

R. Eligehausen
University of Stuttgart, Stuttgart, Germany

ABSTRACT: The transfer of forces from reinforcing bars to surrounding concrete in reinforced concrete
(RC) is influenced by many parameters. Several efforts were made to understand the influence of bond on global
behaviour of RC members. However, the information on bond strength of high strength concrete (HSC) is lacking.
An attempt was made to study the influence of various parameters on bond such as bar diameter, strength of
concrete, lateral confinement and embedment length. The bond lengths were 50 mm and 150 mm with different
bar diameters, strength of concrete and type of confinement. The bar diameters were 16 mm and 20 mm. The bars
were embedded in concrete without confinement and with confinement using spirals and ties. The casting was
done keeping the bars in the horizontal position. The anchorage bond specimens were tested using displacement
control system and the slip of the bars was controlled at a rate of 1.51 mm/minute (0.025 mm/second). The bond
stress-slip response was studied by varying the variables. As the strength of concrete increases the slip at failure
decreases in the descending branch. With smaller bond length, the bond stress was found to be higher. The
bond strength was found to decrease as the bar diameter increased. Splitting failure was observed in unconfined
specimens, whereas pullout failure in confined specimens. The ultimate bond strength ranges between 10.8 MPa
and 19 MPa with spiral confinement, whereas it ranges between 9.2 to 16 MPa with tied reinforcement. The
ductility was found to increase with spiral reinforcement.

1 INTRODUCTION

Bond in reinforced concrete (RC) refers to the resistance of surrounding concrete against pulling out of reinforcing bars. Anchorage bond is developed parallel to the direction of force over a contact surface in order to induce stress in rebars at critical sections. The bond resisting mechanisms in RC members are understood well in normal strength concrete (NSC) after the numerous studies performed in the last thirty years. If the bond resistance is inadequate, slipping of reinforcing bar occurs destroying composite action. In RC members sudden loss of bond between rebars and concrete in anchorage zones causes brittle failure. However, the information on bond strength of high strength concrete (HSC) is scanty (FIB 2000).

Bond is necessary not only to ensure adequate level of safety allowing composite action of steel and concrete, but also to control structural behavior along with sufficient ductility. The bond in RC members depends on a number of factors such as reinforcing unit (bar or multi wire) and stress state in both reinforcing unit and surrounding concrete. Other parameters such as concrete cover, space between rebars, number of

layers and bundled bars, casting direction and bar position play important role. Several research studies have been reported on the influence of deformation patterns and rib geometry on bond (Rehm, 1961; Goto, 1971). For bars with rib face angles, bond behaviour is influenced by the rib face angle. However, when the rib face angle is less than 30 degrees, the bond behaviour is different. In bars with small rib spacing and small rib height the bond strength is reduced.

Mathey and Watstein (1961) reported that the bond strength decreases as the embedment length increases, and decreases as the bar diameter increases. Hansen and Liepins (1962) reported an increase in the bond strength under dynamic loading over static loading. Also progressive bond failure and large slip were expected from large repeated loading. Ferguson and Thomson (1962) reported on development length of rebars and effect of confinement. Bond stress varies as a function of development length rather than bar diameter. Ultimate bond stress varies as a function of $\sqrt{f_c'}$, with other factors being constant, since the bond strength is related to concrete tensile strength. The nature of bond failure and factors influencing splitting, importance of bar spacing and beam width, end anchorage, flexural bond and anchorage bond were reported by *Ferguson et al. (1966).*

Lutz and Gergely (1967) studied the action of bond forces and the associated slip and cracking using rebars with different surface properties. The slip was found to be due primarily to the relative movement between concrete and steel along the surface of the ribs and also due to crushing of mortar. *Goto (1971)* studied the primary and secondary cracking by injecting ink around the deformed rebars in axially loaded tests. *Nilson (1972)* estimated the bond stress from the slope of the steel strain curve. The strain in concrete and steel was measured internally and the bond slip was calculated from the displacement functions obtained by numerical integration of strains. *Jiang et al. (1984)* developed new test method by cutting the reinforcing bars into two halves and placing in two opposite sides of the cross section to study the local slip, secondary cracking and strain distribution in concrete surrounding the interface. A simple one-dimensional analysis predicts the stresses in steel and concrete, local bond-slip, tensile stiffening and total elongation of the reinforcing bar. *Ueda et al. (1988)* studied the beam bar anchorage in exterior beam-column joints. A model has been proposed to predict the load-lead end deformation and anchorage length of rebars extended from beams into exterior columns and subjected to large inelastic loadings.

Effects of anchored bar diameter, confinement of joint and compressive strength of concrete on the hook behaviour in exterior beam-column joints have been studied *(Soroushian, 1988).* An analytical model has been developed for predicting the overall pullout behaviour of rebars, which has been recognised by *ACI–318-83* for development of standard hooks in tension. *Soroushian and Choi (1989)* reported on local bond strength of deformed bars with different diameters in confined concrete. The bond strength decreases as the bar diameter increases. *Soroushian et al. (1991)* studied the influence of strength of concrete with different confinements. Confinement influences local bond of deformed bars. The ultimate bond strength increases as square root of concrete compressive strength.

Abrishami and Mitchell (1992) formulated a new testing technique to simulate uniform bond stress distribution along a rebar to determine bond stress-slip response. *Malvar (1992)* tested specimens with varying confining pressure using confining rings with rebar ribs normal and inclined to the surface and obtained consistent bond-slip response over a short embedded length. Mathematical model for bond-slip behaviour of a reinforcing steel bar embedded in concrete subjected to cyclic loading was reported by *Yankelevsky et al. (1992). Bortolotti (2003)* proposed models to predict the tensile strength of concrete from pullout load.

The confinement improved the bond strength slightly but ductility was improved significantly *(Harajli et al. 2004). Somyaji et al. (1981)* and *Jiang et al. (1984)* conducted several experimental and theoretical studies on bond in NSC. The secondary cracks as well as the distribution of strain in concrete in the vicinity of rebar have been studied. *Darwin et al. (1996)* reported development length criteria for conventional and high relative rib area of reinforcement. On the basis of a statistically based expression, the development length of reinforcement and splice strength in concrete for compressive strength varying between 17 and 110 MPa with and without confinement have been investigated. The effects of cover, spacing, development/spliced reinforcement were incorporated in design equation.

The effects of concrete compressive strength, splice length and casting position on the bond strength of rebars have been studied *(Azizinamini et al. 1993; 1999a).* Increasing the development length in HSC in tension does not seem to increase the bond strength of deformed rebars, when concrete cover is small. Concrete crushing occurred in front of the ribs in NSC, whereas there was no indication of concrete crushing in front of the ribs in HSC with the first few ribs being more active. In HSC with small cover, failure occurred due to splitting of concrete prior to achieving uniform load distribution *Azizinamini et al. (1993). Azizinamini et al. (1999b)* in another study reported that when calculating the development length in HSC for tension splice, a minimum number of stirrups should be provided over the splice region. Statically based on the experimental data an expression has been proposed to calculate the extra number of stirrups required.

Eligehausen et al. (1983) reported comprehensive study on the effect of bar diameter embedded in NSC. The maximum bond capacity decreases slightly with increasing bar diameter. The frictional bond resistance was not influenced by the bar diameter, lug spacing or relative rib area. *Larrard et al. (1993)* investigated the effect of bar diameter on bond strength. The bond strength increases with tensile strength of concrete at a higher rate with smaller bar diameters. A parameter which accounts for the ratios of side cover and bottom faces, and spacing of the spliced bars was introduced. CEB-FIB report *(2000)* presented a general description of the local bond law for tensile forces. Six main stages have been recognized in local bond stress-slip response.

Goto (1971) carried out tests to clarify the propagation of different types of cracks around the tensile reinforcing bars. The internal cracks develop around the reinforcing bars in concrete cylinders as shown in Figure 1. The inclination of internal cracks and the direction of compressive forces on the bar ribs vary between 45 and 80°.

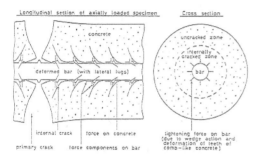

Figure 1. Internal cracks around the reinforcing bar embedded in concrete (Goto, 1971).

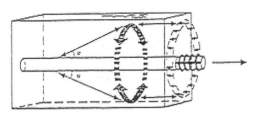

Figure 2. View of tensile ring (Tepfers, 1973).

Tepfers (1973) showed the radial components of bond forces balanced against tensile rings in concrete in Figure 2, using a two dimensional finite element analysis (FEA). The angle "α" is 45 degrees along a perimeter touching the ribs of reinforcing bars independent of rib face angle.

Three mechanisms for bond resistance i.e. (i) chemical adhesion, (ii) friction, and (iii) mechanical interaction between concrete and deformed bars are responsible *(Lutz and Gergely, 1967)*. According to *Rehm (1961)*, and *Lutz and Gergely (1967)*, slip of deformed bars occurs due to (i) splitting of concrete by wedging action, and (ii) crushing of concrete in front of the ribs. For the face angles between 40 and 105 degrees, the slip seems to be not influenced. However, the slip is mostly due to crushing of concrete in front of the ribs. This in effect produces a rib with face angles of 30 to 40 degrees *(Lutz and Gergely, 1967)*.

2 EXPERIMENTAL PROGRAMME

2.1 *Materials*

A 43 grade Portland Pozzolanic Cement (PPC) was used. Specific gravity of cement was 3.12. The fineness of cement was 5.6%. The initial and final setting times were 124 and 300 minutes respectively. Fe 415 grade high strength deformed bars of diameters 16 mm and 20 mm as main reinforcement and 6 mm mild steel (MS) bars for stirrup/spirals as confinement reinforcement were used. Concrete was made from normal

Figure 3. Confining reinforcement in pull-out specimens.

weight black granite aggregate. Specific gravity of cement was 3.12. Specific gravity of coarse aggregate was 2.6. Specific gravity of fine aggregate was 2.6. Two concrete mixes were adopted. In 30 MPa concrete; weight of cement was 300 kg/m^3 and water-cement ratio was 0.45. The concrete mix proportions were 1: 2.30: 4.27: 0.45. In 60 MPa concrete; weight of cement was 400 kg/m^3 and water cement ratio was 0.35. Mix proportions were : 1: 1.64: 3.02: 0.35. Three standard cubes of size 150 mm × 150 mm × 150 mm were used to determine the compressive strength of concrete.

2.2 *Test specimen*

Main bar was embedded in each cube with different confinements such as spirals, ties and without confinement. The test specimen was basically a concrete cube 150 mm × 150 mm × 150 mm with a bar embedded coaxially (IS: 2770-1997). One end of the rebar was projected about 15 mm to measure the free end slip, while the loaded end was jutted out about 750 mm in order to grip the rebar for applying the tensile force. The specimens were cast using 16 and 20 mm diameter rebars in two different concretes with different embedment lengths. By using PVC tubes the required embedment length was achieved. To achieve 50 mm embedment length at the centre of the specimen, PVC tubes were used to unbond the bars from concrete over 100 mm length. The PVC tube neither restrains the slip of the bar nor affects the transfer of bar forces to concrete. The slip was recorded at both loaded and unloaded ends of the bar. The bars were placed in the middle of the specimen horizontally.

2.3 *Test programme*

Tests were carried out on pullout specimen with four different parameters. In a set there were three specimens with different confinements. Only two specimens were tested without confinement. In this study, two concretes of strength 40 and 50 MPa, embedment lengths of 150 and 50 mm, bar diameters 16 and 20 mm and three confinements (spirals, ties and no confinement) were adopted. Typical reinforcements are shown in Figure 3.

Figure 4. Experimental test set-up.

2.4 Fabrication of test specimen

Well seasoned wooden moulds were fabricated to cast 10 specimens in a single batch. Provisions were made at appropriate locations in the moulds to accommodate 3–16 mm, 3–20 mm and 4–25 mm diameter bars. The bars were placed horizontally and concrete was poured vertically. The moulds had provision for fixing the reinforcement cage. Lubricating oil was applied on all sides of the moulds for easy removal of specimens. Concrete was poured carefully from the top and segregation was avoided. Needle vibrator was used to compact the concrete. After twenty four hours the specimens were demolded and cured up to 28 days till testing.

2.5 Experimental set-up and testing

The test setup for testing pullout specimens under controlled displacements is shown in Figure 4. A 250 kN capacity actuator was hung from an existing A frame which could transmit 2000 kN load. The specimen was kept in a frame which hanged from the actuator. To avoid the influence of lateral strains by friction at the bearing plates Teflon sheets were placed between the specimen and base plate of the frame. A steel plate of size 150 × 150 × 12 mm with a central opening of 100 × 100 × 12 mm was placed below Teflon sheet to allow free failure of concrete due to pullout.

The loaded surface was leveled off using gypsum (POP) and a steel plate was placed on top of POP over which an LVDT was fixed to record displacement at unloaded end of the bar. At loaded end two LVDTs were placed to monitor the end displacement. A locking arrangement by means of steel wedges was used. Load cell, which was placed above the wedge along with LVDTs and strain gauges were connected to the data logger which continuously recorded the

respective readings at a frequency of 0.5 Hz. Monotonic load was applied by means of the actuator and the rate of stroke control was maintained at 1.51 mm/min (i.e.0.025 mm/sec). The total displacement of 60 mm in the actuator was set as the end value or complete pullout of the bar whichever occurred first to end the test. Three identical tests were carried out to have a minimal statistical basis.

3 RESULTS AND DISCUSSIONS

Forty pull-out specimens were tested for anchorage bond strength and its variation. The bond stress-slip curves were drawn from the experimentally monitored load vs. slip (at the free and loaded ends) data. The actual materials properties and various parameters obtained from the observations were used. The bond stress (τ) was calculated as the stress developed over an equivalent surface area using the formula

$$\tau = \frac{P}{\pi \, d_b \, l_b} \tag{1}$$

"Where" P = load (N), l_b = embedment length (mm), and d_b = diameter of the rebar (mm).

3.1 Modes of failure

The unconfined concrete at the end of rebar in tension offers the least bond resistance because of early formation of radial splitting cracks caused by high tensile hoop stresses. The failure was caused by the formation of a concrete cone from the concrete block due to bond forces acting on the concrete in front of rebar lugs. Better bond performance was observed with confining reinforcement. Splitting cracks in the plane between bars was developed, but its growth was resisted by the confinement. The bond failure was probably due to shear failure in concrete between the lugs.

An ideal pull-out failure was occurred in all the tested specimens with confinement. However, splitting failure was encountered in the specimens without confinement. Wide longitudinal cracks were formed along the specimen surface. Eventually, the specimen was split into two halves in most of the test specimens as shown in Figure 5. In the tested specimens confined by ties or spirals the splitting cracks were effectively arrested by the confinement, due to which splitting failure was minimized or avoided altogether resulting only in pull-out of rebars from concrete as shown in Figure 6. This was clearly shown by the descending branch of the bond-slip curve which lost its bond stiffness gradually with load. The portion of concrete in between the ribs was sheared off in confined concrete specimens and the bar experienced with stresses much below the yield strength of rebar. The failure of

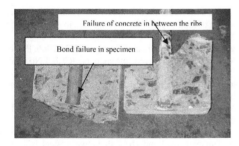

Figure 5.   Splitting failure in unconfined specimen.

Figure 6.   Pullout failure in confined concrete specimen.

concrete in between the ribs was clearly noticed by observing the tested specimens. The unconfined specimens failed due to splitting and the pulled out bar was intact showing that there was no crushing failure and the bar was pulled out from the concrete.

3.2   Bond strength

The maximum bond strength of unconfined concrete ranged between 50 and 60% of that with confinement. The bond stress at the stage of longitudinal splitting was about 8.0 to 11.8 MPa in both 40 MPa and 50 MPa concretes. After attaining the peak value, a sudden drop in the load was observed. The descending branch did not reveal any resistance due to friction. The maximum bond stress in tested specimens with 16 mm diameter bars with different confinements was observed when the slip was between 0.3 and 1.5 mm. The slope of the ascending branch of bond stress-slip curve decreased gradually. It was relatively steeper than those with 20 mm diameter rebars. A horizontal plateau was observed at a maximum bond stress, τ_{max} at a slip of around 0.3 mm and the slip increased further at the same shear stress up to about 1.5 to 4.0 mm. Subsequently the bond resistance dropped to τ_f called frictional bond resistance at a slip of about 10 mm and then was remained constant thereafter up to failure. Coincidentally this value was equal to the clear spacing between the ribs in the bars. The slip continued to increase when the bond stress reached a constant value in the post-peak region of the bond stress-slip curve.

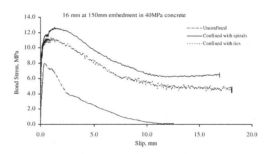

Figure 7.   Bond stress vs. Slip with 16 mm bars at 150 mm embedment in 40 MPa concrete.

3.3   Bond stress-slip response

Slip is defined as the relative displacement of rebar with reference to the surrounding concrete. The relative displacement of the bar is always measured with reference to the undisturbed concrete and consists of relative slip at the interface and shear deformations in concrete. Therefore, displacement occurred due to localized strains in the interface even if there was no slip. Figure 7 shows a typical bond stress-slip response with 16 mm bars at 150 mm embedment length in 40 MPa concrete.

3.4   Effect of various parameters

The bond strength did not change much with diameter of bars. However, the resistance increased slightly with increase in bar diameter in certain cases. The variation in rebar diameter did not influence much the extent of the plateau. The post peak behaviour was better when large bar diameter was used. The maximum bond stress (τ_{max}) was relatively higher at small embedment length, at 50 mm and it was achieved at a slip varying between 0.3 and 1.5 mm. There was no significant change of the slope of the ascending branch as well as the maximum bond stress plateau. But for specimen with 50 mm embedment length, the descending branch was more ductile compared to that with 150 mm embedment length. As the surface area of embedment increased, the maximum bond stress decreased. It shows, however, that there exists a size effect on bond strength of rebars. It has been proved that a splitting failure can be delayed or avoided altogether by providing confining reinforcement. Though the effects of the confining reinforcement (be it spiral or ties) was rather limited, however by providing spirals the bond strength was slightly increased. Usually, it is assumed that once a pull-out failure is initiated, providing a large concrete cover or a sufficiently strong restraining (confining) reinforcement, the value of τ_{max} can not be increased further. The extent of the post-peak curve was significantly increased by spirals showing increase of ductility with spiral reinforcement.

4 CONCLUSIONS

From the present study, the following conclusions can be drawn;

1. The maximum bond stress τ_{max} for unconfined specimens was about 50 to 60% of that of those confined with spirals.
2. The lateral confinement increased the bond strength significantly and the extension of the post-peak curve increased showing improved ductility.
3. The maximum bond stress τ_{max} for specimens confined with spirals was higher and showed significant improved ductility.
4. The influence of bar diameter on the local bond stress-slip relationship was rather small in the tested range ($d_b = 16$ mm and 20 mm).
5. The bond strength also decreased as the embedment length increased. The bond stress varies along the larger embedment lengths while it is more or less uniform in smaller lengths.

REFERENCES

Abrishami, H.H. and Mitchell, D.1992. Simulation of uniform bond stress. *ACI Mat. Jl* 89(2): 161–168.

ACI 318-2005. Building code requirements for structural concrete and commentary, *ACI*, 1995 Farmington Hills, Michigan.

Azizinamini, A.Stark, M. Roller, J. J. and Ghosk, S.K. 1993. Bond performance of reinforcing bars embedded in high strength concrete. *ACI StrJl*, 90(5): 554–561.

Azizinamini, A. Pavel, R. Hatfield, E. and Ghosh, S.K.1999a. Behavior of spliced reinforcing bars embedded in high-strength concrete. *ACI Str Jl* 96(5): 826–835.

Azizinamini, A. Darwin, D. Eligehausen, R. Pavel, R. and Ghosh, S.K. 1999b. Proposed modification to ACI 318-95 tension development and lap splice for high strength concrete. *ACI Str Jl* 96(6): 922–926.

Bortolotti. 2003.Strength of concrete subjected to pull out load. *ASCE Mat, Jl* 15(5): 491–495.

CEB-FIP Report. 2000. Bond of reinforcement in concrete: state of the art report. *FIB Bulletin* 10, Sw.

Darwin, D. Zuo, J. Tholen, M.L. and Idun, E.K. 1996. Development length criteria for conventional and high relative rib area reinforcing bars. *ACI Str Jl* 93(3): 347–359.

De Larrard, F. Schaller, D. and Fuchs, J. 1993. Effect of bar diameter on the bond strength of passive reinforcement in HPC. *ACI Mat Jl* 90(4): 333–339.

Eligehausen, R. Popov, E.G. Bertero, V.V. 1983. Local bond stress-slip relationships of deformed bars under generalized excitations. *R.No.UCB/EERC-83/23,EERC*, Berkeley.

Ferguson, P.M. Robert, I. Thompson J.N. 1962. Development length of high strength reinforcing bars in bond. *ACI Jl* T. No.59-17: 887–922.

Ferguson, P.M. Breen, J.E. and Thompson, J.N. 1966. Pull out tests on high strength reinforcing bars. *ACI Jl*,T.No.62-55, 933–950.

Goto, Y. 1971. Cracks formed in concrete around deformed bars in concrete. *ACI Jl* 68 (2): 244–251.

Hansen, R.J. and Liepins, A.A. 1962. Behaviour of bond in dynamic loading", *ACI Jl*: 563–583.

Harajli, M.H. Hamad, B.S. and Rteil, A.A. 2004. Effect of confinement on bond strength between steel bars and concrete. *ACI Str Jl* 101 (5): 595–603.

IS 2770. 1997. Method of Testing bond in reinforced concrete part i-pullout test. *BIS*, New Delhi.

Jiang, D.H. Shah, S.P. and Andonian, A.T. 1984. Study of the transfer of tensile forces by bond. *ACI Jl* T. No.81-24: 251–258.

Lutz L.A. and Gergely, P. 1967. Mechanics of bond and slip of deformed bars in concrete. *ACI Mat. Jl*. T. No. 64-62: 711–721.

Malvar, L.J. 1992. Bond of reinforcement under controlled confinement. *ACI Mat Jl* 89(6): 593–601.

Mathey, R.G. Watstein, D. 1961. Investigation of bond in beam and pull out specimens with high yield strength deformed bars. *ACI Jl*, T. No.57-50: 1071–1089.

Nilson, A.H. 1972. Internal measurement of bond slips. *ACI Jl* 69 (7): 439–441.

Rehm, G. 1961. Uber die Grundlagen des Verbudzwischen Stahl undBeton. Heft 138, Deutscher Ausschuss fur Stahlbeton, Berlin, 59.

Somayaji, S. and Shah, S.P. 1981. Bond stress versus slip relationship and cracking response of tension members. *ACI Jl* 78(3): 217–225.

Soroushian, P. 1988. Pull out behaviour of hooked bars in exterior beam-column connections. *ACI Str Jl*, 85:269–276.

Soroushian, P. Choi, K.B. Park, G.H. and Aslani, F. 1991. Bond of def. bars to concrete: effects of confinement and strength of concrete. *ACI Mat. Jl* 88(03): 227–232.

Soroushian, P. and Choi, K.B. 1989. Local bond of deformed bars with different diameters in confined concrete. *ACI Str Jl*, 86(02): 217–222.

Tepfers, R.A. 1973. Theory of bond applied to overlapped tensile reinforcement splices for deformed bars. Publ 73:2. Department of Concrete Structures, *Chalmers University of Technology*, Göteborg, p. 328.

Ueda, T. Lin, I. and Hawkins, N.M. 1986. Beam bar anchorage in exterior column-beam connections. *ACI Str. Jl*, T. No. 83-41: 412–422.

Yankelevsky, D.Z. Adin, M.A. and Farhey, D.N. 1992. Mathematical model for bond slip behaviour under cyclic loading. *ACI Str Jl* 89(6): 692–698.

ACI 318-2005. Building code requirements for structural concrete and commentary, ACI, 1995 Farmington Hills, MI.

Fracture Mechanics of Concrete and Concrete Structures – Design, Assessment and Retrofitting of RC Structures – Carpinteri, et al. (eds)
© 2007 Taylor & Francis Group, London, ISBN 978-0-415-44616-7

Realistic model for corrosion-induced cracking in reinforced concrete structures

B.H. Oh & K.H. Kim
Dept. of Civil Engineering, Seoul National University, Seoul, Korea

B.S. Jang
Dam Engineering Research Center, Korea Water Resources Corporation, Daejeon, Korea

J.S. Kim
Dept. of Civil Engineering, Seo Kyeong University, Seoul, Korea

S.Y. Jang
Korea Railroad Research Institute, Euiwang-si, Gyeonggi-do, Korea

ABSTRACT: The mechanism of pressure and thus stress build-up due to corrosion of a reinforcing bar in a reinforced concrete member is a very complex phenomenon because there is a certain pressure-free corrosion strain which has not been explored in the previous studies. The purpose of the present study is to explore the critical corrosion amount which causes the cracking of concrete cover and also to determine the realistic mechanical properties of corrosion layer including the pressure-free corrosion strain and the stiffness. To this end, a comprehensive experimental and theoretical study has been conducted. Major test variables include concrete strength and cover thickness. The corrosion products which penetrate into the pores and cracks around the steel bar have been considered in the calculation of expansive pressure due to steel corrosion. A concept of pressure-free strain of corrosion product layer was devised and introduced to explain the relation between the expansive pressure and corrosion strain. The proposed theory shows good correlation with corrosion test data of reinforced concrete members.

1 INTRODUCTION

The corrosion products of a reinforcing bar in concrete induce pressure to the surrounding concrete due to the expansion of steel. This expansion causes tensile stresses in concrete around the reinforcing bar and eventually induces cracking through the concrete cover (Bazant 1979, Dagher & Kulendran 1992, Jang 2001, Jang 2003, Liu & Weyers 1998, Lundgren 2001, Martin-Perez 1998, Oh et al. 2002, Oh & Jang 2003a, 2003b, 2004, Ohtsu & Yosimura 1997).

The mechanism of pressure and thus stress build-up due to corrosion is a very complex phenomenon because there is a certain pressure-free corrosion strain which has not been explored in the previous studies. The realistic determination of the stiffness of the corrosion layer is also important because it also directly affects the cracking behavior of concrete.

The purpose of the present study is therefore to explore the critical corrosion amount which causes the surface cracking of concrete cover and also to determine the realistic mechanical properties of corrosion layer including the expansion free (pressure-free) corrosion strain and the stiffness of corrosion layer. To this end, a comprehensive experimental and theoretical study has been conducted. Major test variables include concrete strength and cover thickness. Several series of corrosion tests for a steel bar in concrete have been conducted and the strains at the surface of concrete cover have been measured according to the various amount of steel corrosion. The corrosion products which penetrate into the pores and cracks around the steel bar have been considered in the calculation of expansive pressure due to steel corrosion. The critical amount of corrosion, which causes the initiation of surface cracking, was determined from the present test results.

A concept of free expansion (pressure-free) strain of corrosion product layer was introduced to describe the relation between the expansive pressure and corrosion layer strain. A realistic relation between the expansive pressure and average strain of corrosion product layer in the corrosion region has been derived and the representative stiffness of corrosion layer was also determined.

2 BASIC MECHANISM OF DEFORMATION PROCESS DUE TO STEEL CORROSION

2.1 Corrosion and expansion pressure

The corrosion of steel bar in concrete causes the increase of volume. The expansive pressure due to volume increase of a steel bar in concrete generally induces the tensile stresses and strains in the surrounding concrete. The tensile strains of surrounding concrete due to steel expansion increase as the corrosion of steel bar progresses. Further increase of tensile strain will cause cracking in the surrounding concrete and the cracking will also happen at the surface of concrete cover during the expansion process.

In order to analyze the cracking of concrete cover due to steel corrosion, it is necessary to know the relation between the amount of corrosion of a steel bar and the internal pressure arising from corrosion. Therefore, a realistic relation between the amount of corrosion and the internal expansion pressure has been derived in this study.

2.2 Deformation of corrosion layer

The deformation around a steel bar due to corrosion expansion can be described as shown in Figure 1, in which r_b = initial radius of rebar, x_p = loss of radius of rebar due to corrosion, and w_{corr} = the ratio of weight loss due to corrosion to initial weight of rebar, respectively. The loss of rebar area due to corrosion can be expressed as follows.

$$\pi r_b^2 - \pi (r_b - x_p)^2 = \frac{W_{st}}{\rho_{st}} = w_{corr}\pi r_b^2 \qquad (1)$$

where W_{st} = mass of rebar consumed by corrosion and ρ_{st} = density of steel bar.

Equation (1) can be rearranged to obtain the radius loss, x_p as follows.

$$x_p = r_b (1 - \sqrt{1 - w_{corr}}) \qquad (2)$$

The volume of corrosion product made by consumed steel is larger than the volume of consumed steel itself. Therefore, the radius of steel bar after corrosion will increase by Δr_b and the following relation holds.

$$\pi (r_b + \Delta r_b)^2 - \pi r_b^2 = \frac{W_{rust}}{\rho_{rust}} - \frac{W_{st}}{\rho_{st}} = \frac{W_{st}}{\rho_{st}}(\frac{\rho_{st}}{\alpha\rho_{rust}} - 1) =$$

$$\frac{W_{st}}{\rho_{st}}(v_r - 1) = \pi r_b^2 (v_r - 1) w_{corr} \qquad (3)$$

where W_{rust}, ρ_{rust} = mass and density of corrosion products, respectively, and $v_r = \rho_{st}/\alpha\rho_{rust}$ = relative

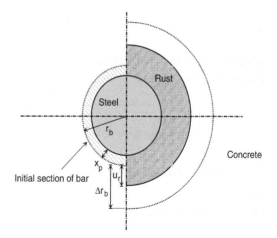

Figure 1. Deformation process due to steel expansion.

ratio of steel density to corrosion product density. The value of α can be obtained from the fact that the mass of steel component among corrosion products is the same as the mass of steel bar consumed by corrosion, i.e. $W_{st} = \alpha W_{rust}$. The value of α is 0.523 for hydrated red rust [Fe(OH)$_3$] and 0.622 for ferrous hydroxides [Fe(OH)$_2$], respectively, because the masses of Fe, O_2, and H_2 per one mol are 55.85 g, 32 g, and 2 g, respectively. The α value ranges from 0.523 to 0.622 depending upon the types of corrosion products. The ratio of steel density to corrosion product density normally ranges from 2 to 4 (Jang 2002). Equation (3) can be rearranged to obtain Δr_b as follows.

$$\Delta r_b = r_b (\sqrt{1 + (v_r - 1)w_{corr}} - 1) \qquad (4)$$

Figure 1 shows the volume expansion of corrosion product layer caused by the radius loss of a steel bar.

Lundgren (2001) considered the compaction effect of corrosion product layer due to surrounding concrete. Therefore, the free-expanded displacement Δr_b is reduced to u_r as shown in Figure 1. The ideal thickness of corrosion product layer, $(x_p + \Delta r_b)$ decreases by $(\Delta r_b - u_r)$ due to pressure force P caused by the restraint of surrounding concrete (see Figure 1).

The strain ε_{rust} of corrosion product layer due to compression effect is here defined as follows.

$$\varepsilon_{rust} = \frac{u_r - \Delta r_b}{x_p + \Delta r_b} \qquad (5)$$

The corrosion product layer is compressed by the strain ε_{rust} due to expansive pressure P and thus the mechanical characteristic of corrosion layer may be expressed as

$$P = K_{rust}\varepsilon_{rust} \qquad (6)$$

in which K_{rust} represents the stiffness of corrosion product layer. This model does not consider the corrosion products that are absorbed into cracks and pores in surrounding concrete. The cracks may happen in the vicinity of reinforcing bar due to circumferential tensile stresses caused by expansion of steel corrosion. Therefore, the corrosion products penetrate into the adjacent cracks in concrete and also into concrete pores. This may be the major drawback of the model previously expressed by Equation (6).

3 TESTS OF CONCRETE CRACKING DUE TO STEEL CORROSION

3.1 Test variables

In the present tests, the surface concrete strains were measured as the corrosion of steel bar in concrete progresses. This will allow the determination of critical corrosion amount which induces the cracking on the surface of concrete cover. To this end, a comprehensive experimental program has been set up to execute the corrosion tests of steel bar in concrete.

The major test variables are the cover thickness and compressive strength of concrete. The concrete cover thicknesses considered were 2, 3, 4, and 5 cm, respectively. In order to consider the effect of compressive strength, the water-cement ratios of concrete were varied from 0.35 to 0.55. The test specimen identification was classified as H, N, and L which represent high strength (H), normal strength (N), and low strength (L) concretes, respectively, depending upon the water-cement (W/C) ratios.

3.2 Test materials

The Type 1 ordinary Portland cement and the river sand with specific gravity of 2.55 were used. The specific gravity of crushed coarse aggregates was 2.6. The slump value of fresh concrete was controlled to be 150 mm and air content was 4.5 percent.

The compressive strengths of concrete for H, N, L series are 45, 40, 28 MPa, respectively. The tensile strengths for H, N, L series are 4.2, 3.9, 3.1 MPa, respectively.

3.3 Test specimens

The size of test specimens was $200 \times 200 \times 200$ mm cube and the cover thicknesses were 20, 30, 40, and 50 mm, respectively. The steel bar of 20 mm diameter was put in concrete specimen at the location of designated cover thickness. Therefore, the ratios of cover thickness to rebar diameter (c/d) are 1.0, 1.5, 2.0, and 2.5, respectively.

The region for steel corrosion was limited to the central portion of steel bar as shown in Figure 2. Both ends

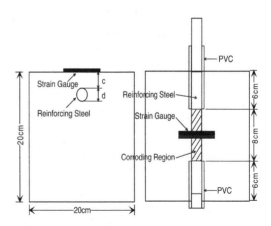

Figure 2. Configuration of test specimen and strain gage attachment.

of steel bar were wrapped with PVC pipes and sealed with epoxy. The cross-hatched region in Figure 2 was exposed to corrosion.

3.4 Test method

The test specimen was immersed in NaCl 3% solution and corrosion circuit was connected using direct-current power supply. The rebar in the specimen plays as an anode and the stainless steel plate as a cathode. The electrical potential between anode and cathode accelerates the penetration of chloride ions into concrete and thus accelerates the corrosion of steel bar. The concrete surface strains and electric potentials were measured every 30 minutes and these measurements were continued until the surface strain increases rapidly and reaches sufficient values. The wide-black line in the Figure 2 indicates the strain gages attached in order to measure the strain increase during the expansion process.

3.4.1 Determination of corrosion amount by faraday's Law

The amount of corrosion can be calculated by Faraday's law. The amount of substance produced or consumed by the electrical quantity of one Faraday (F) is equal to the extracted substance of one chemical equivalent moved by one mol of electron. Equation (7) represents the amount of mol, X, extracted by electrolysis of substance with n electrons.

$$X = \frac{It}{nF} \tag{7}$$

in which I = electric current in ampere, A, n = number of mole participating production reaction, and 1F = electric quantity of one mole of electron = 96,500 C. An electric resistance was installed in the corrosion circuit to measure the electric potential. The electric current was then obtained from the resistance value. The total electric change was also obtained by integrating the electric current values with respect to time and then the number of mol of corroded rebar was determined by using Faraday's law.

3.4.2 Strain measurement

The strain values on the surface of concrete specimen were measured according to time. This is to see the increase of tensile strain due to volume expansion of corroded rebar and to find the time at which the crack occurs on the concrete surface. For this purpose, concrete strain gages were attached on the concrete surface near reinforcing bar as shown in Figure 2. The strain gages were installed in the direction perpendicular to the anticipated crack direction. From the measurement of concrete strains, the critical value of corrosion amount which causes the crack occurrence on the concrete surface has been determined.

4 ANALYSES OF TEST RESULTS

4.1 Surface Strains according to cover thickness

The measured strains on the surface of concrete specimens were plotted. Figure 3 shows the concrete strains according to the amount of corrosion for various specimens with different cover thicknesses. It can be seen from Figure 3 that the development of surface strains is larger and faster as the cover thickness becomes smaller. Namely, much larger strains occur for shallow-thickness specimens at the same corrosion amount. This is because expansive pressure due to same corrosion amount causes much larger surface strains for thinner-covered specimens.

The surface strain increases slowly at lower corrosion amount, but increase rapidly after a certain higher corrosion amount. This may be considered as the start of concrete cracking. The thinner-covered specimens show this rapid increase of strain at relatively lower corrosion amount. The thinner the cover thickness is, the lower the critical corrosion amount is.

4.2 Surface strains according to water-cement ratio

The test results indicate that as the water-cement ratio of concrete decreases the amount of corrosion that induces the same value of surface strain increases. This means that the strength of lower W/C concrete is higher and thus the stiffness is also higher. Higher stiffness of concrete needs higher expansive pressure (and thus higher corrosion amount) in order to develop the same surface strain. However, the water-cement ratio (or strength) of concrete does not affect much the corrosion-induced strains for the specimens with thin cover.

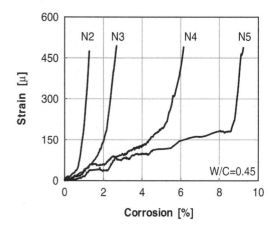

Figure 3. Measured surface strains according to corrosion amount for various cover thicknesses (W/C = 0.45).

4.3 Critical corrosion amount

The point of sudden increase of strain in Figure 3 indicates the start of cracking for each specimen. Therefore, the critical corrosion amounts which cause the initiation of surface cracking were determined from the strain vs. corrosion-amount curves as shown in Figure 3. For example, the critical corrosion amount for N4 series with W/C = 0.45 and 4 cm cover thickness is found to be about 4.25%. The test results indicate that the critical corrosion amount increases with an increase of cover thickness and also increases with an increase of concrete strength.

5 NONLINEAR ANALYSIS FOR CORROSION EXPANSION

5.1 Analysis outline

Finite element analysis has been executed to explore the stress and strain distributions of concrete around the reinforcing bar due to corrosion expansion. The strains on the surface of concrete cover induced from internal pressure due to corrosion expansion have been calculated.

The eight-node plane strain element has been employed to model the test specimens. The material properties obtained from the tests have been used in the analysis. The bilinear stress-deformation relation after tensile strength was employed. The Newton-Raphson method was used for iterative nonlinear analysis.

In order to overcome the difficulties in convergence near peak load, the arc-length method was introduced. The arc-length method follows smoothly the load-displacement curve and enables to predict the behavior after peak load. Finer mesh pattern was used near the reinforcing bar.

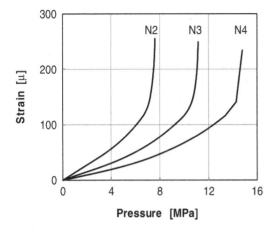

Figure 4. Analysis results for concrete surface strains due to internal pressure increase.

The internal pressure due to corrosion expansion was applied to the radial direction at the outer line of reinforcing bar.

5.2 *Surface strains*

The concrete strains on the surface of cover were obtained from the analysis. Figure 4 shows the relations between the surface strains and internal pressures for normal strength concrete specimens with different cover thicknesses. The surface strains increase almost linearly at low internal pressures and then increase rapidly after certain critical pressure. The overall trend of strain-pressure relation in Figure 4 is very much similar to the relation of experimentally-observed strain-corrosion amount in Figure 3.

6 EXPANSION PRESSURE ACCORDING TO CORROSION AMOUNT

6.1 *Cracks and pores in concrete*

The strains on the surface of concrete cover and the radial displacements at the outer line of rebar due to expansive pressure P were determined from the finite element analysis. The surface strains according to corrosion amounts were also obtained from the present tests. From the values of surface strains, radial displacements u_r, and corrosion amount w_{corr}, the strain ε_{rust} of Equation (5) can be determined with the use of Eqs. (2) ∼ (4). Therefore, the relation between the internal pressure P and the strain of corrosion layer ε_{rust} was obtained.

A major drawback of Lundgren's model is that it does not consider the effects of penetration of corrosion products into the pores and cracks adjacent to the rebar. The cracks occur in concrete around the rebar

due to corrosion expansion. These cracks may play a role to absorb the corrosion products and thus reduce the expansive pressure. In the present study, therefore, the effects of cracks and pores of surrounding concrete have been considered in the modeling of expansive pressure and corrosion layer strain.

The net amount of corrosion product, V_{net} may be summarized as follows.

$$V_{net} = V_{rust} - V_{crack} - V_{pore} \qquad (8)$$

in which V_{net} = net amount of corrosion product, V_{rust} = total corrosion product, V_{crack} and V_{pore} = the amounts of corrosion product which penetrates into the surrounding cracks and pores, respectively. The total corrosion product V_{rust} may be derived as

$$V_{rust} = w_{corr} W_0 \left(\frac{1}{\alpha \rho_{rust}} - \frac{1}{\rho_{st}} \right)$$

$$= \pi r_b^2 (v_r - 1) w_{corr} \qquad (9)$$

The volume of cracks, V_{crack} can be calculated by multiplying the total width of cracks by the depth of cracks.

$$V_{crack} = \frac{1}{2} h \left(\Sigma w_c \right) \qquad (10)$$

in which h = the crack propagation depth which is obtained from the nonlinear finite element analysis at a specified internal pressure, and Σw_c = sum of the crack widths at the outer face of the rebar.

The total crack width Σw_c can be obtained from the circumferential tensile strain ε_θ at the outer line of rebar which arises from expansion pressure. Therefore, Equation (10) can be rewritten as follows.

$$V_{crack} = \frac{1}{2} h \left(\Sigma w_c \right) = \frac{1}{2} h (2 \pi r_b)(\varepsilon_\theta - \varepsilon_0)$$

$$= \pi r_b h(u_r / r_b - \varepsilon_\theta) \qquad (11)$$

in which $\varepsilon_\theta = u_r/r_b$; and ε_0 = the strain at tensile strength = cracking strain of concrete = 0.002. The volume of pore, V_{pore} may be written as

$$V_{pore} = V_{net} \phi_r \qquad (12)$$

In which ϕ_r = the porosity of concrete at the surface location of displaced (expanded) rebar = the porosity at the location $(r_b + u_r)$ from the rebar centroid.

The modified increase of rebar radius Δr_b^* may now be newly derived from Equation (4) and Eqs. (8)∼(9).

$$\Delta r_b^* = r_b \left(\sqrt{1 + (v_r - 1)w_{corr} - (V_{crack} + V_{pore})/\pi r_b^2} - 1 \right) \qquad (13)$$

The strain ε_{rust} is now rewritten as

$$\varepsilon_{rust} = \frac{u_r - \Delta r_b{}^*}{x_p + \Delta r_b{}^*} \qquad (14)$$

The effects of cracks and pores have been considered in the present study as derived in Eqs. (8)~(14). It is generally anticipated that the consideration of cracks and pores around the rebar gives smaller values in radial displacement u_r, the strain ε_{rust} and the stiffness K_{rust} of corrosion layer.

6.2 Variation of porosities from steel interface to distant concrete

The present study indicates that increase of porosity reduces the strain ε_{rust} for same expansive pressure. The porosity here represents that porosity at the displaced location u_r, not the porosity in pure distant concrete. Therefore, it is necessary to derive a porosity relation between the porosity ϕ_i at original steel-concrete interface and the porosity ϕ_0 at far-field distant concrete.

$$\phi_d = \phi_0 + (\phi_i - \phi_0) \exp(-kd) \qquad (15)$$

in which ϕ_d = the porosity at the distance d from the original steel bar. The porosity of normal concrete ϕ_0 may be obtained from the literature and the Powers-Brownyard model which is a function of aggregate content, cement content, water content, and the densities of those ingredients has been used in this study.

The porosity ϕ_i at the initial interface is assumed to be 0.90 and the k value of 200 gives approximately the interface thickness of $20\,\mu m$ (Bourdette et al. 1995, Jang 2003). It is generally known that the interface thickness between steel bar and concrete is about $10\sim30\,\mu m$ (Jang 2003).

The displacement u_r at the first cracking on the surface of concrete cover was obtained from the analysis for each test specimen. The average porosity ϕ_{ave} between ϕ_i at the steel interface and ϕ_r at u_r was obtained from the average area under the porosity distribution curve. These average porosities have been used to derive the relation between P and ε_{rust}.

6.3 Pressure-free strain ε_{fe}

Figure 5 shows the relation between P and ε_{rust} for various values of v_r where v_r represents the relative ratio of steel density to corrosion product density (see Equation (3)). It can be seen that an increase of v_r causes the increase of ε_{rust} which is required to induce the same expansion pressure. This means that larger value of v_r represents the corrosion product of low density

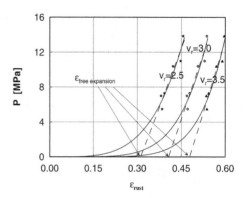

Figure 5. Variation of P according to ε_{rust} for various v_r values.

and thus the compressed strain ε_{rust} of corrosion product layer must be larger to induce the same amount of expansion pressure. The detail examination of Figure 5 enables to determine the so-called free expansion (or pressure-free) strain ε_{fe} which induces no expansion pressure. This was not considered in Lundgren's model in Eqs. (5)~(6).

Therefore, new definition of the strain $\varepsilon_{rust}{}^*$ has been established in this study as follows.

$$\varepsilon_{rust}{}^* = \varepsilon_{rust} - \varepsilon_{fe} = \frac{u_r - \Delta r_b{}^*}{x_p + \Delta r_b{}^*} - \varepsilon_{fe} \qquad (16)$$

It is seen from Figure 5 that linear relations are obtained between P and ε_{rust}^* as follows.

$$P = K_{rust}{}^* \varepsilon_{rust}{}^* \qquad (17)$$

7 COMPARISON OF PROPOSED MODEL WITH TEST DATA

Figure 6 exhibits the comparison of the proposed model with test data on the surface strains and corrosion amounts for cover thicknesses of 4 cm. It can be again seen that the effect of concrete strength is rather small for small cover thickness, but it becomes larger for normally adopted large cover thicknesses. This is important in practice because the concrete structures exposed to sea environments normally have medium or large cover thicknesses to enhance the durability.

The comparisons of proposed analysis with test data were also made on the surface strains and corrosion amounts for various cases. Generally, good correlation between the analysis and test data has been observed. Those comparisons are not included here due to the length limitation of the paper.

786

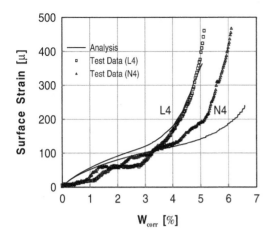

Figure 6. Comparisons of the effects of concrete strength on the surface strains according to corrosion amount w_{corr} for normal cover thickness of 4 cm.

8 CONCLUSION

The purpose of the present study is to explore thecretical corrosion amount which causes the surface cracking of concrete cover and also to determine the realistic mechanical properties of corrosion layer around a rebar which affect the expansion and cracking behavior of concrete structures under sea environments.

It is found that the critical corrosion amount increases greatly with an increase of cover thickness. The concrete strength also affects the critical corrosion amount. The effects of pores and cracks around a rebar have been considered in the modeling of expansion pressure and the strain of corrosion layer. A new concept of free expansion (pressure-free) strain was devised and introduced to derive the modified strain of corrosion product layer. The stiffness of corrosion product layer has been also determined and used to establish a new relation between expansion pressure and the strain of corrosion product layer. The analysis results have been compared with test data and they show generally good agreement.

Future studies may be necessary to develop a realistic method to directly measure the internal expansion pressure due to steel corrosion, which is not an easy task at present.

REFERENCES

Bazant, Z. P. 1979. Physical model for steel corrosion in concrete sea structures-Application. *J. Struct. Div., ASCE*, 105(6): 1155–1166.

Bourdette, B. et al. 1995. Modeling of the transition zone porosity. *Cem Concr Res* 25(4): 741–751.

Dagher, H. J. & Kulendran, S. 1992. Finite element modeling of corrosion damage in concrete structures. *ACI Structural Journal*, 89(6).

Jang, B. S. 2001. Life time estimation method of reinforced concrete structures considering the effects of reinforcements on the chloride diffusion and the non-uniform corrosion distribution. PhD dissertation, Dept. of Civil Engineering., Seoul National University, Seoul. KOREA

Jang, S. Y. 2003. Modeling of chloride transport and carbonation in concrete and prediction of service life of concrete structures considering corrosion of steel reinforcements. PhD dissertation, Dept. of Civil Engineering., Seoul National University, Seoul. KOREA.

Liu, Y. & Weyers, R. E. 1998. Modeling the time to corrosion cracking in chloride contaminated reinforced concrete structures. *ACI Mat. J.*, 95(6): 675–681.

Lundgren , K. 2001. Modeling bond between corroded reinforcement and concrete. *Proceeding of the fourth international conference on fracture mechanics of concrete and concrete structures, Cachan, France, May 28~June 1*, Fracture Mechanics of Concrete Structures, 1: 247–254.

Martin-Perez, B. 1998. Service life modeling of RC highway structures exposed to chlorides. PhD dissertation, Dept. of Civil Engineering, University of Toronto, Toronto, Canada.

Oh, B. H. & Jang, B. S. 2003. Chloride diffusion analysis of concrete structures considering the effects of reinforcements, *ACI Material Journal*, 100(2): 143–149.

Oh, B. H. et al. 2002. Development of high performance concrete having high resistance to chloride penetration. *Nuclear Engineering and Design*, 212(1–3): 221–231.

Oh, B. H. & Jang, S. Y. 2003. Experimental investigation of the threshold chloride concentration for corrosion initiation in reinforced concrete structures. *Magazine of Concrete Research*, 55(2):117–124.

Oh, B. H. & Jang, S. Y. 2004. Prediction of diffusivity of concrete based on simple analytic equations. *Cement and Concrete Research*, 34(3): 463–480.

Ohtsu, M. & Yosimura, S. 1997. Analysis of crack propagation and crack initiation due to corrosion of reinforcement. *Construction and Building Materials*, 11(7–8): 437–442.

Fracture Mechanics of Concrete and Concrete Structures – Design, Assessment and
Retrofitting of RC Structures – Carpinteri, et al. (eds)
© 2007 Taylor & Francis Group, London, ISBN 978-0-415-44616-7

On bond failure by splitting of concrete cover surrounding anchored bars

M.M. Talaat & K.M. Mosalam

Department of Civil and Environmental Engineering, University of California, Berkeley, USA

ABSTRACT: This study develops analytical expressions for the bond stress level surrounding a longitudinal steel bar that causes the cover concrete to fail in a splitting mode, with bar-to-surface cracks forming in the longitudinal direction parallel to the bar. The proposed models use as a foundation the fracture mechanics-based cohesive-elastic ring model, developed by Reinhardt & Van der Veen (1990) based on original work by Tepfers (1979). They propose formulations that better address the assumptions and simplifications made in the original model, namely the biaxial behavior of concrete in tension, the crack-opening displacement profile, and the material law governing post-cracking tension softening. The relative significance of the proposed model enhancements is established through their prediction of bond stress and the associated computational cost for a computational benchmark problem involving a steel bar pullout from a concrete cylinder. Finally, the robustness of the reference and proposed models, measured by their sensitivity to the uncertainty in their respective parameters is evaluated and compared using a deterministic sensitivity analysis. Based on such analysis, recommendations are given on the most suitable and practical enhancements of the original model.

1 INTRODUCTIUON

1.1 *Motivation*

The use of Fracture Mechanics (FM) in estimating concrete behavior and determining the strength of concrete elements has received considerable attention over the past years, primarily because it can describe and interpret size effects in the experimentally-observed behavior of concrete structures. In the field of applying FM to concrete, two approaches came to be widely accepted within the engineering community, both of which adopt the assumptions of Linear Elastic Fracture Mechanics (LEFM). The first is the Fictitious Crack Model (FCM) due to Hillerborg et al (1976). The second approach is the Crack Band Theory model, introduced by Bazant and Oh (1983). Of the two, the former has an advantage of not relying on an empirical parameter such as the crack band width, which makes it more readily adaptable for implementation in Finite Element (FE) codes. The FCM approach has been used to develop the cohesive-elastic bond-splitting model for estimating the cover-splitting strength along a reinforcing steel bar based on bond stress (Tepfers 1979). In this study, this original model is designated A_0 and is briefly reviewed for completeness and its simplifying assumptions identified. Four alternative models are developed and introduced to mathematically address these assumptions and designated A_1 through A_4. The proposed models are compared through their estimation of bond strength in a typical bar pull-out example of a steel bar embedded in a concrete cylinder. After identifying the more significant models among the proposed alternatives, the stability of their predictions with respect to the uncertainty in their individual parameter values is investigated using a deterministic sensitivity analysis, namely the Tornado diagram analysis (Lee & Mosalam 2005) and compared to that of the reference model within the context of the benchmark problem. Besides establishing the sensitivity to modeling assumptions, the discussion of the benchmark problem results addresses the influence and methods of selection for the number of radial cracks in the model under both deterministic and probabilistic contexts.

1.2 *Background – Model A_0*

The reference bond-splitting model assumes the distribution of stresses and deformations shown in Figure 1 for a single bar embedded in concrete. As the bar is being pulled from the concrete, ribs on its perimeter result in an inclined resisting force, which can be resolved into radial "pressure" and longitudinal "bond" components, related through an angle of internal friction α, which reflects the surface conditions. The stress state is assumed axi-symmetric at any radius r. The cylinder enclosed within the cover c surrounding the bar is assumed to have n identical and stable cracks that extend radially to a length e. A polar coordinate

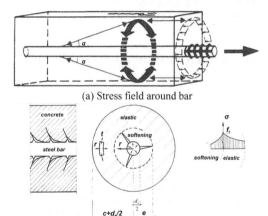

(a) Stress field around bar

(b) Cohesive-elastic ring idealization of cover region

Figure 1. Cohesive-elastic crack model (Tepfers 1979, Reinhardt & Van der Veen 1990).

system is used where the radial and tangential directions are indicated by subscripts r and t, respectively. The hoop (tangential) stress at $r = e$ is equal to the cracking stress f_t, and is assumed to vary elastically in the uncracked region, $r > e$. The model assumes that neighboring longitudinal bars are far enough and that their bond stress fields do not overlap. In the cracked region, $r \leq e$, the tangential stress decreases towards the center with the widening of the crack width w until it vanishes at a crack width w_c, following a power law as follows,

$$\sigma_t(w)/f_t = 1 - (w/w_c)^{k_0} \tag{1}$$

where k_0 = material parameter determined from the tensile fracture energy G_F. The hoop strain at $r = e$, neglecting Poisson's effect, is obtained from

$$\varepsilon_r(e) = f_t/E_c \equiv \varepsilon_{cr} \tag{2}$$

where E_c = concrete modulus of elasticity. This value of the radial strain is assumed constant over the cracked part $r \leq e$. Thus,

$$2\pi e\varepsilon_r(e) = 2\pi r\varepsilon_r(r) + nw(r), \quad \varepsilon_r(r) \approx \varepsilon_r(e) = \varepsilon_{cr} \tag{3a, b}$$

This yields a linear distribution of crack width, i.e.

$$w(r) = 2\pi\varepsilon_{cr}(e-r)/n \text{ for } r \leq e \tag{4}$$

which gives an explicit formula for the hoop stress in the cracked region, namely,

$$\sigma_t(r) = \sigma(w(r)) \text{ for } r_m \leq r \leq e \tag{5}$$

where $r_m = \max\{d_s/2, e - w_c/2\pi\epsilon_{cr}\}$ defines the end of the cohesive zone where softening occurs.

Given the elastic solution (Timoshenko & Goodier 1951) of stresses at radius r, in a thick-walled cylinder of r_i and r_o inner and outer radii, respectively, subjected to internal pressure p_i, i.e.

$$\sigma_r(r) = p_i \frac{r_i^2}{r_o^2 - r_i^2}\left(1 - \frac{r_o^2}{r^2}\right)$$

$$\sigma_t(r) = p_i \frac{r_i^2}{r_o^2 - r_i^2}\left(1 + \frac{r_o^2}{r^2}\right) \quad \text{for } r \geq e \tag{6a, b}$$

The pressure p at the bar surface contributed by both regions, elastic and cracked, can then be calculated by superposition of the equilibrium solution for the elastic region and the integral of $\sigma_t(r)$ over $r \leq e$.

1.2.1 Elastic contribution, p_e
From Equation 6b we can deduce the pressure acting at the bar surface given the pressure applied at the inner wall of the elastic ring. Hence,

$$p_e = \frac{e}{d_s/2}p_i = \frac{2e}{d_s}f_t\frac{(c+d_s/2)^2 - e^2}{(c+d_s/2)^2 + e^2} \tag{7}$$

where $\sigma_t(e) = f_t$ at the onset of cracking, d_s = bar diameter $r = r_i = e$, and $r_o = c + d_s/2$ in Equation 6.

1.2.2 Cohesive contribution, p_c
At any radius $r \leq e$, for the rigid body mechanism assuming Equation 3 holds, using Equations 1, 4, and 5 and finally integrating the hoop stresses over the cracked region leads to

$$p_c = \frac{f_t}{d_s/2}\int_e^{r_m}\frac{\sigma_t(w(r))}{f_t}dr = \frac{2f_t}{d_s}\int_e^{r_m}\left[1 - \left(\frac{w(r)}{w_c}\right)^{k_0}\right]dr$$

$$= \frac{2f_t}{d_s}(e - r_m)\left\{1 - \frac{1}{k_0+1}\left[\frac{2\pi\varepsilon_{cr}}{nw_c}(e - r_m)\right]^{k_0}\right\} \tag{8}$$

Superposing Equations 7 and 8 leads to the total pressure, namely

$$p = p_e + p_c \tag{9}$$

Subsequently, from equilibrium, the shear stress is

$$\tau = p\tan\alpha \tag{10}$$

which represents the bond strength at crack length e. Eligehausen et al. (1983) calculated the average value of the angle of internal friction α, using numerical results of stress distribution around a bar with lateral ribs for a range of cover thickness and bar diameter values. The results were independent of the concrete quality, with values varying between 0.49 and 1.00.

Presently, α is left out by restricting the computation to the radial pressure p due to the ribs on the bar and the pull-out action of the longitudinal stress. Thus, the pressure capacity is given by

$$p_r = \max_e\{p\} \tag{11}$$

where p_r = the maximum pressure value the ring can sustain as the crack propagates radially, after which the crack becomes unstable and runs all the way to the surface causing a longitudinal splitting crack.

This value is dependent on the assumed number of radial cracks n whose increase results in a decrease in the crack opening w per crack at any given crack length e, which leads to an increase in the average tangential stress within the cracked region and a corresponding increase in the cohesive pressure term p_c. The secondary effect of discretely-spaced transverse reinforcement is not explicitly modeled in this model yet can be included in the parameter α.

2 MODEL ENHANCEMENTS

A close investigation of Equation 9 reveals that the resulting pressure is directly proportional to the tensile strength f_t, fracture energy G_F, cover thickness c, number of cracks n, and limiting crack width w_c, while it is inversely proportional to the reinforcing bar diameter d_s. However, this model involves the following assumptions and simplifications:

1. The concrete material is assumed to behave uniaxially. This is reflected in neglecting the effect of radial dilation in calculating ϵ_{cr} in Equation 2, as well as assuming that cracking in tension takes place upon violating a uniaxial stress criterion in the hoop direction which neglects the effect of the radial stress. This radial stress is compressive for the given deformation mode, which should reduce the uniaxial tensile strength and lead to unconservative estimates using model A_0.
2. The crack opening is assumed to vary linearly with the radius in Equation 4. This requires that the tangential strain along the crack length is equal to the cracking strain, which is only valid at $r = e$ (Equation 3). Softening in the region $r < e$ will result in the bulk material between cracks elastically unloading to lower strains and, subsequently, a non-linear crack width distribution along the radius r. This is neglected, with the argument being made that a compensating effect is expected from neglecting radial dilation.
3. The model assumes that the exact shape of the tension softening relationship does not significantly affect the resulting strength, as long as the fracture energy enclosed by the softening curve remains the same. Thus, the model adopts a simple power

softening law (Equation 1), whose coefficient k_0 is determined by equating to G_F the integral of σdw over the range $0 < w < w_c$. Furthermore, while f_t is either directly measured or estimated from the compressive strength, G_F is difficult to measure and is commonly estimated as a function of f_t, and w_c is typically estimated as a multiple of the average aggregate size (Ratanalert & Wecharatana 1989), leading to compounded uncertainty in the model estimation.

The following subsections will present the mathematical formulations developed to address each of the individual assumptions identified above.

2.1 Model A_1 – Biaxial behavior of concrete

The introduction of a concrete biaxial failure criterion explicitly accounts for the effect of radial dilation in the cracked region. The modified cracking stress f_t' is adopted from Gambarova et al. (1994),

$$f_t' = f_t\left(1 + 0.8\frac{\sigma_r(e)}{f_c}\right) \tag{12}$$

where f_c = compressive strength and σ_r = radial (transverse) stress (tension positive). Dividing Equation 6a by Equation 6b and setting $\sigma_t(e) = f_t'$,

$$\sigma_r(e) = -f_t\left[\frac{(c + d_s/2)^2 + e^2}{(c + d_s/2)^2 - e^2} + \frac{0.8f_t}{f_c}\right]^{-1} \tag{13}$$

Substituting Equation 13 in Equation 6 and solving for p_e as in Equation 7 leads to the first term in Equation 14. The second term is analogous to that in Equation 9 after replacing f_t by f_t'. Accordingly, the total pressure sustained in the ring surrounding the bar is given by

$$p = \frac{2f_t}{d_s}(e)\left[\frac{(c + d_s/2)^2 + e^2}{(c + d_s/2)^2 - e^2} + \frac{0.8f_t}{f_c}\right]^{-1}$$
$$+ \frac{2f_t'}{d_s}(e - r_m)\left\{1 - \frac{1}{k_0 + 1}\left[\frac{w_m}{w_c}\right]^{k_0}\right\} \tag{14}$$

where the crack width w_m = the smaller of w_c and the crack width at the steel-concrete interface, and can be expressed as

$$w_m = (2\pi\varepsilon_{cr}(e - r_m))/n \tag{15}$$

Equation 14 explicitly incorporates the effect of the biaxial stress state in the plane of the bar cross-section on the tensile cracking in concrete. The model becomes further complicated upon considering 3-D stress state effects. However, it is argued that splitting failure is investigated in between two existing flexural cracks and that concrete stresses in the longitudinal direction are therefore insignificant. It is also assumed that stress fields surrounding neighboring longitudinal bars are far enough relative to the cover thickness. The effect of radial dilation is accounted for by updating ϵ_{cr} in Equation 16 for each trial crack length e according to

$$\varepsilon_{cr}(e) = \left(f_t - v\sigma_r(e)\right)/E_c \tag{16}$$

where $v =$ Poisson's ratio for concrete.

2.2 Model A_2 – Nonlinear crack width distribution

Accounting for the nonlinear crack width distribution requires the solution of an iterative nonlinear problem. The governing compatibility equation is

$$2\pi r \varepsilon_t(r) + n w(r) = 2\pi r \varepsilon_t(e) \approx 2\pi e \varepsilon_{cr} \tag{17}$$

Thus, along the crack face, according to Equation 1 the tangential stress distribution follows Equation 18; while in the bulk concrete between two cracks, assuming uniform linear-elastic unloading, the tangential stress distribution follows Equation 19. Finally, invoking equilibrium requires equality between Equations 18 and 19.

$$\sigma_t(r) = f_t\left(1 - \left(w(r)/w_c\right)^{k_0}\right) \tag{18}$$

$$\sigma_t(r) = E_c \varepsilon_t(r) \tag{19}$$

Substituting Equation 19 in Equation 17 leads to

$$n w(r) = 2\pi\left(e\varepsilon_{cr} - r\sigma_t(r)/E_c\right) \tag{20}$$

Substituting Equation 20 in Equation 18 yields

$$\sigma_t(r) = f_t\left\{1 - \left(\frac{2\pi}{n w_c}\left[e\varepsilon_{cr} - r\sigma_t(r)/E_c\right]\right)^{k_0}\right\} \tag{21}$$

Equation 21 is nonlinear of the form $x = G(x)$ for $x = \sigma_t$. Convergence requires that $|dG(x)/dx| < 1$. This is indeed satisfied, as illustrated by Equation 22

$$\frac{dG(\sigma_t(r))}{d\sigma_t(r)} = \left(\frac{2\pi}{n w_c}\right)^{k_0}\frac{k_0 r f_t}{E_c}\left[e\varepsilon_{cr} - r\sigma_t(r)/E_c\right]^{(k_0-1)} \tag{22}$$

Numerical investigation of typical parameter ranges suggest that the solution will converge, albeit slowly. After evaluating the hoop stress at sufficient points along the radius, the pressure in the ring is calculated by numerical integration of Equation 8.

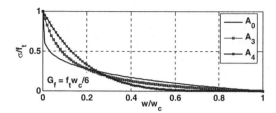

Figure 2. Tension-softening profiles for equal fracture energy.

2.3 Models A_3 and A_4 – Alternate softening laws

Several tensile stress–crack width softening laws have been proposed in the literature. A recent review is available in van Mier (1997). An all-inclusive survey would be prohibitive and beyond the scope of this study. Instead, two alternative laws are considered and shown to have a significant effect on the results. Model A_3 uses the tensile softening law given in Gambarova et al. (1994) by the formula

$$\sigma(w) = f_t\frac{1 - w/w_c}{1 + k_3 w/d_a} \tag{23}$$

where $d_a =$ average aggregate size and $k_3 =$ model parameter. The elastic contribution to the total pressure p_e in Equation 9 is not affected by Equation 23. However, using Equation 23 for calculating the derivative dr/dw and noting that w decreases as r increases, the cohesive contribution p_c becomes

$$p_c = \frac{f_t}{d_s/2}\int_e^{r_m}\frac{\sigma_t(w(r))}{f_t}dr = \frac{2f_t}{d_s}\int_e^{r_m}\left[\frac{1 - w(r)/w_c}{1 + k_3 w(r)/d_a}\right]dr$$
$$= \frac{n d_a f_t}{\pi k_3 w_c \varepsilon_{cr} d_s}\left\{-w_m + \left(1 + \frac{k_3 w_c}{d_a}\right)\left(\frac{d_a}{k_3}\right)\ln\left[1 + \frac{k_3 w_m}{d_a}\right]\right\} \tag{24}$$

Model A_4 uses an alternate form of the power law in Equation 1, according to

$$\sigma_t(w)/f_t = \left(1 - w/w_c\right)^{k_4} \tag{25}$$

where $k_4 =$ model parameter. This leads to the cohesive pressure term in Equation 9 becoming

$$p_c = \frac{2f_t}{d_s}\int_e^{r_m}\left(1 - \frac{w(r)}{w_c}\right)^{k_4}dr = \frac{n w_c f_t}{\pi(k_4+1)\varepsilon_{cr}d_s}\left[1 - \frac{2\pi\varepsilon_{cr}}{n w_c}(e - r_m)\right]^{(k_4+1)} \tag{26}$$

For the purpose of comparing models A_0, A_3 and A_4 and isolating the shape effect of the tensile softening curve, parameters k_3 and k_4 are selected so that the tensile fracture energy G_F under the softening curve in Equations 23 and 25 is maintained equal to that under the softening curve in Equation 1 for a corresponding value of k_0. Figure 2 presents graphic comparison of the alternate softening laws.

3 BENCHMARK PROBLEM

The relative significance of the enhancements described in Section 2 is evaluated using a hypothetical benchmark problem of typical parameter values. This benchmark problem considers the bar-pullout resistance of a single reinforcing steel bar embedded in a concrete cylinder and evaluates the maximum radial pressure developed before bond-splitting failure occurs. The radial pressure is computed and maximized along the crack length for a variable number of radial cracks. The estimated response using model A_0 is compared to estimates using models A_1 through A_4 to investigate the sensitivity to the individual simplifying assumptions. The computational time needed by each model is also compared to assess its practicality. In addition, the sensitivity of the model estimate to the assumed number of cracks is investigated and combined with experimental observations to estimate a value for the parameter n in the absence of experimental data pertinent to the problem of interest. In the presence of such data, a probabilistic approach is outlined to account for randomness in n.

3.1 Problem statement

The benchmark problem geometry is similar to Figure 1. Since the governing equations are highly nonlinear, the use of normalized quantities for force and geometric parameters becomes a matter of form and convenience, because the results remain specific to the neighborhood of the set of geometric and material properties considered. As such, the significance of the different model enhancements, discussed above, is investigated on a problem that represents a commonly encountered set of parameters. The problem data is as given below:

$f_c = 30.0 \, \text{MPa}$, $w_c = 0.2 \, \text{mm}$, $f_t = 3.0 \, \text{MPa}$, $c = 30.0 \, \text{mm}$, $E_c = 22.0 \, \text{GPa}$, $d_s = 10.0 \, \text{mm}$, $G_F = 0.1 \, \text{MPa.mm}$, $d_a = 16.0 \, \text{mm}$, and $v = 0.2$. This data yield the model parameters $k_0 = 0.2$ in Equation 1, $k_3 = 773.0$ in Equation 23, and $k_4 = 5.0$ in Equation 25.

3.2 Comparison results

Figure 3 shows the computational solutions calculated for the benchmark problem using models A_0, A_1, A_3 and A_4. Results for model A_2 are not noticeably different from those of model A_0 and are not shown. The plots demonstrate the variation of the normalized radial pressure capacity p/f_t, versus the normalized radial crack length e/c, for a range of assumed radial cracks n. The number of cracks range from 0 to ∞. The case $n = 0$ corresponds to an assumption of $p_c = 0$ and no cohesion in the cracked concrete, i.e. brittle tensile cracking. The case $n = \infty$ corresponds to a case of an infinitely-rigid confining medium outside the cover

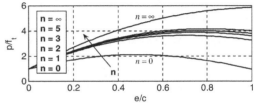

(a) Crack length-radial pressure capacity for model A_0

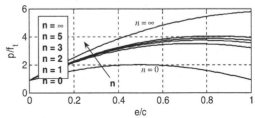

(b) Crack length-radial pressure capacity for model A_1

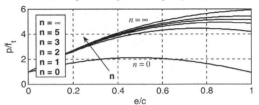

(c) Crack length-radial pressure capacity for model A_3

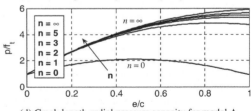

(d) Crack length-radial pressure capacity for model A_4

Figure 3.   Partial results for benchmark problem.

region. Together, both cases define the theoretical bounds on the solution.

It can be observed that the radial pressure capacity is positively correlated to the number of radial cracks, and that it increases as a result of increasing crack length up to a maximum value and then decreases as the crack becomes unstable and splitting failure occurs. It can also be observed that the normalized radial pressure values computed using model A_0 are generally higher than those of model A_1 and lower than those of model A_2.

Given Equation 11, the radial pressure capacity pertinent to the bond-splitting failure is the maximum response in Figure 3 at each n value and can be determined numerically. These values have been computed for $n = 1$, 2, and 3 for models A_0 through A_4 and tabulated in Table 1. The computational time required

Table 1. Comparison between radial pressure capacity of reference and alternative models.

n	Model A_0	Model A_1	Model A_2	Model A_3	Model A_4
1	1.000	0.969	1.060	1.220	1.339
2	1.000	0.972	1.053	1.273	1.372
3	1.000	0.973	1.049	1.293	1.376
Time	1.000	1.333	192.300	0.667	1.000

for completing the solution is also indicated for each model. For ease of comparison, all values in the table have been normalized by the corresponding value (on the same row) for Model A_0.

3.3 Discussion

This section discusses the behavior of the computed response using model A_0 and its dependency on the assumed number of radial cracks, and compares its predictions with those of models A_1 through A_4 to establish the effect of their underlying assumptions. In discussing the model sensitivity to the number of radial cracks in section 3.3.1, the comparison is limited to models A_0, A_3, and A_4. This is because the predicted response of models A_1 and A_2 is not significantly different from that of model A_0.

3.3.1 Effect of number of radial cracks
It is clear in Figure 3 that the influence of n affects primarily the peak pressure capacity, while its effect on the computed response at lower e/c values is less significant. The computed response experiences a major increase upon considering the cohesion contribution (from $n = 0$ to 1), and that the sensitivity decreases quickly as n increases further. A conservative estimate of $n = 1$ is typically assumed in design situations for choosing appropriate cover thickness and bar anchorage lengths. However, experimental observations by Reinhardt et al. (1986) report a consistent observation of three cracks or more. It can also be argued that the assumption of an axi-symmetric stress state, upon which the model formulation is based, is grossly violated by the assumption of one radial crack.

Figure 3 shows that the sensitivity of the model prediction to the uncertainty in the number of radial cracks for values of $n > 1$ is highest in the reference model A_0. This sensitivity decreases significantly in model A_3 and is minimal for model A_4 (11% increase in p_r from $n = 2$ to $n = \infty$, versus 20% and 51% increases for A_3 and A_0, respectively). In the range of commonly-observed values of n between 2 and 5, the sensitivity in prediction for the three models is comparable and is equal to 7%, 11% and 6% for A_0, A_3, and A_4, respectively.

It must be noted that the number of radial cracks is a function of the geometry and boundary conditions surrounding the embedded longitudinal bar, which may differ from the idealized benchmark problem. Therefore, one concludes that an assumption of $n = 2$ or 3 satisfies the model assumptions, results in a stable estimate, and conservatively agrees with experimental observations in cases where experimental data pertinent to the application of interest is not available. In the presence of application-specific experimental data, a probabilistic approach can be followed that mimics the randomness of the process; whereby n can be assigned a discrete probability distribution (e.g. Poisson) whose parameters can be estimated from the data. By simulating n a sufficient number of times from this assigned distribution and substituting in the bond-splitting model, the distribution of the resulting bond strength can be generated and its mean and dispersion estimated.

3.3.2 Effect of biaxial behavior of concrete
It can be observed in Table 1 and Figure 3b that incorporating the biaxial behavior of concrete has no significant effects on the estimated response. The observed effect results in a decrease of the estimated maximum radial pressure capacity and an increase of the e/c value where the maximum capacity is obtained. On average, radial pressure capacity values computed using model A_1 for the benchmark problem are consistently less than those computed using model A_0 by approximately 3%. This results in unconservative estimates but is compensated for by the assumption of linear crack width distribution.

3.3.3 Effect of nonlinear crack width distribution
It can be observed in Table 1 that accounting for nonlinear crack width distribution in model A_2 results in an increase in the estimated radial pressure capacity by approximately 5% and requires approximately 200 times more CPU time for the benchmark problem. The increase in accuracy is deemed infeasible and unjustified for modeling such a local phenomenon in the context of an FE model, especially because the increased accuracy does not render the original model unconservative. Thus, the initial simplifying assumption of neglecting this effect is considered adequate and justified.

3.3.4 Effect of softening law
It can be observed in Table 1 and Figure 3c and d that, counter to commonly assumed, the shape of the softening curve does result in a significant effect both on the radial pressure capacity and the corresponding e/c value. For model A_3, the difference in estimated maximum pressure capacities is approximately 28%, while for model A_4, the difference is approximately 37%; for n ranging between 2 and 5 cracks. This is evidence

that the designation of the fracture energy as a sole parameter – in addition to its being usually empirically assumed rather than directly measured and thus highly uncertain – instead of a more accurate representation of the actual post-cracking behavior is a major source of uncertainty for this model.

4 DETERMINISTIC SENSITIVTY ANALYSIS

It has been established in the previous section that the assumption of alternative material softening laws and the use of cohesive-elastic bond splitting models formulated accordingly leads to a significant variation in the predicted response across these models, namely A_1, A_3, and A_4. Therefore, the adoption of tensile fracture energy as a sole model parameter to characterize the softening behavior is a major source of uncertainty in the model. This uncertainty is further compounded by the uncertainty in the tensile fracture energy value, which is often estimated from indirect measurements. This section individually examines the robustness of the three models A_0, A_3, and A_4 by investigating the sensitivity of the response predicted within each model to the model input parameters. The comparative study is performed using a deterministic sensitivity analysis approach, commonly referred to in the literature as the Tornado diagram analysis method. A summary of this method is described next. An extensive review can be found in (Lee & Mosalam 2005) and a similar application of this approach to the present study can be found in (Binici & Mosalam 2007).

The Tornado diagram analysis is a deterministic method developed to numerically determine the sensitivity of an output quantity of interest to uncertainty in input parameter values, and thus establish the relative importance of the input parameters with respect to their random nature. In this method, a reference point is initially set by computing the predicted model response using the expected mean values of the input parameters. Next, the parameters are individually varied within a given range of uncertainty, typically parameterized by their coefficient of variation (COV). The resulting changes (swings) in the predicted response quantity is computed, compared and sorted across the different parameters. A graphic comparison of the computed sensitivity measures (swings) is used to establish the relative importance of the parameters. A larger sensitivity measure is an indication of a higher relative importance of the associated parameter and, consequently, a larger role for the uncertainty associated with said parameter in determining the outcome and accuracy of the model. In interpreting the results for the comparative study, the following criteria are considered to indicate a higher degree of model robustness: (a) Relatively lower sensitivity to uncertainty in

input parameters, (b) significant sensitivity to only a smaller number of parameters, and (c) especially low sensitivity to parameters in which a higher degree of uncertainty is anticipated.

4.1 Problem statement

The benchmark problem defined in section 3 is used in the present Tornado diagram analysis. The predicted maximum pressure p_r is defined as the output quantity of interest. The list of input parameters being considered and their mean values is composed of the quantities defined at the end of subsection 3.1. The deterministic sensitivity of the model response to parameter uncertainty is calculated using the relative change (swing) in the predicted maximum pressure corresponding to one standard deviation step on either side of the mean value for each input parameter, with an assumed COV of 10%. Since the model sensitivity to parameter uncertainty is generally nonlinear, a tight step size (in terms of COV) is recommended in order to better represent the sensitivity in the neighborhood of the mean response.

4.2 Results

Figure 4 shows the Tornado diagram results for models A_0, A_3, and A_4, where the parameter k represents softening law parameters k_0, k_3, and k_4, respectively. The shown results correspond to the case of $n = 3$. The most important model parameter is the bar diameter, and it has the same sensitivity measure for all models. This is followed by the cover thickness then tensile strength (order reversed for A_1) at approximately equal importance. The fourth parameter on the list is tensile fracture energy, where there is a significant difference in the sensitivity measure across models. Here starts a significant decrease in importance for the remaining parameter in models A_3 and A_4, but not in model A_0. The relative importance of the subsequent variables continues to decrease. The minimum sensitivity measure value is zero for the average aggregate size in models A_0 and A_4, where it is not included in the softening law formulation. Regarding uncertainty in the power law parameters k, model A_0 is most sensitive with parameter k_0 having an importance rank of 4 out of 8. Parameter k_3 has a rank of 2 in model A_3, while k_4 has a rank of 3 in model A_4. However the value of the sensitivity measure indicates that model A_3 is slightly more sensitive to parameter k_3 than model A_4 is to parameter k_4. It is worth noting that the sensitivity of models towards the assumed 10% COV is commeasurable with their sensitivity to the number of radial cracks over their range of expected values (subsection 3.3.1). Thus, for parameters whose expected COV is typically lower (e.g. bar diameter), or typically higher (e.g. fracture energy), the relative importance of the number of cracks can be deduced.

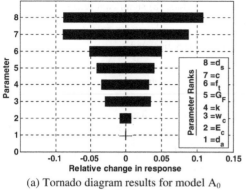

(a) Tornado diagram results for model A_0

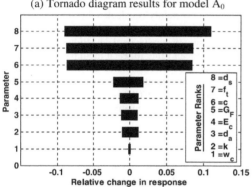

(b) Tornado diagram results for model A_1

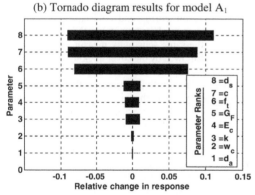

(c) Tornado diagram results for model A_3

Figure 4. Results of Tornado diagram for $n = 3$ radial cracks.

4.3 Discussion

Referring to the criteria of model robustness outlined earlier in section 4, all models display approximately the same sensitivity towards the three highest-ranking parameters. These three parameters are typically easier to estimate or measure directly with limited uncertainty. However, models A_3 and A_4 do not display significant sensitivity towards the remaining parameters, whereas model A_0 does. Moreover, model A_0

is significantly sensitive to the fracture energy, whose estimation typically involves high uncertainty, and to the power law parameters k_θ and w_c, which are highly uncertain owing to perceived randomness in tension-softening response. Therefore, the proposed models A_3 and A_4 are more robust than the reference model A_0. This is further reinforced by the relative sensitivity of the three models to the number of radial cracks previously observed in subsection 3.3.1. More-over, since model A_4 displays less sensitivity towards uncertainty in the remaining parameters than model A_3, is not affected by the uncertainty in estimating the average aggregate size, and shows decreasing sensitivity towards fracture energy and the lower-ranking parameters with increased number of radial cracks (not shown), it is considered the most robust model.

5 CONCLUDING REMARKS

From the previous discussions, the following conclud-ing remarks can be inferred:

1. The cohesive-elastic model for bond failure between concrete and longitudinal steel bars by splitting has been reviewed, and its assumptions and simplifications have been identified for investiga-tion. Four alternate models have been formulated to explicitly address each assumption.
2. A benchmark problem has been introduced to assess the relative significance of improving the individual modeling assumptions.
3. It has been demonstrated that ignoring the biax-ial behavior of concrete in tension results in an insignificant overestimation of bond strength. This effect is counter-balanced by the equally insignifi-cant effect of assuming linear crack width distribu-tion along the crack length.
4. It has been demonstrated that the shape of the tension-softening material law is a significant fac-tor of uncertainty in the analytical model, and that it is important to select a softening law that reliably represents the considered application.
5. The model sensitivity to the number of radial cracks has been investigated. Recommendations for selecting a valid estimate in both deterministic and probabilistic contexts were presented.
6. The relative robustness against parameter uncer-tainty in the reference model as well as two mod-els formulated using alternative tension-softening material laws was assessed using a Tornado dia-gram analysis for the benchmark problem.
7. The reference model was found to be relatively most sensitive to uncertainty in parameters which typi-cally have a highly random nature and thus is the least robust. The two alternate models were found to be significantly less sensitive to highly-random

parameters and should therefore result in a more reliable prediction of bond strength.

ACKNOWLEDGMENT

This study is supported by the Earthquake Engineering Research Centers Program of the NSF under Award No. EEC-9701568 to PEER at UC Berkeley. Opinions and findings presented are those of the authors and do not reflect views of the sponsors.

REFERENCES

Bazant, Z. & Oh, B. 1983. Crack band theory for fracture of concrete. *Materials and Structures*. 16(93): 155–177.

Binici, B. & Mosalam, K. 2007. Analysis of reinforced concrete columns retrofitted with fiber reinforced polymer lamina. *Composites B: Engineering*. 38(2): 265–276.

Eligehausen, R., Popov, E. & Bertero, V. 1983. Local bond stress-slip relationships of deformed bars under generalized excitations.*UCB/EERC-83/23*. Earthquake Eng. Research Center, Univ. of California, Berkeley, USA.

Gambarova, P., Rosati, G. & Schumm, C. 1994. An elasto-cohesive model for steel-concrete bond. In Bazant, Bitnar, Jirasek, & Mazars (eds) *Fracture and Damage in Quasibrittle Structures: Experiment, Modeling, and Computer Analysis*. Chapman and Hall: 557–566.

Hillerborg, A. Modeer, M. & Petersson, P. 1976. Analysis of crack formation and crack growth in concrete by means of fracture mechanics and finite elements. *Cement and Concrete Research*. 6: 773–782.

Lee, T.-H. & Mosalam, K. 2005. Seismic demand sensitivity of reinforced concrete shear-wall building using FOSM method. *Earthquake Engineering and Structural Dynamics*. 34(14): 1719–1736.

Ratanalert, S. & Wecharatana, M. 1989. Evaluation of existing fracture models in concrete. In Li & Bazant (eds) *Facture Mechanics: Application to Concrete*. ACI SP-118: 113–146.

Reinhardt, H. & Van der Vein, C. 1990. Splitting failure of a strain-softening material due to bond stresses. In Carpentieri (ed.), *Application of Fracture Mechanics to Reinforced Concrete*. Elsevier Applied Science: 333–346.

Reinhardt, H., Corneilssen, H. & Hordijk, D. 1986. Tensile tests and failure analysis of concrete. *Journal of Structural Engineering*. ASCE. 112(11): 2462–2477.

Tepfers, R. 1979. Cracking of reinforced concrete cover along anchored deformed reinforcing bars. *Magazine of Concrete Research*. 31(106): 3–12.

Timoshenko, S. & Goodier, J. (2nd Ed.) 1951. *Theory of Elasticity*. McGraw-Hill.

Van Mier, J. 1997. *Fracture Processes of Concrete*. CRC Press.

Fracture Mechanics of Concrete and Concrete Structures – Design, Assessment and Retrofitting of RC Structures – Carpinteri, et al. (eds)
© 2007 Taylor & Francis Group, London, ISBN 978-0-415-44616-7

Influence of cover thickness on the flexural response of fibrous RC beams

G. Campione & M.L. Mangiavillano
Dipartimento di Ingegneria Strutturale e Geotecnica, Università di Palermo, Italy

ABSTRACT: An analytical model is proposed in the present paper aimed at describing the response of simply-supported beams loaded in four-point bending. Several cases concerning normal- and high-strength concrete beams, reinforced with longitudinal bars, stirrups and steel hooked fibers are considered, and their load-deflection curves, as well as their moment-curvature curves and the ultimate deflections at the onset of failure in bending or shear, are obtained by means of the proposed, simplified approach. The satisfactory fitting of the test results confirms the potential of the proposed model, that can accurately predict beam behaviour in bending and shear, including the spalling of the cover and the contribution of the fibers.

1 INTRODUCTION

It is well known that the use of short fibers dispersed in fresh concrete in an adequate percentage and geometry ensures, in the hardened state, effective flexural and shear reinforcement in reinforced-concrete beams and deep elements.

For fibrous reinforced concrete (FRC) also the ultimate shear strength increases with increasing flexural reinforcement ratio and with increasing compressive strength, while not increases with the shear span-to-depth ratio.

From the mechanical point of view several studies presented in the literature refer to the calculus of the shear strength of fibrous reinforced-concrete beams made of normal- or high-strength concretes, normal or lightweight.

The analytical expressions given are often of an empirical nature and they account for the increase in the bearing capacity of the beams, owing to the bridging capacity of fibers across the principal cracks.

It is moreover generally assumed that fibers do not significantly influence the mechanisms observed at failure for beams without fibers (e.g. beam and arch actions) and consequently the expressions given separately consider the strength contributions due to the several parameters governing shear failure.

Referring to the determination of the load-deflection curves, several analytical models based on plane section hypothesis are available for beams affected by flexural failure and implemented in well known computer code (e.g. DRAIN-2DX, 1993).

Analytical simplified flexural models are also available for fibrous concrete beams with and without steel reinforcements, but very few simple models are able to determine the load-deflection curves when shear failure occurs.

Several models for the accurate prediction of load-deflection curves of beams affected by shear or flexural failure are available based on non linear finite element analyses such as the model proposed by Cervenka (2000) utilizing ATENA code or that proposed by Vecchio (2000) the latter (based on the Modified Compression Field Theory) very successfully utilized in many applications.

Very effective computational truss models are also proposed (e.g. Noghabai, 2000).

Most of these models are very effective but are also quite complex.

They require knowledge of the complete constitutive laws of materials including the most relevant parameters governing phenomena, as concrete size effect and strain softening.

On the basis of these considerations, the focus here is on the development of a simplified model able to determine the load-deflection curves of simply-supported, longitudinally- and transversely- reinforced fibrous beams under flexure and shear.

The model proposed is specific for fibrous reinforced concrete beams with steel fibers (most commonly utilised for structural applications) and is based on knowledge of very few mechanical properties of plain concrete, steel bars and the geometrical properties and volume content of steel fibers.

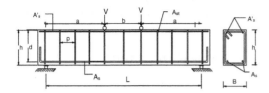

Figure 1. Static scheme of the beam.

2 THE CASE STUDY

The case study considers a prismatic reinforced concrete beam with rectangular cross-section, base B and depth h, which is cast with plain or fibrous concrete. The static scheme adopted consists of simply-supported beams under four-point bending tests, in which the beams are subjected to symmetrical vertical loads, V, acting at distance a from the support, as shown in Figure 1.

The beam is reinforced on the lower side with longitudinal deformed bars having area A_s with a cover thickness c and on the top side with steel bars having area A_s' with a cover thickness c'. The compressed area of longitudinal steel is assumed as being negligible with respect to A_s. Both bars had yielding stress f_y.

In addition, transverse steel stirrups having an area of one leg A_{st} are placed in the beam at pitch p.

The main longitudinal steel bars (bottom bars) are bent at the support with an adequate anchorage length so as to avoid any slippage or premature splitting failure. The beam is reinforced at the bottom with geometrical ratio $\rho = A_s/(B \cdot d)$ with d effective depth. The stirrups had yielding stress f_{yw} and geometrical ratio ρ_{sw} the latter defined as $\rho_{sw} = 2 \cdot A_{st}/(B \cdot p)$.

In the case of fibrous concrete, the fibers utilized have length L_f, equivalent diameter ϕ and volume percentage v_f. In order to take the characteristics of the fibers into account, in the following sections the fiber factor F will be introduced, defined as $F = v_f L_f \beta/\phi$, with β being the shape factor assumed as 1 and 0.5 for deformed and straight fibers respectively.

This assumption for β is in agreement with the results obtained in Banthia and Trottier (1994) from which it appears that the pull-out resistance of straight fibers is less than for hooked or crimped steel fibers (having a similar aspect of ratio) while the values of β are those suggested in Campione et al. (2006).

3 SIMPLIFIED MODEL

In the following sections the evaluation of the bearing capacity of beams and their load-defection curves will be considered in the case of both shear and flexural failure. In the case of flexural failure, under the hypothesis of perfect bond of steel bars and concrete,

the plane section theory will be considered, including also the strength contribution due to the residual tensile strength of fibrous concrete.

No tension stiffening will be considered. Referring to shear failure (cases of shear compression and diagonal tension modes are considered without premature splitting failure) an analytical expression, based on the analysis of beam and arch actions at rupture, will be utilized.

3.1 Flexural strength of fibrous reinforced concrete beams

Using translation and rotational equilibrium conditions it is possible to determine the position of the neutral axis x_c, and the flexural moment at cracking, yielding, compressed cover spalling and failure of compressed zone defined as M_c, M_y M_s and M_u and the corresponding curvatures ϕ_c, ϕ_y, ϕ_s and ϕ_u. The above-mentioned stages are analyzed here referring to the strain and stress distribution shown in Figure 2 (I, II, III, IV) and related to the shear force V by the equilibrium condition V = M/a.

By adopting the above-mentioned model the simplified moment-curvature diagram shown in Figure 3, (stages I, II, III, IV are also indicated), is obtained. A similar approach was also utilised by Rashid and Mansur (2005) for high-strength concrete beams in flexure.

To develop the simplified model the following stress-strain curve, proposed by La Mendola and Papia (2002), under monotonic loading is assumed:

$$\frac{\sigma}{f_{cf}} = \frac{A \cdot \dfrac{\varepsilon}{\varepsilon_{0f}} + (D-1) \cdot \left(\dfrac{\varepsilon}{\varepsilon_{0f}}\right)^2}{1 + (A-2) \cdot \dfrac{\varepsilon}{\varepsilon_{0f}} + D \cdot \left(\dfrac{\varepsilon}{\varepsilon_{0f}}\right)^2} \tag{1}$$

in which $A = E_c/E_0$, with E_0 the secant modulus at peak stress defined as $E_0 = f_{cf}/\varepsilon_{0f}$, being f_{cf} and ε_{0f} the maximum compressive strength and corresponding strain respectively, and E_c the initial modulus of elasticity in compression assumed as in Razvi and Saatciouglu (1999) to be variable with the compressive strength.

The D parameter governs the slope of the descending branch and was the chosen variable with F, according to the expression:

$$D = 0.3136 + 0.175 \cdot F \tag{2}$$

The equation (2) was calibrated on the basis of experimental data available in the literature for fibrous concrete compressed members with hooked steel fibers (data from Fanella and Naaman 1983).

I)

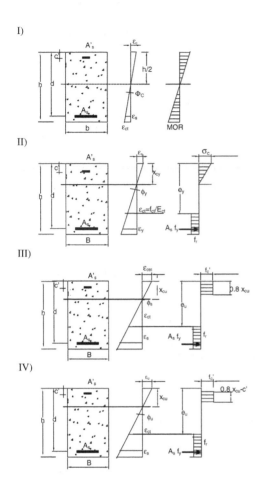

II)

III)

IV)

Figure 2. Design assumptions in the analysis of R/C sections.

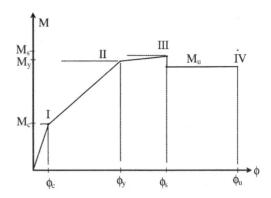

Figure 3. Simplified model for moment-curvature diagram.

The maximum compressive strength f_{cf} and corresponding strain ε_{0f} of normal strength FRC with steel fibers, proposed by Nataraja et al.(1999), can be assumed as:

$$f_{cf} = f_c' + 6.913 \cdot F \qquad (3)$$

$$\varepsilon_{0f} = \varepsilon_0 + 0.00192 \cdot F \qquad (4)$$

being ε_0 the strain at peak stress assumed as in Razvi and Saatciouglu (1999) to be variable with the compressive strength.

It is possible to obtain the value of strain ε_{085} in a closed form by using Eq. (1), assuming that the σ value is equal to 0.85 f_{cf} for $\varepsilon = \varepsilon_{085}$ resulting in the following second order equation:

$$\varepsilon_{0.85}^2 \cdot (1 - 0.15 \cdot D) - \varepsilon_{0.85} \cdot (1.7 + 0.15 \cdot A) \cdot \varepsilon_{of} + \qquad (5)$$
$$+ 0.85 \cdot \varepsilon_{of}^2 = 0$$

This value will be assumed in the following sections for the calculus of M_s and ϕ_s.

Referring to Case I of Figure 2 for the calculus of the cracking moment M_c (if the contribution due to the longitudinal bars is neglected) it is possible to adopt the following relationship:

$$\frac{M_c}{B \cdot h^2} = \frac{MOR}{6} \qquad (6)$$

MOR being the modulus of rupture in flexure of FRC well defined in ACI 544 (1988).

The corresponding curvature is:

$$\phi_c = \frac{2 \cdot MOR}{h \cdot E_c} \qquad (7)$$

in which E_c can be evaluated with the expression $E_c = 4200 \cdot (f_c')^{0.5}$ (MPa) proposed by ACI 318-02 (2002).

Referring to the calculus of the yielding moment M_y reference is made to Case II (Fig. 2) in which the yielding strain $(\varepsilon_y = f_y/E_s)$ is attained in the longitudinal bars in tension and the residual strength f_r of fibrous concrete in tension is reached (this post-cracking tensile strength of FRC is better defined in the following sections). For concrete in compression linear elastic behavior is supposed to be due to the lower deformation reached. To obtain the neutral axis position x_{cy} the stress distribution of Figure 2 (stage II), is assumed. From the translational equilibrium we obtain the following equation:

$$\frac{1}{2} \cdot \frac{E_{cr} \cdot f_y}{E_s} \cdot \frac{x_{cy}}{d - x_c} \cdot B \cdot x_{cy} - f_y \cdot A_s + \qquad (8)$$
$$- f_r \cdot B \cdot (d - e_y) = 0$$

e_y being the distance between the more compressed fiber of the transverse cross-section and the fiber with the maximum tensile strength in concrete, while E_{cr} is a reduced value of initial modulus of elasticity assumed $0.5\ E_c$ as in Rashid and Mansur (2005) to include an average amount of cracking and tension stiffening effects. Its values can be obtained by considering the plane section hypothesis resulting in:

$$e_y = \frac{\dfrac{f_t}{E_{ct}}\cdot(d-x_{cy})+\varepsilon_y\cdot x_{cy}}{\varepsilon_y} \tag{9}$$

E_{ct} being the elastic modulus of concrete in tension assuming half of that in compression.

The values of x_{cy} and e_y are obtained by introducing Eq. (9) in Eq. (8) and producing the second degree equation for position of the neutral axis:

$$\left[\frac{1}{2}\cdot\frac{E_c\cdot f_y}{E_s}\cdot B+f_r\cdot B\cdot\left(\frac{\varepsilon_{ct}-\varepsilon_y}{\varepsilon_y}\right)\right]\cdot x_{cy}^2+ \tag{10}$$

$$\left(f_r\cdot B\cdot c+f_y\cdot A_s\right)\cdot x_{cy}+\left(f_r\cdot B\cdot d^2\cdot\frac{\varepsilon_{ct}-\varepsilon_y}{\varepsilon_y}-f_y\cdot d\cdot A_s\right)=0$$

The internal arm value can be expressed by means of:

$$z=d-\frac{x_c}{3} \tag{11}$$

From the rotational equilibrium it is possible to obtain the yielding moment M_y in the following form:

$$\frac{M_y}{B\cdot d^2}=\rho\cdot f_y\cdot\left(1-\frac{0.33\cdot x_{cy}}{d}\right)+$$
$$+f_r\cdot\left(\frac{h}{d}-\frac{e_y}{d}\right)\cdot\left(\frac{2}{3}\cdot\frac{x_{cy}}{d}-\frac{h-e_y}{2\cdot d}\right) \tag{12}$$

The curvature corresponding to step M_y (first yielding or Stage II in Fig. 2) results in:

$$\phi_y=\frac{f_y}{E_s}\cdot\frac{1}{d-x_{cy}} \tag{13}$$

When referring to the calculus of M_s, corresponding to the cover spalling process (Stage III in Figure 2) and of M_u, at the ultimate state (Stage IV in Figure 2), it was assumed that longitudinal bars are yielded and maximum compressive strength of concrete is reached.

With reference to the symbols in Figure 2 (stage III) and considering the translational equilibrium we obtain:

$$x_{cu}=\frac{d}{0.8}\cdot\frac{\rho\cdot f_y+f_r\cdot\dfrac{h}{d}}{f_c'+f_r\cdot\dfrac{f_t/E_{ct}+\varepsilon_{085}}{0.80\cdot\varepsilon_{085}}} \tag{14}$$

The moment corresponding to the cover spalling process from the rotational equilibrium results in:

$$\frac{M_s}{B\cdot d^2}=\rho\cdot f_y\cdot\left(1-\frac{0.4\cdot x_{cu}}{d}\right)+$$
$$+f_r\cdot\left(\frac{h}{d}-\frac{e_u}{d}\right)\cdot\left(\frac{h}{d}-\frac{h-e_u}{2\cdot d}-\frac{0.4\cdot x_{cu}}{d}\right) \tag{15}$$

e_u being the distance between the more compressed fiber and the fiber at which the maximum tensile strength in concrete is reached.

$$e_u=x_{cu}\cdot\frac{\dfrac{f_t}{E_{ct}}+\varepsilon_{085}}{\varepsilon_{085}} \tag{16}$$

Therefore the curvature at spalling state is:

$$\phi_s=\frac{\varepsilon_{0.85}}{x_{cu}} \tag{17}$$

From the rotational equilibrium the ultimate moment results in:

$$\frac{M_u}{B\cdot d^2}=\rho\cdot f_y\cdot\left(1-\frac{0.4\cdot x_{cu}}{d}-\frac{c'}{2\cdot d}\right)+$$
$$+f_r\cdot\left(\frac{h}{d}-\frac{e_u}{d}\right)\cdot\left(\frac{h}{d}-\frac{h-e_u}{2\cdot d}-\frac{0.4\cdot x_{cu}}{d}+\frac{c'}{2\cdot d}\right) \tag{18}$$

Therefore the curvature at ultimate state is:

$$\phi_u=\frac{\varepsilon_{su}}{x_{cu}} \tag{19}$$

ε_{su} being the maximum strain of bar in tension assumed 0.01.

Moreover, the arm of the internal forces is $j_0 d$ in which j_0 can be expressed by the equilibrium across the compressive centroid in the following form:

$$j_0=\frac{\rho\cdot f_y\cdot\left(1-0.4\cdot\dfrac{x_{cu}}{d}\right)+f_r\cdot\left(\dfrac{h}{d}-\dfrac{e_u}{d}\right)\cdot\left(\dfrac{h+e_u}{2\cdot d}-0.4\cdot\dfrac{x_{cu}}{d}\right)}{\rho\cdot f_y+f_r\cdot\left(\dfrac{h}{d}-\dfrac{e_u}{d}\right)} \tag{20}$$

The previous equations highlight the influence of the main parameters governing the flexural response of the transverse cross-section including the effect of cover thickness also. It is interesting to observe that the cover thickness of the compressed bars plays a role on the ductility of the cross-section also. But, because low values of cover thickness are generally utilized, no significant changes in the overall flexural behavior of the cross-section occurs, while its contribution can be significant in the presence of higher cover thickness that can be required for durability reasons.

In the previous equations, f_t and f_r indicate the maximum and the residual strength of fibrous concrete. For the maximum tensile strength, the same value found in the literature plain concrete was assumed, while for the residual strength the proposal by Marti et al. (1999) was adopted:

$$f_r = 0.375 \cdot F \cdot \left(f_c'\right)^{0.66} \qquad \text{(in MPa)} \qquad (21)$$

To validate the proposed model a comparison with data given in Swamy and Al-Ta'an (1981) is shown. The experimental research refers to a third point bending test on fibrous reinforced concrete beams of 2250 mm length between the two lateral supports. The beams had a rectangular cross-section with B = 130 mm, h = 203 mm, c = 18 mm, the area in tension constituted by two deformed bars of 12 mm diameter, and the area in compression constituted by two 10 mm deformed bars. Plain concrete of $f_c = 30$ MPa and steel bars with yielding stress $f_y = 460$ MPa. The beams were also reinforced with transverse stirrups of 6 mm diameter and pitch p = 125 mm. Hooked steel fibers were utilised with 50 mm length and 0.5 mm diameter at volume percentage of 0, 0.5 and 1%.

Figure 4 shows the experimental and the analytical moment-curvature diagrams for cases of 0 and 0.5% of fibers.

The comparison shows the ability of the simplified model to predict the experimental response including yielding of main bars and crushing of concrete. The model is also able to include the spalling of compressed cover occurring at high curvature.

3.2 Shear strength of fibrous reinforced concrete beams

Bazant and Kim (1984) propose a mechanical model to calculate the flexural capacity in shear of reinforced concrete beams considering the sum of the strength contributions due to the beam and arch actions. These contributions are identified, as shown in Fig. 5, by imposing conditions of equilibrium of the beam enclosed between the support and the loaded section (shear span a). With reference to the symbols shown in Fig. 5, the bending moment M and the shear force V at the generic cross-section can be related to the axial force T in the longitudinal bar and to the internal arm jd by means of:

$$M = V \cdot x = T \cdot jd \qquad (22)$$

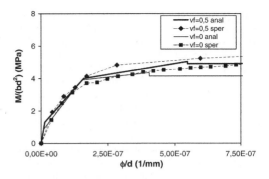

Figure 4. Experimental and analytical moment-curvature curves of beams with $v_f = 0$ and 0.5% (data from Swamy and Al Ta'an, 1981).

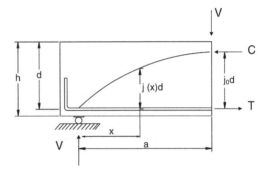

Figure 5. Beam and arch actions.

Moreover, between the shear force V and the bending moment M there is the following, well-known relationship:

$$V = \frac{dM}{dx} = jd \cdot \frac{dT}{dx} + T \cdot \frac{d(jd)}{dx} \qquad (23)$$

obtaining by means of Eq. (22) in Eq. (23) the two fundamental strength contributions well known in the literature as beam effect (jd constant) and arch effect (jd variable).

According to this model, well justified in Bazant and Kim (1984) and here utilized also for fibrous concrete beams, the following simplified hypotheses are assumed: – beam and arch effects are considered separately; – in the evaluation of arch effect it is assumed that the tension force in longitudinal bar remains constant across the shear span;- in the evaluation of beam effect the tension force is supposed to be variable across the shear span. In this was the expression given by Campione et al. (2006) was obtained including in the beam and arch effect the effect of fibers and also considering the contribution made by the fibers through the main cracks. For the beam effect it was assumed that the bond stress q_b is proportional to the tensile strength of the concrete, which in turn is proportional to the square root of the cylindrical compressive strength f'_c. This choice is valid for plain concrete and assumed in ACI 318-02 (2002) and, as suggested in the original work of Bazant and Kim (1984), was also supported experimentally by Harajili et al. (1995) for fibrous concrete. Moreover the expression given by Eq.(20) for the arm of internal forces was assumed also taking into account the presence of fibers by means of the residual tensile strength of the composite. The strength contribution V_2 to the arch action was evaluated, with reference to the mechanism shown in Figure 5, by relating the shear force to the variation in j, T assumed to be constant. In the case of fibrous concrete Campione et al. (2006) to estimate the tensile force in main bars was also included the strength contribution due to the residual strength of composite resulting in a fictitious geometrical ratio of main steel defined as:

$$\rho_t = \rho + \rho_f = \frac{f_r}{\eta \cdot f_y} \cdot \frac{h}{d}\left(1 - \frac{e}{h}\right) \tag{24}$$

in which $\eta = \sigma_s/f_y$, with η (assumed 0.3) a share of the yielding stress assumed in accordance with experimental data available. Finally by considering the sum of the strength contributions due to the beam and arch actions and due to the fibers across the principal cracks we obtain:

$$v_u = \xi \left\{ j_0 \cdot \left[1.3 \cdot \rho^{1/2} \cdot \sqrt{f'_c} + 0.3 \cdot f_y \cdot \varepsilon \cdot \rho_t \cdot \left(\frac{d}{a}\right)^{1.8} \right] + 0.27 \cdot F \cdot \sqrt{f'_c} \right\} \tag{25}$$

where to take the size effect into account the ξ coefficient given by Bazant and Kim (1984) was adopted:

$$\xi = \frac{1}{\sqrt{1 + \frac{d}{25 \cdot d_a}}} \tag{26}$$

where d_a is the maximum aggregate size of the concrete. Eq. (25) was calibrated on the basis of experimental data described in the literature as in Campione et al. (2006).

Figure 6 shows the comparison between the analytical expression here proposed (which has a similar structure to the ACI equation for shear strength) and the expression proposed by CNR-DT 204 (2006) in which the equivalent residual strength is evaluated by Eq. (21). The comparison in terms of shear strength versus a/d ratio referring to beams with and without fibers shows good agreement between the two

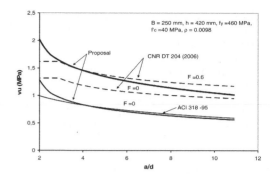

Figure 6. Comparison between analytical expressions.

proposed equations although the origin of the two equations differs.

Several studies show that it is possible to obtain the shear strength of reinforced concrete beams with stirrups by adding the contributions due to beam and arch actions with the strength contribution due to the stirrups bridging the principal crack and not considering the interaction between the single mechanisms.

Moreover, the inclination of the principal cracks is assumed to be 45°. Russo and Puleri (1997) have shown that the stirrups do not always yield at beam rupture and the effective stress can be estimated if the contributions of beam action to the sum of beam and arch action are known. Therefore, they introduce an effectiveness coefficient able to determine the share of yielding stress in the stirrups at beam rupture.

It is possible to include the effect of fibers in the effectiveness function of stirrups originally proposed by Russo and Puleri (1997) and determine the shear strength contribution due to stirrups, expressed as:

$$v_{st} = \Phi_f \cdot \rho_{sw} \cdot f_{yw} \tag{27}$$

where the yielding stress in the stirrups is reduced by Φ_f defined as in Campione et al. (2006). This function which has to be cut at maximum value of one reflects the influence of the beam action (including the effect of fibers also) on the whole strength contribution of the beam and is expressed:

$$\Phi_f = \frac{2.17 \cdot j_0 \cdot \rho^{1/2} \cdot \sqrt{f'_c}}{j_0 \cdot \left[1.3 \cdot \rho^{1/2} \cdot \sqrt{f'_c} + 0.3 \cdot \varepsilon \cdot f_y \cdot \rho_t \cdot \left(\frac{d}{a}\right)^{1.8}\right] + 0.27 \cdot F \cdot \sqrt{f'_c}} \tag{28}$$

Finally, using Eq. (27) and Eq. (28) we obtain an expression of the shear strength in the presence of stirrups:

$$v_u = \xi \cdot \left\{ j_0 \cdot \left[1.3 \cdot \rho^{1/2} \cdot \sqrt{f'_c} + 0.3 \cdot \varepsilon \cdot f_y \cdot \rho_t \cdot \left(\frac{d}{a}\right)^{1.8}\right] + 0.27 \cdot F \cdot \sqrt{f'_c} \right\} + \Phi_f \cdot \rho_{sw} \cdot f_{yw} \tag{29}$$

Campione et al. (2006) also support the proposed expression with available experimental data.

4 LOAD-DEFLECTION CURVES UNDER FLEXURE AND SHEAR

It is possible to determine the load-deflection curves of the simply supported beam knowing the moment curvature-diagrams of the loaded sections, as determined in the previous section, by using Mohr's analogy which considers a fictitious beam simply supported and loaded by the curvature diagrams as shown to provide the simplified load-deflection curves shown in Figure 7.

It emerges from the curve that if overstrength in shear is attained and flexural failure occurs, the response will be that of the simply supported beam in flexure which can be approximated by a four linear diagram (first cracking, yielding, cover spalling and ultimate load). However, if the failure is in shear the diagram can be characterized by a bilinear behaviour constituted by two linear branches up to the maximum load corresponding to shear failure and by a second characterized by residual strength essentially governed by the stirrups and fibers. Consequently the load-deflection curve is that shown in Figure 7 where it must be remembered that instead of V ($V = v_u$ bd) the whole load $P = 2V$ appears for the different values of V. According to Mohr's analogy the deflection of the beam is the moment in the middle section of the fictitious beam. In particular these deflections are expressed in the cases of cracking, yielding, cover spalling and ultimate moments as:

At first cracking:

$$\delta_c = \phi_c \cdot \left[\frac{a^2}{3} + \frac{b^2}{8} + \frac{ab}{2} \right] \qquad (30)$$

At first yielding:

$$\delta_y = \phi_c \cdot \frac{a}{6} \cdot (a + x_f) +$$
$$\phi_y \cdot \left(\frac{a^2}{3} - \frac{a \cdot x_f}{6} + \frac{a \cdot b}{2} + \frac{b^2}{8} - \frac{x_f^2}{6} \right) \qquad (31)$$

with $x_f = (M_c/M_y) \cdot a$.

At the moment corresponding to the cover spalling:

$$\delta_s = \frac{\phi_c}{2} \cdot \left[a \cdot (x_f - x_y) + x_y \cdot \frac{x_y - y_f}{3} \right] +$$
$$+ \frac{\phi_y}{6} \cdot \left[a \cdot (a + x_y) - x_f \cdot (x_y + x_f) \right] +$$
$$+ \phi_s \cdot \left(\frac{a^2}{3} - \frac{a \cdot x_y}{6} + \frac{a \cdot b}{2} + \frac{b^2}{8} - \frac{x_y^2}{6} \right) \qquad (32)$$

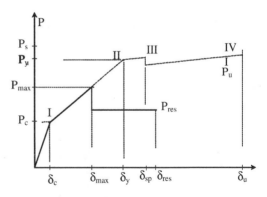

Figure 7. Simplified model for load-deflection curves.

with $x_y = (M_y/M_s) \cdot a$. At the ultimate moment the deflection is expressed by the Eq. (32), with $x_y = (M_y/M_u) \cdot a$.

Should shear failure occur before the yielding of the longitudinal reinforcement, beam capacity would be limited to $M_{max} = (v_u \cdot b \cdot d) \cdot a$.

In this case the corresponding deflection δ_{max} can be obtained as shown in Figure 7 (if we consider for simplicity that up to this stage the response is essentially flexural) as the intersection between the constant value line P_{max} ($2 \cdot v_u \cdot b \cdot d$) and the slope line connecting the first cracking and the yielding stage ultimately state resulting in:

$$\delta_{max} = \delta_c + (v_u \cdot B \cdot d \cdot a - P_c) \cdot \frac{\delta_y - \delta_c}{M_y - M_c} \qquad (33)$$

$$P_c \leq P_{max} \leq P_y$$

Should shear failure occur past the attainment of the yielding moment, the maximum deflection would be as follows:

$$\delta_{max} = \delta_y + (v_u \cdot B \cdot d \cdot a - P_y) \cdot \frac{\delta_u - \delta_y}{M_u - M_y} \qquad (34)$$

Moreover, if it is supposed (as also observed experimentally in Campione et al. 2003) that after shear failure occurs the main contribution to the residual strength is due to the effects of stirrups and fibres (beams and arch effects are not included), it is possible to obtain the residual strength $P_{res} = 2 v_{res} \cdot b \cdot d$, v_{res} being expressed by:

$$v_{ures} = \xi \cdot f_r \cdot \left(\frac{h}{d} - \frac{e_u}{d} \right) + \rho_{sw} \cdot f_y \quad (\text{in MPa}) \qquad (35)$$

To validate the proposed model a comparison with data given in Swamy and Al-Ta'an (1981) mentioned in the previous section and also data from Mansur and

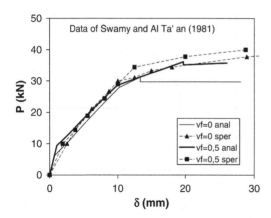

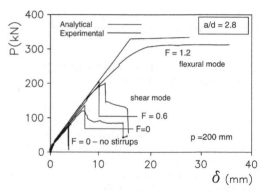

Figure 8. Experimental and analytical load – deflection curves of beams with $v_f = 0$ and 0.5% (data from Swamy and Al Ta'an, 1981).

Figure 10. Experimental and analytical load – deflection curves of beams with data from Campione et al. (2003).

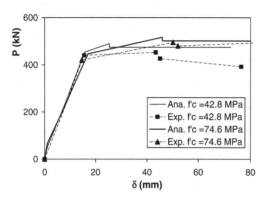

Figure 9. Experimental and analytical load – deflection curves of beams (data from Rashid and Mansur 2005).

Rashid (2005) is shown. The experimental research of Rashid and Mansur (2005) refer to a four point bending test on reinforced concrete beams of 3400 mm length between the two lateral supports. Beams had rectangular cross-section with B = 250 mm, h = 400 mm, c = 20 mm, the area in tension constituted by four deformed bars of 25 mm diameter, and area in compression constituted by two 13 mm deformed bars. Plain concrete were $f'_c = 42.8$ MPa and steel bars were of yielding stress $f_y = 460$ MPa. Beams were also reinforced with transverse stirrups of 10 mm diameter and pitch p = 200 mm. Both researches consider the flexural failure. Figures 8 and 9 show the experimental and analytical load-deflection curves for cases of 0 and 0.5% of fibers.

The comparison shows the ability of the simplified model to predict the experimental response including yielding of main bars and crushing of concrete.

The model is also able to include the spalling of compressed cover occurring at high curvature levels when fibers are added or high strength concrete is utilized.

The results from Campione et al. (2003) refer to four-point bending tests on medium size beams of 2500 mm length and a/d = 2.8 in the presence of longitudinal bars, stirrups and fibers. The stirrup diameter was 6 mm, pitch 200 mm and with yielding stress of 510 MPa. In the case of fibers F = 0, 0.6 and 1.2 were assumed (Fig. 10). In this case shear and flexural failures are obtained depending on fiber amount.

In all cases examined the model shows a good ability to describe the overall flexural behavior of the beams including flexural and shear modes of failure and cover spalling processes.

5 CONCLUSIONS

In this paper an analytical model is proposed and discussed, its primary objective being the description of the load-deflection curve in simply-supported R/C beams, containing stirrups and steel fiber, and subjected to bending and shear.

More specifically, the model: (a) accurately predicts the bearing capacity in bending and shear; (b) takes care of the spalling process ensuing from the compressive stresses acting in the cover of any given section; and (c) makes it possible to evaluate the residual capacity in shear, after shear failure, taking into account stirrups and fiber contributions.

REFERENCES

ACI Committe 318, 2002. Building code requirements for reinforced concrete. *American Concrete Institute*, Detroit, Michigan.
ACI Committe 544, 1988. Design considerations for steel fiber reinforced concrete. *ACI Structural Journal* 85(5): 563–580.

Banthia, N. & Trottier, J.F. 1994. Concrete reinforced with deformed steel fibers, part I: bond-slip mechanism. *ACI Material Journal* 91(5): 435–446.

Bazant, Z.P. & Kim, J.K. 1984. Size effect in shear failure of longitudinally reinforced beams. *ACI Structural Journal* 81(5): 456–468.

Campione, G., Cucchiara, C. & La Mendola, L. 2003. Role of fibres and stirrups on the experimental behaviour of reinforced concrete beams and flexure and shear. *Int. Conf. on Composites in Construction, September 2003*, Rende, Italy.

Campione, G., La Mendola, L. & Papia, M. 2006. Shear strength of fiber reinforced beams with stirrups. *Structural Engineering and Mechanics* 24(1): 107–136.

Cervenka, V. 2000. Simulating a response. *Concrete Engineering International* 4(4): 45–49.

CNR-DT 204/2006 Instructions for design, execution and control of fibrous reinforced concrete structures. Consiglio Nazionale Ricerche ROMA. (only available in Italian).

Fanella, D.A. & Naaman, A.E. 1983. Stress-strain properties of fiber reinforced mortar in compression. *ACI Journal,* Proceedings no. 82-41:475–483.

Harajli, M., Hout, M., & Jalkh, W. 1995. Local bond stress-slip behavior of reinforcing bars embedded in plain and fiber concrete. *ACI Materials Journal* (4) : 343–353.

La Mendola, L. & Papia, M. 2002 General stress-strain model for concrete or masonry response under uniaxial cyclic compression. *Structural Engineering and Mechanics* 14(4): 435–454.

Marti, P., Pfyl T., Sigrist, V. & Ulaga T. 1999. Harmonized test procedures for steel fiber-reinforced concrete. *ACI Materials Journal* 96 (6): 676–685.

Nataraja, M.C., Dhang, G N. & Gupta A.P. 1999. Stress-strain curves for steel fiber reinforced concrete under compression. *Cement & Concrete Composite* 21: 383–390.

Noghabai, K. 2000. Beams of fibrous concrete in shear and bending: experiment and model. *ASCE J. of Structural Engineering* 125(2): 243–251.

Prakash, V., Powell, G.H, & Campbell, S. 1993. DRAIN-2DX Inelastic dynamic response of plane structures. Berkley U.S.

Rashid, M.A. & Mansur, M.A. 2005. Reinforced high-strength concrete beams in flexure. *ACI Structural Journal* 102(3): 462–471.

Razvi S, and Saatcioglu M. 1999. "Confinement model for high-strength concrete". Journal of Structural Engineering ASCE, Vol. 125, N. 3, pp. 281–288.

Russo, G. & Puleri, G. 1997. Stirrup effectiveness in reinforced concrete beams under flexure and shear. *ACI Structural Journal* 94(3): 227–238.

Swamy, R.N. & Al-Ta'an, S.A. 1981. Deformation and ultimate strength in flexure of reinforced concrete beams made with steel fibers. *ACI Structural Journal* 78(3): 395–405.

Vecchio, F.J. 2000. Analysis of shear-critical reinforced concrete beams. *ACI Structural Journal* 97(1): 102–110.

Fracture Mechanics of Concrete and Concrete Structures – Design, Assessment and Retrofitting of RC Structures – Carpinteri, et al. (eds)
© 2007 Taylor & Francis Group, London, ISBN 978-0-415-44616-7

Fracture behavior of the concrete cover due to rebar corrosion

Y. Kitsutaka
Tokyo Metropolitan University, Tokyo, Japan

K. Arai
Taisei Construction Co., Tokyo, Japan

ABSTRACT: Evaluation of the performance recovery of reinforced concrete (RC) members by repair is an important issue in recent years. In this paper, first, fracture behavior of concrete RC cover with cracks due to the rebar corrosion was analyzed by using fracture mechanics approach and the diagrams to evaluating the degree of rebar corrosion by a concrete surface crack width and rebar diameter were proposed. Next, effects of the repairing material properties of polymer cement mortar (PCM) on fracture properties of RC cover were evaluated by a bending test of center notched small cubic RC specimen. Strength recovery by repairing is satisfactory regardless of the rebar corrosion. Toughness is also significantly improved by the repairing. The degree of improvement to toughness becomes greater as the volumetric content of fibers of PCM.

1 INTRODUCTION

Extending the service lives of buildings has become increasingly important from the standpoint of global environmental protection and economic efficiency. Maintenance of reinforced concrete buildings, which are in stock in large quantities, is particularly of social significance. In order to improve the durability of reinforced concrete structures, it is essential to select a repair method suitable for the identified deterioration phenomena of existing structures. Deterioration phenomena critical to reinforced concrete structures are interpreted as reductions in the reinforcement strength causing reductions in the structural capacity of members due to reinforcing steel corrosion, which is aggravated by cracking in cover concrete under expansive and tensile stresses resulting from reinforcement corrosion and/or aggregate reaction. In other words, it is vital for the durability of concrete that the concrete be sound and resistant to expansive and tensile forces causing cracking and that the cover including reinforcement develop sufficient strength performances. To this end, it is important to be able to evaluate the strength of cover concrete for reinforcement when deteriorative actions have caused corrosion on reinforcement and reduced its adhesion. It is also crucial to be able to evaluate the strength restoration of such cover concrete after repair and estimate the degree of reinforcement deterioration. Nevertheless, few studies have dealt with these subjects.

In this study, a method of estimating the amount of reinforcement corrosion from the crack width on the surfaces is proposed based on fracture mechanics approach. Also the authors investigated methods of evaluating the fracture properties of cover concrete including reinforcement and experimentally examined the effects of the degree of reinforcement corrosion and the repair methods on such fracture properties.

2 ESTIMATION OF THE DEGREE OF CORROSION FROM SURFACE CRACK WIDTH

2.1 *Analysis method*

The changes in the surface crack width due to reinforcement corrosion were estimated by nonlinear analysis using techniques used in fracture mechanics. The CMOD under mode I (tensile) deformation due to reinforcement corrosion was modeled by applying stresses perpendicular to the crack surfaces (mode I). Two models representing before and after the crack reached the surface were formulated as shown in Figure 1 (a) and (b).

Crack propagation is modeled by means of the fictitious crack concept with cohesive forces. The relationship of expansion load and displacement is obtained by solving the equilibrium equations for stress intensity factors, the crack opening displacement (COD) at

a) before surface cracking b) after surface cracking

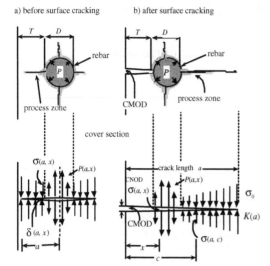

Figure 1. Models of cracking due to rebar corrosion.

the crack surface, and constitutive function of cohesive stress and COD. These equations are as follows:

$$K(a) = K_p(a) + K_r(a) \qquad (1)$$

$$\delta(a,x) = \delta_p(a,x) + \delta_r(a,x) \qquad (2)$$

where, a is the crack length, x is the point on crack surface in which the COD is calculated, $K(a)$ is the total stress intensity factor on crack tip, $K_p(a)$ is the stress intensity factor due to rebar expansion load, $K_r(a)$ is the stress intensity factor due to cohesive stress, $\delta(a,x)$ is the COD on x, $\delta_p(a,x)$ is the COD on x due to rebar expansion load, $\delta_r(a,x)$ is the COD on x due to cohesive stress and $\sigma(a,x)$ is the cohesive stress on x.

Expansion stress due to rebar corrosion $P(a, x)$ is modeled by assuming the pressure on the rebar surface is uniform as $P_0(a)$ and acting only the crack opening direction.

$$P(a,x) = P_0(a) \cdot I(x);$$
$$I(x) = \sqrt{1 - \left(1 - \frac{2(x-T)}{D}\right)^2}, \quad (T \le x \le T+D) \qquad (3)$$

where, T is the concrete cover, D is the rebar diameter, $I(x)$ is the function of stress component.

The relationship between $\delta(a,x)$ and $\sigma(a,x)$ shown in Figure 1 is the tension softening diagram considered a material property. Applying the bi-linear

approximation for a tension softening diagram, we obtain

$$\sigma(a,x) = m(\delta,x) \cdot \delta + n(\delta,x);$$
$$\delta = \delta(a,x), \qquad (4)$$
$$m(\delta,x) = \begin{cases} m_1 = (f_t - \sigma_1)/\delta_1, & (0 < \delta \le \delta_1) \\ m_2 = \sigma_1/(w_c - \delta_1), & (\delta_1 < \delta \le w_c) \end{cases}$$

where, $m(\delta, x)$ and $n(\delta, x)$ are coefficients of a linear simple equation of multilinear softening curve, δ is COD, σ is the cohesive stress, and w_c is the critical value of COD.

The stress intensity factor in the Figure 1. (a) and (b) cases by mode I stresses on crack surface are given by (Tada 1985)

$$K_1 = \frac{2f}{\sqrt{2d}} G(a,c,d), \quad K_1 = \frac{2f}{\sqrt{\pi a}} G(a,c,d) \qquad (5)$$

where, f is node force, d is the width of the body, c is the co-ordinate indicating the point on crack surface where cohesive stress is acting, and $G(a, c, d)$ is the weight functions (APPENDIX).

So stress intensity factors in Equation (2) are normalized as

$$K_p(a) = P(a) \int_0^a I(c) g_1(a,c) dc \qquad (6)$$

$$K_r(a) = \int_0^a \sigma(a,c) g_1(a,c) dc \qquad (7)$$

Castigliano's theorem can be applied for the calculation of COD for uniform materials. Displacement of the cracked body can be expressed as

$$d\delta = \delta_0 + \frac{2}{E} \int_x^a K(z) \left[\frac{\partial K_F(z)}{\partial F} \right]_{F=0} dz \qquad (8)$$

where, dy is the displacement, d_0 is the displacement of the uncracked body, z is the co-ordinate indicating the crack length for the integration, f is the fictitious force acting on the point z, $E^*(z)$ is the generalized equivalent elastic modulus of the material on z (i.e., E for plane stress, and $E/(1-v2)$ for plane strain, where E and n is Young's modulus and Poisson's ratio respectively), and $K_F(z)$ is the stress intensity factor due to fictitious force f. From (7), crack opening displacements, $\delta_p(a,x)$ and $\delta_r(a,x)$ can be obtained as

$$\delta_p(a,x) = \frac{P(a)}{E} \int_0^a I(c) g_2(a,x,c) dc \qquad (9)$$

$$\delta_r(a,x) = \frac{1}{E} \int_0^a \sigma(a,c) g_2(a,x,c) dc \qquad (10)$$

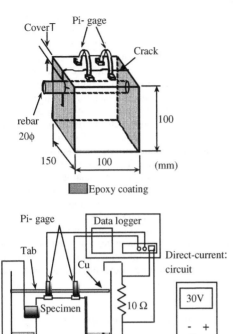

Figure 2. Outline of corrosion accelerated test.

Substituting (6) and (7) into (1), also (9) and (10) into (2), and with cancellation of $P(a)$, the fundamental simple crack integral equation is obtained (Kitsutaka 1997) as

$$\delta(a,x) = \frac{1}{E} \int_0^a \sigma(a,c) H^e(a,x,c) dc;$$

$$H^e(a,x,c) = g_2(a,x,c) - g_1(a,c) \frac{\int_0^a I(c) g_2(a,x,c) dc}{\int_0^a I(c) g_1(a,c) dc} \quad (11)$$

where, $H^e(a,x,c)$ is the weight function (called as H-function) of the specimen and does not depend on the external forces and the cohesive stresses.

Substituting $\sigma(a,x)$ of (4) into the crack integral equation of (11) and expressing in the matrix form for the total number of nodes (=n) on the crack surface, we get the simultaneous crack equations. COD of each nodes are calculated linearly by these equations. This problem can be solved by several iterations (Kitsutaka 1997). From the solution of COD, $\sigma(a,x)$ is calculated from (4). Then, $P(a,x)$ are calculated from (3).

The advantage of this analysis method is that the singular solution of COD distribution can be calculated directly by using the simultaneous crack equations with boundary conditions of H-function. Because of

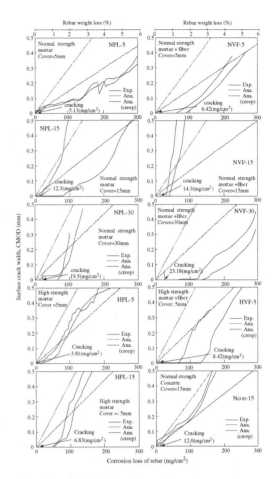

Figure 3. Results of corrosion accelerated test.

this singularity, we do not need to assume the crack surface profile, and the relationship between COD and the expansion stress with different types of constitutive law on each point of crack surface can be obtained by only iterating on the inclination of tension softening curves.

2.2 Analysis results

Figure 3 shows the analysis results related to the results of accelerated corrosion tests on specimens other than those described in the previous paper (Kitsutaka 1998). Figure 2 sows the outline of accelerated corrosion test. The tests were conducted on five mortars/concrete: normal strength plain mortar (NPL, Fc = 55 MPa), normal strength fiber-reinforced mortar (NVF, Fc = 58 MPa, Vinylon-Fiber 1%), high strength plain mortar (HPL, Fc = 120 MPa), high strength fiber-reinforced mortar (HVF, Fc = 120 MPa, Vinylon-Fiber 1%) and normal concrete (Ncon,

Fc = 55 Mpa). Two levels of cover depths, 5 and 15 mm, were selected for reinforcing bars 10 mm in diameter. The tension softening curve for each specimen was estimated in a bilinear form. The corrosion loss was calculated from the analysis results of rebar expansion (CMOD of rebar areas in the adhesion model). The elastic modulus and volumetric expansion coefficient of rust were assumed to be 250 MPa and 2.5, respectively, referring to the literature, such as reference (Kitsutaka 1998). Analysis was also conducted considering the creep deformation of concrete near the rebar. In this case, the CMOD of the rebar area for the same expansive pressure was calculated in consideration of the creep coefficient. The creep coefficients were assumed to be 1.0 and 0.5 for series N and H, respectively. Figure 3 reveals that the analysis results incorporating creep generally agree better with the test results, presumably because the plastic deformation near the rebar under expansive pressure can be expressed by the creep coefficient. Also, the analysis results of the high strength series agree better with the test results than those of the normal strength series. Whereas cracking in normal strength concrete is apt to disperse around rebars, that of high strength concrete tends to occur singly in a brittle manner. This may be the reason for the good agreement with the adherence model, which assumes a single crack.

Figure 4 shows the relationship between the crack width and section loss ratio under various conditions based on the present analysis. The compressive strength is in three levels: 20, 50, and 100 MPa, each with a creep coefficient (ϕ) of 1.5, 1.0, and 0.5. The rebar diameter is in three levels: 6, 10, and 20 mm, and the cover thickness is in five levels: 10, 20, 30, 40, and 50 mm. An approximate section loss ratio of reinforcement can be estimated from the surface crack width using this diagram by selecting the concrete strength, rebar diameter, and cover depth. The section loss ratio of reinforcement increases as the crack width increases. When the crack width is the same, the section loss ratio decreases as the cover depth increases and as the bar diameter increases.

3 EVALUATION OF FRACTURE PROPERTIES OF COVER REPAIRED CONCRETE

3.1 Test procedure

Geometry of the specimen is same as the one showed in Figure 2. A single reinforcing bar (SD295A-D10) was placed in each prismatic specimen measuring 100 by 100 by 100 mm with a cover depth of 10 mm, assuming an insufficient cover in a reinforced concrete structure, and subjected to electrical charges to accelerate its corrosion. The protruding ends of the bar were sealed with epoxy to prevent water seepage during electrical corrosion.

The method of accelerating reinforcement corrosion is same as the method showed in Figure 2. Each specimen was immersed in a 3% NaCl solution and subjected to an electrical current from a stabilized power source through the rebar serving as the anode and copper foil serving as the cathode. This caused both anodic reaction, in which ferrous ions were dissolved while generating electrons, and cathodic reaction, in which the generated electrons were consumed for reduction of dissolved oxygen, simultaneously on the reinforcement surfaces, thereby accelerating corrosion. The cumulative amperage was continuously monitored using dataloggers, based on which the section loss ratio of reinforcement due to corrosion was changed. The relationship between the corrosion loss and the cumulative amperage was determined beforehand by conducting preliminary tests. A portable rebar corrosion meter was used to grasp the relationship between the degree of reinforcement corrosion and the half-cell potential during accelerated testing.

3.2 Evaluation of fracture properties of cover concrete

A slab having corroded lowermost bars under normal loading subjected to partial flexural and shear forces was specifically assumed for this study, in order to evaluate the fracture properties of cover concrete for reinforcement. In other words, short-span three-point loading tests were conducted on the prismatic specimens to apply both bending and shear forces, with the axis being the direction of reinforcement as shown in Figure 5. Also, a notch to a depth of 10 mm perpendicular to the bar was made in the tension edge to pinpoint the location of fracture. A closed-loop servo-control hydraulic testing machine manufactured by MTS Systems Corporation was used for the tests, with the load being controlled by the crack mouth opening displacement (CMOD) at the notch. Sensitive clip gauges were used for measuring the CMOD. The load point displacements were also measured using two displacement transducers attached to the sides of each specimen.

3.3 Test conditions

Four concrete mixtures were proportioned: normal strength with and without fibers and high strength with and without fibers. Normal portland cement and crushed sand of hard sandstone from Tama were used as the cement and fine aggregate, respectively. The chemical admixture was an air-entraining and high-range water-reducing admixture of a polycarboxylic ether type. The fibers and reinforcement were polyvinyl alcohol (PVA) fibers 400 μm in diameter and 12 mm in length and deformed bars SD295A-D10,

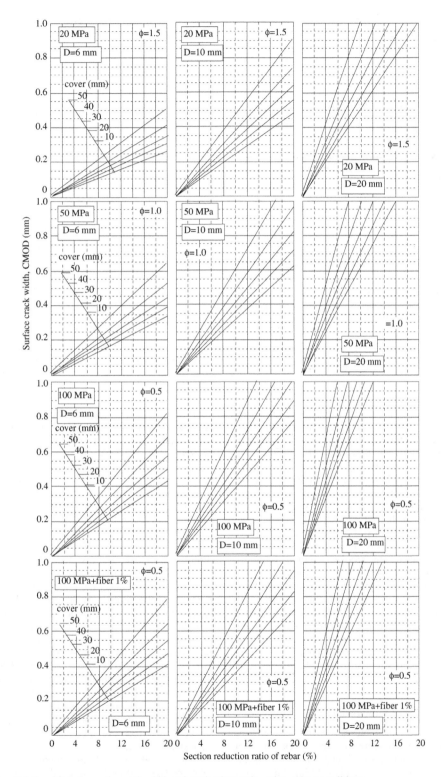

Figure 4. Relationship between the crack width and section loss ratio under various conditions.

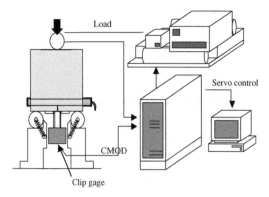

Figure 5. Test method to evaluate the cover concrete strength.

respectively. The factors and levels of experiment were as follows: two levels of water-cement ratios (W/Cs) (30% with the chemical admixture and 60%), two levels of fiber content (0% and 2%), and five levels of section loss ratios of rebars in specimens by electrical corrosion (0%, 5%, 10%, 15%, and 30%). These levels totalled 20.

3.4 Repair methods and conditions

Specimens with section loss ratios of 0%, 5%, 10%, 15%, and 30% were prepared by accelerating corrosion deterioration, to which two types of repair, patching and crack injection, were applied. Repaired specimens were further subjected to accelerated corrosion to final section loss ratios of 0%, 5%, 10%, 15%, and 30%. For patching, a polymer cement patch material made by mixing a premixed powder comprising fibers and cement and a liquid containing polymer was used. After chipping off the deteriorated area with an impact hammer to form a V-shaped groove of a right triangular cross section, rust and concrete on rebar surfaces were removed using a wire brush, and the patch material was filled in the groove. For crack injection, a two-liquid low-viscosity acrylic crack injection compound was used. It is a product designed to retain its reactivity even when applied to cracking where water is present. After diking the area along the cracking above the rebar using a silicone sealant, 6 ml of the injection compound was allowed to seep into the cracking using a dropper. The excess compound on the concrete surface was removed after confirming the hardening of the compound.

3.5 Test results and discussion

From the results of compression and tension tests on concrete standard-cured for four weeks in water at 20°C. The compressive strength increases as the W/C decreases (80 Mpa and 40 MPa) but a little

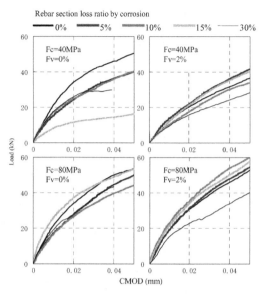

Figure 6. Results of three-point loading tests with different degrees of deterioration before repairing.

bit decreases when fibers are included. Fibers are expected to resist cracking induced by corrosive expansion pressure and cover concrete deterioration.

The section loss ratio of rebars were calculated from the cumulative amperage on the horizontal axis and the half-cell potential on the vertical axis. This reveals that the half-cell potential is in a stable state ($E < -200\,\text{mV}$) with no corrosion (0%) in all mixtures. With a section loss ratio of 5% to 15%, the half-cell potential is plotted in both the uncertain ($-200\,\text{mV} \le E \le 350\,\text{mV}$) and corrosion ($-350\,\text{mV} < E$) regions, whereas nearly all plots fall in the corrosion region when the section loss ratio is 30%.

Figure 6 shows the results of three-point loading tests on specimens with different degrees of deterioration before repair. The CMOD of normal strength specimens with no fibers increases as the section loss ratio increases under the same loads, presumably due to the losses in the adhesion of rebars. No clear changes related to the section loss ratio are observed in high strength specimens with no fibers, presumably because the high strength of concrete prevents the adhesion of rebars from decreasing even under the corrosion-induced expansive pressure. In regard to specimens containing fibers, no marked increase in the displacement is observed with a section loss ratio of up to 15%, though the displacement is significant with a section loss ratio of 30%. Fibers are therefore expected to produce an effect of resisting the corrosion-induced expansive pressure.

Figures 7 and 8 shows the maximum loads and maximum road recovering ratio during three-point loading

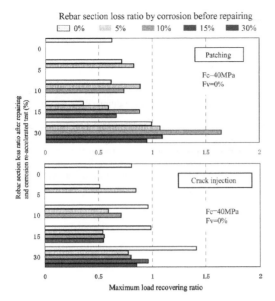

Figure 7. Maximum loads on repaired specimens and specimens subjected to accelerated rebar corrosion.

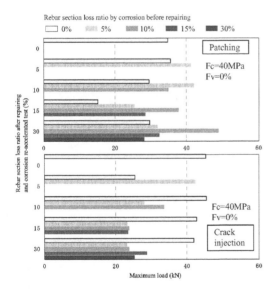

Figure 8. Maximum loads recovering ratio on repaired specimens and specimens subjected to accelerated rebar corrosion.

tests on repaired specimens and specimens subjected to accelerated rebar corrosion to the specified degrees after repair. When repaired by patching, the strength of repaired specimens with no corrosion is lower than unrepaired specimens with no corrosion. The strength of specimens with a higher degree of corrosion after repair is lower. When repaired by crack injection, the strength of specimens with a low degree of corrosion is higher than those repaired by patching, and the strength loss due to re-deterioration is also smaller than those repaired by patching. This can be attributed to increases in the strength of cover concrete impregnated with the injection resin. The strength after repair tends to decrease as the degree of corrosion before repair increases and as the degree of corrosion after repair increases, regardless of the repair method.

4 CONCLUSIONS

Conclusions of this study are as follows.

A method of analyzing the relationship between crack width and corrosion loss of reinforcement based on fracture mechanics was presented. By incorporating creep deformation near reinforcement, the analysis results agreed relatively well with the test results. A diagram for estimating the section loss ratio of internal reinforcement from the surface crack width was also presented.

The strength of cover concrete evaluated by three-point loading testing decreases as the section loss ratio of reinforcement increases, presumably due to reductions in the adhesion of reinforcement as corrosion develops. Increased strength of concrete and inclusion of fibers are effective measures against losses in the evaluated strength of cover concrete associated with reinforcement corrosion. When the degree of reinforcement corrosion is low, crack injection is more effective than patching in restoring the losses in the evaluated strength of cover concrete. Both patching and crack injection do not significantly contribute to the restoration of the evaluated strength losses of cover concrete. These are therefore considered to be of significance in inhibiting reinforcement corrosion.

5 APPENDIX

$$K_I = \frac{2f}{\sqrt{2d}} G(a,c,d) \qquad (A.1)$$

$$G(a,c,d) = \left[1 + 0.297 \sqrt{1 - \left(\frac{c}{a}\right)^2} \left(1 - \cos\frac{\pi a}{2d}\right) \right] F(a,c,d) \quad (A.2)$$

$$K_I = \frac{2f}{\sqrt{\pi a}} G(a,c,d) \qquad (A.3)$$

$$G(a,c,d) = \frac{G'(a,c,d)}{(1-A)^{3/2}\sqrt{1-B^2}}; \quad (A = a/d, B = c/a) \quad (A.4)$$

$$G'(a,c,d) = g_1(A) + g_2(A) \cdot B + g_3(A) \cdot B^2 + g_4(A) \cdot B^3$$

$$g_1(A) = 0.46 + 3.06A + 0.84(1 - A)^5 + 0.66A^2(1 - A)^2$$

$$g_2(A) = -3.52A^2$$

$$g_3(A) = 6.17 - 28.22A + 34.54A^2 - 14.39A^3 - (1 - A)^{1.5}$$
$$-5.88(1 - A)^5 - 2.64A^2(1 - A)^2$$

$$g_4(A) = -6.63 + 25.16A - 31.04A^2 + 14.41A^3 + 2(1 - A)^{1.5}$$
$$+5.04(1 - A)^5 + 1.98A^2(1 - A)^2$$

ACKNOWLEDGEMENT

This study was conducted as a part of the 21st Century COE Program for Tokyo Metropolitan University, "Development of Technology for Activation and Renewal of building stocks in Megalopolis," and as a part of the Ministry of Education, Science, Sports and Culture, Grant-in-Aid for scientific Research (B).

REFERENCES

Tada H, Paris PC, Irwin GR. 1985. The stress analysis of crack handbook, Second Edition. Paris Productions Incorporated.

Kitsutaka Y. 1997. Fracture parameters by poly-linear tension softening analysis. *Journal of Engineering Mechanics, ASCE* 123(5): 444–450.

Nakamura, N. & Kitsutaka, Y. 1998. Crack opening of high-strength concrete surface due to corrosion of reinforcing bar. *Proceedings of Japan Concrete Institute* 20(2): 871–876.

Part VI
Practical problems in RC structural applications

Part VI
Practical problems in RC structural applications

Fracture Mechanics of Concrete and Concrete Structures – Design, Assessment and Retrofitting of RC Structures – Carpinteri, et al. (eds)
© 2007 Taylor & Francis Group, London, ISBN 978-0-415-44616-7

FEM model for analysis of RC prestressed thin-walled beams

M. Bottoni, C. Mazzotti & M. Savoia

DISTART – Structural Engineering, University of Bologna, Bologna, Italy

ABSTRACT: This paper deals with service behavior of prestressed cracked thin-walled beams. A non linear finite element model for thin-walled beams has been developed. Deflections in the non-linear range are obtained by adopting a smeared crack model for concrete in tension. Contribution of steel bars in reducing deformation has been considered. An example of a thin-walled prestressed girder is presented.

1 INTRODUCTION

Microcracking and cracking have a strong influence on instantaneous and long-term behaviour of concrete (Ghali & Favre 1994). Adoption of constitutive law for concrete in his cracked state is essential and mandatory to describe structural deformability under service loads. Another important phenomenon for structures during their service life is creep deformation, which can cause deflections two or three times greater then instantaneous counterpart (Bažant 1988, Favre et al. 1997).

Service behaviour of reinforced concrete is particularly important for some kinds of structures such as bridges or roof girders, due to their large spans and small thicknesses, increasing deformation and creep effects. For these structural elements, warping effects may be very significant and traditional beam models do not apply properly; theories for thin-walled beams are much more appropriate (see Laudiero & Savoia 1990, Capuani et al. 1998, Prokič 1996, Prokič 2002).

In the present paper, a non-linear finite element model is presented for the analysis of deformability of reinforced concrete thin-walled beams.

A finite element formulation for thin-walled beams is firstly introduced (see also Bottoni et al. 2006); the kinematic model is described and equations derived from FE discretization are given. Main assumption is that of cross-sections rigid in their own planes; this hypothesis reduces the total number of dofs involved in the analysis and, correspondingly, computational effort. Furthermore, it is acceptable for analysis of beams under service loadings. On the contrary, longitudinal displacements are obtained through dofs in the section plane, so allowing for warping or shear-lag effects. Longitudinal displacement is considered constant within the thickness. The resulting finite element is essentially a membrane element with null deformation along the transverse axis. The presence of ordinary or prestressed steel reinforcement is considered through homogenization over the thickness.

As far as material modelization is concerned, a smeared crack approach is adopted for concrete, providing the average behaviour under tension after cracking. Ordinary steel reinforcement is modelled as layers of different materials in the finite element, whereas prestressed steel strands are modelled as separate panels.

Solution algorithm in the non-linear range is developed, adopting a displacement-control method as proposed by Batoz & Dhatt (1979).

Finally, an example of a roof precast and prestressed RC element is reported. Results are given in term of load-deflection curves, strains and stresses distributions as well as concrete cracking maps.

2 THE FINITE ELEMENT MODEL

2.1 Kinematic model

The kinematic model is defined according to the reference system shown in Figure 1, where a prismatic reinforced concrete thin-walled beam is depicted; each branch of the cross-section can be constituted by different materials through-the-thickness, in order to consider the presence of reinforcement steel.

A global right-handed orthogonal coordinate system $(O; x, y, z)$ is adopted, where x- and y- are two general axes belonging to the cross-section plane. Location of the reference system origin O and direction of x- and y-axes are general, i.e. they do not necessarily coincide with centroid (or shear centre)

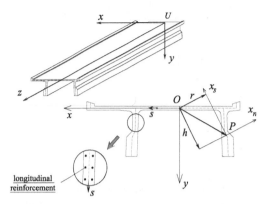

longitudinal
reinforcement

Figure 1.   Thin-walled beam and its general cross-section.

and principal axes of the cross section. Coordinates of general point P can be expressed as:

$$x = x(s), \quad y = y(s) \tag{1}$$

where s is a curvilinear coordinate lying on the profile centerline. At any general point P of section contour, a local right-handed orthogonal coordinate system $(P; x_s, x_n, z)$ is also defined (Fig. 1), where x_s- and x_n- axes, lying in the section plane, are tangential and orthogonal to the centerline, respectively.

Together with the definition of a reference system, the kinematic model is based upon the following main assumptions:

(a) Thickness of various branches constituting the cross-section is small with respect to overall cross-section dimensions;
(b) Cross-sections are rigid in their own planes;
(c) Displacements along z-direction are independent from cross-section in-plane displacements;
(d) No variation of kinematic or static variables across the thickness is considered.

According to hypothesis (b), in-plane displacements of point P are defined through rigid movements of the cross-section itself; therefore, they can be expressed as a function of displacement components $\xi(z)$ and $\eta(z)$ of O along x- and y-axes and of $\theta(z)$, cross-section rotation around O (Fig. 2).

According to hypothesis (c), $w(s,z)$ is the displacement along the longitudinal axis z and it is independent from other displacement components. Accordingly, displacements of general point $P(s,z)$ along x-, y- and z-directions are, respectively (Bottoni et al. 2006):

$$u(s,z) = \xi(z) - y\theta(z) \tag{2a}$$

$$v(s,z) = \eta(z) + x\theta(z) \tag{2b}$$

$$w = w(s,z) \tag{2c}$$

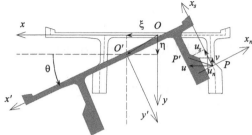

Figure 2.   Displacement components in global and local reference systems.

By geometrical considerations, displacements of point $P(s,z)$ along local coordinate axes x_n, x_s can be expressed as:

$$u_n(s,z) = \xi(z)\frac{dy(s)}{ds} + \eta(z)\frac{dx(s)}{ds} - h(s)\theta(z) \tag{3a}$$

$$u_s(s,z) = \xi(z)\frac{dx(s)}{ds} + \eta(z)\frac{dy(s)}{ds} + r(s)\theta(z) \tag{3b}$$

respectively, where $r(s)$ and $h(s)$ are components along x_n, x_s of vector OP, i.e. (see Fig 1):

$$r(s) = x(s)\frac{dy(s)}{ds} - y(s)\frac{dx(s)}{ds} \tag{4a}$$

$$h(s) = x(s)\frac{dx(s)}{ds} + y(s)\frac{dy(s)}{ds} \tag{4b}$$

2.2   Strain and stress components

From Equations 3, strain components ε_z, γ_{zs}, γ_{zn} can be written as:

$$\varepsilon_z(s,z) = \frac{\partial w(s,z)}{\partial z} \tag{5a}$$

$$\gamma_{zs}(s,z) = \frac{\partial w}{\partial s} + \frac{\partial u_s}{\partial z} = \frac{\partial w}{\partial s} + \left(\frac{dx}{ds}\frac{d\xi}{dz} + \frac{dy}{ds}\frac{d\eta}{dz} + r\frac{d\theta}{dz} \right) \tag{5b}$$

According to hypothesis (d) of section 2.1, strain components are constant across wall thickness. Moreover, due to small wall thickness, strain component γ_{zn} can be neglected. Finally, strain components ε_s, ε_n, $e\gamma_{ns}$ are zero due to hypothesis (b) of rigid cross-sections.

Shear strain γ_{zs}^{SV}, corresponding to Saint Venant torsion and linearly varying across wall thickness with null value on the centerline, is then superimposed to shear strain defined in Equation 5b. This strain

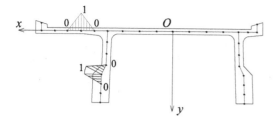

Figure 3. Nodes on transverse cross-section and linear approximation for axial displacements.

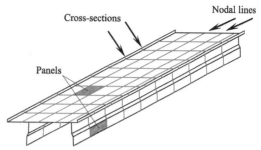

Figure 4. Discretization of the beam. Longitudinal and cross-sectional subdivision.

component cannot be derived rigorously from displacement field described by Equations 2, due to the assumption of constant axial displacement through-the-thickness. However, Saint-Venant stiffness is essential for open thin-walled beams to withstand uniform torsion. Corresponding strain γ_{zs}^{SV} is assumed to be proportional to the derivative of the torsion angle θ with respect to longitudinal axis z, according to the expression (Laudiero & Zaccaria 1988):

$$\gamma_{zs}^{SV} = 2\frac{d\theta(z)}{dz}x_n \tag{6}$$

The model can be further simplified by making assumptions on stress components. First of all, due to the assumption of rigid cross-section, in-plane normal and shear stresses (σ_s, σ_n, τ_{ns}) cannot be determined explicitly, but they are negligible with respect to other components.

Moreover, due to small thickness of thin-walled beam, shear stress τ_{zn} is also negligible. Hence, only stresses σ_z and τ_{zs} are considered in the present model, together with shear stress τ_{zs}^{SV}. Stress components are related to corresponding strains through constitutive laws, i.e.:

$$\sigma_z^i = E_z^i \varepsilon_z \tag{7a}$$

$$\tau_{zs}^i = G_{zs}^i \gamma_{zs} \tag{7b}$$

$$\tau_{zs}^{SV,i} = G_{zs}^i \gamma_{zs}^{SV} \tag{7c}$$

where superscript i refers to i-th layer and E_z and G_{zs} are secant values (in non linear range) of longitudinal Young modulus and transverse shear modulus, respectively. Considering different materials through the wall thickness, corresponding stress components, σ_z, τ_{zs} and τ_{zs}^{SV}, are obtained through individual constitutive laws.

3 NUMERICAL MODEL

3.1 *The finite element model*

The application of the finite element method requires a proper discretization of geometric domain. In the

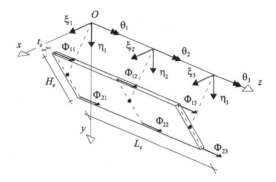

Figure 5. Finite element of *panel* type and corresponding dofs.

proposed model, discretization is performed following two main phases: first, thin-walled beam is divided along the longitudinal direction in sub-elements, by using a number of reference cross sections; secondly, general transverse cross-section is discretized into a number of straight segments defined between couples of nodes (Fig. 3). Nodes are forced to be located where some kind of discontinuity arises: geometrical discontinuity (first derivative of curvilinear coordinate and wall thickness), material discontinuity and where three or more walls intersect each other. Additional intermediate nodes are also considered in order to improve the solution. Joining the corresponding nodes of different transverse sections, a nodal line is obtained. The single finite element, called *panel*, is defined as the portion of wall between two adjacent sections and two adjacent nodal lines (Fig. 4). A linear variation of longitudinal displacement $w(s,z)$ is considered along the cross-section curvilinear coordinate s, whereas quadratic approximation is adopted along z-direction (Bottoni et al. 2006).

Moreover, quadratic functions are used for in-plane cross-sectional displacements $\xi(z)$, $\eta(z)$, $\theta(z)$. The resulting finite element has 6 nodes and 15 dofs (Fig. 5): 6 for axial displacements and 9 for

821

rigid body transverse displacements of three involved cross-sections.

All finite elements whose edges are located in a common transverse section share the same rigid body displacements. Hence, displacement functions providing rigid body displacements of cross-sections can be written in the form:

$$\xi(z) = \mathbf{N}^T(z)\boldsymbol{\xi} \qquad (8a)$$

$$\eta(z) = \mathbf{N}^T(z)\boldsymbol{\eta} \qquad (8b)$$

$$\theta(z) = \mathbf{N}^T(z)\boldsymbol{\theta} \qquad (8c)$$

where vectors:

$$\boldsymbol{\xi}^T = [\xi_1, \xi_2, \xi_3], \quad \boldsymbol{\eta}^T = [\eta_1, \eta_2, \eta_3],$$

$$\boldsymbol{\theta}^T = [\theta_1, \theta_2, \theta_3] \qquad (9)$$

contain rigid body dofs while vector $\mathbf{N}^T = [N_1, N_2, N_3]$ contains quadratic shape functions. Moreover, axial displacement is given by:

$$w(s, z) = \mathbf{N}_\phi^T(s, z)\boldsymbol{\Phi} \qquad (10)$$

where vector:

$$\boldsymbol{\Phi}^T = [\phi_{11}, \phi_{12}, \phi_{13}, \phi_{21}, \phi_{22}, \phi_{23}] \qquad (11)$$

contains axial dofs, while shape functions collected in vector $\mathbf{N}_\phi^T(s, z)$ are defined by the composition of $\mathbf{N}(z)$ and linear Lagrangean interpolation functions $\mathbf{M}^T(s) = [M_1(s), M_2(s)]$ according to expression:

$$\mathbf{N}_\phi^T(s, z) = [M_1(s) \cdot \mathbf{N}^T(z), M_2(s) \cdot \mathbf{N}^T(z)] \qquad (12)$$

From Equations 5, 8, strain components can then be written as:

$$\varepsilon_z(s, z) = \frac{\partial w(s, z)}{\partial z} = \mathbf{N}_\phi'^T(s, z)\boldsymbol{\Phi} \qquad (13a)$$

$$\gamma_{zs}(s, z) = \frac{\partial u_s(s, z)}{\partial z} + \frac{\partial w(s, z)}{\partial s} =$$
$$= \alpha_x \mathbf{N}'^T(z)\boldsymbol{\xi} + \alpha_y \mathbf{N}'^T(z)\boldsymbol{\eta} + \qquad (13b)$$
$$+ r\mathbf{N}'^T(z)\boldsymbol{\theta} + \mathbf{N}_{\phi,s}^T(s, z)\boldsymbol{\Phi}$$

where prime stands for partial derivative with respect to z and coefficients $\alpha_x = dx/ds$, $\alpha_y = dy/ds$, $r = xdy - ydx$ are constant for each finite element.

Finally, Saint Venant shear strains are given by (see Equation 6):

$$\gamma_{zs}^{SV}(s, z) = 2\theta'(z)x_n = 2\mathbf{N}'(z)\boldsymbol{\theta}\,x_n \qquad (14)$$

3.2 Equilibrium conditions for the finite element

Equilibrium conditions for the single finite element can be written by the principle of virtual displacements. Making use of Equations 5, 6 and 7, inner virtual work can be written as:

$$\delta L_{vi}^e = \int_{V_e} (\sigma_z \,\delta\varepsilon_z + \tau_{zs}\,\delta\gamma_{zs} + \tau_{zs}^{SV}\,\delta\gamma_{zs}^{SV})\,dV_e =$$
$$= t_e \int_{H_e} \int_{L_e} (\delta\varepsilon_z E_z^{eq}\varepsilon_z + \delta\gamma_{zs}G_{zs}^{eq}\gamma_{zs})\,ds dz + \qquad (15)$$
$$+ \frac{1}{3}t_e^3 \int_{H_e} \int_{L_e} (\delta\vartheta' \cdot G_{zs}^{SV,eq}\,\vartheta')\,ds dz$$

where E_z^{eq}, G_{zs}^{eq} and $G_{zs}^{SV,eq}$ are equivalent moduli, taking into account single contributions of each material through the finite element thickness:

$$E_z^{eq} = \frac{1}{t_e}\sum_{i=1}^m E_z^i \cdot t_i \qquad (16a)$$

$$G_{zs}^{eq} = \frac{1}{t_e}\sum_{i=1}^m G_{zs}^i \cdot t_i \qquad (16b)$$

$$G_{zs}^{SV,eq} = \frac{4}{t_e^3}\sum_{i=1}^m G_{zs}^i\left(x_{n,i}^3 - x_{n,i-1}^3\right) \qquad (16c)$$

Substitution of Equations 13, 14 into Equation 15, yields:

$$\delta L_{vi}^e = t_e E_z^{eq}\int_{H_e}\int_{L_e}(\delta\boldsymbol{\Phi}^T\mathbf{N}_\phi')\,(\mathbf{N}_\phi'^T\boldsymbol{\Phi})\,ds\,dz +$$
$$+ t_e G_{zs}^{eq}\int_{H_e}\int_{L_e}(\delta\boldsymbol{\Phi}^T\mathbf{N}_{\phi,s} + \alpha_x\delta\boldsymbol{\xi}^T\mathbf{N}' + \alpha_y\delta\boldsymbol{\eta}^T\mathbf{N}' + r\delta\boldsymbol{\theta}^T\mathbf{N}')\cdot$$
$$\cdot (\mathbf{N}_{\phi,s}^T\boldsymbol{\Phi} + \alpha_x\mathbf{N}'^T\boldsymbol{\xi} + \alpha_y\mathbf{N}'^T\boldsymbol{\eta} + r\mathbf{N}'^T\boldsymbol{\theta})\,ds\,dz + \qquad (17)$$
$$+ \frac{1}{3}t_e^3 G_{zs}^{SV,eq}\int_{H_e}\int_{L_e}(\delta\boldsymbol{\theta}^T\mathbf{N}')\,(\mathbf{N}'^T\boldsymbol{\theta})\,ds\,dz = \delta\mathbf{U}_e^T\mathbf{K}_e\,\mathbf{U}_e$$

where vector $\mathbf{U}_e^T = [\boldsymbol{\xi}^T, \boldsymbol{\eta}^T, \boldsymbol{\theta}^T, \boldsymbol{\Phi}^T]$ collects all dofs of the element and $\mathbf{K}_e$ is the corresponding stiffness matrix. Moreover, external virtual work is given by:

$$\delta L_{ve}^e = \int_{L_e}(q_x\delta\xi + q_y\delta\eta + m_t\delta\theta)\,dz +$$
$$+ \delta\boldsymbol{\xi}^T\mathbf{Q}_x + \delta\boldsymbol{\eta}^T\mathbf{Q}_y + \delta\boldsymbol{\theta}^T\mathbf{M}_t \qquad (18)$$

where:

- q_x, q_y, m_t are external distributed loads corresponding to rigid body cross-section displacements $\xi(z)$, $\eta(z)$, $\theta(z)$,
- $\mathbf{Q}_x^T = [Q_{x1}, Q_{x2}, Q_{x3}]$, $\mathbf{Q}_y^T = [Q_{y1}, Q_{y2}, Q_{y3}]$, $\mathbf{M}_t^T = [M_{t1}, M_{t2}, M_{t3}]$ are vectors of generalized applied forces associated with dofs ξ, η, θ.

Substituting Equations 8 in Equation 18 yields:

$$\delta L_{ve}^e = \delta\boldsymbol{\xi}^T \mathbf{t}_\xi + \delta\boldsymbol{\eta}^T \mathbf{t}_\eta + \delta\boldsymbol{\theta}^T \mathbf{t}_\theta + \\ + \delta\boldsymbol{\xi}^T \mathbf{Q}_x + \delta\boldsymbol{\eta}^T \mathbf{Q}_y + \delta\boldsymbol{\theta}^T \mathbf{M}_t = \delta\mathbf{U}_e^T \mathbf{F}_e \tag{19}$$

where:

$$\mathbf{t}_\xi = \int_{L_e} \mathbf{N}(z)q_x dz, \quad \mathbf{t}_\eta = \int_{L_e} \mathbf{N}(z)q_y dz,$$

$$\mathbf{t}_\theta = \int_{L_e} \mathbf{N}(z)m_t dz \tag{20}$$

and:

$$\mathbf{F}_e^T = [\mathbf{t}_\xi + \mathbf{Q}_x, \ \mathbf{t}_\eta + \mathbf{Q}_y, \ \mathbf{t}_\theta + \mathbf{M}_t, \ 0] \tag{21}$$

Setting $\delta L_{vi}^e = \delta L_{ve}^e$, under the assumption of arbitrary values of variations $\delta\xi$, $\delta\eta$, $\delta\theta$, for $0 \le z \le L_e$, Equations 17 and 19 yield:

$$\mathbf{K}_e \mathbf{U}_e = \mathbf{F}_e \tag{22}$$

stating equilibrium condition for the general finite element.

4 SOLUTION METHOD IN THE NON-LINEAR RANGE

When a reinforced concrete structure in the cracked range is considered, Equation 22 gives a non linear system of equations due to non linear behavior of concrete. A Newton-Raphson method has then been adopted to solve the non-linear FE problem. The solution algorithm originally proposed by Batoz & Dhatt (1979) has been used. This method enables step-by-step controlling of a single, monotonically growing dof, while shape of external load distribution is defined separately. All unconstrained dofs are unknown, except for the displacement component adopted as the control parameter; increment of load multiplier λ is also unknown. With this method, softening branches in load-displacement response (as in the case of crack formation in concrete under tension) can be correctly followed without imposition of the deformed shape to the structural element.

First, equation system stating equilibrium for the linearized Newton-Raphson problem is written. Then, in the usual way, partitioning of the system is operated, thus imposing boundary conditions on displacements. The following reduced equation system of order m can be obtained, being m the number of unconstrained dofs and with i indicating general Newton-Raphson iteration:

$$\mathbf{K}_i d\mathbf{U}_i = d\lambda_i \hat{\mathbf{F}} + \mathbf{R}_i + \mathbf{C}_i \tag{23}$$

where $\mathbf{K}_i$ is the structural tangent matrix, $\hat{\mathbf{F}}$ a constant vector determining the shape of external load distribution, $\mathbf{R}_i$ the vector of residual forces and $\mathbf{C}_i$ a vector originating from system partitioning. Moreover, $d\mathbf{U}_i$ and $d\lambda_i$ are variations of dofs vector and load multiplier, respectively. For each Newton-Raphson iteration i, the following equation system can be derived from Equation 23:

$$d\mathbf{U}_i = d\lambda_i \mathbf{U}^a + \mathbf{U}^b \tag{24}$$

where:

$$\mathbf{U}^a = \mathbf{K}_i^{-1}\hat{\mathbf{F}}, \quad \mathbf{U}^b = \mathbf{K}_i^{-1}(\mathbf{R}_i + \mathbf{C}_i) \tag{25a,b}$$

Equation system 24 must be solved for $d\lambda_i$ and $d\mathbf{U}_i$, for a given step increment $\Delta\bar{u}$ of the controlled dof. This is done easily, if j-th equation (corresponding to controlled dof) is extracted, stating:

$$d\lambda_i U_j^a + U_j^b = \Delta\bar{u} \quad \text{for iteration} \quad i = 1 \\ d\lambda_i U_j^a + U_j^b = 0 \quad \text{next} \quad (i > 1) \tag{26}$$

Increment $\Delta\bar{u}$ is imposed at the first iteration, as evidenced by Equation 26.

5 MODELIZATION OF PRESTRESSED REINFORCED CONCRETE

5.1 Constitutive laws for concrete

In the present study, concrete behavior in tension is considered linear up to tensile strength with softening branch after the peak stress. Shear deformability is considered purely elastic; beam is then supposed not to crack under shear stresses for service load levels.

A smeared crack law (Reinhardt & Yankelewsky 1989) has then been used for concrete in tension. The law is linear with elastic modulus E_c up to cracking occurring at point $[\varepsilon_{ct}, f_{ct}]$; softening branch is described by the following relationship between concrete stress σ_c and strain ε_c:

$$\sigma_c = f_{ct}\left(1 - \frac{\varepsilon_c}{\varepsilon_{c0}}\right)^2 \tag{27}$$

where ε_{c0} is value of strain where stress becomes zero. The complete constitutive law in tension (Fig. 6) is obtained by shifting Equation 27 to the right of the quantity ε_{ct}. Strain with zero value for stress becomes $\varepsilon_u = \varepsilon_{c0} + \varepsilon_{ct}$.

In compression a linear law with elastic modulus E_c is adopted, since service loads are considered.

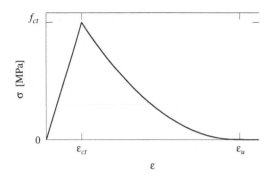

Figure 6. Reinhardt & Yankelewsky law (1988) for concrete in tension.

Table 1. Material properties.

Property	Symbol	Value	Unit
Elastic modulus of concrete	E_c	40	GPa
Shear modulus of concrete	G_c	17.4	GPa
Tensile strength of concrete	f_{ct}	3.7	MPa
Ultimate strain of concrete	ε_u	0.5‰	GPa
Elastic modulus of steel strands	E_{sp}	196	GPa
Elastic modulus of steel bars	E_s	206	GPa
Shear modulus of steel strands	E_{sp}	75.4	GPa
Shear modulus of steel bars	E_s	79.2	GPa

5.2 Modelization of bars and pretensioning strands

For both pretensioning strands and steel bars, a linear elastic behavior is assumed. In the finite element model, the presence of ordinary steel bars is modeled as layers made of a different material. On the contrary, strands are modeled as a unique layer with different width.

Pretensioning is then introduced in the form of concentrated forces applied on strands at beam extremities.

6 RESULTS

Results of the non linear analysis on a simply supported, prestressed thin-walled beam are reported. Beam length is 24 m and cross-section is depicted in Figure 7. For beam geometry, see also Di Prisco et al. (1990). A uniformly distributed load q along beam length is applied. Overall pretensioning force is set to 2180 kN. Material parameters are listed in Table 1.

Modelization of the cross-section is shown in Figure 8; thickness of branches and main nodes of discretization are reported. Additional nodes have then been placed for bars and strands, as well as along section branches to improve the solution (see Fig. 10).

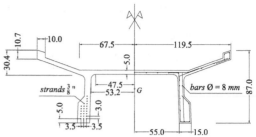

Figure 7. Beam cross-section with prestressing and ordinary reinforcement.

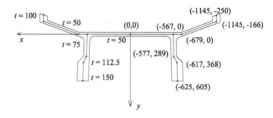

Figure 8. Modeling of cross-section: coordinates of main points (right) and thickness of branches (left).

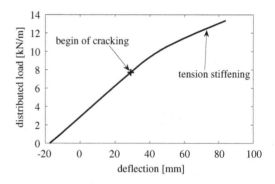

Figure 9. Distributed load vs. deflection.

Prestress load has been introduced from the beginning together with beam weight.

In a second stage, vertical load has been incremented by controlling increase of beam deflection, following the method described in Section 4; analysis has been carried out up to the attainment of compression strains (0.8‰) corresponding to about 40% of compressive strength (conventional limit of linear elastic behaviour). Cracking first occurs for a distributed load of about 8 kN/m. By increasing load level, beam stiffness in bending reduces significantly, as shown by slope decrease of the load-deflection curve in Figure 9. After an intermediate phase with cracking extending from middle-span, curve becomes approximatively linear. Due to the presence of prestress load, cracking is revealed by a very gradual

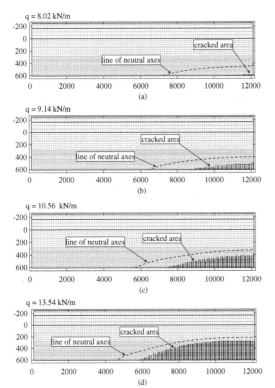

Figure 10. Cracking configuration and line of neutral axes at different loading levels.

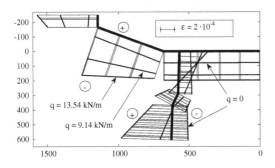

Figure 11. Strains close to middle-span for different load levels.

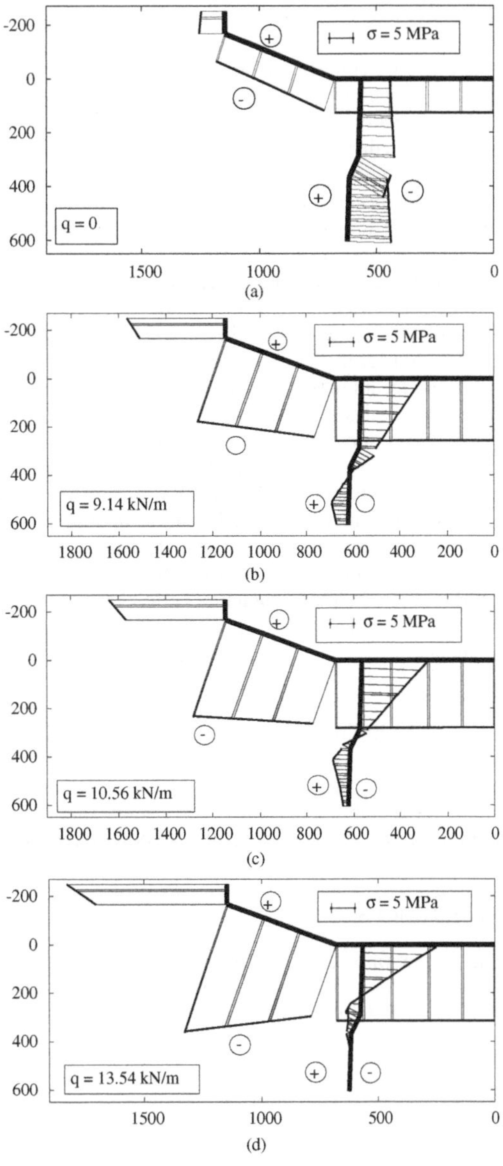

Figure 12. Stress distributions close to middle-span for different load levels.

change in stiffness, without softening branches as in the case of ordinary steel reinforcement only.

Figures 10a–d show the evolution of cracked region with load increase, expanding from middle-span. Black points indicate Gauss points where strain overcomes the cracking limit ε_{ct}; moreover, black line indicates the neutral axis position (null axial strain). Having adopted a local model, strain localization in

tension could be expected. Nevertheless, Figures 10 indicate a well distributed cracking state.

Shows strain profile for three different load levels at the cross-section 48 cm far from middle-span. Moreover, for the same section, in Figure 12 normal stresses in concrete are reported, for different load levels. Softening behaviour of concrete in tension due to cracking in the bottom portion of the beam is shown.

7 CONCLUSIONS

A new finite element model is developed for non linear behaviour of concrete under service loads.

Cross-sections are assumed rigid in their own planes, so reducing the number of dofs. Shear elastic deformability is considered into the model, though inelastic deformability due to cracking is only taken into account for normal strains. A smeared cracked model describes the average damaging behaviour.

In a future study, a non local constitutive model will be used to model cracking behaviour of concrete (Pijaudier-Cabot & Bažant 1986, Bažant & Cedolin 1991). Non local models are typically used as regularization techniques when strain localization occur in numerical analysis.

Moreover, the model will be extended to cover the case of long-term loadings, including creep deformation of concrete and steel relaxation of prestressing strands. Since membrane deformation of individual panels only are considered (no out-of-plane bending), introduction of creep deformation according to solidification theory is much more effective and computationally efficient with respect to ordinary shell elements.

ACKNOWLEDGEMENT

The financial support of (Italian) MIUR (PRIN 2006 Grant: "Structural behaviour of self compacting fibre reinforced concrete") is gratefully acknowledged.

REFERENCES

Batoz, J.L. & Dhatt, G. 1979. Incremental displacement algorithms for non linear problems. *International Journal for Numerical Methods in Engineering* 14: 1262–1267.

Bažant, Z.P. & Cedolin, L. 1991. *Stability of Structures.* Oxford University Press.

Bažant, Z.P. (Ed.) 1988. *Mathematical modeling of creep and shrinkage in concrete.* John Wiley and Sons.

Bottoni, M., Mazzotti, C. & Savoia, M. 2006. A F.E. model for viscoelastic behaviour of composite pultruded shapes. *Proceedings of ECCM 2006*, Lisbon, Portugal.

Capuani, D., Savoia, M. & Laudiero, F. 1988. Dynamics of multiply connected perforated cores. *Journal of Engineering Mechanics, ASCE*, 124:622–629.

Di Prisco, M., Gambarova, P., Toniolo, D. & Failla, C. 1990. Sperimentazione su prototipo di un componente prefabbricato in c.a.p. per coperture di grande luce (in Italian). *Proceedings of CTE 1990*, Bologna, Italy.

Favre, R., Jaccoud, J.P., Burdet, O. & Charif, H. 1997. *Dimensionnement des structures en béton. Aptitude au service et éléments de structures* – Vol. 8 (in French). Presses Polytechniques et Universitaires Romandes.

Ghali, A. & Favre, R. 1994. *Concrete structures, stresses and deformations*, London: E & FN Spon Publisher.

Laudiero, F. & Zaccaria, D. 1998. A consistent approach to linear stability of thin-walled beams of open sections. *International Journal of Mechanical Sciences*, 30(8): 543–557.

Laudiero, F. & Savoia, M. 1990. Shear strain effects in flexure and torsion of thin-walled beams with open or closed cross-section. *Thin-Walled Structures*, 10: 87–119.

Prokič, A. 1996. New warping function for thin walled beams, I: Theory. *Journal of Structural Engineering, ASCE*, 122(2):1437–1442.

Prokič, A. 2002. New finite element for analysis of shear lag. *Computers & Structures*, 80:111–1024.

Pijaudier-Cabot, G. & Bažant, Z.P. 1986. Nonlocal damage theory, *Journal of Engineering Mechanics, ASCE*, 113(2): 1512–1533.

Reinhardt, H.W. & Yankelewsky, D.Z. 1989. Uniaxial behavior of concrete in cyclic tension, *Journal of Structural Engineering, ASCE*, 115(1): 166–182.

Fracture Mechanics of Concrete and Concrete Structures – Design, Assessment and
Retrofitting of RC Structures – Carpinteri, et al. (eds)
© 2007 Taylor & Francis Group, London, ISBN 978-0-415-44616-7

Crack width evaluation in post-tensioned prestressed concrete bridge-deck slabs

Y.C. Choi & B.H. Oh
Dept. of Civil Engineering, Seoul National University, Seoul, Korea

S.L. Lee
Dept. of Civil Engineering, Mokpo National University, Muan-gun, Jeonnam, Korea

S.W. Yoo
Dept. of Civil Engineering, Woosuk University, Wanju-gun, Jeonbuk, Korea

M.K. Lee
Dept. of Civil Engineering, Jeonju University, Jeonju-si, Jeonbuk, Korea

B.S. Jang
Dam Engineering Research Center, Korea Water Resources Corporation, Daejeon, Korea

ABSTRACT: The purpose of the present study is to investigate the cracking behavior and crack width of transversely post-tensioned concrete decks in PSC box-girder bridges. For this purpose, full-scale box-girder members were made and tested. Major test variables include the amount of prestressing and prestressing steel ratios. The crack width, strains and deflections of deck slabs were measured automatically according to the increase of applied loads. The bond effectiveness of prestressing steel was determined from the test data and it was found to be about 0.468 compared with non-prestressed ordinary rebar. A realistic formula for predicting crack width of PSC deck slabs has been derived. The comparison of proposed equation agrees very well with test data while the existing Gergely-Lutz equation and CEB-FIP code equation show large deviation from the test data. The proposed equation can be efficiently used for the realistic analysis and design of prestressed concrete decks in box-girder bridges.

1 INTRODUCTION

The transverse prestressing of bridge decks was first introduced in box-girder bridges mainly to maximize the length of cantilever overhangs and to reduce the number of webs (Almustafa 1983, Oh et al. 2005). The post-tensioning system is generally used in the transverse prestressing of bridge decks in box-girder bridges (Figure 1). Recently, it was reported that many serious longitudinal cracks occurred at the bottom of top slab of prestressed concrete (PSC) box-girder bridges which are not prestressed laterally (Almustafa 1983, Oh et al. 2005). The transverse prestressing reduces definitely the widths of longitudinal cracks. The crack widths must be limited to certain allowable values especially in PSC structures under service loads. It is therefore necessary to have a realistic prediction equation for crack width under applied

Figure 1. Transverse post-tensioning system of bridge deck in box girder bridges.

loading for rational and safe design of PSC box-girder bridges.

In the post-tensioned prestressed members, prestressing steel stresses are transferred from the prestressing steel via the surrounding grout to the duct, containing the prestressing tendon, and from the duct to the structural concrete (Almustafa 1983). It is generally difficult to evaluate the local bond behavior of prestressing tendon because the tendons are partly in contact with the duct and partly with the surrounding concrete. It is therefore necessary to identify the effective circumference of multi-strands in which the strands are bonded to concrete. The purpose of the present study is first to investigate the bond behavior of prestressing steel and then to propose a realistic formula for predicting maximum crack width of PSC deck slabs. The maximum crack width is derived in terms of steel stress which represents the magnitude of applied loadings. For this purpose, four full-scale box girder members were fabricated and tested. The major objective of the present tests was to see the transverse behavior of PSC box-girders including the cracking behavior of upper deck slabs.

Many strain gages were installed on transverse rebars and prestressing steels to measure the strain changes under applied loads. The crack width gages were also installed to monitor the variation of crack widths at the bottom surface of top slab during the loading process. From the comparisons of the strains of PS steels with the strains of nonprestressed rebars at the same locations, the bond characteristic of posttensioned steel was evaluated. The proposed equation for maximum crack width is a function of steel stress after decompression, effective tensile area, bond characteristic parameter of PS steel, and diameter of steel bar. To verify the applicability of proposed equation, the proposed formula is compared with test data, Gergely and Lutz equation (1968) and CEB-FIP model code (1990).

2 BOND MECHANISM OF PRESTRESSED CONCRETE MEMBERS

The cracking of prestressed concrete members is influenced by the interaction between steel and concrete. Between adjacent cracks the tensile forces are transferred form the steel to the surrounding concrete by bond stress. The distribution of the bond stress along the bar is generally nonlinear. Since the distance between the cracks varies, it is difficult to calculate the exact distributions of bond stresses between steel and concrete. In order to simplify the calculation, the analysis is based on an average constant bond stress (CEB-FIP 1990, Alvarez & Marti 1996) along the bar and it is also assumed that the tensile

strength of the concrete does not vary over the length of the bar. Alvarez & Marti (1996) analyzed the cracked prestressed concrete tension members with the approximate distributions of bond stresses for the reinforcing bar and prestressing steel. The basic equilibrium in tension requires that the total tensile force is equal to the sum of the tensile forces of the reinforcing bar and prestressing steel.

The CEB-FIP model code (1990) considers that the bond behavior of prestressing tendons is different form that of deformed reinforcing bars. The different magnitude of steel stress develops in prestressing steel and reinforcing bar, respectively. The CEB-FIP code (1990) considers two stages separately for crack formation, i.e., single crack formation phase and stabilized crack formation phase. For single crack formation phase, the different transmission lengths l_s and l_p for bond stresses are calculated for reinforcing bar and prestressing steel, respectively, with the assumption of equal crack widths for prestressing and reinforcing steels. The increase of stress after cracking is much less in prestressing steel than in reinforcing steel. For stabilized multiple cracking phase, the transmission length is considered to be equal for prestressing steel and reinforcing bar, which is calculated by considering the different diameters and different bond behavior of reinforcing bars and prestressing steels, respectively. However, the effect of multiple strands in a duct on the bond behavior of prestressing steel was not considered in these calculations in the code (CEB-FIP model code 1990). As can be seen in the next section of present paper, however, the effects of multi-strands in a duct must be considered correctly in order to obtain realistic bond behavior and thus reasonable crack width equation for prestressed concrete members.

3 EXPERIMENTAL INVESTIGATION

3.1 Outline of test members

Four full-scale concrete box girder members have been designed and made to investigate the crack width and crack spacing of deck slab in PSC box-girder bridges. The transverse width, longitudinal length and total height of test members of concrete box girders are 6,500 mm, 2,300 mm and 1,600 mm, respectively as shown in Figure 2. The main design variables include the magnitude of precompression, diameter of steel bars, non-prestressed steel ratio and prestressing steel ratio in the top slab of box-girder bridges. Figure 2 shows the detailed section A-A' of top slab of test members. The magnitudes of precompression due to lateral prestressing are 0, 2.35, 3.52, and 4.70 MPa for test member LP0, LP1, LP2, and LP3, respectively. This is to see the effect of prestressing on the control of crack width under service loads.

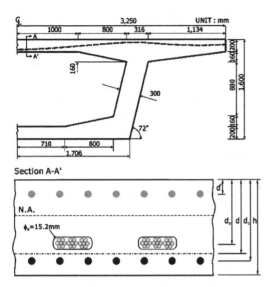

Figure 2. Schematic diagram for cross section properties.

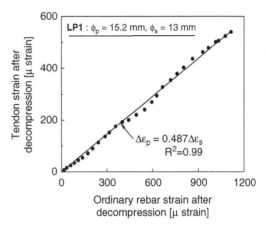

Figure 3. Relationship between tendon and reinforcing bar strains after decompression for LP1 test members.

3.2 *Material properties*

The regular deformed bars have been used in the present test members. The yield strength of mild reinforcement was 398 MPa from material tests in the laboratory. The steel strain gages were installed at the transverse reinforcements. Those gages were protected by covering them with proper waterproof materials. In this study, all prestressing steels were uncoated, low-relaxation 15.2 mm seven-wire strands and the flat duct (70 × 20 mm) were used to locate the tendons inside. The yield and ultimate strength of the strands were 1,600 MPa and 1,892 MPa, respectively. To evaluate the strain increase of tendon after cracking, the steel strain gages were also installed with adequate protection. The concrete was carefully placed in the form so as not to hurt the attached steel strain gages and the maximum-size of aggregate was 25 mm (1 in.). The concrete cylinders with size of ø100 × 200 mm were cast and cured under similar curing condition of test members. The average compressive strength concrete was 40 MPa.

3.3 *Measurement*

Transverse post-tensioning was conducted by using the mono-hydraulic jack. The jacking stress of prestressing steel was $0.75f_{pu}$. The test members were loaded at two points on the upper deck, which simulate two lateral wheel loads. The concrete surfaces were painted white with emulsion paint to facilitate the measurement of crack propagation. The load was applied to the test members using the automatically controlled actuators. The linear variable differential transducers (LVDTs) were installed to measure the deflection profiles of the top slab of the box-girder.

In order to monitor the flexural crack width under applied load, the crack width gages were attached over the cracks after reading the initial values of crack width with microscopes. All strain values of ordinary bars and prestressing tendons and the crack width values at concrete surface were automatically measured and stored in the computer during the loading process. The number of cracks and crack spacings were also measured at each loading step.

4 TEST RESULTS AND ANALTYSIS

4.1 *Bond characteristics of post-tensioned steel*

When prestressed steels and ordinary rebars are simultaneously used, different amounts of steel stresses are expected to develop in prestressed steel and ordinary rebar because the bond behavior of prestressing tendons is different from that of reinforcing bars.

Figure 3 shows the relationship between the increases of tendon stress and rebar stress after decompression for test member LP1. It can be seen that fairly good linear relations are obtained between the increments of tendon stress and ordinary rebar stresses. The increment of tendon stress is about 46.8 percent (0.468) on the average of that of ordinary rebar stress. This indicates that the effectiveness of bond property of prestressing steels is much lower than that of ordinary ribbed bars. The bond effectiveness ratio may be set as about 0.468 between prestressing steel and ordinary rebar from the present test results.

For single crack formation stage in a prestressed member, the relationship between the bond stresses of prestressing steel and ordinary rebar may be written as follows according to CEB-FIP model code (1990).

$$\Delta f_p = \sqrt{\frac{\tau_{ap}}{\tau_{as}} \frac{\sum_{op} A_s}{\sum_{os} A_p}} f_s = \sqrt{\frac{\tau_{ap} \phi_s}{\tau_{as} \phi_p}} f_s \qquad (1)$$

where Δf_p, f_s = increments of tendon and reinforcing bar stresses after decompression, ϕ_p, ϕ_s = nominal diameters of strand and reinforcing bar, τ_{ap}, τ_{as} = average bond stresses of tendon and rebar, respectively, $\sum_{op}$, $\sum_{os}$ = effective circumferential perimeter of multi-strands in a flat duct and reinforcing bar, and A_p, A_s = total area of multiple strands in a flat duct and the area of reinforcing bar, respectively.

Equation 1 neglects the effect of multi-strands in a duct. In flat duct systems, the tendons are partly in contact with the duct and partly with concrete. In order to identify the bond property between concrete and tendon, it is necessary to determine the circumferential perimeter of multi-strands which are directly bonded to surrounding grout materials. In this study, the effective circumferential perimeter of multi-strands is derived as Equation 2.

$$\sum_{op} = \pi \phi_p + (n-1) \phi_p \qquad (2)$$

where n is the number of strand in a flat duct.

Equation 1 is now then modified by using the effective circumference of multiple strands in a duct.

$$\Delta f_p = \sqrt{\frac{\tau_{ap}}{\tau_{as}} \frac{\sum_{op}}{A_p} \frac{\phi_s}{4}} f_s = \sqrt{\frac{\tau_{ap}}{\tau_{as}} \frac{\pi + (n-1)}{n\pi} \frac{\phi_s}{\phi_p}} f_s = \xi f_s \qquad (3)$$

The relation between Δf_p and f_s of Equation 3 can be obtained from the test data of Figure 3. The average value of these ratios between Δf_p and f_s was obtained as 0.468. Now, the value of τ_{ap}/τ_{as} can be obtained from Equation 3 with the test results of Figure 3. The average value of τ_{ap}/τ_{as} was found to be 0.465 from the present test data. The CEB-FIP code (1990) specifies the value of τ_{ap}/τ_{as} to be 0.4 for post-tensioned strands, which is somewhat lower than the average value from the present test results. The present test data indicate that CEB-FIP model code (1990) overestimates the circumferential perimeter for multiple strands within a duct.

4.2 Crack occurrence

The transverse flexural cracks were firstly appeared in the top slab at an approximately applied load of 500 kN for all test members. The first longitudinal flexural cracks of top slabs were formed at the applied load of 550 kN, 564 kN, 643 kN and 916 kN (where 1 kN = 220.46 lb) for LP0, LP1, LP2 and LP3, respectively. This indicates that test member LP3 which has largest precompressive stress exhibits much higher initial cracking load. This can be seen clearly from the present test results which show less cracking at the same applied load levels as the precompression increases.

In pure bending region, the crack spacing was investigated for various increasing load levels. The average crack spacing decreases with increase of load level which means that the number of cracks increases as the applied load increases. The increase of prestressing force (LP3) exhibits rather large crack spacing. The average crack spacing may be a function of prestressing steel ratio.

4.3 Derivation of crack width formula

The phenomenon of cracking of concrete is complex because of the tensile weakness of concrete. The crack width may depend on many factors such as steel stress, effective concrete area in tension, nonprestressed reinforcement type and ratio, prestressing steel type and ratio, and strength of concrete.

If time-dependent effects are considered in the PSC members, the stress of ordinary rebar will increase due to creep deformation of concrete.

In this study, the increase of ordinary steel stress due to time-dependent effects after prestressing has been taken care of and only the increment of steel stress after decompression has been considered to compare with crack width increase according to load increase.

The relationship between rebar stress after decompression and maximum crack width can be assumed to be linear (Bazant & Oh 1983, Gergely & Lutz 1968, Oh & Kang 1997). The value of crack width is also generally assumed to increase in proportion to the diameter of rebar (Oh & Kang 1997). The effective concrete area in tension accounts for the non-uniform normal stress distribution arising from bond forces into the concrete cross-section at the end of the transmission length.

In this study, the effective concrete area in tension is written as Equation 4 which is given by CEB-FIP model code (1990).

$$A_{t,eff} = \min[2.5(c + \phi_s/2)b, (h-x)b/3] \qquad (4)$$

where c is concrete clear cover depth, x is the neutral axis depth of fully cracked section and b is the width of member.

Figure 4 shows the nonlinear relationship between the ratio $w_{max}/f_s \phi_s$ and the ratio ρ_{eff} [$= (A_{st} + \xi A_{pt})/A_{t,eff}$] which are obtained from the test results. The value of ξ represents the bond effectiveness between the prestressing steel and ordinary rebar. The value of ρ_{eff} [$= (A_{st} + \xi A_{pt})/A_{t,eff}$] may be defined as the

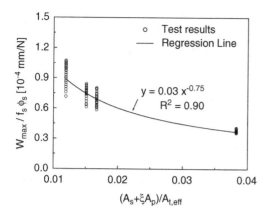

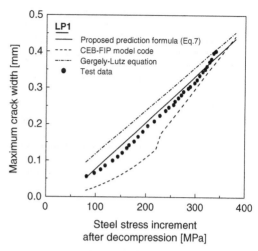

Figure 4. Correlation between effective reinforcement ratio and observed crack width.

effective combined reinforcement ratio of ordinary rebar plus effective prestressing steel with reference to effective tensile concrete area.

Figure 4 shows that the crack width decreases with an increase of effective combined reinforcement ratio ρ_{eff} and is inversely proportional to $\rho_{eff}^{0.75}$. From this relation of ρ_{eff} with crack width w_{max}, the maximum crack width equation is reasonably derived as follows.

$$w_{max} = 3 \times 10^{-6} (f_s - f_0) \phi_s \left(\frac{A_{t,eff}}{A_{st} + \xi A_{pt}} \right)^{0.75} \frac{h-x}{d-x} \quad (5)$$

$$\xi = \sqrt{\frac{\tau_{ap}}{\tau_{as}} \frac{\pi + (n-1)}{n\pi} \frac{\phi_s}{\phi_p}} \quad (6)$$

where w_{max} = predicted maximum crack width in mm, A_{st}, A_{pt} = total area of prestressing steels and reinforcing bars, x = neutral axis depth of cracked section in mm, h = height of cross section in mm, d = effective depth in mm, ϕ_s = diameter of rebar in mm, ϕ_p = diameter of PS strand in mm, f_s = increment of rebar stress after decompression in MPa, f_0 = steel stress at the initial occurrence of crack in MPa, and τ_{ap}/τ_{as} is 0.465 for post-tensioned tendons which was obtained from this study. Here, $(h-x/d-x)$ was introduced to represent the crack width at the bottom face of flexural members.

Generally, the cracking starts at the tensile strain of about 0.0002 in concrete. Therefore, the reinforcement stress after decompression at the first occurrence of crack is 40 MPa (5800 psi) because the cracking strain of concrete is about 0.0002. Therefore, the realistic prediction equation of crack width is now reasonably suggested as follows.

$$w_{max} = 3 \times 10^{-6} (f_s - 40) \phi_s \left(\frac{A_{t,eff}}{A_{st} + \xi A_{pt}} \right)^{0.75} \frac{h-x}{d-x} \quad (7)$$

Figure 5. Comparison of proposed crack width equation with experimental data.

5 COMPARISON WITH TEST DATA

The proposed crack width formula has been compared with the present test results. Figure 5 shows the comparisons of proposed crack width equation with the experimental data. Figure 5 also shows the comparisons with the simplified crack width equation of Gergely and Lutz (1968) and maximum crack width equation of CEB-FIP Model Code (1990) which consists of two parts, i.e., single crack formation phase and stabilized crack formation phase.

Gergely and Lutz equation overestimates the observed maximum crack widths for all test members. This is probably due to the fact that Gergely and Lutz equation is basically based on the test data of reinforced concrete members. On the other hand, the CEB-FIP model code equation generally underestimates the maximum crack width especially in the lower stress range of reinforcing bar. This may be due to the fact that the CEB-FIP code (1990) considers two different stages in the crack formation separately, i.e., single crack formation phase and stabilized crack formation phase.

However, it can be clearly seen from the test data of Figure 5 that the crack width increases almost linearly form the initial stage. The proposed equation agrees very well with test data for all stress ranges. Therefore, the suggested formula may well be used for the realistic analysis and design of deck slabs in prestressed concrete box-girder bridges.

6 CONCLUSION

The purpose of the present study is to explore the cracking behavior of laterally prestressed bridge decks in box-girder bridges. For this purpose, a series of full-scale box girders have been fabricated and tested.

The bond characteristics of prestressing steels were identified from the measurements of strains for both prestressing steels and ordinary rebars. The bond effectiveness of prestressing steels was obtained from the present test results and it was found to be about 0.468 compared with ordinary rebars.

The effective circumferential perimeter of multi-strands in a duct which is bonded to concrete is derived and used to determine the crack width under applied loads. A realistic prediction equation for crack widths was derived in terms of effective combined reinforcement ratio, diameter of rebar, and the rebar stress after decompression.

The proposed equation correlates very well with test data of crack width under applied loads, while the existing formulae of CEB-FIP and Gergely & Lutz equation deviate greatly from the test data for prestressed concrete flexural members.

REFERENCES

Almustafa, R.A. 1983. *The analysis of transverse prestressing effects in bridge decks*: 6–7. PH.D. thesis. The university of Texas at Austin.

Bažant, Z.P. & Oh, B.H. 1983. Spacing of cracks in reinforced concrete. *Journal of Structural Engineering, ASCE* 109(9): 2066–2085.

Alvarez, M. & Marti, P. 1996. Experimental bond behavior of reinforcing steels at plastic strains. *Institut für Baustatik, ETH*, Zurich, Switzerland: 134.

Comite-Euro International du Beton/Federation Internationale de la Precomtrainte. *CEB-FIP Model Code 1990*: 274–255. London: Thomas Telford.

Gergely, P. & Lutz, L.A. 1968. Maximum crack width in reinforced concrete flexural members.*Causes, Mechanism, and Control of Cracking in Concrete*. SP-20: 87–117. Detroit: American Concrete Institute.

Oh, B.H et al. 2005. Flexural behavior of PSC box girders according to haunch size. *Journal of KSCE*, 25(2-A): 217–280.

Oh, B.H. & Kang, Y.J. 1997. New formulas for maximum crack width and crack spacing in reinforced concrete flexural members. *ACI Journal*, March–April: 103–112.

Fracture Mechanics of Concrete and Concrete Structures – Design, Assessment and Retrofitting of RC Structures – Carpinteri, et al. (eds)
© *2007 Taylor & Francis Group, London, ISBN 978-0-415-44616-7*

Crack width prediction in RC members in bending: A fracture mechanics approach

S. Sakey & D. Binoj

B. M. S. College of Engineering, Bangalore, India

ABSTRACT: Cracking is a very common occurrence in reinforced concrete (RC) structures. Cracks in RC structures are characterized by crack width and crack spacing. In the present study an expression is developed using a cohesive crack model having a bilinear strain softening relationship to predict crack widths in RC beams. One of the assumptions made in the development of the model is that there is complete loss of bond between the bar and the concrete. However, this crude assumption leads to too conservative values for crack width, since the frictional forces at the bar-concrete interface limit bar slip and consequently crack width. Therefore, it is necessary to introduce a restraining force to model bond and to find crack-closing displacement. The crack width values so obtained from the proposed model are compared with code predictions and with experimental results available in literature. The results show that the proposed approach is sound, consistent and realistic.

1 INTRODUCTION

The occurrence of cracks in reinforced concrete structures is inevitable because of the low tensile strength of concrete. Cracks form when the tensile stress in concrete exceeds its tensile strength. Cracking in reinforced concrete structures has a major influence on structural performance, including tensile and bending stiffness, energy absorption capacity, ductility, and corrosion resistance of reinforcement. Cracking at the service load should not extend to such a limit that it spoils the appearance of the structure or leads to excessive deformation of the members. This may be achieved by specifying an allowable limit on crack width values. In order to assure a satisfactory performance of the structure even under service loads, an important limit state i.e., the limit state of serviceability (cracking) is introduced into the limit state design procedure. This limit state is assumed to be satisfied if crack widths in a concrete member are within a maximum allowable limit. While the need for a crack limit state has been universally agreed on, the formulae for predicting the crack width extensively vary in the various codes of practice. Inspection of crack width prediction procedures proposed by various investigators indicates that each formula contains a different set of variables. A literature review also suggests that there is no general agreement among various investigators on the relative significance of different variables affecting the crack width, despite the large number of experimental work carried out during the past few decades. Taking all the parameters into account in a single experimental program is not normally feasible due to the large number of variables involved, and the interdependency of some of the variables.

In this paper, an attempt is made to predict an expression for crack width by incorporating a bilinear strain softening function and all the variables which influence crack widths. The proposed formulas are also compared comprehensively with the test results available in the literature (Hognestad, 1962; Kaar and Mattock, 1963; Clark, 1956). To access the relative performance of the proposed crack width equation, it is compared with the international codes of practice.

2 CRACK WIDTH EXPRESSION

Gerstle et al. (1992) developed simplified assumptions that allow analytical solutions for flexural cracks in singly reinforced beams in bending while retaining the significant features of the fictitious crack model (FCM) which was introduced by Hilllerborg et al. (1976). The FCM has the potential of being very useful in understanding the fracture and failure of concrete structures. It assumes that the fracture process zone at the crack tip is long and infinitesimally narrow. The fracture process zone is characterized by a normal stress versus a crack opening displacement curve which is considered as a material property. The shape of this stress-crack opening displacement (softening) curve can be either linear/bilinear/tri-linear or a power law. Gerstle et al. (1992) assumed a linear softening relation in his formulation and predicted the crack width as a product of a constant C (a function of brittleness of concrete or a function of reinforcement) and critical crack opening displacement COD_{cr} (a function of the softening curve or the fracture energy). This expression for crack width does not explicitly include such parameters as the diameter/perimeter of reinforcement which influences the values of crack widths. In the literature, bilinear softening seems to describe the behavior of concrete in tension more appropriately than the linear softening. An attempt is made here to work out an expression for crack width based on bond mechanics (bar slip included), and formulated as the product of the crack spacing times the mean strain in the reinforcement by incorporating the bilinear softening function.

2.1 Significance of the strain softening curve

The stress-crack opening law for concrete in tension is found to have a descending branch in the post-peak region. The simplest idealization for this behavior is a linear softening relation, but it is more realistic to consider a bilinear softening relationship. A typical bilinear shape of the softening curve is shown in Figure 1. Linear softening seems to be an obvious choice when the data describing the actual material behavior is limited. However, linear softening proved to overestimate structural capacity. Therefore, bilinear curves have been accepted as reasonable approximation of the softening curve for concrete, although there seems to be no agreement about the precise location of the kink point. In the literature, several researchers have given the kink positions (break points) on the basis of experiments, and there are quite a few simple methods to identify any bilinear softening to fit particular experimental data (Guinea et al. 1994). Brincker and Dahl (1989) reformulated the substructure method introduced by Petersson (1981) for the three point bending specimen in order to obtain complete load displacement relations. From the sensitivity analysis of their method using linear, bilinear and tri-linear models, it is evident that the shape of the stress crack opening displacement relation has significant influence on the results. However, tri-linear approximation does not seem to deviate significantly from the bilinear approximation indicating the sufficiency of the bilinear approximation. In this study, the crack width is calculated considering specific kink positions (break points) as suggested by Brincker and Dahl (1989), having the value of $k_1 = 0.308$ and $k_2 = 0.161$.

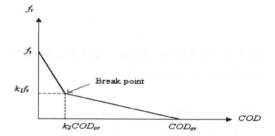

Figure 1. Typical bilinear stress versus crack opening displacement curve.

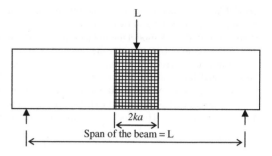

Figure 2. Schematic diagram of the beam showing central elastic band width.

2.2 Proposed methodology using a flexural cracking model which incorporates the bilinear softening function in tension with non linearity of concrete in compression

The main assumptions of the model are as follows:

1. Plane sections remain plane before and after deformation within the central elastic band.
2. The beam is considered rigid outside the central elastic band.
3. Fictitious crack surfaces remain plane after deformation.
4. The stress versus crack opening displacement curve is assumed as bilinear softening in tension.
5. Concrete is homogeneous, isotropic and non-linear elastic.
6. The steel has a perfectly plastic material model.
7. The reinforcement can slip with respect to concrete within the central elastic band (2 ka).
8. The centroid of the steel is located at the bottom of the beam and the concrete cover below the steel level is deliberately neglected for simplicity in the derivation of the expressions.

As shown in Figure 2, this model considers a varying central elastic band whose width varies k times the length of crack from the crack surface. The width of the central elastic band considered is $2ka$, i.e., at a distance of ka on either side from the crack surface, where k is a constant and a is the crack length. The beam is

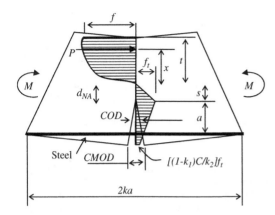

Figure 3. Schematic of stress variation (Case I – Stage I).

considered rigid outside the central elastic band. The model is capable of predicting the flexural behavior of concrete beams.

Figure 3, shows an idealization of the deformed shape (greatly magnified) of a crack in a reinforced concrete beam, together with normal stress distribution considering a bilinear stress crack opening displacement relationship.

Two cases are considered and they are as follows:

(i) Case I, the fictitious crack is not sufficiently open to relieve the normal stress at its mouth i.e., ($CMOD < COD_{cr}$). Case I is further subdivided into two stages, viz. Stage I (just before the kink), and Stage II (just after the kink), and
(ii) Case II, in which the fictitious crack is sufficient open to relieve the normal stress at its mouth ($CMOD > COD_{cr}$).

2.3 Normalization of parameters

- Crack mouth opening displacement $C = \frac{CMOD}{COD_{cr}}$
- Crack length $= A = \frac{a}{h}$
- Distance from crack tip to neutral axis $S = \frac{s}{h}$
- Distance from neutral axis to top fiber of beam $T = \frac{t}{h}$
- Material-scale parameter for concrete $= \beta = \frac{f_t h}{E_c COD_{cr}}$
- Material parameter(for reinforcement), $\alpha = \rho n$

where, m is the applied moment, $n = E_s/E_c$ is the modular ratio, ρ is the geometric reinforcement ratio and α is the mechanical reinforcement ratio.

Deriving on similar lines to that suggested by Gerstle et al. (1992) the expression for the crack mouth opening displacement ($CMOD$) is obtained at various stages of loading as equal to constant C multiplied by critical crack opening displacement (COD_{cr}). The $CMOD$ is nothing but the crack width at any point of loading in the reinforced concrete flexural member at the level of steel.

$$CMOD = C*COD_{cr}$$

From the derivation of the flexural cracking model with bilinear softening in tension and nonlinearity in compression, we have the strain at the bottom of the beam which is also equal to the strain at the level of steel.

$$\varepsilon_s = \varepsilon_b = \frac{k_1 f_t (1-C)}{E_c (1-k_2)} \tag{2}$$

Rearranging the above equation,

$$\left[C = 1 - \frac{E_c (1-k_2)}{k_1 f_t} \varepsilon_s \right] \tag{3}$$

Substituting (3) in (1),

$$CMOD = \left[1 - \frac{E_c (1-k_2)}{k_1 f_t} \varepsilon_s \right] COD_{cr} \tag{4}$$

$$= \left[\frac{1}{\varepsilon_s} - \frac{E_c (1-k_2)}{k_1 f_t} \right] COD_{cr} \varepsilon_s$$

$$= \left[\frac{E_s}{f_s} - \frac{E_c (1-k_2)}{k_1 f_t} \right] COD_{cr} \varepsilon_s$$

$$= \left[\frac{n E_c}{f_s} - \frac{E_c (1-k_2)}{k_1 f_t} \right] COD_{cr} \varepsilon_s$$

$$= \left[\frac{n}{f_s} - \frac{(1-k_2)}{k_1 f_t} \right] E_c COD_{cr} \varepsilon_s \tag{5}$$

As we know,

$$\beta = \frac{f_t h}{E_c COD_{cr}}, \alpha = \rho n, \rho = \frac{A_s}{bh} = \frac{\varepsilon.\pi.\phi.\phi}{4.b.h},$$

ε = number of bars

and ϕ = Diameter of the reinforcing bar.

Using the above relationships, we get

$$CMOD = \left[\frac{\alpha}{\rho f_s} - \frac{(1-k_2)}{k_1 f_t} \right] \frac{f_t h}{\beta} \varepsilon_s \tag{6}$$

$$= \left[\frac{\alpha f_t h}{\rho f_s \beta} - \frac{(1-k_2)h}{\beta k_1} \right] \varepsilon_s$$

$$CMOD = \left[\frac{\left(4bh^2\alpha f_t\right)}{\left(\varepsilon\pi\phi\phi\beta f_s\right)} - \frac{\left(1-k_2\right)h}{k_1\beta} \right]\varepsilon_s \qquad (7)$$

where, $CMOD$ = Maximum crack width at the level of steel & k_1 k_2 = Kink Positions (Break points). Therefore, Maximum crack width in the above expression is a function of nine variables viz: reinforcement ratio (α), brittleness of concrete (β), tensile strength of concrete (f_t), stress in steel (f_s), diameter of bar (ϕ), perimeter of bar ($\varepsilon\pi\phi$), cross section dimensions of the beam (b, h) and the strain in steel (ε_s), which directly influence the crack width prediction and hence exhibits a physically realistic expression. The crack width in question is at the level of the reinforcement. The crack width at the bottom of the beam with the reinforcement cover shall be equal to $[(h-x)/(d-x)]$ times the crack width at the level of steel, where, h and d are overall depth and effective depth respectively and x is the neutral axis position from the top fiber. In the above expression for crack width Eq. 7 is derived from a flexural cracking model considering bilinear strain softening in tension and non-linearity of concrete in tension and one of the assumptions is that there is a complete slip between the crack face and the steel. However, this need not be the case and the predictions made using Eq. 7 are likely to be conservative estimates of the crack width. Further, it can be understood physically that there cannot be a complete loss of bond between the steel and concrete in ordinary bond conditions. Therefore, in order to account for this anomaly, a restraining bond force is introduced against complete loss of bond between the steel and concrete, to predict crack width values that are realistic. The effect of this crack closing force on the predicted values of the crack width is discussed in the following section.

3 CRACK WIDTH CORRECTION

Assuming total loss of bond at the bar-concrete interface within the distance $\pm kA$ from the cracked plane is too conservative, since there is the restraining action due to bond. Therefore, in order to realistically model cracking and to avoid any crack-width overestimation, it is necessary to introduce a restraining force at the level of the reinforcement and to find the crack-closing displacement at that level. This is done using the expression as given in the hand book stress intensity factors (Gdoutos 2003),

$$U_\theta = \frac{K_1^{total}}{4\mu}\left(\frac{r}{2\pi}\right)^{1/2}\left[-(2k+1)\sin\frac{\theta}{2}+\sin\frac{3\theta}{2}\right] \qquad (8)$$

where, $k = \left[\frac{3-\nu}{1+\nu}\right]$ for plane stresses and
$k = (3-4\nu)$ for plane strains
$\mu = E/2(1+\nu) =$ Rigidity Modulus, $\nu_{concrete} = 0.15$,
$r =$ distance to the crack tip.

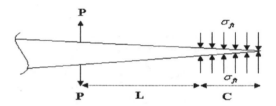

Figure 4. Schematic of the restraining force and the crack closing stresses.

K_1^{total} = Stress intensity due to applied load 'P' and tensile stress 'f_t' of concrete.

The stress intensity factor K_I is calculated from Dugdale's model (Gdoutos 2003). According to this model there is a fictitious crack equal to the real crack (L) plus the length of fracture process zone (C). The crack is loaded by a restraining force (P) at the level of steel and an additional crack closing stress which is equal to the tensile strength of concrete. Therefore, the stress intensity factor $K_1^{(total)}$ acting at tip of the fictitious crack is expressed as:

$$K_1^{(P)} + K_1^{(f_t)} = K_1^{(total)} \qquad (9)$$

$K_1^{(P)}$ = Stress intensity factor due to applied loads P is

$$K_1^{(P)} = \frac{2*P}{\left[2*\pi*(C+L)\right]^{1/2}} \qquad (10)$$

where,

$C =$ length of fictitious crack obtained from the model.
$L =$ Real crack obtained from the model.
$P =$ Restricting force
$P = B*(D/2)*f_t*0.09$ when D/B = 1.
  $= B*(D/2)*f_t*0.2$ when D/B = 2.
  $= B*(D/2)*f_t*0.26$ when 2 < D/B < 3.
$P = B*(D/2)*f_t*0.13$ when 3 < D/B < 3.5.
  $= B*(D/2)*f_t*0.15$ when 3.5 < D/B < 4.
  $= B*(D/2)*f_t*0.41$ when D/B = 4.
$B =$ Width of the beam in mm.
$D =$ Overall depth of the beam, mm.
$f_t =$ Tensile strength of concrete in N/mm^2.

$K_1^{(f_t)}$ = The stress intensity factor due to tensile stress (f_t) acting along the length of the fracture process zone.

$$K_1^{(f_t)} = -\frac{4*f_t*C^{1/2}}{\left(2*\pi\right)^{1/2}} \qquad (11)$$

Here, f_t = Cohesive force acting over the fictitious region.

836

Substituting Eq.10 and Eq.11, in Eq. 9

$$K_1^{(total)} = \frac{\left[(2P) - \left(4 * f_t * \sqrt{C} * \sqrt{(C+L)}\right)\right]}{\sqrt{2\pi} * \sqrt{(C+L)}} \quad (12)$$

Substituting $\theta = \pi$, in Eqn. 8, we obtain the expression for the crack closing displacement at the level of steel as

$$U_\theta = \frac{-K_1}{2\mu}(k+1)\left(\frac{r}{2\pi}\right)^{1/2} \quad (13)$$

where, $r = (C+L) =$ Total crack, $k = \left(\frac{3-v}{1+v}\right)$ for plane stresses & $(3-4v)$ for plane strains

$$\mu = \frac{E}{2(1+v)} = \text{Rigidity Modulus, } v(concrete) = 0.15$$

Substituting $K_1^{(total)}$ from Eqn. 12 into Eqn. 13. Therefore, $U_\theta =$

$$= -\left\{\left[\frac{(2*P) - \left(4*f_t*\sqrt{C}*\sqrt{(C+L)}\right)}{2*\mu*\sqrt{2\pi}*\sqrt{(C+L)}*\varepsilon_s}\right]*\left(\sqrt{C+L}\right)*(k+1)\right\}*\varepsilon_s \quad (14)$$

The restricting force P used in the expression is calculated as the product of the effective area of concrete A_{ct} around the reinforcement contributing to this effect and the maximum tensile stress f_t of concrete. The restricting force P for various combinations of depth to width ratio (D/b) is calculated and it is substituted in Eq (3.10) to obtain the stress intensity factor due to an applied load. Therefore the crack is loaded by a restricting force (P) at the level of steel and an additional crack closing stress which is equal to the tensile strength of concrete and the stress intensity factor $K_1^{(total)}$ acting at the tip of the fictitious crack will be the combination of the stress intensity factor due to the applied load and stress intensity factor due to tensile stress (f_t) acting along the length of the fracture process zone. Therefore, the maximum crack width at the level of steel is computed as the difference in the crack width values with a complete loss of bond and the crack closing displacement due to the restricting force.

i.e., $(CMOD)_{wcf} = CMOD_{wocf} - U_\theta \quad (15)$

where, $(CMOD)_{wcf} =$ Maximum Crack width with closing force

$CMOD_{wocf}$ = Maximum Crack width without closing force

U_θ = Crack closing displacement due to closing force

4 RESULTS AND DISCUSSION

In order to assess the soundness of the proposed expression (Eqn 7 and 15), they are compared with the test results available in literature (Kaar and Mattock, 1963; Hognestad, 1962; Clark, 1956) and also with the expressions adopted in the international codes of practice.

4.1 Test results of Kaar and Mattock

Kaar and Mattock (1963) of the Portland Cement Association (PCA) modified the CEB equation (1959) to express the maximum crack width at the level of reinforcement on the concrete surface. Two full scale T-beams and a half and quarter scale model of one of these beams were tested. The T-beam specimens were loaded by hydraulic rams under the center diaphragm and were restrained by tie rods near the beam ends. This loading arrangement was used to simulate a negative moment region in a continuous T-beam. A 40-power microscope graduated in thousandths of an inch hydraulic actuators placed at mid-span was used to measure crack width.

4.2 Test results of Hognestad

Hognestad (1962) tested reinforced concrete members with high-strength deformed bars and concluded that (i) the mechanism of crack formation is such that a wide experimental scatter must inherently occur. (ii) both maximum and average crack width are essentially proportional to the stress in steel and (iii) the crack width that developed in the case of beams reinforced with state-of-the-art deformed bars was less than one half of that for plain bars. He reported crack widths at the centroid of reinforcement for steel stresses ranging from 20000 lb/in² (137.9 N/mm²) to 50000 lb/in² (344.7 N/mm²) for every 10000 lb/in² (68.9 N/mm²) increments.

4.3 Test results of Clark

Clark (1956) tested 54 specimens and reported maximum crack width and spacing for steel stresses ranging from 15000 lb/in² (103.4 N/mm²) to 45000 lb/in² (310.2 N/mm²) at every 5000 lb/in² (34.5 N/mm²) increment. Crack widths on the tensile face were determined by the use of Tuckerman optical strain gages, strains in the tensile reinforcement were measured with electrical resistance strain gages. The location and extent of cracks were observed and recorded. A number of R/C Slabs and beams with different geometries and bar arrangements were tested in 4-point bending.

4.4 Consolidated test results

The proposed method for predicting the maximum crack width is compared using the test results reported

Table 1. Statistical comparison of the proposed expression at kink positions (break points) of bilinear curve ($K_1 = 0.308$ $K_2 = 0.161$) with the reported test results.

| Source | No. of observation | (Crack width ratio)W_{cal}/W_{exp} for Bilinear ($K_1 = 0.308$ $K_2 = 0.161$) | | |
		Avg.	Std. Dev.	C.O.V
Clark	15	1.154	0.284	24.614
Kaar & Matock	6	1.093	0.226	20.714
Hognestad	27	1.081	0.246	22.708

Table 2. Statistical comparison of various codes with the proposed method.

| Source | No. of observation | (Crack width ratio)W_{cal}/W_{exp} for Bilinear ($K_1 = 0.308$ $K_2 = 0.161$) | | |
		Avg.	Std. Dev.	C.O.V
BS 8110 equation	32	0.726	0.214	29.46
Model code equation	32	0.620	0.270	43.55
Gergely and Lutz equation	32	0.892	0.210	23.57
Chinese code equation	32	0.833	0.200	24.02
Bilinear	27	1.081	0.246	22.708

in the literature (Hognestad 1962), Kaar and Mattock (1963), (Clark 1956). For each beam, the theoretical crack width obtained by means of the proposed expression is divided with the corresponding experimental crack width i.e., (W_{cal}/W_{exp}) and the average ratio is obtained at steel stress of 275.8 N/mm^2 (40000 Psi). The respective standard deviation and coefficient of variation are also obtained as shown in the Table 1.

4.5 Comparison of the crack widths from the proposed expression along with the Codes of Practice with reference to the test results of Hognestad

In order to assess the relative performance of the proposed expression (Eqn 7 and 15), the average crack width ratios, standard deviation and coefficient of variation are obtained for the test results of Hognestad (1962) and compared with the corresponding values obtained from the expression adopted for crack width prediction in the international codes of practice.

4.6 Discussion of the test results

From the results obtained (Tables 1 & 2) the following points are noted:

From Table 1, it is observed that an average crack width ratio of 1.081 and the coefficient of variation of 22.71% is obtained at kink position of ($K_1 = 0.308$ $K_2 = 0.161$) for the test results of Hognestad (1962), indicating that theoretical values of crack width obtained from the proposed expression is closer to the experimental results.

For the test results of Kaar and Mattock, the proposed expression produces a crack width ratio of 1.093 with a standard deviation 0.226 and a coefficient of variation of 20.714% at kink position of ($K_1 = 0.308$ $K_2 = 0.161$) These values indicate that the crack width predicted by the proposed expression is consistent and reliable, and that the coefficient of variation is lower.

For the test results of Clark, the proposed expression provides a crack width ratio of 1.154 at kink position of

($K_1 = 0.308$ $K_2 = 0.161$). The deviation of theoretical crack width from experimental crack width was 0.284 with a coefficient of variance of 24.614%.

From Table 2, it can be observed that the BS 8110 equation underestimates the crack width by 27.4% for the test results of Hognestad at a steel stress of 275.8 N/mm^2 (40000 Psi) with an average crack width ratio of 0.726 and a coefficient of variation of 29.46%.

The Model code equation 1990 also underestimates the values of crack width by 38% for the test results of Hognestad at a value of steel stress equal to 275.8 N/mm^2 with an average crack width ratio (W_{cal}/W_{exp}) and coefficient of variation as 0.620 and 43.55% respectively.

The Gergely and Lutz equation which is based on a statistical analysis provides an average crack width ratio of 0.892 with a coefficient of variation of 23.57% at a steel stress of 275.8 N/mm^2 (40000 Psi) for experimental values of Hognestad. It can be observed that even though the coefficient of variation is lower, the average crack width is still underestimating by 10.8%.

For the test results of Hognestad, the Chinese code underestimates the values of crack width by 16.7% with a coefficient of variation of 24.02% at a steel stress of 275.8 N/mm^2 (40000 Psi). The average crack width ratio is 0.833, which clearly shows that even though the coefficient of variation is lower, the crack width ratio still underestimates.

From Table 2, it is also observe that the proposed expression provides better crack width ratio (1.081) and coefficient of variation (22.708%). These statistics indicate that this proposed expression is able to predict consistent crack width values with a significantly lower coefficient of variation as compared to the crack width values provided by the codes.

The graphical illustration of the statistical comparison of the proposed expression at kink position (break

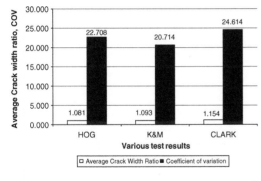

Figure 5. Graphical representation of the average crack width ratio and coefficient of variation with reference to test results of various investigators.

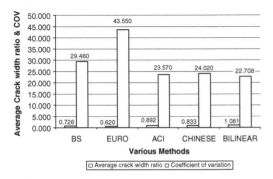

Figure 6. Comparison of Average crack width ratio and coefficient of variation of proposed expression along with various codes of practice using the test results of Hognestad (1962).

point) of the bi-linear curve ($k_1 = 0.308$ & $k_2 = 0.161$) as given in Table 1 is presented in Figure 5 and the graphical illustration of the statistical comparison of the various codes with the proposed method as given in Table 2 is presented in Figure 6.

5   CONCLUSION

In the present study an expression is developed to predict crack width in R/C beams, taking advantage of the cohesive-crack model. This expression is a function of the brittleness of concrete (a function of the tensile strength of concrete, beam depth, elastic modulus of concrete and the fracture energy), reinforcement ratio, crack length, bar diameter, stress in steel and Young's modulus of steel. To assess the validity of the expression, it was compared with other test data on crack width and crack spacing and also with various international codes of practice. The results show that the proposed approach obtained from the model using the bi-linear softening makes it possible to evaluate more accurately the crack width, compared to other formulations. Furthermore, the proposed approach has a rational and mechanically-sound basis, since it is rooted in concrete fracture mechanics.

REFERENCES

Brincker, R. and Dahl, H. 1989, "Fictitious crack model of concrete fracture" Magazine of Concrete Research, 41, No. 147, 79–86.

Clark, A.P. 1956, "Cracking in reinforced concrete flexural members", *ACI Journal, proceedings* V. 52, No. 8, 851–862.

Dugdale, D.S. 1960 "Yielding of steel sheets containing slits." *J. Mech. Phys. Solids*, 8, 100–108.

Gergely, P. and Lutz, L.A. 1968, " Maximum crack width in reinforced concrete flexural members", causes, mechanism and control of cracking in concrete, SP – 20, *American Concrete Institute*, Detroit, pp. 87–117.

Gerstle, W.H. Partha, P.D. Prasad, N.N.V. Rahulkumar, P. and Ming Xie 1992, "Crack growth in flexural members – A fracture mechanics approach" *ACI Journal*, Vol. 89, No. 6, pp. 617–625.

Gdoutos 2003, "Fracture mechanics – An introduction".

Hognestad, E. Jan 1962, "High strength bars as concrete reinforcement – Part 2: Control of flexural cracking," *Journal, PCA Research and Development Laboratories*, V. 4, No. 1, pp. 46–63.

Kaar, P.H., and Mattock, A.H. Jan 1963, "High strength bars as concrete reinforcement – Part 4: Control of cracking," *Journal, PCA Research and Development Laboratories*, V. 5, No. 1, pp. 46–63.

Edward G. Nawy Oct 1968 " Crack control in reinforced concrete structures", *ACI Journal*, pp. 825–836.

Fracture Mechanics of Concrete and Concrete Structures – Design, Assessment and Retrofitting of RC Structures – Carpinteri, et al. (eds)
© 2007 Taylor & Francis Group, London, ISBN 978-0-415-44616-7

Crack widths in reinforced cement-based structures

A.P. Fantilli & P. Vallini
Politecnico di Torino, Torino, Italy

H. Mihashi
Tohoku University, Sendai, Japan

ABSTRACT: The theoretical approaches used for the evaluation of crack width in reinforced concrete (RC) structures, are generally based on the hypothesis of parallel crack surfaces. In this way, crack width measured on the concrete cover should be equal to that on the bar surface. The results of several experimental analyses do not justify this assumption. Therefore, to better define the effective crack profile of RC structures, a new model, able to analyze the whole structural response of reinforced concrete ties, is here presented. In the proposed approach, all the physical phenomena involved in the cracking process are taken into account. A good agreement between numerical results and experimental data is found both in case of steel rebar and ordinary Fiber Reinforced Cementitious Composites (R/FRCC), and in case of steel rebar and High Performance Fiber Reinforced Cementitious Composites (R/HPFRCC).

1 INTRODUCTION

The demand of improving and increasing durability of reinforced concrete (RC) structures, which are generally deteriorated by the corrosion of steel reinforcement, has driven several researchers to analyzing the cracking phenomenon of concrete in tension. As a first approach, crack pattern evolution has been experimentally investigated in RC members in tension since the second half of the past Century (Broms, 1965; Goto, 1971; Watstein & Mathey, 1959). Such elements are generally composed of a steel reinforcing bars (Φ is the diameter of its cross-section) covered by a concrete cylinder of thickness R_1-$\Phi/2$ (Fig. 1a). During the test, when the normal load N is applied to the ends of the element, a crack pattern composed by different types of cracks appears. Precisely, the cracking process produces both main cracks, which are observed on the surface of the tie, and internal (or hidden) cracks (Fig. 1b), which are not visible to an external viewer. They have been detected in the experimental analyses of Broms (1965) and Goto (1971) by injecting ink into the specimen during the loading. After the ink had hardened, the tested member was cut open and the internal cracks were measured both in width and in length.

The cracking phenomenon mainly affects RC structures subjected to bending actions. Anyway, it is a widespread opinion that a RC tie is a satisfactory model for reproducing the tensile regions between two consecutive cracks of a flexural member. In other words, modeling RC members in tension can be considered a reasonable way to compute all the phenomena involved in the cracking of RC structures. Therefore, the first theoretical researches on the evolution of main cracks have regarded RC ties. In these elements, according to the linear elastic fracture mechanics theory, the growth of a single crack can be analyzed as a stability problem (Bianchini et al., 1968). For instance, it is possible to investigate the evolution of a crack profile, or the function $w(R)$ (that is, the crack width w measured at the distance R from the reinforcement axis), in the portion depicted in Figure 1b. If a crack is supposed to start from the steel – concrete interface (where $R = \Phi/2$), its propagation is initially stable, and it does not reach the surface of the member (where $R = R_1$). This is the condition of internal cracks in Figure 1b, for which the stress intensity factor K_I is lower than the critical value K_{IC}. By increasing of the normal force N, K_I becomes higher than K_{IC}, and crack propagation is unstable. In this case, crack tip rapidly reaches the external surface (main cracks in Fig. 1b). As is well known, the value of K_I is strictly connected to the geometry of the problem and to the mechanical properties of concrete.

Therefore, in the case of Figure 1b, only when the distance l_c between main cracks is much longer than the thickness of the concrete cover ($l_c \gg R_1 - \Phi/2$), can the crack tip rise to concrete surface. On the contrary, if $l_c < (R_1 - \Phi/2)$, the propagation of internal

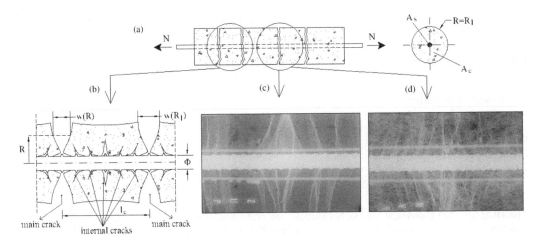

Figure 1. Main cracks and internal cracks in reinforced concrete structures: (a) steel reinforced members subjected to axial loads; (b) a portion of a RC tie (Goto, 1971); (c) a portion of a R/FRCC tie; (d) a portion of a R/HPFRCC tie (Mihashi et al. 2003).

cracks remains stable, and their tips do not reach the external surface.

Therefore, main cracks can be considered as the final stage of the unstable propagation of internal cracks, which are initially stable. This assumption is in accordance to the tests of Broms (1965) and Goto (1971), and to the numerical investigations of Lutz (1970), which confirmed the presence of main cracks and internal cracks at all the loading stages.

Starting from these theoretical and experimental observations, many formulae for the evaluation of crack width, and crack distance, have been proposed (ACI Committee 224, 1986; CEB, 1993). However, as recently pointed out by Beeby (2004), in these approaches crack width on the surface of the concrete cover, $w(R_1)$, is assumed to be equal to that on the steel-concrete interface, $w(\Phi/2) = w_0$. This assumption disagrees with the results of several tests, where crack shapes [i.e. the crack profile $w(R)$] have been measured. Such investigations have been initially conducted by Broms (1965), who has observed and measured the evolution of main cracks (Fig. 1b). At low axial loads N, crack widths are nearly the same, independently of the distance R from the reinforcement [$w(R) \cong$ constant in Fig. 1b]. In this situation, crack surfaces are approximately plane and perpendicular to the reinforcement axis. Subsequently, with the increase of N, crack widths are narrower on the reinforcement, where w_0 becomes approximately 1/2 to 1/3 of $w(R_1)$. Similar conclusions have been drawn by Beeby (1972), who has tested several tension members with different dimensions and reinforcement ratios.

The distribution of tensile stresses σ_{ct} in the concrete cover is the fundamental cause of different crack widths along a main crack. Since $\sigma_{ct}(R) \neq$ constant on crack surfaces, shear stresses, originated by the bond-slip between steel and concrete, can be detected in the concrete surrounding the main cracks. According to Watstein & Mathey (1959), the shear components of stress give rise to the crack profile $w(R)$ shown in Figure 1b. Thus, it is not sufficient to take into account only the bond-slip mechanism between steel and concrete and the fracture mechanics of concrete, as proposed by Fantilli et al. (1998). To evaluate the crack profile of a main crack, the shear deformability of concrete cannot be neglected (Walraven & Reinhardt, 1981).

In order to take into account all the mechanisms involved in the cracking phenomenon, a new model, able to compute the structural response of a member in tension, is presented in this paper. Not only the tensile members made of classical concrete, but also those made by ordinary Fiber Reinforced Cementitious Composite (FRCC) and by High Performance Fiber Reinforced Cementitious Composite (HPFRCC) are taken into consideration. In these new cement-based composites, crack profiles appear different from those depicted in Figure 1b. As observed in the tests by Mihashi et al. (2003) and Otsuka et al. (2003), even for high steel strains, crack widths away from the bars, of reinforced FRCC (R/FRCC) and reinforced HPFRCC (R/HPFRCC) members in tension, are narrower than those directly measured over the reinforcement surface. This is shown by the two X-ray images reported in Figures 1c–1d, which have been taken at yielding of steel rebar in tensile members made, respectively, of R/FRCC (Fig. 1c) and of R/HPFRCC (Fig. 1d).

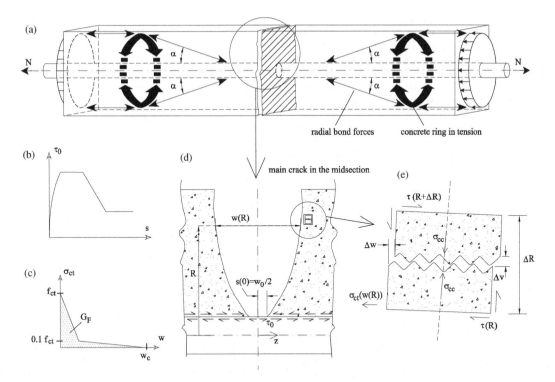

Figure 2. First crack in the midsection of a RC tie: (a) the strut and tie model of Tepfers (1979); (b) bond-slip τ_0-s relationship (CEB, 1993); (c) fictitious crack model $\sigma_{ct} - w$ (CEB, 1993); (d) crack profile of the main crack in the midsection of the element; (e) kinematical and static variables of the Rough Cracks model (Bazant & Gambarova, 1980).

2 A MODEL FOR THE EVALUATION OF CRACK PROFILE

The evaluation of crack profiles of a main crack cannot be separated from the structural analysis of the tension member depicted in Figure 1a.

The mechanical response of RC ties can be computed by referring to the well-known strut and tie model introduced by Tepfers (1979), which is schematically shown in Figure 2a. It consists of radial components of the bond forces, inclined of α with respect to reinforcement axis and produced by the ribs of the rebar, which are balanced against tensile stress rings. Such a mechanism correctly reproduces an anchorage zone, but it vanishes in the midsection of the RC member in tension depicted in Figure 1a. In this zone, concrete is less confined than elsewhere (Fig. 2a), due to the absence of a tensile stress ring near the external surface. Thus, the first main crack usually develops in the middle of the element.

For the sake of simplicity, the definition of crack profiles $w(R)$ will regard a RC tie with a single main crack in its midsection. The effects produced by other cracks, such as the internal cracks (Fig. 1b), are indirectly considered by adopting the phenomenological

relationships τ_0-s depicted in Figure 2b, which reproduces the bond-slip behavior between steel and concrete. Since this mechanism generates bond stresses τ_0 around the concrete-rebar interface of the cracked cross-section, shear stresses can be detected in the surrounding concrete. According to several experimental and theoretical observations, shear stresses (and shear strains) produce different crack widths along R. In particular, crack width w_0 on the bar surface, whose magnitude is related to the bond stress by the τ_0-s relationship of Figure 2b (in which $s = w_0/2$ has to be considered), is different from $w(R)$ measured at distance $\Phi/2 < R \leq R_1$ (Fig. 2d). However, not only shear stresses τ affect the concrete around the cracks. In an element closer to the crack surfaces (Fig. 2e), the axial tensile stresses σ_{ct} and radial compressive stresses σ_{cc}, produced by the bond mechanism in the Tepfers' model, can be also detected. According to the Fictitious Crack model introduced by Hillerborg et al. (1976), σ_{ct} is a function of the crack width $w(R)$, as shown by the cohesive relationship of Figure 2c. The single discrete crack can be divided into two different zones: the process zone, where $w(R) \leq w_c$ (and $\sigma_{ct} \geq 0$), and the macro-cracked zone, where $w(R) > w_c$ (and $\sigma_{ct} = 0$).

Shear stresses τ and compressive stresses σ_{cc} are governed by the so-called aggregate interlock mechanism. It has been investigated for a long time, and different models have been proposed to reproduce it theoretically (Bazant & Gambarova, 1980). An aggregate interlock model is generally found on the definition of the relationship between the kinematical variables (i.e., the increment of the longitudinal displacement, or crack width, Δw, and the increment of the radial displacement Δv) and static variables (i.e., shear stresses τ and compressive stresses σ_{cc}). The Rough Cracks model introduced by Bazant & Gambarova (1980) is here adopted for the aggregate interlock mechanism. It has been widely used to reproduce effectively the shear resistance of RC beams (Dei Poli et al., 1986). In particular, in the present paper, the Rough Cracks model is considered in the form of shear stress increment $\Delta \tau$ within the element of finite length ΔR (Fig. 2e) (Gambarova, 1980):

$$\Delta\tau = \tau(R+\Delta R) - \tau(R) = \tau_a \cdot \frac{a_0}{a_0 + (\Delta v/d_a)^2} \cdot \frac{\Delta w}{\Delta v} \cdot \frac{a_2 + a_3 |\Delta w/\Delta v|^3}{1 + a_3 (\Delta w/\Delta v)^4} \quad (1)$$

where $\tau_a = 0.25 f_c$ ($f_c =$ compressive strength of concrete); $a_0 = 0.111$; $a_1 = 0.435\text{E-3 N/mm}$; $a_2 = 2.45/\tau_a$ MPa; $a_3 = 2.44$ $(1 - 4/\tau_0)$ MPa; $d_a =$ maximum aggregate size. If the function of radial displacement increment Δv and the function of shear stress $\tau(R)$ are known, the increment of the crack profile $\Delta w(R)$ can be obtained by solving Eq. (1).

2.1 The shear stress function $\tau(R)$

The shear stress function $\tau(R)$ can be defined by imposing the equilibrium condition in the longitudinal direction. More precisely, referring to the concrete cylinder of length dz depicted in Figure 3, whose cross-section is a circular crown defined by the radius R_0 and R, it is possible to write:

$$-2\pi R \tau(R)dz + 2\pi R_0 \tau(R_0)dz + \int_{R_0}^{R} \frac{\partial\sigma_{ct}}{\partial z} dz dA_c = 0 \quad (2)$$

where $\tau(R) =$ shear stress on the external surface of the cylinder; $\tau(R_0) =$ shear stress on the internal surface and of the cylinder; $\sigma_{ct} =$ longitudinal tensile stress in the concrete (due to symmetry the radial stresses are assumed to be equal to zero).

Neglecting small quantities of higher order, the cross-sectional area dA_c can be evaluated with the following formula:

$$dA_c = 2\pi R dR \quad (3)$$

By assuming $R_0 = \Phi/2$, the function $\tau(R)$ can be found by substituting Eq. (3) into Eq. (2):

$$\tau(R) = \frac{\tau_0 \cdot \Phi}{2 \cdot R}[1 + \gamma(R)] \quad (4)$$

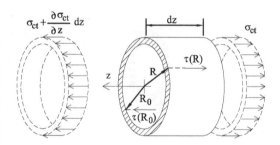

Figure 3. The state of stress in a concrete cylinder of length dz.

where $\tau_0 =$ bond stress on the bar surface, whose magnitude depends on the crack width w_0 (according to the τ_0-s relationship of Fig. 2b). The function γ may be written as:

$$\gamma(R) = \frac{2}{\tau_0 \Phi} \int_{R_0}^{R} \frac{\partial\sigma_{ct}}{\partial z} R dR \quad (5)$$

To evaluate $\gamma(R)$, the function $\partial\sigma_{ct}/\partial z$ should be defined. It is important to clarify that the crack of Figure 2d, as well as those considered in this paper, are located in the symmetrical cross-section of a tensile element. Thus, shear stresses τ are not admitted on crack surfaces, while possible interactions between the Fictitious Crack Model of Figure 2c and the Rough Crack model of Figure 2e are completely neglected. Shear stresses, originated by the bond-slip mechanism on the bar surface, are only transmitted to the upper layers by shear transfer mechanism.

This mechanical behavior is schematically reproduced in Figure 4a. In this way, the shape of crack profile suggests the shape of $\tau(R)$. For instance, it is possible to analyze the crack profile shown in Figure 4a, where the distances of the concrete surface and crack tip from the reinforcement axis are, respectively, R_1 and R_B. From the concavity of $w(R)$, the sign (i.e. the direction) of shear stresses in each point of the crack surface can be immediately defined. It can be concordant or discordant to the direction of τ_0, which is produced by the bond slip mechanism on steel surface. Obviously, $\tau = 0$ in the point of the profile where $w = 0$, and in the point where the derivative dw/dR changes its sign (the radius R_A is assumed to be the distance between this point and the reinforcement axis). In this way, a qualitative shape of $\tau(R)$ is therefore obtained (Fig. 4a).

According to the tests of Broms (1956) and Goto (1971), and to the theoretical results of Bianchini et al. (1968), all the possible crack profiles are shown in Figures 4b–4d. For low values of N, crack is extended within the concrete thickness (internal crack) and it does not appear on the concrete surface (Fig. 4a). Increasing N, crack grows (Fig. 4b) and its tip

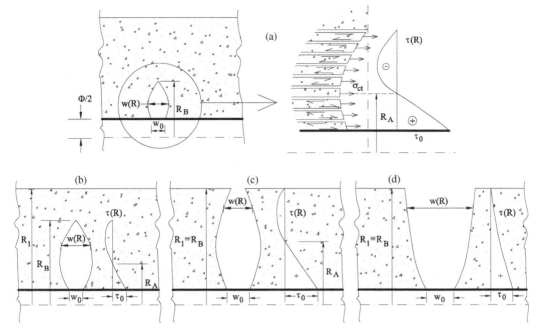

Figure 4. Evolution of crack profile: (a) directions of shear stress $\tau(R)$; (b) internal crack; (c–d) evolution of a main crack.

reaches and overcomes the external surface ($R_B = R_1$ in Fig. 4c). If N remains low, $w(R) \cong$ constant and crack surfaces can be considered plane and vertical. On the contrary, for higher values of the applied loads, crack width is shorter on the reinforcement than on the concrete surface (Fig. 4d).

In Figure 4, shear stress distributions $\tau(R)$, corresponding to the crack profiles of a main crack, are also depicted. In the cases of Figures 4b–4c, when $R_A < R_1$, the values of shear stresses are zero in two different points, which are located at two different depths from the reinforcement axis (R_A and R_B).

More exactly, in the zone where $R_A < R < R_B$, shear stresses are discordant to the direction of bond stress τ_0. When $R_A = R_B$ (Fig. 4d), $\tau(R)$ is monotonic and shear stresses have the same sign of τ_0 for all the values of R. In this situation, $\tau(R)$ seems to be independent of crack width and, consequently, of the cohesive law σ_{ct}-w (Fig. 2c).

Such a consideration is confirmed by the tests of Scott & Gill (1987), who have measured the displacement profile of the concrete in one of the ends of a RC tie. In this cross-section, where $\sigma_{ct} = 0$, they found a displacement field similar to the shape of crack profile $w(R)$ reported in Figure 4d. Thus, the situation $R_A = R_1 = R_B$ (Fig. 4d) corresponds to the condition $w_0 \geq w_c$ in the cohesive model of Figure 2c (in which w_c is the maximum crack width with non-zero stresses). In the other cases (Figs. 4b–4c), a linear variation of R_A, within the domain $[\Phi/2, R_1]$, can be

related to the variation of w_0 in the domain $[0, w_c]$. In this way, R_A is given by:

$$R_A = \Phi/2 + (R_1 - \Phi/2)\frac{w_0}{w_c} \qquad \text{if } w_0 < w_c$$

$$R_A = R_1 \qquad\qquad\qquad \text{if } w_0 \geq w_c$$

(6)

Similarly, the definition of function γ (Eq. [5]) is ruled by the value of w_0. In particular, when $w_0 \geq w_c$, $\tau(R)$ is a monotonic function and consequently $\gamma \geq -1$. In the case $w_0 < w_c$, $\tau(R)$ must be equal to zero at the distances R_B and R_B from the reinforcement axis (that is, $\gamma < -1$ in the domain $[R_A, R_B]$, and $\gamma > -1$ in the domain $[R_0, R_A]$). Under these conditions, it is assumed that the function $\partial\sigma_{ct}/\partial z$ takes the following form:

$$\frac{\partial\sigma_{ct}}{\partial z} = (R_1 - R_A)\cdot(a + b\cdot R) + c \qquad \text{if } w_0 < w_c$$

$$\frac{\partial\sigma_{ct}}{\partial z} = c \qquad\qquad\qquad \text{if } w_0 \geq w_c$$

(7)

where a, b, c = coefficients. Substituting Eqs. (6) and Eqs. (7) into Eq. (5), the function $\gamma(R)$ becomes:

$$\gamma(R) = \frac{1}{\tau_0 \cdot R_0}\left[\frac{R^2 - R_0^2}{2}c + (R_1 - R_0)\cdot\left(1 - \frac{w_0}{w_c}\right)\cdot\left(\frac{R^2 - R_0^2}{2}a + \frac{R^3 - R_0^3}{3}b\right)\right] \quad \text{if } w_c < w_0 \quad (8)$$

$$\gamma(R) = \frac{1}{\tau_0 \cdot R_0}\cdot\frac{R^2 - R_0^2}{2}c \qquad\qquad\qquad \text{if } w_c \geq w_0$$

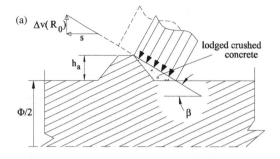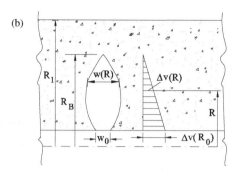

Figure 5. Definition of $\Delta v(R)$: (a) concrete displacements around the ribs; (b) possible distribution of $\Delta v(R)$ around the cracked cross-section.

For a given value of R_B ($\leq R_1$), the coefficients a, b and c can be evaluated by imposing, simultaneously, the following boundary conditions:

$$w_0 < w_c \Rightarrow \begin{cases} \gamma = -1 & \text{if} \quad R = R_A \\ \gamma = -1 & \text{if} \quad R = R_B \end{cases}$$

(9)

$$w_0 \geq w_c \Rightarrow \{\gamma = -1 \quad \text{if} \quad R = R_B\}$$

When a, b and c are known, both $\gamma(R)$ and $\tau(R)$ can be evaluated by means of Eqs. (4) and Eq. (8), respectively.

2.2 *The radial displacement $v(R)$ function*

Due to the longitudinal slip $s(z)$ between steel and concrete, a radial displacement $v(R)$ can be also observed around the ribs of deformed bars. The description of the complete steel-concrete interaction, introduced by Tepfers (1979), is schematically reproduced in Figure 5a. More precisely, the slips $s(z)$ produce crushing of the concrete in front of the ribs. Sliding planes, inclined of $\beta = 30° \div 40°$ with respect to the reinforcement axis, are generated by this phenomenon. The displacement is therefore independent of the rib face angle, when it exceeds $40°$.

The increment of radial displacements Δv, which must be introduced in Eq. (1), is therefore a function of the longitudinal slip $s(z)$. Consequently, on the reinforcing bar of the cracked cross-section, where $s = w_0/2$, it is possible to obtain $v(\Phi/2) = \Delta v(\Phi/2) = \tan \beta w_0/2$. Since the vertical concrete ring in tension (Fig. 2a) resists to the vertical displacement, the condition $\Delta v(R_B) = 0$ is here assumed. On the contrary, in the range $\Phi/2 \leq R \leq R_B$ (Fig. 5b), the following linear variation of $\Delta v(R)$ is considered:

$$\Delta v(R) = \frac{w_0}{2} \cdot \tan \beta \cdot \left(1 - \frac{R - \Phi/2}{R_B - \Phi/2}\right) \leq h_a$$

(10)

where h_a = rib depth of reinforcing bars.

Since ribs on deformed rebars play a fundamental role in the definition of $\Delta v(R)$ (Eq. [10]), in the present model the shape of the crack profile $w(R)$ (Eq. [1]) seems to be strictly connected to the type of reinforcement. This is in accordance with the theoretical and experimental observations of Walraven & Reinhardt (1981). In their tests, $w(R)$ appears nearly constant, independently of the external load N, if smooth steel reinforcement is adopted.

3 A POSSIBLE SOLUTION OF THE PROBLEM

The problem previously formulated can be divided into two parts. In a first step, the crack profile $w(R)$ can be evaluated from Eq. (1). Afterward, the computed $w(R)$ can be used in the evaluation of the structural response of cracked RC members in tension. A similar problem has been solved under the hypothesis of parallel crack surfaces [$w(R) = $ constant] (Fantilli et al., 1998; Fantilli at al., 2005). However, both this hypothesis and the assumption of plain and vertical strain profile in each cross-section of a RC tie are here removed.

In the evaluation of $w(R)$, instead of N, the crack width w_0 at steel-concrete interface (Fig. 2d) is considered as the independent variable. This choice depends on the relationship between $w(R)$ and the applied load N that can be obtained from a test, during which the displacements in the end sections of a RC tie are generally controlled. There is not a one-to-one correspondence between $w(R)$ and N, because of the softening behavior subsequent to the formation of the main cracks. In other words, crack growth produces a temporary unstable behavior of the member, during which an increase of the average elongation and a reduction of the applied load N can be measured (Fantilli et al., 1998). Since the width of a single crack increase monotonically (Bosco et al., 1990) , for a given w_0, it is possible to define univocally the crack profile $w(R)$ and the corresponding applied load N.

846

To define the crack width w at a distance R from the reinforcement axis, all the increments $\Delta w(R)$, measured by means of Eq. (1) in the domain $[\Phi/2, R]$, must be added to w_0. If the τ_0-s relationship (Fig. 2b) and the mechanical properties of concrete and steel are known, the function $w(R)$ can be computed by applying an iterative procedure. When $w(R)$ is known, the state of stress $\sigma_{ct}(w)$ on the crack surfaces can be defined by the cohesive model σ_{ct}-w of Figure 2c. If $R_B < R_1$, in the ligament of internal cracks a uniform stress (equal to the tensile strength f_{ct}) is considered. For the evaluation of steel stress in the cracked cross-section $\sigma_s(z = 0)$, and of the applied load N, it is necessary to formulate a tension stiffening problem. According to Fantilli et al. (1998), it consists of a system of equilibrium and compatibility equations, as in a classical structural problem.

4 ANALYSIS OF R/FRCC AND R/HPFRCC MEMBERS IN TENSION

The Broms' method of measuring crack widths cannot be easily applied to fiber reinforced concrete, because of narrow crack widths. In case of thick concrete covers, it is very difficult, or totally impossible, to measure crack profiles around the steel-concrete surface, even at the failure of the structure. For this reason, to evaluate $w(R)$ in fiber reinforced concrete structures, a X-ray technique has been adopted by Mihashi et al. (2003).

Two ties tested, named R/FRCC and R/HPFRCC, are here considered. They consist of a single deformed bar of diameter $\Phi = 16$ mm embedded in the center of a rectangle cross-section prism made by FRCC, containing only polyethylene fibers, or by HPFRCC containing steel cords with polyethylene fibers. The geometrical properties of the specimens, the test equipment, and the mechanical properties of the cement-based composites are shown in Mihashi et al. (2003). In these elements, the observation of crack pattern was possible by means of the Otsuka's technique (1989), which consists of injecting a contrast medium into holes embedded in the specimen, and taking radiographs at certain stages of loading. Figure 1c and Figure 1d show the X-ray photographs taken at yielding of rebar ($\sigma_s = 420$ MPa), respectively in the R/FRCC and in the R/HPFRCC ties (Mihashi et al., 2003). In both the specimens, the X-ray photographs seem to show qualitatively the stress distribution in the cement-based composite around the main crack. In fact, where the higher values of σ_{ct} have been reached, there is higher concentration of micro-cracks and, consequently, a greater presence of white parts in the radiographs.

In correspondence of a certain value of N, a qualitative comparison between the stress distribution in the cement-based composite around the main crack,

obtained from the proposed model, and the X-ray photograph can be shown. This is possible by assuming that the bond-slip relationship and the aggregate inter-lock mechanism (Fig. 2) can be extended to fiber reinforced composites.

Unfortunately these relationships, regarding FRCC and HPFRCC, are still unknown. However, since the comparison has to be only qualitative, the bond-slip relationship and the Rough Cracks model valid for ordinary concrete are here adopted. At steel yielding, Figure 6 shows the comparison between the σ_{ct} distributions in the two cement-based composites (FRCC in Fig. 6a, and HPFRCC in Fig. 6b) and the X-ray photographs. It is possible to observe a distinct similarity between the shape and the position of the contour lines, which represent the curves at the same tensile stress in the concrete, and the white part of the X-ray images.

Both the X-ray radiographs and the σ_{ct} distributions show the different cracking process in the two specimens. This is due to the different behavior of the cement based composite under tensile actions. In the case of R/FRCC members in tension (Fig. 6a), the decrement of tensile stress affects a wide area of concrete around the main crack. In this zone, the white parts in the X-ray image also show the localized damage of composite. In this case, the crack is entirely cohesive $[w(R) < w_c]$, although the steel reinforcing bar is yielded. As a matter of fact, stresses on the crack surface, and on the surrounding cement-based composite, are different from zero (Fig. 6a). This is also true for R/HPFRCC ties (Fig. 6b), where the reduction of tensile stresses is localized in a restricted area around the main crack. The contour lines do not show a great variation of the tensile stress magnitude, which remains almost equal to the tensile strength of the composite. This is shown by the X-ray photograph of Figure 6b, where the white and the black parts are not clearly separated as in the R/FRCC element (Fig. 6a).

5 CONCLUSIONS

A new model, able to define the crack profile in a cementitious composite of reinforced elements in tension, is proposed. The bond-slip mechanism, the fracture mechanics of concrete in tension and the shear resistance of cracked concrete are taken into account. Regarding to fiber reinforced cementitious composites, a good agreement between the distribution of tensile stresses in the composites and the X-ray images is also found in R/FRCC and R/HPFRCC members in tension. In these ties, made of cement-based composites having a higher fracture energy, or a strain hardening, cracks are much narrower than in RC ties. Moreover, crack widths are larger on the interface between steel rebar and concrete than on the surface of concrete cover, even after yielding of the rebar.

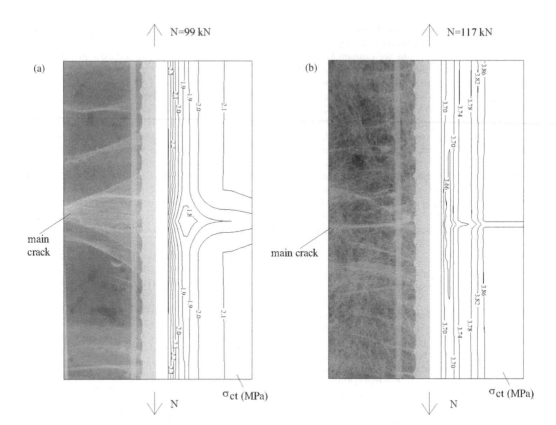

Figure 6. Comparison between the X-ray photographs (Mihashi et al., 2003; Otsuka et al., 2003), and σ_{ct} distributions obtained with the proposed model (contour lines on the right): (a)R/FRCC specimen at steel yielding; (b) R/HPFRCC specimen at steel yielding.

ACKNOWLEDGEMENTS

This work has been financially supported by the Italian Ministry of Education, University and Research (PRIN 2004–2005).

REFERENCES

ACI Committee 224, 1986. Cracking of Concrete Members in Direct Tension. *ACI Journal* 83(1): 3–13.

Bazant, Z.P., and Gambarova, P., 1980. Rough Cracks in Reinforced Concrete. *Journal of Structural Division ASCE* 106(4): 819–842.

Beeby, A.W., 1972. A study of cracking in reinforced concrete members subjected to pure tension. *Cement and Concrete Association, Technical Report n.42.468*, London, U.K.

Beeby, A.W., 2004. The influence of the parameter ϕ/ρ_{eff} on crack widths. *Structural Concrete* 5(2):71–83.

Bianchini, A.C., Kesler, C.E., and Lott, J.L., 1968. Cracking of Reinforced Concrete Under External Load. *In ACI-SP 20, Causes, mechanism and control of cracking in concrete*: 73–85.

Bosco, C., Carpinteri, A., and Debernardi, P.G., 1990. Fracture of Reinforced Concrete: Scale Effect and Snap-Back

Instability. *Engineering Fracture Mechanics* 35(4/5): 665–677.

Broms, B.B., 1965. Crack Width and Crack Spacing in Reinforced Concrete Members. *ACI Journal Proceedings*, 62(10): 1237–1256.

Comitè Euro-International du Bèton (CEB), 1993. CEB-FIP Model Code 1990. *CEB Bulletin d'Information n. 213/214*, Lausanne, Switzerland.

Dei Poli, S., Gambarova, P.G., and Karakoc, C., 1986. Aggregate Interlock Role in R.C. Thin-Webbed Beams in Shear. *Journal of Structural Engineering ASCE* 113(1): 1–19.

Fantilli, A.P., Ferretti, D., Iori, I., and Vallini, P., 1998. Behaviour of R/C elements in bending and tension: the problem of minimum reinforcement ratio. *In A. Carpinteri (ed) Minimum Reinforcement in Concrete Members, ESIS Publication 24*, Elsevier, Amsterdam, The Netherland: 99–125.

Fantilli, A.P., Mihashi, H., and Vallini, P., 2005. Strain Compatibility between HPFRCC and Steel Reinforcement. *Materials and Structures* 38(5): 495–503.

Gambarova, P.G., 1980. On Aggregate Interlock Mechanism in Reinforced Concrete Plates with Extensive Cracking. *Studi e Ricerche, Corso di Perfezionamento per le Costruzioni in Cemento Armato F.lli Pesenti, Politecnico di Milano*, Vol. 2: 43–102 (in Italian).

Goto, Y., 1971. Crack Formed in Concrete Around Deformed Tension Bars. *ACI Journal Proceedings* 68(4): 244–251.

Hillerborg, A., Modéer, M., and Petersson, P.E., 1976. Analysis of crack formation and crack growth in concrete by means of fracture mechanics and finite elements. *Cement and Concrete Research* 6(6): 773–782.

Lutz, L.A., 1970. Analysis of Stresses in Concrete Near a Reinforcing Bar Due to Bond and Transverse Cracking. *ACI Journal Proceedings* 67(10): 778–787.

Mihashi, H., Kawamata, A., Kiyota, M., Otsuka, K., and Suzuki, S., 2003. Bond Crack and Tension Stiffening Effect in High-Performance Cementitious Composite. *In Japan Concrete Institute JCI Symposium on Ductile Fiber Reinforced Cementitious Composites (DFRCC)*, Tokyo, Japan: 1–8.

Otsuka, K., 1989. X-ray technique with contrast medium to detect fine cracks in reinforced concrete. *In Mihashi et al. (Eds) Fracture Toughness and Fracture Energy*, Sendai, Japan: 521–534.

Otsuka, K., Mihashi, H., Kiyota, M., Mori, S., and Kawamata, A., 2003. Observation of multiple cracking in hybrid FRCC at micro and macro levels. *Journal of Advanced Concrete Technology* 1(3): 291–298.

Scott, R.H., and Gill, P.A.T., 1987. Short-term distributions of strain and bond stress along tension reinforcement. *The Structural Engineer* 65B(2): 39–48.

Tepfers, R., 1979. Cracking of concrete cover along anchored deformed reinforcing bars. *Magazine of Concrete Research* 31(106): 3–12.

Walraven, J.C., and Reinhardt, H.W., 1981. Theory and Experiments on the Mechanical Behaviour of Cracks in Plain and Reinforced Concrete Subjected to Shear Loading. *Heron* 26(1A): 1–68.

Watstein, D., and Mathey, R.G., 1959. Width of Cracks in Concrete at the Surface of Reinforcing Steel Evaluated by Means of Tensile Bond Specimens. *ACI Journal Proceedings* 55(7): 47–56.

Fracture Mechanics of Concrete and Concrete Structures – Design, Assessment and Retrofitting of RC Structures – Carpinteri, et al. (eds)
© 2007 Taylor & Francis Group, London, ISBN 978-0-415-44616-7

Concrete structures under severe loading: A strategy to model the response for a large range of dynamic loads

A. Rouquand
Centre d'Etudes de Gramat, France

C. Pontiroli
Communications & Systems, Merignac, France

J. Mazars
Laboratoire Sols, Solides, Structures, Risques and VOR research network, Grenoble, France

ABSTRACT: Realistic dynamic description and modelling of material failure is one of the actual problems in structural mechanics. Analyses of failure processes require the use of complex FE analyses and advanced constitutive models. For modeling of concrete it is necessary to capture several important phenomena such as damage and ductility (possibly softening), rate effects ... This implies the description of the concrete behaviour with constitutive equations as refined as possible. In this work a coupled damage and plasticity model including the effective stress concept is used to solve time dependent problems. This is done using an explicit procedure contributing to a reduction of the computational time. Such a procedure requires no iterations and no tangent stiffness matrix. Stability is automatically assured by using small time increments. This strategy has been successfully applied during the last years to model a large range of severe loadings on complex reinforced concrete structures. The mean model concepts are presented in this paper and some examples of numerical simulations are given and compared to experimental data.

1 INTRODUCTION

The simulation of the failure process in complex reinforced structures is a big challenge. Several physical phenomena must be considered. For example, for high velocity impacts and explosion events near concrete structures, high pressures and high strain rate loading occur locally around the projectile and around the explosive charge. Such phenomena generate pore collapse mechanisms that dissipate a large amount of energy. Irreversible shear strains under high pressure can also be observed driving a significant part of the material response. Under a high pressure regime in porous material, the elastic response becomes non linear and pressure dependent. For soils, rocks and concrete, the water content inside the open voids is very important. This parameter can control the pressure volume relationship and heavily influences the shear material response.

At some distance from the projectile or the explosive charge, the physical phenomena change progressively to become structure oscillations at moderate strain rate levels. The material response is now driven by an increase of concrete damage due to crack opening mechanisms, crack closure effects and friction phenomena related to differential displacements at the crack tip level. The material model has to account for all of these effects such as stiffness deterioration, recovery of stiffness due to crack closure, or permanent strains and frictional stresses that generate hysteretic loops during unloading and reloading paths. All these mechanisms must be implemented together in a unique material model able to simulate a large range of dynamic problems.

Different kinds of models are proposed to simulate the behaviour of concrete structures including plasticity (Ottosen 1979), damage (Mazars 1986, 1989, Jirasek 2004) or fracture based approaches (Bazant et al. 1996). Nevertheless, very few are able to simulate crash tests (Krieg 1978, Van Mier et al 1991).

The ability of the constitutive model to reproduce the real material behaviour is not the only challenge. Numerical aspects, related to the algorithm used to compute the stress tensor at the local level or related to the computation of structural displacements at the global level in a finite element analysis are also very important. At each level, the computational procedure has to be numerically efficient and robust.

This paper gives some details of the numerical procedure used to perform numerical simulations of concrete structures under severe loadings. Examples are given and finite element results are compared to experimental data.

2 DAMAGE AND PLASTIC MODEL FOR CONCRETE: PRM CRASH MODEL

2.1 The scalar damage model (PRM model)

2.1.1 Constitutive relations
To simulate the behaviour of concrete at a moderate stress level, a two scalar damage model has been proposed from works by J. Mazars (1986), C. Pontiroli (1995), A. Rouquand (1995 & 2005). The named PRM model simulates the cyclic behaviour of concrete. This model distinguishes the behaviour under tension and the behaviour under compression. Between theses two loading states a transition zone is defined by $(\sigma_{ft}, \varepsilon_{ft})$. Where σ_{ft} and ε_{ft} are respectively the crack closure stress and the crack closure strain. The main equations of the PRM model for a uniaxial loading are:

under traction: $(\sigma - \sigma_{ft}) = E_0 \cdot (1-D_t) \cdot (\varepsilon - \varepsilon_{ft})$

under compression: $(\sigma - \sigma_{ft}) = E_0 \cdot (1-D_c) \cdot (\varepsilon - \varepsilon_{ft})$

E_0 is the initial Young's modulus. D_t evolves as well in tension as in compression through the variable

$$\tilde{\varepsilon} = \sqrt{\sum_i \langle x_i \rangle_+^2} \ [1], <x_i>_+ = x_i \ \text{if} \ x_i > 0 \ \text{and} \ <x_i>_+ = 0$$

if not; $x_i = \varepsilon_i$ are the principal strain components in compression and $x_i = (\varepsilon - \varepsilon_{ft})_i$ in tension. $\tilde{\varepsilon}$ is an indicator of the local state of extension (positive strain state), responsible of damage. The general evolution of damage is an exponential form driven by $\tilde{\varepsilon}$: $D_t = fct$ $(\tilde{\varepsilon}, \varepsilon_{0t}, A_t, B_t)$, $\varepsilon_{0t}, A_t, B_t$ are material parameters. ε_{0t} is the tensile damage threshold. D_c is driven by the same variable $\tilde{\varepsilon} = \sqrt{\sum_i \langle \varepsilon_i \rangle_+^2}$ and evolves through the same function: $D_c = fct \ (\tilde{\varepsilon}, \varepsilon_{0c}, A_c, B_c)$.

Initially $\varepsilon_{ft} = \varepsilon_{ft0}$ is a material parameter. Afterwards ε_{ft} is directly link to Dc. $\sigma_{ft} = f(\varepsilon_{ft}, D_c)$ gives the stage where the transition between the two kinds of damage occurs. The corresponding response for a uniaxial cyclic loading is given Figure 1.

We can observe that the behaviour can be described by the classical equation: $\sigma_d = E_0 (1 - D_i) \varepsilon_d$ with i = t, c, $\varepsilon_d = \varepsilon - \varepsilon_{ft}$ and $\sigma_d = \sigma - \sigma_{ft}$.

The general 3D constitutive equation of the model relating strain and stress tensors (in bold) is reported below:

$$(\mathbf{\sigma} - \mathbf{\sigma}_{ft}) = \Lambda_0 (1-D) (\mathbf{\varepsilon} - \mathbf{\varepsilon}_{ft}) \quad \text{or} \tag{1}$$

$$(\mathbf{\sigma} - \mathbf{\sigma}_{ft}) = (1-D) [\lambda_0 \ \text{trace}(\mathbf{\varepsilon} - \mathbf{\varepsilon}_{ft})\mathbf{1} + 2\mu_0 (\mathbf{\varepsilon} - \mathbf{\varepsilon}_{ft})]$$

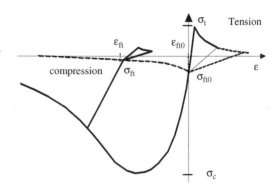

Figure 1. Stress strain curve for a tensile – compressive loading.

where σ_{ft} and ε_{ft} are the crack closure stress and strain tensors used to manage permanent effects; Λ_0 is related to the initial mechanical characteristics of the material. D, the damage remains a scalar and is issued from a combination of the two modes of damage:

$$D = \alpha_t D_t + (1- \alpha_t)D_c \tag{2}$$

α_t evolves between 0 and 1 and the actual values depend on $(\varepsilon - \varepsilon_{ft})$. For more details see Mazars (1986).

This formulation is an explicit one. It has been implemented into "ABAQUS explicit" and is used for dynamic structural simulations. In order to avoid depending mesh size solutions, a Hillerborg method has been used (Hillerborg 1976) which allows to control the dissipation of energy in each element.

2.1.2 Strain rate effects – internal friction damping
It is well known that concrete is strain rate dependent particularly by pure tensile loading. This effect is accounted for using dynamic thresholds $(\varepsilon_{0t}^d \ \text{and} \ \varepsilon_{0c}^d)$ instead of static one's $(\varepsilon_{0t}^s \ \text{and} \ \varepsilon_{0c}^s)$. Dynamic thresholds are deduced from the static ones through a dynamic increase factor $R = \varepsilon_0^d/\varepsilon_0^s$. Its value for a compressive dynamic loading takes the following form:

$$R_c = \min(1.0 + a_c \dot{\varepsilon}^{b_c}, \ 2.50) \tag{3}$$

And for a dynamic tensile loading:

$$R_t = \min[\max(1.0 + a_t \dot{\varepsilon}^{b_t}, \ 0.9\dot{\varepsilon}^{0.46}), \ 10.0] \tag{4}$$

a_c, b_c and a_t, b_t are material coefficients defined by the user. For a high strain rate, the tensile dynamic increase factor is supposed to follow an empirical formula: $0.9\dot{\varepsilon}^{0.46}$ that agrees very well with the experimental data obtained by Brara & Klepaczko (1999)

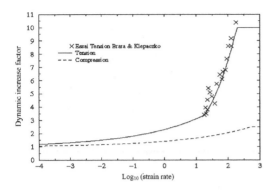

Figure 2. Strain rate effects (PRM damage model).

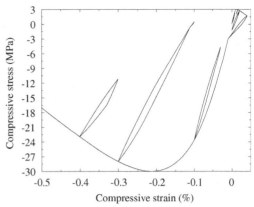

Figure 3. Cyclic loading including damping stresses.

on a particular micro concrete. Figure 2 illustrates the evolution of the compressive (dashed line) and tensile (continuous line) dynamic increase factors versus the strain rate.

For cyclic loading, as the one encountered during an earthquake loading, friction stresses induce significant dissipated energy during unloading and reloading cycles. To account for this important phenomenon an additional damping stress is introduced in the model:

$$\underline{\underline{\sigma}} - \underline{\underline{\sigma}}_{ft} = (\underline{\underline{\sigma}} - \underline{\underline{\sigma}}_{ft})^{damage} + \underline{\underline{\sigma}}^{damping} \qquad (5)$$

The damping stress generates a hysteretic loop during the unloading and the reloading cycle. This stress is calculated from the damping ratio ζ classically defined as the ratio between the area under the closed loop and the area under the linear elastic-damage stress curve:

$$\varsigma = \frac{A_h}{E_0(1-D)(\overline{\varepsilon}_{max} - \overline{\varepsilon}_{ft})^2} \qquad (6)$$

A_h is the loop area under the stress strain curve, $E_0(1-D)$ is the current material stiffness. ε_{max} is the maximum strain before unloading, ε_{ft} is the closure strain that defines the transition point between compression and tension.

The damping stresses are computed in such a way that the damping ratio ζ is related to the damage D according to the relation:

$$\zeta = (\beta_1 + \beta_2 D) \qquad (7)$$

β_1 is a damping ratio for an undamaged and perfectly elastic material. $\beta_1 + \beta_2$ is the damping ratio for a fully damaged material. β_1 and β_2 are material parameters. Usually β_1 can be chosen equal to 0.02 and β_2 can be chosen equal to 0.05. Figure 3 shows, for cyclic tensile or compressive loading, the strain stress curve including damping stresses.

2.2 Plastic model with effective stresses

The previous damage model is very efficient to simulate the behaviour of concrete for unconfined or low confined cyclic loading (Rouquand 2005). For very high dynamic loads leading to a higher pressure level, an elastic plastic model is more appropriate. For example, the impact of a projectile striking a concrete plate at 300 m/s induces local pressures near the projectile nozzle of several hundred MPa. The previous damage model cannot simulate the pore collapse phenomena rising at this pressure level. It also cannot model the shear plastic strain occurring in this pressure range. To overcome these limitations, the elastic and plastic model proposed by Krieg (1978) has been chosen to simulate this kind of problem. From this simple elasto-plastic model a first improvement has been introduced in order to simulate the non linear elastic behaviour encountered during an unloading and reloading cycle under a high pressure level. A second improvement has been made to account for the water content effects introducing an effective stress theory as described by C. Mariotti (2002). This effect induces change on the pressure volume curve and on the shear plastic stress limit.

2.2.1 The modified Krieg model (dry material)

The Krieg model can be applied to describe the behaviour of a dry material. The improvements made here concern the elastic behaviour which is now non linear and pressure dependent. This non linearity increases as the pore collapse phenomena progresses. Figure 4 shows a typical pressure volume curve used in the modified Krieg model. For pressure values under P_1, the behaviour between pressure and volume is linear and elastic. For a pressure greater than P_1, the pore collapse mechanism becomes effective. During the loading process, the pressure-volume response follows a curve identified from experiments. During the

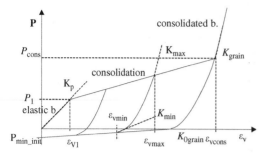

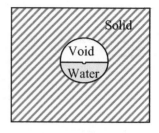

Figure 5. Simplified geologic media of a partially saturated material.

Figure 4. Pressure volume behaviour in the modified elastic and plastic Krieg model.

unloading, the behaviour is elastic but non linear. The bulk modulus becomes pressure dependent. It is equal to K_{max} at the first unloading point and decreases to K_{min} when the tensile pressure cut-off P_{min} is reached (this value is generally negative, which means that traction is necessary to recover the initial volume). This pressure cut-off becomes smaller and smaller as the maximum pressure P_{max} increases. When P_{max} is close to P_1, K_{max} is close to K_{min} and also closed to the initial bulk modulus K_p. When P_{max} reaches P_{cons}, K_{max} becomes equal to K_{grain} and K_{min} becomes equal to K_{0grain}. So the non linearity becomes more and more important as the pore collapse phenomena progresses.

When P_{max} becomes greater than P_{cons}, the pore collapse phenomena is achieved because all the voids are removed from the material. At this pressure level the material is consolidated and the behaviour becomes purely elastic and non linear.

2.2.2 Improvement of the Krieg model for partially saturated materials

2.2.2.1 Pressure volume behaviour

Many concrete and geologic media have an open porous structure. The water can move through the porous media from one void to another. Consequently, the void can be partially or totally filled with water. This induces significant changes in the material response and particularly on the relation between pressure and volume.

To understand more easily the water effect on a geologic medium, the material structure can be studied as a mixture of a solid medium with a void partially filled with water as shown in Figure 5.

For high dynamic loads, the time scale is very low (few milliseconds or less) so the water has no time to move inside the material and undrained conditions can be considered. Figure 6 shows the generic response of a partially saturated material. This response is given in terms of pressure versus the volume change. *For a dry material*, the pressure volume response follows the solid curve shown on figure 6. When the pressure is sufficient to remove all the voids, the response is

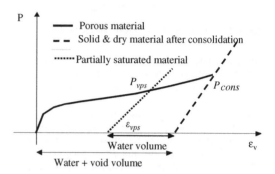

Figure 6. Water content effect on pressure volume relationship.

given by the thick dashed curve. *In case of a partially saturated material*, the relation between pressure and volume is given by the response of the dry material until all the voids (part of the pores without water) are removed from the medium. Thereafter, the thin dashed curve gives the response of the solid and water mixture. The intersection of the large dashed curve with the horizontal axis gives the porosity of the dry material. The intersection of the dashed light curve with the horizontal axis gives the "free porosity" ε_{vps} of the partially saturated material. Consequently, when the material becomes more and more dry, the thin dashed curve moves to the right. In the modified plastic model presented here, the knowledge of the water content ratio η (water volume divided by the total volume) is sufficient to deduce all the improvements of the material behaviour.

When the pressure reaches the particular value P_{vps} corresponding to the intersection of the solid line with the thin dashed line, all the voids of the partially saturated medium are removed, so the medium becomes a two phase mixture of liquid and solid. To define the behaviour of this solid and water mixture (thin dashed curve) the pressure is assumed to increase in the same way in the two phases (solid and liquid phases). So an iterative procedure as to be run in order to find the relative volume changes of each phase. This procedure gives a pressure difference equal to the consolidation pressure of the partially saturated material P_{vps} when the total volume change of the two phases $(\varepsilon_v - \varepsilon_{vps})$ is known. Liquid behaviour is described using the Mie Gruneisen equation of state and the solid phase behaviour is the non linear elastic model briefly described in §2.2.1.

2.2.2.2 Shear behaviour

Water content has an effect on the shear behaviour. In the Krieg model, the plastic shear strength q_0 (computed as the Von Mises stress) is pressure dependent (see Figure 7). As the pressure increases, the shear yield stress increases too. This effect is the consequence of the porous structure of the material. During the pore collapse phenomena the void volume decreases, the pressure increases so the contact area of the solid grains inside the material matrix increases and the shear forces inducing sliding motions between the solid grains also increase. When all the voids are removed, the shear strength remains constant and becomes pressure independent because the contact area cannot increase any more. The material becomes "homogeneous" and the shear strength reaches a limit that is material dependent. For a partially saturated material, the behaviour remains similar to the behaviour of a dry material until all the voids are removed. Thereafter we suppose that water pressure and solid grain pressure increase together in the same way. So the pressure difference between the two phases remains constant, contact forces and contact areas at the micro scale level maintain constant and the shear strength remains also constant.

At this point, the effective stress concept can be introduced. The shear strength is related to the effective pressure and this effective pressure is taken equal to the interstitial pressure. For a dry material, the effective pressure is always equal to the total pressure. But for a partially saturated material, the effective pressure is the total pressure like in dry material until all the voids are removed. *After consolidation the interstitial (or the effective) pressure does not increase any more and consequently the shear yield strength remains constant.* As the water contents increase, the pressure level P_{vps} decreases and then the shear strength q_0 also decreases. Figure 7 illustrates the effect of the effective pressure concept. The solid line gives the shear yield strength versus the pressure for a dry material. For a partially saturated one, the shear strength follows the solid line until the pressure P_{vps} is reached. Afterwards the shear strength does not increase and it follows the dashed horizontal line.

2.3 Coupling procedure for the damage and the modified Krieg model: PRM crash model

The scalar damage model has been coupled with the modified Krieg model. The coupling procedure ensures a perfect continuity between the two model

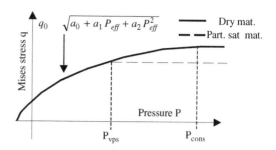

Figure 7. Shear yield strength versus pressure for a dry and a partially saturated material (q is the Von Mises stress).

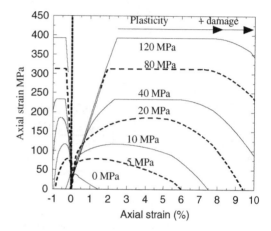

Figure 8. Stress strain response on a concrete given by the coupled damage and plastic model (tri-axial tests with increasing lateral pressure).

responses. The predicted stresses correspond to the damage model response if the maximum pressure is too low to start the pore collapse phenomena or if the shear stress is too low to reach the shear yield stress. If not, the plastic model is activated and pilots the evolutions until the extensions sufficiently increase to lead to a damage failure.

Figure 8 shows the static response obtained on a cylindrical specimen for tri-axial tests with increasing lateral pressure. Tests performed on the GIGA machine at 3S-R Grenoble prove the pertinence of these results (Gabet 2006).

3 NUMERICAL SIMULATIONS

This model has been implemented in the ABAQUS explicit finite element code and it has been extensively used to simulate a lot of complex problems. The PRM damage model can be used with most of the available finite elements (1D truss elements, beam elements, 2D plane stress and plane strain elements, 2D

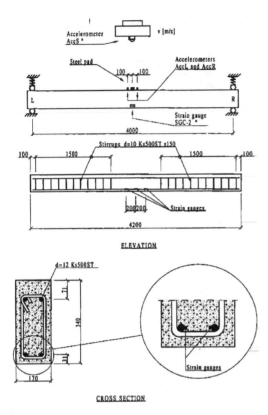

Figure 9. Experimental device and beam characteristics (dynamic three points bending tests).

Figure 10. Computed and observed crack pattern.

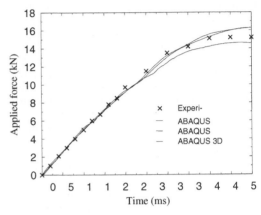

Figure 11. Measured and computed dynamic loads (three points bending test).

axisymmetric elements, shell elements, 3D solid elements, etc.). The coupled damage and plastic model (PRM crash model) can be used with 2D plane strain elements, 2D axisymmetric and 3D solid elements. In order to show the capabilities of the coupled model some applications are presented here and numerical results are compared to experimental data.

3.1 Dynamic three points bending test on a reinforced concrete beam

Figure 9 shows the experimental device and the beam characteristics (in mm). These tests have been conducted by Agardh, Magnusson & Hanson (1999) in Sweden on a high strength reinforced concrete beam.

Beam elements, 2D plane stress elements and 3D solid elements are used to model the reinforced concrete beam. A single element is used in the depth direction with the 3D model. Taking advantage of the symmetry, only a half part of the beam is modelled. The reinforcement material model is the classical Johnson Cook plasticity model. The concrete and the steel reinforcement are supposed to be perfectly bonded. Figure 9 details the experimental apparatus. When

the deflection becomes greater than 90 mm, shock absorbers damp the central part of the beam.

Figure 10 shows, at the end of the dynamic test, the tensile damage contours on the 3D beam model (upper part of the figure). The lower part shows the corresponding observed crack pattern. The computed cracks are mainly concentrated in the central part of beam like in the experiment.

In figure 11 the measured force (cross points) is compared to the three computed forces resulting from the three different meshes. The beam model gives the lower force. The 2D and 3D models give very similar results.

3.2 Impact on a T shape reinforced concrete structure

This study is related to the analysis of the vulnerability of concrete structures under intentional actions. More specifically, the effect of a projectile of about 80 kg striking a reinforced concrete plate is studied. Such an experiment has been done by E. Buzaud et al. (2003). The 35NCD16 steel projectile has an ogival nozzle. Its diameter is 160 mm and its length is 960 mm (Figure 13). An accelerometer recorder system is mounted inside the projectile to measure the axial and lateral accelerations during the tests. Figure 12 shows the test configuration with a T shape concrete structure. The size of each reinforced concrete square plate composing the target is 3 m. The thickness of the front

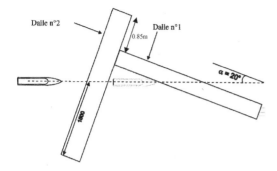

Figure 12. Test configuration, impact on T structure.

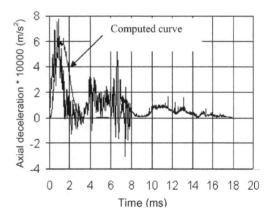

Figure 13. Projectile mesh.

part of the concrete target is 400 mm and the thickness of the rear part is 300 mm. Reinforcement is composed of two steel layers (one on each side of the concrete plate) with 16 mm diameter bars. Other 10 mm diameter bars link each reinforcement mesh node of the face to face layers. The distance that separates each bar is 100 mm. The distance between the reinforcement layer and the top (or the bottom) plate surface is 50 mm.

3D numerical simulations have been done using the ABAQUS explicit finite element code. The total number of the finite elements is about 530 000 for the entire model. The projectile material (Figure 13) is simulated using an elastic and perfectly plastic model with a plastic yield stress of 1300 MPa. The reinforcement is also modelled with an elasto-plastic model with isotropic hardening. The initial yield stress is 600 MPa and reaches 633 MPa for a failure strain $\varepsilon = 0.13$. The concrete behaviour is simulated with the coupled plastic and damage model.

Table 1 gives the concrete material data used in the simulation. Most of these values are taken from literature data relative to similar materials.

On Figure 14, the measured deceleration is compared to the computed value. Some differences can be seen but the overall deceleration shape is correctly predicted.

Figure 15 shows the tensile damage contours at the end of the numerical simulation (T = 20 ms). The first part of the target is perforated and a rebound off the rear part is observed. This has been observed experimentally. The projectile velocity, at the exit of the first impacted plate, is also close to the measured one.

Table 1. Concrete material data.

Symbol	Parameter	Value S.I. units
E_0	Young modulus	$3.5 \ 10^{10}$
ν_0	Poisson ratio	0.2
σ_c	Compressive strength	$-41 \ 10^6$
σ_t	Tensile strength	$3.3 \ 10^6$
G_f	Fracture energy	120
a_0	1st coefficient (shear strength)	$1.8 \ 10^{15}$
a_1	2nd coefficient (shear strength)	$2.4 \ 10^8$
a_2	3rd coefficient (shear strength)	0.6
n	No. of points (compaction curve)	2
P_1	Pressure	$60 \ 10^6$
ε_{v1}	Volume	-0.00308
P_{cons}	Consolidation pressure (last point)	$2 \ 10^9$
ε_{vcons}	Corresponding volume (last point)	-0.1284
K_{grain}	Bulk modulus at consolidation	$3.9 \ 10^{10}$
K_{0grain}	Bulk modulus unloaded material	$3.9 \ 10^9$
η_{eau}	Water contents ratio	0
ρ_0	Density	2300

Figure 14. Measured and computed projectile decelerations.

4 CONCLUSION

A general constitutive model for a concrete structure submitted to extreme loading (high velocity and high confinement) has been developed and implemented into the "ABAQUS explicit" code in the framework of damage and plasticity mechanics. The resulting coupled damage and plasticity model (*PRM crash model*) can simulate a lot of physical mechanisms like crack opening and crack closure effects, strain rate effects, material damping induced by internal friction, compaction of porous media, shear plastic strains under high pressure, water content effects on the pressure volume behaviour and on the shear strength.

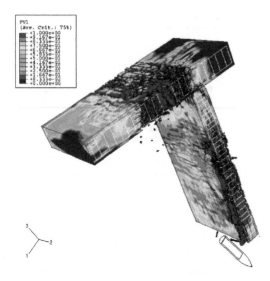

Figure 15. Tensile damage contours at 20 ms. The projectile has perforated the upper part and penetrated the right part after a rebound.

To validate this particular coupling of plasticity and damage, an extensive experimental program has been performed at 3S-R Grenoble using the GIGA machine which allows high confinement up to 1 GPa (Gabet 2006), and a new program is in progress on the large Hopkinson bar at JRC Ispra to complete the data base under high velocity loading.

The new model has been extensively used and can advantageously simulate a large panel of problems going from quasi-static simulations on concrete structures to high dynamic problems related to the effect of high velocity impacts.

The examples presented here and during the conference, demonstrate the efficiency of the proposed numerical procedure.

REFERENCES

Agardh L., Magnusson J., Hansson H., 1999, High strength concrete beams subjected to impact loading, an experimental study, FOA Defence Research Establishment, FOA-R-99-01187-311—SE.

Bazant Z.P., 1994, "Nonlocal damage theory based on micromechanic of crack interaction". *Journal Engineering Mech.* ASCE 120, pp. 593–617.

Brara A., 1999, Etude expérimentale de la traction dynamique du béton par écaillage, *thèse de l'université de Metz – France*

Buzaud E. et al., 2003, An experimental investigation of corner effects resulting from vertical attack on hardened structures, *proceedings of 11th ISIEMS,* Mannheim, Germany.

Gabet T, 2006, Comportement triaxial du béton sous fortes contraintes: Influence du trajet de chargement, Phd thesis, Université Joseph Fourier, Grenoble.

Hillerborg A., Modeer M., Petersson P.E., 1976, Analysis of crack formation and growth in concrete beams of fracture mechanics and finite elements, *Cement and Concrete Research*, Vol. 6, pp. 773–782.

Jirásek M., 2004, "Non-local damage mechanics with application to concrete". *Revue française de génie civil*, 8 (2004), pp. 683–707.

Krieg R.D., 1978, A simple constitutive description for soils ans crushable foams, Sandia National Laboratories, SC-DR-72-0833, Albuquerque, New Mexico.

Mariotti C., Perlat J.P., Guerin J.M., 2002, A numerical approach for partially saturated geomaterials under shock, CEA/DAM Bruyères le Châtel, *International Journal of Impact Engineering* 28 (2003) 717–741.

Mazars J., 1986. A description of micro and macro scale damage of concrete structures, *Engineering Fracture Mechanics*, V. 25, n° 5/6.

Mazars J., Pijaudier-Cabot G., 1989, "Continuum damage theory – application to concrete. *Journal of Engineering Mechanics*". 115(2), pp. 345–365.

Ottosen N.S., 1979, "Constitutive model for short time loading of concrete". *Journal of Engineering Mechanics,* ASCE, Vol 105, pp. 127–141.

Rouquand A., Pontiroli C., 1995, Some considerations on explicit damage models including crack closure effects and anisotropic behaviour, *Proceedings FRAMCOS-2,* Ed.F.H.

Wittmann, AEDIFICATIO Publisher, Freiburg.

Rouquand A., 2005, Presentation d'un modèle de comportement des géomatériaux, applications au calcul de structures et aux effets des armes conventionnelles, Centre d'Etudes de Gramat, rapport technique T2005-00021/CEG/NC.

Van Mier J.G., Pruijssers A., Reinhardt H.W., Monnier T., 1991, Load time response of colliding concrete bodies, J. of Structure Engineering, vol. 117, p. 354–374.

Fracture Mechanics of Concrete and Concrete Structures – Design, Assessment and Retrofitting of RC Structures – Carpinteri, et al. (eds)
© 2007 Taylor & Francis Group, London, ISBN 978-0-415-44616-7

Nonlinear sub-structuring applied to pseudo-dynamic tests on RC structures

F. Ragueneau, A. Souid, A. Delaplace & R. Desmorat
LMT-Cachan, ENS Cachan/CNRS/Univ. Paris 6, Cachan, France

ABSTRACT: This paper aims at giving some insights within the pseudo-dynamics tests on reinforced concrete structures. Nonlinear substructuring is used making benefits of simplified finite element analysis based on multifibers beams theory approach. Continuum damage mechanics is used for the modelled structure allowing creating a realistic dynamic environment for the tested substructure.

1 INTRODUCTION

The comprehension of the ultimate behaviour of Civil Engineering structures subject to natural or industrial risks such as shocks, impacts, earthquake, explosions, etc can be handled in two ways: experimental testing and numerical modelling. In earthquake engineering, the major drawback remains in the experimental work. One has to deal with large scale structures subject to dynamic and complex loading. Classical tests are performed on shaking tables, allowing reproducing real or artificial earthquake, but with reproducibility difficulties and physical measurement limitations. To overcome these difficulties, the pseudo-dynamic or hybrid testing are under developments (Pegon & Pinto 2000). A combination between the numerical modelling (into which one can introduce the suitable model of material behaviour) and a test on parts of the structures can be made to better understand the structure response while benefiting from the substructuring technique. To numerically determine the inertia forces for performing static tests instead of dynamic ones leads to the so-called pseudo-dynamics (PSD) modelling. We describe and present the results for such PSD tests with sub-structuring technique carried out in nonlinear range. In fact, taking into account the nonlinear behaviour of the modelized structure is of major importance since the concomitant stiffness or eigenfrequency decrease change the whole response of the structure and so, the boundary conditions and loadings of the tested structure.

The tests were conducted for a two level reinforced concrete frame which is subjected to a real two components earthquake (horizontal and vertical). We present the use of computations for the modelized and tested sub-structures in the PSD tests. Both an implicit and an explicit time integration schemes are used for the simulated parts in parallel with an explicit one used for the tested part (Souid *et al.* 2005). Tests results are used to identify and validate the nonlinear constitutive equations. For instance, a three dimensional damage model with induced damage anisotropy is described and used for quasi-brittle materials such as concrete. The quasi-static condition of the tests allows performing refined field measurements using the digital images correlation techniques. At the scale of a reinforced concrete beam, one can distinguish for different geometries and steel reinforcement ratio, the rupture kinematics and make easier numerical model identification.

2 PSEUDO-DYNAMIC TESTING

2.1 *General scheme*

Evaluation of the seismic response of a structural system is usually conducted using a shaking table. However, shaking-table experiments for large-scale structures are difficult, for instance due to table capacity limitations. An alternative way of testing full or large scale structures is the PSD testing (Shing and Mahin, 1984, Takanashi and Nakashima, 1987). The PSD testing is an experimental technique developed to evaluate the seismic performance of structure samples in a laboratory by means of computer-controlled simulation. It is an hybrid method, in which the structural displacements due to the earthquake are computed by using a stepwise integration procedure. Let us consider the dynamic equilibrium equation under seismic external acceleration $a(t)$:

$$M\ddot{u}(t) + C\dot{u}(t) + r(t) = f(t) = -M\{1\}a(t) \tag{1}$$

where M, C are the mass and damping matrices, $\dot{u}(t)$ and $\ddot{u}(t)$ are respectively the relative velocity and acceleration vectors at time t. Knowing variables at time t_n one can compute displacement and velocity at time t_{n+1} by using a numerical scheme. Only the $r(t)$ forces are experimentally measured. Numerical time discretization schemes belong to the Newmark family. Within the framework of experiment-computation interaction (Shing et al. 1991), an efficient and pragmatic choice consists in implementing an Operator Splitting algorithm, allowing for a direct integration without iteration in the linear range (Nakashima et al. 1993). In nonlinear regime up to rupture, it becomes necessary to damp the high frequencies, sources of numerical instabilities when an explicit procedure is adopted. In that purpose, the OS technique can be coupled to the HHT algorithm (Hilber et al. 1977, Combescure & Pegon 1997). Knowing the accelerogram and so the acceleration vector $\ddot{u}^{n+1}$, the displacements and velocities are predicted as following

$$u_{trial}^{n+1} = u^n + \Delta t \dot{u}^n + \frac{1}{2}\Delta t^2(1-2\beta)\ddot{u}^n \tag{2}$$

$$\dot{u}_{trial}^{n+1} = \dot{u}^n + \Delta t(1-\gamma)\ddot{u}^n \tag{3}$$

and corrected using:

$$u^{n+1} = u_{trial}^{n+1} + \Delta t^2 \beta \ddot{u}^{n+1} \tag{4}$$

$$\dot{u}^{n+1} = \dot{u}_{trial}^{n+1} + \Delta t \gamma \ddot{u}^{n+1} \tag{5}$$

with $\beta = (1-\alpha)^2/4$ et $\gamma = (1-2\alpha)/2$. For $\alpha = 0$, we recover the classical Newmark scheme (1/2, 1/4) and for $\alpha \in [-1/3\,;0[$, the numerical scheme dissipates energy. To get an explicit solution, one may approximate the stiffness forces by:

$$r^{n+1}\left(u^{n+1}\right) \approx K^I u^{n+1} + \left(r_{trial}^{n+1}\left(u_{trial}^{n+1}\right) - K^I u_{trial}^{n+1}\right) \tag{6}$$

K^I is a stiffness matrix (from the initial virgin one to the tangential one). Using the equation of motion at time $n+1$ shifted of α, one may obtain the acceleration vector by solving the linear algebraic system:

$$\widehat{M}\ddot{u}^{n+1} = \widehat{f}^{n+1} \tag{7}$$

with $M = M + \gamma\Delta t(1+\alpha)C + \beta\Delta t^2(1+\alpha)K^I$ and:

$$\widehat{f}^{n+1} = (1+\alpha)f^{n+1} - \alpha f^n + \alpha r_{trial}^{n+1} - (1+\alpha)r_{trial}^n + \alpha C\dot{u}_{trial}^n \\ - (1+\alpha)C\dot{u}_{trial}^{n+1} + \alpha\left(\gamma\Delta t C + \beta\Delta t^2 K^I\right)\ddot{u}^n \tag{8}$$

2.2 Sub-structuring

The PSD testing with substructuring can significantly reduce the cost of the tests to get the seismic performance of the structures (Chung et al. 1999). In substructuring technique, a physical model is built

only on the part or parts where nonlinearity is expected (the physical substructure), with the remaining parts modeled computationally (the numerical substructure). This method initially developed by (Takanashi and Nakashima 1987), Mahin and Shing 1985) has been considerably extended by researchers at the JRC, (Buchet and Pegon 1994). The numerical part is simulated by using a finite element code in a computer connected through a network with other computers that realize the experimental procedures of the PSD test. The displacement at the interface between the physical and numerical substructures is obtained and applied to the test specimen by hydraulic actuators. The resulting resistance forces are measured by load cells and fed back to the numerical model, together with the next increment of earthquake ground motion. A new interface displacement is then calculated and applied to the tested specimen, and the loop is repeated until the test is completed, (Pegon & Pinto 2000, Chang 2001, Williams and Blakeborough 2001).

We denote by the subscripts S and T the matrices corresponding respectively to the simulated and the tested substructure. The i and j indices correspond to the internal nodes of the simulated substructure, the I and J indices to the internal nodes of the tested substructure and δ and θ to the interface nodes between the two substructures. Based on the equation 7, the system to solve sums up to:

$$\begin{bmatrix} {}^S\widehat{M}_{ij} & {}^S\widehat{M}_{i\theta} & 0 \\ {}^S\widehat{M}_{\delta j} & {}^S\widehat{M}_{\delta\theta} + {}^T\widehat{M}_{\delta\theta} & {}^T\widehat{M}_{\delta J} \\ 0 & {}^T\widehat{M}_{I\theta} & {}^T\widehat{M}_{IJ} \end{bmatrix} \begin{bmatrix} \ddot{u}_j^{n+1} \\ \ddot{u}_\theta^{n+1} \\ \ddot{u}_J^{n+1} \end{bmatrix} = \begin{bmatrix} {}^S\widehat{f}_i \\ {}^S\widehat{f}_\delta + {}^T\widehat{f}_\delta \\ {}^T\widehat{f}_I \end{bmatrix} \tag{9}$$

A static condensation applied to interface nodes allows to treat only two systems: the first one for the simulated substructure and the second one for the tested substructure.

In order to account for the diffused cracking in the whole concrete structure, the nonlinear behaviour of materials has to be introduced in the simulated substructure as well. In order to ensure efficiency and robustness, the framework of simplified multifibres analysis has been chosen.

2.3 Numerical implementation and multifibres analysis

For a simple reason of excessive computational costs, complete 3D approaches to structural dynamics in civil engineering are not commonly used. Nonlinear dynamic analysis of complex civil engineering structures based on a detailed finite element model requires large scale computations and handles delicate solution techniques. The necessity to perform parametric studies due to the stochastic characteristic of the input accelerations imposes simplified numerical modeling which will reduce the computation cost. In classical

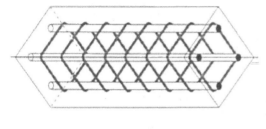

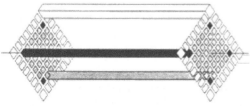

Figure 1. Multifibre mbeam for concrete structures (after Guedes *et al.* 1994).

multifibre analysis (Bazant *et al.* 1987, Spacone *et al.* 1996) the latter is achieved by selecting the classical Euler-Bernoulli beam model for representing the global behavior of the structural components of a complex civil engineering structure. With respect to the large spreading of the zone with nonlinear behavior it is further seek to limit the model complexity (and resulting computational costs) by limiting the diversity of possible deformation global patterns which is achieved in a multifibre beam model with fibres restricted to beam kinematics and with each one employing its own constitutive model (see figure 1). The main advantage of using a multifibre type finite element concerns the possibility to use a simple uniaxial behavior which allows for a very efficient implementation of nonlinear constitutive equations. This is no longer possible for thick beams where shear strains play a major role (Dubé 1994).

The multifibre beam element developed herein employs the standard Hermite polynomial shape functions to describe the variation of the displacement field along the beam. For the Euler-Bernoulli element, the shear forces are computed at the element level through the equilibrium equations. Reinforcement bars are introduced as special fibres, whose behaviour is obtained as a combination of those for concrete and steel (mixture law). The difference with "classical" beam elements concerns the cross section behaviour, i.e. the relation between the generalized strains **e** and the generalized stresses **s**. In the general 3D case the latter includes:

$$\mathbf{s} = \begin{pmatrix} N & M_x & M_y & M_z \end{pmatrix}^T \text{ and } \mathbf{e} = \begin{pmatrix} \varepsilon & \theta_x & \chi_y & \chi_z \end{pmatrix}^T \quad (10)$$

where N is the normal force, M_x the torque, M_y and M_z are the bending moments, ε the axial strain, θ_x

the twist, χ_y and χ_z the curvatures. The cross section behaviour is expressed with the constitutive matrix:

$$K = \begin{bmatrix} K_{11} & 0 & K_{13} & K_{14} \\ & K_{22} & 0 & 0 \\ & & K_{33} & K_{34} \\ sym & & & K_{44} \end{bmatrix} \quad (11)$$

where the coefficients are obtained by integration over the cross section (y and z axes):

$$K_{11} = \int_S E dS, K_{13} = \int_S E z dS;$$

$$K_{14} = -\int_S E y dS; K_{22} = \int_S G(y^2 + z^2) dS \quad (12)$$

$$K_{33} = \int_S E z^2 dS; K_{34} = -\int_S E y z dS; K_{44} = \int_S E y^2 dS$$

where E and G are Young's and shear moduli which vary in y and z. The chosen moduli can be initial, secant or tangent, depending upon the iterative algorithm used to solve the global equilibrium equations. The components of the constitutive matrix are computed by means of numerical integrations, often with one Gauss point per fibre. For the Euler-Bernoulli element, the shear forces are computed at the element level through the equilibrium equations (included in the Hermite polynomial shape functions).

When dealing with structures such as shear walls, which posses the slenderness ratio far from the classical beam ones, a more reliable representation of shear deformations and shear stresses has to be provided. One possibility in that respect is to use the classical Timoshenko beam model, which can describe the constant shear strain. The main difficulty of developing the finite element implementation of the Timoshenko beam model concerns the so-called shear locking phenomena, or inability of the standard finite element approximations to represent pure bending vanishing shear modes. A number of different remedies to shear locking problem has been proposed, ranging from selective or reduced integration, assumed shear strain, enhanced shear strain or hierarchical displacement interpolations. A recent work of Kotronis (2000) extends these ideas in order to construct shear locking remedies for a mulifibre Timoshenko beam.

3 CONCRETE MODELLING

Concerning the concrete constitutive equations, a refined modelling within the earthquake engineering scope should account for decrease in material stiffness as the microcracks open, stiffness recovery as crack closure occurs, inelastic strains concomitant to damage and induced anisotropy. The latter is obtained by an anisotropic damage model based on Continuum Damage Mechanics. The model is written within the thermodynamics framework and introduces only one damage 2nd order tensor variable. To describe the damage evolution, a damage criterion of Mazars (Mazars 1984) type is used. It introduces an equivalent strain computed from the positive part of the strain tensor. The numerical scheme used for the implementation in a F.E. code is implicit, with all the advantages of robustness and stability. However, the constitutive equations of the anisotropic damage can be solved in an exact way on an integration time step. The calculation of the damage and of the stress is then completely explicit from a programming point of view.

3.1 Elasticity-damage coupling

The damage state is represented by the 2nd order tensor D and there is one known thermodynamics potential $\rho\psi^*$ (Ladevèze 1983) from which derives a symmetric effective stress $\tilde{\sigma}$ independent from the elasticity parameters (Lemaitre & Desmorat 2000):

$$\rho\psi^* = \frac{1+\nu}{2E}Tr\left[(I-D)^{-1/2}\sigma_+^D(I-D)^{-1/2}\sigma_+^D + \left\langle\sigma^D\right\rangle_-^2\right] + \frac{1-2\nu}{6E}\left[\frac{\langle Tr\sigma\rangle_+^2}{1-D_H} + \left\langle-Tr\sigma\right\rangle_+^2\right] \quad (13)$$

with E, ν the Young modulus and Poisson ratio of initially isotropic elasticity and $\sigma^D = \sigma - 1/3 Tr[\sigma]\,I$ is the deviatoric stress and where D_H the hydrostatic damage $D_H = 1/3 TrD$.

Quasi-brittle materials such as concrete exhibit a strong difference of behavior in tension and in compression due to damage. This micro-defects closure effect usually leads to complex models when damage anisotropy is considered (Ladevèze 1983, Chaboche 1993, Dragon et Halm 1996) and the purpose here is to show that it is important for cyclic applications to consider damage anisotropy with a quasi-unilateral effect acting on the hydrostatic stress and on the deviatoric one. The thermodynamics potential writes:

$$\rho\psi^* = \frac{1+\nu}{2E}Tr\left[(I-D)^{-1/2}\sigma_+^D(I-D)^{-1/2}\sigma_+^D + \left\langle\sigma^D\right\rangle_-^2\right] + \frac{1-2\nu}{6E}\left[\frac{\langle Tr\sigma\rangle_+^2}{1-TrD} + \left\langle-Tr\sigma\right\rangle_+^2\right] \quad (14)$$

$\langle.\rangle_-$ corresponds to the negative part of a tensor, expressed in its eigen-coordinates. In order to keep differentiability properties of the Gibbs free energy, the positive part σ_+^D of σ^D has to be carefully built (Ladevèze 1983). The following eigenvalue problem (eigenvalues λ^I and corresponding eigenvectors $\vec{T}^I$) has to be solved :

$$\sigma_+^D\vec{T}^I = \lambda^I(I-D)^{1/2}\vec{T}^I \quad (15)$$

The norm being defined as: $\vec{T}^{I^T}(1-D)^{1/2}\vec{T}^J = \delta_{IJ}$. The positive part of the deviator is then expressed as:

$$\sigma_+^D = \sum_{I=1}^{3}\left[(I-D)^{1/2}\vec{T}^I\right]\left[(I-D)^{1/2}\vec{T}^I\right]^T\left\langle\lambda^I\right\rangle_+ \quad (16)$$

The elasticity law reads

$$\varepsilon = \rho\frac{\partial\psi^*}{\partial\sigma} = \frac{1+\nu}{E}Tr\left[\left((I-D)^{-1/2}\sigma_+^D(I-D)^{-1/2}\right)^D + \left\langle\sigma_-^D\right\rangle^D\right]$$
$$+ \frac{1-2\nu}{3E}\left[\frac{\langle Tr\sigma\rangle_+}{1-TrD} - \left\langle-Tr\sigma\right\rangle_+\right]I \quad (17)$$
$$= \frac{1+\nu}{E}\tilde{\sigma} - \frac{\nu}{E}Tr\tilde{\sigma}I$$

and defines the symmetric effective stress $\tilde{\sigma}$ independent from the elasticity parameters:

$$\tilde{\sigma} = \left[(I-D)^{-1/2}\sigma_+^D(I-D)^{-1/2}\right]^D + \left\langle\sigma^D\right\rangle_-^D + \frac{1}{3}\left[\frac{\langle Tr\sigma\rangle_+}{1-TrD} - \left\langle-Tr\sigma\right\rangle_+\right]I \quad (18)$$

The notation $\langle x\rangle_+$ stands for the positive part of a scalar, $\langle x\rangle_+ = x$ if x > 0, $\langle x\rangle_+ = 0$ else.

3.2 Damage threshold function

As for plasticity, the elasticity domain can be defined through a criterion function f such as the domain $f < 0$ corresponds to elastic loading or unloading. Many criterion can be used, written in terms of stresses such as plasticity criteria, strains, or strain energy release rate density leading or not to dilatancy in compression. The purpose here is to built a constitutive model with a restricted number of material parameters, robust and easy to implement in Finite Element computer codes. Dilatancy will not be taken into account and one will accept an open criterion for the tricompression states. These remarks lead us to the simple choice of Mazars criterion, function of the positive extensions $\langle\varepsilon_I\rangle$ of the I_{th} principal strain ε_I,

$$f = \hat{\varepsilon} - \kappa(trD) \text{ with } \hat{\varepsilon} = \sqrt{\langle\varepsilon\rangle_+ : \langle\varepsilon\rangle_+} = \sqrt{\sum\langle\varepsilon_I\rangle^2} \quad (19)$$

where $\hat{\varepsilon}$ is the equivalent strain for quasi-brittle materials and κ is the elastic strain limit in tension. Different expressions for the equivalent strain may be adopted, allowing dealing with biaxial behaviour in a more appropriate way. For example, one can consider the de Vree (de Vree et al. 1995) formulation :

$$\hat{\varepsilon} = \frac{k-1}{2k(1-2\nu)}I_1 + \frac{1}{2k}\sqrt{\frac{(k-1)^2}{(1-2\nu)^2}I_1^2 - \frac{12k}{(1+\nu)^2}J_2} \quad (20)$$

with $I_1 = Tr\varepsilon$ and $J_2 = 1/6I_1^2 - 1/2\varepsilon : \varepsilon$. The biaxial response of anisotropic modelling using Mazars or Vree equivalent strain is given in figures 2 and 3.

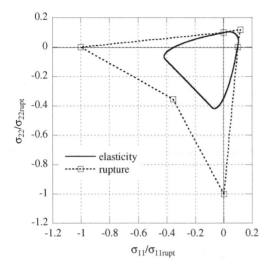

Figure 2. Elasticity and rupture for the Mazars equivalent strain.

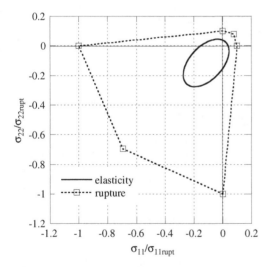

Figure 3. Elasticity and rupture for the de Vree equivalent strain.

3.3 Damage evolution laws

To propose a damage model written in the thermodynamics framework, consider a damage pseudo-potentiel $F = \mathbf{Y} : \langle \boldsymbol{\varepsilon} \rangle_+$ where $\boldsymbol{\varepsilon}$ acts as a parameter so that the damage evolution law is derived from the normality rule as

$$\dot{\mathbf{D}} = \dot{\lambda} \frac{\partial F}{\partial \mathbf{Y}} = \dot{\lambda} \langle \boldsymbol{\varepsilon} \rangle_+ \tag{21}$$

The damage multiplier $\dot{\lambda}$ is determined from the consistency condition $f = 0, \dot{f} = 0$. with $f = \hat{\varepsilon} - \kappa(\xi)$, and, $\xi = \mathbf{D} : \langle \boldsymbol{\varepsilon} \rangle_+ / \max(\varepsilon_I)$.

Concerning the consolidation function κ to account for damage increase in tension as well as in compression, a possible choice is: $\kappa = \kappa(\xi)$. One has still to define the function κ. The simplest choice is to consider a linear function introducing two parameters only (law 1): the damage threshold $\kappa_0 = \kappa(0)$ and a damage parameter A as

$$\kappa(\mathbf{D}) = \frac{1}{A}(\xi) + \kappa_0 \tag{22}$$

Damage anisotropy is different in tension and in compression. It affects differently the elasticity law and a strong difference in tension and in compression is finally obtained with the quite simple damage evolution law 1 (figure 1). Important point, this feature is gained with the consideration of one (tensorial) damage variable only in accordance with the thermodynamic definition of a state variable: if one degradation mechanism is observed, only one damage variable shall represent the micro-cracks or micro-defects pattern, whatever the material is in tension or in compression. The dissymmetry is nevertheless not sufficient with the linear κ-function with a too high damage rate in compression leading to a non physical snapback. One prefers then to consider as damage evolution law:

$$\dot{\mathbf{D}} = \frac{d\kappa^{-1}}{d\hat{\varepsilon}} \frac{\langle \boldsymbol{\varepsilon} \rangle_+}{\hat{\varepsilon}} \dot{\hat{\varepsilon}} = A \left[1 + \frac{\hat{\varepsilon}^2}{a^2} \right]^{-1} \frac{\langle \boldsymbol{\varepsilon} \rangle_+}{\hat{\varepsilon}} \dot{\hat{\varepsilon}} \tag{23}$$

with a a material parameter of the order of magnitude the value of the strain reached in compression. This defines κ^{-1} and κ as

$$\kappa(\xi) = a . \tan \left[\frac{\xi}{aA} + \arctan \left(\frac{\kappa_0}{a} \right) \right] \tag{24}$$

3.4 Model responses

Using the following material parameters ($E = 42$ GPa, $v = 0.2$, $\kappa_0 = 5 \; 10^{-5}$, $A = 5 \; 10^3$, $a = 2.93 \; 10^{-4}$), the uniaxial response of the model is given in figure 4 for compression loading. The cyclic behaviour is presented in figure 5. The unilateral behaviour as well as the damage deactivation when passing from tension to compression are recovered.

3.5 Numerical implementation

The anisotropic damage model is in fact quite simple to implement in a FE code. A global resolution of the equilibrium equations gives the displacements at time

863

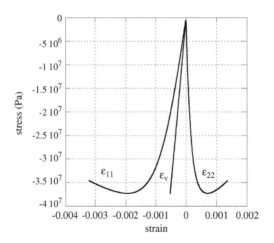

Figure 4. 3D model response in compression.

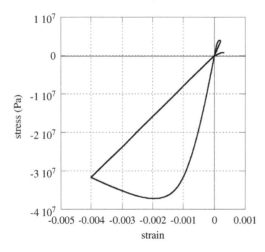

Figure 5. Cyclic uniaxial model response Tension – Compression – Tension.

t_{n+1} with the internal damage variable $D = D_{n+1}$ kept unchanged from the last computed increment t_n. The strains ε_{n+1} at each Gauss point are calculated from the elements interpolation functions. To integrate the constitutive equations means to determine the stress σ_{n+1} and the damage D_{n+1} at time t_{n+1}. An iterative process, not described here, made of global equilibrium resolutions followed by local time integration of the constitutive equations often takes place. One focuses here on the numerical scheme for the local integration of the damage law.

Compute the equivalent strain:

$$\hat{\varepsilon}_{n+1} = \sqrt{\langle \varepsilon_{n+1} \rangle_+ : \langle \varepsilon_{n+1} \rangle_+} \qquad (25)$$

Make a test on the criterion function:

$$f = \hat{\varepsilon}_{n+1} - \kappa(D_n) \qquad (26)$$

If $f \leq 0$, $D_{n+1} = D_n$ (material behaves elastically), else the damage must be corrected by using the damage evolution law discretized as

$$\Delta D = D_{n+1} - D_n = \Delta \lambda \langle \varepsilon_{n+1} \rangle_+ \qquad (27)$$

The damage multiplier is determined from the consistency condition numerically written $f_{n+1} = \hat{\varepsilon}_{n+1} - \kappa(D_n) = 0$ so that

$$\Delta \lambda = \frac{D_{n+1} : \langle \varepsilon_{n+1} \rangle_+ - D_n : \langle \varepsilon_n \rangle_+}{\hat{\varepsilon}_{n+1}^2} \qquad (28)$$

with $D_{n+1} : \langle \varepsilon_{n+1} \rangle_+ = |\max \langle \varepsilon_I \rangle_+| \kappa^{-1}$ being known and the exact actualisation of D:

$$D_{n+1} = D_n + \Delta \lambda \langle \varepsilon_{n+1} \rangle_+ \qquad (29)$$

Stresses computation : Using the elasticity law allow for the computation of the effective stress tensor (E is the elastic Hooke tensor),

$$\tilde{\sigma}_{n+1} = E : \varepsilon_{n+1} \qquad (30)$$

The stress tensor is obtained through the relation between the effective stress tensor and the stress tensor in an anisotropic framework (equation 18).

The numerical scheme is fully implicit, therefore robust, but it has the main advantage of the explicit schemes: there is no need of a local iterative process as the exact solution of the discretized constitutive equations can be explicited (Desmorat et al. 2004).

4 EXPERIMENTAL TESTS ON RC STRUCTURES

A full nonlinear pseudo-dynamic test with substructuring is proposed. A reinforced concrete structure, shown in figure 6 is considered. The clamped frame is loaded with a dynamic seismic signal applied on its foundation. The damage model is applied for concrete material, and a nonlinear plastic behaviour is chosen for steel bars. Computational time is limited in the following by taking into account a multifibre model with the Finite Element Code CAST3M. A distributed loading mass m is applied on the upper beam and on the right middle one. A concentrated loading mass M is applied in the middle of the left beam, with $M \gg m$. The frame failure is supposed to occur after the rupture of the left beam, as a low level of damage occurs in the rest of the structure. Then, just the left beam is tested

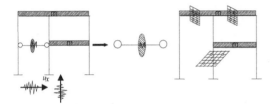

Figure 6. Substructuring decomposition for the nonlinear testing.

Figure 7. Experimental set-up for cyclic three points bend tests on RC beams.

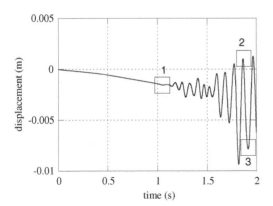

Figure 8. Experimental mid-span deflection.

and its stiffness is obtained experimentally (Laborderie 1991) as the response of the rest of the structure is computed. An additional condition is applied on the structure: horizontal displacements of the beam ends are equal. By the way, just a single degree of freedom is controled on experimental setup. This assumption is valid in that case, where horizontal stiffness is much greater than the vertical one. The experimental set-up for testing RC beams under cyclic three points bend test is presented in figure 7.

A first experimental result is presented in figure 8, it shows the vertical displacement of the centre of the beam during the first seconds of the signal. The realization of quasi-static tests enables us to perform

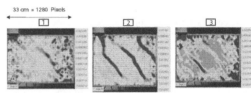

Figure 9. Results of digital image correlation analysis performed on the tested RC beams.

more precise measurements than in the case of tests in dynamics. At critical states of the seismic structural response, it is possible to carry out field measurements via digital images correlation (Hild 2002). First results of this type are presented for three moments of the response: at the beginning of the earthquake, for a maximum negative moment and for a maximum positive moment. Placed at one third of the beam, the camera makes it possible to observe cracks openings in shear. Figure 9 presents the field of horizontal displacement. Openings of cracks from 50 to 700 microns thus could be observed at various instants.

5 CONCLUSIONS

Pseudo-dynamic tests with substructuring allow dynamic studies of large structures with a moderate experimental setup. Inertia forces are computed and the critical part of the structure is statically tested, while the non-critical parts of the structure are modelized. Based on this approach, our work focuses particularly on the damage model used for the modelized parts of the structure, and on the experimental measurements of damage on the tested structure. After a recall of the PSD test basis, we introduce an anisotropic damage model, based on a second order damage tensor, allowing describing induced anisotropy and crack closure with just five parameters. This model is used in the study of a reinforced concrete clamped frame, loaded with a seismic signal. One overloaded beam of the frame is experimentally tested, as the rest of the structure is modelized. Digital image correlation technique is used to study crack apparition and closure, allowing a fine identification and validation of the damage model. Experimental results validate the necessity to take into account the low damage level of the non-critical parts.

REFERENCES

Bazant, Z.P., Pan J.-Y. & Pijaudier-Cabot, G., 1987, Softening in reinforced concrete beams and frames, ASCE J. of Struct. Engrg. 113(12), pp. 2333–2347.
Buchet, P. & Pegon, P., 1994, PSD testing with substructuring implementation and use, specialpublication, 1.94.25 JRC ISPRA.

Chaboche, J.L., 1993, Development of continuum Damage Mechanics for elastic solids sustaining anisotropic and unilateraldamage, *Int. J. Damage Mechanics*, Vol. 2, pp. 311–329.

Chang, S.Y., 2001, Application of the momentum equation of motion to pseudo-dynamic testing, Phil.Trans. R. Soc. London, 359, 1801–1827.

Chung, J., Yun, C.B., Kim, A.S. & Seo, J.W., 1999, Shaking table and pseudo dynamic tests for the evaluation of the seismic performance of base-isolated structure, *Engng. Struct.*, 21, 365–379.

Combescure, D. & Pegon, P., 1997, α–Operator splitting time integration technique for pseudo-dynamic testing – error propagation analysis, *Soil Dynamics and Earthquake Engineering*, Vol. 16, pp. 427–443.

de Vree J., Brekelmans W. & van Gils M., 1995. Comparison of nonlocal approaches in continuum damage mechanics, *Comp. Struct.* 55: 581–588.

Desmorat, R., Gatuingt, & F. Ragueneau, F., 2004. Explicit evolution law for anisotropic damage : application to concrete structures, *NATO conf.*, Bled, Slovénie, juin 2004.

Dragon, A. Halm, D., 1996 Modélisation de l'endommagement par mésofissuration: comportement unilatéral et anisotropie induite, *C. R. Acad. Sci.*, t. 322, Série IIb, p. 275–282.

Dubé, J.F., 1994 Modélisation simplifiée et comportement visco-endommageable des structures en béton, Ph. D. thesis: E.N.S.- Cachan.

Hilber, H.M., Hughes, T.J.R. & Taylor, R.L., 1977 Improved numerical dissipation for time integration algorithms in structural dynamics, *Earthquake Engineering Structural Dynamics*, vol. 5, pp. 283–292.

Hild F., 2002, CORRELI^{LMT}: a software for displacement field measurements by digital image correlation, Internal report N° 254. LMT-Cachan.

Kotronis, P., 2000. Cisaillement dynamique de murs en béton armé. Modèles simplifiés 2D et 3D. Ph. D. thesis: ENS-Cachan.

Laborderie C., 1991. Phénomènes unilatéraux dans un matériau endommageable, PhD thesis University Paris 6.

Ladevèze, P., 1983. On an anisotropic damage theory, Proc. CNRS Int. Coll. 351 Villars-de-Lans, Failure criteria of structured media, Edited by J. P. Boehler, pp. 355–363.

Lemaitre, J. & Desmorat, R., 2005 Engineering Damage Mechanics: Ductile, Creep, Fatigue and Brittle Failures, Springer.

Mahin, S. & Shing, P.B., 1985. Pseudodynamic method for seismic testing, Struct. Eng., 111, 1482–1503.

Mazars, J., 1984. Application de la mécanique de l'endommagement au comportement non linéaire et 'a la rupture du béton de structure, Thèse d'état Université Paris 6.

Nakashima, M., Akazawa, T. & Sakaguchi, O., 1993. Integration method capable of controlling experimental error growth in substructure pseudo-dynamic test. *Journal of Structural and Construction Engineering AIJ*, 454, pp. 61–71.

Pegon, P. & Pinto, A.V., 2000. Pseudo-dynamic testing with substructuring at the elsa laboratory. *Earthquake Engng. Struct.Dyn* 29 905.925.

Shing, P.B. & Mahin, S.A., 1984. Pseudodynamic method for seismic performance testing: theory and implementation. *UCB/EERC-84/01, Earthquake Engineering Research Centre*, University of California, Berkeley.

Shing, P.B., Vannan, M.T. & Cater, E. 1991. Implicit time integration for pseudodynamic Pegon tests. *Earthquake Engineering Structural Dynamics*, Vol. 20, pp. 551–576.

Souid, A., Delaplace, A., Ragueneau, F. & Desmorat, R., 2005. Pseudodynamic tests and substructuring of damageable structures, Sixth European Conference on Structural Dynamics, Paris, France.

Spacone, E., Filippou, F.C. & Taucer, F.F. 1996. Fiber Beam-Column Model for Nonlinear Analysis of R/C Frames. I: Formulation. *Earthquake Engineering and Structural Dynamics*, Vol. 25, N. 7, pp. 711–725.

Takanashi, K. & Nakashima, M. 1987, Japanese activities online testing, *Engng. Mech.*, 113, 1014–1032.

Williams, M.S. & Blakeborough, A., 2001. Laboratory testing of structures under dynamic loads: an introductory review., Phil. Trans. R. Soc. Lond., 1651–1669.

Fracture Mechanics of Concrete and Concrete Structures – Design, Assessment and Retrofitting of RC Structures – Carpinteri, et al. (eds)
© 2007 Taylor & Francis Group, London, ISBN 978-0-415-44616-7

Behavior of RC beams under impact loading: Some new findings

S.M. Soleimani
Associated Engineering Ltd., Burnaby, BC, Canada

N. Banthia & S. Mindess
The University of British Columbia, Vancouver, BC, Canada

ABSTRACT: The load recorded by the striking tup has been used to study the impact behavior of reinforced concrete beams. It was noted that this load could not be considered as the bending load experienced by the concrete beam. A portion of this load is used to accelerate the beam, and therefore, finding the exact bending load versus time response has been one of the most challenging tasks for impact researchers. To capture a true bending load versus time response a special test setup was designed and built. Tests showed a time lag between the maximum load indicated by the instrumented tup and the maximum load indicated by the instrumented supports. This time lag has confirmed that the inertial load effect must be taken into account. It was also found that beyond a certain impact velocity, the flexural load capacity of RC beams remained constant; further increases in stress rate did not increase their load carrying capacity.

1 INTRODUCTION

Material properties change under high strain rates. As a result, reinforced concrete (RC) beams made of reinforcing bars and concrete will respond differently at different loading rates. Since the compressive (and tensile) strength of concrete and the yield strength of steel increase when loaded at high strain rates, increasing the strain rate will generally increase the flexural capacity of RC beams (Takeda and Tachikawa 1971, Bertero et al. 1973, Kishi et al. 2001).

2 IMPACT TEST SETUP

2.1 Drop weight impact machine

A drop weight impact machine with a capacity of 14.5 kJ was used in this research study. A mass of 591 kg (including the striking tup) can be dropped from as high as 2.5 m. During a test, the hammer is raised to a certain height above the specimen using a hoist and chain system. At this position, air brakes are applied on the steel guide rails to release the chain from the hammer. On releasing the brakes, the hammer falls and strikes the specimen. Three load cells were designed and built at the University of British Columbia for this project. During preliminary tests, it was discovered that if the specimen was not prevented from vertical movements at the supports, within a very short period after first contact of the hammer with the

specimen, contact with the supports was lost and as a result, loads indicated by the support load cells were not correct. For instance, the loads recorded by the support load cells for two identical tests were totally different. This phenomenon was further verified by using a high speed camera. To overcome this problem, the vertical movement of RC beams at the supports was restrained using two steel yokes (Figure 1). In order to assure that the beams were still simply supported, these yokes were pinned at the bottom to allow rotation during beam loading. To permit an easier rotation, a round steel bar was welded underneath the top steel plate where the yoke touched the beam.

Figure 1. Impact test setup with steel yokes.

2.2 Beam design and testing procedure

A total of 12 identical RC beams were cast to investigate the behavior of RC beams under impact loading. These beams contained flexural as well as shear reinforcement. They were 1 m in total length and were tested over an 800 mm span. Load configuration and cross-sectional details are shown in Figure 2.

Seven beams were tested under impact with different impact velocities ranging from 2.8 m/s to 6.26 m/s, and three beams were tested under quasi-static, 3-point loading. The remaining two beams were strengthened by GFRP fabric; one was tested under quasi-static and the other under impact loading

(impact velocity = 3.43 m/s). Table 1 shows the beam designations and configuration.

Based on the Canadian Concrete Design Code, the capacity of this beam under quasi-static loading is 51 kN, when the tension reinforcement starts to yield. It is worth noting that the beam was designed to fail in a flexural mode, since enough stirrups were provided to prevent shear failure.

Under quasi-static loading conditions, all of the beams (i.e. BS-1, BS-2, BS-3 and BS-FRP) were tested in 3-point loading using a Baldwin 400 kip Universal Testing Machine. Under impact loading conditions, all of the beams were tested using an instrumented drop-weight impact machine as explained in 2.1.

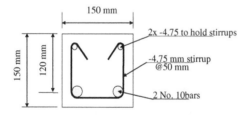

Figure 2. Load configuration and cross-sectional details of RC beams.

Table 1. RC beams designations.

Beam No.	Quasi-static Loading	Impact Loading Drop Height, h (mm)					GFRP Fabric
		400	500	600	1000	2000	
BS-1	✓	–	–	–	–	–	–
BS-2	✓	–	–	–	–	–	–
BS-3	✓	–	–	–	–	–	–
BS-FRP	✓	–	–	–	–	–	✓
BI-400	–	✓	–	–	–	–	–
BI-500-1	–	–	✓	–	–	–	–
BI-500-2	–	–	✓	–	–	–	–
BI-500-3	–	–	✓	–	–	–	–
BI-600	–	–	–	✓	–	–	–
BI-1000	–	–	–	–	✓	–	–
BI-2000	–	–	–	–	–	✓	–
BI-600-FRP	–	–	–	✓	–	–	✓

3 RESULTS AND DISCUSSION

3.1 Quasi-static loading

The results of the three beams loaded quasi-statically (i.e. BS-1, BS-2 and BS-3) were quite consistent. The load vs. deflection curve for beam BS-1, shown in Figure 3, represents a typical flexural failure mode of RC beams. Load vs. deflection responses for the other two beams (BS-2 and BS-3) were very similar to that of beam BS-1. Initially, the beam was uncracked (i.e. from the beginning of the curve till Point A). The cross-sectional strains at this stage were very small and the stress distribution was essentially linear. When the stresses at the bottom side of the beam reached the concrete tensile strength, cracking occurred. This is shown as Point A in Figure 3. After cracking, the tensile force in the concrete was transferred to the steel reinforcing bars (rebars). As a result, less of the concrete cross section was effective in resisting moments and the stiffness of the beam (i.e. the slope of the curve) decreased. Eventually, as the applied load increased, the tensile reinforcement reached the yield point shown by Point B in Figure 3. Once yielding had occurred, the midspan deflection increased rapidly with little increase

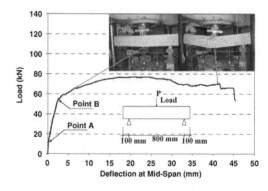

Figure 3. Load vs. deflection curve for RC beam with a flexural failure mode.

in load carrying capacity. The beam failed due to crushing of the concrete at the top of the beam. The experimental test result showed 54 kN capacity for this beam, corresponding to Point B in Figure 3. Thus there is a good agreement between theoretical and experimental values for the load carrying capacity of this RC beam.

3.2 Impact loading

For all impact tests using the drop-weight machine, PCB Piezotronics™ accelerometers were employed. These accelerometers were screwed into mounts which were glued to the specimens prior to testing. Vertical accelerations at different locations along the beam were recorded with a frequency of 100 kHz using National Instruments™ VI Logger software. Locations of the accelerometers are shown in Figure 4.

The velocity and displacement histories at the location of the accelerometers were obtained by integrating the acceleration history with respect to time using the following equations:

$$\dot{u}_o(t) = \int \ddot{u}_0(t)dt \qquad (1)$$

$$u_o(t) = \int \dot{u}_0(t)dt \qquad (2)$$

where $\ddot{u}_0(t)$ = acceleration at the location of the accelerometer; $\dot{u}_0(t)$ = velocity at the location of the accelerometer; and $u_0(t)$ = displacement at the location of the accelerometer.

The contact load between the specimen and the hammer is not the true bending load on the beam, because of the inertial reaction of the beam. A part of the tup load is used to accelerate the beam from its rest position. D'Alembert's principle of dynamic equilibrium can be used to write equilibrium equations in dynamic load conditions. This principle is based on the notion of a fictitious inertial force. This force is equal to the product of mass times its acceleration and acting in a direction opposite to the acceleration. D'Alembert's principle of dynamic equilibrium states that with inertial forces included, a system is in equilibrium at each time instant. As a result, a free-body diagram of a moving mass can be drawn and principles of statics can be used to develop the equations of motion. Thus, one can conclude that in order to obtain the actual bending load on the specimen the inertial load must be subtracted from the observed tup load. It is also important to note that the tup load throughout this paper is taken as a point load acting at the mid-span of the beam, whereas the inertial load of the beam is a body force distributed throughout the body of the beam. This distributed body force can be replaced by an equivalent inertial load, which can then be subtracted from the tup load, to obtain the true bending load, which acts at the mid-span. Therefore, at any time t, the following equation can be used to obtain the true bending load that the beam is experiencing (Banthia et al., 1989):

$$P_b(t) = P_t(t) - P_i(t) \qquad (3)$$

where $P_b(t)$ = true bending load at the mid-span of the beam at time t; $P_t(t)$ = tup load at time t; and $P_i(t)$ = point load representing the inertial load at the mid-span of the beam at time t equivalent to the distributed inertial load.

In this research program, support anvils in addition to the tup were instrumented in order to obtain valid and true bending loads at any time t directly from the tests. Therefore, the true bending load at time t, $P_b(t)$, which acts at the mid-span can also be obtained by adding the reaction forces at the support anvils at time t:

$$P_b(t) = R_A(t) + R_C(t) \qquad (4)$$

where $R_A(t)$ = reaction load at support A at time t; and $R_C(t)$ = reaction load at support C at time t as shown in Figure 2. This has been verified experimentally by Soleimani (2006).

Three identical beams (i.e. BI-500-1, BI-500-2 and BI-500-3) were tested under a 500 mm drop height and steel yokes were used to prevent upward movement of beams at the support locations at the instant of impact. Load vs. time histories of these beams are shown in Figure 5 (a) to (c). There are two important points to mention here: (1) true bending load, $P_b(t)$, obtained from support load cells (load cell A + load cell C) are pretty much the same for all three beams; (2) maximum tup load (denoted as load cell B) recorded by the striking hammer is not consistent and is in the range of 158 kN to 255 kN. It is also worth mentioning that the results obtained from the two support load cells are quite similar to each other and the peak load in both load cells occurred at the same time as expected. This phenomenon can be seen in Figure 6 for the case of beam B1-500-2.

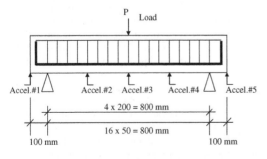

Figure 4. Location of the accelerometers in impact loading.

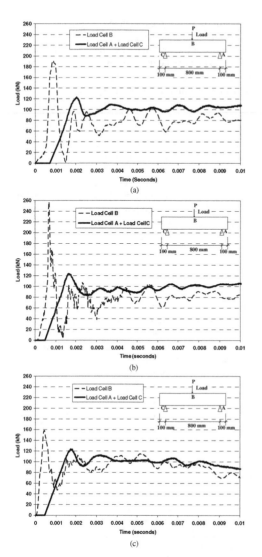

Figure 5. Load vs. time for beam (a) BI-500-1; (b) BI-500-2; and (c) BI-500-3.

Equations 1 and 2 were used to calculate the displacement of the RC beam at the locations of the accelerometers. For beam BI-500-1, the displacement curve along half of the beam's length is shown in Figure 7. Note that the deflection distribution is essentially linear, consistent with many earlier studies on beams with various types of reinforcement (e.g., Banthia 1987; Bentur et al. 1986). Since the beam failed in flexure, the displacement on the other half of the beam was symmetrical to the displacement shown in this Figure. The diamond-shaped points in this Figure show the actual displacement of the beam. The best-fit line is drawn and its equation along with its

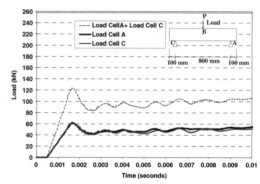

Figure 6. Load vs. time for support load cells in beam BI-500-2.

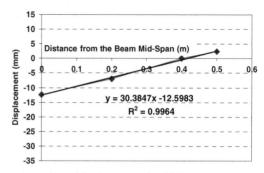

Figure 7. Displacement of beam BI-500-1 at t = 0.005 s.

R^2 value is given. The displacements shown in Figure 7 were recorded at 0.005 seconds after the impact.

Therefore, one can conclude that the deflected shape for a simply supported RC beam at any time t under impact loading produces a linear deflection profile that can be approximated by a V-shape consisting of two perfectly symmetrical lines.

At the instance of impact, the hammer has a velocity V_h given by:

$$V_h = \sqrt{2gh} \qquad (5)$$

where V_h = velocity of the falling hammer at the instance of impact in m/s; g = acceleration due to gravity (=9.81 m/s^2); and h = drop height in m.

The impact velocities at the instant of impact for the hammer with a mass of 591 kg for different drop heights, calculated using Equation 5, are given in Table 2.

As an example, the velocity vs. time calculated by Equation 1 for beam BI-500-2 is shown in Figure 8. Interestingly, the velocity of the hammer at the instant of impact (3.13 m/s from Table 2) and the maximum velocity of the beam (which occurred 0.001 s after the impact as shown in Figure 8) are very similar to each other. This, at least to some extent, can explain why

Table 2. Impact velocity for different drop height.

Drop height (mm)	Velocity (m/s)
400	2.80
500	3.13
600	3.43
1000	4.43
2000	6.26

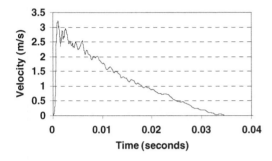

Figure 8. Velocity vs. time at the mid-span, beam BI-500-2.

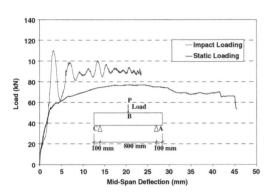

Figure 9. Load vs. mid-span deflection, beam BI-400.

the tup load at the very beginning of impact decreased almost to zero, after a very rapid increase to a maximum value (see Figure 5). In other words, the beam was accelerated by the hammer and reached its maximum velocity while at the same time (i.e. $t = 0.001$ s) the tup load (load cell B) decreased to zero as the beam sped away from the hammer and lost contact. The hammer was back in contact with the beam after some time (in the case of BI-500-2, after about 0.0005 s) and the load rose again. Some time after impact started (in the case of BI-500-2, after 0.035 s) the velocity of both (i.e. hammer and beam) decreased to zero.

The true bending load vs. mid-span deflection curves for beams BI-400, BI-500-1, BI-500-2, BI-500-3, BI-600, BI-1000 and BI-2000 are shown in Figures 9 to 13. The numbers 400, 500, 6000, 1000 and 2000 refer to the drop height in mm (see Table 1).

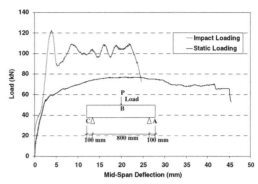

Figure 10. Load vs. mid-span deflection, beam BI-500-1.

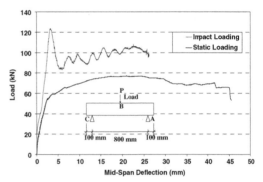

Figure 11. Load vs. mid-span deflection, beam BI-500-2.

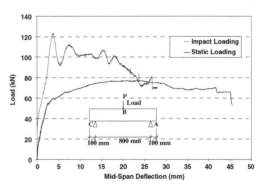

Figure 12. Load vs. mid-span deflection, beam BI-500-3.

Equation 4 was used to find the true bending load and Equations 1 and 2 were used to find the deflection at mid-span from the acceleration histories of mid-span accelerometers (accelerometer #3 in Figure 4) in each case. To provide a meaningful comparison, mid-span deflections are shown out to 50 mm in all cases.

Load vs. mid-span deflection of the same beam tested under static loading is also included in each

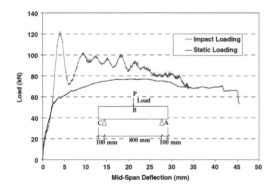

Figure 13.   Load vs. mid-span deflection, beam BI-600.

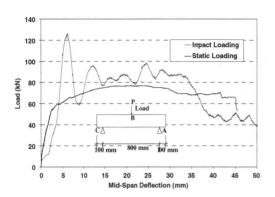

Figure 14.   Load vs. mid-span deflection, beam BI-1000.

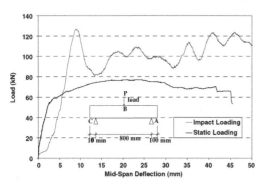

Figure 15.   Load vs. mid-span deflection, beam BI-2000.

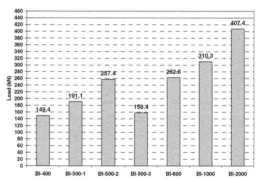

Figure 16.   Maximum recorded tup load for different beams/drop height.

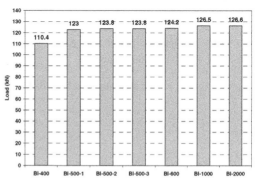

Figure 17.   Maximum recorded true bending load for different beams/drop height.

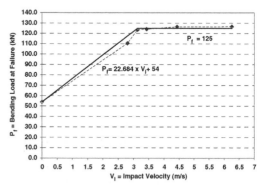

Figure 18.   Bending load at failure vs. impact velocity.

graph to show the differences in beam responses to different loading rates.

Maximum recorded tup loads for beams tested under different drop heights are compared in Figure 16. Maximum recorded true bending loads (summation of support load cells) are shown in Figure 17.

It is clear that while the recorded tup load in these beams, in general, increased with increasing drop height, at a constant drop height (i.e. 500 mm), the

maximum values for tup load were not the same. However, beyond a certain drop height, the maximum true bending load (i.e. load cell A + load cell C) did not change with further increases in drop height.

The bending load at failure vs. impact velocity is shown in Figure 18. Bending load at failure is defined as the maximum recorded true bending load for impact

872

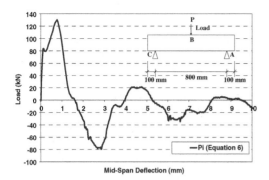

Figure 19. Inertia load for beam BI-400.

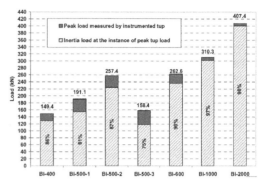

Figure 20. Inertia load at the peak of tup load.

loading. This is also the load at which, presumably, the steel rebars in tension start yielding under static loading.

It may be seen that bending load at failure increased with increasing velocity of the impact hammer until it reached a velocity of about 3 m/s. After this point, the bending load at failure was independent of impact velocity and remained constant. It is very important to note that for this hammer, with a mass of 591 kg, a minimum drop height is needed to make the RC beam fail. For example a drop height of only 100 mm of this hammer most probably would not break the beam, but failure might occur if a heavier hammer was employed. Since the impact velocity is directly related to hammer drop height, one can conclude that for a given hammer mass, there exists a certain threshold velocity (or drop height) after which the bending load at failure will not increase with increasing velocity. This threshold velocity for the hammer used in this research was found to be ~3 m/s. Figure 18 also shows that the impact bending capacity of this RC beam is about 2.3 times its static bending capacity. Therefore, an impact coefficient of 2.3 can be used to estimate the impact bending capacity of this RC beam from its static bending capacity.

Equation (7.8) can be rewritten as:

$$P_i(t) = P_t(t) - P_b(t) \qquad (6)$$

Therefore, the inertial load at any time t is the difference between the tup load and the true bending load. As an example, the inertial load for beam BI-400 calculated by Equation 6 is shown in Figure 19. The values obtained by this equation are the most accurate values coming from a fully instrumented test setup.

A large portion of the peak load measured by the instrumented tup is the inertial load. This is shown in Figure 20. At the peak load measured by the instrumented tup, the inertial load, used to accelerate the beam from its rest position, may account for 75% to 98% of the total load.

3.3 Beams strengthened by GFRP fabric

The Wabo®MBrace GFRP fabric system was used to strengthen two RC beams for flexure and shear. One layer of GFRP fabric with a thickness of about 1.2 mm, length of 750 mm and width of 150 mm was applied longitudinally on the tension (bottom) side of the beam for flexural strengthening and an extra layer with fibers perpendicular to the fiber direction of the first layer was applied on 3 sides (i.e. 2 sides and bottom side) for shear strengthening.

One of these beams was tested under quasi-static loading, while the other was tested under impact with a 600 mm hammer drop height (i.e. impact velocity of 3.43 m/s). Load vs. mid-span deflections of these RC beams are shown in Figure 21 (a) and (b). It is important to note that while the control RC beam (i.e. when no GFRP fabric was used) failed in flexure, the strengthened RC beam failed in shear indicating that shear strengthening was not as effective as flexural strengthening; perhaps more layers of GFRP were needed to overcome the deficiency of shear strength in these beams.

In general, these tests showed that GFRP fabric can effectively increase an RC beam's load capacity under both quasi-static and impact load conditions.

Load carrying capacities of these beams are compared in Table 3. While an 84% increase in load carrying capacity was observed in quasi-static loading, the same GFRP system was able to increase the capacity by only 38% under impact loading. It is also worth mentioning that while the maximum bending load under impact loading for the un-strengthened RC beam was 2.26 times its static bending capacity, the ratio of maximum impact load to static load for the strengthened RC beam was only 1.69. This difference can certainly be explained by the change in failure mode from bending to shear when GFRP fabric was applied to these RC beams. The area under the load-deflection curve in Figure 21 (b) was measured and it was found that about 86% of the input energy was absorbed by the strengthened RC beam during the impact.

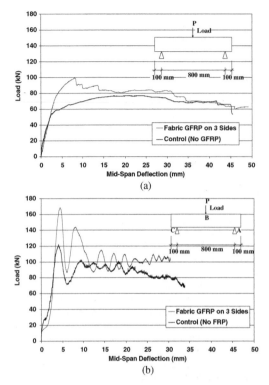

(a)

(b)

Figure 21. Load vs. mid-span deflection for RC beam strengthened in shear and flexure using fabric GFRP; (a) Quasi-static loading, (b) Impact loading (velocity = 3.43 m/s).

Table 3. Load carrying capacity of RC beams strengthened by fabric GFRP.

Loading type	Load carrying capacity [kN]	Increase in load carrying capacity [%]
Quasi-static	99.4 (54)*	84%
Impact	168.4 (122.2)*	38%

* Numbers in brackets are the load carrying capacity of un-strengthened RC beams.

4 CONCLUSIONS

Based on the results and discussions reported above, the following conclusions can be drawn:

1. Load carrying capacity of RC beams under impact loading can be obtained using instrumented anvil supports.
2. The use of steel yokes at the support provides more reliable and accurate results.
3. Loads measured by the instrumented tup will result in misleading conclusions due to inertia effect.

4. There is a time lag between the maximum load indicated by the instrumented tup and the maximum load indicated by the instrumented supports. This lag is due to the stress pulse travel time from the centre to the support. This time lag shows that the inertial load effect must be taken into account.
5. Inertial load at any time t can be obtained by subtracting the summation of the support load cells (i.e. true bending load) from the load obtained by the instrumented tup.
6. Bending load capacity of an RC beam under impact loading can be estimated as 2.3 times its static capacity for the conditions and details of tests performed here. Similar coefficient was reported for different types of RC beams (different cross-sectional areas and different reinforcement ratios) tested under impact loading (Kishi et al. 2001).
7. After a certain impact velocity, the flexural load capacity of RC beams remains constant, and increases in stress (or strain) rate will not increase their load carrying capacity.
8. GFRP fabric can increase the load carrying capacity of RC beams in both static and impact loading conditions.
9. The use of fabric GFRP may change the mode of failure, and as a result, the load carrying capacity of an RC beam strengthened by fabric GFRP under impact loading can be much lower than the anticipated 2.3 times its static capacity (see conclusion 6 above).

REFERENCES

Banthia, N. 1987. Impact resistance of concrete. PhD Thesis, University of British Columbia, Vancouver, Canada.
Banthia, N., Mindess, S., Bentur, A. and Pigeon, M. 1989. Impact testing of concrete using a drop-weight impact machine. *Experimental Mechanics* 29(2): 63–69.
Bentur, A., Mindess, S. and Banthia, N. 1986. The behaviour of concrete under impact loading: experimental procedures and method of analysis. *Materiaux et Constructions*, 19(113): 371–378.
Bertero, V.V., Rea, D., Mahin, S. and Atalay, M.B. 1973. Rate of loading effects on uncracked and repaired reinforced concrete members. *Proceedings of 5th world conference on earthquake engineering, Rome*. Vol. 1: 1461–1470.
Kishi, N., Nakano, O., Matsuoka, K.G. and Ando, T. 2001. Experimental study on ultimate strength of flexural-failure-type RC beams under impact loading. *Proceedings of the international conference on structural mechanics in reactor technology. Washington, DC*. Paper # 1525, 7 pages.
Soleimani, S.M. 2006. Sprayed glass fiber reinforced polymers in shear strengthening and enhancement of impact resistance of reinforced concrete beams. PhD Thesis, University of British Columbia, Vancouver, Canada.
Takeda, J. and Tachikawa, H. 1971. Deformation and fracture of concrete subjected to dynamic load. *Proceedings of the international conference on mechanical behaviour of materials, Kyoto, Japan*. pp. 267–277.

*Fracture Mechanics of Concrete and Concrete Structures – Design, Assessment and
Retrofitting of RC Structures – Carpinteri, et al. (eds)
© 2007 Taylor & Francis Group, London, ISBN 978-0-415-44616-7*

Fracture and explosive spalling of concrete slabs subjected to severe fire

F.A. Ali, A. Nadjai, D. Talamona & M.M. Rafi
FireSERT, University of Ulster, Jordanstown, UK

ABSTRACT: The paper presents an experimental investigation on explosive spalling and fracture-induced
deformation of 6 full-scale simply supported reinforced concrete slabs subjected to conventional fire curve
(BS476) and severe hydrocarbon fire curve, performed at the Fire Research Centre, University of Ulster, UK.
Each slab was loaded with 14% of its ultimate load and was heated from the bottom side only. Temperature
profile was recorded at 3 depths within the slabs and the moisture content was also measured before and after
the tests. The deflection of the slabs was recorded at the middle of the 3 meters span. All slabs suffered from
explosive spalling at different degrees. The study showed that explosive spalling took place much earlier and was
more violent in slabs subjected to hydrocarbon curve in comparison with BS476. The paper ends with analysis
of interaction between the parameters involved in the tests.

1 INTRODUCTION

The majority of the research studies on behaviour of
concrete elements which were performed in the last
decades have mainly focused on beams and columns.
Some of these studies investigated small scale slabs El-
Hawary, *et. al.* (1996), Shuttleworth (2001), but a very
small number of investigations have involved large
scale specimens. Most of these slabs were tested under
standard normal heating rates BS476 or ASTME119
(Shirly *et. al.* 1988). However, the effect of more
severe fires (hydrocarbon fire curve for example) on
the performance of structural elements is gathering
momentum following the tragic events of 9/11 in New
York. The research performed in the BRE (Cooke
(2001)) is among the rare works, which involved the
effect of hydrocarbon curve on the fire resistance and
deflection of concrete slabs. However, no explosive
spalling of concrete was reported in this study. Previous research (Cooke (2001), Ali. *et. al.* (2004), Gamal
et. al. (1995), Selih *et. al.* (1991), Chung *et. al.* (2005)
and others) has shown that the probability of explosive
spalling of concrete increases under higher heating
rates. It is well established now that heating a concrete element from one side creates two moving fronts:
heat and moisture front. The two fronts move away
from the heated face of the concrete towards the cold
unheated side. The speed of the two fronts depends on
several factors including the heating rate where the two
fronts could meet at a specific distance inside the concrete. This causes the water within the moisture clog
to transform to vapour. The build up of high vapour
pressure which may reach 3 to 5 MPa (Chuang *et. al.*

2003) can cause explosions in the concrete. In addition, recent research suggests that the presence of steel
reinforcement impedes moisture movement and produces quasi-saturated moisture clog zones that could
lead to the development of significant pore pressure
(Chuang *et. al.* 2003).

The objective of this paper is to represent
the outcomes of an experimental study performed on 6 large simply supported concrete slabs
$3300 \times 1200 \times 200$ mm. The slabs were subjected to
conventional (BS476) and hydrocarbon severe heating
rates. The temperatures were measured inside the slabs
at three depths: – surface, 40 mm (steel reinforcement),
100 mm. The mid-span deflection of the concrete slabs
was also measured is presented. The explosive spalling
observed during the tests and an assessment criterion of spalling are discussed. The paper includes also
an analysis of the interaction between the parameters
involved in the tests.

2 THE EXPERIMENTAL PROGRAMME

The experimental work involved testing 6 large
scale $3300 \times 1200 \times 200$ mm normal-strength reinforced concrete slabs with an average concrete strength
of 42 N/mm^2 at 28 days. During the tests the concrete
slabs were mounted on the top of the furnace with clear
span of 3000 mm (see Figure 1). Each slab was reinforced with 6 T12 steel bars, embedded longitudinally
along the slab at a spacing of 220 mm centre to centre,
as illustrated in Figure 1.

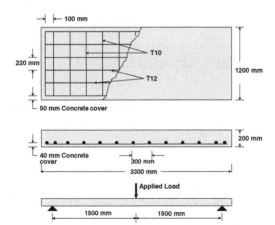

Figure 1. Loading and reinforcement details of concrete slabs.

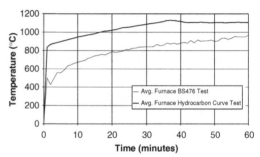

Figure 2. Average temperatures recorded in the furnace.

Table 1. Tests results.

Fire Curve	Slab Ref.	Spalling Degree %	Moisture Content %		Max. Defl. mm
			Top	Bottom	
BS476	S1	1.0	3.6	5.5	29.7
	S2	2.7	3.7	5.7	32.1
	S3	2.2	3.9	6.0	34.2
Hydro.	S4	2.1	3.7	5.2	44.8
	S5	4.3	3.8	4.9	44.7
	S6	3.2	3.5	4.8	43.9

Secondary reinforcement comprising of 13 T10 steel bars were placed perpendicular to the main reinforcement at a spacing of 300 mm c/c. The concrete cover of the steel reinforcement was 40 mm vertically and 50 mm on slab sides (see Figure 1).

2.1 Test parameters

All slabs were tested under the same loading level = 0.14 of the design load of BS8110. This load was applied at the mid-span point of each slab as shown in Figure 1. Two heating regimes were used during the experimental programme, BS476 and hydrocarbon fire curves. Figure 2 shows the average experimental fire curves achieved during the tests. Three slabs S1, S2 and S3 were tested using the standard temperature-time curve BS 476, while the remaining slabs S4, S5 and S6 were tested under the severe hydrocarbon fire curve (see Table 1). All the tests were performed in a $4 \times 3 \times 3$ m combustion chamber, with the slab specimen situated on top of the furnace and heated from beneath.

2.2 Experimental data measurement

In order to measure the temperature distribution within the specimens, each slab was fitted with five 1.5 mm sheathed thermocouples. Three of the thermocouples were used to measure the bottom surface temperature (one located at the mid span of the slab and the other two were located 400 mm from both supported ends of the slab). The remaining two thermocouples were cast within the slab. The first one was located at the centre of the slab, while the second was touching the 7th reinforcement bar (mid-span of the slab) at a depth of approximately 40 mm from the bottom of the slab surface. Slab deflection was measured using a Linear

Variable Differential Transformer (LVDT), located at the mid span of the slab. Temperature distribution and slab deflection data was recorded using a data logging system. The moisture content was measured on both of the top and bottom surfaces of each slab using Tramex CRH concrete moisture content measuring device.

These measure–ments were then repeated in the same locations following the test process. Audio and visual observations of explosive spalling were made during the tests through the quartz windows situated in the door of the combustion chamber.

2.3 Test methodology

The slabs were set on top of the combustion chamber and the necessary measuring devices were attached. The load was applied at a constant rate until the desired loading level was reached. Then the combustion chamber burners were ignited and controlled to achieve the required time-temperature curve. The applied load was kept at a constant level throughout the test. The duration of each test was 60 minutes.

2.4 Spalling assessment

The primary criterion used to assess spalling was the degree of spalling. Following the test, the concrete lost due to spalling was collected and weighed. The degree of spalling (S_d) was then measured using the following formula:

$$S_d = W_L / W_C \qquad (1)$$

where W_L is the weight of concrete lost due to explosive spalling and W_C is the weight of the concrete slab before testing. The depth of spalling and percentage surface area lost due to explosive spalling were approximately measured after the test. This enabled to estimate the volume of concrete that has disintegrated from the slab due to explosive spalling.

3 TESTS RESULTS

The main results from the series of experiments undertaken are shown in Table 1. All slabs were subjected to the same loading level = 0.14 of the design load of BS8110 represented in one concentrated load of 27 kN at mid span of the slab as shown in Figure 1. All of the six slabs experienced explosive spalling during testing, with more violent spalling of slabs exposed to the hydrocarbon fire curve where spalling started after 2 minutes of heating. Slabs subjected to the BS476 fire curve did not experience explosive spalling until the 15th minute of heating.

3.1 *Slabs tested under BS476 fire curve*

3.1.1 *Explosive spalling*
All slabs have experienced explosive spalling. The spalling was violent with distinctive noises. Figure 3 shows explosive spalling of the three slabs S1, S2 and S3. The degree of spalling experienced by slabs S1–S3 is presented in Table 1. The degree of spalling of the three slabs was reasonably close to each other. The depth of the spalled areas varied along the surface of the slabs, with the greater depths noted towards both ends of the slabs. Slab S2 experienced the largest degree of spalling at 2.7%; with a maximum spalling depth of 25 mm. Slabs S1 and S3 had maximum spalling depth of 20 mm and 15 mm respectively. The first occurrence of explosive spalling was observed approximately after 15 minutes at a slab surface temperature around 750°C. This explosive spalling phase lasted for approximately 20 minutes, after which only minor isolated incidents of explosive spalling were noted.

3.1.2 *Temperature profile along slab thickness*
A significant temperature gradient was recorded along the slab thickness during heating. Figure 4 shows the development of temperature of slab S2 at three point:– surface, at reinforcement level and at the mid height of the slab. Figure 4 clearly shows a high thermal gradient of around 880°C between the surface and the center

Figure 3. Explosive spalling of slabs S1–S3.

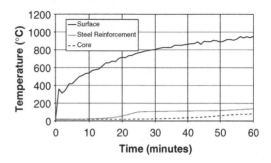

Figure 4. Temperature development within slab S2.

of the slab. It is important to note that slab S2 has experienced the highest degree of spalling when subjected to the BS476 fire curve. Due to violent explosive spalling, the surface thermocouple of slab S1 has dislodged from the slab surface between the 25th and 26th minute. For this reason the surface temperature recorded after the 25th minutes for slab S1 will be disregarded in this analysis. All slabs showed a similar temperature development, with slab 2 achieving a slightly higher maximum temperature of 138°C after 60 minutes. The thermocouple utilized to measure the temperature at the steel reinforcement is situated at a depth of 40 mm from the exposed surface of the concrete slab.

3.1.3 *Slab deflection*
Slab deflection was measured at the central point of the slab span. Figure 5 shows the development of the deflection of slabs S1, S2 and S3 subjected to BS476 fire curve.

It can be seen from Figure 5 that the three slabs have experienced reasonably similar rates and values of deflection (Table 1). From Figure 5 it can also be seen that slab's deflection is relatively small, up to a temperature of approximately 450°C. Beyond this temperature the deflection increases at higher rate up

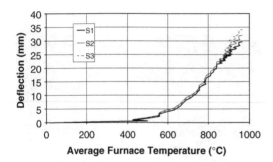

Figure 5. Deflection recorded for slabs S1–S3.

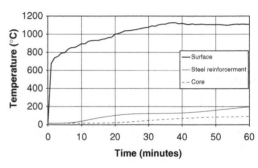

Figure 6. Temperature development within slab S5.

to a temperature of approximately 800°C. After that the rate of deflection increases again until the end of the test. This increase appears to coincide with the onset of severe explosive spalling and the reduction in the slabs cross section. In general, slabs have achieved an average maximum displacement of 32 mm at a furnace temperature approaching 965°C, with slab S3 showing a marginally greater maximum deflection of 34.2 mm.

3.2 Slabs tested under hydrocarbon fire curve

3.2.1 Explosive spalling
In an apparent difference from BS476 slabs tested under Hydrocarbon fire have experienced more violent explosive spalling which happened on noticeably early time, after 2 minutes of the start of the test. The degree of spalling experienced by slabs S4–S6 is presented in Table 1. Slab S5 experienced the highest degree of spalling of 4.3% with a maximum spalling depth of 20 mm. Slabs S4 and S6 both have a maximum spalling depth of approximately 15 mm. Also it is important to emphasis that most of the violent explosive spalling occurred within the first 12 minutes. After that only occasional occurrences of mild spalling were noted.

3.2.2 Temperature profile along slab thickness
As expected the temperature gradient in slabs subjected to hydrocarbon fire was more evident and higher in values (1020°C). In particular Figure 6 presents the temperature profile of slab S5, which has experienced a degree of spalling of 4.3% (Table 1). From Figure 6, it can be seen that a similar temperature gradient exists within slab S5 to that of slab S2 tested under the BS476 curve. However, the temperatures recorded within slab S5 are higher which indicates the increased severity of the hydrocarbon fire curve. After 30 minutes of fire exposure, the surface of slab S5 has reached a temperature of approximately 1100°C, while temperature at the steel reinforcement was approximately 130°C. Even the centre of the slab has experienced a slight temperature increase to 50°C only. All three slabs showed a very similar temperature profile to that

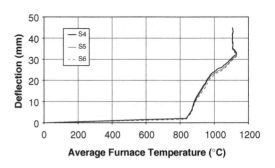

Figure 7. Deflection recorded for slabs S4–S6.

of the hydrocarbon fire curve displayed in Figure 2, reaching a maximum temperature of approximately 1100°C after 60 minutes.

Figure 6 does not only highlight the large temperature gradient that exists in normal strength concrete but also underline the enhanced temperature profile experienced by normal strength concrete when exposed to hydrocarbon fire conditions as opposed to the conventional BS476 time-temperature curve.

3.2.3 Slabs deflection
Figure 7 shows the development of the deflection of slabs S4–S6 when exposed to the hydrocarbon fire curve. All 3 slabs showed almost identical rates of deflection. Initially the slabs deformed at gradual rate, until a temperature of approximately 840°C was reached after 2 minutes. At this point, the gradient of the curves in Figure 7 shows a sudden and sharp increase, indicating a rapid increase in the deflection of the slabs. This point coincides with the moment when explosive spalling has started and caused a reduction in the slab cross section. This increase continues for the remainder of the experiment, with the slabs achieving an average maximum displacement of 44.5 mm at a temperature approaching 1100°C. Slab S4 displays the largest deflection of 44.8 mm.

4 ANALYSIS OF RESULTS

4.1 *Effect of fire severity on slab deflection*

From Figure 8, it can be seen that all slabs showed a slow deformation rate during the early stages of fire exposure. After 2 minutes, at which point the slab deformation is almost negligible, the average furnace temperature under the BS476 and Hydrocarbon fire curves has reached approximately 450°C and 840°C respectively.

At this stage, slabs subjected to both heating regimes showed a sudden and sharp increase in deflection, with the gradient of the curves in Figure 8 indicating that the slabs exposed to the hydrocarbon fire experienced a more rapid rate of deflection.

Under the BS476 curve, the slab deflection increases as the temperature in the furnace (and therefore the surface temperature of the concrete) continues to rise. However, it is noticeable that even when the temperature of the hydrocarbon fire reaches a 'ceiling' of approximately 1100°C, the slab deflection continues to increase. In summary, although both heating regimes appear to induce a similar rate of deflection on the concrete slabs, the faster temperature development of the hydrocarbon fire ensures that slabs exposed to the more severe fire conditions will experience a more rapid rate of deflection, and ultimately a greater deformation will be recorded for those slabs.

4.2 *Effect of fire severity on explosive spalling*

From Table 1, it can be seen that the specimens exposed to the hydrocarbon fire curve experienced a greater degree of explosive spalling. The maximum degree of spalling exhibited by slab S2 under the conventional BS476 fire curve was 2.7%, whereas 4.3% of slab S5 was removed by explosive spalling when exposed to the hydrocarbon fire. Although the slabs subjected to the hydrocarbon fire curve exhibited the greater amount of spalling, the actual depth of the spalling was quite similar for all slabs, measuring between 15 mm and 25 mm, regardless of the fire severity. Also, while explosive spalling commenced on all slabs in the same temperature range of 750°C–850°C, the more advanced temperature profile of the hydrocarbon fire ensured that explosive spalling occurred much earlier under hydrocarbon fire conditions than under the BS476 regime. To summarize the analysis, an increase in fire severity resulted in much earlier onset and a greater degree of explosive spalling. However, from the experiments undertaken, the depth of spalling was not noticeably affected by the increase in fire severity but the area affected by spalling was larger under the hydrocarbon fire curve.

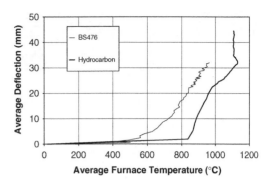

Figure 8. Average deflection of slabs under BS476 and Hydrocarbon fire curves.

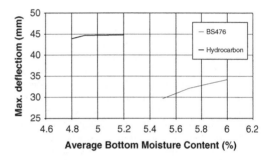

Figure 9. Effect of Average Moisture Content on maximum slab deflection.

4.3 *Effect of moisture content on slab deflection and explosive spalling*

Figure 9 demonstrates that a relationship appears to exist between the moisture content and the maximum deflection recorded for the slabs. This relationship was more obvious in the slabs exposed to the conventional BS476 time-temperature curve, as the maximum deflection of the normal strength concrete slabs clearly increased with an increase in average moisture content. Although the maximum deflection of the slabs also increased slightly with the increase of moisture content under hydrocarbon fire conditions, this increase was not as marked for hydrocarbon fire as that experienced under the BS476 heating regime.

4.4 *Evidence of moisture front movement through observation*

Observations were made of the moisture front movement of slab S6, which was exposed to the hydrocarbon fire curve. Prior to testing, the average moisture content of the slab was recorded both on the top surface of the slab and on the bottom surface (Table 1) exposed to the heating regime, with moisture content of 3.5%

Figure 10. Top surface of slab S6 at minute 50 during the test.

and 4.8% obtained for the respective surfaces. After 8 minutes of exposure to the hydrocarbon fire curve, water started to appear on the top surface of the slab at the four lifting hooks (embedded for slab lifting). As the temperature within the slab continues to rise throughout the test, the moisture is driven away from the advancing heat front towards the top surface. During the fire test and after 50 minutes, when even the slab core temperature was only 100°C, large pools of expelled water were visible on the surface of the slab (Figure 10). Following the completion of the experiment, the average moisture content of the surface exposed to the fire was measured, and a value of 0% was recorded – as opposed to 4.8% prior to testing.

5 CONCLUSIONS

1. All the normal-strength reinforced-concrete slabs experienced explosive spalling, regardless of the heating regime utilised during the experiments.
2. An increase in fire severity resulted in earlier occurrences of explosive spalling, with slabs exposed to the hydrocarbon and BS476 fire curves experiencing explosive spalling after 2 and 15 minutes respectively.
3. It was noticed that all the normal strength reinforced concrete slabs exhibited a large thermal gradient between the slab surface, the steel reinforcement and the slab core.
4. Concrete slabs exposed to more severe fires experienced more rapid deflection rate.

5. Slabs exposed to the severe hydrocarbon fire have experienced a higher degree of explosive spalling; therefore lower heating rates minimized, but did not eliminate, the risk of explosive spalling.
6. From the tests undertaken, the actual depth of explosive spalling was not noticeably affected by an increase in fire severity.
7. The maximum deflection of normal strength reinforced concrete slabs appears to increase with higher moisture content, both under high and low heating rates.
8. From the tests undertaken, the moisture front within the concrete slabs appears to move away from the heated surface during fire exposure. This movement appears to be accelerated with the increase of fire severity.
9. A hard evidence in the form of pictures, confirming the moisture clog movement away from the heated concrete face was obtained.

REFERENCES

Ali, F. A. Nadjai, A., Slikock, G. Outcomes of a Major research on high strength concrete in Fire. Fire Safety Journal, Volume 39, Issue 6, September 2004, p 433–445.

Chung, J.H., Consolazio, G.R. Numerical modeling of transport phenomena in reinforced concrete exposed to elevated temperatures. Cement and Concrete Research, v35, n 3, March, 2005, p 597–608.

Cooke, G.M.E. Behaviour of precast concrete floor slabs exposed to standardised fires. Fire Safety Journal, v 36, n 5, July, 2001, p 459–475.

El-Hawary, M.M., Ragab, A.M., Osman, K.M., Abd El-Razak, M.M. Behavior investigation of concrete slabs subjected to high temperatures. Computers and Structures, v 61, n 2, Oct, 1996, p 345–360.

Gamal, A. Hurst, J. Modeling the thermal behavior of concrete slabs subjected to the ASTM E119 standard fire condition. Journal of Fire Protection Engineering, v 7, n 4, 1995, p 125–132.

Huang, C.L.D, Gamal N.A. Influence of slab thickness on responses of concrete walls under fire Numerical Heat Transfer. An International Journal of Computation and Methodology; Part A: Applications, v 19, n 1, Jan–Feb, 1991, p 43–64.

Selih, J., Sousa, A.C.M., Bremner, T.W. Moisture and heat flow in concrete walls exposed to fire. Journal of Engineering Mechanics, v 120, n 10, Oct, 1994, p 2028–2043.

Shirley T., Burg R. G., Fiorato A. E. Fire Endurance Of High-Strength Concrete Slabs. ACI Materials Journal V.85, No. 2, March-April 1988, p 102–108.

Shuttleworth P. Fire Protection of Concrete Tunnels Linings. The Third International Conference on Tunnel Fires and Escape from Tunnels, 9–11 October 2001, Washington DC, USA, p157–165.

Fracture Mechanics of Concrete and Concrete Structures – Design, Assessment and Retrofitting of RC Structures – Carpinteri, et al. (eds)
© 2007 Taylor & Francis Group, London, ISBN 978-0-415-44616-7

Reinforced and prestressed concrete beams subjected to shear and torsion

H. Broo, M. Plos, K. Lundgren & B. Engström

Department of Civil- and Environmental Engineering, Structural Engineering, Concrete Structures, Chalmers University of Technology, Gothenburg, Sweden

ABSTRACT: Today, the nonlinear finite element method is commonly used by practicing engineers, although design and assessment for shear and torsion in reinforced concrete structures are still made using methods based on sectional forces. By modelling the shear behaviour, using 3D nonlinear FEM, higher load carrying capacity and more favourable load distribution was shown, compared to conventional analysis. A modelling method using four-node curved shell elements with embedded reinforcement was evaluated in this study. Tests of reinforced and prestressed beams loaded in bending, shear and torsion were simulated. The increase in shear capacity, in addition to the reinforcement contribution, was modelled with a relationship for concrete in tension according to the modified compression field theory and compared with the use of a relationship related to the fracture energy of plain concrete. The results show that evaluations of the load-carrying capacity or crack width will be on the safe side, if only the fracture energy is used to define the concrete in tension.

1 INTRODUCTION

For structural design and assessment of reinforced concrete members the non-linear finite element (FE) analysis has become an important tool. However, today, design and assessment for shear and torsion are still made using simplified analytical or empirical design methods based on sectional forces and moments. The current calculation method for reinforced concrete members subjected to combined shear and torsion, in the European Standard EC2 CEN/TC250/SC2 (2004), adds stresses from shear and from torsion linearly without taking into account deformations and compatibility within the member. However, earlier research indicates that there is a redistribution even within concrete members without transverse reinforcement, see Gabrielsson (1999) and Pajari (2004), and that this could be modelled with non-linear FE analyses, see Broo *et al.* (2005) and Broo (2005). Nonlinear FE analyses of concrete members with transverse reinforcement subjected to shear have been studied by several researchers, for example Ayoub & Filippou (1998), Yamamoto & Vecchio (2001), Vecchio & Shim (2004) and Kettil *et al.* (2005). In recent research projects, failures due to shear and torsion was successfully simulated with nonlinear FE analyses, also for members with transverse reinforcement, see Plos (2004). A higher load carrying capacity compared to conventional analysis was shown. Thus, a more favourable load distribution, compared to conventional analysis, was found when the structure was analysed in three dimensions and by including the fracture energy

associated with concrete cracking. Modelling R/C and P/C members subjected to shear and torsion needs to be further studied and verified in order to be reliable and practically applicable.

The aim of this study is to work out a modelling method to simulate the shear-induced cracking and the shear failure of R/C and P/C members. The modelling method should be possible to use for analyses of more complicated structures, for example box-girder bridges, subjected to bending, shear, torsion and combinations of these load actions. Engineers using commercial nonlinear FE programs, not especially designed for shear analysis, should be able to use the modelling method in their daily practice. Further aims are to determine the most important parameters that need to be accounted for in the material model or in the material properties used.

The different contributions to the nonlinear response of shear are concisely presented and discussed, as well as the different approaches to shear modelling. The proposed FE approach to R/C and P/C modelling is also shown to fit quite well some available test results concerning concrete beams loaded in bending, torsion and shear.

2 SHEAR BEHAVIOUR

2.1 *The non-linear response in shear*

Both shear forces and torsional moments cause shear stresses that could result in cracks in a concrete member. Cracks due to shear stresses are usually inclined compared to the direction of the reinforcement. To satisfy the new equilibrium after shear cracking, longitudinal reinforcement and transversal reinforcement or friction in the crack is required. After cracking the shear stresses are transmitted by compression in the concrete between the inclined cracks, by tension in the transverse reinforcement crossing the inclined cracks, by tension in the longitudinal reinforcement, by compression in the compressive zone and by shear transfer in the crack. The visual shear cracks are preceded by the formation of micro-cracks. The micro-cracking reduces the stiffness of the member and a redistribution of stresses can occur resulting in strut inclinations smaller than 45°, Hegger et al. (2004). Due to the rotation of the struts more transverse reinforcement can be activated. The rotation of the compressive struts can continue until crushing of the concrete between the inclined cracks occurs, Walraven & Stroband (1999). Possible failure modes due to shear cracks are either crushing of the concrete between two shear cracks or sliding along a shear crack. It is well-known that the shear capacity is larger than what can be explained by the reinforcement contribution determined from a truss model. This increase in shear capacity is due to tension stiffening, dowel action, and aggregate interlock, and is also known as the concrete contribution.

After cracking, concrete can transmit tensile stresses due to tension softening and for reinforced concrete also due to tension stiffening. Tension softening is the capability of plain concrete to transfer tensile stresses after crack initiation. In a reinforced concrete member subjected to tensile forces, the concrete in between the cracks carry tensile stresses transferred from the reinforcement through bond, thus contributing to the stiffness of the member. This is known as the tension stiffening. The tension-stiffening effect increases the overall stiffness of the reinforced concrete member in tension compared to that of a bare reinforcing bar. Due to the bond action there are still high transverse tensile stresses in the compressive struts. Cracked concrete subjected to high tensile strains in the direction normal to the compression is softer and weaker than concrete in a standard cylinder test, Vecchio & Collins (1986), Vecchio & Collins (1993) and Belarbi & Hsu (1995).

The complex behaviour of reinforced concrete after shear crack initiation has been explained in several papers, for example ASCE-ACI Committe 445 on Shear and Torsion (1998), Vecchio & Collins (1986), Pang & Hsu (1995), di Prisco & Gambarova (1995), Walraven & Stroband (1999), Zararis (1996), Soltani et al. (2003). Several mechanisms contribute to the non-linear response in shear: bridging stresses of plain concrete (tension softening), interaction between reinforcement and concrete due to bond (tension stiffening), aggregate interlocking, dowel action, and

reduction of concrete compressive strength due to lateral cracking. The stress equilibrium can be expressed in average stresses for a region containing several cracks or in local stresses at a crack. The local stresses normal to the crack plane are carried by the reinforcement and by the bridging stresses of plain concrete (tension softening). Along the crack plane, the shear stresses are carried by friction due to aggregate interlocking and dowel action. The stresses will depend on the crack width, the shear slip, the concrete mix-design (strength, grading curve and maximum aggregate size) and of course the reinforcement (type, diameter and spacing), fib (1999).

2.2 Modelling of the non-linear shear response

Several analytical models that are capable of predicting the nonlinear response in shear has been presented. for example the modified compression field theory (MCFT), Vecchio & Collins (1986), the distributed stress field model (DSFM), Vecchio (2000), the cracked membrane model (CMM), Kaufmann & Marti (1998), the rotating-angle softened truss model (RA-STM), Pang & Hsu (1995), the fixed-angle softened truss model (FA-STM), Pang & Hsu (1996), and the softened membrane model (SMM), Hsu & Zhu (2002). All these models are based on the smeared approach, i.e. the influence of cracks is smeared over a region and the calculations are made with average stresses and average strains. Stress equilibrium, strain compatibility and constitutive laws are used to predict the shear force for chosen strains. Some models use a rotating crack concept and thus no relationship between shear stress and shear strain is needed. Others are based on a fixed crack concept including a relationship for average shear stresses and average shear strains. Most of the models are also implemented in finite element programs. Soltani et al. (2003) propose a model that calculates local stresses and strains at the crack plane, separating the contribution from tension softening, tension stiffening, aggregate interlock and dowel action, to predict the nonlinear shear response.

If the shear-induced cracking and shear failure is modelled with a nonlinear FE program, not especially designed for shear analysis, parts of the concrete contribution needs to be accounted for by modifying the constitutive relationships used. The required modifications depend on the modelling philosophy, on crack representation (fixed or rotating cracks) and on how the interaction between reinforcement and concrete is modelled.

Modelling the reinforcement and the interaction between reinforcement and concrete can be more or less detailed. When modelling larger structures, i.e. box-girder bridges, the reinforcement can be modelled as embedded in the concrete elements. The embedded reinforcement adds stiffness to the FE model, but the

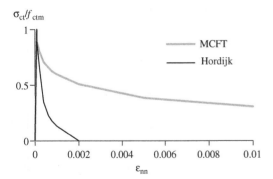

Figure 1. Different tension-softening relations used, for the four-point bending beam. For the curve by Hordijk the fracture energy is smeared over a length of 107 mm (the crack band with, h), which corresponds to the mean crack spacing obtained in the test.

reinforcement has no degree of freedom of its own. Hence, the reinforcement is perfectly bonded to the surrounding concrete and no slip can occur. Embedded reinforcement can be applied to any type of finite element that represents the concrete. In this case the above-mentioned effects of the concrete contribution, must be taken into account in the constitutive relations describing the materials behaviour, e.g. in the concrete tensile response or in the reinforcement response. Different ways of doing this for the tension stiffening effect has been proposed by Lackner & Mang (2003) and Kaufmann & Marti (1998). Relationships for tensile stresses versus crack openings in plain concrete are based on fracture mechanics and related to the fracture energy, G_f; an example is the relation proposed by Hordijk, as described in TNO (2002). In Figure 1 the curve by Hordijk is compared with the expression by Collins & Mitchell (1991) as described below.

For reinforced concrete members subjected to shear also the contribution from dowel action and friction due to aggregate interlock can be accounted for in the constitutive relations. Such relationships that link average tensile stresses to average tensile strains for orthogonally reinforced cracked concrete have been established trough shear panel tests, Vecchio & Collins (1986) and Pang & Hsu (1995). The relationship

$$\sigma_{c1} = \frac{f_{ctm}}{1 + \sqrt{500 \cdot \varepsilon_1}} \tag{1}$$

is the one used in the modified compression field theory (MCFT), Collins & Mitchell (1991). Here σ_{c1} is the mean principal tensile stress, f_{ctm}, the mean tensile concrete strength and ε_1 is the mean principal tensile strain. This relationship has later been modified by Bentz (2005).

The relationship should be limited so that no concrete tensile stress is transmitted after the reinforcement has started to yield. This is a problem when modifying the relationship for the concrete tensile response in a FE program since there is no obvious link between the steel strain in the reinforcement direction and the concrete strain in the principal stress direction. Hence, the cracked concrete can transfer tensile stresses in the principal stress direction even after the reinforcement in any direction yields.

The relationships by Collins & Mitchell (1991), Pang & Hsu (1995) and Bentz (2005) were established for analysis of orthogonally reinforced concrete specimens subjected to shear. However, more general applicability for members with deviating reinforcement or specimens subjected to, for instance, bending or tension is not shown and is likely doubtful. In Broo et al. (2006) some of the shear panels tested by Vecchio & Collins (1986) and Pang & Hsu (1995) were analysed with the FE program Diana, TNO (2002), and the use of the tension softening curve by Hordijk and a curve according to MCFT, Equation 1, were compared. The results showed that – by merely considering plain-concrete fracture energy – the capacity was underestimated and the average strains, i.e. the crack width, were overestimated. On the other hand, if the concrete contribution was modelled with the expression from MCFT, the capacity was overestimated and the average strains underestimated for most specimens, except for the panels tested by Vecchio & Collins (1986). It should be mentioned that results from these panel tests are included in the test results, used to calibrate the expression in the MCFT. This means that the concrete contribution to shear capacity can be accounted for by modifying the constitutive relationship used for concrete in tension. However, caution is recommended in order not to overestimate the capacity. If no modification of the tension-softening curve is done, the shear capacity will at least not be overestimated. Moreover, it was found that it is important to include the reduction of compression strength due to lateral cracking if the failure mode is crushing of the concrete between the shear cracks. In the analyses presented here, the results obtained by using Hordijk's tension-softening curve and MCFT's curve are compared for a reinforced concrete beam subjected to bending and shear and a prestressed box-beam subjected to bending, shear and torsion.

3 MODELLING TECHNIQUES

To investigate the general applicability of the modelling method worked out for shear panels in Broo et al. (2006) for members with non-orthogonal reinforcement and subjected to mixed loading, two beam tests were modelled. In all analyses, the concrete was

Table 1. Material properties for concrete used in the analyses.

Test	f_{cm} MPa	f_{ctm} MPa	E_c GPa	G_f Nm/m²	h m
NSC 3	27.3	2.16	30.05	88.9	0.107
Beam 5	24.9	1.97	29.14	47.3	0.050

Table 2. Material properties for reinforcement and pre-stressing strands used in the analyses.

Test	Dim and Qual.	f_y MPa	f_u MPa	ε_{sy} ‰	ε_{s2} ‰	ε_{su} ‰	E_s GPa
NSC 3	ϕ 8 K500 ST	574	670	3.1	29.3	99	199.8
	ϕ 20 K500 ST	468	600	2.4	21.5	132	195.6
Beam	5 ½″ St 150/170	1840	–	–	–	–	–
	ϕ 8 Ks40s	456	600*	2.09	–	150*	218
	ϕ16 Ks60	710	900*	3.05	–	110*	233

*Values taken as mean value of test values from several other reports using same kind of reinforcement from the same time period.

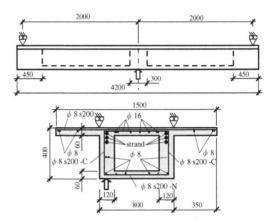

Figure 2. Test set-up, geometry and reinforcement of the prestressed box-beam, Beam 5, Karlsson & Elfgren (1976).

modelled with four-node curved shell elements. For the specimens modelled here, plane-stress elements or even beam elements would have been more appropriate, but the aim was to work out a modelling method that can be used also for more complicated structures, for which curved shell elements are more suitable. Full interaction was assumed between the reinforcement and the concrete, which was modelled with a constitutive model based on nonlinear fracture mechanics. A rotating crack model based on total strain, see TNO (2002), was used in all analyses. In the analyses of the beams, concrete compressive stresses localise into one element, whose size does not correspond to the size of the specimens used to calibrate the compression relationship by Thorenfeldt as described in TNO (2002). Hence, if the relationship by Thorenfeldt was used the model could not predict the response. This disadvantage was overcome by modelling the concrete in compression with an elastic-ideal plastic relationship. The reduction of the hardening in compression due to lateral cracking was modelled according to Vecchio and Collins, as described in TNO (2002). For the tension softening, two approaches to account for the concrete contribution from tension stiffening, dowel action and friction due to aggregate interlock were compared. The curve by Hordijk, see TNO (2002), merely based on plain-concrete fracture energy is compared with the expression in Equation 1, taken from the MCFT, Collins & Mitchell (1991), which attempts to take also the concrete contribution into account, see Figure 1. The concrete material properties for the beams analysed are presented in Table 1. The concrete tensile strength, f_{ct}, the concrete modulus of elasticity, E_c, and the fracture energy G_f were calculated according to CEB (1993), from the mean cylinder compressive strength, f_{cm}, reported from the tests. For the curve by Hordijk the fracture energy is smeared over a length, h, the crack band with that corresponds to the mean crack spacing obtained in the test.

The constitutive behaviour of the reinforcement and the prestressing steel was modelled by the von Mises yield criterion, with an associated flow law and isotropic hardening. In Table 2 the material properties of the reinforcement used for the beam analyses are presented. No hardening parameters were presented for the reinforcement used in the box-beam test; the values presented in Table 2 are mean values taken from several other test reports using the same kind

of reinforcement, from the same laboratory and the same time period.

4 ANALYSES OF P/C BOX-BEAM

4.1 FE model

A prestressed box-beam (Beam 5) provided also with ordinary longitudinal and transverse reinforcement tested by Karlsson & Elfgren (1976) was analysed. The box-beam was subjected to bending, shear and torsion and the final failure was due to large opening of a shear and torsion crack in the loaded web. Figure 2 shows the dimensions and support conditions of the simulated box-beam.

Due to symmetry only half the beam was modelled, as shown in Figure 3, using curved shell elements and material properties as described above. The box-beam was reinforced as shown in Figure 2. The prestressing strands and the longitudinal reinforcement with dimension 8 mm were modelled as embedded bars, TNO (2002). All other reinforcements were modelled as embedded grids, TNO (2002).

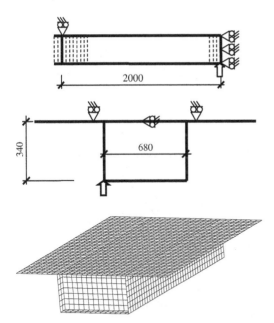

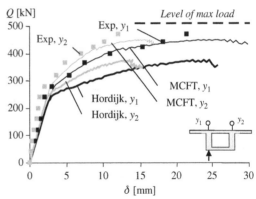

Figure 3. Restraints and finite-element mesh used to model the box-beam tested by Karlsson & Elfgren (1976). Dashed lines mark the cross-sections that are stiffened and tied to keep the cross-section in plane.

The support conditions of the model are shown in Figure 3. In the test the box-beam was supported on roller bearings with a load distributing support plate. In the analyses, the nodes in the centre of the supports were fixed in the vertical direction. The nodes on each side of this node were forced to have the same vertical displacement but in opposite direction, thus enabling a rotation and simulating a free support with a distribution length equal to the support plate in the test.

Stiffeners at the support and at the mid-span where the load was applied were taken into account in two ways. All shell elements for the box wall and flanges in the area of the stiffeners were given a thickness twice the thickness of the elements outside these parts. The density of the concrete was also modified to maintain the correct self-weight of the box-beam. Furthermore, all nodes in each cross-section of the stiffened areas were tied so that each cross-section remained plane.

In the box-beam test the load was applied in steps of 40 kN up to 320 kN. Thereafter, the load was increased by controlling the mid-deflection in steps of 1–2.5 mm. In the analyses, the load was applied as a prescribed deformation of the loading node, i.e. the bottom corner node in the symmetry section. The box-beam analysis had to be performed in two phases. In the first phase the loading node was not supported; here the prestressing force (110 kN) was released and the self-weight was applied. In the second phase, the loading node was

Figure 4. Comparison of results from test and analyses of a prestressed concrete box-beam subjected to bending, shear and torsion; applied load versus mid-span displacement.

supported vertically at the location obtained in the first phase. Thereafter, the loading was applied by increasing the vertical displacement of the loading node. An implicit solving method was used. Iteration was made with constant deformation increments of 0.1 mm. For each increment equilibrium was found using the BFGS secant iteration method, TNO (2002). The analysis was continued if the specified force, energy or displacement convergence criterion was fulfilled, according to default values, see TNO (2002). If the convergence criterion was not fulfilled within twenty iterations, the analysis was continued anyhow. Afterwards the convergence criteria ratios were checked.

4.2 Results

The applied load versus vertical displacements from the analyses and the test are compared in Figure 4. The results show, as expected, that if only the fracture energy of plain concrete was taken into account, the capacity was underestimated and the vertical deflections were overestimated. However, when the concrete contribution was modelled with the expression from MCFT, the capacity was still underestimated but the fitting of the results was satisfactory for the vertical deflections.

In the test, the first crack, going in transverse direction across the top flange, occurred at a load of 240 kN due to bending. This crack propagated downwards in the most loaded web at a load of 280 kN. At a load of 320 kN the first shear and torsion crack appeared near the support. The final failure, at a load of 510 kN, was due to large opening of a shear and torsion crack in the loaded web. The angle of the cracks in the most loaded web varied between 45 and 60 degrees, while they remained vertical in the other web. The crack propagation and the crack pattern from both analyses agree well with those observed in the test.

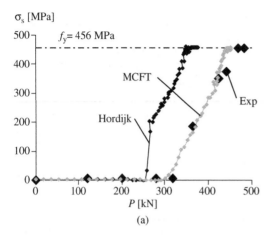

σ_s [MPa]

f_y = 456 MPa

MCFT

Hordijk

Exp

P [kN]

(a)

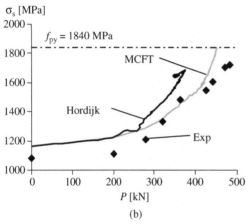

σ_s [MPa]

f_{py} = 1840 MPa

MCFT

Hordijk

Exp

P [kN]

(b)

Figure 5. Comparison of results from test and analyses of a prestressed concrete box-beam subjected to bending, shear and torsion; applied load versus steel stresses in; (a) a stirrup in the loaded web and 400 mm from the load (b) the top prestressing strand in the loaded web and the most loaded section.

In Figure 5, the load versus steel stresses for one strand and one stirrup, from the test and the analyses, are compared. The steel stresses increases first when the box-beam starts to crack. In the analysis with the tension softening modelled according MCFT the steel stresses increase is slower which corresponds better with the steel stresses measured in the test.

5 ANALYSES OF P/C BEAM

5.1 FE model

A R/C beam loaded in four-point bending, NSC3, tested by Magnusson (1998) was modelled with curved shell elements. The beam was subjected to bending and

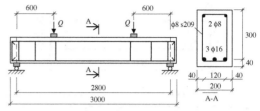

Figure 6. Test set-up, geometry and reinforcement of the R/C beam loaded in four-point bending, NSC3, Magnusson (1998).

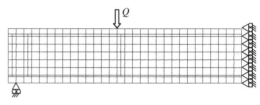

Figure 7. FE mesh used to simulate the R/C beam loaded in four-point bending.

shear and failed in bending due to yielding of the longi-tudinal reinforcement and crushing of the concrete in the compressive zone in the mid-span part of the beam. Figure 6 shows the geometry and support conditions of the simulated bending beam.

Due to symmetry only half the beam was modelled, as shown in Figure 7, using curved shell elements and material properties as described in Section 3. The beam was reinforced as shown in Figure 6. The longitudi-nal reinforcement and the stirrups between the support and the load were modelled as embedded bars, TNO (2002). The stirrups in the middle part of the beam were modelled as an embedded grid, TNO (2002).

In the test the beam was supported on roller bearings with a load distributing steel plate. In the analyses, the node in the centre of the support was fixed in the vertical direction. The nodes on each side of this node were forced to have the same vertical displacement but in opposite direction, thus enabling a rotation and simulating a free support with a distribution length equal as the support plate in the test.

The loading of the bending beam was controlled by displacement both in the test and in the analysis. In the analysis the loading was applied by increasing the vertical displacement of the loading node in steps of 0.1 mm. In the test the load was distributed by a loading plate. In the analyses, this was simulated by locking the nodes on each side of the loading node to the loading node, in such a way that their vertical displacement remained in a plane.

In the bending beam analyses, large compressive strains localised in one element, which was also subjected to large lateral tensile strains due to a

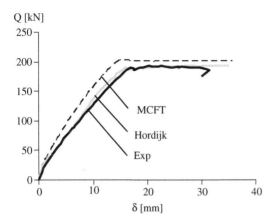

Q [kN]

δ [mm]

Figure 8. Comparison of results from test and analyses of the R/C beam loaded in four-point bending; applied load versus mid-span displacement.

flexural shear crack. This flexural shear crack was also observed in the test, but there it did not go into the compressive zone. Consequently, reducing the compressive strength due to lateral strains resulted in an unreasonable response. Therefore, for these analyses, the compressive strength was not reduced. In these analyses, the loading was applied by increasing the vertical displacement of the loading node. An implicit solving method was used. Iterations were made with constant deformation increments of 0.1 mm. For each increment, equilibrium was found using the BFGS secant iteration method, TNO (2002). The analysis was continued if the specified energy convergence criteria was fulfilled, according to default value, see TNO (2002). However, if the convergence criterion was not fulfilled within twenty iterations, the analysis was continued. Afterwards the convergence criteria ratios were checked.

5.2 Results

The applied load versus vertical displacements from the analyses and the test are compared in Figure 8. These results show that, if tension softening is modelled according to MCFT, the behaviour was too stiff and the capacity was overestimated. Hence, the cracked concrete transfers tensile stresses even if the longitudinal reinforcement yields.

With only the fracture energy of plain concrete taken into account, the capacity is very well estimated and the behaviour is just a little bit to stiff. The conclusion is that if a tension softening curve including the concrete contribution to shear capacity is used, it needs to be modified, so that no tensile stresses are transferred when the reinforcement yields. Otherwise the capacity will be overestimated for the parts of a member, which are subjected to tension or bending.

6 CONCLUSIONS

It is well-known that the shear capacity is larger than what can be explained by the reinforcement contribution determined from a truss model. This increase in shear capacity is due to tension stiffening, dowel action, and friction due to aggregate interlock, and is also known as the concrete contribution. If the shear-induced cracking and shear failure is modelled nonlinear with a FE program, not especially designed for shear analysis, parts of the concrete contribution needs to be accounted for by modifying the constitutive relationships used.

In the analyses presented here the use of the tension-softening curve by Hordijk and a curve according to MCFT are compared for a prestressed box-beam subjected to bending, shear and torsion, and for a reinforced concrete beam subjected to bending and shear.

The commercial FE program Diana was used to model a test of a box-beam that failed in shear. It was shown that four-node curved shell elements with embedded reinforcement can describe the nonlinear shear response also for P/C members loaded in bending, shear and torsion. The results show that – by merely considering plain-concrete fracture energy – the capacity is underestimated and the vertical deflections are overestimated. However, when the concrete contribution was modelled with the expression from MCFT, the capacity was still underestimated but the fitting of the results was satisfactory for the vertical deflections.

By modelling a test of a R/C beam loaded in four-point bending that failed in bending it was found that, when the tension softening was modelled according to MCFT, the behaviour was to stiff and the capacity was overestimated. Hence, the cracked concrete transferred tensile stresses even when the longitudinal reinforcements yield. This implies that an analysis of a concrete member subjected to shear, torsion, and bending will be on the safe side when evaluating the load-carrying capacity or crack width, if only the fracture energy is used to define the unloading branch of the concrete in tension.

REFERENCES

ASCE-ACI Committe 445 on Shear and Torsion (1998). Recent approches to shear design of structural concrete. *Journal of Structural Engineering* 124 12: 1375–1417.

Ayoub A. & Filippou F. C. (1998). Nonlinear Finite-Element Analysis of RC Shear Panels and Walls. *Journal of Structural Engineering* 124 3: 298–308.

Belarbi A. & Hsu T. T. C. (1995). Constitutive laws of softened concrete in biaxial tension-compression. *ACI Structural Journal* 92 5: 562–573.

Bentz E. C. (2005). Explaining the Riddle of Tension Stiffening Models for Shear Panel Experiments. *Journal of Structural Engineering* 131 9: 1422–1425.

Broo H. (2005). *Shear and torsion interaction in prestressed hollow core slabs*. Thesis for the degree of licentiate of engineering Department of civil and environmental engineering Chalmers University of Technology, Göteborg.

Broo H., Lundgren K. & Engström B. (2005). Shear and torsion interaction in prestressed hollow core units. *Magazine of Concrete Research* 57 9: 521–533.

Broo H., Plos M., Lundgren K. & Engström B. (2006). Simulation of shear type cracking and failure with non-linear finite element method. *Submitted to Magazine of Concrete Research*.

CEB (1993). *CEB-FIP Model Code 1990*. Lausanne, Switzerland: Bulletin d'Information 213/214.

CEN/TC250/SC2 (2004). *Eurocode 2: Design of concrete structures Part 2: Concrete bridges Design and detailing rules Stage 49*. Brussels: European Committee for Standardization.

Collins M. P. & Mitchell D. (1991). *Prestressed Concrete Structures*. Englewood Cliffs, New Jersey: Prentice Hall.

di Prisco M. & Gambarova P. G. (1995). Comprehensive Model for Study of Shear in Thin-Webbed RC and PC Beams. *Journal of Structural Engineering* 121 12: 1822–1831.

fib (1999). *Structural concrete. Textbook on behaviour, Design and performance updated knowledge of the CEB/FIP model code 1990*. Lausanne, Switzerland: International Federation for Structural Concrete (fib).

Gabrielsson H. (1999). *Ductility of High Performance Concrete Structures*. Ph. D. Thesis Division of Structural Engineering Luleå University of Technology 1999:15, Luleå, Sweden.

Hegger J., Sherif A. & Görtz S. (2004). Investigation of Pre- and postcracking Shear Behavior of Prestressed Concrete Beams Using Innovative measuring Techniques. *ACI Structural Journal* 101 2: 183–192.

Hsu T. T. C. & Zhu R. R. H. (2002). Softened membrane model for reinforced concrete elements in shear. *ACI Structural Journal* 99 4: 460–469.

Karlsson I. & Elfgren L. (1976). *Förspända lådbalkar belastade med vridande och böjande moment samt tvärkraft*. Rapport 76:10 Göteborg: Chalmers tekniska högskola, Institutionen för konstruktionsteknik, Betongbyggnad.

Kaufmann W. & Marti P. (1998). Structural concrete: Cracked membrane model. *Journal of Structural Engineering* 124 12: 1467–1475.

Kettil P., Ródenas J. J., Aguilera Torres C. & Wiberg N.-E. (2005). Strength and deformation of arbitrary beam sections using adaptive FEM. *Submitted to Computers & Structures*.

Lackner R. & Mang H. A. (2003). Scale transition in steel-concrete interaction. I:Model. *Journal of Engineering Mechanics* 129 4: 393–402.

Magnusson J. (1998). *Anchorage of deformed bars over end-supports in high-strength and normal-strength concrete beams: an experimental studie*. Report 98:7 Göteborg: Division of concrete structures.

Pajari M. (2004). *Shear-torsion interaction tests on single hollow core slabs*. VTT Research Notes 2275 Espoo: Technical Research Centre of Finland, VTT Building and Transport.

Pang X.-B. D. & Hsu T. T. C. (1995). Behavior of reinforced concrete membrane elements in shear. *ACI Structural Journal* 92 6: 665–679.

Pang X.-B. D. & Hsu T. T. C. (1996). Fixed angle softened truss model for reinforced concrete. *ACI Structural Journal* 93 2: 197–207.

Plos M. (2004). *Structural Assessment of the Källösund bridge using Finite Element Analysis – Evaluation of the load carrying capacity for ULS*. Rapport 04:1 Göteborg: Concrete Structures, Department of Structural and Mechanical Engineering, Chalmers University of Technology.

Soltani M., An X. & Maekawa K. (2003). Computational model for post cracking analysis of RC membrane elements based on local stress-strain characteristics. *Engineering Structures* 25: 993–1007.

TNO (2002). *DIANA Finite Element Analysis, User's Manual release 8.1*. TNO Building and Construction Research.

Walraven J. & Stroband J. (1999). The behaviour of cracked in plain and reinforced concrete subjected to shear. *5th International Syposium on Utilization of High Strength/ High Performance Concrete*, Rica Hotel Sandefjord, Norway, 1999.

Vecchio F. J. (2000). Disturbed Stress Field Model for Reinforced Concrete: Formulation. *Journal of Structural Engineering* 126 9: 1070–1077.

Vecchio F. J. & Collins M. P. (1986). The modified compression-field theory for reinforced concrete elements subjected to shear. *Journal of the American Concrete Institute* 83 2: 219–231.

Vecchio F. J. & Collins M. P. (1993). Compression response of cracked reinforced concrete. *Journal of Structural Engineering* 119 12: 3590–3610.

Vecchio F. J. & Shim W. (2004). Experimental and Analytical Reexamination of Classical beam Tests. *Journal of Structural Engineering* 130 3: 460–469.

Yamamoto T. & Vecchio F. J. (2001). Analysis of Reinforced Concrete Shells for Transvers Shear and Torsion. *ACI Structural Journal* 98 2: 191–199.

Zararis P. D. (1996). Concrete Shear Failure in Reinforced-Concrete Elements. *Journal of Structural Engineering* 122 9: 1006–1015.

Fracture Mechanics of Concrete and Concrete Structures – Design, Assessment and Retrofitting of RC Structures – Carpinteri, et al. (eds)
© 2007 Taylor & Francis Group, London, ISBN 978-0-415-44616-7

Assessment of the residual fatigue strength in RC beams

T. Sain & J.M. Chandra Kishen

Dept. of Civil Eng Indian Institute of Science, Bangalore, India

ABSTRACT: In conventional analysis and design procedures of reinforced concrete structures, the ability of concrete to resist tension is neglected. Under cyclic loading, the tension-softening behavior of concrete influences its residual strength and subsequent crack propagation. The stability and the residual strength of a cracked reinforced concrete member under fatigue loading, depends on a number of factors such as, reinforcement ratio, specimen size, grade of concrete, and the fracture properties, and also on the tension-softening behavior of concrete. In the present work, a method is proposed to assess the residual strength of a reinforced concrete member subjected to cyclic loading. The crack extension resistance based approach is used for determining the condition for unstable crack propagation. Three different idealization of tension softening models are considered to study the effect of post-peak response of concrete. The effect of reinforcement is modeled as a closing force counteracting the effect of crack opening produced by the external moment. The effect of reinforcement percentage and specimen size on the failure of reinforced beams is studied. Finally, the residual strength of the beams are computed by including the softening behavior of concrete.

1 INTRODUCTION

Reinforced concrete members subjected to cyclic loading may exhibit both stiffness and strength degradation depending on the maximum amplitude and the number of cycles experienced by the member. Most of the models available currently simulate the cycle dependent stiffness loss that is observed in experiments. The well known Park and Ang (1985) model defines a damage index which is expressed as a linear relation with displacement ratio and absorbed cyclic energy, to describe the hysteretic damage in reinforced concrete members. Garstka et al. (1993) have defined damage indicator in terms of energy ratios, for computing the stiffness loss due to inelastic deformation under earthquake loading. These models are based on elastic-plastic response of RC members, henceforth do not discuss on cracking behavior of concrete. In general, it is accepted that highly reinforced beams that fail by steel yielding are mostly fracture-insensitive. So, structures of these type have not been much investigated from the viewpoint of fracture mechanics. However, there are situations in which fracture plays a role, e.g. failure of normally and lightly reinforced beams (Bazant and Planas 1998). In particular, if crack forms within the tensile zone of RC beams due to fatigue loading, it provokes unstable behavior, which may introduce snap back response in the post peak region. For avoiding such a situation, the criterion to compute the minimum reinforcement for concrete members under flexure, is determined through fracture mechanics approach

(Bosco et al. 1990), as the condition for which first concrete cracking and steel yielding are simultaneous. All LEFM models that are currently available in the literature have roots in the model proposed by Carpinteri (1981). In the LEFM based models, to compute the fracture moment, it is assumed that the $K_I = K_{Ic}$, and steel yielding occurs simultaneously. In the present approach, the assumption of steel yielding is not considered. The limiting criterion is assessed in terms of the tip opening displacement, hence the presence of the process zone is incorporated in the formulation through softening laws. Further, the stability of the crack propagation is also determined using crack extension based approach.

2 FRACTURE MECHANICS BASED MODEL OF REINFORCED CONCRETE BEAM

Based on Carpinteri's (1981) seminal work on LEFM models of reinforced concrete beam, the member with a crack of length a subjected to bending is approximated by a beam subjected to the bending moment and to the steel force applied remotely from the crack plane as shown in Figure 1(a), (b) and (c). Next the steel action is decomposed in a standard way into a bending moment and a centric force. Under the action of applied moment M, the steel force is a statically undetermined reaction. Carpinteri assumed that the crack remains closed while the steel is in elastic regime. Therefore, the crack growth takes place only when the

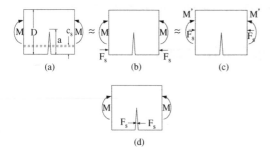

(a)   (b)   (c)

(d)

Figure 1. (a), (b), (c), Carpinteri's LEFM approximation for RC beam; (d) Bosco and Carpinteri's Modification.

steel yields and simultaneously $K_I = K_{Ic}$. Using this condition, the unknown steel force can be computed as described by Carpinteri (1984). With these conditions, the parametric equations of the moment rotation curves could be obtained easily. The limitation of this model in terms of crack closing in elastic regime of the reinforcement was removed by Baluch et al. (1992) and by Bosco and Carpinteri (1992). In the present work, the model proposed by Bosco and Carpinteri (1992) is used. They have modified an earlier model by letting the force of the reinforcement act on the crack faces rather than remotely from the crack plane as shown in Figure 1(d). Hence, it is no longer necessary to assume that the crack is closed everywhere while the steel is elastic; it is enough to assume that the crack is closed at the point where the reinforcement crosses it. Hence the condition for obtaining the unknown steel force in the elastic regime can be written as,

$$w_s = (w_s)_M - (w_s)_S = 0 \tag{1}$$

where w_s is the crack opening at the level of steel bar; $(w_s)_M$ and $(w_s)_S$ are the crack opening due to bending moment and closure forces exerted by the reinforcement respectively. These are computed using the following relations (Alaee and Karihaloo 2003);

$$(w_s)_M = \lambda_{SM} M \tag{2}$$

and

$$(w_s)_S = \lambda_{SS} F_S \tag{3}$$

where λ_{SM} and λ_{SS}, the compliance coefficients due to unit moment M and unit steel force F_S, can be written as,

$$\lambda_{SM} = \frac{2}{BDE} \int_{C_s}^{a} Y_M\left(\frac{x}{a}\right) F_1\left(\frac{x}{a}, \frac{a}{D}\right) dx \tag{4}$$

and

$$\lambda_{SS} = \frac{2}{BE} \int_{C_s}^{a} F_1\left(\frac{x}{a}, \frac{a}{D}\right)^2 dx \tag{5}$$

where C_s is the clear cover to the steel bar. In the above Equations, $Y_M\left(\frac{x}{a}\right)$ and $F_1\left(\frac{x}{a}, \frac{a}{D}\right)$ are the geometry factors expressed as follows:

$$Y_M(\alpha) = \frac{6\alpha^{1/2}(1.99 + 0.83\alpha - 0.31\alpha^2 + 0.14\alpha^3)}{(1-\alpha)^{3/2}(1+3\alpha)} \tag{6}$$

where $\alpha = a/D$ is the relative crack depth;

$$F_1\left(\frac{x}{a}, \alpha\right) = \left\{ \frac{3.52(1-\frac{x}{a})}{(1-\alpha)^{3/2}} - \frac{4.35 - 5.28\frac{x}{a}}{(1-\alpha)^{1/2}} + \left[\frac{1.3 - 0.3(\frac{x}{a})^{3/2}}{\sqrt{1-(\frac{x}{a})^2}} + 0.83 - 1.76\frac{x}{a} \right] \left[1 - \left(1 - \frac{x}{a}\right)\alpha \right] \right\} \frac{2}{\sqrt{\pi\alpha}} \tag{7}$$

Once the unknown steel force is computed using the above relation, based on the principle of superposition the stress intensity factor can be expressed as a summation of K_{IM} and K_{IF} as,

$$K_I = K_{IM} - K_{IF} \tag{8}$$

where K_{IM} and K_{IF} are the stress intensity factors produced by the bending moment and steel force, which can be written as follows:

$$K_{IM} = \frac{M}{BD^{3/2}} Y_M(\alpha) \tag{9}$$

and

$$K_{IF} = \frac{F_S}{BD^{1/2}} F_1\left(\frac{C_s}{D}, \frac{a}{D}\right) \tag{10}$$

In the present study, the above described model is used for determining the condition for unstable crack propagation based on the crack extension resistance approach. The tension softening behavior of concrete in the post peak region is also incorporated in the analysis. The detailed description is given in the subsequent section.

3 DETERMINATION OF FRACTURE STABILITY CRITERION: CRACK EXTENSION RESISTANCE BASED APPROACH

The residual strength assessment of a reinforced concrete member essentially involves the determination of critical condition with respect to fracture failure. In the present study, the condition for unstable crack propagation is found out based on the crack extension resistance based approach. The crack extension resistance based approach originally proposed by Reinhardt and Xu (1998) for plain concrete specimen, in which the crack extension resistance is computed considering the effect of cohesive forces within the process zone. The basic principle of the approach is that the crack extension resistance is composed of two parts. One part is the inherent toughness of the material, which resists the initial propagation of an initial crack under loading, and is denoted as K_{Ic}^{ini}. The cohesive force distributed on the fictitious crack during crack propagation gives another part of the extension resistance. Therefore, it is a function of the cohesive force distribution $f(\sigma)$, tensile strength f_t of the material and the length a of the propagating crack, which can be written as follows,

$$K_R(\Delta a) = K_{Ic}^{ini} + K^c(f_t, f(\sigma), a) \qquad (11)$$

The inherent initiation toughness K_{Ic}^{ini} for a standard three-point bending beam can be computed using

$$K_{Ic}^{ini} = K(P_{ini}, a_0) = \frac{3P_{ini}L}{2BD^2}\sqrt{\pi a_0}g_1\left(\frac{a_0}{D}\right) \quad (12)$$

where P_{ini} is the initial cracking load; a_0 is the initial notch length; L, B, D is the span, width and depth of the beam respectively and $g_1(a_0/D)$ is the geometric factor.

Similarly, the general expression of the crack extension resistance due to cohesive force is given by Reinhardt and Xu 1998),

$$K^c(f_t, f(\sigma), a) = \int_{a_0}^{a} 2\sigma(x)F_1\left(\frac{x}{a}, \frac{a}{D}\right)/\sqrt{\pi a}dx \quad (13)$$

where F_1 is the geometry factor as defined in Equation 7. In the above Equation $\sigma(x)$ is the assumed stress distribution within the fracture process zone. In the present study, the three idealizations for the traction-separation law are considered in order to determine the crack extension resistance and the corresponding critical crack length for which unstable fracture takes place. In available literature, the post-peak softening behavior has been mathematically modeled by different investigators using linear, bilinear, power-law and other relationship depending on the trend followed by experimental results. In this work, we consider the effect of linear, bilinear and power law softening behavior on the fatigue strength of reinforced concrete beams. Amongst these, the simplest approximation is the linear softening relation as proposed by Hillerborg et al. (1976), and stress at any point in the process zone is considered to be a function of the crack opening only. Mathematically, the

linear softening relation can be written as (Hillerborg et al. 1976),

$$\sigma = f_t\left(1 - \frac{w}{w_c}\right) \qquad (14)$$

where f_t is the tensile strength, w the crack opening displacement and w_c the critical crack opening displacement.

Similarly, the bilinear softening behavior can be mathematically expressed as,

$$\sigma = \begin{cases} f_t - (f_t - \sigma_1)w/w_1 & w \leq w_1 \\ \sigma_1 - \sigma_1(w - w_1)/(w_c - w_1) & w > w_1 \end{cases} \quad (15)$$

where w_1 is the opening displacement when the softening curve changes slope due to bi-linearity and the corresponding stress is σ_1.

The power function suggested by Reinhardt (1984) is given by,

$$\sigma = f_t\left[1 - \left(\frac{w}{w_c}\right)^n\right] \qquad (16)$$

where n is an index which is assumed to be 0.248 based on experimental calibrations. After a crack starts from a notch, the size of the fracture process zone grows as the crack advances. The consequence is that the crack resistance K_R to propagation increases. The condition for crack propagation within a member is considered when $K_R(\Delta a)$ equals K_I. K_{IP} is the mode I stress intensity factor under the external loading P, which for a RC beam under three-point bending is given by Equation 8 together with 9 and 10. The important point to be noted here is that, the effect of reinforcement is not considered in terms of resistance, instead it is incorporated while evaluating the stress intensity factor. This is done because, the steel force depends on the applied external moment, therefore not an inherent property of the material.

Further, to determine the instability condition, the well known concept of fracture equilibrium is used (Bazant and Cedolin 1998). If the fracture equilibrium state is unstable, the crack will propagate by itself. Formally, these conditions can be stated as follows:

$$K_R'(\Delta a) - K_{IP}' > 0 \Rightarrow \quad stable$$

$$K_R'(\Delta a) - K_{IP}'(P) = 0 \Rightarrow \quad critical$$

$$K_R'(\Delta a) - K_{IP}'(P) < 0 \Rightarrow \quad unstable \quad (17)$$

where the primes in $K's$ indicate the slope of the quantities concerned. From the above conditions it turns out that, when the slope of the resistance curve is lesser than the slope of the K_I curve, unstable crack propagation takes place. In the present study,

Table 1. Details of the RC beam.

Depth	150 mm
Width	100 mm
Length	1200 mm
Steel area	113.09 mm²
Yield stress	544 MPa
E	35.6e3 MPa
f_{ck}	45 MPa
f_t	3.75 MPa
G_F	0.0725 N/mm
w_c (Linear)	0.037 mm
w_c (Bi-linear)	0.094 mm
w_c (Linear)	0.067 mm

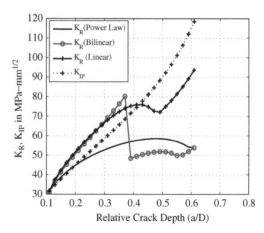

Figure 3. K_R, K_{IP} for Linear, bilinear and Power-law Softening.

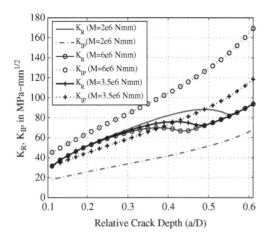

Figure 2. K_R, K_{IP} for Linear Softening.

a reinforced concrete beam is considered for numerical validation of the above method. The details of the specimen geometry and fracture properties are listed in Table 1. The specimen was originally used by Alaee and Karihaloo (2003) for their study on CARDIFRC.

The proposed method of obtaining the instability condition is applied for the RC beam under consideration, for three different values of external moment as $M = 2E6, 3.5E6$ and $6E6$ Nmm as shown in Figure 2. In this particular case, only linear softening is assumed. It is seen that, for the lowest value of M, the slope of the resistance curve remains higher than the slope of K_{IP} curve throughout the assumed crack length regime; hence the crack propagation is stable for this case. For $M = 3.5e6$ Nmm, the resistance curve intersects the stress intensity factor curve at a point $\alpha = 0.418$, and K'_R is lesser than K'_{IP} for α greater than 0.418. Hence, according to the stability condition stated above, the crack propagation remains stable upto relative crack depth of 0.418, becomes critical at that particular value of α, and the unstable region follows in case of further crack propagation. For the M value of 6e6 Nmm, the resistance is always lesser than the stress intensity factor throughout the crack

propagation region resulting in an unstable fracture phenomenon. It can be concluded that as the applied moment value increases the α value corresponding to fracture instability decreases. The study is further extended to determine the influence of different softening approximations as described by Equation 14, 15 and 16, on the computation of instability limits. Figure 3 shows the K_R, K_{IP} curves obtained considering the three softening laws under the external moment of 3.5e5 Nmm. Since, the external load remains constant, and the softening does not take part in K_{IP} computation, the K_{IP} curves are the same for all the three cases. Only the resistance curve will depend on the softening approximations. The comparative study reveals that the linear softening predicts highest value of $\alpha_C = 0.418$ corresponding to unstable condition, followed by bilinear ($\alpha_C = 0.375$) and power-law ($\alpha_C = 0.195$). Therefore, one has to correlate the experimental data of unstable crack propagation with the numerical predictions and conclude about the ideal softening approximation, which would result into realistic prediction on stability condition. Once the critical value of α is determined, the residual strength of the RC beam has to be computed as a function of increasing crack length in the stable region. Before entering into the discussion of fatigue behavior, it is to be noted here, that a parametric study is performed on the steel percentage and the size of the specimen, to find out the influence of these factors on fracture stability issue. Figure 4 represents the K_{IP} and K_R curves for three different percentages of steel and considering linear softening. The resistance will be same for all the three cases and the SIF due to applied loading also does not vary with p unless steel yielding occurs. Therefore, the curves coincide with each other and intersect at a specific point. Hence, it can be concluded, that the reinforcement percentage does not have any

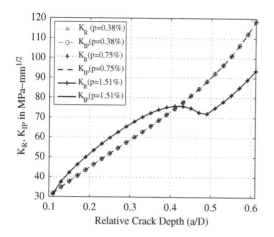

Figure 4. K_R, K_{IP} for three different percentage reinforcement.

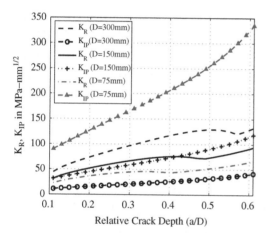

Figure 5. K_R, K_{IP} for three different depth of specimen.

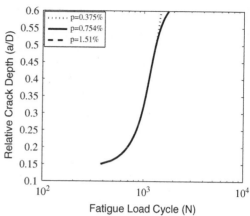

Figure 6. Fatigue crack propagation in RC beam.

perceivable effect on the stability criterion. The critical value of α remains unchanged for increasing as well as decreasing value of steel percentage. Whereas, a strong effect of specimen size is observed on the instability condition as predicted in Figure 5. Here, three different values of specimen depth are considered ($D = 75, 150$ and 300 mm). It is observed that as the specimen size decreases, the zone corresponding to stable crack propagation almost vanishes as observed for $D = 75$ mm; where the resistance is always lesser than the SIF value.

4 FATIGUE CRACK PROPAGATION IN REINFORCED CONCRETE

In the present study, to analyze the fatigue behavior of reinforced concrete, the LEFM based fatigue law as proposed by Slowik et al. (1996) is used, with suitable modifications to incorporate the effect of reinforcement. The fatigue law proposed by Slowik et al. (1996) for describing the complex phenomenon of crack propagation in concrete is given by:

$$\frac{da}{dN} = C\frac{K_{Imax}{}^m \Delta K_I{}^n}{(K_{IC} - K_{Isup})^p} + F(a, \Delta\sigma) \qquad (18)$$

where C is a parameter which gives a measure of crack growth per load cycle, K_{Isup} is the maximum stress intensity factor ever reached by the structure in its past loading history, K_{IC} the fracture toughness, K_{Imax} is the maximum stress intensity factor in a cycle, N is the number of load cycles, a is the crack length, ΔK is the stress intensity factor range, and m, n, p, are constants. These constant co-efficients are determined by Slowik et al. through an optimization process using the experimental data and are 2.0, 1.1, 0.7 respectively. In an earlier work, the above law is modified by the authors (Sain and Chandra Kishen 2004), to incorporate the effect of frequency of the cyclic loading, and the effect of overloads, the details of which are not repeated here.

In case of reinforced concrete beams, the stress intensity factor is considered to be a combined effect of applied loading and the tensile reinforcement as mentioned in Equation 8. The presence of reinforcement introduces a negative SIF (resistance to crack opening), which essentially reduces the crack propagation rate. When the steel is in the elastic regime, the unknown reaction force in steel is computed using the method as described in earlier section. The crack propagation procedure is applied for the considered specimen, for three different values of reinforcement percentage ($p = 0.36, 0.75, 1.51\%$). Figure 6 shows the $a - N$ curve obtained using the proposed method, for three different steel area under constant amplitude

893

fatigue load with maximum moment of 3.5E6 Nmm and a minimum value of zero. It is observed that the rate of crack propagation remains same for all the three p values, if the steel does not yield. The steel yielding occurs only for $p = 0.36\%$, at $\alpha = 0.5$, under the given loading condition. Hence, after $\alpha = 0.5$, the crack propagation rate differs from the other two case as shown in the Figure.

5 RESIDUAL STRENGTH ASSESSMENT

The tensile strength and toughness of concrete are usually disregarded in the strength assessment of reinforced concrete member. In the present study, the post-peak behavior is considered in terms of the tension-softening law, as described earlier, for computing the residual capacity of a cracked RC beam. The available methods for determining the fracture moment, either assume the limiting condition as yielding of reinforcement or assume the length of the process zone. These assumptions are relaxed in the foregoing analysis. The criterion used for computing the ultimate moment capacity is the crack tip opening displacement, w at the tip of each incremental crack length reaching the critical crack tip opening displacement, w_c, which is a material parameter. Hence, the reinforcement does not necessarily reach yielding corresponding to all the crack length values. The following assumptions are made in the analysis regarding the stress-strain distribution along the cracked section:

1. Strain varies linearly across the depth of beam during bending.
2. The crack opening profile is linear.
3. The softening behavior is known in terms of cohesive force-crack opening law. Alternatively, an average strain ϵ_t on the continuum scale may be defined as representative of the opening displacement of the microcracks within an effective softening zone width h_s. In this way, an effective stress-strain constitutive relationship can be adopted in the spirit of the nonlocal continuum concept (Bazant and Oh 1983). The crack opening displacement in the discrete crack model and the post-peak strain in the continuum model are related by $w = h_s \epsilon_t$. In the present study, h_s is taken as $0.5D$, where D is the beam depth.

By fixing the limiting tip opening displacement, corresponding equivalent strain is calculated following assumption (3). The ultimate tensile strain corresponding to $w = w_c$ is denoted as ϵ_{tu}, and the strain corresponding to elastic limit (in other words ϵ for $w = 0$ or $\sigma = f_t$) is represented as ϵ_{tp}. The fracture process zone of length l_p is assumed to form in front of the crack tip. It comprises of the zone starting from the crack tip where ($w = w_c$) or equivalently $\epsilon_t = \epsilon_{tu}$ and extending until $w = 0$ or $\epsilon_t = \epsilon_{tp}$.

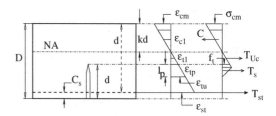

Figure 7. Stress-Strain distribution of the RC beam.

To compute the moment carrying capacity for the assumed strain distribution, an incremental procedure as proposed by Raghuprasad et al. (2005), is adopted. The method is based on the fundamental equilibrium equation for the progressive failure of concrete beams. The uncracked ligament portion $(d - \alpha d - l_p)$ as shown in Figure 7 is divided into a number of segments (say 10,000); each having a segment of depth $\delta x = [(1 - \alpha)d - l_p]/10,000$. To calculate the neutral axis depth factor k, a trial and error procedure is adopted. Knowing k, by the linearity assumption, l_p can be computed as,

$$l_p = \left(1 - \frac{\epsilon_{tp}}{\epsilon_{tu}}\right)(1 - k - \alpha)d \qquad (19)$$

Hence, the resistance provided by the softening zone, (assuming linear softening behavior) can be expressed as,

$$T_s = \frac{1}{2}Bl_pf_t \qquad (20)$$

Next in the uncracked ligament portion, the stresses are calculated at each segment for the compressive strains $\epsilon_{c1}, \epsilon_{c2} \ldots \epsilon_{cm}$ and for the tensile strains $\epsilon_{t1}, \epsilon_{t2} \ldots \epsilon_{tn}$; where m = number of segments in compression zone, and n = number of segments in tension zone. Then the compressive forces $f_{c1}, f_{c2}, \ldots f_{cm}$ and the tensile forces $f_{t1}, f_{t2}, \ldots f_{tm}$ are calculated incrementally. The stress and strain in the tensile reinforcement can be computed as;

$$\epsilon_{st} = \epsilon_{tu}\frac{1 - k}{1 - \alpha - k} \qquad (21)$$

and the stress, can be written as,

$$\sigma_{st} = E_s\epsilon_{st} \quad \leq f_Y \qquad (22)$$

where, E_s is the elastic modulus of steel and f_Y is the yield stress. Steel is assumed to behave in an elastic perfectly plastic manner. Hence, the resistance provided by the reinforcement is expressed as,

$$T_{st} = \sigma_{st}A_{st} \qquad (23)$$

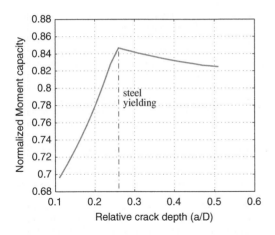

Figure 8. Normalized moment capacity computed through present method.

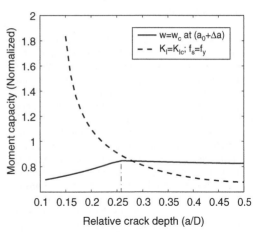

Figure 9. Normalized moment capacity (Comparison between proposed method and LEFM theory).

where, A_{st} is the area of reinforcement. The depth of the neutral axis is calculated such that the total compressive force ($C = f_{c1} + f_{c2} + \ldots + f_{cm}$) equals the total tensile force ($T = T_s + f_{t1} + f_{t2} + \ldots + f_{tm} + T_{st}$). Once the neutral axis depth factor is known the moment carrying capacity can be computed for that equilibrium configuration. The moment of resistance can be computed as,

$$M_R = M_{soft} + M_{UT} + M_{st} \qquad (24)$$

where, M_{soft} is the moment of resistance provided by the softening zone and equals,

$$M_{soft} = T_s[(1 - \alpha - k/3)D - 2/3l_p] \qquad (25)$$

and M_{UT} is the moment of resistance provided by the uncracked tension concrete, which is given by,

$$M_{UT} = T_{Ut}[(1 - \alpha - k/3)D - xx/3 - l_p] \qquad (26)$$

where T_{Ut} is the tensile resistance provided by the uncracked concrete and xx is the length of the corresponding uncracked portion. Finally, the moment of resistance due to the reinforcement M_{st} is computed as,

$$M_{st} = T_{st}\left(1 - \frac{k}{3}\right)d \qquad (27)$$

The procedure is repeated for different crack lengths $a_1, a_2 \ldots a_n$ as long as equilibrium is satisfied. The proposed method is applied to determine the moment carrying capacity for the RC beam specimen as considered above, as a function of increasing crack length. Figure 8 shows the normalized moment value obtained for the considered specimen as a function of increasing

crack length. It is observed that, before steel yielding, the moment carrying capacity increases along with increase in crack length, and the value starts decreasing once the steel undergoes plastic deformation, which is reasonable in case of reinforced specimen. The normalization is done with respect to ($K_{Ic}BD^{3/2}$).

Figure 9 shows the normalized moment value for the specimen, computed using the above method as well as assuming the LEFM criteria, which considers the failure condition to be ($K_I = K_{Ic}$) together with the assumption of steel yielding. In the second case the ultimate moment can be expressed as follows;

$$M_F = \frac{K_{Ic}BD^{3/2}}{Y_M(\alpha)} + \frac{F_P D}{Y_M(\alpha)}$$

$$\left[F_1\left(\frac{x}{a}, \alpha\right) + Y_M(\alpha)\left(\frac{1}{2} - \frac{C_s}{D}\right)\right] \qquad (28)$$

It is seen from the results, that before steel yielding, the second condition predicts higher value than the first case. Whereas, after steel yielding it is lower than the present method. The moment value computed using Equation 24 becomes greater than those computed using Equation 28. Hence, it can be concluded that the assumption of steel yielding for throughout the crack propagation regime, overestimates the capacity of the member in the initial crack propagation stage. However, a situation may arise, when the member fails due to crack propagation even though the steel may not have yielded. Therefore, the present method relaxes the assumption of steel yielding, instead it considers the formation of full process zone at the crack tip. The failure is to be governed by the crack tip opening displacement as mentioned earlier, which is more reasonable in case of residual strength assessment of cracked member.

6 CONCLUSIONS

In the first part of the present study, a method is proposed to determine the condition for unstable crack propagation in a RC beam, considering the post-peak softening response of the concrete. The crack extension resistance based approach is followed to determine the critical value of relative crack depth. Three standard approximations namely, linear, bilinear and power-law are used for describing the softening zone behavior. A parametric study is performed over the reinforcement percentage and the depth of the beam, to find out the influence of each on the stability criterion. It is observed that the percentage reinforcement does not effect the stability phenomenon, unless the steel yields, whereas the depth of the specimen has a strong influence on fracture instability. As the depth increases, the critical crack length corresponding to instability decreases. In the second part of the analysis, the residual strength of a cracked RC beam is assessed by considering the critical tip opening displacement as the governing parameter. The numerical example shows, that the prediction through proposed method is reasonable in the sense, it relaxes the assumption of steel yielding for each and every crack length, which is commonly followed in LEFM based analysis. It is observed that the capacity of the member increases along with crack length when the reinforcement is within the elastic regime, whereas after steel yielding the value reduces with further propagation of crack.

REFERENCES

Alaee, F. and B. Karihaloo (2003). Fracture model for flexural failure of beams retrofitted with cardifrc. *Journal of Engg. Mech.,ASCE 129*(9), 1028–1038.

Baluch, M., A. Azad, and W. Ashmawi (1992). Fracture mechanics application to reinforced concrete members in flexure. In A. Carpinteri (Ed.), *Applications of Fracture Mechanics in Reinforced Concrete*, pp. 413–436. Elsevier Applied Science.

Bazant, Z. and L. Cedolin (1998). *Stability of structures: Elatsic, Inelastic, Fracture and Damage theories*. Dover Publications.

Bazant, Z. and B. Oh (1983). Crack band theory for fracture of concrete. *Materials and Structures 16*, 155–177.

Bazant, Z. and J. Planas (1998). *Fracture and size effect in concrete and other quasibrittle materials*. CRC Press.

Bosco, C. and A. Carpinteri (1992). Fracture mechanics evaluation of minimum reinforcement in concrete structures. In *Applications of Fracture Mechanics in Reinforced Concrete*, pp. 347–377. Elsevier Applied Science.

Bosco, C., A. Carpinteri, and P. Debernardi (1990). Minimum reinforcement in high strength concrete. *ASCE Journal of Struc. Engg. 116*(2), 427–437.

Carpinteri, A. (1981). A fracture mechanics model for reinforced concrete collapse. In *Proc. IABSE Colloquium on Advanced Mechanics of Reinforced Concrete*, pp. 17–30.

Carpinteri, A. (1984). Stability of fracturing process in rc beams. *ASCE Journal of Struc. Engg. 110*, 2073–2084.

Garstka, B., W. Kratzig, and F. Stangenberg (1993, JUNE). Damage assessment in cyclically loaded reinforced concrete members. In *Proceedings, Second EURODYNE*, pp. 121–128.

Hillerborg, A., M. Modeer, and P. Petersson (1976). Analysis of crack formation and crack growth in concrete by means of fracture mechanics and finite elements. *Cement and Concrete Research 6*, 773–782.

Park, Y. and A. Ang (1985). Mechanistic seismic damage model for reinforced concrete. *ASCE Journal of Struc. Engg. 11*(4), 722–757.

Raghuprasad, B., B. Bharatkumar, D. Ramachandra Murthy, R. Narayanan, and S. Gopalakrishnan (2005). Fracture mechanics model for analysis of plain and reinforced high-performance concrete beams. *Journal of Engg. Mechanics, ASCE. 131*(8), 831–838.

Reinhardt, H. (1984). Fracture mechanics of an elastic softening material like concrete. *HERON 29*, 1–44.

Reinhardt, H. and S. Xu (1998). Crack extension resistance based on the cohesive force in concrete. *Engg. Frac. Mech. 64*, 563–587.

Sain, T. and J. Chandra Kishen (2004, April). Damage and residual life asessment using fracture mechanics and inverse method. In V. C. Li, C. Leung, K.William, and S. Billington (Eds.), *Proc., Fracture mechanics of concrete and concrete structures*, Vail, Colorado,USA, pp. 717–724.

Slowik, V., G. Plizzari, and V. Saouma (1996). Fracture of concrete under variable amplitude loading. *ACI Materials Journal 93*(3), 272–283.

Fracture Mechanics of Concrete and Concrete Structures – Design, Assessment and Retrofitting of RC Structures – Carpinteri, et al. (eds)
© *2007 Taylor & Francis Group, London, ISBN 978-0-415-44616-7*

On the cracking stress of RC elements subjected to pure shearing loads

T. Uno
Eight Consultants, Okayama, Okayama, Japan

I. Yoshitake, H. Hamaoka & S. Hamada
Yamaguchi University, Ube, Yamaguchi, Japan

ABSTRACT: The purpose of the present study is to investigate the influence of re-bars on the fracture behavior of RC elements subjected to pure shearing stress in order to evaluate the shear strength on a member level or higher. In particular, in this paper, we discuss the influence of internal re-bars by comparing an RC element and a plain concrete element under cracking load (stress). Furthermore, in order to investigate the influence of shrinkage of concrete on the cracking load (stress), we report the results of pure shearing tests on concrete employing a shrinkage reducing admixture and expansive material.

1   INTRODUCTION

Shearing fracture of concrete structures is characterized by brittleness, and the quantification of shearing fracture behavior is an important factor in securing the safety of structures. Therefore, re-bars for shear reinforcement are more often than not placed in many concrete structures. However, the influence of such re-bars on the shear strength of concrete has not proven to be necessarily satisfactory.

In previous studies, not a few experimental and analytical researches have been conducted on reinforced concrete beam and slab members. In such researches targeted at a member level, it would be difficult to directly evaluate the influence of internal re-bars on the shear strength of concrete. The present work aims to obtain the pure shearing property of reinforced concrete (RC) on an element level, facilitating the evaluation of the relative influence of the re-bars.

In previous studies by the authors, a simplified apparatus for a pure shearing test was developed, and the mechanical properties of a plain concrete element subjected to pure shearing stress, of carbon fiber sheets reinforced concrete element, etc. were reported. However, we have so far carried out pure shearing tests of an RC element targeting the most common RC structures. Common concrete structures have an RC structure, and when shearing stress acts on it, the influence of re-bars on fracture behavior should not be negligible. Consequently, we attempted in the present study the experimental evaluation of a pure shearing fracture property of an RC element in comparison with a plain concrete element which had posed cleavage fracture (mode I) in the previous studies. In particular, in

this paper, the influence of the presence or absence of re-bars on cracking load is reported regarding the concrete element subjected to pure shearing stress.

2   PURE SHEARING TEST PROCEDURE

2.1   *Loading apparatus and loading method*

The medium-sized apparatus for the pure shearing test (Figure 1) used in the present study can be installed in

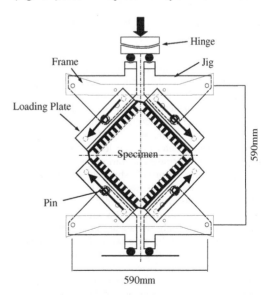

Figure 1.   Pure shearing test.

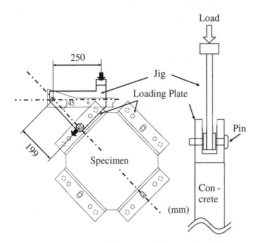

Figure 2. Providing method of pure shearing stress.

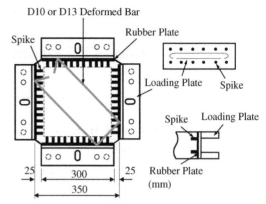

Figure 3. Detail of specimen.

an Amsler type testing machine. As shown in Figure 2, a (vertical) uniaxial load is distributed in 45° directions via the rotating loading jig so that pure shearing stress can be applied on a concrete test specimen. The reaction force in the horizontal direction is structurally received by the steel frames installed on the top and bottom. This method can be considered very economical because it uses a relatively simple jig and does not require multiple pressing devices.

Each loading plate used for this medium-sized pure shearing test is channel-shaped as shown in Figure 2, which shows that one rotating loading jig is pin-coupled structurally.

2.2 Test specimens and experimental parameters

As shown in Figure 3, test specimens for medium-sized pure shearing tests were prepared by attaching steel bolts (M12, embedded 25 mm in length) to each of the steel loading plates on the four sides, with the role of spikes. Here, the spike-embedded areas in the concrete are considered to be parts conveying shearing stress to the interior concrete element.

Additionally, a long aperture (15 × 265 mm) was provided at the center of each loading plate so that re-bars could be arranged at various angles. For pure shearing tests in the present study, a 1 mm-thick rubber plate was installed between the loading plate and concrete so as to avoid excessive deformation to the concrete test specimen due to the rigid loading plate.

In the present study, re-bars having relatively small diameters (D13, D10) were worked into a channel shape and also provided with a hook on both ends in order to secure sufficient fixity of re-bars in the concrete element. The spacing between re-bars was 150 mm. The bar arrangement condition before casting is shown in Figure 4.

Figure 4. Re-bar set up.

Generally, it is known that the (bending) cracking load of reinforced concrete is low in comparison with plain concrete because a tensile force is potentially generated in the concrete due to partial constraint of the free volume change (shrinkage) of the concrete excited by the re-bars (Niwa, Hidaka, Tanabe, 1996). Also, according to the previous study, it is reported that shear reinforcing bars influence the shear cracking load F_{ps} of RC beams employing high-strength concrete, but the fact is that there is not sufficient information about the influence of re-bars on the shear cracking of general-strength-level concrete.

With that, we investigated in the present study the influence of re-bars with respect to the pure shear cracking load F_{ps} by using normal-strength (W/C = 60%) concrete together with a shrinkage reducing admixture and expansive material. The

Table 1. Experimental parameters.

No.	Specimen	Re-bar diameter	Expansive material (Ex) kg/m³	Shrinkage reducing admixture (SRA) kg/m³	Curing method
1	NS-M0	—	—	—	Wet
1′	D13(0°)-M0	D13	—	—	Wet
2	NS-M20	—	20	—	Wet
2′	D13(0°)-M20	D13	20	—	Wet
3	NS-M0	—	—	—	Wet
3′	D10(0°)-M0	D10	—	—	Wet
4	NS-M0	—	—	—	Dry
4′	D10(0°)-M0	D10	—	—	Dry
5	NS-M0+	—	—	10	Wet
5′	D10(0°)-M0+	D10	—	10	Wet
6	NS-M20	—	20	—	Wet
6′	D10(0°)-M20	D10	20	—	Wet
7	NS-M30	—	30	—	Wet
7′	D10(0°)-M30	D10	30	—	Wet
8	NS-M40	—	40	—	Wet
8′	D10(0°)-M30	D10	30	—	Wet

Table 2. Mix proportions of concrete.

Mix No.	W/(C + Ex) %	W/C %	W kg/m³	C kg/m³	S kg/m³	G kg/m³	Ad kg/m³	Ex kg/m³	SRA kg/m³
M0	60	60	160	267	790	1092	C × 1.0%	—	—
M0+	60	60	160	267	790	1092	C × 1.0%	—	10
M20	60	65	160	247	790	1092	C × 1.0%	20	—
M30	60	68	160	237	790	1092	C × 1.0%	30	—
M40	60	71	160	227	790	1092	C × 1.0%	40	—

experimental parameters employed in the present study are given in Table 1.

2.3 Materials used and mixing conditions

The mixing proportions of the prepared concrete are shown in Table 2. The experiments were conducted by setting the water-powder ratio (W/(C + Ex)) constant at 60% for the purpose of investigating the pure shearing property of normal strength concrete.

Also, due to the use of early-strength Portland cement, we conducted experiments at the age of 7–8 days in Tests No.1, No.2, and No.4 to No.8 (15 days in Test No.3 only). In these tests, drying effects were minimized by sufficient compress curing until the tests performed.

3 PURE SHEAR CRACKING AND CRACKING LOAD (PURE SHEARING STRENGTH) OF RC ELEMENT

3.1 Pure shear cracking pattern

Cracking pattern caused by the pure shearing tests and cracking loads (pure shearing strengths f_{ps}) are

collectively shown in Figure 5. In every experiment using the medium-sized apparatus for the pure shearing test, the plain concrete element posed cleavage fracture (mode I) in which one vertical crack (pure shear cracking) occurred and developed as the principal tensile stress component was prominent, as show in the previous studies using the small-sized apparatus.

On the other hand, the cracking patterns differed with experimental parameters in the pure shearing tests of the RC element. Pure shear cracking patterns are discussed below for the various experimental parameters (re-bar diameter and mix proportion).

3.1.1 Diameter of re-bar

On the D13(0°)-M0 test specimen and D13(0°)-M20 test specimen employing D13 re-bars, cracking occurred along the re-bar axis and along the interface between the loading plate and concrete (tips of spikes) after the occurrence of pure shear cracking. On the other hand, on the D10(0°)-M0 test specimen employing D10 re-bars, pure shear cracking occurred, and no cracking was found in other directions.

The reason for this can be explained as follows: since the (shearing) reinforcing effects of the D13 re-bars were sufficiently high with respect to the

No.	1		2	
	NS-M0	D13(0°)-M0	NS-M20	D13(0°)-M20
Crack sketch	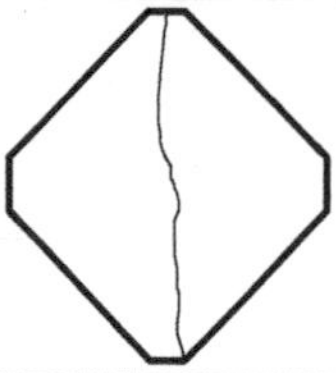	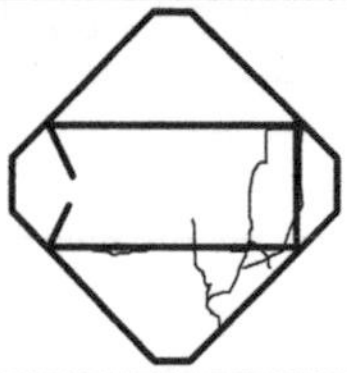	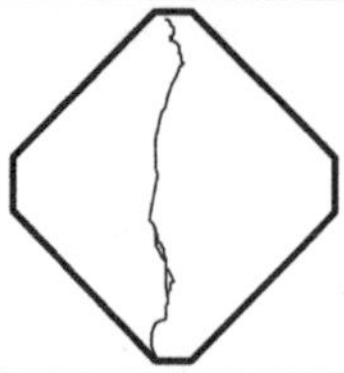	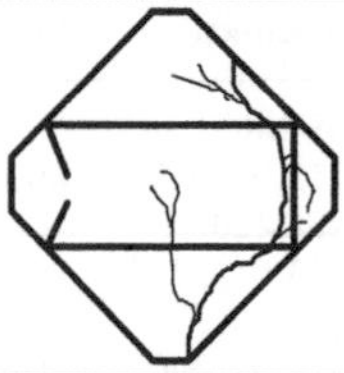
F_{ps}	97.8kN	79.7kN	79.5kN	78.7kN
f_{ps}	2.04N/mm^2	1.66N/mm^2	1.66N/mm^2	1.64N/mm^2
RC/NS	0.81		0.99	
No.	3		4	
	NS-M0	D10(0°)-M0	NS-M0	D10(0°)-M0
Crack sketch	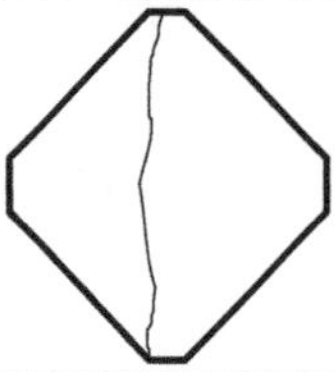	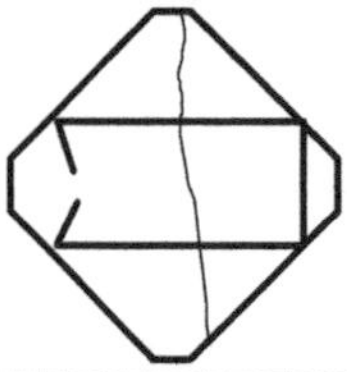	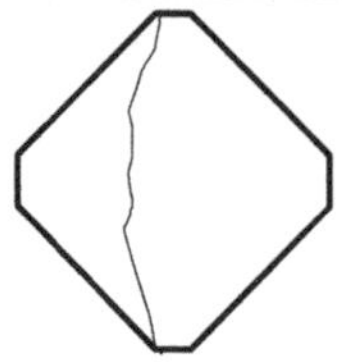	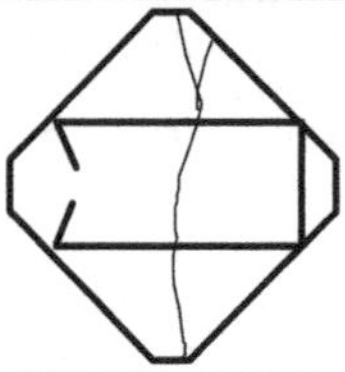
F_{ps}	92.4kN	83.3kN	121.1kN	104.0kN
f_{ps}	1.93N/mm^2	1.74N/mm^2	2.53N/mm^2	2.17N/mm^2
RC/NS	0.90		0.86	
No.	5		6	
	NS-M0+	D10(0°)-M0+	NS-M20	D10(0°)-M20
Crack sketch	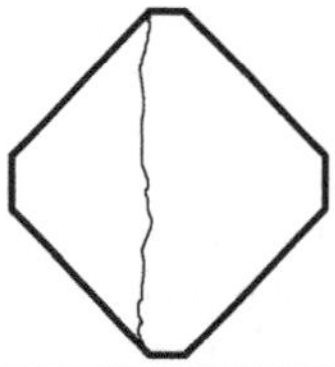	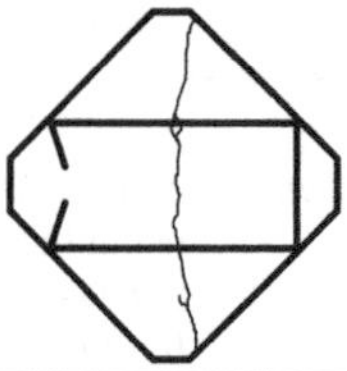	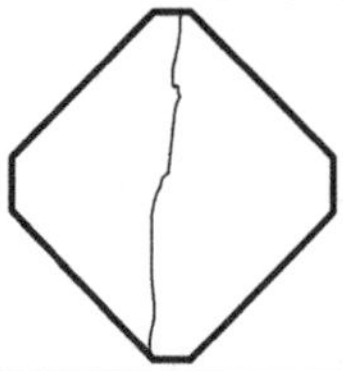	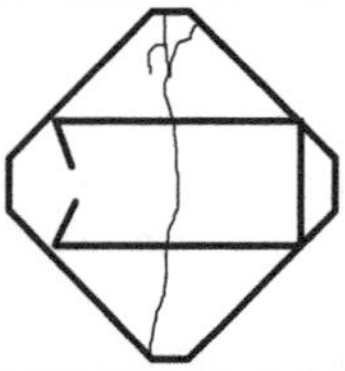
F_{ps}	108.6kN	108.8kN	74.5kN	83.3kN
f_{ps}	2.27N/mm^2	2.27N/mm^2	1.56N/mm^2	1.74N/mm^2
RC/NS	1.00		1.12	
No.	7		8	
	NS-M30	D10(0°)-M0+	NS-M40	D10(0°)-M40
Crack sketch	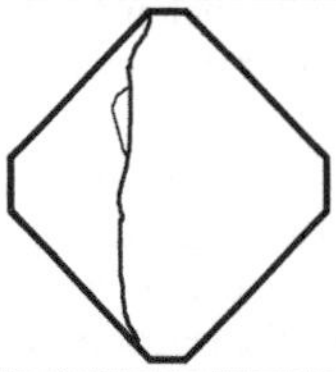	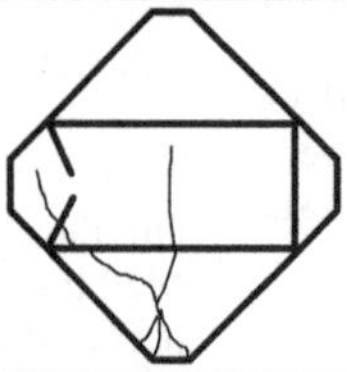	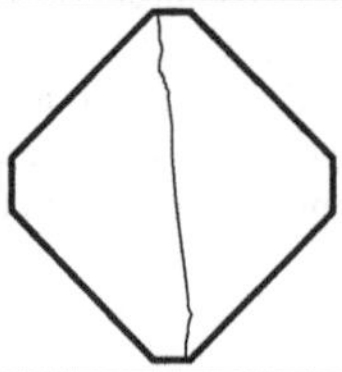	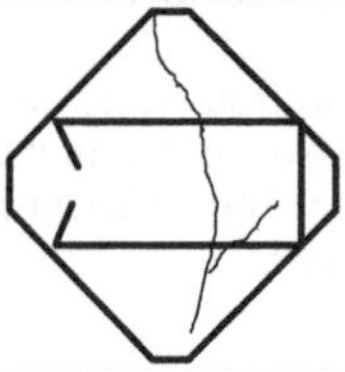
F_{ps}	44.4kN	51.7kN	40.5kN	39.2kN
f_{ps}	0.93N/mm^2	1.08N/mm^2	0.85N/mm^2	0.82N/mm^2
RC/NS	1.17		0.97	

Figure 5.   Crack sketch and cracking load (f_{ps}, F_{ps}).

dimensions of this test specimen, cracking occurred at the interface between the loading plate and concrete and near the re-bar axis, which presented relatively low strength since the force was applied against the principal tensile stress component of the pure shearing stress state. In the present study, which concerns experiments on the shearing properties of an element of the RC structure, the fracture behavior obtained by using the D13 re-bars may not be a representative. Therefore, in the following experiments, we discussed the pure shear cracking property by using the D10 re-bars.

3.1.2 Amount of expansive material and shrinkage reducing admixture

On the D10(0°)-M20 test specimen employing expansive concrete with the amount of expansive additive 20 kg/m^3 (standard amount added), pure shear cracking developed in the vertical direction developed. At the same time, on the test specimens (D10(0°)-M30 and D10(0°)-M40) with the amounts of expansive material 30 kg/m^3 and 40 kg/m^3, respectively, cracking occurred along the interface between the loading plate and concrete (tips of spikes) after the occurrence of pure shear cracking. It is presumed that the chemical pre-stressing (strain) effects of expansive concrete limited the principal tensile stress component that acted on the cross sections at the center of the test specimens, and cracking developed in the areas in which they had little influence.

The D10(0°)-M0+ test specimen employing the shrinkage reducing admixture showed cleavage fracture (mode I) in which one pure shear cracking occurred as in the D10(0°)-M0 and D10(0°)-M20 test specimens.

3.2 Cracking load (pure shearing strength)

The purpose of the present study is to investigate the influence of internal re-bars in the concrete element on the first cracking (shear cracking) load. Therefore, we decided to perform comparative evaluation of cracking loads F_{ps} by preparing test specimens with and without re-bars in No.1 to No.8 concurrently. The cracking loads, pure shearing strengths f_{ps}, and strength ratios (RC/NS) associated with the existence and non-existence of the re-bars of the respective test specimens are shown in Figure 5 as mentioned above. In the present study using the RC element, pure shearing strengths fps were obtained from the cracking load Fps, because the purpose of the present study is to evaluate the strength characteristics of concrete element. As in the previous section, pure shear cracking loads F_{ps} (pure shearing strengths f_{ps}) are shown below as systematically arranged by experimental parameters.

3.2.1 Re-bar diameter

Focusing on the results of Tests No.1 (D13) and No.3 (D10), we note that the strength ratios associated with the existence and non-existence of the re-bars are 0.81 and 0.90, respectively, and the cracking load F_{ps} of the RC element is smaller than that of the plain concrete element in either Test No. In Test No.4 on condition of 15 days of air curing after casting (conditions other than curing were the same as in Test No.3), the strength ratio was the smallest among Tests No.3 ~ No.8 using D10 re-bars.

Also, when the results of No.2 (D13) and No.6 (D10) using expansive concrete with the amount of expansive material 20 kg/m^3 are compared, the strength ratios associated with the existence and non-existence of the re-bars were 0.99 and 1.12, respectively. With the use of expansive concrete, the cracking load F_{ps} of the RC element is comparatively improved, but the cracking load F_{ps} of the D13(0°)-M20 test specimen employing the D13 re-bars remains almost equal to that of the NS-M20 test specimen of the plain concrete element.

3.2.2 Shrinkage reducing admixture

From the above experiment, since the action of potential initial stress associated with the shrinkage of concrete was theorized, similar experiments were conducted on concrete employing a shrinkage reducing admixture (Test No.5). The result was that the cracking loads F_{ps} of the NS-M0+ test specimen and D10(0°)-M0+ test specimen were almost the same, approximately 109 kN (RC/NS = 1.00). By comparing this result with the results of Tests No.3 and No.4, it can be seen that the initial stress acting on the concrete could also greatly influence the shear cracking load F_{ps} in the element-level experiments used in the present study.

3.2.3 Amount of expansive material

Based on the above results, we conducted similar comparative experiments on concrete having different amounts of expansive material. In Test No.6 (M20), the strength ratio associated with the existence and non-existence of the re-bars was 1.12 as mentioned above, while the same ratio in Test No.7 (M30) was 1.17, and furthermore the strength ratio associated with the existence and non-existence of the re-bars in Test No.8 (M40) was 0.97, which indicates that there was not much improvement of the strength as a result of the expansion. The experimental results indicate that reinforcing effects against the occurrence of shear cracking may be obtained by adding appropriate amounts of expansive additive even with re-bars having relatively small diameters, such as stirrups.

4 CONCLUSIONS

In the present study, we discussed the influence of internal re-bars by comparing the cracking loads (stresses) of an RC element and plain concrete element subjected to pure shearing stress. Conclusions obtained in the present study are listed below.

1) Even with the use of normal-strength concrete, the cracking load of the RC element under the influence of potential initial stresses caused by re-bars was reduced to approximately 80–90% compared with the plain concrete element. From this, emerges the possibility that the influence of internal re-bars may not always be negligible in the evaluation of a proof strength of concrete members on which shearing stress acts on the element level.
2) If the reinforcing effects of re-bars are large and if the chemical pre-stressing (strain) effects of expansive concrete are large, only the (pure) shear cracking of the RC element is prominent, and does not lead to fracture, and cracking therefore develops at other brittle zones.

3) With the concrete employing a shrinkage reducing admixture, the cracking load was approximately equal irrespective of the presence or absence of re-bars. Also, with the concrete employing appropriate amounts of expansive material, the cracking load of the RC element can be improved compared to that of the plain concrete element.

REFERENCES

Niwa, J., Hidaka, S. & Tanabe, T. 1996. The influence of initial stresses on the size effect of concrete flexural strength. *Journal of materials, concrete structures and pavements* 550 V-33: 85–94.
Tanaka, H., Yoshitake, I. & Hamada, S. 2003. Experimental study on the strength of concrete element subjected to pure shearing stress. *Journal of materials, concrete structures and pavements* 746 V-61: 205–214.
Yoshitake, I., Honjo, K., Hisabe, N., Tanaka, H. & Hamada, S. 2006. Experimental study on fracture behavior of concrete element subjected to pure shearing stress. *Journal of materials, concrete structures and pavements* 62(1): 29–37.

Fracture Mechanics of Concrete and Concrete Structures – Design, Assessment and Retrofitting of RC Structures – Carpinteri, et al. (eds)
© *2007 Taylor & Francis Group, London, ISBN 978-0-415-44616-7*

Cracking and deformation of RC beams strengthened with Reactive Powder Composites

I. Ujike, H. Sogo & D. Takasuga
Ehime University, Graduate school of science and engineering, Matsuyama, Ehime, Japan

Y. Konishi & M. Numata
Aikyo Corporation, Toon, Ehime, Japan

ABSTRACT: A reinforced concrete beam without cracking under serviceability imit state has been developed. A part of the tension zone in the reinforced concrete beam was fortified with Reactive Powder Composite (RPC). Flexural load tests on the beam were carried out in this study. The cracking moment of the beam reinforced with RPC could be estimated by the elastic theory, provided that the stress due to the restraint of reinforcing bar against the autogenous shrinkage of RPC must be taken into consideration. After the generation of cracking, RPC has little effect on the deformation of beam reinforced with RPC due to the increase of bending moment. However, the flexural capacity of beam fortified with RPC is larger than that of the reinforced concrete beam without RPC and increases with the increase in the reinforced area of RPC.

1 INTRODUCTION

Cracks developed in a reinforced concrete member speed up the deterioration of the reinforced concrete member, because cracks make the invasion of harmful substances easy. It is difficult to evaluate the geometrical properties of the crack quantitatively by the irregularity of the crack. Therefore, an estimate method for durability of reinforced concrete members in consideration of the effect of the crack hasn't been established. Although crack width is controlled to stay within the permissible crack width by the usual design, the crack is a weak point even if it is narrow. In order to give the reinforced concrete member high durability, it is necessary to avoid cracking.

Recently, Ductile Fiber Reinforced Cementitious Composites (DFRCC) which destruction energy and ductility improve greatly under tensile stress condition have been developed (JCI 2002). Reactive Powder Composite (RPC) is one of DFRCC and the paste of RPC is made super high strength by the use of reactive powder, fine granulating of aggregate and tightest filling of powder (Musha et al 2002). From this, the stress which a crack occurs in first after elastic deformation by the action of tensile force, first cracking strength, is very high about 8 N/mm².

Authors gave attention to super high first cracking strength of RPC. A reinforced concrete beam without cracking under serviceability limit state has been developed (Ujike et al. 2005). The part of tension zone

in the reinforced concrete beam was strengthened with RPC. In this study, flexural loading tests on the beams were carried out and cracking moment, curvature, reinforcement strain and flexural capacity of the beams fortified by RPC were investigated.

2 EXPERIMENTS

2.1 Materials

RPC used in this study consists of premixed powder in a carefully selected combination of cement, silica particles and siliceous sand and steel fibers, without coarse aggregate. Special water reducing agent was added to mixing water. For conventional concrete, high early strength portland cement (density 3.14 g/cm³) was used. Crushed sand (density of surface-dry 2.57 g/cm³, absorption 1.33%) and crushed gravel (density of surface-dry 2.62 g/cm³, absorption 0.88%) were used as a fine and coarse aggregate, respectively. Water reducing agent was also used.

2.2 Specimen

Beam specimens were produced for a flexural loading test. Specimens have the height of 200 mm, the width of 150 mm and overall length of 1800 mm. Details of cross sections are illustrated in Fig. 1. In figure, a gray part shows RPC. Deformed bars with a bar

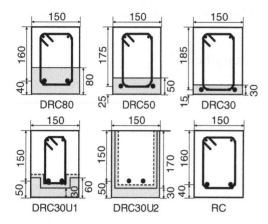

Figure 1. Cross–sections of specimens.

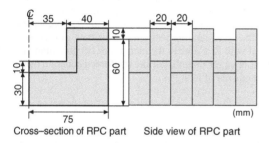

Figure 2. Outline of irregularity on RPC (DRC30UI).

Table 1. Specified mix proportion of RPC

Water powder ratio (%)	Unit content (kg/m³)			
	Water	Premix	Steel powder	SP* fiber
8.0	180	2254	157	27

∗ Special water reducing agent.

Table 2. Specified mix proportion of concrete.

W/C (%)	s/a (%)	Unit content (kg/m³)			
		Water	Cement	Fine aggregate	Coarse aggregate
45.0	35.0	165	367	645	1219

Table 3. Mechanical properties of RPC and concrete.

Specimen	RPC (N/mm²)			Concrete (N/mm²)	
	f'_D	f_{cr}	E_D	f'_c	E_c
DRC80	193.0	14.1	53830	32.4	28130
DRC50	193.0	14.1	53830	45.9	31680
DRC30	193.0	14.1	53830	45.9	31680
DRC30U1	179.3	11.8	52270	35.4	25420
DRC30U2	213.5	11.1	55510	44.8	29050
RC	—	—	—	45.9	31680

f'_D: Compressive strength of RPC
f_{cr}: First crack strength of RPC
E_D: Elastic modulus of RPC
f'_c: Compressive strength of concrete
E_c: Elastic modulus of concrete

diameter of 16 mm were used as a primary tension reinforcement. For DRC80, DRC50 and DRC30, reinforcing bars were arranged in the center of height direction of the RPC part. Deformed bars with a bar diameter of 10 mm were used as a stirrup and were arranged along the entire length of the specimen at 100 mm spacing. These stirrups also have the function to prevent the slip between concrete and RPC in addition to the function of shear reinforcement. For DRC30U1 and DRC30U2, the bottom and the side sections except for reinforcing bars in the beam were fortified in the concavity-shaped. This arrangement of steel bars, as described later, has the function to decrease the influence of the autogenous shrinkage of RPC on the cracking moment of the beams. Furthermore, convex parts with height of 1 cm and width of 2 cm were formed on the RPC surface in contact with concrete at intervals of 2 cm, as shown in Fig. 2.

The specified mix proportions of RPC and concrete are tabulated in Table 1 and Table 2, respectively. The beams reinforced with RPC were produced with the following process. Assembled steel bars were fixed within a form. For DRC30U1 and DRC30U2, the concrete part in the beam was made from polystyrene foam. The polystyrene foam was turned upside down and was placed in the form. RPC having the flow

value of 270 mm was mixed and RPC was cast into the form with vibration. After 48 hours from casting of RPC, the polystyrene foam was removed from RPC and RPC specimens were steam-cured at 98 degrees C for 48 hours. Concrete was constructed on RPC part after 3 days from the finish of steam curing and a reinforced concrete beam without RPC was also constructed at the same time, and subsequently left motionless in the laboratory until a loading test. Table 3 shows the mechanical properties of RPC and concrete used for each specimen. The mechanical properties were measured at the time of the loading test.

2.3 Loading test

Flexural loading tests were performed after 7–10 days from the cast of concrete. Figure 3 shows the side view of a specimen and the loading points. As shown in Fig. 3, the load was applied at two positions which divide the span into three equal and gradually

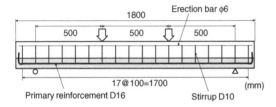

Figure 3. Side view of specimen and loading point.

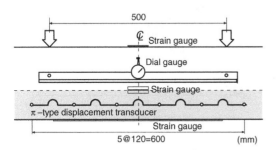

Figure 4. Outline of measurement.

increased in 2 kN steps. Reinforcement strain, concrete strain at upper fiber, RPC strain at bottom fiber, strains at the adjacent joint between RPC and concrete, deflection and crack width were measured in the tested zone with the length of 500 mm subjected to pure bending. Fig. 4 shows the outline of measurement. The reinforcement strain was measured by strain gauges fixed at the center of a steel bar. The deflection was measured with dial gauges having the accuracy of 1/1000 mm. For measurement of the RPC strain, nine strain gauges with the length of 60 mm were affixed at the bottom surface of the specimen. In the cases of DRC80, DRC50 and DRC30, strains adjacent at joint between RPC and concrete were measured. Crack widths were measured with a π-type displacement transducer having the accuracy of 1/2000 mm.

3 EXPERIMENTAL RESULTS AND DISCUSSION

3.1 Cracking moment

In this study, the serviceability limit state for cracks is investigated. In order to evaluate the performance on crack prevention of the beam reinforced with RPC, the moment corresponding to the permissible crack width is calculated as the external moment to act under the serviceability limit state. The moment corresponding to the permissible crack width is calculated based on Standard Specification for Design and Construction of Concrete Structure (JSCE 2002) (hereafter referred as Specification).

According to Specification, when environmental condition regarding the corrosion of reinforcement is normal environment, the permissible crack width w_a for the corrosion of reinforcement is given as a function of cover c:

$$w_a = 0.005c \quad (1)$$

As cover of RC shown in Fig. 1 is 32 mm, we obtain $w_a = 0.16$ mm. The following equation for crack width w has been proposed by Specification.

$$w = 1.1k_1k_2k_3\{4c+0.7(C_s-\phi)\}(\sigma_s/E_s+\varepsilon'_{cs}) \quad (2)$$

where, k_1 is a constant to take into account the difference in surface geometry of reinforcement, $k_1 = 1.0$ as deformed bar was used in this study. k_2 is a constant to take into account the influence of concrete strength f'_c, $k_2 = 0.93$ as k_2 is calculated by $k_2 = 15/(f'_c + 20) + 0.7$. k_3 is a constant for arrangement of reinforcement, $k_3 = 1.0$ as reinforcing bars are arranged in one layer. And C_s = center to center distance of reinforcing bar; ϕ = bar diameter; σ_s = reinforcement stress; E_s = modulus of elasticity of reinforcement. ε'_{cs} is the strain to take into account the influence of creep and drying shrinkage on crack width, $\varepsilon'_{cs} = 150 \times 10^{-6}$ is used in general. From Eq. (2), the reinforcement stress $\sigma_s = 159$ N/mm² is obtained. Furthermore, the relationship between moment M and reinforcement stress σ_s is given from elastic analysis with the assumption that tensile stress in concrete is negligible as follow:

$$\sigma_s = n(M/I_i)(d-x) \quad (3)$$

where, n = elastic modular ratio of steel bar to concrete; d = effective depth; x = neutral axis depth; I_i = moment of inertia of transformed cross section about neutral axis. Substituting $\sigma_s = 159$ N/mm² into Eq. (4), $M = 8.82$ kNm is obtained.

Figure 5 shows an example of the change of strains measured by the strain gauges affixed at a bottom fiber of the beam with the increase of moment. The strains of RPC at the bottom fiber increase linearly until the generation of cracking. There are an increase and a decrease in strain of RPC rapidly due to the generation of cracking. In this study, the moment just before the strain changed remarkably is decided with the cracking moment. Table 4 shows the cracking moment of specimens reinforced with RPC. The cracking moments of the beams fortified with RPC except for DRC30 are larger than the moment corresponding to permissible crack width of RC, that is, it means that a crack doesn't occur in the beam appropriately reinforced with RPC under serviceability limit state.

The calculated values of the cracking moment for specimens reinforced with PRC are also shown in

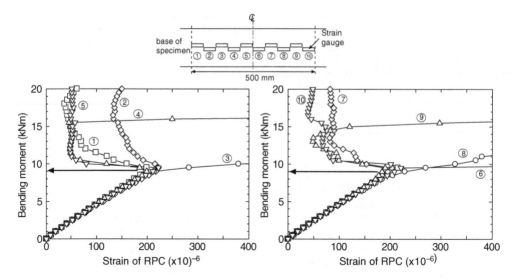

Figure 5.   Strains of RPC at bottom fiber of DRC50.

Table 4.   Cracking moment of specimens reinforced with RPC.

| Specimen | Cracking moment (kNm) | | | |
	Measure-ment	Calcula-tion I	Calcula-tion II	$\sigma_{c,as}$ (N/mm²)
DRC80	10.15	13.13	10.31	3.04
DRC50	9.00	13.18	8.85	4.63
DRC30	6.00	12.68	6.29	7.12
DRC30U1	10.00	9.11	—	—
DRC30U2	8.89	8.99	—	—

$\sigma_{c,as}$:stress due to restraint of reinforcing bar against autogenous shrinkage of RPC.

Table 4. The cracking moments of calculation I are obtained from the following equation based on elastic theory under the assumption that a crack generates when the tensile stress at the bottom fiber of the beam reaches the first crack strength of RPC.

$$M_{cr} = (f_{cr}I_g)/(h-y) \qquad (4)$$

where, M_{cr} = cracking moment, f_{cr} = first crack strength of RPC, h = height of beam. y and I_g are depth of centroid and the moment of inertia of transformed cross section about centroid as already shown in Table 4. The calculations on DRC80, DRC50 and DRC30 overestimate the cracking moment of specimen reinforced with RPC.

It has been reported that the autogenous shrinkage of RPC is large (Katagiri et al 2002). The effect of autogenous shrinkage of RPC is taken into consideration

by the calculation on the cracking moment. $\sigma_{c,as}$ in Table 4 is the stress due to the restraint of a reinforcing bar against the autogenous shrinkage of RPC produced by the end of steam curing. It is assumed in the calculation of restraint stress that both RPC and a steel bar change in a body. From previous study (Katagiri et al 2002), the autogenous shrinkage strain used for the calculation is set 500×10^{-6} for RPC containing the steel fiber. The relaxation of the restraint stress by creep isn't taken into consideration, because creep of RPC is very small. In calculation II, the restraint stress was deducted from the first cracking strength, and except for this, it is the same way as the calculation I. The calculation II is comparatively well in agreement with the measurement. The cracking moment of the beam reinforced with RPC can be estimated by the elastic analysis in consideration of the restraint stress due to autogenous shrinkage of RPC.

From this fact, if the restraint stress can be made small, it expects that the resistance against cracks for the beam reinforced with RPC can be improved more. In DRC30U1 and DRC30U2, in order to reduce the restraint stress developed in RPC, the reinforcing bar isn't arranged in RPC. The cracking moments of DRC30U21 and DRC30U2 are equal or greater than that of DRC50, although the reinforcement area of RPC in the tension zone of the beam is small. And the calculated values of DRC30U1 and DRC30U2 are almost in agreement with the measured values.

3.2   Cracking

Figure 6 shows the crack width of RC and the displacement of RPC at the depth of the reinforcing bar of DRC80 and DRC30. About RC, one crack generated to

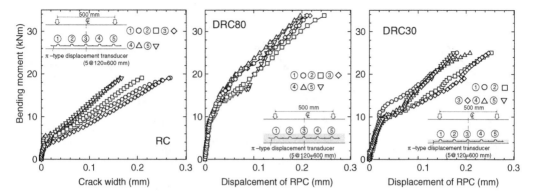

Figure 6. Crack width of RC and displacement of RPC at depth of reinforcing bar (DRC80, DRC30).

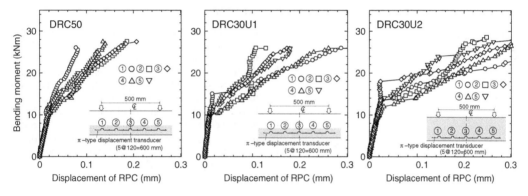

Figure 7. Displacement of RPC at depth of 25 mm from bottom fiber (DRC50, DRC30U1, DRC30U2).

the extent of measurement of one displacement transducer. The value shown in Fig. 6 is the average of the value measured at the both sides of beam. About DRC80 and DRC30, The values shown in Fig. 6 are also the average of the measured value of the displacement transducer of both sides of beam. However, several cracks were generated in the one measurement section. Therefore, at the failure of the beam, the width of one crack that was generated on the beam fortified with RPC is 0.1 mm or less. The increase ratio of displacement of RPC to the increase in bending moment for DRC30 is larger than that for DRC80. The reason for this is that the effective depth of the reinforcing bar is large and the area of reinforcement with RPC is small.

Figure 7 shows the displacement of RPC at the depth of 25 mm from the bottom fiber for DRC50, DRC30U1 and DRC30U2. About these beams, several cracks were also generated in the one measurement section. Although the cracking moment of DRC30U1 and DRC30U2 equals or is greater than that of DRC50, the increase ratio of displacement of RPC to the increase in bending moment for DRC30U1 and DRC30U2 is larger than that for DRC50. This may be

because the reinforcing bar isn't arranged inside RPC in DRC30U1 and DRC30U2.

3.3 Reinforcement strain

Figure 8 shows the strain of a reinforcing bar in the beams fortified with RPC. The broken line in Fig. 8 is the calculation by elastic analysis in consideration of full section of the beam. The contribution of RPC is evaluated by the use of elastic modular ratio of RPC to concrete, like the case of steel bars in reinforced concrete. Before the generation of cracking, experimental values agree well with the calculated values. The dot-dash line is the calculation neglecting the tension zone of the beam. The measurement values are smaller than the calculation values due to the contribution of RPC after crack generating.

In order to consider the contribution of RPC, the softening stress – crack opening displacement (COD) relationship shown in Fig. 9 is used in this study. The relationship was established by referring to the guidelines for design of DFRCC (JSCE 2004). From the measurement shown in Fig. 6 and Fig. 7, it is considered that the stress equivalent to first crack

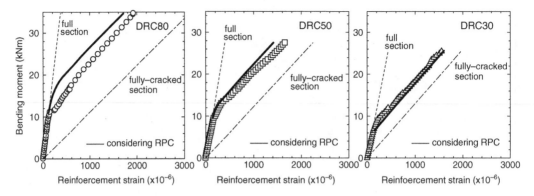

Figure 8. Reinforcement strain of beams fortified with RPC (DRC80, DRC50, DRC30).

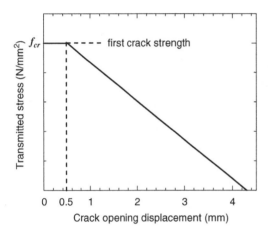

Figure 9. Tension softening diagram of RPC.

strength is developed in RPC after generating crack, because the crack width of RPC is as small as 0.1 mm or less. Furthermore, based on this, the distributions of strain and stress at arbitrary section of the beam fortified with RPC are assumed as shown in Fig. 10. We have two equations for equilibrium of axial force and bending moment, therefore, concrete strain and reinforcing strain can be obtained by solving both equations simultaneously. The solid line in Fig. 8 is the calculation value which takes the effect of RPC into consideration, as mentioned above. In calculation of DRC80, when bending moment is not less than 20.3 kNm, the stress corresponding to first crack strength is distributed over the entire cross-section of the RPC part. About DRC50, it is 12.9 kNm or more and it is simultaneous with the generation of the crack about DRC30. The calculation value is comparatively well in agreement with the measurement value, although the calculated value tends to underestimate the measured value as the area of RPC in beam increases.

3.4 Curvature

Figure 11 and Fig. 12 show comparisons of experimental and calculated changes of curvature with the increase of moment. About the calculation shown in the figures, I_g is the moment of inertia of transformed cross section about centroid. I_{cr} is the moment of inertia of fully cracked section transformed to concrete. It is assumed that tension of concrete and RPC is neglected when I_{cr} is calculated. I_e is the effective moment of inertia, and assuming cross section stiffness is constant all over the member length, I_e is computed by following equation (JSCE 2002).

$$I_e = \left(\frac{M_{cr}}{M_{max}}\right)^3 I_g + \left\{1 - \left(\frac{M_{cr}}{M_{max}}\right)^3\right\}I_{cr} \qquad (5)$$

where, M_{max} is the moment in computation of curvature and is larger than M_{cr}. The measured values shown in Table 4 are used for the cracking moment M_{cr} in this study.

About RC, the calculated value using the effective moment of inertia well corresponds to measured value to the moment of about 10 kNm, however, from this point, the calculation becomes smaller than the measurement. On the other hand, the calculated values for the beams fortified with RPC are also in fair agreement with measured values until around the destruction except for DRC30U2. Figure 11 and Fig. 12 indicate that RPC has little effect on the increase in curvature with the increase of bending moment after the generation of cracking. As already mentioned in the explanation on the behavior of reinforcement strain in the beam fortified with RPC, from the generation of cracking to the destruction of the beam, the width of crack developed in RPC is narrow. The constant stress corresponding to the first crack strength will be generated in RPC. Therefore, the internal force produced in RPC which resists the external moment may not change.

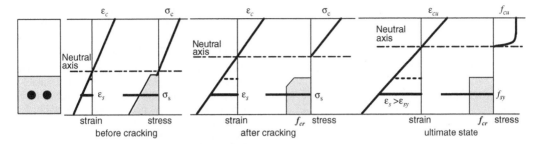

Figure 10. Distributions of strain and stress in beam fortified with RPC.

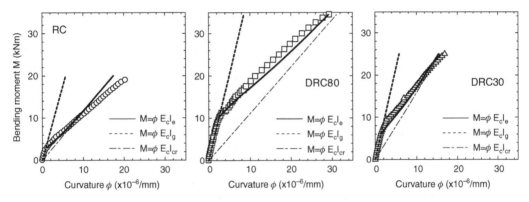

Figure 11. Comparison measured and calculated moment-curvature relationship of RC, DRC80 and DRC30.

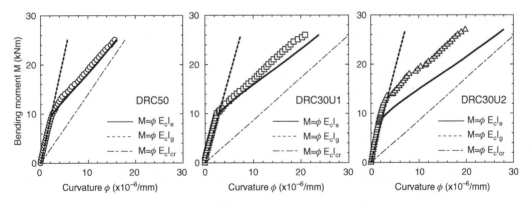

Figure 12. Comparison measured and calculated moment-curvature relationship of DRC50, DRC30U1 and DRC30U2.

About DRC30U2, it is confirmed from the value of strain gauges affixed at the bottom surface of the beam that DRC30U2 beam cracked at the bending moment of 8.89 kNm. But, the crack does not affect the increase of curvature. In this connection, if 13.5 kNm of bending moment is used for cracking moment M_{cr} in Eq.(5), the calculation is in agreement with the measurement.

3.5 Capacity of beam

Table 5 shows the measured flexural capacity of specimen. The failure type of all beams was the tension failure that the yield of reinforcing bar precedes the failure of concrete, because the reinforcing bars in all specimens were designed as the under-reinforcement, In addition, although the diagonal cracks which developed from the flexural cracks occurred in shear span

Table 5. Flexural capacity of specimens

| Specimen | Flexural capacity (kNm) | | | $\sigma_{c,as}$ (N/mm^2) |
	Measure- ment	Calcula- tion I	Calcula- tion II	
DRC80	34.5	37.81	34.16	3.04
DRC50	27.5	36.01	31.22	4.63
DRC30	25.5	31.72	27.04	7.12
DRC30U1	26.0	27.80	—-	—-
DRC30U2	28.5	27.12	—	—
RC	20.5	19.15		

$\sigma_{c,as}$:stress due to restraint of reinforcing bar against autogenous shrinkage of RPC.

of all specimens, that crack didn't influence the failure of the beam because of the arrangement of stirrups except for DRC30U2. Generally, diagonal cracking is the crack that flexural cracking is sloped by bending moment and shear force. The cracking moment of DRC30U2 is high and the cracking doesn't develop easily by the steel fiber in RPC. Therefore, the load on the initiation of diagonal cracks of DRC30U2 becomes high and DRC30U2 may have high shear capacity although stirrups aren't arranged at shear span in the beam.

The flexural capacity of the beams reinforced with RPC is larger than that of RC beam and increases with the increase of the area reinforced with RPC. This increase of flexural capacity may be due to RPC, however, the increasing ratio of flexural capacity isn't necessarily in proportion to the reinforced area of RPC. The calculation I is made according to the assumption on the distribution of stress shown at the right side of Fig. 10. The calculated values for DRC30U1 and DRC30U2 are well in agreement with the measured values. However, the calculated values for DRC80, DRC50 and DRC30 are overestimated the measured values. In order to solve this overestimation, like the case of the evaluation for the cracking moment of the beam fortified with RPC, the stress due to the restraint of the reinforcing bar against the autogenous shrinkage of RPC is also taken into consideration when calculating the flexural capacity of the beam reinforced with RPC. In the calculation II, the restraint stress generated in RPC is subtracted from the first crack strength of RPC. The calculated value of the flexural capacity of the beam reinforced with RPC is improved by the consideration of the restraint stress of RPC as shown in Table 5.

4 CONCLUSIONS

The reinforced concrete beam without cracking under serviceability limit state has been developed by

authors. Flexural load tests of the beam fortified with RPC were carried out and the mechanical behaviors of the beams were investigated. The following conclusions are drawn within the scope of this study.

1) The cracks don't occur in the reinforced concrete beam fortified the tension zone with RPC suitably, when the moment corresponding to the generation of cracks below permissible crack width for corrosion of steel bar acts.
2) The cracking moment of the beam fortified with RPC can be evaluated by using the elastic analysis, provided that it is necessary to take the restraint stress into consideration when autogenous shrinkage of RPC is restrained.
3) The crack generated in the beam fortified with RPC is distributed by the steel fiber in RPC and the width of the crack is as small as 0.1 mm or less from the generation of cracking to the destruction of the beam.
4) It is necessary to take into consideration that RPC shares the stress equivalent to first cracking strength of RPC for the evaluation of strain of reinforcing bar in the beam fortified with RPC after cracking.
5) The curvature of the beam fortified with RPC can be evaluated by the calculation using the effective moment of inertia like the case of the reinforced concrete beam.
6) The flexural capacity of the beam fortified with RPC is larger than that of the reinforced concrete beam and increases with the increase of the area reinforced with RPC.

REFERENCES

JCI 2002 Research committee report on performance evaluation and structural utilization of ductile fiber reinforced cementitious composites. Tokyo: Japan Concrete Institute.
JSCE 2002. Standard specification for design and construction of concrete structures Structural performance verification edition). Tokyo: Japan Society of Civil Engineers.
JSCE 2004. Recommendations for Design and Construction of Ultra High Strength Fiber Reinforced Concrete Structures. Concrete Library 113. Tokyo: Japan Society of Civil Engineers.
Katagiri, M., Maehori, S., Ono, T., Shimoyama, Y. and Tanaka, Y. 2002. Physical properties and durability of reactive powder composite material (Ductal). Proceedings of the first fib congress 2002 Osaka, pp.133–138.
Musha, H. et al. 2002. Design and construction of SAKATA-MIRAI bridge using of reactive powder composite. Bridge and Foundation Engineering 36(11): pp.5–15.
Ujike, I et al. 2005. A study on crack prevention at service state of reinforced concrete member fortified tension zone by Reactive Powder Composite. Journal of Society of Material Science 54(8): pp.855–860.

Fracture Mechanics of Concrete and Concrete Structures – Design, Assessment and Retrofitting of RC Structures – Carpinteri, et al. (eds)
© *2007 Taylor & Francis Group, London, ISBN 978-0-415-44616-7*

Finite-element simulations of the punching tests on shear-retrofitted slab-column connections

A. Negele, R. Eligehausen & J. Ožbolt
Institute of Construction Materials, Universitaet Stuttgart, Germany

M.A. Polak
Department of Civil and Environmental Engineering, University of Waterloo, Canada

ABSTRACT: A new type of shear reinforcement, shear bolts, are used for retrofit and strengthening of existing previously built concrete slabs to increase the punching capacity. The bolts are installed in holes drilled in a slab in concentric perimeters around the column. The test results showed a load increase up to flexural failure of the slabs and a significantly improved ductility when using shear bolts. The punching tests were investigated using the 3D nonlinear finite-element code MASA which was developed at the Institute of Construction Materials of the Universitaet Stuttgart. The paper discusses the modeling of slabs with shear reinforcement. A proper idealization of the shear reinforcement is essential for the accuracy of the finite-element simulations. The finite-element simulations show good agreement with the test results and give a more detailed insight in the failure mechanism. The models build the basis for an extensive parametric study on the punching behavior of flat slabs.

1 INTRODUCTION

Flat slabs are widely used in reinforced concrete constructions. The connections between columns and slabs are subject to high stresses which might lead to a brittle punching shear failure. To increase the punching shear capacity and the ductility of the slab-column connections shear reinforcement can be used. Conventional shear reinforcements such as stirrups, bent up bars and shear studs are installed before casting of the concrete. Shear studs have been proved to be the most effective and easy to install shear reinforcement (e.g. Dilger & Ghali 1981).

Changes in the building use, the need of installing new services and construction or design errors might require strengthening of existing slabs. Therefore, a simple and effective method for subsequent installation of shear reinforcement was developed. These shear bolts consist of a rod with a head at one side and a thread at the other side. The shear bolts are slid in drilled holes and fixed with a washer and nut system at the threaded end. With this system the bolts are supplied with a comparatively stiff anchorage system as shear studs.

Tests on slab column connections of interior and edge slab column connections (Adetifa & Polak 2005, El-Salakawy et. al. 2003) show a load increase up to flexural failure of the slab column connections and a considerable increase in ductility.

For a better understanding of the punching failure and as a basis for parametric studies, the tests on interior slab column connections were modeled using the finite element method. In the following the modeling of the slabs and the results of the finite element simulations with the program MASA are presented.

2 PUNCHING TESTS

A series of four punching tests on interior slab-column connections was used for the finite element study. A detailed description of the tests can be found in Adetifa & Polak (2005).

2.1 *Test setup and parameters*

All tests were performed on square slabs with side lengths of 1800 mm and a thickness of 120 mm. The loading was applied on square column stubs with cross sections of 150×150 mm. The test parameters can be found in Table 1.

The main test parameter was the number of shear bolts. A slab reinforced with two rows of shear bolts is shown in Figure 1. The distance of the first shear bolt to the column was $s_0 = 50$ mm all other rows were spaced at $s_i = 80$ mm (Table 1). The bolts were placed in a cross shape with two bolts on each column face. They were installed in drilled holes with a diameter of

Table 1. Test parameters.

Spec.	SB rows	s_o	s_i	bolt Ø [mm]	thick-ness h [mm]	depth d [mm]	flexural reinf. ρ [%]
SB1	0	–	–	–	120	90	1.25
SB2	2	0.56d	0.89d	9.5	120	90	1.25
SB3	3	0.56d	0.89d	9.5	120	90	1.25
SB4	4	0.56d	0.89d	9.5	120	90	1.25

SB: shear bolts

Figure 1. Strengthened slab with shear-bolts.

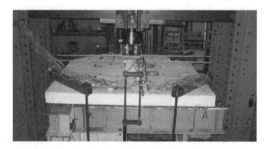

Figure 2. Experimental setup with test specimen.

16 mm. Washers with a diameter of 44 mm and thicknesses of 10 mm were used. The nuts were torqued hand tight before testing. The flexural tension reinforcement consisted of 10 M bars ($A = 100 \, mm^2$) with 90 mm and 100 mm spacing for the top and bottom layers, respectively.

All slabs were simply supported along the edges at a distance of 1500 mm; the corners were restraint to avoid lifting up (Figure 2). The simple support was realized through 40 mm wide steel plates with Neoprene strips to allow rotations at the supports.

2.2 Test results

The slab without shear reinforcement failed in punching. Specimen SB2 failed in a combined failure mode of punching outside the shear bolts and flexure. All other slabs failed in flexure. The slabs show a distinct flexural crack on the tension side of the slab between the column face and the first row of shear bolts. After a considerable yield plateau a shear crack occurred

Table 2. Test results.

Spec.	SB rows	$f_{c,cyl}$ [MPa]	Failure load [kN]	Displ. [mm]	Mode
SB1	0	44	253	12	P
SB2	2	41	364	28	P/F
SB3	3	41	372	33	F
SB4	4	41	360	48	F

SB: shear bolts
Displ.: Column displacement at failure load
Mode: P – Punching, F – Flexural Failure

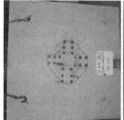

a) compression side   b) tension side

Figure 3. SB3 –crack development after testing.

outside the shear reinforced zone. The shear crack is clearly visible on the compression side of the slab. On the tension side the shear crack is partly visible. This is due to the fact that the shear crack runs along the tensional flexural reinforcement and does not directly protrude through the slab. The test results are summarized in Table 2, typical crack development is shown in Figure 3. More detailed results are discussed in chapter 4.

3 FINITE-ELEMENT CODE AND MODELING

The finite-element simulations were performed with the three-dimensional nonlinear finite-element code MASA. This code was developed at the Institute of Construction Materials at the Universitaet Stuttgart. The finite element code MASA is designed for the nonlinear analysis of quasi-brittle materials such as concrete.

3.1 MASA fundamentals

The finite element code MASA is based on the mircroplane material model with relaxed kinematic constraint (Ožbolt et al. 2001). In numerous investigations it has been shown that MASA is able to predict the behavior of reinforced concrete structures and the punching failure of flat slabs realistically (Ožbolt et al.

1999, Beutel 2002). The program uses a smeared crack approach in combination with the crack band method as a localization limiter (Ožbolt & Bažant 1996).

3.2 Modeling with three-dimensional elements

Concrete or massive steel members are modeled with three dimensional elements. Hexahedra elements result in simple meshes with relatively few elements. Tetrahedral elements allow the meshing of arbitrary geometries and are therefore used more often. Principally bar reinforcement can also be modeled with three dimensional elements. However, in large scale structures the reinforcement needs to be simplified with one dimensional truss elements.

3.3 Modeling of reinforcement with bar elements

Reinforcement bars can be modeled with one dimensional bar elements. Bar elements have a real length and a defined cross section. An additional virtual length that is used for calculating the bending stiffness in the finite element program allows controlling the flexural rigidity of the elements. Bar elements share their end nodes with the adjacent three-dimensional concrete elements, which results in a fixed connection between both elements.

3.4 Modeling of reinforcement with bond elements

In all cases were the bond behavior of the reinforcement influences the load bearing behavior considerably. The use of bond elements gives more realistic results of the finite element simulations. A bond element is an additional two node spring element that connects the end node of a bar element with the adjacent node of the concrete elements (Figure 4). The bond element is defined by a bond-slip curve and realizes the load transfer between the bar element and the concrete elements according to this definition. The development of the bond element and the implementation of bond-slip curves for standard deformed reinforcement bars can be found in Lettow et al. (2004) and Lettow (2006).

3.5 Modeling of anchorage elements

When bar or bond elements are used to model headed bars, stirrups or hooked bars the anchorage element must also be idealized by bar or bond elements. The anchorage element can be idealized applying two different methods. The first option is to arrange the bar elements in a cross or hook shape (Figure 5a, b). The anchorage slip of the idealized head is controlled by introducing a virtual element length for the calculation of the internal flexural element stiffness. The stiffness can be calibrated on finite-element pull-out tests.

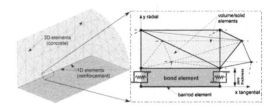

Figure 4.   Definition of the bond element.

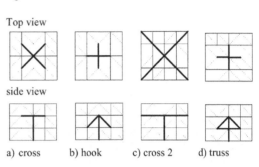

Top view

side view

a) cross   b) hook   c) cross 2   d) truss

Figure 5.   Anchorage of bar elements.

A representative slip is computed using equation (1) for head slip; it depends on the head pressure (Furche 1994):

$$s = \frac{k_A}{c} \cdot \left(\frac{\sigma}{f_c} \right)^2 \qquad (1)$$

$$k_A = 0.5 \cdot \sqrt{d_s^2 + 9 \cdot (d_k^2 - d_s^2)} - d_k / 2 \qquad (2)$$

s: head slip
d_k: head diameter
d_s: shaft diameter
k_A: factor for influence of head-shaft ratio
σ: head pressure
c = 600 for uncracked concrete
c = 300 for cracked concrete
f_c: concrete strength, 200 mm cube

It must be noted that the forces that can be transmitted when a bar element is anchored in cracked concrete elements are very small. Therefore, in such case the number of elements for the head must be increased (tension side of a flat slab, Figure 5c). However, elements with a large flexural stiffness might influence the flexural behavior of the whole structure or lead to local damage if they are used in regions with high local deformations. Therefore, the second modeling option is to use bar elements with a flexible joint and arrange them in a truss shape to provide the anchorage (Figure 5d). With this option a rotation of the head is possible and the damage is minimized. The disadvantage of this solution is that the slip behavior of the

913

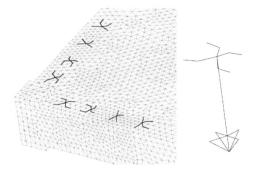

a) Slab section with column stub   b) Shear bolt

Figure 6.   Model of the slab.

Table 3.   Material properties.

Spec	SB1	SB2-4
Concrete		
$f_{c,cyl}$ [MPa]	44	41
$f_{c,t}$ [MPa]	2.2	2.1
G_f [N/mm]	0.07	0.07
E_c [MPa]	28000	28000
Steel		
f_y [MPa]	455	455
f_t [MPa]	650	650
E_S [MPa]	197000	197000

$f_{c,cyl}$: concrete compressive cylinder strength
$f_{c,t}$: concrete tensile strength
G_f: concrete fracture energy
E_c: concrete modulus of elasticity
f_y: steel yield strength
f_t: steel ultimate strength
E_S: steel modulus of elasticity

anchorage element cannot be directly controlled. Furthermore, the modeling depends on a regular finite element mesh as shown in Figure 5d.

Preliminary studies have shown that for anchorage in severely cracked concrete, as in the case of the bending cracks of a slab, a large cross as shown in Figure 5 c needs to be applied to ensure adequate anchorage. In regions with large deformations and without flexural cracks, as can be found on the bottom side of the slab, especially close to the column, truss elements as shown in Figure 5 d should be used.

4 FINITE-ELEMENT SIMULATIONS

In the finite-element simulations, a symmetrical slab can be modeled by a quarter of a slab. All nodes along symmetry planes need to be fixed in the direction orthogonal to the plane to account for symmetry.

4.1 *Model of the slab*

The slab is modeled with column stubs on both sides of the slab. The concrete is modeled with tetrahedral elements with an element size of approximately 14 mm. A section of the quarter of the slab is shown in Figure 6a. It shows a view on the tension side of the slab column connection. The modeling of a shear bolt is shown in Figure 6b. The bolt is modeled according to section 3.5 with bond elements with a defined flexural stiffness on the tension side of the slab, and bond elements as a truss on the compression side of the slab. The sum of the mechanical and frictional bond is set to 0.2 N/mm² to simulate the post installed shear bolts with no friction along the drilled hole. The flexural reinforcement is modeled with bar elements; the layout is according to the reinforcement layout in the tests.

The load is applied on the column stub, in displacement controlled mode.

4.2 *Material properties*

The concrete and steel material properties were used according to the material test results of the punching tests, as shown in Table 3. The stress-strain curve for steel was modeled as a trilinear function with yield at 2.3‰ strain reaching ultimate stress at 5% strain.

The concrete at the highly stressed compressive zone at the column face is loaded in a multiaxial state of stress, which results in an increased compressive strength. For slabs with shear reinforcement this strength is further increased by the additional restraint of the compressive zone by the heads of the first studs. This mechanism contributes the ultimate load increase in tested slabs. However, the modeling of shear reinforcement with two dimensional bar or bond elements in finite element simulations does not result in an increased concrete strength at the column face. This needs to be compensated in the simulations by introducing a region with increased concrete strength in the compressive zone at the column face. To account for the increase in the concrete strength, the concrete in that region is modeled with a reduced Poisson's ratio.

4.3 *Modeling of support conditions*

The vertical supports are realized by restraing of the nodes at a support point or along a support line. In the following simulations the line support of the test is simplified to a radial point support. This better represents the actual test support conditions which are never perfectly restrained and allow some movement.

4.4 *Load-displacement curves*

The punching tests with slabs without and with shear reinforcement show a considerable load increase when

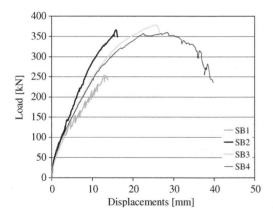

Figure 7. Load displacement curves tests.

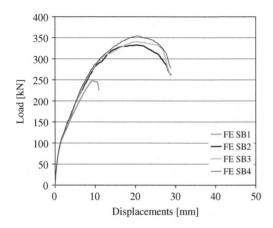

Figure 8. Load displacement curves simulations.

shear reinforcement is used (Figure 7). Furthermore the ductility is increased with the number of rows of shear bolts.

The results of the finite-element simulations (Figure 8) show a good agreement with the experimental values of the ultimate loads. Simulation SB1 shows a brittle failure as in the test. For all other tests the ductility is increased considerably. However, the simulations do not show the same large influence of the number of bolts on the ductility as the tests.

4.5 Cracking

The crack development in the finite-element simulations can be visualized through the principal tensile strains in the concrete. The cracking is shown at two sections through the slab. Section 1 is parallel to the y-axis at a distance of 45 mm from the axis. The section cuts one row of shear bolts. Section 2 lies in the slab diagonal where no shear bolts are used. The black

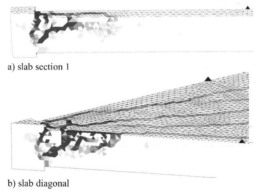

a) slab section 1

b) slab diagonal

Figure 9. Cracking at ultimate load: SB1.

shaded areas with maximum strains of 0.03 represent a crack width of 0.4 mm for elements with a side length of 14 mm. The different strain values are shown at steps of 0.005. The support points are visualized by black triangles.

The crack development of Slab SB1, without shear reinforcement, is shown in Figure 9. The sections show an inclined shear crack of about 37° at section 1 and of about 46° , 32° and 20° at section 2. A direct comparison to the crack development in the test is not possible since the test specimens were not saw cut after testing. A variation of the failure crack angle along the column perimeter has also been reported by Clauss & Birkle (2002). A flexural crack at the face of the column is clearly visible.

Figure 10a shows the crack development for a slab with shear reinforcement, SB4, along section 1. The crack development shows a distinct bending crack at the column face beginning on the tension side of the slab as observed in the tests. Outside the shear reinforcement there is a visible shear crack, which was the failure crack in the test. The region between the first and second row of studs is cracked. This might be caused by the influence of the stiff anchorage elements of the shear bolts. This additional cracking influences the ductility of the load displacement curves and leads to a second post failure shear crack between the first two rows of bolts.

The finite-element simulation gives the opportunity to visualize vertical and horizontal strains corresponding to shear and flexural cracking, respectively. Figure 10b shows the horizontal strains along the axis parallel to the slab section. The main crack is the flexural crack at the column face; minor flexural cracking occurs between the first two rows of studs. Figure 10c shows the vertical cracking. The vertical part of shear crack outside the shear reinforcement is clearly visible. The shear crack angle is about 40°. Additionally there are minor vertical strains between the first two shear bolts adjacent to the column.

a) Principal tensile strains

b) Tensile strains in x-direction - flexural cracks

c) Tensile strains in z-direction – shear cracks

Figure 10.   Cracking at ultimate load: SB4 section 1.

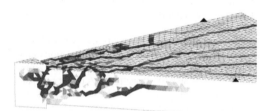

a) Principal tensile strains

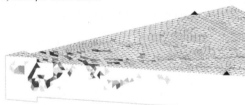

b) Tensile strains in x-direction - flexural cracks

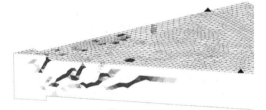

c) Tensile strains in z-direction – shear cracks

Figure 11.   Cracking at ultimate load – SB4 section 2.

a) slab section 1

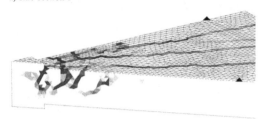

b) slab diagonal

Figure 12.   Cracking at 80% of ultimate load, SB4.

Section 2 through the slab diagonal shows consider-ably more shear cracking than the section parallel to the y-axis. This is due to the orthogonal distribution of the shear bolts. The shear crack angles are about 48°, 41°, 25° and 15°. Comparing the crack develop-ment on the slab diagonal with the cracking of the slab without shear reinforcement, SB1, it can be observed that the first three cracks are slightly steeper but other-wise similar to the cracks of slab SB1. The additional shallow crack develops at about 90% of the ultimate load. This is the continuation of the crack that devel-ops outside the shear reinforced zone in section 1 and becomes the failure crack in the test.

The crack development before ultimate load is illus-trated in Figure 12. The flexural cracks at the column face and the shear cracks in the shear reinforced zone have developed. The failure crack outside the shear reinforcement is formed at ultimate load.

4.6   Flexural reinforcement

The flexural reinforcement in the tests was designed to reach yield before the shear failure occurred in the slab SB1. Figure 13 shows the good agreement of the activation of the flexural reinforcement in the test and simulation before and after reaching the yield point. In the test the strain gage was damaged when the strains reached 6.5‰, so test values for ultimate load are not comparable.

4.7   Activation of the shear reinforcement

The activation of the shear reinforcement gives infor-mation on the formation of the shear cracks in the slab. In the punching tests, the shear bolts were very little stressed until ultimate load was reached (Figure 14 Bolt1-4). The only exception was bolt one which was subjected to small stresses at loads higher than 200 kN and was strongly activated starting at a load of 300 kN. The beginning of the bolt activations coincides with the ultimate load of a slab without shear reinforcement, SB1. This shows that shear cracking begins at the same time for slabs without and with shear reinforcements.

916

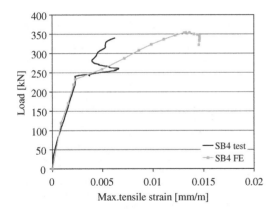

Figure 13.   Flexural reinforcement strains.

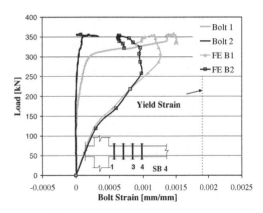

a) Bolts 1,2

However, the bolts can retard the opening of the shear crack considerably. Before ultimate load, the first bolt is highly stressed which shows that considerable shear cracking occurres at the column face. Furthermore the two outer rows of shear bolts are suddenly stressed at ultimate load. This shows that the shear crack outside the shear reinforcement occurs together with shear cracks that cross the outer rows of bolts.

In the simulation, the activation of all the shear bolts starts from the beginning of the slab loading were no cracking has occurred (Figure 14 FE B1-B4). This is due to the fact that the modeling of the anchorage elements of the shear bolts are connected to the concrete elements at their common nodes which allows a force transfer between them in compression and tension, while in the tests the bolt heads can only transfer load in compression. When the slab is loaded and starts deforming the part of the stud head that is away from the column face lifts from the slab surface and in the simulation tension forces in the bolts are generated. Therefore, the bolt activation alone cannot give the direct information on the crack development in the simulation. However, a stronger increase in the bolt activation shows the beginning of shear cracking as well as the shear cracking can be directly observed through the concrete strains.

4.8   Summary of the finite element simulations

The finite-element simulations show very good agreement with the test results in respect to ultimate loads, crack development and activation of the flexural reinforcement. It is clearly shown how shear bolts allow to avoid extensive shear cracking and therefore increase the punching shear load of flat slabs. From the crack development in the slab sections it can also be seen that this influence on the crack development is only valid for the regions adjacent to the shear bolts. The cracking on the slab diagonal with no shear reinforcement is

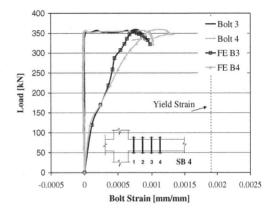

b) Bolts 3,4

Figure 14.   Bolt strain for SB4.

almost identical to slabs without shear reinforcement. The modeling of the shear bolts and their anchorage elements is essential for the load bearing behavior of the slab. Inappropriate modeling of the anchorage elements might lead to concrete damage at the column face or no activation of the shear reinforcement which both result in ultimate loads that are too low. Even a most skillful modeling of the anchorage elements results in a bolt activation that is not entirely the same as in the tests. However, this early bolt activation can be accepted as long as the overall load bearing behavior of the slab is not negatively influenced.

5   CONCLUSIONS

The simulations of the punching tests show a good agreement with the test results. Therefore, the developed model can be used as a basis for parametric studies on the influence of the slab thickness, the reinforcement ratio and of openings in the slab adjacent to the column. The simulations can be used to gain information on the ultimate punching shear loads, the activation of the flexural reinforcement and the formation of the flexural and shear cracks.

REFERENCES

Adetifa, B., Polak, M. A. 2005, Retrofit of interior slab column connections for punching using shear bolts, *ACI Structural Journal* 102(2): 268–274.

Beutel, R. 2002, Durchstanzen schubbewehrter Flachdecken im Bereich von Innenstützen, *Dissertation*, RWTH Aachen.

Clauss, A., Birkle, G., Durchstanzen an Innenstützen – Die Auswirkung der Anordnung der Doppelkopfbolzen. In: Beiträge aus der Befestigungstechnik und dem Stahlbetonbau, *Festschrift zum 60. Geb. von Prof. Dr.-Ing. R. Eligehausen*. Stuttgart: ibidem-Verlag, 2002, S. 45–55.

Dilger, W. H., Ghali, A. 1981, Shear reinforcement for concrete slabs, *Journal of Structural Division*, ASCE, V. 107, No. ST12, 2403–2420.

El-Salakawy, E. F., Polak, M. A., Soudki, K.A., 2003, New shear strengthening technique for concrete slab-column connections, *ACI Structural Journal*, 100(3): 297–304.

Furche, J., 1994, Zum Trag- und Verformungsverhalten von Kopfbolzen bei zentrischem Zug, *Dissertation*, Universität Stuttgart.

Lettow, S. Mayer, U., Ožbolt, J., Eligehausen, R. 2004, Bond of RC memebers using nonlinear 3D FE analysis, *Proceedings of the Fifth International Conference on Fracture Mechanics of Concrete and Concrete Structures. FraMCoS 5*, Eds.: V.C. Li, C.K.Y. Leung, K.J. Willam, S.L. Billington , 12–16 April, Vail, Colorado, USA: 861–868. – ISBN 0- 87031-135-2.

Lettow, S., 2006, Ein Verbundelement für nichtlineare Finite Elemente: Analysen – Anwendung auf Übergreifungsstöße, *Dissertation*, Institut für Werkstoffe im Bauwesen, Universität Stuttgart.

Ožbolt, J., Bažant, Z. P. 1996, Numerical smeared fracture analysis: Nonlocal microcrack interaction approach, *International Journal for Numerical Methods in Engineering*, Jg. 39, Nr. 4: 635–661.

Ožbolt, J.,Li, Y.-J., Kožar, I. 2001, Microplane model for concrete with relaxed kinematic constraint, *International Journal* of Solids and Structures, 38: 2683–2711.

Ožbolt, J., Mayer, U., Vocke, H., Eligehausen, R. 1999, Das FE-Programm MASA in Theorie und Anwendung, *Betonund Stahlbetonbau*, Jg. 94, Nr. 10: 403–412.

Fracture Mechanics of Concrete and Concrete Structures – Design, Assessment and Retrofitting of RC Structures – Carpinteri, et al. (eds)
© 2007 Taylor & Francis Group, London, ISBN 978-0-415-44616-7

The effects of transverse prestressing on the shear and bond behaviors of R/C columns

Y. Shinohara
Structural Engineering Research Center, Tokyo Institute of Technology, Tokyo, Japan

H. Watanabe
Department of Architecture, Nagasaki Institute of Applied Science, Nagasaki, Japan

ABSTRACT: Experiments and 3-D FEM analyses were performed on reinforced concrete columns laterally prestressed by the shear reinforcement to study the influence of the active confinement upon shear crack behaviors and bond splitting strength. Many strain gauges were attached to main bars and hoops, and the width of each crack over transverse hoops was measured by a digital microscope. Transverse prestressing increases the bond splitting strength as well as the shear capacity at first cracking and the ultimate shear strength, and decreases the width values of cracks. The FEM analyses can evaluate the shear capacity at first cracking, ultimate shear strength and the bond splitting strength with a fair degree of precision, and provide valuable information about the effect of the active confinement on shear and bond behaviors by evaluating the intensity of confinement in tri-axial state of stress with minimum principal stress and equivalent confining pressure.

1 INTRODUCTION

Prestressing in concrete structures is generally aimed at controlling the flexural cracks by the arrangement of tendons in an axial direction of any given member. On the other hand, in order to delay the onset of shear cracking and to reduce its width, not to control a flexural crack, experimental studies have been conducted on the reinforced concrete (RC) columns laterally prestressed by the shear reinforcement with high strength (Watanabe et al. 2004). The results of the flexure-shear tests have indicated that the shear capacity at first diagonal cracking is increased and the width values of shear cracks, especially their residual opening are remarkably reduced by transverse prestressing. This reduction of the crack opening has improved not only durability but also earthquake resistance since the ability to transmit shear force across a rough crack increases dramatically by reducing its width (Shinohara et al. 1999). The three dimensional finite element (FEM) analyses were also performed on RC columns mentioned above to investigate the effect of the lateral confinement using the equivalent confining pressure and the degree of damage in compressive zone as the gauges that evaluate active confinement and compressive-shear failure quantitatively (Shinohara et al. 2004). These studies have revealed that an increase in the resistance against shear failure as well as shear cracking with increasing prestress in the shear reinforcement could be explained by the triaxial state of stress in the core concrete.

The primary purpose of this study is to investigate how transverse prestressing in RC columns would affect the shear behaviors and bond behaviors on the basis of the triaxial state of stress, to clarify the relationship between the width of a shear crack and the strain of a shear reinforcement, and to see the extent to which the FEM analysis with a smeared crack model and a bond-slip model can evaluate the actual shear crack behavior and the bond strength.

2 OUTLINE OF TEST AND ANALYSIS

2.1 Test specimens

The details of the test specimen are shown in Figure 1. The main characteristics of the specimens are summarized in Table 1. The flexure-shear tests have been performed on four columns which were laterally prestressed (LPRC) and not prestressed (RC). The specimens were designed to cause a shear failure before the longitudinal reinforcement yield by Architectural Institute of Japan (1999). Two columns of them (B-series) were designed to cause a bond failure first by removing reinforcement (D13 in Fig. 1) that prevent bond splitting failures. The test specimens had a square cross section of 340 mm × 340 mm and a height of 900 mm. The lateral prestress was introduced into concrete as follows: (1) the high strength transverse hoops (U6.4 in Fig. 1) were pretensioned to about 40% of the

yield stress using the rigid steel molds and special jigs shown in Figure 1, (2) concrete was vertically placed into the molds and cured until the strength of concrete increased adequately, (3) the core concrete was laterally pre-stressed by removing the steel molds. The product of the ratio (p_w) and the stress (σ_{wp}) of the pretensioned transverse reinforcement is defined as average lateral prestress σ_L ($=p_w\sigma_{wp}$) to indicate the intensity of lateral prestress. The mix proportion of concrete used in the test specimens is given in Table 2. The coarse aggregate used in the mix is natural round sea gravel with a maximum aggregate size of 25 mm. The bond splitting and shear strength are calculated in accordance with AIJ design guideline (1999) and other researchers, summarized in Table 3. The effect of transverse prestressing on bond splitting strength is counted by adding the average prestress to the tensile strength of concrete.

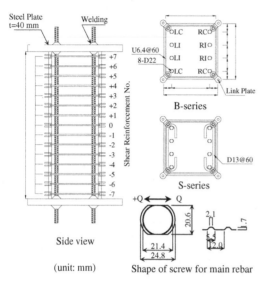

Figure 1.   Details of test specimen and reinforcement.

Table 1.   List of test specimens.

Test Series	M/QD	% p_w	p_{wb}	σ_B N/mm²	σ_{wp}	σ_L
B-RC	1.3	0.29	0.29	47.7	0	0
B-LPRC	1.3	0.29	0.29	47.1	587	1.7
S-RC	1.3	0.29	1.54	50.8	0	0
S-LPRC	1.3	0.29	1.54	46.5	536	1.6

M/QD = shear span-depth ratio, p_w = ratio of transverse hoop, p_{wb} = ratio of transverse reinforcement including hook against bond split, σ_B = compressive strength of concrete, σ_{wp} = introduced prestress in transverse hoop, σ_L = lateral prestress ($=p_w\sigma_{wp}$)

2.2   Test set-up and instrumentation

The loading apparatus is shown in Figure 2. The vertical force on the test specimen was supplied by the 2 MN-hydraulic jack, and the ratio of axial load to axial strength was kept constant at 0.3 during a test. The horizontal force was supplied by two hydraulic jacks with the capacity of 500 kN, and controlled in displacement. The cyclic horizontal load was applied in the way to produce an antisymmetric moment in a column. The horizontal load was turned back when the rotation angle of member, R reached ±1/400, ±1/200, ±1/133 (B-series only), ±1/100, ±1/67, ±1/50 and ±1/33, until after the peak load. The width of each shear crack close to the shear reinforcement was measured using two digital microscopes with a resolution

Table 2.   Mix proportion.

Proportion, by wieght				Admixture	Slump
Cement	Sand	Gravel	Water	Super	cm
1	2.04	2.53	0.5	plasticizer	21

Table 3.   Bond splitting strength and shear strength.

Test Series	Bond strength N/mm² $\tau_{bu}F$	$\tau_{bu}M$	$\tau_{bu}D$	$Q_{bu}F$ kN	$Q_{bu}M$	$Q_{bu}D$	$_{cal}Q_{sc}$	$_{cal}Q_{su}$
B-RC	4.2	5.4	3.4	445	511	399	470	629
B-LPRC	5.9	7.1	5.1	580	647	536	612	730
S-RC	–	–	8.7	–	–	694	496	648
S-LPRC	–	–	10.1	–	–	809	606	725

$\tau_{bu}F = \tau_{bu}$ by Fujii (1983), $\tau_{bu}M = \tau_{bu}$ by Maeda (1994), $\tau_{bu}D = \tau_{bu}$ by AIJ Design guidelines (1999), $Q_{bu}F$, $Q_{bu}M$ and $Q_{bu}D$ = shear strength using $\tau_{bu}F$, $\tau_{bu}M$ and $\tau_{bu}D$ $_{cal}Q_{sc}$, $_{cal}Q_{su}$ = shear crack and ultimate strength by Watanabe (2004).

of 0.01 mm every cycle three times in loading and two times in unloading. The crack width used in this paper is defined as a distance normal to the direction of a crack, as illustrated in Figure 3. Three strain gauges were glued on each leg of all transverse hoops, and their locations and designations are shown in Figure 3. Seven strain gauges were also glued on each longitudinal reinforcement of B-series specimens to evaluate the bond stress.

2.3 Analytical models and idealizations

The finite element mesh and boundary condition of the analytical model are shown in Figure 4. The mechanical properties of concrete and reinforcement are shown in Figures 5 and 6 together with their idealizations in analyses. Due to the symmetry, only one half of the column was analyzed. The stiff elements were attached at the top and bottom of a column to idealize steel stubs. The top nodes were constrained to move uniformly in the vertical direction and not to allow the upper stiff elements to rotate, so that a column deformed in an antisymmetric mode. The bond-slip between concrete and reinforcement was not considered in analysis of S-series because an additional reinforcement was installed to avoid a bond splitting failure. On the other hand, the bond-slip relation shown in Figure 7 was assumed in analysis of B-series. When loading, the prescribed prestress was first introduced into the shear reinforcement, and then the axial load was applied in load control, finally the shear load was applied in displacement control. The maximum-tensile-stress criterion of Rankine was adopted as a failure criterion in the tension zone of concrete. Smeared cracking and bi-linear tension softening shown in Figure 5 are adopted in this analysis. The shear stiffness of cracked concrete is generally dependent on the crack width. This phenomenon is taken into account by decreasing the shear stiffness with an increase of the normal crack strain. Drucker-Prager criterion was used for a failure criterion in the compressive zone of concrete. The formulation is given by

$$f(I_1, J_2) = \alpha I_1 + \sqrt{J_2} - k = 0 \qquad (1)$$

$$\alpha = \frac{2\sin\phi}{\sqrt{3}(3 - \sin\phi)} \qquad (2)$$

$$k = \frac{6\cos\phi}{\sqrt{3}(3 - \sin\phi)}c \qquad (3)$$

$$I_1 = \sigma_1 + \sigma_2 + \sigma_3 \qquad (4)$$

$$J_2 = \left[(\sigma_1 - \sigma_2)^2 + (\sigma_2 - \sigma_3)^2 + (\sigma_3 - \sigma_1)^2\right]/6 \qquad (5)$$

Where ϕ is the internal-friction angle, c is the cohesion; σ_1, σ_2 and σ_3 are the principal stresses (see Chen

1982). The internal-friction angle of Drucker-Prager was determined based on experimental results performed on concrete cylinders with different strengths and hoops to study the effect of lateral confinement by

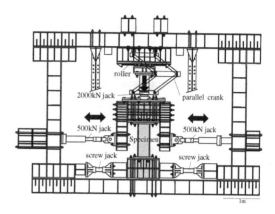

Figure 2. Loading apparatus and test specimen.

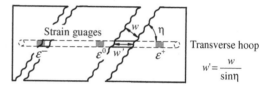

Figure 3. Definition of crack width and designations of tree strain gauges.

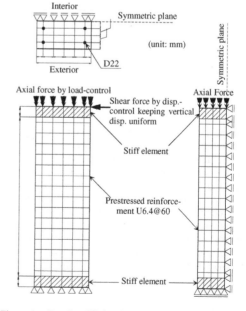

Figure 4. Details of finite element model of B-series.

921

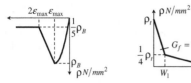

Compressive stress-strain curve   Tensile stress-crack width curve

Test	ρ_B	ε_{max}	E_c	σ_t	W_1	W_2
Series	N/mm²	%	N/mm²	N/mm²	mm	mm
B-RC	47.7	-0.2	3.38E+4	3.0	0.030	0.15
B-LPRC	47.1	-0.2	3.56E+4	3.0	0.030	0.15
S-RC	50.8	-0.2	3.51E+4	2.9	0.031	0.15
S-LPRC	46.5	-0.2	3.45E+4	2.9	0.031	0.15

Poisson's ratio $\nu=0.2$

Figure 5. Mechanical properties and idealizations for concrete.

Bar	ρ_y	ρ_{max}	E_s	
Type	N	/mm²		
D22	1196	1281	1.92E+5	
U6.4	1459	1499	2.04E+5	
D13	344	488	1.92E+5	

Stress-strain curve

Figure 6. mechanical properties and idealizations for rebars.

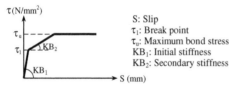

S: Slip
τ_1: Break point
τ_u: Maximum bond stress
KB_1: Initial stiffness
KB_2: Secondary stiffness

Test	Bar	KB_1	KB_2	τ_1	τ_u
Series	Position	N/mm³		N/mm²	
B-RC	Corner	88.2	5.88	1.76	4.95
B-RC	Intermediate	88.2	5.88	1.76	4.24
B-LPRC	Corner	88.2	5.88	1.76	7.88
B-LPRC	Intermediate	88.2	5.88	1.76	5.41

Figure 7. Analytical model for bond-slip relationship.

Takamori et al. (1996). According to their test results, the strength of concrete confined by lateral reinforcement similar to our specimen increases to $(\sigma_B + 2.0\sigma)$, where σ_B is the compressive strength of plain concrete and σ is the averaged lateral confining stress. An increasing rate of 2.0 to σ is under half 4.1 proposed by Richart (1928) due to the partial confinement by hoops. From this equation, a set of principal stresses that corresponds to the strength of concrete in a triaxial state of stress with confining pressure is determined as: $\sigma_1 = \sigma_2 = -\sigma, \sigma_3 = -(\sigma_B + 2.0\sigma)$. The minus refers to compression. By substituting these principal stresses into Equation 1

$$f(I_1, J_2) = (1 - \sqrt{3}\alpha)\sigma_B + (1 - 4\sqrt{3}\alpha)\sigma - \sqrt{3}k = 0 \quad (6)$$

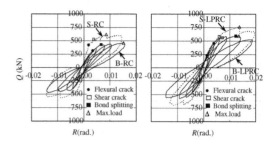

Figure 8. Comparisons of Q-R between S series and B series specimens (Envelop curves for S-series.

Because Takamori's tests (1996) showed a constant coefficient of 2.0 for any value of the averaged confining stress, Equation 6 must be valid regardless of σ, as well. Therefore, the multiplication factor of second term in Equation 6 must be zero:

$$(1 - 4\sqrt{3}\alpha) = 0 \quad \Rightarrow \quad \alpha = \frac{1}{4\sqrt{3}} \quad (7)$$

By substituting Equation 7 into Equation 2, the internal-friction angle of 20° is estimated to be suitable for triaxial state of stress confined laterally by reinforcement in a similar way to this specimen.

3 RELATION BETWEEN SHEAR BEAHAVIOR AND LATERAL CONFINEMENT

3.1 Shear load-rotation angle curves

The shear load Q-rotation angle R curves obtained from the tests of B-series and S-series are shown for comparison in Figure 8, and figure 9 shows $Q - R$ curves, compared with the results of analysis. For the typical crack behavior observed during S-series tests, flexural cracks appeared first, and they extended into flexural shear cracks near the both end of the specimen, and finally shear cracks occurred with increasing shear load. The maximum shear loads for S-RC and S-LPRC were 617 kN when $R = 1/100$ and 762 kN when $R = 1/67$ respectively. The shear loads for both S-RC and S-LPRC were gradually reduced without any reinforcement's yielding by crushing concrete in the compressive zone at the top and bottom ends. For B-series specimens designed to cause bond failure, on the other hand, the maximum shear loads were 25 % lower than those of S-series due to the bond splitting, and the stiffnesses deteriorated slightly to reach the peak loads at $R = 1/67$. The flexural cracks for both B-RC and B-LPRC column appeared first at $R = 1/200$. The bond splitting cracks in B-RC were observed at $R = 1/133$ and extended over the longitudinal reinforcement with sequent loading cycles. For B-LPRC, the shear cracks preceded the bond splitting

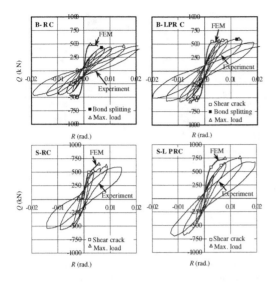

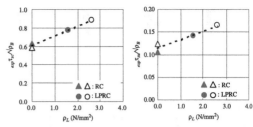

Figure 10. Increase of strength with increasing lateral prestress.

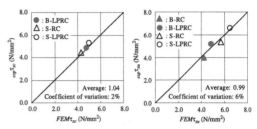

Figure 9. Comparisons between analytical and experimental Q-R curves.

Table 4. Shear cracking, bond splitting and ultimate strength.

Test	$_{exp}Q_{sc}$	$_{exp}Q_{bu}$	$_{exp}Q_{su}$	$_{FEM}Q_{sc}$	$_{FEM}Q_{bc}$	$_{FEM}Q_{su}$
Series	kN					
B-RC	–	461	–	–	493	–
B-LPRC	562	595	–	547	565	–
S-RC	515	–	617	495	–	655
S-LPRC	611	–	762	577	–	747

$_{exp}Q_{sc}$ and $_{FEM}Q_{sc}$ = shear capacity at first cracking by test and FEM

$_{exp}Q_{bc}$ and $_{FEM}Q_{bc}$ = bond splitting strength by test and FEM

$_{exp}Q_{su}$ and $_{Fem}Q_{su}$ = ultimate shear strength by test and FEM

Figure 11. Comparison of strength from experiment and FEM.

with Watanabe's data (2004) marked by solid-white ($\sigma_B = 35$ N/mm^2). The difference in strength of concrete was adjusted by dividing them by the characteristic strength to determine their failure modes. It can be seen from Figure 10 that the shear crack and ultimate strength have increased proportionally with increasing lateral pressure. Furthermore, Figure 11 shows a comparison of the shear strengths obtained from experiment and analysis in both test series. The predictions of FEM analysis are consistent with all existing experimental data for the shear capacity at first cracking and the ultimate shear strength.

cracks when $R = 1/100$ at the bottom of specimen and extended along the longitudinal reinforcement. The shear cracking, bond splitting and ultimate shear strength obtained from experiments and FEM analyses are compared in Table 4. The shear capacity at first cracking provided by the analysis is defined as the shear load which causes a strain in shear reinforcement to increase rapidly. Compared with the bond strength in Table 3, the test results are 10% higher than AIJ design guidelines (1999) and 10% lower than Maeda's equation (1994). It should be noted that the longitudinal rebars are not directly supported by transverse rebars (see Fig. 1) and this may reduce the effect that controls bond splitting.

As for S-series tests, the relations between shear crack stress $_{exp}\tau_{sc}(=_{exp}Q_{sc}/bD)$ and lateral prestress, and between ultimate shear stress $_{exp}\tau_{su}(=_{exp}Q_{su}/bD)$ and lateral prestress are plotted in Figure 10 together

3.2 Triaxial state of stress by FEM analysis

The effect of the lateral confinement on the shear behavior is evaluated based on the triaxial state of stress in the concrete. Figure 12 shows the distributions of the minimum principal stress in the center of S-RC and S-LPRC specimens. For S-RC specimen, the compressive strut formed by a large compressive stress was revealed at the shear load of 550 kN, thereafter, the width of the strut reduced slightly and localized in a diagonal direction at the maximum load. For S-LPRC specimen, on the other hand, the compressive strut appeared at the shear load roughly similar to the maximum load of RC and the width of the strut increased gradually up to the maximum load. The degree of damage for compressive failures and the equivalent confining pressure were introduced as the gauges to evaluate the effect of active confinement on

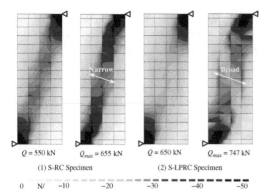

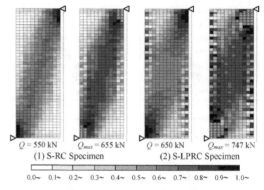

| Q = 550 kN | Q_{max} = 655 kN | Q = 650 kN | Q_{max} = 747 kN |
| (1) S-RC Specimen | | (2) S-LPRC Specimen | |

0 N/ −10 −20 −30 −40 −50

Figure 12.   Compression struit based on minimum principal stress.

| Q = 550 kN | Q_{max} = 655 kN | Q = 650 kN | Q_{max} = 747 kN |
| (1) S-RC Specimen | | (2) S-LPRC Specimen | |

0.0~ 0.1~ 0.2~ 0.3~ 0.4~ 0.5~ 0.6~ 0.7~ 0.8~ 0.9~ 1.0~

Figure 15.   Comparison of equivalent confining pressure at maximum shear load of S-series specimens.

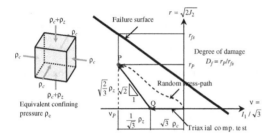

Figure 13.   The degree of damage and equivalent confining pressure for triaxial state of stress in meridian plane of Drucker-Prager criterion.

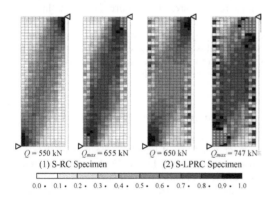

| Q = 550 kN | Q_{max} = 655 kN | Q = 650 kN | Q_{max} = 747 kN |
| (1) S-RC Specimen | | (2) S-LPRC Specimen | |

0.0 • 0.1 • 0.2 • 0.3 • 0.4 • 0.5 • 0.6 • 0.7 • 0.8 • 0.9 • 1.0

Figure 14.   Comparison of damage in compression zone.

the stress state in the core concrete quantitatively, as shown in Figure 13. The degree of damage for compressive failures is defined using the deviated part of the stress state in principal stress space. Figure 14 shows the degree of damage for compressive failures in the surface of S-RC and S-LPRC specimens at the same shear load as Figure 12. It can be seen from Figure 12 and Figure 14 that the distributions of the

degree of damage correspond roughly with those of the minimum principal stress. This degree of damage for S-LPRC is lowered compared with RC specimen. The red parts dotted with a white dot at the top and bottom ends represent the post-peak softening zone of concrete, so that the failure mode of analysis is quite similar to that of experiment. The post-peak softening of concrete for S-RC specimen occurred when the shear load is about 600 kN, and the softening zone was limited to the small area. This softening for S-LPRC specimen, on the contrary, occurred at the shear load of 700 kN, and the softening zone was gradually expanded up to the maximum load. The equivalent confining pressure is defined as the lateral pressure when converting the stress state on a random stress-path into that on the stress-path according to the triaxial compressive test with a constant lateral pressure, as shown in Figure 13. Consequently, the equivalent confining pressure increases with increasing hydrostatic component and decreasing deviated component of the stress state in principal stress space. The ratio of the equivalent confining pressure to the strength of concrete is shown in Figure 15 to compare S-RC with S-LPRC at maximum shear loads. As for S-LPRC specimen, the active confinement that is over ten times higher than the passive confinement was produced after applying the axial load, and it covered wider parts of the specimen than RC specimen until the maximum shear load.

4   SHEAR CRACK BEHAVIOR

4.1   Shear crack patterns

Figure 16 shows the contours of the crack strain obtained by analyses and the diagrams of shear cracks observed by S-series tests. The lateral confinement in S-LPRC specimen restrained greatly shear cracks from propagating at the shear load similar to the peak

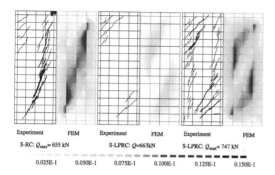

| Experiment | FEM | Experiment | FEM | Experiment | FEM |

S-RC: Q_{max}= 655 kN   S-LPRC: Q=663kN   S-LPRC: Q_{max}= 747 kN

0.025E-1   0.050E-1   0.075E-1   0.100E-1   0.125E-1   0.150E-1

Figure 16. Crack patterns from experiment and crack strain from FEM analysis in S-series tests.

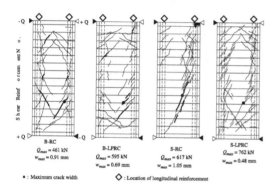

B-RC
Q_{max} = 461 kN
w_{max} = 0.91 mm

B-LPRC
Q_{max} = 595 kN
w_{max} = 0.69 mm

S-RC
Q_{max} = 617 kN
w_{max} = 1.05 mm

S-LPRC
Q_{max} = 762 kN
w_{max} = 0.48 mm

• : Maximum crack width   ◇ : Location of longitudinal reinforcement

Figure 17. Crack patterns at maximum shear load.

load of S-RC specimen, and the final crack pattern of S-LPRC differed drastically from that of S-RC. The shear cracks developed scatteringly in the upper and lower side of S-LPRC specimen, whereas they developed intensively in the center of S-RC specimen. The FEM analyses using smeared crack model cannot evaluate an individual crack but can detect the difference in crack patterns between S-RC and S-LPRC specimen.

Figure 17 shows the crack patterns observed in B-series and S-series tests at the maximum loads. Cracks with the width of 0.5 mm and over are emphasized using thick lines. In B-series columns, more bond splitting cracks appears along the longitudinal reinforcement and the crack angle is smaller than S-series columns. Small crack angle weakens the effect of shear reinforcement and brings wider cracks. The distributions of crack width over the transverse reinforcement where the total width of cracks is the largest are shown in Figure 18 to compare B-series with S-series. Transverse prestressing can disperse shear cracks along the depth of columns in S-series, whereas bond splitting cracks in B-series columns tend to concentrate near the longitudinal reinforcement regardless of the lateral prestress.

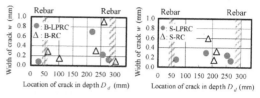

Figure 18. Width and distribution of shear crack.

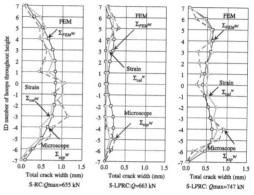

S-RC:Q_{max}=655 kN   S-LPRC:Q=663 kN   S-LPRC: Q_{max}=747 kN

Figure 19. Plots of crack width measured by means of microscope and of strain in hoops, and comparison with analysis.

4.2 Estimations of shear damage by FEM analysis

To investigate how accurate the FEM analysis with a smeared crack model can evaluate the extent of actual shear crack damage, the total crack width by FEM analysis, $\Sigma_{FEM}w$, which is estimated from the nodal displacements at both ends of a shear reinforcement by neglecting the strains of concrete, is shown in Figure 19 together with $\Sigma_{exp}w'$ by microscopes and $\Sigma_{cal}w$ calculated by integrating strains of hoops. Although $\Sigma_{exp}w'$ is observed on the surface of concrete while $\Sigma_{cal}w$ and $\Sigma_{FEM}w$ are estimated by a reinforcement, these three crack width values exhibit broadly similar behavior due to the small concrete cover of 9 mm. The difference between the crack width values of S-RC and S-LPRC specimen is basically consistent with that of shear crack behaviors shown in Figure 16 since the total crack width faithfully reflects their behaviors. Especially, the experimental and analytical crack width, $\Sigma_{cal}w$ and $\Sigma_{FEM}w$ are similar along the height of columns.

5 RELATION BETWEEN BOND BEAHAVIOR AND LATERAL CONFINEMENT

5.1 Strains in longitudinal reinforcement

The identification of longitudinal reinforcement and zonings divided by strain gauges to calculate bond

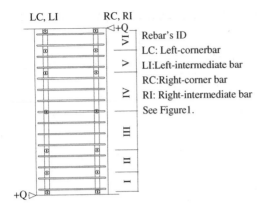

Rebar's ID
LC: Left-cornerbar
LI:Left-intermediate bar
RC:Right-corner bar
RI: Right-intermediate bar

See Figure1.

Figure 20. Identification of longitudinal reinforcement and zoning for bond stress.

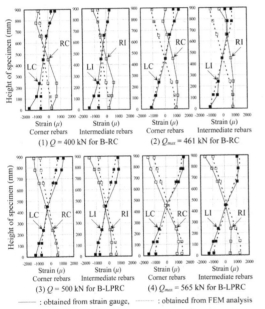

—— : obtained from strain gauge, ········· : obtained from FEM analysis

Figure 21. Strain distribution in longitudinal reinforcements.

stress is shown in Figure 20. After axial loading, compressive strain of each longitudinal rebar was about 300 to 400 μ and no bar yielded until shear loading was terminated. Figure 21 shows the distributions of strains in longitudinal reinforcement of B-series specimens, compared with the results of FEM analysis. The strains of tension zones damaged by shear cracks become constant as it called tension-shift. The analytical results of left rebars agree with the experimental data much better than those of right rebars. This is mainly because the experimental data of bond strength for left rebars are adopted in bond-slip relations. However, this analysis

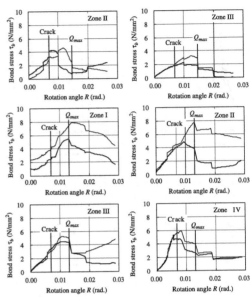

Figure 22. Relations between local bond stress of left side rebar and rotation angle.

Table 5. List of maximum local bond stress at each zone.

Test B-Series	Bar ID	Maximum local bond stress τ_b (N/mm^2)					
		Zone I	Zone II	Zone III	Zone IV	Zone V	Zone VI
RC	LC	–	4.95	3.24	2.64	–	–
RC	LI	–	4.24	2.31	2.10	–	–
RC	RC	–	–	3.79	4.00	3.80	2.40
RC	RI	–	–	3.29	3.11	2.31	4.17
LPRC	LC	7.88	7.75	5.27	5.91	–	–
LPRC	LI	5.41	4.87	4.52	5.14	–	–
LPRC	RC	–	–	4.64	4.70	5.83	6.90
LPRC	RI	–	–	3.10	3.47	4.44	5.64

that considered only the bond strength cannot estimate the effect of lateral confinement on the bond stiffness and the difference between corner and intermediate rebars. It is necessary to improve further the analytical model.

5.2 Bond behaviors in longitudinal reinforcement

The local bond stress is defined as an average bond stress of zonings shown in Figure 20. The average bond stress is calculated by dividing the differential force between zonings by the product of the zone's length and rebar's perimeter. The relations between local bond stress τ_b and rotation angle R for left rebars in each

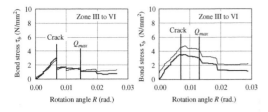

Figure 23. Relations between average bond stress of right side rebar and rotation angle.

zone are in Figure 22 where bold lines indicate corner bars and thin lines intermediate bars. Each local bond stress apart from two zones damaged by shear cracks is summarized in Table 5. The bond strength of corner rebars is higher than that of intermediate rebars and this trend is stronger for the compressive zones of B-LPRC (zone I and II). Figure 23 shows the relation between average bond stress of right rebars for zone III to VI and rotation angle. As can be predicted from the strain distributions in Figure 21, the bond stress of B-RC specimen decreases rapidly when bond cracks occurred at the shear load of 415 kN and less than half of the maximum bond stress at the peak shear load irrespective of the position of the rebar. The bond stress of B-LPRC specimen, on the other hand, retains most of the maximum bond stress until the peak shear load and decreases slowly after that. This reason is that transverse prestressing prevents the bond splitting crack along the longitudinal rebars from developing afterwards into a side splitting failure.

6 CONCLUSIONS

The following conclusions are obtained from experiments and 3-D FEM analyses performed on reinforced concrete columns laterally prestressed to investigate the effect of the active confinement upon shear and bond behaviors.

(1) The shear capacity at first cracking and ultimate shear strength have increased proportionally with increasing lateral pressure.

(2) The FEM analyses have revealed that an increase in resistance against shear failure as well as shear cracking with increasing lateral prestress could be explained by the triaxial state of stress in the core concrete.

(3) The shear crack patterns have changed appreciably, and the spacing and width of cracks have decreased drastically by transverse prestressing.

(4) The FEM analyses using smeared crack model cannot evaluate a localized crack accurately but can provide valuable information about the total damage along the depth of a column.

(5) Removing reinforcement to prevent bond splitting failures reduces the ultimate shear strength by 25 % and leads to bond failure mode.

(6) The FEM analyses using bond-slip model can evaluate the bond splitting strength and the distributed strains of longitudinal reinforcement.

(7) The bond strength after cracking has been improved by transverse prestressing.

REFERENCES

Architectural Institute of Japan (1999). Design Guidelines for Earthquake Resistant Reinforced Concrete Buildings Based on Inelastic Displacement Concept (in Japanese)

Chen, W.F. 1982. Plasticity in Reinforced Concrete. McGraw-Hill Book Company

Fuji, S. & Morita, S. 1983. Splitting bond capacities of deformed bars. *Transactions of the architectural institute of Japan* (AIJ) No. 324: 45–53 (in Japanese)

Maeda, M. & Otani, S. 1994. An equation for bond splitting strength based on bond action between deformed bars and concrete. *Summaries of technical papers of annual meeting*, AIJ,Structures II: 655–658 (in Japanese)

Richart, F.E., Brandtzaeg, A. & Brown, R. L. 1928. A Study of the Failure of Concrete under Combined Compressive Stresses. Univ. of Illinois, *Engineering Experiment Station*, Bulletin, No.185

Shinohara, Y. & Kaneko, M. 1999. Compressive shear behavior in fracture process zone of concrete. *Journal of Structural and Construction Engineering*, AIJ, No. 525: 1–6 (in Japanese)

Shinohara, Y., Miyano, K., Watanabe, H. & Hayashi, S. 2004. Active confining effect and failure mechanism for RC columns prestressed laterally. *Journal of Structural and Construction Engineering*, AIJ, No. 578: 115–121 (in Japanese)

Takamori, N., Benny Benni Assa, Nishiyama, M. & Watanabe, F. 1996. Idealization of Stress-Strain Relationship of Confined Concrete. *Proceedings of the Japan Concrete Institute*, Vol.18, No. 2: 395–400 (in Japanese)

Watanabe, H., Katori, K., Shinohara, Y. & Hayashi, S. 2004. Shear crack control by lateral prestress on reinforced concrete column and evaluation of shear strength. *Journal of Structural and Construction Engineering*, AIJ, No. 577: 109–116 (in Japanese)

Fracture Mechanics of Concrete and Concrete Structures – Design, Assessment and Retrofitting of RC Structures – Carpinteri, et al. (eds)
© 2007 Taylor & Francis Group, London, ISBN 978-0-415-44616-7

Test up to ultimate limit state and failure of innovative prebended steel-VHPC beams for railway bridges in France

S. Staquet
BATir, Université Libre de Bruxelles, Brussels, Belgium

F. Toutlemonde
Structures Laboratory, French Public Works Research Institute LCPC, Paris, France

ABSTRACT: Experimental validation of a possibly attractive extension of the pre-bended composite beams technique, first applied in Belgium as "Preflex", was carried out. Very high performance concrete (C80/95) was used for potential advantages of low delayed strains and high tensile strength. A major issue of the programme consisted in validating a design method for the serviceability limit state (SLS) linked to the cracking of the concrete and for the ultimate limit state (ULS) corresponding to an instability (warping) or to yielding of the steel girder. Detailed results of the global response of two 13 m-long beams up to ultimate limit state and failure are analyzed in conjunction with deflection measurements. The ultimate limit state computation based on the strength and elastic stability is always controlled by the serviceability limit state for this kind of structures.

1 INTRODUCTION

1.1 *The preflex beam*

The Belgian engineer, A. Lipski, with assistance of Professor L. Baes from the Université Libre de Bruxelles (Baes and Lipski, 1957) invented the system of the precambered composite steel-concrete beam, initially known by the patent name "Preflex beam". The first project using this type of beams dates back to 1951. The two best-known structures are the Southern Tower (Tour du Midi) and the Berlaymont Building, both in Brussels (Belgium). The typical construction sequence of a prebended beam is as follows (see Figure 1):

a. In the plant, setup a steel I-girder, with a precamber, supported at each end.
b. Prebend the steel girder by applying two concentrated loads at one-quarter and three-quarters of the span.
c. Cast the 1st phase concrete (C50/60) at the level of the bottom flange of the steel girder while keeping in place the loads of the prebending phase of the girder.
d. Two days after casting the concrete, remove the prebending loads. As a result, the beam goes up, the precamber becomes smaller than the original precamber and the concrete is now subjected to compression.
e. Cast the 2nd phase concrete on site.

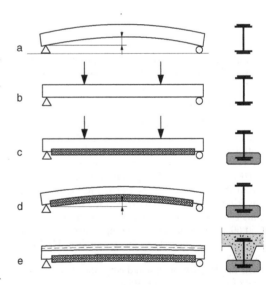

Figure 1. Construction steps of a prebended beam.

Due to steel-concrete bond, which makes it possible to consider the cross-section as composite, the concrete of 1st and 2nd phases significantly increase the stiffness of the composite beam as compared to the stiffness of the steel girder alone. Provided the concrete of the bottom flange remains subjected to compression even during the application of service loads, the

929

Figure 2. Prebending of the steel girder (step b and form-work assembly before step c).

requirement of no cracking in the concrete is satisfied. The design principles for the service loads are partly based on the class II type of BPEL (BPEL, 1999) implying no decompression in concrete under frequent service load combinations and a maximum concrete tensile stress limited to the actual concrete tensile strength under rare service load combinations. The design criteria for the ultimate limit state (ULS) correspond to an instability (warping) of the steel profile or to yielding of the steel girder upper flange. The Preflex system has been particularly successful in Belgium because it allows long spans with a minimal construction depth and offers excellent fire resistance (Staquet et al., 2004). It is presently mainly used in Belgium for railway bridges due to its high fatigue performance (first verified experimentally by Verwilst, 1953).

1.2 Research significance

The aim of the research program that has been carried out at LCPC in the framework of the French National Project MIKTI was to extend the system of the Preflex beam to Very High Performance Concrete (VHPC) and to give background for further updating of the Eurocode 4 (EN 1994-2) to the design of this kind of structure. Actually, the concrete grade which has been used until now is lower or equal to C50/60 whereas the average compressive strength at 28 days of the concrete in this research is 110 MPa. In the comparison to the present realizations coming from the Belgian precast industry, the main advantage of using VHPC with silica fume is to reduce the prestressing losses of the system thanks to a significant decrease of the creep deformations, as predicted by Eurocode (EN 1992-2:2005) together with the possibility to optimize the beam weight and its serviceability domain. Applications might concern French Railways Bridges, where stiffness has to be maintained or increased, due to high speed train requirements, with maintained clearance and reduced beam inertia (accounting for ballast between the rails and the

bridge structure). A first theoretical step developed the updated background of these beams design, relatively to delayed strains of concrete and their effects (Mannini, 2001). This development was confirmed by research work conducted in Belgium for prebended U-shaped Bridges (Staquet & Espion, 2005). Then, the interest of VHPC in optimizing the 1st phase concrete deadweight and the global stiffness was confirmed. Consequently, it was decided to carry out an experimental investigation of long-term performance under dead and fatigue loads and behaviour up to ultimate limit state and failure of VHPC-prebended beams, mainly aiming to validate the scientific method for taking concrete creep into account. Beside data of concrete delayed effects inducing specific monitoring (Staquet et al. 2006), the fatigue performance and safety of fatigue design provisions were validated by this experimental program (Toutlemonde & Staquet, 2007). The present paper focuses on validating a design method for the serviceability limit state (SLS) linked to cracking of the concrete and for the ultimate limit state (ULS) corresponding to an instability (warping) or to yielding of the steel girder.

2 EXPERIMENTAL INVESTIGATIONS

2.1 Steps of the experimental program

The construction stages a to d (Figure 1) were carried out at LCPC on two factory-made steel girders HEB 360, 13 m-long, using a specially designed self levelling (Roussel et al., Staquet & Toutlemonde, 2006) VHPC. The cross section of the beams illustrated on Fig. 3 shows that the thickness of the concrete situated below the bottom flange of the steel girder was only 55 mm. Moreover, in addition to the longitudinal passive reinforcements (diameter 12 mm), ribbed stirrups (diameter 8 mm) were disposed every 15 cm along the beam and steel square ribs (25 × 5 × 200 mm) were welded to the bottom flange of the steel girder every 45 cm (Fig. 4). Dimensions of the ribs were defined as usual for taking all the design shear stresses at steel lower flange – concrete interface. As usually realized, stirrups are fixed across the steel profile web, which also favours regular stress transfer between concrete and steel, ensuring the desired composite behaviour.

Two days after concrete casting, the prestressing was transferred by releasing the prebending loads (145.2 kN for beam P1 and 137.5 kN for beam P2) initially applied by two jacks at one quarter and at three quarters of the span (Fig. 5). Two months after, the beams were submitted to permanent loads (40 kN applied by lead masses, representing the load of the upper concrete deck, superstructures and ballast). After 4 months, live loads representative of railway traffic were applied: 1000 cycles representing the effect of trains possibly transiting once a year (half UIC

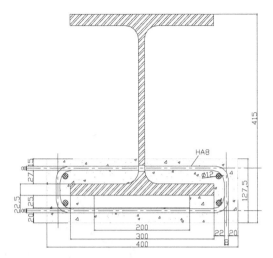

Figure 3. Cross section of the prebended beams tested (dimensions in mm).

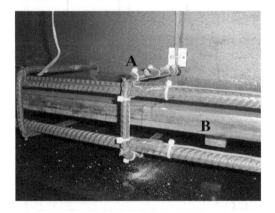

Figure 4. Ribbed stirrups (A) every 15 cm along the beam and steel square ribs (B) every 45 cm.

conveys), then one or two million cycles corresponding to more frequent heavy trains. The deflection was monitored for more than 8 months as well as numerous complementary strain measurements. Mechanical characterization (Young's modulus), creep and shrinkage tests were also performed so that a correct analysis of the structural behaviour of the prebended beams can be done, focusing on the steel-concrete behaviour, and taking advantage of low delayed deformations of VHPC. Finally the beams were tested up to failure for quantifying their safety margin under service and ultimate loads, and checking the accuracy of the design predictions.

2.2 First phase VHPC mechanical characteristics

Optimization of the cross-section required using VHPC (C80/95) for the 1st phase concrete. Moreover,

Figure 5. Transfer of prestressing to the composite beam at 48 hours of concrete age.

Table 1. VHPC mechanical properties of the cylinders 11/22 exposed to 40% RH and 20°C after 1 day of concrete age

Age (d)	f_c (MPa)	f_t (MPa)	E (GPa)
2	71.5	–	38.2
7	87.8	4.4	40.6
28	99.3	4.5	42.8
56	103.1	–	43.4
100	103.3	–	43.5
168	106.1	5.6	43.3

transfer of the prebending only 48 hours after casting required strength at least equal to 55 MPa at 2 days, leading to a ratio of about 40% between the maximal compressive stress applied at 2 days and the average compressive strength, which corresponds to the practical current ratio. Self compacting properties were required for the beam realization, due to congested areas under the lower steel flange. The feasibility of this optimization was demonstrated (Staquet & Toutlemonde 2006). Mechanical characteristics of VHPC were measured carefully, under different standard and realistic conditions, for further analysis of the rigidity evolution of the beams. It is noticeable that, even though the cement paste content of this self-levelling VHPC is quite important, the delayed strains amplitude is rather small, due to the silica fume content. These strains are correctly estimated by CEB-FIP 1999 model for creep of specimens loaded at 2 days, and by EN 1992-2 model for creep of specimens loaded at 28 days. Table 1 shows the evolution of the VHPC mechanical properties: f_c, average compressive strength; f_t, average splitting tensile strength; E, average modulus of elasticity, determined on cylinders 11/22 cm exposed to 40% of relative humidity and 20°C at one day of age. The E value at 168 days is

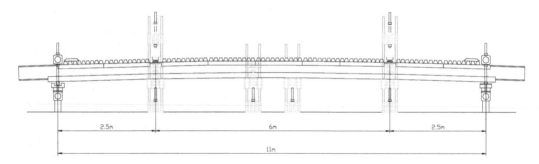

Figure 6. Static loading up to failure for the two beams.

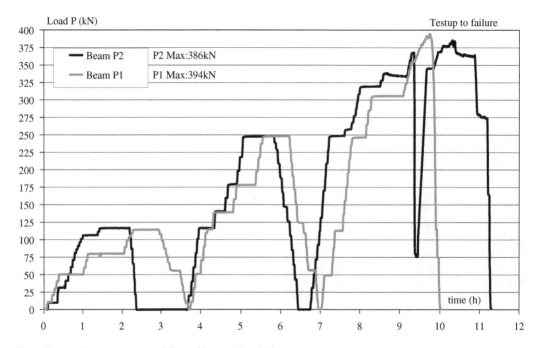

Figure 7. Loading sequence up to failure of Beams P1 and P2.

slightly lower than the one at 100 days, explained by an effect of rather severe drying conditions.

2.3 Loading up to failure

The estimation of the cracking load (average stress equal to about 1.5 f_t in the whole 1st phase concrete: 8.4 MPa), the warping load for a free span of 6 m (between application points of the loads by the jacks, Fig. 6) and the yield load of the upper flange of the steel girder (fy: 420 N/mm²) by neglecting the concrete participation, provided the following theoretical values of the load at each jack (EN 1993): 93 kN, 336 kN and 410 kN respectively. The tests were carried out in 3 phases (Fig. 7 and 8): up to 115 kN then unloading;

up to 250 kN (3/4 of the warping theoretical load or 60% of the theoretical yield load) then unloading; and loading up to failure.

The value of 115 kN is considered as an excess value of the design service load which is related to the damage occurrence in the concrete flange (stabilized cracking pattern) and the theoretical value of 336 kN corresponds to the ultimate limit state (ULS) loading value. The table 2 shows the extreme first phase concrete stress values under permanent loads computed by using the CEB MC 90 model code in its version 99 (fib, 1999) for the prediction of creep and shrinkage of the VHPC. The stress values for the beam P1 are put in italic in Table 2 because, due to an extra-severe loading leading to an irrecoverable cracking that was applied

Figure 8. View of loading of Beam P2.

Table 2. Extreme stress computed values in concrete before the test up to failure (negative value in compression)

Stress value in the 1st phase concrete (MPa)	Bottom fiber	Upper fiber	Average
Beam P1	−2.9	+1.8	−0.55
Beam P2	−2.8	+1.7	−0.6

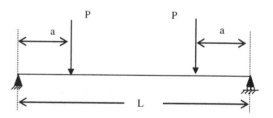

Figure 9. Loading configuration during test up to failure.

on Beam P1 before the fatigue test and the test up to failure, the stress state of concrete is different from these computed values in some parts of this beam.

2.4 Results and discussion

According to the loading configuration shown in the figure 9, the apparent bending stiffness can be determined from the deflection f at mid-span and the global applied load 2P, using the following equation:

The static bending stiffness EI of the steel girder alone is 90,700 kN.m^2 whereas the static bending stiffness of the composite beam corresponding to a full concrete contribution (with instantaneous modular ratio) is 152,700 kN.m^2, the concrete corresponding to one third of the global stiffness. The values of the stiffness k reported in Table 3 were obtained from a linear regression of the experimental relation load-deflection at mid-span for each loading and unloading phase in the range of 0 to 20 kN. As the P1 beam had been submitted to an extra-severe loading leading to irrecoverable cracking, its static bending stiffness at the beginning of the test was lower than the one of the P2 beam.

Table 3. Static bending stiffness of the tested beams during the test up to failure (L:loading; U: unloading).

P1	k (kN/mm)	EI (kN.m^2)	Variation
L 0-115 kN	7.12	125,000	
U 115 kN-0	6.04	106,000	−15.2 %
L 0-250 kN	6.96	123,000	−2.2 %
U 250 kN-0	5.56	98,000	−21.9 %
P2	k (kN/mm)	EI (kN.m^2)	Variation
L 0-115 kN	7.58	133,000	
U 115 kN-0	6.28	111,000	−17.2 %
L 0-250 kN	7.30	129,000	−3.7 %
U 250 kN-0	5.48	96,000	−27.7 %

Figure 10. View of the cracking pattern.

After having reached the SLS load (115 kN), the bending stiffness loss which was obtained from the experimental relation force-deflection in the loading phase up to 250 kN, was only 2% for Beam P1 and 4% for Beam P2. The reduction of the static bending stiffness after having loaded the beam up to 250 kN and determined in the unloading phase was 22% for Beam P1 and 28% for Beam P2 and is less than 10% higher than stiffness of the steel alone. A significant reduction of the bending stiffness in the range of 15% to 17% was also obtained in the unloading phase from 115 kN. The bending stiffness was subsequently fully recovered after having completely unloaded the beam because the level of compressive stresses was high enough to close the cracks in the concrete flange of the two beams. The observation of the first cracking visible on the bottom surface of the concrete flange around mid-span appeared at a load of 40 kN for Beam 1 and 85 kN for Beam 2 or at a bending moment of 100 kN.m and 212 kN.m respectively. As the load increased up to the SLS value, the crack network extended until it reached a pattern in close relation with the steel mesh of reinforcement: ribbed stirrups every 15 cm, visible on lateral and upper surfaces along the beam and steel square ribs welded to the bottom flange of the steel girder, visible on the bottom surface every 45 cm along the beam (Fig. 10). The maximal crack opening reaches 0.15 mm under 115 kN and the crack pattern was stabilised, uniformly distributed between the two concentrated loads at this loading value.

At a loading value of 50 kN for Beam P1 and 115 kN for Beam P2, a longitudinal crack was opened in the central part, corresponding to the lateral boundary of the steel square ribs.

During the second loading, the maximal crack opening reached 0.3 mm under 140 kN and 0.4 mm under 180 kN for Beam P1 whereas 0.2 mm under 180 kN and 0.3 mm under 250 kN for Beam P2.

The non-linearity of the load-deflection curve was observed since 40 kN for the two beams (Fig. 11). After the first loading phase up to 115 kN with a maximal deflection at mid-span of 38.3 mm for Beam P1 and 38.2 mm for Beam P2, the residual deflection was 0.6 mm and 1.4 mm respectively. After the second loading phase up to 250 kN with a maximal deflection at mid-span of 89.5 mm for Beam P1 and 91.2 mm for Beam P2, the residual deflection was 1.7 mm and 2.6 mm respectively. The global behaviour of the beam, relatively linear until 250 kN (Fig. 11) became very non linear from 310 kN for the two beams (Fig. 12) due to the development of cracks and slidings between cross-sections. During the third loading, the maximal crack opening under 350 kN reached 0.7 mm at the bottom surface of the concrete flange. The warping appeared at 394 kN for Beam P1 and 386 kN for Beam P2 with a deflection at mid-span equal to 24 cm at this stage and a typical diagonal cracking in concrete flange around mid-span due to the rotation of the sections (Fig. 13).

The failure corresponds firstly to the instability of the steel upper flange as predicted because the maximum experimental loading value of about 390 kN is a little bit lower than the theoretical value of 410 kN, corresponding to the yield load of the upper flange of the steel girder (taking into account a safety factor according to EN-1993). The presence of VHPC does not modify the failure mode linked to the steel girder but provides important reserve with respect to an ultimate limit state of warping instability type by delaying its onset: about 390 kN instead of 336 kN. Moreover, its decreases the suddenness of the warping thanks to the energy which is dissipated by diagonal cracking of the concrete flange in torsion. After the final unloading, an elastic recovery still existed. The figure 14 displays the VHPC-steel beam P2 after loading up to failure. Figure 15 shows the lateral displacement between the position of the upper flange of the Beam P2 after failure due to the warping and its initial position. The maximal value of the lateral displacement occurred around mid-span and was equal to 9 cm (Fig. 15).

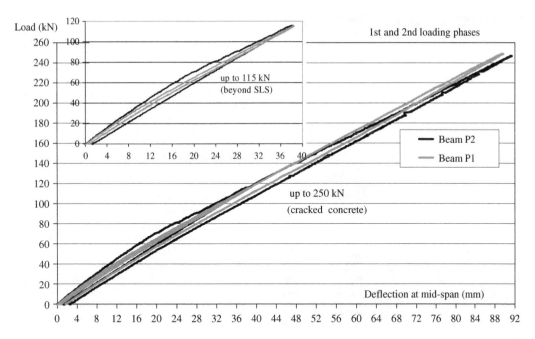

Figure 11. Loading of Beams P1 and P2 up to 250 kN.

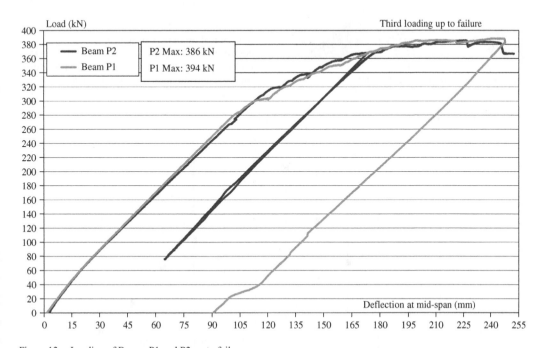

Figure 12. Loading of Beams P1 and P2 up to failure.

935

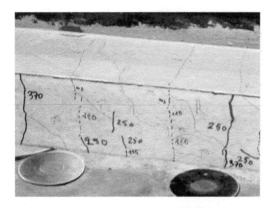

Figure 13. View of the cracking pattern after failure.

Figure 14. View of Beam P2 after failure.

3 CONCLUSIONS

The experiments realised on the two prebended steel-VHPC beams allow concluding to the following points:

– low evolutions of the deflection after 6 to 8 weeks confirm the interest to use VHPC;
– feasability to optimize a composite prebended beam with VHPC is demonstrated;
– the evolution of the structural stiffness under rare or fatigue loads lower than the cracking limit of concrete (design of class II type of BPEL) is negligible;
– the presence of VHPC does not modify the failure mode linked to the steel girder but provides important reserve with respect to an ultimate limit state of warping instability type by delaying its onset and reducing its suddenness.
– The remarkable repeatitivity of the behaviour of the two beams brings a great credibility to the results. The ultimate limit state computation based on the strength and the stability of the single girder appearsas safe. This is even safer in case of 2nd phase concrete for a multi-beams structure.
– Practically, the design of prebended beams is always controlled by the serviceability limit state verifications due to the following criteria:
– control of the delayed effects of concrete by limiting the compressive stress during the prestressing, which implies to have accurate data on creep and mechanical properties of the concrete;
– control of cracking implying to keep compression under permanent and even frequent loads and to limit the tension under rare service loads.

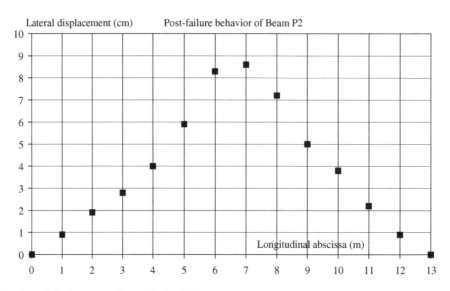

Figure 15. Lateral displacement of Beam P2 after failure.

Provided these criteria are met, composite behaviour is normally ensured under service loadings and up to ultimate limit state, and corresponding design methods are valid.

4 ACKNOWLEDGEMENTS

The authors are pleased to thank number of colleagues from BCC, MI and SFIOA units at LCPC who participated to this study. The support of the National Project 'MIKTI' (sponsored by the Ministry for Public Works – Civil Engineering R&D funds) is also gratefully acknowledged.

REFERENCES

Baes L. & Lipski A. 1957. La poutre Preflex, la décompression du béton enrobant l'aile tendue, le problème du retrait et du fluage. *Revue C (Gent)* I-4 : 29–49 (in French).

BPEL 1999. Règles BPEL 91 modifiées 99 (Règles techniques de conception et de calcul des ouvrages et constructions en béton précontraint suivant la méthode des états limites) fascicule n° 62 titre 1er section II du CCTG applicable aux marchés publics de travaux, JO du 16 février 1999.

EN 1992-2 : 2005 Eurocode 2 – Calcul des structures en béton – Partie 2 : Ponts en béton – calcul et dispositions constructives, *CEN*.

EN 1993 Eurocode 3 – Design of steel structures, *CEN*.

EN 1994-2 Eurocode 4 – Design of composite steel and concrete structures – Part 2: Bridges, *CEN*.

fib-CEB-FIP 1999, Structural Concrete Volume 1, *Bulletin fib n° 1, fib.*

Mannini C. 2001. Etudes des effets différés dans les poutres Preflex, mémoire de projet de fin d'études, ENPC, Paris, 157 pp.

Roussel N., Staquet S., D'Aloia L., Le Roy R., Toutlemonde F. SCC casting prediction for the realization of prototype VHPC-precambered composite beams, accepted for publication in *Materials and Structures*.

Staquet S., Rigot G., Detandt H., Espion B. 2004. Innovative Composite Precast Precambered U-shaped Concrete Deck for Belgium's High Speed Railway Trains. *PCI Journal* 49 (6): 94–113.

Staquet S. & Espion B. 2005. Deviations from the principle of superposition and their consequences on structural behaviour. *Shrinkage and Creep of Concrete ACI SP* 227: 67–83.

Staquet S. & Toutlemonde F. 2006. Innovation for the railway bridge decks in France: a precambered composite beam using VHPC. *La Technique française du Béton, Special Publication for the 2nd fib Int. Symposium, Naples, Italy, AFGC, Paris.*

Staquet et al. 2006. Détermination expérimentale, au moyen d'un contraintemètre actif, de l'évolution des contraintes dans une poutre mixte acier-BTHP préfléchie. *Journées des Sciences de l'Ingénieur (JSI), Marne-la-Vallée, 5–6 décembre.*

Toutlemonde F. & Staquet S. 2007. Fatigue response of VHPC for innovative composite prebended beams in railway bridges. *Concrete under Severe Conditions: Environment & Loading,* submitted for Publication *in the Proc. of the Int. Conference CONSEC07, Tours, France.*

Verwilst Y. 1953. Essais pulsatoires et essai statique à la rupture d'une importante poutre de 14,50 m sur l'installation G.I.M.E.D. de l'Association des Industriels de Belgique (A.I.B.), *revue «l'Ossature Métallique»(O.M.)* 3 : 165–169.

Fracture Mechanics of Concrete and Concrete Structures – Design, Assessment and Retrofitting of RC Structures – Carpinteri, et al. (eds)
© 2007 Taylor & Francis Group, London, ISBN 978-0-415-44616-7

Delamination and structural response of RC thin shell in nuclear shield buildings with unanticipated construction openings*

S.C. Mac Namara
Syracuse University, Syracuse, NY, USA

M.M. Garlock & D.P. Billington
Princeton University, Princeton, NJ, USA

ABSTRACT: This paper examines the structural response in a thin shell concrete dome with construction openings. The weight of the dome is carried in axial compression along the hoops and meridians of the dome. The openings interrupt the hoops and meridians and the dome's weight must be redistributed around the openings; resulting in zones of increased compression near the opening. In the affected areas there is a significant difference between the compression on the top surface of the dome, and that on the bottom surface. The dome was cast in two layers. This study examines the potential for fracture (and thus delamination) as a result shear stress at the interface of the two layers. The stresses are largest around the opening. The shear capacity of the interface is determined by previous empirical studies of composite beams. The shear stresses in the dome with openings are not large enough to cause delamination.

1 INTRODUCTION

1.1 Introduction

As Nuclear Power Plants age many require steam generator replacement. There is a nickel alloy in the steam generator tubes that is susceptible to stress cracking and although these cracks can be sealed the generator becomes uneconomical without 10%–15% of the tubes (Chernoff & Wade, 1996). The steam generator in a typical nuclear power plant is housed in the containment structure next to the reactor. The equipment hatch is not big enough to facilitate steam generator replacement, thus construction openings in the dome of the containment structure are required. Where both the walls and the dome of the structure have been post-tensioned such openings are generally made in the walls, and where only the walls are post-tensioned it is easier to put the openings in the dome. This paper examines the effects of such openings in the dome.

The prototype nuclear containment shield building is made up of a 0.6 m (2 ft) thick dome atop 0.91 m (3ft) thick and 51.9 m (170 ft) high cylindrical walls, radius 20 m (65.5 ft), with a tension ring (15 ft) high and 2.4 m (8 ft) thick in between. The dome of the building is cast in two layers; a lower 23 cm (9 in) layer that serves as the formwork for an upper 38 cm (15 in) layer.

The aim is to evaluate the stresses through the 0.6 m depth of the dome roof of the shield building. The finite element model of the dome is made from a series of layers of solid-elements. In the model the hoop and meridian stresses are not uniform through the depth at any point on the dome. The stresses vary linearly from the top surface of the dome to the underside of the dome. This variation is indicative of bending, and where this bending stress itself varies there will be a horizontal shear stress in the dome. This paper aims to establish the extent and the magnitude of these horizontal shear stresses in the dome with openings and in the dome without openings (if any).

Should the shear stresses at the interface between the two layers of the dome prove to be significant there is the potential for cracking leading to delamination.

1.2 Previous research

There is little published research on the structural response of nuclear containment structures to any unanticipated construction openings.

Mac Namara et al, 2007 examines a model (with the same geometry as that considered in this paper) made from shell elements and demonstrates the redistribution of the weight of the dome around construction openings leading to zones of increased meridian compression either side of the opening, and zones of increased hoop compression above and below the opening.

Bennett, 2005 studies construction openings (of a similar size to those considered here) made in the *walls* of a shield building for steam generator replacement. Bennett concluded that the impact of the opening on the stresses due to self weight was local, and though significant in comparison to the stresses in the structure without openings, insignificant in terms of material strength.

For the question of delamination, the available references are empirical studies of delamination at the interface of two layers of concrete in *beams* only. Patnaik, 2001 gives horizontal shear strengths for beam interfaces of between ~1700 kN/m² and ~6200 kN/m² (~250 psi and ~900 psi), and cites a series of previous studies that give similar results. The study concluded that the horizontal shear strength at the interface of the beams was independent of concrete strength and depth of beam to tie spacing ratio.

1.3 *Description of the Model*

This paper examines a generic nuclear containment shield building as described in section 1.1.

The analysis of a model made from shell elements (Mac Namara et al, 2007) showed that the effect of the boundary conditions of the dome had only a local impact on the stresses and forces in the dome. That analysis also established that the openings in the dome had no significant effect on the stresses in the walls or the tension ring. For this reason and for computational simplicity the analysis in this paper was limited to the dome roof of the structure. The dome is modeled as a series of layers of solid-elements with fixed connections at the base. For further ease of computation and to allow for a finer mesh only one quarter of the dome is modeled. The boundary conditions along the x and y axes replicate the symmetry of the dome. The full dome and the final model are shown in Figure 1.

The model is analyzed under dead load only, the only load that acts on the shield building while the steam generator replacement is carried out (and the construction openings are in place).

2 RESULTS

2.1 *Axial and bending stresses in the dome without openings*

We first examine the distribution of stresses in the structure without openings. These stresses are compared both to the expected stresses under membrane theory (to verify the model) and to the stresses in the model with openings (to fully examine the impact of the openings on the structure).

Thin-shell concrete domes are assumed to carry all loads as axial forces in two orthogonal directions. Meridian forces and stresses act along the meridians of the dome (the meridians of a dome are like lines

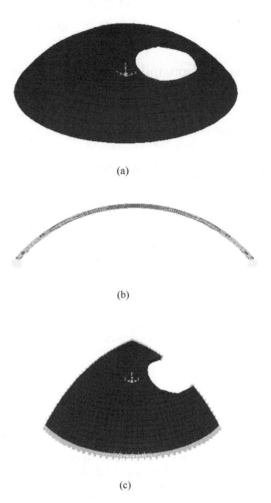

(a)

(b)

(c)

Figure 1. The finite element model; (a) the full dome, (b) the cross section, and (c) The quarter section.

of longitude on a globe). Hoop forces and stresses act along a series of parallel hoops (the hoops of a dome are like lines of latitude on a globe)

In Membrane Theory, the meridian stress resultant;

$$N_\varphi = aq\left(\frac{1}{1+\cos\varphi}\right) \quad (1)$$

and the hoop stress resultant;

$$N_\theta = aq\left(\frac{1}{1+\cos\varphi} - \cos\varphi\right) \quad (2)$$

where a is the radius of curvature of the dome; 26.8 m (88ft), q is the load on the dome (self weight), and ϕ is the angle between the radii of the dome at the crown and at the point in question (Billington, 1982).

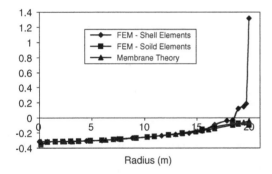

Figure 2. Hoop stresses for the solid-element and shell-element models compared with membrane theory.

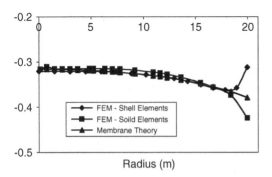

Figure 3. Meridian stresses for the solid-element and shell-element model models compared with membrane theory.

Figures 2 and 3 show the membrane theory values for hoop and meridian stresses respectively and compare those to the *average* value for stresses at the same point returned by the finite element model. The hoop and meridian stresses are in compression throughout and are, as expected, the same at the crown, 320 kN/m^2 (46psi) of compression. Hoop stress decreases from the crown to the base where it is approximately 48.3 kN/m^2 (7psi). The meridian stress increases from the crown to the base where it is approximately 390 kN/m^2 (56psi). The stresses in the finite element model agree closely with those predicted by membrane theory with small deviations close to the base of the dome. These deviations are easily explained by the different boundary conditions assumed by membrane theory and by the model.

The boundary condition assumed by membrane theory at the base of the dome is that of a roller perpendicular to the tangent of the dome at that point. The dome is constrained the direction of the tangent at the base, it is free to move in the direction perpendicular to the tangent, and free to rotate. The finite element model assumes a fixed connection at the base of the dome (Billington, 1982).

Figures 2 and 3 also compare the hoop and meridian stresses found for the dome in the shell-element model in Mac Namara et al, 2007 to those of the solid-element model. The only significant difference is that the shell model analysis has hoop tension close to the base that is not evident in the solid-element model. This is also due to a difference in boundary conditions. Where the shell-element model has the ring and wall attached to the dome, the solid-element model has fixed boundary conditions. A separate analysis of a shell model with fixed boundary conditions confirms that we do not expect to see hoop tension at the base of a fixed dome, made of either shell or solid-elements (Mac Namara, 2007).

The stresses for the solid-element model in Figures 2 and 3 are the *average* stresses through the depth of the dome. The model does indicate variation of hoop and meridian stresses through the depth of the dome. For the most part these variations are small, indicating little bending. In most of the dome the portion of total stress that is a bending stress is on the order of 10%. This indicates that although the model is made up of solid-elements, it is exhibiting shell behavior and is an appropriate approach to examine the stresses in the structure

There is more significant bending associated with the meridian stress at the base of the dome. This part of the dome is expected to deviate from membrane theory of shell behavior (much of the self weight is carried by out of plane shear). This deviation is also a result of the difference in boundary conditions discussed above.

2.2 Axial stresses in the dome with openings

Figures 4 and 5 show the hoop and meridian stresses in the dome with openings compared to the dome without openings. The stresses in the dome with openings are plotted for two axes, Axis 90 is the axis of symmetry through the dome that bisects the openings and Axis 0 is the axis of symmetry through the dome perpendicular to Axis 90. Note that Figures 4 and 5 plot the *average* stresses through the depth of the dome at any point.

As expected the hoop stresses along axis 90, the axis that bisects the opening, show zones of increased hoop compression with respect to the dome without openings between the crown and the opening. The stress is 480 kN/m^2 (70psi) in the dome with openings compared to 311 kN/m^2 (45psi) in the dome without). There is another zone of increased hoop compression between the opening and the base. The stress is 360 kN/m^2 (52psi) in the dome with openings compared to 191 kN/m^2 (27 psi)in the dome without.

The meridian stresses along axis 90 show the expected decrease in meridian compression both above and below the opening and 0 kN/m^2 immediately above the opening (where there is no material below to support self weight) and 0 kN/m^2 immediately below

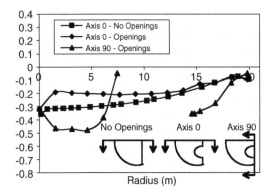

Figure 4. Hoop stresses for the solid-element models with and without openings.

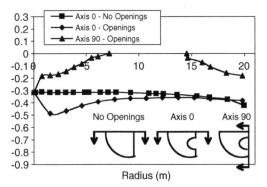

Figure 5. Meridian Stresses for the solid-element models with and without openings.

the opening (where there is no material above to cause meridian stress).

Generally the stresses in the solid-element mode compare well with shell-element model in Mac Namara et al, 2007. The shell-element mode does display a small hoop tension stress immediately above and below the opening that is not present in the results for this solid-element model. However, the solid-element model shows considerable bending stress at these points.

the solid-element model has fixed boundary conditions. A separate analysis of a shell model with fixed boundary conditions confirms that we do not expect to see hoop tension at the base of a fixed dome, made of either shell or solid-elements (Mac Namara, 2007).

Figures 4 and 5 show that on Axis 0 (the axis furthest from the openings) the hoop and meridian compression in the dome with openings is close to that of the dome without openings. In the top half of the dome the hoop *and* meridian stresses, in the dome with openings, are within 40% of those in the dome without, this difference falls to 10% and less in the lower half

of the dome. This behavior is the same as that of the shell model, and illustrates the local and limited nature of the influence of the openings on the stresses in the dome.

The values for hoop and meridian stress across the dome (and not just along the axes shown in Figures 4 and 5) show the redistribution of hoop and meridian compression around the openings. The hoop and meridian compression that would have been taken by the missing material is redistributed around the opening. Thus, there is increased hoop compression between the opening and the crown and between the opening and the base, and increased meridian compression on the either side of the openings. There is a corresponding decrease in hoop compression on either side of the openings, and a decrease in meridian compression between the opening and the crown and between the opening and the base.

The maximum hoop stress at any point through the depth of the model with openings is 703 kN/m^2(102psi) of compression and is found at the bottom corner of the opening. The maximum meridian stress is 1110 kN/m^2 (161psi) and is found in the region of increased compressive meridian stress alongside the opening.

These maximum stresses are between 2 and 3 times larger than the maximum stresses in the dome without openings. However magnitude of the stresses is insignificant when compared to material strength ($\sim$20,000 kN/m^2, $\sim$3000psi). This is not surprising considering the large factor safety such a structure would have under dead load alone.

2.3 Bending stresses in the dome with openings

There is more variation of hoop stresses (i.e. more bending) in the dome with openings as compared to the dome without, even along axis 0 away from the openings. Bending constitutes up to $\sim$30% of total hoop stress along axis 0 as opposed to $\sim$10% in the dome without openings. Along this axis there is less hoop bending near the crown and the base than in the rest of the dome. The pattern and magnitude of bending is same along axis 90, except immediately above and below the opening where the variation of hoop stress through the depth is even more significant. Bending constitutes $\sim$60% of the total hoop stress immediately above the opening and $\sim$40% immediately below.

The impact of the openings on the variation of meridian stress is less significant than the impact on hoop stress. Along axis 0, bending accounts for $\sim$10% of the total meridian stress, the same as in the dome without openings. Along axis 90 the variation of meridian stress through the depth is significant in comparison with average stress through the depth. However, as the total meridian stresses along this axis are very small (due to the opening – see Figure 5) the magnitude of this bending is insignificant.

Near the base, along both axes, there is bending on the order of 40–80% of total stress, this is not however an impact of the opening and is present in the dome with openings and is a function of the boundary conditions of the dome.

2.4 *Shear stresses in the dome with openings*

The bending evident in the solid-element model of the dome with openings is accompanied by shear stresses. The shear stresses are largest where the magnitude of the bending changes most rapidly. The maximum shear stresses in the model are immediately adjacent to the opening.

In general the shear stresses associated with the hoop stress (the S13 stresses, where the one direction is the hoop direction and the three direction is the local vertical axis for the element) are larger than those shear stresses associated with the meridian stresses (the S23 shear stresses where, the two direction is along the meridians of the dome, and the three direction is the local vertical axis for the element).

The maximum S13 shear stresses are found in an area where the hoop stresses are increased relative to the dome without openings; immediately above the opening. Incidentally, this is not the location of maximum hoop stress which is below the opening. The maximum S13 shear stress is 86 kN/m² (12.5 psi) and is found at the top corner of the opening.

The maximum S23 shear stresses are also found where the associated (meridian) stresses are largest. Although the S13 stresses are generally the larger, the maximum S23 shear stress is 95 kN/m² (13.8 psi), and is found at the side of the opening.

The shear stresses in the rest of the dome, away from the opening and away from the base of the dome, are even smaller, with maximums of ~ 20 kN/m² (3psi).

All of the shear stresses in the dome with openings are trivial and they are not significant enough to represent any threat of delamination.

3 CONCLUSIONS

3.1 *Conclusions*

Text The solid-element dome without openings displays little bending and insignificant shear stresses and the average stresses through the depth of the dome are consistent both with the stresses in the shell-element model model and with membrane theory. Thus, the use of such a model made of solid-elements to examine shear stresses and the potential for delamination in a dome with openings is valid.

The pattern and magnitude of average hoop and meridian stresses through the depth of the solid-element model of the dome with openings also compares well with the hoop and meridian stresses in the shell-element model model. The analysis of the solid-element dome with openings shows considerably more bending than the dome without openings (by a factor of 6). The analysis also shows significantly more shear stress in the dome with openings. These shear stresses are largest along the edges of the opening; however the magnitude of the shear stresses is very low and will not cause delamination (even locally).

REFERENCES

Bennett, J.L. 2005. An Analysis of Pre-stressed Concrete Containment Buildings with an Unanticipated Construction Opening: Steam Generator Replacement at Duke Power's Oconee Nuclear Station. *Princeton University Department of Civil and Environmental Engineering, Senior Thesis.*

Billington, D.P. 1982. Thin Shell Concrete Structures. New York: McGraw-Hill.

Chernoff, H. & Wade, K. 1996. Steam Generator replacement overview Part 4. *Power Engineering* 100 (1): 25–28

Mac Namara, S.C., Garlock, M.M, Billington, D.P. 2007. Structural Response of Nuclear Containment Shield Buildings with Construction Openings. *ASCE Journal of Performance of Constructed Facilities.* In press.

Mac Namara, S.C. 2007. Structural Response of Thin Shell Concrete in Nuclear Containment Shield Buildings to Unanticpated Construction Openings. *Princeton University Ph.D. Thesis, 2007.*

Patnaik, A. 2001. Behavior of Concrete Composite Beams with Smooth Interface *Journal of Structural Engineering* 127 (4): 359–366

Fracture Mechanics of Concrete and Concrete Structures – Design, Assessment and Retrofitting of RC Structures – Carpinteri, et al. (eds)
© *2007 Taylor & Francis Group, London, ISBN 978-0-415-44616-7*

Stiffness requirements for baseplates

R. Eligehausen & S.Fichtner

Institute of Construction Materials, Fastening Technology Department, Universitaet Stuttgart, Germany

ABSTRACT: In Europe, the resistance of fastenings with cast-in-place and post-installed anchor systems is calculated according to the CC-Method (Concrete Capacity Method). The actions on the fasteners are calculated according to the theory of elasticity assuming a rigid baseplate. A sufficient stiffness of the baseplate has to be ensured and therefore certain criteria must be fulfilled. Because of some investigation showing that the design method for the baseplate does not lead to the estimated ultimate load, further numerical and experimental research was done and is described in this paper. It results in a new design approach, which takes the previous method into account and extends it with further parameters to get a baseplate thickness which ensures a safe and economic fastening.

1 INTRODUCTION

The CC-method is very useful for calculating the ultimate load of fastenings in concrete because of its simplicity and good predictability. Considering the concrete parameters, embedment depth and distances to other fasteners and edges, it is possible to determine the ultimate loads.

The CC-Method is also used to calculate complex arrangements of fasteners, which are connected by baseplates to a single fastening point.

Usually one has different loading situations on such a fastening. While applied normal and shear forces are usually distributed equal to the fasteners, eccentrically positioned forces result in a bending moment. This moment takes loading from some anchors and puts it on others. But the quantity of this unequal loading not only depends on the relation of the normal/shear force to the bending moment. It is also influenced by the stiffnesses of all parts of the fastening – usually one would think about the baseplate thickness at first.

2 STATE-OF-THE-ART SYNTHESIS

In the past a few approaches were made to determine the distribution of forces under the baseplate to predict a reliable ultimate load. Following the method used for steel constructions (e.g. T-Stubs), one could

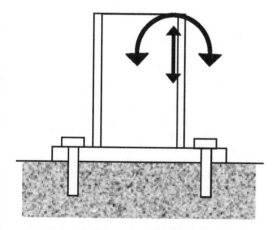

Figure 1. Example of a baseplate construction.

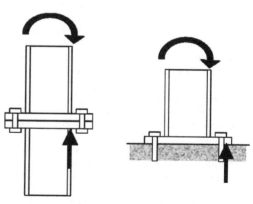

Figure 2. Approaches for the location of the compression force.

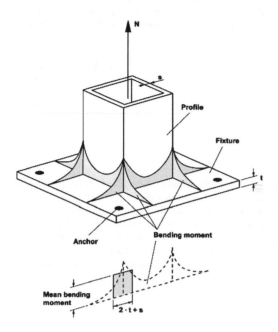

Figure 3. Stress distribution in the baseplate and calculation of mean values (Mallée, 1999).

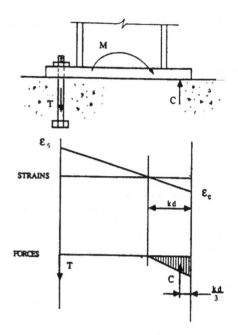

Figure 4. Static model of an elastic calculation of a baseplate.

assume the compressive force under the baseplate near or under the attachment. Several investigations stated this approach to be very conservative, because of the very different stiffnesses using steel bolts/screws on the steel construction side and anchors in the fastening technology.

The other, more realistic, approach is to get the distribution from the concrete methods. Usually the concrete constructions do not have large deformations and so the cross-section is supposed to remain plain. The so-called Bernoulli-Hypothesis, which is applied here, allows the use of the theory of elasticity.

Using a baseplate the concrete surface would stay plain and so it is only necessary to take care of the baseplate to do the same – till it reaches the ultimate load. The method developed by Mallée (1999) is about limiting the stress in the baseplate (stress limit approach). These stresses – mean values over an area of $2*t+s$ (Figure 3) – have only to be lower than the uniaxial yield limit of the baseplate steel. Mallée concludes that the deformation of the baseplate then is only elastic and small enough to ensure a plane surface, too.

If the baseplate is assumed to remain plane then one has only to consider the – mostly elastic - deformation that the baseplate brings to the concrete surface in the area of compression (Figure 4).

At least it is – compared to the first assumption – a little more complex, but still a linear elastic problem. Like many of the companies of the fastening

technology sector and at first the IWB (Institute of Construction Materials) have shown, it is possible to calculate this FE model in a few seconds on usual PC configurations.

3 PROBLEM AND INVESTIGATION

Of course this method is not quite suitable for all of the constructions one can think of. Mainly it never takes into account all the differences in the load-displacement curves that exist for fastening components.

This problem was first observed by the state office for structural engineering Baden-Württemberg, Schneider (1999) who carried out a few finite element studies. In this investigation 3 different assemblages were tested, which mainly differed in their loading direction and position of attachment while all baseplates had four anchors connected.

The results discovered problems especially with structures with eccentrically applied attachment.

Mallée (1999) carried out numerical and experimental tests to state his theory to be safe. Like Schneider he calculated the distribution of forces in the anchors while reaching the design load. His conclusion was quite different from Schneider's.

But for the quantification of safety it is not only necessary to look at the design load, because it mainly

Table 1. Investigated parameters and values.

No.		Count	Value
1	Stiffness of fastener	3	30–160 kN/mm
2	Embedment depth of fastener (h_{ef})	3	80–240 mm
3	Eccentricity of attachment	2	0-XXX mm
4	Eccentricity of loading	3	−5000, 0, 5000 mm
5	Type of loading	2	normal force w&w/o bending moment
6	Number of fasteners	3	4, 6, 9
7	Size of baseplate	3	1, 2, 3*h_{ef}

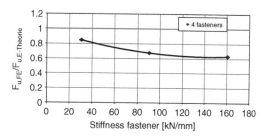

Figure 6. Influence of stiffness of fasteners.

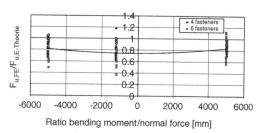

Figure 7. Influence of the type of applied loads.

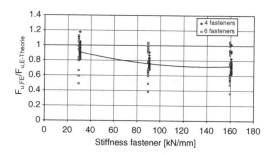

Figure 5. Influence of the fastener stiffness.

takes into account of the deformation. One has to observe the ultimate load of the construction, too.

The author of this paper did further numerical and experimental research to describe the most relevant parameters. The parameters are summed up in Table 1. In column 2 the varied number of each parameter is shown and in column 3 the applied values.

Because not all combinations make sense for describing the problems of baseplate constructions, not every possible model was created. In total over 200 simulations were carried out. The baseplate thickness was calculated using the stress limit approach by Mallée.

4 RESULTS

4.1 Stiffness of fasteners

The first parameter to explain is the stiffness of each fastening. Figure 5 shows the influence of the stiffness (in KN/mm) on the ultimate load of the baseplate.

All results given next are in comparison to the predicted values using the theory of elasticity.

If the stiffness of the fastener increases, the ultimate load decreases up to 50% to 70% of the predicted value

using the theory of elasticity. If the fastener stiffness is below 30 kN/mm, the decrease is less 20% or even less. The spread is very large.

Of course, if one takes out a few series of simulations with absolutely comparable conditions, the diagram looks like Figure 6.

The decrease in ultimate loading does not differ that much using the highest and lowest stiffness, so this parameter cannot be the only one, that affects the load-displacement curve of such a construction.

4.2 Influence of type of loading

Figure 7 shows the influence of the load type. The large values on the X-axis indicate a huge eccentricity of the tension or compression normal force and therefore a relatively high bending moment.

In the middle (that is near the Y-axis) the related normal force is much higher and produces less bending moment. The loading on the baseplate is mainly done by the high compression reaction forces under the baseplate.

The average-value curve in Figure 7 shows that a larger external lever arm yields decreasing ultimate loads of the construction.

4.3 Influence of the profile position

The next parameter is the eccentricity of the attached profile. The focus is on small attachments compared to the baseplate size. It's obvious to conclude in case of

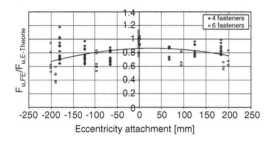

Figure 8. Influence of the position of the profile on the baseplate.

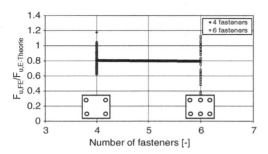

Figure 9. Influence of the number of fasteners.

large attachments that the baseplate would not deform that much because of its strong bracing.

In Figure 8 shows the effects that eccentricity of the attachment has on its ultimate capacity.

The drawing shows that with mid-positioned attachments the calculated values according to the theory of elasticity are not essentially decreasing. But if the profile is moved out of the centre of the baseplate the baseplate seems overstressed and the ultimate load falls.

4.4 Influence of the number of fasteners

The last view on results presented here is on the number of fasteners. In Figure 9 only the simulations with 4 and 6 fasteners are shown.

As one can see, the mean value nearly stays constant, but the spread is larger while taking 6 anchors. An explanation is the generally larger dimensions of a baseplate in case of 6 anchors and the additional 2 fasteners in the middle of the plate.

5 COMPARISON OF SIMULATIONS TO EXPERIMENTAL STUDIES

The evaluation of the numerical studies was done by simulating experimental tests with the finite element model. Therefore all parameters were considered and

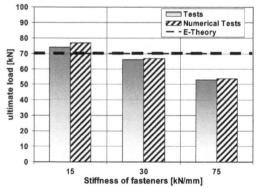

Figure 10. Parameter fastener stiffness.

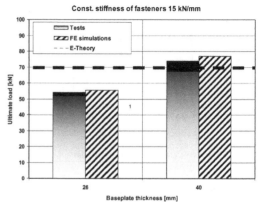

Figure 11. Parameter thickness of baseplate.

taken as far as known. While the conditions of constraints do not affect the whole test – the edge distances and concrete body dimensions are large enough – some parameters like concrete, anchor and baseplate type were known and modeled in detail.

Like in the tests, all simulations were loaded until a concrete cone failure occurred. The ultimate load was taken at the highest point of the load-displacement curve.

In Figure 10 the values for the ultimate load are compared. On the X-axis the stiffness of the fastener is written and the dashed horizontal line describes the ultimate load, which is calculated using the theory of elasticity.

In all three cases the simulations are very close to the test results. They all are in a range of 1–4% compared to the tests. In the following diagram the ultimate loads according to the plate thickness is shown. The difference between tests and simulation is as small as in the first figure.

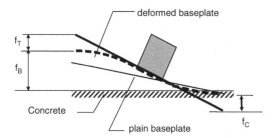

Figure 12.   Schematic drawing; relation of the deformations.

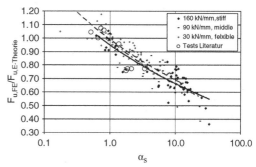

Figure 13.   Correlation between the stiffness parameter and the ultimate loads of a fastening point.

6   DESIGN APPROACH

While the results show, that some parameters have more influence on the load-displacement behavior of a construction, there are other – smaller – boundaries which are worth to be considered, too.

Since the new design approach takes into account the stiffnesses of the different substructures coming into play in the behavior of a baseplate, the proposed approach can be shortly called "stiffness criterion". The design should first be made with the stress limit approach as suggested by Mallée. With the results of the elastic calculation the following formula is entered:

$$\alpha_s = \frac{f_C}{f_B + f_T} \leq 1{,}0$$

$$f_B = \frac{N_u}{K} \qquad \text{deformation of anchor}$$

$$f_T = \frac{N_u * l_T^3 * 12}{E * b\, t_{fix}^3} \qquad \text{deformation in tension area}$$

$$f_C = \frac{C * l_C^3 * 12}{E * b\, t_{fix}^3} \qquad \text{deformation in compression area}$$

The use of this approach should be done – for difficult constructions (e.g. loading in two directions) – by implementing it into the already existing finite element programs. The advantage is that the values f_C and f_T can be calculated automatically.

The following diagram shows, that the Parameter α_S is a good indicator whether the assemblage will reach the ultimate which was calculated using the theory of elasticity.

7   CONCLUSIONS

Baseplates are very often used to connect steel constructions to concrete foundations or walls. Because of its heterogeneity those assemblies need to be designed carefully.

The approach by limiting the stresses in the baseplate below the yield strength of the steel material by choosing an appropriate thickness is in many cases a sufficient method which leads to fast, safe and economic solutions.

But not all baseplate constructions with their various parameters fit into this scheme. Sometimes the stiffnesses of all parts of the connection point are very different and have to be investigated further.

With the presented results and approach, all influences of the connection are considered and will lead – in addition to the stress limit approach – to a safe solution for a much larger application range.

REFERENCES

European Organisation for Technical Approvals (EOTA) 1997: *Guideline for metal anchors in concrete.* DIBt, Nr. 16, Appendix C: Design guide for fastenings

CEB Design Guide 1997: *Design of Fastenings in Concrete,* Comite Euro-International du Beton, Thomas Telford

Eligehausen, R.; Mallée, R. 2000: *Befestigungen im Beton- und Mauerwerkbau (Fastenings in concrete and masonry),* Bauingenieur-Praxis, Ernst & Sohn

Eligehausen, R.; Fichtner, S. 2003: *Steifigkeit von Ankerplatten (Stiffness of baseplates),* Universität Stuttgart, Institut für Werkstoffe im Bauwesen

Mallée, R.; Burkhardt, F. 1999: *Befestigungen von Ankerplatten mit Dübeln (Connection of baseplates with post-installed anchors),* Beton- und Stahlbetonbau 94, Heft 12, S. 502–511, Ernst & Sohn Verlag

Schneider, H. 1999: *Zum Einfluss der Ankerplattensteifigkeit auf die Ermittlung der Dübelkräfte bei Mehrfachbefestigungen (Influence of the baseplate stiffness on the load distribution of group fastenings),* Landesgewerbeamt Baden-Württemberg, Landesstelle für Bautechnik

Mallée, R. 2004: *Versuche mit Ankerplatten und unterschiedlichen Dübeln (Tests with baseplates and different fastening systems),* fischerwerke

Ožbolt, J. 1999: *"MASA- Macroscopic Space Analysis",* Stuttgart: Institut für Werkstoffe im Bauwesen, Internal report

Fracture Mechanics of Concrete and Concrete Structures – Design, Assessment and Retrofitting of RC Structures – Carpinteri, et al. (eds)
© 2007 Taylor & Francis Group, London, ISBN 978-0-415-44616-7

Behaviour of a support system for precast concrete panels

G. Metelli
University of Brescia, Italy

P. Riva
University of Bergamo, Italy

ABSTRACT: Even though a large amount of precast concrete panels are produced every year in Europe, there is a lack of studies on the seismic behaviour of connection systems. In order to understand the interaction between the concrete panels and the structure, a numerical and experimental study on a connection system subjected to seismic action has been faced. Non-linear finite element analyses of a concrete panel portion connected to a concrete column by a steel system have been conducted with FE program DIANA. The non linear behaviour of the materials and of the contact surface between the panel and the support system are considered. Experimental tests have been carried out on prototype specimens representing a column portion linked by the connection system to a precast panel portion. The specimens have been subjected to cyclic horizontal displacement histories by imposed transverse displacement. The tests are an effective tool to validate the numerical results and to define an accurate force-displacement constitutive law of the connection.

1 INTRODUCTION

Precast concrete panels are often used for the façades of modern warehouses and commercial malls.

In Italy, the precast concrete structures are generally designed neglecting the interaction between the structure and the cladding panels. These elements are hung to the columns or the beams and are considered to be mass, contributing only to the dynamic properties of the structural skeleton of the building, having no influence on the lateral stiffness. The connection systems of the cladding panels should be designed in order to provide the displacement demand during serviceability earthquake without any damage occurring in the panel and in the structure. At the same time, they should be designed also to transmit forces to the structure due to ultimate limit state earthquakes (Eurocode 8, 2003). Furthermore, this assumption, which neglects the structural behaviour of the external cladding, is presumed to be conservative in the design of seismic resistant concrete frames. Linear static and linear dynamic analyses of reinforced concrete frames have shown a reduction of the lateral drift and a change in the natural frequency and member force distribution, by considering external precast concrete wall claddings as opposed to the bare frame (Henry and Roll, 1986). These numerical results indicate that neglecting the structural role of the external concrete walls might not be conservative.

Although several works have been focused on the seismic behaviour of RC frames with masonry infills (Biondi et al., 2000; Mehrabi et al., 1996; Mehrabi and Shing, 1997) or moment resisting steel frames (De Matteis, 2005; Dogan et al., 2004; Pinelli et al., 1995), the dynamic and non linear behaviour of pre-cast structures under seismic actions, with pre-fabricated RC panels used as curtain walls, is still not well known due to the lack of studies on the interaction of the RC frame and the external pre-cast concrete panels. This aspect has been widely studied for moment resisting steel frames: a recent study (De Matteis, 2005) proves, by non-linear dynamic time-history analyses, that light sandwich panels bolted to edge members of the frame can improve the lateral stiffness of the structure, allowing a remarkable reduction in size of the MR steel bare frame. The hysteretic behaviour of the panels has been experimentally studied and has been taken into account in the numerical analyses. In (Pinelli et al., 1995) an advanced connection between architectural cladding panels and steel frames has been designed in order to dissipate energy in the engineered connection elements (Figure 1). The experimental behaviour of the connection is described in (Pinelli et al., 1996). The connection behaves like a passive dissipater, which provides lateral stiffness to the main steel structure because of the bracing effects of the cladding panels. This way, the cladding walls are not only mass hung to the structure, but, at the same time, by providing

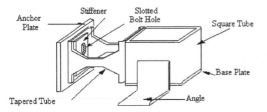

Figure 1. Ductile advanced cladding connection (Pinelli et al., 1995).

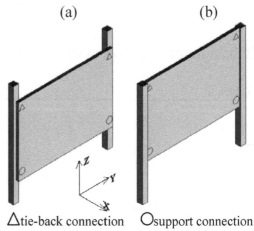

Δtie-back connection   Osupport connection

Figure 2. Typical arrangement of connections for precast concrete panels.

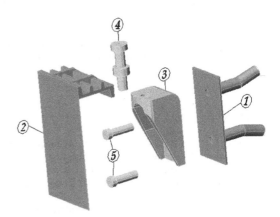

Figure 3. Components of the support system: exploded view.

a bracing action to the frame, they change the fundamental frequency of the building, in comparison with the bare resisting frame. The numerical results shown in (Pinelli et al., 1995) point out that the effectiveness of the cladding connection in the reduction of lateral displacements depends significantly on the ratio of the modified fundamental frequency of the structure and the critical frequency of the earthquake. The combination of the advanced cladding connections with a base isolator system should be more effective in high-rise steel frames. Some preliminary numerical studies show that this hybrid passive energy dissipation system provides a reduction of the base shear, earthquake input energy, and ductility demand in frame members (Dogan et al., 2004).

In precast concrete industrial buildings, the concrete cladding panels are often fixed to the structure by means of two support connections at the bottom, carrying the gravity loads, and two tie-back connections at the top of the panels, avoiding the out of plane movement of the panel (Figure 2). The four panel connections are designed to carry the same amount of the horizontal (parallel or normal to the panel) loads due to wind or earthquake. The cladding panels can be placed outside the structure, thus hiding it (case (a) in Figure 2), with the support system statically loaded in a plane perpendicular to the panel, or between the columns with the connection system, loaded in the plane of the façade, placed on the lateral edges of the panel (case (b) in Figure 2).

As previously mentioned, one of the problems related to precast RC panels, is that the behaviour of the supporting systems under earthquake actions is not well known, both with respect to their strength and ductility.

This paper aims at presenting finite element analyses (FEA) and the results of a wide experimental program of a particular steel support system connecting pre-cast concrete panels to concrete structures. This support system, called MT, has been designed and used since the '90s to carry only the concrete panel weight.

The numerical and experimental results presented in this paper provide a useful indications on the local behaviour of connection system loaded by a horizontal action and on diffusion phenomena in the concrete which cause the damage in the connected members. Furthermore the results can define an accurate force-displacement constitutive law of the connection, which should be an effective tool to study the seismic response of precast RC buildings.

2 THE CONNECTION SYSTEM

As shown in Figure 3, the MT support system is composed of three main components: the anchor steel plate with welded inserts embedded in the column (1); the anchor steel distributing plate (2), built in the precast panel; the steel bracket supporting the panel (3); a leveling bolt to adjust the vertical position of the panel during installation (4); two bolts transferring the dead load of the panel to the column (5). Figure 4 shows a

Figure 4. Typical application of the support system.

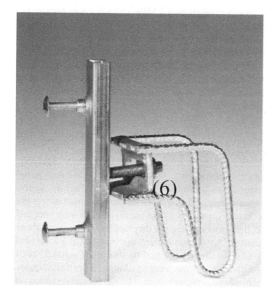

Figure 5. The tie-back connection.

typical application of the support connection system. Figure 5 shows the tie-back system, which is composed of an anchor channel built in the column (6) and a steel bracket embedded in the panel (7). Furthermore, the head of the levelling bolt and the distributing plate, built in the precast panel, are characterised by a saw-toothed surface in order to improve the friction between the bolt and the distributing plate.

3 NUMERICAL ANALYSIS

3.1 *FE model*

A three dimensional Finite Element (FE) model of the MT6 support system has been carried out by adopting

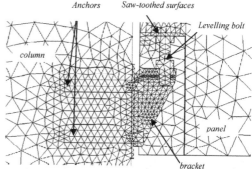

Figure 6. FE model of the assembled components of the support system.

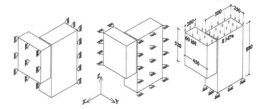

Figure 7. FE model: restraints and dimensions.

8-node tetrahedral elements. Over 12,000 elements have been used, with a dimension varying from 5 mm to 50 mm away from the diffusion zone (Figure 6). The restraints and the geometry of the model are shown in Figure 7: the panel restraints avoid out of plane movements, while the column is assumed to be fixed. The column is loaded by a vertical pressure equal to 8 MPa, in order to simulate the load of the roof of a typical one storey industrial precast building. A weight of 120 kN has been assigned to the panel, which is carried by two support systems. A horizontal displacement parallel (Y axis in Fig. 2) or normal (X axis in Fig. 2) to the building façade has been applied to the precast panel in order to evaluate the behaviour of the system loaded by seismic actions. All the details of the FE model can be find in the work of Metelli and Riva (2006).

The non linear analyses have been carried out with the FE program DIANA V.9.1. The non linear behaviour of the materials and of the contact surface between the panel and the support system have been considered. A rotating smeared crack approach has been used to model the concrete cracking zone around the anchors of the support system.

The friction between the levelling bolt head and the distributing plate has been modelled by linking the symmetrical nodes of the two elements with two springs parallel to the contact surface (x and y directions in Figure 8), thus allowing the support system

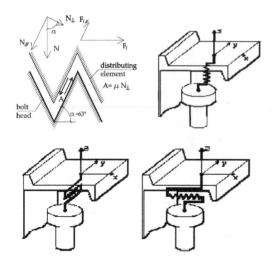

Figure 8. FE modeling of the friction between the distributing plate and the head of the leveling bolt.

Table 1. Mechanical properties of the materials.

	f_c [MPa]	E_c [GPa]	v	G_f [N/mm]	f_t [MPa]	ε_{c1} [%]	ε_{cu} [%]
Concrete	35	35	0.2	0.15	3.2	0.22	0.34

		E_s [GPa]	v	f_y [MPa]	f_u [MPa]	ε_{su} [%]
Steel						
Bolts	210	0.3	640	800	11
Brackets and plates	210	0.3	350	520	29
Anchors	210	0.3	500	500	12

E: modulus of elasticity; v: Poisson's ratio; f_c: compressive strength; f_t: tensile strength; ε_{c1}: strain at the peak compressive stress; G_f: fracture energy; ε_{cu}: ultimate strain; ε_{su}: ultimate plastic strain; f_y: yield strength; f_t: ultimate strength.

strength f_t equal to 3.2 MPa and a fracture energy G_f equal to 0.15 N/mm. The behaviour of the concrete far from the connection system has been assumed as linear elastic. For the steel components of the system, an elasto-plastic with linear hardening constitutive model and a Von Mises yield criterion has been assumed. The material properties are summarised in Table 1.

3.2 Numerical results

As previously mentioned, the analyses have been conducted considering the case of the cladding panel external to the structure (case (a) in Fig. 2), applying a horizontal displacement to the panel along the Y axis (δ_y, parallel to the panel) or X axis (δ_x, normal to the panel). For each displacement direction, three analyses have been carried out by assuming a vertical load N in the panel equal to 100%, 66% and 33% of the panel weight in order to simulate a vertical component of the seismic action. The numerical results with normal displacement δ_x are representative also of the case with the cladding panel placed between the two columns (case (b) in Figure 2), being the imposed displacement parallel to the bracket and to the anchors embedded in the column.

The main results of the numerical analyses are summarized in Table 2 and Table 3, where the failure load $F_{y,u}$ or $F_{x,u}$, the lateral displacement $\delta_{y,u}$ or $\delta_{x,u}$ at failure, and the tangent stiffness k_y or k_x are shown for six loading configurations. The load-displacement curve is illustrated in Figure 9 for a vertical load equal to 60 kN and a longitudinal displacement δ_y. In order to evaluate the contribution of each component to the lateral deformation of the support system, the lateral absolute displacement of four points has been plotted: the displacement imposed to the panel (δ_1), the bolt head (δ_2) and the bracket (δ_3 and δ_4). The results of all other studied cases are discussed and reported in details by the work of Metelli and Riva (2006). Based

to carry the horizontal loads. A no tension gap element allowed the vertical load to be transferred from the panel to the column. As shown in Figure 8, the stiffness k of the longitudinal spring depends on the geometrical characteristic of the toothed surfaces and has been calculated considering the force F_f causing the panel to slide along the toothed surface. Considering a slip metal to metal friction coefficient μ equal to 0.3 and the equilibrium along the tooth face, it is possible to define an equivalent interlocking coefficient ξ between the distributing plate and the levelling bolt (Fig. 8):

$$F_f \cos \alpha = \mu N \cos \alpha + N \sin \alpha \quad (1)$$

$$F_f = (\mu + \tan \alpha) N = \xi N \quad (2)$$

This way, by assuming a tooth angle α equal to 63°, the horizontal force F_f causing the panel slip is 2.3 times the vertical load N carried by the support system. Hence, the stiffness k of the longitudinal springs is given by equation (3):

$$k = \frac{N \xi}{n} \frac{1}{\delta_{0.1}} \quad (3)$$

where $n =$ number of the linked nodes on the head of the bolt; $\delta_{0.1} =$ conventional slip of 0.1 mm corresponding to a longitudinal force equal to $2.3N$ (slip onset).

A C35/45 concrete has been assumed with the compressive stress–strain relationship provided by EC2 (Eurocode 2, 2004). In tension, a linear cohesive crack model has been assumed with a tensile

Table 2. Numerical results with an applied displacement in the plane of the panel (Y axis).

Case MT6	N [kN]	$F_{y,u}$ [kN]	$\delta_{y,u}$ [mm]	k_y [kN/mm]	Panel $\delta_1-\delta_2$ [mm]	Bolt $\delta_2-\delta_3$ [mm]	Bracket $\delta_3-\delta_4$ [mm]	Column δ_4 [mm]	Failure
ay1	60	50.7	26.0	27.5	0.10	5.20	12.94	7.75	Bolt
					0.4%	20%	49.8%	29.8%	yield
ay2	40	50.7	26.0	27.5	0.10	5.20	12.94	7.75	Bolt
					0.4%	20%	49.8%	29.8%	yield
ay3	20	46.0	7.00	25.5	0.81	1.19	3.24	1.75	Panel
					11.6%	17.0%	46.4%	25.0%	slip

Table 3. Numerical results with an applied displacement in the plane of the panel (Y axis).

Case MT6	N [kN]	$F_{x,u}$ [kN]	$\delta_{x,u}$ [mm]	k_x [kN/mm]	Panel $\delta_1-\delta_2$ [mm]	Bolt $\delta_2-\delta_3$ [mm]	Bracket $\delta_3-\delta_4$ [mm]	Column δ_4 [mm]	Failure
ax1	60	95.1	3.80	49.6	0.21	3.05	0.44	0.10	Bolt
					5.5%	80.3%	11.6%	2.6%	yield
ax2	40	92.0	2.90	51.2	0.24	2.20	0.37	0.09	Panel
					8.3%	75.9%	12.7%	3.1%	slip
ax3	20	46.0	0.90	51.0	0.13	0.57	0.15	0.04	Panel
					14.6%	63.8%	16.9%	4.7%	slip

on the results, the following main observations may be made:

– the value of the vertical load N governs the failure and the resistance of the support system due to the friction provided by the interlocking of the bolt head and the distributing plate: for longitudinal displacement with a vertical load reduced to the 33% of the panel weight (Table 2 – case ay3) or for normal displacement with a vertical load reduced to the 66% of the panel weight (Table 2 – case ax2 and ax3) the failure of the system is due to the panel slip. In all other cases, the failure is due to the bolt yield;

– the support system loaded in the direction normal to the panel shows a better performance than the one loaded in the longitudinal direction, both in term of resistance and stiffness. The resistance of the system varies from 46.0 kN for case ay3 to 95.1 kN for case ax1, while the stiffness varies from 27.5 kN/mm to 51.2 kN/mm;

– by representing the experimental results of each analysis by means of a bilinear curve, it is possible to define a ductility factor q as the ratio between the ultimate lateral displacement $\delta_{x,u}$ or $\delta_{y,u}$ and the yield displacement $\delta_{x,y}$ or $\delta_{y,y}$,

$$q = \frac{\delta_{x,u}}{\delta_{x,y}} \quad \text{or} \quad q = \frac{\delta_{y,u}}{\delta_{y,y}} \quad (4)$$

The ductility factor q varies from 1.45 in case ax2 to 13.5 in case ay1. The support system loaded in the

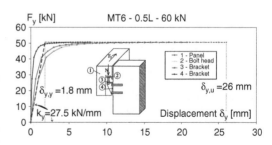

Figure 9. Longitudinal Load F_y – Lateral Displacement δ_y with a dead load equal to 60 kN.

longitudinal direction shows a good ductility due to the low torsional resistance of the bracket. The main lateral deformation δ of the system in the case of the support system loaded in the longitudinal direction is caused by the bracket torsion, which accounts for approximately 50% of the total imposed displacement δ_y, while in the case of the support system loaded along the normal direction it is mostly due to the bolt deformation, accounting for 63.8% to 80.3% of the total imposed displacement δ_x.

– The damage of the concrete is limited to the zone around the anchors as shown by the crack pattern of the column at failure: the average crack width around the anchors is approximately equal to 0.30 mm.

Table 4. Summary of the experimental results.

	N [kN]	k_y [kN/mm]	$F_{y,u}$ [kN]	$\delta_{y,u}$ [mm]	Panel sliding $\delta_1-\delta_2$ [mm]	Bolt $\delta_2-\delta_3$ [mm]	Bracket $\delta_3-\delta_4$ [mm]	E [kJ]	Failure
MT6_1.0L	60	10	18.8	34	12.5 (37%)	18.0 (53%)	2.9 (8%)	1.43	Bolt yield
MT6_0.0L	60	23	51.0	26	17.5 (68%)	4.4 (17%)	4.8 (18%)	2.79	Panel slip*)
MT6_0.5L	60	11	41.0	26	11.6 (45%)	5.8 (22%)	8.6 (33%)	2.53	Bolt eld
MT6_0.5L	40	15	43.3	29	16.2 (56%)	5.7 (20%)	7.1 (24%)	2.21	Bolt yield
MT6_0.5L	20	15	37.5	29	21.4 (74%)	2.2 (8%)	5.4 (18%)	1.83 1.83	Panel slip*)
MT9_0.5L	90	30	43.8	35	11.2 (32%)	15.5 (44%)	8.0 (23%)	2.71	Bolt yield
MT9_0.5L	60	25	46.8	35	8.7 (25%)	14.1 (40%)	15.1 (43%)	2.19	Bolt yield
MT9_0.5L	30	20	31.4	30	13.7 (45%)	8.0 (26%)	14.1 (47%)	1.67	Panel slip*)
MT4_0.5L	40	12	30.5	12	3.1 (26%)	4.3 (36%)	4.7 (40%)	0.71	Panel slip*)
MT4_0.5L	27	13	20.3	12	6.2 (50%)	3.8 (30%)	1.6 (13%)	0.68	Panel slip*)
MT4_0.5L	13	11	19.3	16	9.4 (59%)	3.0 (19%)	2.9 (18%)	0.58	Panel slip*)

4 EXPERIMENTS

4.1 Test set-up

The aims of the experimental program are to evaluate the stiffness, the ductility and the energy dissipation characteristic of each type of MT support system as well as to investigate the role of the vertical load and of the extension length of the levelling bolt on the lateral behaviour of the support system. The experimental tests concern 11 prototype specimens representing a column portion linked by the support system to a precast panel portion. Three batches of tests have been carried out at the P.Pisa Laboratory of the University of Brescia in order to investigate the behaviour of different support system, called MT4, MT6 and MT9, where the number refers to the nominal vertical load transferred by the support system to the column (half weight of the concrete panel) (see Table 4). The MT4 and the MT9 series consist of 3 specimens, each one with a different vertical load, while the MT6 series consists of five specimens with a varying vertical load and different extension length of the levelling bolt. All the specimens tested had a 250×250 mm column cross section and a column length equal to 1250 mm. The panel portions were 200 mm depth, 300 mm wide and $480 \div 580$ mm high. The geometry of the tested specimens and the mechanical characteristic of concrete

and reinforcing steel are shown in Figure 10, while Figure 11 and Figure 12 show the experimental setup adopted. The bench consists of a 1800×2150 mm steel ring frame (a) designed in order to allow the panel to move in its plane and to avoid any movement of the column portion. For all the tests, the vertical load N, varying from 13 kN to 90 kN, was applied and kept constant during the test by means of two tendon rods (b). The horizontal displacement was applied at the top of the panel by means of threaded steel bars (c) (16 mm in diameter), which are instrumented with strain gauges to measure the applied load F. As in the numerical analysis, the column is loaded by a vertical pressure equal to 8 MPa, by means of four 24 mm diameter steel bars (d).

The specimens were subjected to eight cyclic horizontal displacement histories by imposed transverse displacement δ of increasing amplitude (Figure 13) with a step increment of about 0.1 mm/s. In order to evaluate the contribution of each component to the lateral deformation of the support system, the lateral absolute displacement of four points were measured: the displacement imposed to the panel (δ_1), the bolt head and bottom displacements (δ_2, δ_3), and the bracket displacement (δ_4) (Figure 13). In order to verify that the in plane panel rotation was correctly controlled by the two vertical tendons, the horizontal

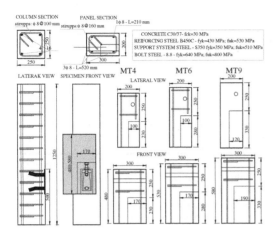

Figure 10. Dimensions of the specimens and mechanical characteristic of the materials.

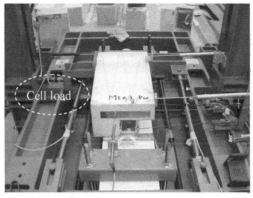

Figure 12. Picture of the bench with MT6 support system.

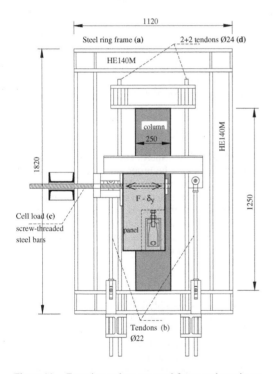

Figure 11. Experimental set-up: steel frame and specimen.

displacement δ_5 of the panel bottom (b) was also measured.

4.2 Experimental results

The experimental horizontal load F – displacement δ_1 curves of five tested specimens are shown in Figure 16

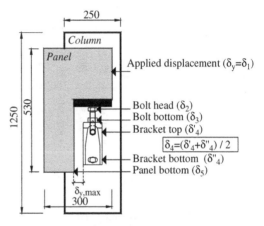

Figure 13. Measured points.

while Table 4 show in details the main results of each test. For the MT6 support system with 60 kN vertical load and leveling bolt at the middle extension (0.5 L), the comparison between the numerical and experimental results is also shown. On the basis of the results, the following main observations can be drawn:

– the shear resistance $F_{y,u}$ of the support system and failure are strongly affected by the leveling bolt extension length. For the MT6 support system the failure load decreases from 51.0 kN in the case of completely screwed leveling bolt (MT6_0.0L_60kN) to 18.8 kN (MT6_1.0L_60kN) in case of completely unscrewed leveling bolt. At the same time, longer extension length of the leveling bolt allows the system to develop larger lateral deformation $\delta_{y,u}$ due to the bolt plastic deformation (see pictures in Figure 16);

– in case of short leveling bolt extension or small vertical load, the failure is due to the panel slip with a severe damage of the saw-toothed surfaces of bolt

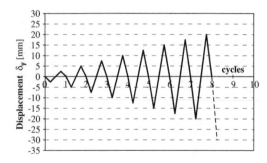

Figure 14. Loading history.

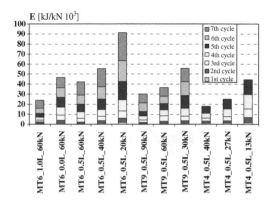

Figure 15. Dissipated energy.

and plate. In the case of completely screwed bolt (MT6_0.0L_60kN) the lateral deformations of the bolt and the panel sliding correspond respectively to the 17% and 68% of the total lateral deformation. On the contrary the completely unscrewed bolt allows a lower panel slip (37%) and a larger bolt deformation (53%);

– the vertical load N affects the behaviour of the support system both in terms of components deformation and ultimate load. A decrease of 67% of the vertical load N causes a reduction of shear resistance F_u equal to 9% for MT6 support system, 28% for MT9 and 36% for MT4. As lower the vertical load is, as lower the friction between the distributing plate and the head of the leveling bolt becomes, causing the damage of the saw-toothed surfaces and a failure by panel slip. The teeth damage and the slip panel are pointed out by the horizontal path in the F-δ curves;

– the support systems showed good energy dissipation properties with very high lateral displacements and without loosing the capability to sustain the vertical load. The specimen Mt6_0.0L_60kN shows a larger energy dissipation capacity than the other

specimens. Furthermore most of the energy is dissipated by friction, so that it increases with the vertical load. Figure 15 shows the comparison of the specific dissipated energy, normalized with respect to the vertical load N, among all of tested specimens;

– in Figure 14 the comparison between numerical monotonic and hysteretic experimental curves is presented for the MT6_0.5L_60kN specimen. It is worth pointing out that the numerical analyses catch the same failure mode than the tests with a shear ultimate load equal 1.20 time the experimental value. Nevertheless, the numerical initial stiffness k of the system is double than the experimental one because a conventional infinitely high stiffness (given by eq. (3)) of the longitudinal springs which connect the bolt head and the distributing plate of the panel is assumed in the FE model;

– the tests pointed out that the damage in the supporting system is localized either in the saw-toothed surfaces or in the plastic deformation of the leveling bolt. Very little damage pattern was observed in concrete elements: in the MT6_0.0L_60kN, MT6_0.5 L_60 kN, MT9_0.5 L_90 kN specimens a crack may be observed, departing from the right top of the bracket toward the side column with an angle approximately equal to 50°.

5 DESIGN CONSIDERATIONS

On the basis of the experimental results, some design considerations may be proposed concerning the MT support system behaviour. According to Eurocode 8 (§4.3.5.2), the non structural elements, as well as their connections and attachments or anchorages, shall be verified for the seismic design situation by applying a horizontal force F_a, acting at the centre of mass of the non-structural element, which is defined as follows:

$$F_a = \left(S_a \cdot W_a \cdot \gamma_a\right)/q_a \qquad (5)$$

where W_a is the weight of the element, S_a is the seismic coefficient applicable to non-structural elements, (defined by the eq. (6)), γ_a is the importance factor of the element assumed equal to 1, q_a is the behaviour factor of the element assumed equal to 2. The seismic coefficient S_a may be calculated using the following expression:

$$S_a = \alpha \cdot S \cdot \left|3\left(1+z/H\right)/\left(1+\left(1-T_a/T_1\right)^2\right)-0.5\right| \qquad (6)$$

where α is the ratio of the design ground acceleration (PGA) on type A ground, a_g, to the acceleration of gravity g; S is the soil factor; T_a is the fundamental vibration period of the non-structural element; T_1 is the fundamental vibration period of the building in the

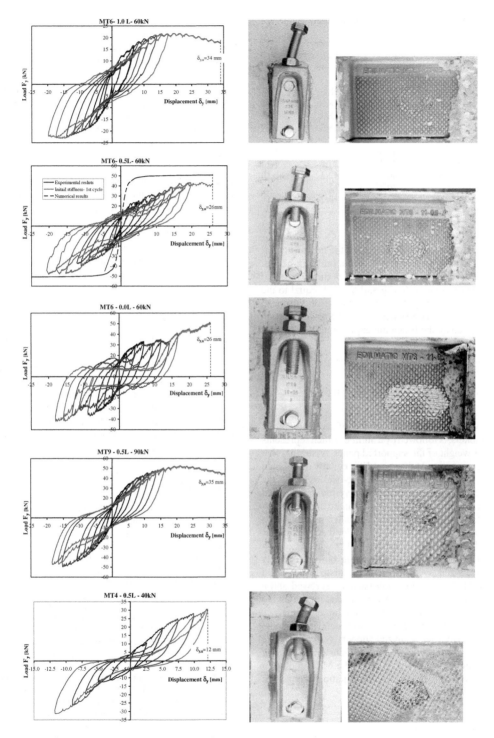

Figure 16. Force F_y –displacements δ_y curves and pictures at failure.

Table 5. Verification of the MT support system at ultimate limit state.

Support system	F_u	F_{as}	ψ
MT6-0.5L-60 kN	41.0	18.3	2.24
MT6-0.5L-40 kN	43.3	18.3	2.36
MT6-0.5L-20 kN	37.5	18.3	2.05
MT6-1.0L-60 kN	18.8	18.3	**1.03**
MT6-0.0L-60 kN	51.0	18.3	2.79
MT9-0.5L-90 kN	43.8	27.5	1.59
MT9-0.5L-60 kN	46.8	27.5	1.70
MT9-0.5L-30 kN	31.4	27.5	**1.14**
MT4-0.5L-40 kN	30.5	12.2	2.50
MT4-0.5L-27 kN	20.3	12.2	1.66
MT9-0.5L-13 kN	19.3	12.2	1.58

relevant direction; z is the height of the non-structural element above ground; and H is the building height. Assuming the ratio T_a/T_1 equal to 0 and z/H equal to 1, a PGA equal to 0.35 g and $S = 1.4$, equal to the most severe PGA and soil conditions for European sites, the seismic coefficient S_a and the seismic action F_a are given by the following expressions:

$$S_a = 0.35 \cdot 1.4 \cdot \left[3(1+1)/\left(1 + (1-0)^2\right) - 0.5 \right] = 1.225 \quad (7)$$

$$F_a = (S_a \cdot W_a \cdot \gamma_a)/q_a = \frac{1.23 \cdot W_a \cdot 1}{2} \cong 0.61 W_a \quad (8)$$

As previously mentioned, the label of each type of the MT brackets, is followed by a number which represents the half weight of the supported panel. Assuming that the seismic action of the panel F_a is equally distributed among the two brackets and two tie-back systems, the seismic action on each connection system F_{as} is equal to $F_a/4$. So that it is possible to calculate the safety factor ψ for each tested bracket, as shown in Table 5. In all cases, the seismic action is smaller than the bracket strength, and the safety factor $\psi > 1.5$, with the exception of the MT6-1.0L-60 (levelling bolt completely unscrewed and a vertical load equal to 60 kN) and MT9-0.5L-30 (levelling bolt partially screwed and a vertical load equal to 30 kN) specimens for which the safety factor is smaller than 1.5, but larger than 1.

Regarding the serviceability limit state, it is not possible to verify each analyzed support system because it depends on the concrete panel dimension. Assuming a panel height h equal to 2.50 m, the MT6 and MT9 support system can develop deformations consistent with an interstory drift of $0.01h$.

6 CONCLUSIONS

The behaviour of a typical support system for precast panels has been analysed both numerically and

experimentally to asses its capacity to resist seismic (i.e. transverse) forces.

Although the support system was not originally designed to carry lateral loads, the results show a good performance of the support bracket, both in terms of resistance, energy dissipation and ductility with a very limited damage in the concrete elements.

The transverse force transmission of the supporting system is ensured by the enhanced friction due to the saw toothed surface of the levelling bolt head and of the corresponding plate.

Further numerical analyses involving the FE modelling of the saw-toothed surfaces should be carried out in order to fine-tune the constitutive model of the levelling bolt-to-plate interface.

Finally, based on the FE analysis and experimental results, a simplified constitutive model of the support system has to be developed in order to allow the study of the seismic response of industrial buildings taking into account the interaction between the concrete structure and the cladding panels.

ACKNOWLEDGMENTS

The authors gratefully acknowledge the support of Edilmatic srl (Pegognaga (MN), Italy) for financing this research project on concrete panel support systems. The cooperation of Gino Antonelli and Luca Scartozzoni in carrying out the numerical analysis and Andrea Facchinetti in carrying out the tests are gratefully acknowledged.

REFERENCES

Biondi, S., Colangelo, F., Nuti, C. 2000 La Risposta Sismica dei Telai con Tamponature Murarie, *CNR- Gruppo Nazionale pe la Difesa dei Territori*.

De Matteis, G. 2005. Effect of Lightweight Cladding Panels on the Seismic Performance Of Moment Revisiting Steel Frames, *Engineering Structures*, 27, n° 11, 1662–1676.

Cohen, J. M., Powell, G. H. 1993. A Design Study of an Energy-Dissipating Cladding System, *Earthquake Engineering and Structural Dynamics*, 22, 617–632.

Diana v. 8.1.2. 2003. User's manual. TNO DIANA BV, Delft.

Dogan, T., Goodno, B. J., Craig, J. I. 2004. Hybrid Passive Control In Steel Moment Frame Buildings, *Proceedings of the 13th World Conference on Earthquake Engineering, Paper 2387*.

Eurocode 2: Design of concrete structures – Part 1-1: General Rules, and Rules for Buildings, EN 1992-1-1:2004, *European Committee for Standardization*, December 2004.

Eurocode 8: Design of structures for earthquake resistance – Part 1: General Rules, Seismic Actions and Rules for Buildings, PrEn 1998-1, *European Committee for Standardization*, December 2003.

Henry, R. M., Roll, F. 1986. Cladding-Frame Interaction, *Journal of Structural Engineering*, 112, n° 4, 815–834.

Goodno, B. J., Craig, J. I., Dogan, T., and Towashiraporn, P. 1998. Ductile Cladding Connection Systems for Seismic Design, *Building and Fire Research Laboratory, NIST, Report GCR 98-758*.

Pinelli, J. P., Craig, J. I., Goodno, B. J. 1995. Energy-Based Seismic Design of Ductile Cladding Systems, *Journal of Structural Engineering*, 121, n° 3, 567–578.

Pinelli, J. P., Moor, C., Craig, J. I., Goodno, B. J. 1996. Testing of Energy Dissipating Cladding Connections, *Earthquake Engineering and Structural Dynamics*, 25, 129–147.

Mehrabi, A. B., Shing, B. P., Schuller, P., Noland, J. 1996. Experimental Evaluation of Masonry-Infilled RC Frames, *Journal of Structural Engineering*, 122, n° 3, 228–237.

Mehrabi, A. B., Shing. 1997. Finite Element Modelling of Masonry-Infilled RC Frames, *Journal of Structural Engineering*, 123, n° 5, 604–613.

Metelli, G. and Riva, P. 2006. Numerical analyses of a support system for pre-cast concrete panels. *Proceedings of the First European Conference on Earthquake Engineering and Seismology*, Geneve, Switzerland, 3–8 September 2006, paper n° 612.)

Paulay, T. and Priestley, M.J.N. 1992. Seismic Design of Reinforced Concrete and Masonry Buildings, *John Wiley & Sons*, New York.

Fracture Mechanics of Concrete and Concrete Structures – Design, Assessment and Retrofitting of RC Structures – Carpinteri, et al. (eds)
© 2007 Taylor & Francis Group, London, ISBN 978-0-415-44616-7

Preliminary computations for a Representative Structural Volume of nuclear containment buildings

L. Jason
CEA Saclay, DEN/DM2S/SEMT/LM2S, Gif sur Yvette, France

S. Ghavamian
EDF R&D, Clamart, France

A. Courtois
EDF SEPTEN, Villeurbanne, France

ABSTRACT: The behaviour of a Representative Structural Volume for containment buildings of 1300 and 1450 MWe nuclear power plants is studied. First computations are considered to investigate the fracture mode of the structure and especially the role of the prestressing tendons. Cables initiate the development of a localized mechanical damage and help its propagation. The preliminary computations also underline the need for an experimental validation to further investigate the issue of the mechanical degradation. That is why an experimental device is proposed and described. As the new structure is supposed to represent the containment building of a 1450 MWe nuclear power plant, new computations are carried out with the finite element code Cast3M. They confirm the key role of the prestressed tendons in the development of damage.

1 INTRODUCTION

In high-power French nuclear power plants (1300 and 1450 MWe), the concrete containment vessel represents the third passive barrier after the fuel cladding and the containment vessel of the reactor core (Fig. 1). It is responsible for the safety of the environment as it is supposed to prevent leakage in case of accidents. That is why it is carefully monitored. Integrity tests, consisting in an internal pressure inside the structure, are carried out every ten years to check the leakage rate (if any). As the gas transfer through concrete is directly influenced by the mechanical degradation (Picandet et al., 2001) and as experiments can be hardly carried out because of the difficult environmental conditions, numerical studies remain the most convenient way to understand the degradation process. That is why, for the last decade, Electricité de France (EDF) has launched several important civil engineering research and development programs, dedicated to the analysis of prestressed pressure containment vessel (PCCV) reactor building. These concern the elaboration of new constitutive laws for concrete, techniques of modelling and resolution algorithms. The validity of the models (damage (Mazars, 1984) and/or plasticity (Jason et al., 2006), fracture (Ngo & Scordelis, 1967)...) and more generally the methodology for non-linear calculation

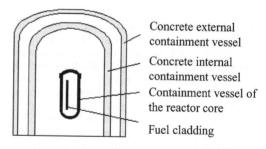

Figure 1. Barriers of a 1300 or 1450 MWe nuclear power plant.

must be obtained by comparing their performances with experimental results (benchmarking (Ghavamian & Delaplace, 2003) for example).

The validation is generally obtained from simple tests on small size specimen, where elementary features of models are qualified. But more complex tests are also essential to determine the capacity of the calculations to predict the structural behaviour of more realistic and industrially representative cases.

In the field of nuclear containment buildings, a Representative Structural Volume (RSV) was designed. As the dimensions of containment vessels were not suitable for detailed studies, only one part was modelled.

The RSV was designed to reproduce, as accurately as possible with acceptable computational costs, the behaviour of the typical part of the internal vessel. It contained concrete, passive steel bars and pretensioned cables. Preliminaries computations on this volume are presented in the section 2. At first the structure was considered as a mere numerical tool, to investigate the ability of constitutive laws to be applied on an industrial test. But unknowns about the role of the pretensioned tendons and mesh-dependency problems finally proved clearly that an experimental validation was necessary. That is why an experimental device was designed to validate, first qualitatively, then quantitatively, the tendencies that were observed through the parametric studies. Dimensions of the new structure and experimental processes are depicted in section 3.

As the new structure is supposed to represent the containment building of a 1450 MWe nuclear power plant, its dimensions and components are different from those of the preliminary RSV. That is why new computations have been carried out with the finite element code Cast3M at CEA (Cast3M, 2006) to evaluate the expected mechanical behaviour. The mode of fracture is especially highlighted in section 4 and a parametric study is also proposed to emphasize the mesh-dependency problem.

2 PRELIMINARY REPRESENTATIVE STRUCTURAL VOLUME

The application presented in this part has been proposed by Electricité de France in 2002. The test, named PACE 1300, is a Representative Structural Volume (RSV) of a PCCV of a French 1300 MWe nuclear power plant. Figure 2 illustrates the location of the RSV within the entire PCCV. The model incorporates almost all components of the real structure: concrete, vertical and horizontal reinforcement bars, transversal reinforcements, and pretensioned tendons in both horizontal and vertical directions. The size of the RSV is chosen to respect three conditions: large enough to include a sufficient number of components (and especially prestressed tendons) and to offer a significant observation area in the centre, far enough from boundary conditions, while remaining as small as possible to ease computations. The RSV includes 11 horizontal and 10 vertical reinforcement bars (on both internal and external faces), 5 horizontal and 3 vertical prestressed tendons, and 24 reinforcement hoops uniformly distributed in the volume. The geometry of the problem is given in figure 3. Figure 4 provides information about the steel distribution and properties.

The behaviour of the RSV needs to be as close as possible to the in situ situation. The following boundary conditions have thus been chosen: face SB blocked along the vertical direction, on face SH all nodes are

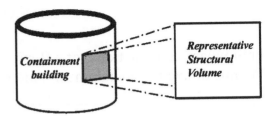

Figure 2. Position of the extracted Representative Structural Volume (RSV).

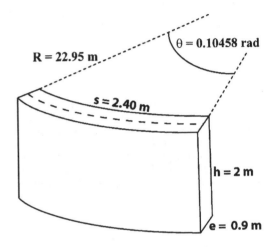

$R = 22.95$ m

$\theta = 0.10458$ rad

$s = 2.40$ m

$h = 2$ m

$e = 0.9$ m

Figure 3. Geometry of the Representative Structural Volume (RSV).

restrained to follow the same displacement along Oz and no rotations are allowed for faces SG and SD (see figures 5 and 6).

In order to model the effect of the pretensioned tendons, bar elements are anchored to faces SG and SD for horizontal cables and to faces SB and SH for vertical tendons, then prestressed using internal forces.

Then, these elements are restrained to surrounding concrete elements to represent the prestressing technology applied in French PPCVs.

The integrity test loading is represented by a radial pressure on the internal face SI and the bottom effect applied on face SH (tensile pressure proportional to the internal pressure to simulate the effect of the neighbouring structure). The self weight of the RSV and that of the surrounding upper-structure are also taken into account. With these conditions, a mesh containing 16,500 Hexa20 elements for concrete and 1200 bar element for reinforcement and tendons is used in this contribution.

Figure 7 provides the internal pressure applied on the volume as a function of the radial displacement of a point located at the bottom right of the internal face, using the elastic damage law developed by

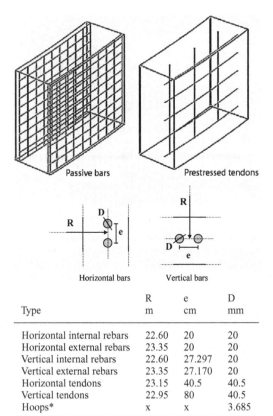

Passive bars | Prestressed tendons

Horizontal bars | Vertical bars

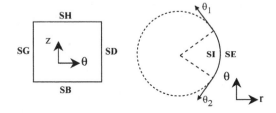

Type	R m	e cm	D mm
Horizontal internal rebars	22.60	20	20
Horizontal external rebars	23.35	20	20
Vertical internal rebars	22.60	27.297	20
Vertical external rebars	23.35	27.170	20
Horizontal tendons	23.15	40.5	40.5
Vertical tendons	22.95	80	40.5
Hoops*	x	x	3.685

* Hoops are uniformly distributed in the RSV

Figure 4. Steel geometries.

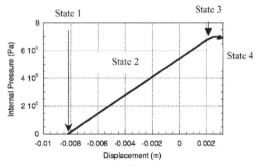

Figure 5. Definition of the Finite Element model indicating the boundary SG, SD, SH and SB.

Mazars (1984). This curve can be divided in four parts. The initial state corresponds to the application of the prestress on the tendons.

This yields a compaction of the volume and due to the boundary conditions (no normal displacement on the lateral face), it imposes an initial negative radial displacement (see figure 8). Then, upon application of the internal pressure, there is a zone of linear behaviour where the compaction is reduced and the structure returns towards its initial rest position before undergoing tension for higher values of internal pressure

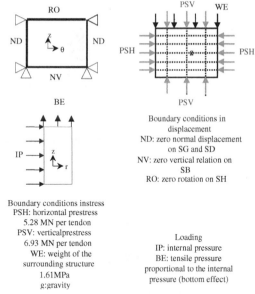

Boundary conditions in displacement
ND: zero normal displacement on SG and SD
NV: zero vertical relation on SB
RO: zero rotation on SH

Boundary conditions instress
PSH: horizontal prestress 5.28 MN per tendon
PSV: verticalprestress 6.93 MN per tendon
WE: weight of the surrounding structure 1.61MPa
g:gravity

Loading
IP: internal pressure
BE: tensile pressure proportional to the internal pressure (bottom effect)

Figure 6. Boundary conditions and loading for the RSV.

Figure 7. Displacement–pressure curve for the RSV.

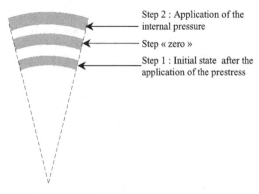

Step 2 : Application of the internal pressure

Step « zero »

Step 1 : Initial state after the application of the prestress

Figure 8. Radial deflection of the RSV through different steps (schematic). View from the top of the volume.

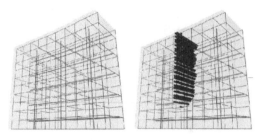

Figure 9. Development of a damage band in the RSV. Initial state (on the left) and damage band (on the right).

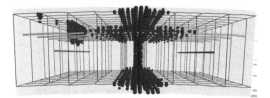

Figure 10. Development of the damage band. View from the top.

(almost 0.7 MPa). Damage does not evolve during state 2 but increases during state 3. Finally, a partial unloading of the volume appears (state 4) due to heavy cracking of the structure.

Figure 9 illustrates the evolution of the damage distribution during the simulation. Black points correspond to zones where the damage reaches a value up to 0.7. A first localization band appears in the middle of the volume along the vertical prestress tendon. Figure 10 presents the damage evolution with a view from the top of the RSV. The internal variable initiates from the middle of the structure along the vertical tendon then propagates along the horizontal pretensioned cables.

Moreover, a "localization" of a small damage can also be observed along the horizontal pretensioned tendons, just after the prestress and before the application of the internal pressure (figure 11) (Jason, 2004). It is probably due to the boundary conditions, the steel-concrete interface (supposed perfect here) and/or the choice in the modelling (one dimensional bar elements in a three dimensional concrete volume).

These preliminaries computations show that the pretensioned tendons seem to play a key role in the development of the mechanical degradation. They are responsible for the initiation of the damage and help its propagation. But as these observations are up to now only based on numerical observations, further studies are needed. They clearly prove that an experimental validation is necessary to validate, first qualitatively, then quantitatively, the tendencies that have been observed through the computations.

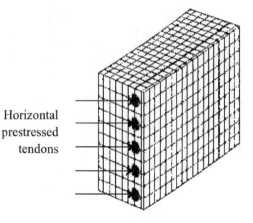

Horizontal prestressed tendons

Figure 11. Damage distribution just after the prestress and before the application of the internal pressure.

3 EXPERIMENTAL PROGRAM

To validate the above-mentioned structural behaviour for prestressed reinforced concrete elements, an experimental program has been set up in association between EDF and the University of Karlsruhe. The test named PACE 1450 is quite similar to the PACE 1300 RSV described previously; the difference concerns the dimensions, reinforcement ratios and prestressing level. Since 1450 MW are obviously more than 1300 MW, thicker walls with more prestressing are required.

The precise purpose of the test is to validate computational models and techniques. To do so, within the specimen we shall create mechanical conditions as similar as possible to those of the real structure, and simulate a scenario where the containment vessel undergoes an uprising internal pressure. The aim is to follow the evolution of stress developed within the structure, and localise crack initiations and progression, which eventually could lead to microcracking of concrete and the decay in leakage tightness of the vessel. That is why the specimen will be equipped with sound detection sensors and strain field measuring devices, both in the volume and on the surface of concrete.

To do so, the following experimental setup was imagined by the university of Karlsruhe (fig. 12):

- The specimen lays horizontally with its curvature pointing upwards.
- Hoop tendons apply their prestressing through steel abutments ('ears') to which they are anchored.
- A slight vertical prestressing is also applied.
- Internal vessel pressure is applied over the top surface of the specimen.
- The 8 hydraulic jacks stretch the specimen along hoop direction.

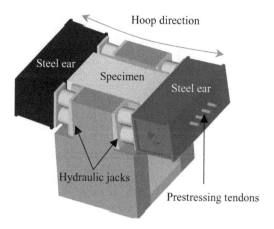

Figure 12. Experimental setup for the RSV.

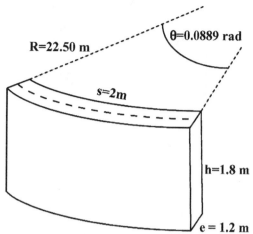

Figure 14. Dimensions of the new PACE 1450 RSV.

Computational simulations will then be carried out by constraining the model to follow the displacements recorded at 8 corners of the model while prestressing and internal pressures are also applied. Then results will be compared to evaluate the possibility of the modelling to reproduce experimental observations.

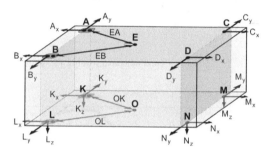

Figure 13. Position of displacement measurement devices.

Loadings applied to the specimen will produce both radial displacement and membrane stretching of the specimen as in the real structure.

The structure will undergo several consecutive loading sequences, always applying the same internal pressure using dry air with normal humidity and temperature conditions. To encounter the aging effect of a real containment vessel, the prestressing is gradually reduced from 100% in RUN 1, down to 60% for RUN 4. This should produce more and more microcracks in concrete, and result in slight changes in structural permeability.

During the tests different kind of measures are recorded (fig. 13):

– Displacements of all 8 corners of the observation volume
– Volume strain field observation using Bragg fibre optic network cast in concrete
– Surface strain field observation using optical camera
– 3D localisation of cracking using sound detection sensors
– Gas leakage through a collecting chamber

4 NEW PACE 1450 RSV

The dimensions, geometries, boundary conditions and loading are adapted from the definition of the experimental program. The new RSV includes 9 horizontal and 11 vertical reinforcement bars (on both internal and external faces), 4 horizontal and 1 vertical prestressed tendons. Hoops are not represented in the new RSV in order to simplify the model. The new geometry of the problem is given in figure 14. Figure 15 provides information about the steel distribution and properties.

The same boundary conditions as those chosen in the preliminary RSV are applied: face SB blocked along Oz, on face SH all nodes are restrained to follow the same displacement along Oz and no rotations are allowed for faces SG and SD. Only horizontal cables are prestressed as the vertical tendon is only cast to maintain the rigidity. Instead, a compressive homogeneous pressure of 1 MPa is applied during the prestressing on SH. The horizontal tension force is taken equal to 4.15 MN, which corresponds to a loss of 40 % of the initial 6.93 MN prestressing tension (simulation of the ageing process). For the sake of simplicity, the effect of gravity is not considered. Contrary to the preliminary computations, the bottom effect and the weight of the surrounding structure are not taken into account. Boundary conditions and loading are summarised in figure 16.

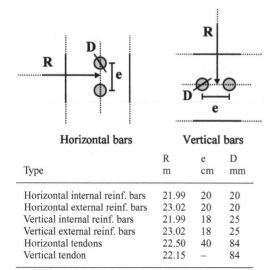

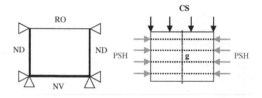

| Type | Horizontal bars | Vertical bars | |
	R m	e cm	D mm
Horizontal internal reinf. bars	21.99	20	20
Horizontal external reinf. bars	23.02	20	20
Vertical internal reinf. bars	21.99	18	25
Vertical external reinf. bars	23.02	18	25
Horizontal tendons	22.50	40	84
Vertical tendon	22.15	–	84

Figure 15.   Steel geometries of the new RSV.

Boundary conditions in displacement
ND : zero normal displacement on
SG and SD
NV : zero vertical relation on SB
RO : zero rotation on SH

Boundary conditions in stress

PSH : horizontal prestress
4.15 MN per tendon
CS : compressive stress
1 MPa

Loading

IP : internal pressure

Figure 16.   Boundary conditions and loading for the new RSV.

Figure 17 illustrates the radial displacement – pressure curve. It is still divided into four parts with an initial negative displacement due to the prestress, a zone of linear behaviour where damage does not evolve, then the apparition of a non linear response, followed by a partial unloading. Due to a change in the boundary and loading conditions, the admissible internal pressure is lower than in the initial RSV. The pressure – displacement curve is associated with different distributions of damage, as illustrated in figure 18. Two main phenomena are highlighted. First, a development of damage is observed from the inner surface of the volume.

It was expected since the loading is an internal pressure inside part of a cylinder. More interesting is the initiation and development of damage along the horizontal prestressed tendons. The internal variable indeed appears along the cables then propagates from the cables to the rest of the structure. It shows that the tendons play a key role in the damage distribution.

In comparison with the preliminary RSV, the mechanical mode of degradation is totally different, as we do not observe the localisation of strains or damage any more (due to a change in the loading conditions and especially the tensile bottom effect which is not taken into account any more). Moreover, in the present case, the level of prestress does not impose enough compression to create damage without internal pressure, as observed in figure 11 for the first RSV. Nevertheless, some similarities are also noticed about the role of the tendons. In both cases, they initiate, partly (new RSV) or totally (preliminary RSV) the evolution of damage and help its propagation.

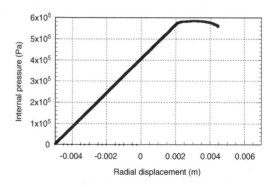

Figure 17.   Radial displacement – internal pressure curve for the new RSV.

The last step of these preliminary computations consists in investigating the mesh-dependency of the response. To this aim, two meshes, with different densities, are considered (figure 19). The value of the maximum of the equivalent strain is compared for elastic and nonlinear computations.

The equivalent strain, defined by Mazars (1984) quantified the tensile strains that are responsible for the development of the mechanical damage.

$$\varepsilon_{eq} = \sqrt{\sum_{i=1}^{3} (<\varepsilon_i>_+)^2} \qquad (1)$$

$<\varepsilon_i>_+$ represents the positive principal values of the strain tensor.

968

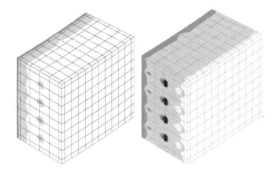

Figure 18. Evolution of the damage distribution in the new RSV after the application of an internal pressure.

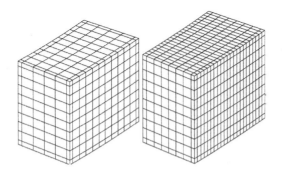

Figure 19. Meshes for the mesh-dependency study.

Table 1. Mesh-dependency studies. Difference between the maximum of the equivalent strain for the coarse mesh and the medium mesh

	Prestress 6.93 MN	Prestress 4.15 MN
Elastic computation	16 %	0.03 %
Non linear computation	19 %	2.46 %

To emphasize the role of the prestress, a computation with a tension of 6.93 MN is also carried out (initial tension in the cables if the ageing process is not taken into account).

The mesh-dependency problem that appears in the non linear computation can be partly explained by the strain and damage localization. It is an usual effect of softening laws (Crisfield, 1982) and could need the use of a regularization technique (Pijaudier-Cabot & Bazant, 1987), that will not be discussed in this contribution. More interesting is the dependency observed during the elastic computation. As the constitutive law is elastic, only the material components can explain this effect. If we compare the calculation with 6.93 MN

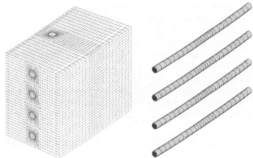

Figure 20. Mesh for future computations. Discretization of the cables with full 3D elements.

and 4.15 MN prestresses, it can be noticed that the higher the pretension is, the more important the mesh-dependency becomes. The application of prestress thus triggers a local mesh-dependency, probably due to the one-dimensional representation of the cables into a three dimensional volume (localization of the application of the tension force).

5 CONCLUSIONS

Different preliminary computations have been carried out to characterize the mechanical behavior of a Representative Structural Volume of a containment building of 1300 and 1450 MWe nuclear power plant. The role of the prestress tendons is especially underlined. They are partly responsible for the initiation of damage and help its propagation. Moreover, from a numerical point of view, they seem responsible for a mesh-dependency problem that even appears for elastic computations.

These observations require an experimental validation, that will start soon, to validate the role of the prestress and also the need for further numerical studies to understand the influence of the one-dimensional modelling of the pretensioned cables. This issue will be investigated by comparing the present results with a full three dimensional analysis (discretization of the cables with 3D elements, figure 20).

REFERENCES

Cast3M. 2006. Description of the finite element code Cast3M. http://www-cast3m.cea.fr.
Crisfield M.A. 1982. Local instabilities in the non linear analysis of reinforced concrete beams and slabs, *Proceedings of the Institution of Civil Engineers* 73:135–145
Ghavamian, S. & Delaplace, A. 2003. Modèles de fissuration de béton. Projet MECA. *Revue Française de Génie Civil*, 7, 5
Jason, L. Huerta, A. Pijaudier-Cabot, G. & Ghavamian, S. 2006. An elastic plastic damage formulation for concrete: application to elementary and comparison with an

isotropic damage model. *Computer Methods in Applied Mechanics and Engineering.* 195, 52:7077–7092

Jason, L. 2004. *Relation endommagement perméabilité pour les bétons. Application aux calculs de structures.* PhD Thesis, Nantes, Université de Nantes et Ecole Centrale de Nantes, France

Mazars, J. 1984. *Application de la mécanique de l'endommagement au comportement non linéaire et à la rupture du béton de structure.* PhD thesis. Paris, Université Paris VI, France

Ngo, D. & Scordelis, A.C. 1967. Finite element analysis of reinforced concrete beams. *Journal of the American Concrete Institute.* 64: 152–163

Picandet, V. Khelidj, A. Bastian, G. 2001. Effect of axial compressive damage on gas permeability of ordinary and high performance concrete. *Cement and Concrete Research.* 31:1525–1532.

Pijaudier-Cabot, G. & Bazant Z.P. 1987. Nonlocal damage theory. *Journal of Engineering Mechanics.* 113: 1512–1533.

Part VII
Monitoring and assessment of RC structures

Fracture Mechanics of Concrete and Concrete Structures – Design, Assessment and Retrofitting of RC Structures – Carpinteri, et al. (eds)
© 2007 Taylor & Francis Group, London, ISBN 978-0-415-44616-7

The creative response to concrete cracking

D.P. Billington & P. Draper
Princeton University, Princeton, New Jersey, USA

ABSTRACT: Cracking in concrete has served as an important influence in the development of structural design. Careful observation and insightful responses to concrete cracking by designers have led to a series of innovative new structural solutions. This progression is illustrated through several specific historical design examples.

1 INTRODUCTION

Judging from the proceedings of your last conference in 2004, we can see that there is little that we can contribute to the many interesting developments taking place in the main subject of your organization. Our contribution will thus be of a very different sort that will treat concrete cracking as an integral part of the history of structural design. It is our hope that the more scientific work that makes up most of your research will prove to be not just useful to designers but will stimulate them to make better designs and to benefit by future collaboration.

Our thesis is that major advances in concrete structures have arisen through designers' contemplation of observed cracking in their own works as well as in the works of their predecessors. The collected record of a series of such contemplations and the resulting improved or new designs will become, we believe, a necessary part of the education both of students and practitioners. The cracking events we will describe are not the results of shoddy workmanship or incompetent designers but in all cases were the surprising defects that well trained and careful engineers made usually following state-of-the-art ideas.

2 THE CLASSICAL PERIOD

The Pantheon in Rome is undoubtedly the greatest concrete structure of the classical period (Figure 1). It is also the first such structure to exhibit substantial cracking which led to a creative design addition as well as to the form of a similar dome nearly half a millennium later (Mark and Robinson, 1993).

Constructed between A.D. 118 and 128, the Pantheon is a concrete dome far larger than any earlier

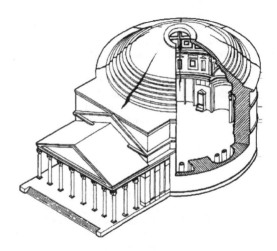

Figure 1. Drawing of the Pantheon.

such structure. Its 43 meter clear span would not be exceeded in any concrete dome until the 20th century. Made of Roman pozzolan concrete with a lightweight aggregate, the form of the monumental structure had been misunderstood until the pioneering studies of Robert Mark explained it and its influence on later forms. Designed as a hemisphere, the dome was taken to be fully monolithic with concrete rings encircling the dome near its junction with the cylindrical wall. Many students of the Pantheon thought that these rings, clearly expressed above the concrete shells, acted as hoops in reinforcing the shell against circumferential tension stresses. As Mark demonstrated convincingly, these are not rings at all but rather they are extra loading on arches that the Roman builders used to prevent arch bending. Thus the lower part of the dome is really not a dome but a series of arches separated from each other by radial cracks that extend upwards over half

Figure 2.   Cracks in the Pantheon.

the vertical distance between the hemispherical base and the crown (Figure 2).

One can observe these cracks even today nearly 1900 years after completion. The Romans clearly saw them and piled on the extra weight to keep the arches from bending and cracking in the circumferential direction. So the striking ring stiffened form of the Pantheon was a creative response to major radial cracking and the structure has proven durable ever since. The extent of the crack is easily predicted by the simple membrane-theory formula for hoop stresses.

$$N_\theta^1 = aq\left(\frac{1}{1+\cos\phi} - \cos\phi\right)$$

where a is the dome radius in meters, q the dead load in Kg per meter squared, and ϕ is the angle between the axis of rotation and a radius defined by any point down the meridian. Thus when ϕ is greater than $51^0 50^1$ the N_θ^1 will be positive hence tension so that the crack could extend up the shell to a vertical distance of greater than $a/2$, which we can observe even today. In that case it is more than likely that the builders of the Hagia Sophia in Constantinople would have studied the Pantheon as the only extinct dome of nearly the same size as their planned work to the east.

The structure was built quickly between A.D. 532 and 537 but its great dome fell in 558 following the two earthquakes of 553 and 557. Right away a new and higher dome began and was completed between 558 and 562. Despite subsequent partial destruction in earthquakes, that 562 dome is what can be seen today (Mark and Robinson, 1993).

A most striking feature of the Hagia Sophia is the array of windows around the base of the dome which from inside gives the impression that the dome floats on light. While this ideal may have been in the minds of the architects it is nearly certain that the builders intended these openings also to reflect the fact that in a non reinforced dome there would be meridional

cracks which the window openings replace. In fact these openings extend above the hemispherical plane to an angle just over 50° between the axis of rotation and the meridional point when the windows end. Thus the remarkable floating feeling is in reality a creative response to the meridional cracking at Rome. The windows are formed by arches that support the great dome so that the defect at Rome became the stimulus for a glorious form at Constantinople 430 years later.

3   THE GOTHIC PERIOD

There are at least two examples of cracking in the great churches of northern France, one of which led to a clear creative response by the designers. Again this is the work of Robert Mark in discovering this example and illustrating it in engineering terms in reaction to a deep misunderstanding by an art historian who studied the same problem (Mark, 1982). This question about the justification for the pinnacles on pier buttresses raises a much more fundamental issue, one that arose in the 19th century with the renewed cultural interest in gothic cathedrals.

The story begins with the gothic revival in the late 18th century – largely a literary movement famous for the section in Victor Hugo's *Hunchback of Notre Dame* where he describes the sad state of that great Paris cathedral and begins to make a case for restoration of these works. The architect Eugene Viollet-le-duc took up this task as head of the commission for ancient monuments and proceeded to write profusely on the engineering and architectural features of these structures. One example will illustrate the issue of cracking.

In the cathedral at Amiens there are main massive pier buttresses that carry roof loads as well as other loads from the interior. Viollet wanted to show that all main parts of Gothic form were essential to the structural design or the construction process and one set of examples he took was the small pinnacle atop each pier buttress. Were they useful or were they merely decoration? Viollet answered this with a resounding yes in favor of useful because as he explained it there are vertical forces on the pier and horizontal forces. These former ones, he claimed, served to compress the masonry and the masonry within the buttress on the foundation and therefore the weight of the pinnacle would keep the pier buttress from sliding or bending outward.

This reasonable sounding defense of the utility of the pinnacle was attacked by the architectural historian Pol Abraham who ridiculed Viollet's logic by showing first that the small stone weight was too small to have any significant influence on the immense buttress with its large forces. Second, he argued that even if the pinnacle were of a helpful weight it was clearly in the wrong place (at the outer edge of the buttress) where

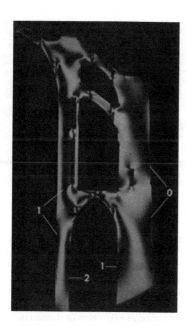

Figure 3.   Photoelastic analysis of Amiens.

it would contribute to its outward bending instability. Clearly, said Abraham, the pinnacle was a decorative feature to improve the appearance of an otherwise dull flat upper terminus of the pier buttress. Abraham used many more examples to illustrate what he considered to be the general fallacy that Gothic designers based their overall forms on structural and constructional ideas. This controversy between the so-called rationalism of Viollet and the illusionism of Abraham became an important debate in the 20th century where some architects designed to express structural or constructional ideals while others believed that those engineering features were irrelevant to the visible expression of form. But most agreed that the Gothic form was beautiful so Abraham's argument was essential to an architecture that strove to be separated fully from any expression of engineering. Was Abraham correct?

Robert Mark entered that debate with a series of examples for which the pinnacle is the simplest to describe. By means of photoelastic model analysis, he showed that without the pinnacle at Amiens there would have existed in one predominate place on the buttress a region of tension stress which could easily have led to cracking near the top outside edge (Mark, 1982) (Figure 3). He then reanalyzed the structure with the pinnacle on top and that tension disappeared. The strong supposition is that Gothic builders saw such tension because the structure is of course stone on stone and the interface could therefore easily open up. The pinnacles are therefore of just the right weight and in just the right location to eliminate that cracking. Once decided upon, as in much gothic architecture,

the element was shaped elegantly so that the form is rational and the choice of detail was aesthetic. Here the response to cracking led to an elegant addition to the already striking exterior form of high Gothic structure.

4   THE PERIOD OF EARLY ADVANCES IN REINFORCED CONCRETE

A number of entrepreneurs began to design and build reinforced concrete structures before 1900 and major advances came soon thereafter. Probably the most widespread system of design-build was that pioneered by the French engineer, François Hennebique, one of whose most famous structures was the bridge over the Vienne River at Châtellerault in 1899. Engineers admired the lightness of its three arch spans but the more perceptive ones recognized that the arches exhibited cracking at both supports and at the crown. This cracking led one of Hennebique's employees in 1900 to propose an arch design with hinges built in at each support and at the crown. This three-hinged arch could expand or contract as the ambient temperature changed without cracking because the hinges allowed freedom of movement (Billington, 1976).

This young engineer, Robert Maillart (1872–1940), made the design for a single-span arch bridge over the Inn River at Zuoz in eastern Switzerland; it was completed the following year. Maillart used that opportunity to design a completely new form in concrete – the hollow box – in which arch, side walls, and deck were all built together in one monolithic form that was far stronger and substantially lighter than Hennebique's bridge and other concrete arch bridges of that time (Billington, 1976).

Hennebique's bridge cracking suggested to Maillart the three-hinged form but he was concerned that the increased flexibility owing to the hinges would make the structure too light and subject to higher stresses and more serious vibrations under traffic load. Maillart therefore, by overcoming the cracking problem, went on to overcome the extra flexibility through a greatly stiffened arch achieved by connecting it to walls and the roadway deck. This led him to his great innovation, the concrete hollow box – still a major form for bridge design in the 21st century.

But here Maillart was misled by his knowledge of classical forms, especially that of Roman arch bridge design that had been transferred to keep the Pantheon from dangerous arch bending. In their bridges the Romans had used circular forms to simplify construction by cutting the stones all to the same wedge shapes. They were good engineers and knew that the circular shape was wrong so they corrected for it by piling extra weight near the abutments (what we would call changing the pressure line to keep it inside the kern). This weight was normally rubble masonry and being loose

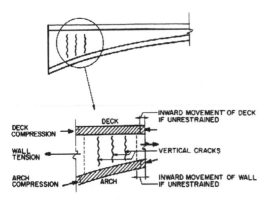

Figure 4.   Zuoz Bridge and cracks.

Figure 5.   Tavanasa Bridge.

required spandrel walls on either side of the arch. These containment walls reached from arch to deck and gave the visual impression of a haunched beam-arch that by the 19th century was an accepted aesthetic form for water crossings. Michelangelo's famous bridge over the Arno River at Florence is a fine example of such a bridge.

So Maillart designed the walls for his hollow box to extend all the way to the abutments even though the support hinges were placed in the arch well below the walls. The result was a set of cracks in the spandrel walls that arose from differential movements between the arch (wet from the river) and the deck (dry and hot from the sun of the Oberengadine) (Figure 4). The owners of the bridge called Maillart to the site several years later to explain the problem and determine its danger. Maillart realized that there was no danger of failure but rather of gradual deterioration through heavy weather – he recommended whitewashing it for protection. The bridge lasted well for 65 years; it was then rehabilitated and is in good shape today.

But the cracks set Maillart thinking deeply about his form and he realized that the spandrel wall was unnecessary near the abutments so that in his next new project, the Tavanasa Bridge of 1905, he removed that part and produced a clear expression of the three-hinged form (Figure 5). This was the first example of a great concrete work of structural art; it was also efficient in using a minimum of material and economical in being built by Maillart himself as the least expensive in a design-build competition. The high art world of Switzerland was, however, hostile to this completely unprecedented shape and for 25 years Maillart could not complete a bridge in that form. He therefore turned to other bridge forms and once again cracking came to the rescue of creativity.

In 1910 the cantons of Aargau and Solothurn asked for design-build bids for a bridge over the Aare River near the town of Aarburg and three companies submitted bids of which Maillart was second lowest in price. Because of his strong reputation for high quality construction, Maillart gained the contract, designed a traditional arch and completed construction in 1912. A few years later the owners called him back to the site to explain cracks in the deck above the arch (Figure 6). This defect surprised Maillart because his design was thoroughly conventional – almost Roman in concept – where he designed the deck to carry loads to the columns which in turn transferred those loads to the arch which carries all the load to the abutments. There was no interaction; he designed each part to carry all its loads with no help from any other part.

As Maillart thought about this problem which, as at Zuoz, did not signal a danger of collapse, he began to realize that the location and extent of the cracks were expressive of a structural behavior that he had not anticipated and which reminded him of an idea presented in class almost 30 years earlier by his professor at the ETH, Wilhelm Ritter (Ritter, 1883).

The lesson was that a monolithic structure will act as a single unit even if the analysis imagines it to be individual elements acting separately as was the standard approach. He saw that when the arch deflected under traffic loads over half the span, the deck must also deflect that way and not, as he had assumed, as if it were deflected over the far shorter length between columns. The cracks illustrated that behavior clearly and this problem stimulated Maillart to invent his second major bridge form beginning with a small 1923 bridge over the Flienglibach. Here he used the parapet as a stiffener to carry the half load bending and as a consequence he could design the arch about one-third the thickness of the arch at Aarburg. This new idea found full expression two years later in the Valtschielbach Bridge (Figure 7). As with the 1905 Tavanasa, this 1925 bridge signaled a major new design using the potential of monolithic concrete and resulting from an earlier design that exhibited highly visible and extensive cracking. Neither of these two bridge innovations, however, reached the highest point of structural art that those cracking experiences promised. Maillart would

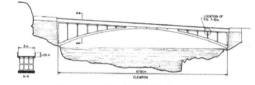

Figure 6. Aarburg Bridge and cracks.

Figure 7. Valtschielbach Bridge.

achieve that point during the last decade of his life from 1930 to his death in 1940.

5 PRESTRESSING AND CRACKLESS CONCRETE

In the meantime another more sweeping innovation appeared. While cracks can stimulate creativity it is also true that the possibility of completely eliminating cracks can inspire hopes for a new world of structural design. Such was the vision of a French engineer who first imagined such a world in the same year that Maillart first saw his cracked Zuoz Bridge. This Frenchman, Eugene Freyssinet (1879–1962), developed the idea, patented it and began using it only after the age of 50. Later he would express his vision that prestressed concrete held much more promise than just one more development because, as he wrote in 1949, "in itself the idea of prestressing is neither complicated nor mysterious; it is even remarkably simple,

but it does belong to a universe unknown to classical structural materials and the difficulty for those first coming to the idea of prestressing is to adapt themselves to this new universe." Even Hennebique had not claimed so much for reinforced concrete, although its properties could have rightly been called unknown to the universe of stone, wood, and iron structures. Yet both Frenchmen proclaimed a new era in building based on the union of metal and concrete, and both sought what Maillart had called the lightness of metal and the permanence of stone (Billington, 1985).

Because this was a "new universe", Freyssinet claimed that "the fields of prestressed and reinforced concrete have no common frontier". Either a structure was fully prestressed or it was not prestressed concrete. Either the design was to be fully crack free or it was not part of the "new universe". His idea was revolutionary, it was stimulating, it had productive results, but it was wrong.

Nevertheless, in the United States it became the rule that prestressed concrete was distinctly separate from reinforced concrete and to this day almost no civil engineering undergraduate, while taking a course in concrete structures, has been exposed to prestressed concrete. In standard texts prestressing is usually relegated to a single separate chapter or left out entirely. There is no doubt that prestressing was a true revolution in structural engineering and it would lead eventually to beam bridge spans of 1000 feet and to the economical use, as Freyssinet predicted, of high strength steel and high strength concrete. But as a new universe of crack free forms it would be more a questionable idea than a sweeping change. I can illustrate an example of the dangers inherent in the early idea of crack free design by taking the concept which, although not strictly speaking prestressing, nevertheless shows the problem.

In the 1960s there began to be built large power plants, some for nuclear power but others using fossil fuel. One striking visual component of many such plants was the natural draft cooling tower reaching above 300 ft. in height and eventually over 500 ft. high. Originally these towers were designed quite simply for dead weight and wind load where the base stresses due to each counteracted (compression for dead load and tension on the windward side for wind load). Because they were made to balance, very little vertical reinforcing steel was used. As an example, if the dead weight gave a maximum of 2000 psi compression and the wind load (assumed here as 100 mph) gave a maximum of 2000 psi tension, the design result would be zero stress at one point and some expression elsewhere around the circumference of the thin shell tower. What happened to a set of such towers in 1965 was that there occurred a wind higher than the design by only about 12% (112 mph). Since the pressure is proportional to velocity squared that meant the maximum tension

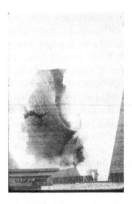

Figure 8.   Ferrybridge cooling tower failure.

increased by about 25% to roughly 2500 psi, leaving a net tension of 500 psi and hence cracking. Three towers failed and the primary lesson was that there needed to be an ultimate load analysis as well as the correct working load analysis that was used (Central Electricity Generating Board, 1966) (Figure 8).

Thus the concept of crack free design did not initially include a concern for overloads. Freyssinet was far too good an engineer to have made a mistake like that. But his enthusiasm for prestressing did require, for the profession as a whole, the additional emphasis on the potential for a cracked state and subsequently for the assurance of appropriate reinforcing still to prevent danger.

Fortunately three engineers almost immediately after World War II, when Freyssinet's ideas began to receive serious attention, spoke out strongly in favor of a more balanced view, one which still recognized the extraordinary potential for prestressing but one which recognized the great significance of recognizing cracking as an integral part of design. This recognition led to a new concept for prestressing that would bring in a new era in understanding concrete structural engineering – not the new universe postulated by Freyssinet, but still a new universe that was inclusive rather than exclusive.

6   STRUCTURAL CONCRETE AND THE CENTRALITY OF CRACKING

The three pioneering structural engineers who took issue with Freyssinet saw that prestressing was really a new idea that should be seen as improving reinforced concrete structures rather than supplanting them. The first was Paul Abeles, an Austrian who fled from fascism and brought his ideas to Great Britain. His idea which he named partial prestressing was to provide enough prestressing to control cracking in reinforced beams rather than to completely remove it. His designs

aimed first at railway beams where the high live load caused the greatest tensile stresses. If these were to be fully counteracted by prestressing there would remain, with only dead load, large stresses (both compression and tension) due to the prestressing and also the possibility of large upward displacements. Abeles argued that since reinforced concrete practice permitted cracking, why not allow some and use reinforcing steel to control cracking and carry live load. The prestressing would carry dead load so that most of the time – when the heaviest trains were not running over the elements – there would be no cracking or, more properly put, the cracks due to live load would be closed (Abeles, 1962).

Abeles found a sympathetic colleague in the second major figure of the post war era, the Belgium professor of structures Gustave Magnel (1889–1955). Magnel was one of the two greatest teachers of structures that I know about during this post war time and he carried out numerous tests in his well equipped laboratory at the University of Ghent. His tests on Abeles' designs materially helped the Austrian make his point about partial prestressing but even more importantly Magnel had the background and depth of understanding that allowed him to openly dispute Freyssinet (Magnel, 1954).

An ally of Magnel's in combating the extreme position of Freyssinet was the Swiss professor, Pierre Lardy (1903–1958). Like Magnel, Lardy produced (with his predecessor Max Ritter) a pioneering text on prestressed concrete in the 1940s and like Magnel, Lardy was a superlative teacher who taught prestressed concrete to his students in the late 1940s. Lardy's two most famous students both took up the issue of cracking but from different perspectives (Billington, 2003). Christian Menn (b1927), the most gifted bridge designer of the last half century, quickly followed the ideas that Magnel and Lardy had emphasized that prestressing and reinforcing should not be separated. In Menn's bridges he often used prestressing to counteract dead loads and sometimes part of the live load and then employed reinforcing steel to carry the remaining load.

One clear example is the transverse reinforcement in the hollow box deck for the 1974 Felsenau Bridge (Figure 9). Here the wide cantilever overhangs are prestressed for dead load and reinforcing steel added to carry the live (traffic) loads. This allowed Menn to have unusually long cantilevers which provided a substantial shadow on the web of the box and contributed to the light appearance of the haunched main spans (the longest in Switzerland at the time). Had he tried to follow Freyssinet's crack free universe the cantilevers would surely have been overstressed when no live load was present and probably deflected upward as well. Menn thus achieved the high goal of the structural artist by allowing for cracks and creating thereby a more elegant design (Billington, 2003).

Figure 9. Felsenau Bridge and wide overhanging wings.

Figure 10. Sicli Building with no waterproofing.

Lardy's other most famous student, Heinz Isler (b1926), actually realized Freyssinet's goal of a new universe but without the prestressing that the French engineer believed essential. Isler's new universe is truly a revolution in concrete design that the profession as a whole has only barely recognized. Isler achieves crack free roof spans in reinforced concrete entirely by creating unprecedented shapes that avoid almost entirely any tension stresses and hence any cracking. In this way Isler's thin shell concrete roofs, with spans sometimes above 150 ft., are entirely waterproof and are built with no roofing or other water proofing materials (Figure 10). They are bare concrete usually only three inches thick and created by his hanging membrane reversed process. In this process, Isler invents a form by suspending a cloth between the desired number and location of supports. He then pours on the cloth a fluid plastic which causes the cloth to assume a shape dictated solely by gravity. When the plastic hardens, Isler overturns the model (of the order of three feet in span) and measures the ordinates precisely so that the full scale structures will be built of the same form (Billington, 2003).

These shells are designed so that the model supports and the real structural supports are usually at ground level and then, because the shells are curved, there will be both vertical and horizontal reactions needed for stability. Isler ties the supports together by prestressing tendons normally below grade. The secret of Isler's crackless concrete, therefore, is not direct prestressing but rather the creation of a new form, largely loaded by gravity, that will by its shape avoid cracks and lead to a new vision for concrete.

All of these events have led to a new designation, structural concrete, which is a material that makes no exclusive claims for either prestressing or reinforcement but rather considers them each as part of the designer's education and practice. Two major figures in concrete structural engineering of the 20th century are credited with advancing this term: John Breen, a distinguished professor at the University of Texas at Austin and Jörg Schlaich, both a professor at Stuttgart but also head of one of the most creative design firms.

7 CONCLUSION

This paper has sought to illustrate, through specific design examples, the way in which the contemplation of concrete cracking has led to new ideas and new designs. In a more general way, this brief study also seeks to emphasize the centrality of historical cases to the education of students as well as practitioners. Isolated or disconnected historical case studies are usually of much less significance than a set of such cases that are connected by a central idea. That is what makes history both important and of interest in education. This present sketch attempts to develop the idea of cracking as an important part of concrete studies not only as scientific analysis but also as design insight.

ACKNOWLEDGEMENTS

Material for this paper was developed as part of a teacher-scholar program partly funded by the National Science Foundation Grant No. 0095010. All the examples from the classical and Gothic periods are taken directly from the pioneering publications of Robert Mark. We are also indebted to Kathy Posnett who ably typed the manuscript.

REFERENCES

Abeles, P. W. 1962. *An Introduction to Prestressed Concrete.* Concrete Publications Ltd. See especially Paul Abeles

"Comments," *PCI Journal*, Vol. 22, No. 3 (May/June 1977): 109–115.

Billington, D. P. 1976. *Robert Maillart's Bridges*. Princeton University Press: 13, 19–26.

Billington, D. P. 1985. *The Tower and the Bridge*. Princeton University Press: 194.

Billington, D. P. 2003. Chapter 4: From Mathematics to Aesthetics. *The Art of Structural Design: A Swiss Legacy*. Princeton University Art Museum and Yale University Press: 112–127, 146, 178.

Central Electricity Generating Board, London, England. 1966. Report of the Committee of Inquiry into Collapse of Cooling Towers at Ferrybridge, Monday, November 1, 1965.

Magnel, G. 1954. *Prestressed Concrete*, 3rd Ed. Concrete Publications Ltd.: 134, 149–151.

Mark, R. 1982. *Experiments in Gothic Structure*. MIT Press: 13, 32, 52–55, and 102.

Mark, R. and Robinson, E. C. 1993. Vaults and Domes. *Architectural Technology up to the Scientific Revolution*. MIT Press: 141–152.

Ritter, W. 1883. Statische Berechnung der Versteifungsfachwerke der Hängebrücken. *Schweizerische Bauzeitung*. Vol. 1, No. 1.

*Fracture Mechanics of Concrete and Concrete Structures – Design, Assessment and
Retrofitting of RC Structures – Carpinteri, et al. (eds)
© 2007 Taylor & Francis Group, London, ISBN 978-0-415-44616-7*

Damage evaluation and corrosion detection in concrete by acoustic emission

M. Ohtsu & Y. Tomoda
Kumamoto University, Kumamoto, Japan

T. Suzuki
Nihon University, Kanagawa, Japan

ABSTRACT: Acoustic emission (AE) techniques have been extensively studied and applied to nondestructive testings (NDT) of concrete and concrete structures. For damage evaluation, AE behavior of concrete under compression could be analyzed, applying the rate process theory. Based on Loland's model in damage mechanics, a relation between AE rate and the damage parameter is correlated. By quantifying intact moduli of elasticity of concrete from the database on the relationship, relative damages of concrete in existing structures are successfully estimated by the compression test of concrete samples. The technique is applied to estimate concrete samples of recycled aggregate. The deteriorated degree of recycled concrete is reasonably estimated. For corrosion detection, continuous monitoring of AE signals is useful for earlier warning of corrosion in reinforcement. It is demonstrated that the onset of corrosion in reinforcement and the nucleation of corrosion cracking in concrete could be clearly identified by AE parameter analysis.

1 INTRODUCTION

Concrete and concrete structures could deteriorate due to the environmental effects. Consequently, evaluation of deteriorated degree or damage in concrete has been in so great demand that it is necessary to develop quantitative techniques for damage evaluation in concrete.

In the case of diagnostic inspection of concrete structures, mechanic properties of concrete are normally evaluated by taking core samples. However, properties obtained from the compression test have not been directly applied to damage evaluation. In this concern, acoustic emission (AE) is known to be promising to evaluate the degree of damage. The authors have proposed to evaluate AE activity under compressions by introducing the rate-process analysis (Ohtsu, 1992). By calculating an intact modulus of elasticity from the database based on a relation between AE rate and the damage parameter, a procedure to estimate the relative damage of concrete is implemented as Damage evaluation of Concrete by AE raTe-process analysis (DeCAT) (Suzuki and Ohtsu, 2004). In this study, the DeCAT system is applied to estimate a deteriorated degree of concrete samples made of recycled aggregate.

Corrosion of reinforcing steel bars (rebars) is one of critical deteriorations in reinforced concrete. When chloride concentration at rebar exceeds a range of values with a probability for corrosion initiation, a passive film on the surface of rebar is destroyed and corrosion is initiated. Then electrochemical reaction continues with available oxygen and water. Corrosion products on rebar surfaces grow with time and nucleate microcracks in concrete. According to the Japanese standard specifications on maintenance (JSCE, 2001), deterioration process is specified as illustrated in Figure 1.

There exist two transition periods at onset of corrosion and at nucleation of cracks. The former is

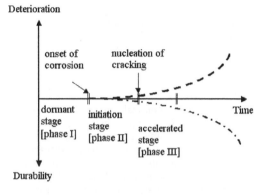

Figure 1. Corrosion process of reinforced concrete.

associated with the transition from the dormant stage (phase I) to the initiation stage (phase II). Corrosion-related damage could begin at this point, which is normally defined as the initiation time of corrosion. The latter is the transition from the initiation stage (phase II) to the accelerated stage (phase III), and is critically important to assess the durability of reinforced concrete structures. For nondestructive evaluation (NDE) of corrosion, electrochemical techniques of half-cell potential and polarization resistance have been widely employed. Yet, it is known that these techniques can not provide precise information on the two transition periods. As a result, these periods are normally defined by chloride concentration levels at cover thickness in concrete. Here, continuous AE measurement is applied to identify the transition periods at the onset of corrosion and at the nucleation cracking.

2 ANALYSIS OF AE SIGNALS

2.1 AE rate-process analysis

AE behavior of a concrete sample under compression is associated with generation of micro-cracks. These micro-cracks are gradually accumulated until final failure. The number of AE events, which correspond to the generation of these cracks, increases acceleratedly along with the accumulation of micro-cracks. It appears that the process is dependent on the number of cracks at a certain stress level as to be subjected to a stochastic process. Therefore, the rate process theory is introduced to quantify AE behavior under compression.

The following equation of the rate process is introduced to formulate the number of AE hits dN due to the increment of stress from V to $V + dV$,

$$f(V)dV = \frac{dN}{N} \tag{1}$$

where N is the total number of AE events and $f(V)$ is the probability function of AE at stress level V(%). For $f(V)$ in Equation 1, the following hyperbolic function is assumed,

$$f(V) = \frac{a}{V} + b \tag{2}$$

where a and b are empirical constants. Here-in-after, the value 'a' is called the rate, which reflects AE activity at a designated stress level. At a low stress level the probability varies, depending on whether the rate 'a' is positive or negative. In the case that the rate 'a' is positive, the probability of AE activity is high at a low stress level, suggesting that concrete is damaged. In the case of the negative rate, the probability is low

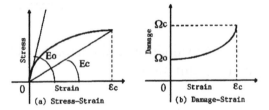

Figure 2. Stress-strain-damage relations under compression.

at a low stress level, revealing that the structure could be in stable condition. Substituting Equation 2 into Equation 1, a relationship between total number of AE events N and stress level V is obtained as the following equation,

$$N = CV^a \exp(bV) \tag{3}$$

where C is the integration constant.

A damage parameter Ω in damage mechanics can be defined from a relative ratio of the moduli of elasticity (Loland, 1980),

$$\Omega = 1 - \frac{E}{E^*} \tag{4}$$

where E is the modulus of elasticity of concrete and E^* is the ideal modulus of elasticity, which is assumed to be intact or completely undamaged. Assigning Ω_0 is the initial damage at the onset of the compression test, the following equation is derived,

$$E_0 = E^*(1 - \Omega_0) \tag{5}$$

In the compression test of a concrete sample, a relation between stress and strain is typically plotted as shown in Figure 2 (a). According to Equation 5, the initial modulus of elasticity E_0 is associated with the initial degree of damage Ω_0. Corresponding to the damage Ω_c at the ultimate strain ε_c, the scant modulus of elasticity, E_c, is defined. In this study, the modulus of elasticity, E_0, was estimated as a tangential modulus, after approximating the stress-strain relation by a parabolic function.

As given in Equation 5, the initial damage Ω_0 is an index of damage. Still, it is fundamental to know the intact modulus of elasticity of concrete, E^*. But it is not easy to determine the modulus E^* of concrete taken from an existing structure. Consequently, it is attempted to estimate the modulus E^* from AE measurement.

A correlation between the decrease in the moduli of elasticity, $\log_e(E_0 - E_c)$, and the rate 'a' derived from

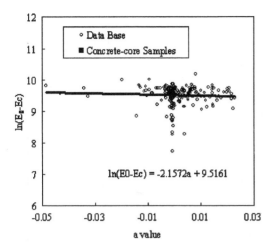

Figure 3. Database on rate 'a' and the modulus.

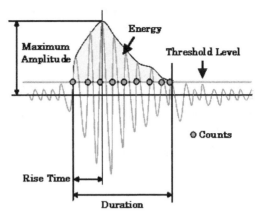

Figure 4. AE wavefrom parameters.

AE rate-process analysis is shown in Figure 3. Results of concrete samples previously tested are plotted by open circles and results of core samples recently tested (Suzuki and Ohtsu, 2004) are denoted by solid circles. A linear correlation between $\log_e(E_0 - E_c)$ and the rate 'a' value is reasonably observed. The decrease in the moduli of elasticity, $E_0 - E_c$, is expressed,

$$E_0 - E_c = E*(1 - \Omega_0) - E*(1 - \Omega_c).$$
$$= E*(\Omega_c - \Omega_0) \tag{6}$$

Based on a linear correlation in Figure 3,

$$\log_e(E_0 - E_c) = \log_e[E*(\Omega_c - \Omega_0)]$$
$$= xa + c \tag{7}$$

Here, it is assumed that $E_0 = E*$ when $a = 0$. This allows us to estimate the intact modulus of modulus of concrete, $E*$, from AE rate-process analysis as,

$$E* = E_c + e^c \tag{8}$$

Each time when we conduct the compression test, AE data are added to Figure 3 and the intact modulus is calculated by Equation 8. Thus, Figure 3 works as the database of the DeCAT system.

2.2 AE parameter analysis

Characteristics of AE signals were estimated by using two indices of RA value and average frequency. These are defined from such waveform parameters as rise

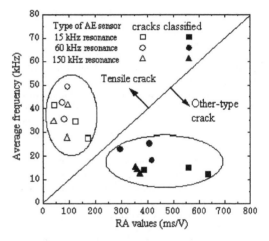

Figure 5. Crack classification by AE indices.

time, maximum amplitude, counts and duration shown in Figure 4, as follows;

$$RA \; value = Rise \; time \; / \; Maximum \; amplitude, \tag{9}$$

$$Average \; frequency = Counts \; / \; Duration. \tag{10}$$

According to the Japan code (JCMS, 2003), AE sources of active cracks are classified into tensile cracks and other-type cracks, based on the relationship between the RA values and the average frequencies, as shown in Figure 5. When the RA value is small and the average frequency is high, AE source is classified as a tensile crack. In the other case, AE source is referred to as a crack other than a tensile crack, including a shear crack. This criterion is applied to classify AE events detected in the corrosion process.

In addition, to evaluate sizes of AE sources, the amplitude distribution of AE events was applied.

A relationship between the number of AE events, N, and the amplitudes, A, is statistically represented as,

$$\text{Log}_{10} N = a - b\text{Log}_{10}A, \tag{11}$$

where a and b are empirical constants. The latter is called the b-value, and is often applied to estimate the size distribution of AE sources (Shiotani et al., 2001). In the case that the b-value is large, small AE events are mostly generated. In contrast, the case where the b-values become small implies active nucleation of large AE events.

3 EXPERIMENTS

3.1 Concrete of recycled aggregate

Cylindrical samples of 10 cm in diameter and 20 cm in height were made, which were made of four types of coarse aggregate. These are listed in Table 1. Two types of aggregate were commercially available. One was crushed aggregate and the other was heated and milled. We have recently developed a technique to take coarse aggregate out of concrete, applying the pulse-power, where 100 pulses of 400 kV and 6.4 kJ/shot were discharged in water. Thus, recycled aggregate was taken out of cylindrical concrete samples made of original aggregate, and then was applied to recast concrete samples. In all the types, the maximum gravel size of coarse aggregate was 20 mm, and the water-to-cement ratio was 55%. Air-entrained admixture was added to control the slump vales at around 7 cm and air contents at 6%. In the table, densities and the absorption coefficients f these aggregates are given. It is realized that densities decrease and the absorptions increases in recycled aggregate, compared with the concrete of original aggregate. Among the recycled aggregates, that of pulse-discharged has the higher density and the lower absorption.

For each aggregate, 3 cylindrical samples were made and tested after 28 day-standard curing. During the compression test, AE measurement was conducted. Silicon grease was pasted on the top and the bottom of the specimen, and a Teflon sheet was inserted to reduce AE events generated by friction. MISTRAS-AE system (PAC) was employed to count AE hits. AE hits were detected by using an AE sensor (UT-1000: 1 MHz resonance frequency). The frequency range was set from 60 kHz to 1 MHz. An experimental set-up is shown in Figure 6.

For event counting, the dead time was set as 2 msec.. It should be noted that AE measurement was conducted at two channels as well as the measurement of axial and lateral strains. AE hits and strain of the two channels were averaged and estimated as a function of stress level.

Table 1. Properties of coarse aggregate.

Type of recycled aggregate	Saturated density (g/cm³)	Dried density (g/cm³)	Water absorption (%)
Crushed	2.53	2.49	2.71
Heated and milled	2.59	2.54	2.10
Pulse disharged	2.95	2.90	1.42
Original	3.06	3.04	0.49

Figure 6. AE measurement in the compression test.

3.2 Corrosion tests

Reinforced concrete slabs tested were of dimensions 300 mm × 300 mm × 100 mm. Configuration of specimen is illustrated in Figure 7. Reinforcing steel-bars (rebars) of 13 mm diameter are embedded with 15 mm cover-thicknesses for longitudinal arrangement. When making specimens, concrete was mixed with NaCl solution. In order to investigate the threshold limit of chloride concentration for corrosion, the lower-bound threshold value (chloride amount 0.3 kg/m³ of concrete volume, 0.088% mass of cement) prescribed in the code (JSCE, 2001) was taken into consideration. After the standard curing for 28 days in 20°C water, chloride content was measured and found to be 0.125 kg/m³ in concrete volume (0.036% of mass of cement). Mixture proportion of concrete was the same as that of recycled concrete, but the maximum size of aggregate was 10 mm. A compressive strength at 28 days of the standard curing was 35.0 MPa. Following the standard curing, all surfaces of the slab specimen were coated by epoxy, except the bottom surface for one-directional diffusion as illustrated in Figure 7.

An accelerated corrosion test and a cyclic wet-dry test were conducted. In the accelerated corrosion test, the specimens were placed on a copper plate in a container filled with 3% NaCl solution as shown in Figure 8. Between rebars and the copper plate, 100 mA

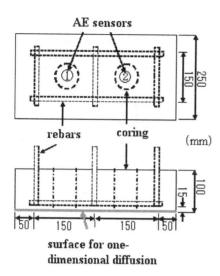

surface for one-
dimensional diffusion

Figure 7. Sketch of a slab specimen.

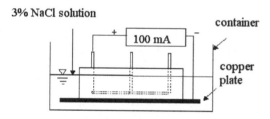

Figure 8. Test set-up for an acceleration test.

electric current was continuously charged. In the cyclic wet-dry test, the specimens were cyclically put into the container in the figure without charge for a week and subsequently dried under ambient temperature for another week.

AE measurement was continuously conducted, by using LOCAN 320 (PAC). Two AE sensors were attached to the upper surface of concrete at the center of coring locations shown in Figure 7. In order to detect AE signals, a broad-band sensor (UT-1000, PAC) was employed. Frequency range of the measurement was 10 kHz – 1 MHz and total amplifications was 60 dB gain. For event counting, the dead-time was set to 2 msec. with 40 dB threshold.

Half-cell potentials at the surface of the specimen were measured by a portable corrosion-meter, SRI-CM-II (Shikoku Soken, Japan). In the accelerated corrosion test, the measurement was conducted twice a day until the average potentials reached to −350 mV (C.S.E.), which gives more than 90% possibility of corrosion (ASTM, 1991). In the cyclic test, the specimen was weekly measured until the average potentials in dry condition reached to −350 mV (C.S.E.). During

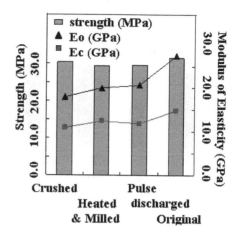

Figure 9. Strength and modulus of elasticity.

the half-cell potential measurement, AE measurement was discontinured.

Chloride concentrations were measured at several stages. At first, the initial concentration was measured by using a standard cylinder sample after 28-day moisture-curing. At other stages, two core samples of 5 cm diameter were taken from the specimens, of which locations are illustrated in Figure 7. Slicing the core into 5 mm-thick disks and crushing them, concentrations of total chloride ions were determined by the potentiometric titration method.

4 RESULTS AND DISCUSSION

4.1 Damage evaluation of recycled concrete

Strengths and moduli of elasticity in four-types of concrete are shown in Figure 9. These are averaged values of three samples for each aggregate. It is found that the strengths of recycled aggregate and the moduli of elasticity are lower than that of the original. Among the recycled concrete, concrete of crushed aggregate has the poorest properties, because a thin mortar-layer is still stuck to the interface with aggregate.

From Equation 8, the intact modulus of elasticity, E^*, for each aggregate was evaluated as an averaged value of three samples. Then, as relative damage the ratios of initial moduli E_0 to intact moduli E^* are evaluated. Results are given in Figure 10.

As mentioned before, experimental values are only available for concrete of pulse-discharged and original aggregates. In the figure, a mechanical property of recycle aggregate is evaluated as the relative damage. The poorest property is observed in concrete of crushed aggregate. The property of concrete of recycled aggregate by the pulse-discharged method is better than that of heated and milled, but a slightly lower

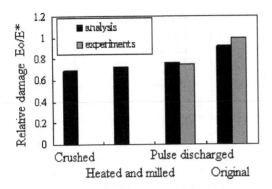

Figure 10. Relative damages of recycles aggregate.

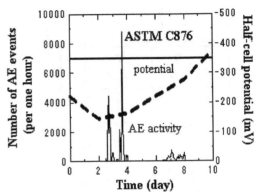

Figure 11. AE activities and half-cell potentials in the accelerated corrosion test.

than that of original aggregate. Agreement between results of AE rate process analysis and actual values in the experiments is remarkable. Although the experimental value of concrete of recycled aggregate by the pulse-discharged method was estimated in comparison with that of original aggregate, analytical value was estimated by AE rate process analysis without knowing that of original aggregate. Thus, it is demonstrated that an application of DeCAT system to evaluate the property of recycled concrete is quantitatively effective.

4.2 Corrosion detection

4.2.1 Accelerated corrosion test
A relation between AE activity and half-cell potentials measured are shown in Figure 11. The number of AE events is plotted as a sum of AE events at two channels counted for one hour. Two periods of high AE activities are observed at around 3 days elapsed and 7 days elapsed. It is noted that the half-cell potentials start to decrease after the 1st activity, but are still higher than −350 mV around at the 2nd activity. Because the half-cell potential lower than −350 mV is prescribed as more than 90% probability of corrosion (ASTM, 1991), results suggest that the corrosion in rebars is detected by AE activity more confidently than the half-cell potential.

Total chloride ions were determined in depths, and chloride concentrations at cover thickness were analytically estimated by,

$$C(\mathbf{x}, \mathbf{t}) = C_0 (1 - erf[\frac{x}{2\sqrt{Dt}}]) \qquad (12)$$

During the test, core samples were taken at four stages, testing all of three specimens. Initially, chloride concentration was measured by using a standard cylindrical sample for the compression test. Following two periods of high AE activities and at the final stage, concrete cores were taken out from the three specimens.

Then, chloride concentrations at cover thickness were measured. These experimental values were analyzed by Equation 12 and are compared in Figure 12 (b), where total AE hits observed during the test are compared with chloride concentration at the rebar.

According to a phenomenological model of reinforcement corrosion (Melchers and Li, 2006), typical corrosion loss is reported as illustrated in Figure 12 (a). At phase 1, corrosion is initiated. The rate of the corrosion process is controlled by the rate of transport of oxygen. As the corrosion products build up on the corroding surface of rebar, the flow of oxygen is eventually inhibited and the rate of the corrosion loss decreases as illustrated as phase 2 in Figure 12 (a). This is a nonlinear corrosion-loss-time relationship for corrosion under aerobic conditions. The corrosion process involves further corrosion loss as phases 3 and 4 due to anaerobic corrosion. Accordingly, two stages of active corrosion loss are modeled. In Figure 12 (b), total number of AE hits are plotted, which were obtained from Figure 11. AE activity is compared with chloride concentration at rebar of measured and analyzed values by Equation 12.

It is observed that the increase in total AE hits during the acceleration test is in remarkable agreement with the typical corrosion loss in the phenomenological model. This implies that AE activity clearly corresponds to corrosion activity on the rebar surface. In the figure, two-threshold values of chloride concentration are denoted. One is the lower-bound threshold for the onset of corrosion (0.3 kg/m³ in concrete volume, 0.088% mass of cement volume) and the other is the threshold value for the performance-based design. According to the code (JSCE, 2001), concrete with chloride content over 1.2 kg/m³ in concrete volume (0.35% mass of cement volume) is not accepted for construction to prevent the corrosion. Comparing AE activity with chloride concentration in Figure 12 (b), it is found that chloride concentration becomes higher

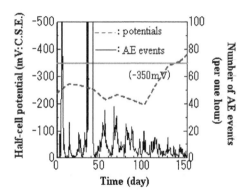

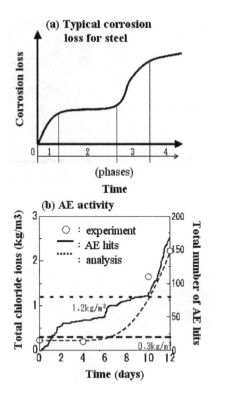

Figure 12. Total number of AE hits and chloride concentration.

Figure 13. AE activities and half-cell potentials in the cyclic wet-dry test.

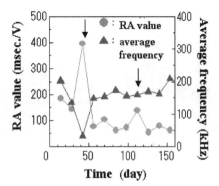

Figure 14. RA values and the average frequencies in the cyclic wet-dry test.

than 0.3 kg.m³ after the 1st AE activity, and it reaches over 1.2 kg/m³ following the 2nd AE activity. This implies that chloride concentrations for corrosion phenomena at rebars reasonably correspond to two-stage high AE activities for the initiation of corrosion and the nucleation of corrosion cracking.

4.2.2 AE activities in the cyclic tests

The number of AE events and the half-cell potentials during the cyclic test are shown in Figure 13. The number of AE events for one hour is again plotted. AE events are periodically observed along with cycles of wet and dry. The 1st high AE activity is observed at 40 days elapsed, while the 2nd activity is not obvious. According to the half-cell potentials, the values start to decrease at around 100 days elapsed.

In order to identify the period of the 2nd AE activity, the RA values and the average frequency were determined from Equations 9 and 10, averaging the data during two weeks of wet-dry cycles. Results are shown in Figure 14. Corresponding to the 1st period as denoted by an arrow symbol, the RA value becomes large and the average frequency is low. According to Figure 5, AE sources are classified as other than tensile cracks. Toward 100 days elapsed, the increase in

the RA value is observed. Thus, the 2nd period is reasonably identified around at 100 days elapsed, where the RA values are low and the average frequencies are fairly high. From Figure 5, tensile cracks are to be nucleated at this period. By analyzing two AE indices of the RA value and the average frequency, the 1st and the 2nd periods of high AE activities are reasonably identified.

The b-value was also determined for each wet-dry cycle as the average value. Results are shown in Fig. 15. The b-value becomes large at the 1st period (upward arrow symbol) and then the b-values keep fairly low. This result implies generation of small cracks of other than a tensile crack around at the 1st period. Then, nucleation of fairly large tensile cracks follows, leading to the 2nd period (downward arrow symbol).

It was found that chloride concentration at rebar reached to the lower-bound threshold of 0.3 kg/m³ after 40 days, and at around 100 days elapsed, the concentration became higher than 1.2 kg/m³. After coring the specimens at 42 days elapsed and 126 days elapsed, rebars were removed and visually inspected.

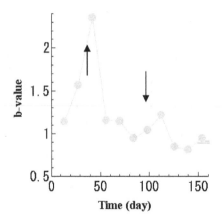

Figure 15.   Variation of b-values in the cyclic wet-dry test.

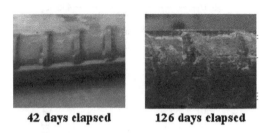

42 days elapsed   **126 days elapsed**

Figure 16.   Visual observation of rebars.

As given in Figure 16, no corrosion was observed after 42 days, while rebar was fully corroded after 126 days. These results imply that AE activities after 100 days could reasonably result from concrete cracking due to expansion of corrosive products in rebars, as chloride concentration reached over $1.2\ kg/m^3$ in concrete.

In order to investigate the condition of rebar at the 1st AE activity in detail, a rebar skin in the left figure of Figure 16 was examined by the scanning electron micrograph (SEM). Distributions of ferrous ions at the initial as received and at 42 days elapsed are compared in Figure 17. At the initial stage, homogeneous distribution of ferrous ions is observed, while they disappear at some regions on the surface at 42 days. This implies that onset of corrosion occurs in rebars at the 1st period of AE activities. Although no corrosion was observed visually, rebar were actually corroded as found in the SEM observation.

Eventually, it is realized that the 1st high AE activity corresponded to the onset of corrosion in the rebars, and the 2nd high AE activity could be generated due to concrete cracking.

Comparing these findings with the deterioration process in Figure 1, it is reasonably demonstrated that two key periods in the corrosion process can be identified by AE monitoring. At the onset of corrosion in rebar, small AE events of the other-type cracks are

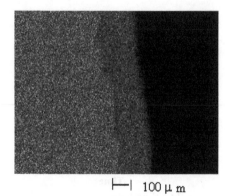

├─┤ 100 μ m

(a) Distribution of ferrous ions (grey region) prior to the test.

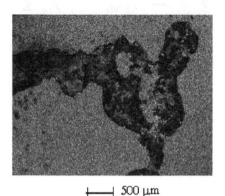

├──┤ 500 μm

(b) Distribution of ferrous ions (grey region) after the 1st period.

Figure 17.   Distributions of ferrous ions on rebar surface.

actively observed. At nucleation of cracking in concrete, tensile cracks are generated as fairly large AE events, which result from cracking in concrete due to expansion of corrosive products.

5   CONCLUSIONS

In the first half of the paper, the quantitative evaluation of damage in concrete is proposed by applying the AE rate-process analysis and damage mechanics. The technique is developed as the DeCAT software. We have the following conclusions.

(1) The relative damage is evaluated by the ratio of the initial modulus of elasticity to the intact modulus, obtained from the DeCAT analysis.
(2) The analysis is applied to evaluate the mechanical property of recycled concrete. Recycled aggregate were obtained by crushing, heated and milled,

and pulse-discharged. Making cylindrical concrete samples of similar mix proportions, the compression test of the samples were conducted. Relative damages evaluated are in reasonable agreement with actual deterioration degrees of recycled concrete.

(3) The experimental value of concrete of recycled aggregate by the pulse-discharged method was compared with that of original aggregate as the relative damage. It is confirmed that the relative damage of the analytical value, which was estimated by the DeCAT analysis without knowing that of original aggregate, is in remarkable agreement with the experimental value.

In the second half, the application of continuous AE monitoring in the corrosion process is studied. Because high AE activities are observed in the corrosion process, AE parameters of the RA value and the average frequency were estimated, along with the b-value of AE amplitude distribution Ingress of chloride ions was measured and analyzed, comparing with AE results. We have the following conclusions.

(4) At the 1st period of high AE activities, the RA values become high, the average frequencies are low and the b-value is large. This implies that small shear cracks are actively generated as AE sources. Approaching to the 2nd period, the RA values become low, the average frequencies are getting higher and the b-values are small. The fact reasonably suggests that fairly large tensile cracks are generated due to expansion of corrosive products.

(5) Compared with AE results, it is found that the onset of corrosion starts, when the chloride concentration exceeds the lower-bound threshold. Nucleation of corrosion cracking is observed, when the chloride concentration becomes over the threshold level specified in the codes.

(6) Removing rebars from the specimen, it is confirmed that rebars could corrode after chloride concentration reaches over the specified threshold, and AE activities after 100 days result from concrete cracking due to expansion of corrosive products in rebars.

(7) In order to investigate the condition of rebars at the 1st activity, a rebar skin was examined by the scanning electron micrograph (SEM). It is found that ferrous ions have disappeared. This suggests that the onset of corrosion occurs in rebars at the 1st period of high AE activities.

(8) These findings demonstrate that two key periods in the corrosion process can be identified both at the onset of corrosion and at the nucleation of cracking by continuous AE monitoring.

REFERENCES

ASTM. 1991, *Standard Test Method for Half-Cell Potentials of Uncoated Reinforcing Steel in Concrete*, ASTM C876.

JCMS. 2003, *Monitoring Methods for Active Cracks in Concrete by A E*, JCMS-II B5706.

JSCE. 2001. *Standard Specifications for Concrete Structures – Version of Maintenance.*

Loland, K.E. 1980. "Continuous Damage Model for Load – Response Estimation of Concrete", *Cement and Concrete Research* , Vol.10, 395–402.

Melchers, R. E. and Li, C. Q. 2006. Phenomenological Modeling of Reinforcement Corrosion in Marine Environments, *ACI Materials Journal*, Vol. 103, No. 1, 25–32.

Ohtsu, M. 1992. "Process Analysis of Acoustic Emission Activity in Core Test of Concrete", *Proc. of JSCE*, No.442/V-16, 11–217.

Shiotani, T. Ohtsu, M. and Ikeda, K. 2001. Detection and Evaluation of AE Waves due to Rock Deformation, *Construction and Building Materials*, Nos. 5–6, 235–246.

Suzuki, T. and Ohtsu, M. 2004, Quantitative Damage Evaluation of Structural Concrete by a Compression Test based on AE Rate Process Analysis, *Construction and Building Materials*, 18, 197–202.

Fracture Mechanics of Concrete and Concrete Structures – Design, Assessment and Retrofitting of RC Structures – Carpinteri, et al. (eds)
© 2007 Taylor & Francis Group, London, ISBN 978-0-415-44616-7

Cracking mechanisms of diagonal-shear failure monitored and identified by AE-SiGMA analysis

K. Ohno, S. Shimozono & M. Ohtsu
Graduate School of Science and Technology, Kumamoto University

ABSTRACT: The maintenance of concrete structures has become a serious problem, because concrete is to be realized as no longer maintenance-free. Recently, diagonal shear failure of concrete structures draws a great attention because of disastrous damages due to earthquakes. Accordingly, structural monitoring and assessment of failure or damage by nondestructive evaluation (NDE) is in remarkable demand. Acoustic emission (AE) is known to be promising for NDE of concrete structures for diagnostics and health monitoring. It is known that fracture mechanisms are identified by AE wave form analysis. As a quantitative waveform analysis of AE signals, SiGMA (simplified Green's functions for moment tensor analysis) procedure has been developed. Based on the moment tensor analysis, crack location, crack type and crack orientation are readily identified. In the present study, diagonal shear failure in reinforced concrete (RC) beams is investigated, applying the SiGMA analysis. Thus, cracking mechanisms are clarified and an application to structural monitoring is discussed.

1 INTRODUCTION

The mechanisms of diagonal shear failure in reinforced concrete (RC) beams have not been completely clarified yet. The failure type of RC beams depends on the ratio of the shear span to the effective depth (a/d). Generally, in the case where the ratio a/d is large than 3.0, diagonal tensile failure occurs in RC beams as generated cracks lead to the ultimate state in the beams.

AE method is one of nondestructive testings for concrete structures for diagnostics and health monitoring. AE phenomena are theoretically defined as elastic waves emitted due to microfracturing or faulting in a solid. Emitted AE waves of feeble amplitudes are characterized by high-frequency components in the ultrasonic range. Because the detected AE waves associated with the sources, information on the source mechanisms are contained in AE waves. As a quantitative inverse analysis of AE waveforms, SiGMA (simplified Green's functions for moment tensor analysis) procedure has been developed (Ohtsu, 1991). Kinematics of AE source, such as crack location, crack type and crack orientation can be analyzed from recorded AE waveforms.

In the present paper, AE method is applied to diagonal shear failure of RC beams. Prior to bending tests of RC beams, theoretical waveforms were calculated in order to determine proper location of AE sensors. Theoretical waveforms were synthesized by applying the dislocation model and Green's functions in a half space. Then, the mechanisms of internal cracks due to bending fracture were identified by SiGMA analysis. In three-dimensional (3D) massive body of concrete, the applicability of SiGMA analysis has been confirmed (Ohtsu et al., 1998). Here AE sources due to diagonal shear failure are located and classified of crack type from recorded AE waveforms.

2 SIGMA ANALYSIS

2.1 Theory of moment tensor

As formulated in the generalized theory (Ohtsu and Ono, 1984), AE waves are elastic waves generated by dynamic-crack (dislocation) motions inside a solid. As AE waves are generated by microcracks, wave motion $u_i(\mathbf{x},t)$ can be represented,

$$u_i(x,t) = \int_F T_{ik}(x,x',t) * b_k(x',x) dS ,\tag{1}$$

where T_{ik} is Green's function of the second kind and * denotes the convolution integral. b_k is the crack motion.

In case of an isotropic elasticity,

$$T_{ik} = \lambda G_{ij,j} n_k + \mu G_{ik,j} n_j + \mu G_{ij,k} n_j ,\tag{2}$$

where λ and μ are Lame constants. G_{ik} are the Green's functions. n_k is the crack normal vector.

Substituting Equation 2 into Equation 1, and introducing moment tensor, M_{pq}, $u_i(\mathbf{x},t)$ can be represented as,

$$u_i(x,t) = \int_F T_{ik}(x,x',t) * b_k(x',t)dS$$

$$= G_{ip,q}(x,x',t)M_{pq} * S(t) \tag{3}$$

Here, $G_{ip,q}(\mathbf{x},\mathbf{x}',t)$ are spatial derivative of Green's functions and $S(t)$ represents the source kinetics (the source-time function). Inverse solutions of Equation 3 contain two-fold information of the sources. Source kinetics are determined from the source-time function $S(t)$ by a deconvolution procedure. Source kinematics are represented by the moment tensor, M_{pq}. In order to perform the deconvolution and to determine the moment tensor, the spatial derivatives of Green's functions or the displacement fields of Green's functions due to the equivalent force models are inherently required. Consequently, based on the far-field term of the P-wave, a simplified procedure was developed (Ohtsu, 1991). The procedure is implemented as the SiGMA (Simplified Green's functions for Moment Tensor Analysis) code.

Mathematically, the moment tensor in Equation 3 is defined by the tensor product of the elastic constants, the normal vector $\mathbf{n}$ to the crack surface and the crack-motion (dislocation or Burgers) vector $\mathbf{l}$.

$$M_{pq} = C_{pqij}l_i n_j \Delta V \tag{4}$$

The elastic constants C_{pqij} have a physical unit of $[N/m^2]$ and the crack volume ΔV has a unit of $[m^3]$. The moment tensor has the physical unit of a moment, $[Nm]$. This is the reason why the tensor M_{pq} was named the moment tensor. The moment tensor is a symmetric second-rank tensor and is comparable to the elastic stress in elasticity as,

$$[M_{pq}] = \begin{bmatrix} m_{11} & m_{12} & m_{13} \\ m_{12} & m_{22} & m_{23} \\ m_{13} & m_{23} & m_{33} \end{bmatrix} \Delta V \tag{5}$$

All elements of the moment tensor are illustrated in Figure 1. In a similar manner to stress, diagonal element represent normal components and off-diagonal elements are shown as tangential or shear components.

2.2 Equivalent Force Models

AE sources can be represented by equivalent force models, such as a monopole force, a dipole force and a couple force. Relations among crack (dislocation) models, equivalent force models and moment tensors are straightforward. From Equation 4, in an isotropic material we have

$$M_{pq} = (\lambda l_k n_k \delta_{pq} + \mu l_p n_q + \mu l_q n_p)\Delta V \tag{6}$$

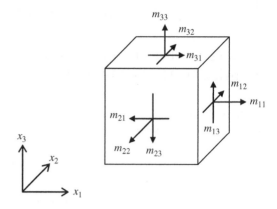

Figure 1. Elements of the moment tensor.

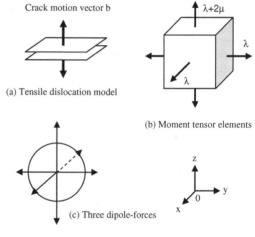

(a) Tensile dislocation model

(b) Moment tensor elements

(c) Three dipole-forces

Figure 2. (a) Tensile dislocation model, (b) related moment tensor elements and (c) three dipole-force.

In the case that a tensile crack occurs on a crack surface parallel to the x-y plane and opens in the z-direction as shown Figure 2, the normal vector $\mathbf{n} = (0,0,1)$ and the crack vector $\mathbf{l} = (0,0,1)$. Substituting these into Equation 6, the moment tensor becomes,

$$M_{pq} = \begin{bmatrix} \lambda & 0 & 0 \\ 0 & \lambda & 0 \\ 0 & 0 & \lambda+2\mu \end{bmatrix} \Delta V \tag{7}$$

Only diagonal elements are obtained, which are shown in Figure 2(b). Replacing these diagonal elements as dipole forces, three dipole-forces are illustrated in Figure 2(c). This implies that combination of three dipoles is necessary and sufficient to model a tensile crack.

In Figure 3, the case of a shear crack parallel to the x-y plane is shown with the normal vector $\mathbf{n} = (0,0,1)$.

992

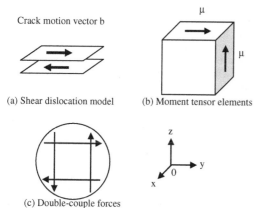

Crack motion vector b

(a) Shear dislocation model

(b) Moment tensor elements

(c) Double-couple forces

Figure 3. (a) Shear dislocation model, (b) related moment tensor elements and (c) double-couple forces.

Shear motion occurs in the y-direction with the crack vector $\mathbf{l} = (0,1,0)$. From Equation 6, we have,

$$M_{pq} = \begin{bmatrix} 0 & 0 & 0 \\ 0 & 0 & \mu \\ 0 & \mu & 0 \end{bmatrix} \Delta V \qquad (8)$$

As seen in Figure 3(c), the double-couple force model is comparable to off-diagonal elements of the moment tensor in Equation 8.

2.3 SiGMA Code

Taking into account only P-wave motion of the far field (1/R term) and considering the effect of reflection at the surface, the amplitude of the first motion is derived from Equation 3. The reflection coefficient Ref($\mathbf{t,r}$) is obtained as $\mathbf{t}$ is the direction of sensor sensitivity and $\mathbf{r}$ is the direction vector of distance R from the source to the observation point, as $\mathbf{r} = (r_1, r_2, r_3)$. The time function is neglected in Equation 3, and the amplitude of the first motion A($\mathbf{x}$) is represented,

$$A(x) = Cs \cdot \frac{\mathrm{Re}f(t,r)}{R} \cdot (r_1, r_2, r_3) \begin{bmatrix} m_{11} & m_{12} & m_{13} \\ m_{12} & m_{22} & m_{23} \\ m_{13} & m_{23} & m_{33} \end{bmatrix} \begin{bmatrix} r_1 \\ r_2 \\ r_3 \end{bmatrix} \qquad (9)$$

where Cs is the calibration coefficient of the sensor sensitivity and material constants. Since the moment tensor is a symmetric tensor of the 2nd rank, the number of independent elements is six. These are represented in Equation 9 as m_{11}, m_{12}, m_{13}, m_{22}, m_{23}, and m_{33}.

These can be determined from the observation of AE waves at more than six sensor locations. In the

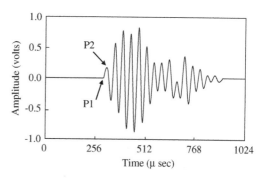

Figure 4. Detected AE waveform.

SiGMA procedure, two parameters of the arrival time (P1) and the amplitude of the first motion (P2) are visually determined from AE waveform as shown in Figure 4. In the location procedure, the source (crack) location $\mathbf{x'}$ in Equation 3 is determined from the arrival time differences t_i between the observation points $\mathbf{x}_i$ and $\mathbf{x}_{i+1}$, by solving equations,

$$R_i - R_{i+1} = |\mathbf{x}_i - \mathbf{y}| - |\mathbf{x}_{i+1} - \mathbf{y}| = v_p t_i \qquad (10)$$

Here v_p is the velocity of P-wave.

After solving Equation 10, the reflection coefficient Ref($\mathbf{t,r}$), the distance R, and direction vector $\mathbf{r}$ are readily obtained to solve Equation 9. The amplitude of the first motions P2 in Figure 4 at more than six channels are substituted into Equation 9, and all the elements of the moment tensor are determined. Since the SiGMA code requires only relative values of the moment tensor elements, the relative coefficients Cs are sufficient.

2.4 Eigenvalue analysis of the moment tensor

In order to classify a crack into a tensile or shear type, a unified decomposition of the eigenvalues of the moment tensor was developed (Ohtsu, 1991). In general, crack motion on the crack surface consists of slip motion (shear components) and crack-opening motion (tensile components), as illustrated in Figure 5.

Thus, it is assumed that the eigenvalues of the moment tensor are the combination of those of a shear crack and of a tensile crack, as the principal axes are identical. Then, the eigenvalues are decomposed uniquely into those of a shear crack, the deviatoric components of a tensile crack and the isotropic (hydrostatic mean) components of a tensile crack. In Figure 5, the ratio X represents the contribution of a shear crack. In that case, three eigenvalues of a shear crack become X, 0, −X. Setting the ratio of the maximum deviatoric tensile component as Y and the isotropic tensile component as Z, three eigenvalues of a tensile crack are denoted as Y + Z, −Y/2 + Z, and −Y/2 + Z. Eventually the decomposition leads to relations,

$$1.0 = X + Y + Z,$$

the intermediate eigenvalue/the maximum eigenvalue

$$= 0 - Y/2 + Z,$$

the minimum eigenvalue/the maximum eigenvalue

$$= -X - Y/2 + Z. \tag{11}$$

It should be pointed out that the ratio X becomes larger than 1.0 in the case that both the ratios Y and Z are negative (Suaris and van Mier, 1995). The case happens only if the scalar product $l_k n_k$ is negative, because the eigenvalues are determined from relative tensor components. Making the scalar product positive and re-computing Equation 11, the three ratios are reasonably determined. Hereinafter, the ratio X is called the shear ratio.

In the present SiGMA code, AE sources with shear ratios less than 40%, are classified as tensile cracks. The sources with X > 60% are classified as shear cracks. In between 40% and 60%, the cracks are referred to as mixed-mode.

From the eigenvalue analysis, three eigenvectors **e1**, **e2**, **e3** are also obtained. Theoretically, these are derived as,

$$
\begin{aligned}
\mathbf{e1} &= \mathbf{l} + \mathbf{n} \\
\mathbf{e2} &= \mathbf{l} \times \mathbf{n} \\
\mathbf{e3} &= \mathbf{l} - \mathbf{n}
\end{aligned}
\tag{12}
$$

Here × denotes the vector product, and the vectors **l** and **n** are interchangeable. In the case of a tensile crack, the vector **l** is parallel to the vector **n**. Thus, the vector **e1** could give the direction of crack-opening, while the sum **e1** + **e3** and the difference **e1** + **e3** give the two vectors **l** and **n** for a shear crack.

To locate AE sources, at least 5-channel system is necessary for 3-D analysis. Since 6-channnel system is the minimum requirement for the moment tensor, 6-channel system is required for the SiGMA-3D analysis.

3 THEORETICAL AE WAVEFORMS

3.1 *AE source models*

In order to determine AE sensor locations, the theoretical waves are analyzed. Based on the location and moment tensors, elastic waves due to a tensile crack, an in-plane shear crack and an out-of-plane shear crack in a half-space were calculated theoretically at the sensor locations. The basic code for computation was already published (Ohtsu & Ono, 1984, Ohtsu & Ono, 1988, Ohtsu & Ohno & Hamstad, 2005).

RC beams of dimensions 250 mm × 150 mm × 2000 mm with 400 mm shear span were tested. The compressive strength and the tensile strength of concrete at 28-day standard curing were 29.7 MPa and 3.03 MPa, respectively. The velocity of P-wave was 4230 m/s and the modulus of elasticity was 28.2 GPa. Poisson's ratio was 0.2. P-wave velocity and Poisson's ratio were applied to SiGMA analysis.

At the origin of the coordinates system, three cracks were considered as source models. Cracks that are a

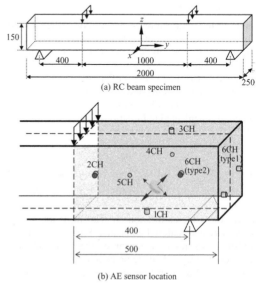

(a) RC beam specimen

(b) AE sensor location

Figure 6. (a) RC beam specimen, (b) AE sensor location (unit:mm).

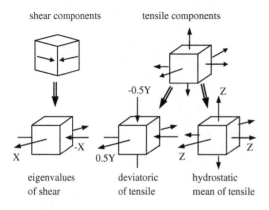

shear components   tensile components

eigenvalues   deviatoric   hydrostatic
of shear   of tensile   mean of tensile

Figure 5. Unified decomposition of eigenvalues of the moment tensor.

tensile crack, an in-plane shear crack and an out-of-plane shear crack are assumed to be generated in the shear span (Figure 6).

A tensile crack, of which the normal vector $\mathbf{n} = (0, 1/\sqrt{2}, 1/\sqrt{2})$ and the crack vector $\mathbf{l} = (0, 1/\sqrt{2}, 1/\sqrt{2})$, occurs inclined 45° to y-axis. The moment tensor of a tensile crack is represented as,

$$M_{pq} = \begin{bmatrix} \lambda & 0 & 0 \\ 0 & \lambda+\mu & \mu \\ 0 & \mu & \lambda+\mu \end{bmatrix} \Delta V . \tag{13}$$

The moment tensor of an in-plane shear crack model the crack vector of which is $\mathbf{l} = (0, -1/\sqrt{2}, 1/\sqrt{2})$ is represented as,

$$M_{pq} = \begin{bmatrix} 0 & 0 & 0 \\ 0 & -\mu & 0 \\ 0 & 0 & \mu \end{bmatrix} \Delta V \tag{14}$$

The moment tensor of an out-of-plane shear crack model is obtained as setting $\mathbf{l} = (1, 0, 0)$

$$M_{pq} = \begin{bmatrix} 0 & \dfrac{1}{\sqrt{2}}\mu & \dfrac{1}{\sqrt{2}}\mu \\ \dfrac{1}{\sqrt{2}}\mu & 0 & 0 \\ \dfrac{1}{\sqrt{2}}\mu & 0 & 0 \end{bmatrix} \Delta V \tag{15}$$

Since six sensors were assumed as one group, two types of sensor sets were located in the specimen. The coordinate of these AE sensor locations are shown in Figure 6 and Table 1. The origin of X and Y coordinates is set at the center of the specimen and Z origin is at the bottom in the specimen. Three types of cracks

Table 1. The coordinate of AE sensor location.

	x(m)	y(m)	z(m)
Type1			
1CH	0.030	0.700	0.000
2CH	0.075	0.600	0.120
3CH	0.000	0.800	0.250
4CH	−0.075	0.750	0.200
5CH	−0.075	0.650	0.050
6CH	−0.025	1.000	0.125
Type2			
1CH	0.030	0.700	0.000
2CH	0.075	0.600	0.120
3CH	0.000	0.800	0.250
4CH	−0.075	0.750	0.200
5CH	−0.075	0.650	0.050
6CH	0.075	0.900	0.105

were modeled in the shear span (Figure 6(b)). At five locations in the shear span, these cracks are nucleated for each crack model. Elastic waves generated due to three types of crack models in the specimen were detected at the surface of specimen by two types of sensor locations.

3.2 Results of SiGMA analysis

In Figure 7, examples of waveforms computed are given. Here, the rise time of the source-time function

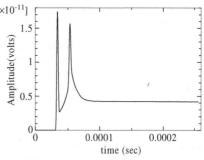

(a) Detected AE waveform due to a tensile crack

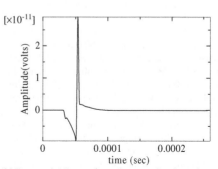

(b) Detected AE waveform due to an in-plane shear crack

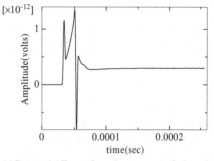

(c) Detected AE waveform due to an out-of-plane shear crack

Figure 7. Example of theoretical waveforms.

Table 2. Results of SiGMA analysis.

Crack mode	Tensile-mode	Mixed-mode	Shear-mode
Assumed crack	Results of Type1		
Tensile	1	1	3
In-plane-Shear	3	1	1
Out-of-plane-Shear	2	1	2
Assumed crack	Results of Type 2		
Tensile	1	4	0
In-plane-Shear	2	0	3
Out-of-plane-Shear	3	0	2

was set to 7 μsec. The SiGMA analysis was applied to these theoretical waves.

Results are given in Table 2. In the both types, location errors between the crack model and the results of SiGMA analysis do not exist. From the number of classified cracks, it is found that the difference of results between Type 1 and Type 2 does not appear clearly.

4 BENDING TEST OF RC BEAM

4.1 Experimental procedure

The bending test was carried out in RC beam specimens in the laboratory. The specified mix of concrete is shown in Table 3. The effective depth of reinforcing bar is 203.5 mm and shear span is 400 mm (a/d = 1.97). AE activities were detected by AE sensors of 150 kHz resonance (R15, PAC) and the sampling frequency for recording waveforms is 1 MHz (DiSP, PAC). AE hits were amplified with 40 dB gain in a pre-amplifier and 20 dB gain in a main amplifier.

Based on results of AE source models, AE sensors were arranged, following type 1 and type 2. To detect as many as possible of AE events due to cracks, 8-channel system extending type 1 and type 2 were employed for AE measurement. The tested RC beam and AE sensor configuration are shown in Figure 6 and Figure 8. To monitor diagonal shear failure in the RC beam with AE sensors, stirrups were arranged in a half portion of the specimen. 8 AE sensors covered the whole area of the shear span without stirrups. For other measurements, displacements on two sides were measured with displacement-transducers.

4.2 Results of AE parameter analysis

In the bending test, the ultimate load of the RC beam was 96.1 kN. Load and displacement during the test are shown in Figure 9. The displacement varied linearly with stress. As the load increased, the number of AE hits increased and the hits were observed frequently near the ultimate load (Figure 10). The similar

process of fracture of the asbestos-cement pipe samples is reported in the previous paper (Suzuki et al., 2006).

The flexural crack was observed in middle of the specimen visually at about 46.0 kN. After the occurrence of these cracks, the number of AE hits increased. The shear cracks were observed in shear span at about 84.8 kN.

AE parameter analysis can classify easily crack into two types of tensile mode and shear mode (JCMS-III B5706-2003 code). Two AE parameters which are the RA value and the average frequency are applied to classification of cracks generated. The results of AE parameter analysis are shown in Figure 12. AE hits of 69.14 % of total is classified into the tensile mode, the ratio of the shear mode is 30.86 %. Here, paying attention to variation of AE hits, the testing period is divided into three stages. The stage 1 is the period where the number of AE hits is a few until 53 minutes. In stage 2 until 68 minutes, as the load increased, frequent AE generation is observed. In the final stage, the number of AE hits became the largest and the diagonal failure was occurred.

The results of the three stages are summarized in Table 4. It is clearly found that almost over 60 % AE hits were classified into tensile mode. In the stage 2, as the flexural crack was observed in the middle of the specimen, it is thought that the ratio of tensile mode increased. In addition, in the stage 3, as many cracks grew from bottom of specimen to top of specimen in the shear span, it is thought that the ratio of shear mode increased.

4.3 Results of SiGMA analysis

During the bending test, 160 AE events were detected by 8 AE sensors and have been analyzed. Results of the SiGMA analysis are shown in Figure 12. Kinematics of AE sources are found on the plane of diagonal shear failure. Here, a shear mode is indicated with the cross symbol, a mixed-mode is the triangle symbol and a tensile mode is the bar symbol.

It is noted that positions of shear cracks are plotted higher than positions of tensile cracks. AE sources of three types are mostly concentrated in around 0.8 m from the center of the specimen. In the SiGMA analysis, the event definition time (EDT) is set to 95 μsec. EDT is uses to recognize waveforms occurring within the specified time from the first-hit waveform and to classify them as part of the current event. Therefore, this time might have influence on AE source locations.

The results of three stages are given in Table 5. In all of three stages, shear cracks are distinguished in these events. The results between Table 4 and Table 5 are different. On the other hand, it is realized that the ratio of tensile cracks increases from stage 1 to stage 2. This result is similar to result of AE parameter

Table 3.   Mix proportion and properties of concrete.

Weight per unit volume (kg/m³)

W/C (%)	Water	Cement	Fine aggregate	Coarse aggregate	Admixture (cc)	Slump (cm)	Air (%)	Maximum gravel size (mm)
55	175	318	717	1178	132	8	6.0	20

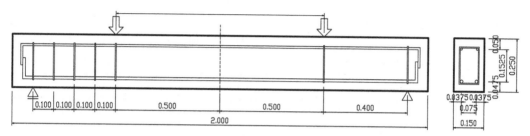

Figure 8.   Sectional view of the specimen (unit:m).

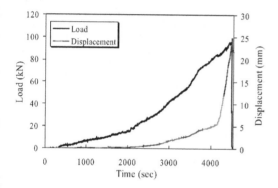

Figure 9.   The relation between load and displacement.

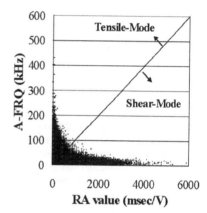

Figure 11.   Results of AE parameter analysis.

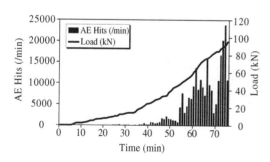

Figure 10.   AE generation behavior in the bending test.

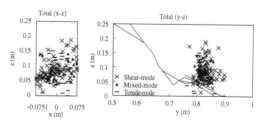

Figure 12.   The result of SiGMA analysis.

analysis. From stage 2 to stage 3, the tendency of results of SiGMA analysis is different from results of AE parameter analysis.

The difference between the AE parameter analysis and the SiGMA analysis could result from the fact that AE parameter analysis is carried out based on all AE hits, but SiGMA analysis is applied to only AE events. AE parameter analysis might be included AE events generated in middle of the specimen. This could be the reason why the ratios of the tensile mode are always high in the AE parameter analysis.

997

Table 4. The ratio of each mode according to three stages. (AE parameter analysis)

	Tensile-mode	Shear-mode
Stage 1	66.81%	33.19%
Stage 2	75.04%	24.95%
Stage 3	61.40%	38.60%
Total	69.14%	30.86%

Table 5. The ratio of each mode according to three stages. (SiGMA analysis)

	Tensile crack	Mixed-mode	Shear crack
Stage1	20.00%	20.00%	60.00%
Stage2	28.81%	15.25%	55.93%
Stage3	29.73%	16.22%	54.05%
Total	28.75%	15.63%	55.63%

5 CONCLUSION

In this paper, the bending test was carried out in the RC beam specimen and diagonal shear failure process was monitored by AE.

Theoretical waveform analysis was applied to decide the optimal arrangement of AE sensor. As a result, the difference of arrangement of AE sensor between type 1 and type 2 is not clear. Therefore, to detect AE events as many as possible due to cracks, 8-channel system was employed for AE measurement.

It is confirmed that there are three stages in AE generating behaviors. It is found that dominant motions of diagonal shear failure are of the tensile model by AE parameter analysis. The results of SiGMA analysis, however, dominant source motions in shear span are of the shear mode.

AE parameter analysis carried out based on detected all AE hits, but SiGMA analysis is applied to only AE events. As a result, the ratio of tensile mode is larger than the ratio of shear mode in the AE parameter analysis. This could be the reason why the ratios of the tensile mode are always high in the AE parameter analysis.

REFERENCES

Ohtsu, M. & Ono, K. 1984. A Generalized Theory of Acoustic Emission and Green's Function in a Half Space. *Journal of AE*. 3(1): 124–133.

Ohtsu, M. & Ono, K. 1988. AE Source Location and Orientation Determination of Tensile Cracks from Surface Observation: *NDT International*. Vol.21. No.3: 143–150.

Ohtsu, M. 1991. Simplified Moment Tensor Analysis and Unified Decomposition of AE Source: Application to in Situ Hydrofracturing Test. *J.Geophys. Res.* 95(2): 6211–6221.

Ohtsu, M., Okamoto, T. & Yuyama, S. 1998. Moment Tensor Analysis of Acoustic Emission for Cracking Mechanisms in Concrete. *ACI Structural Jounal* Vol.95(2): 87–95.

Ohtsu, M., Ohno, K. & Hamstad, M. 2005. Moment Tensors of In-Plane Waves analyzed by SiGMA-2D. *Journal of AE*. Vol.23: 47–63.

Suaris, W. & van Mier, J.G.M. 1995. Acoustic Emission Source Characterization in Concrete under Biaxial Loading. *Materials and Structures*, No.28: 444–449.

Suzuki, T., Ohno, K. & Ohtsu, M. 2006. Damage Evaluation of Existing Asbestos-Cement Pipe by Acoustic Emission. *Proceedings of The 18th International Acoustic Emission Symposium*: 257–262.

Fracture Mechanics of Concrete and Concrete Structures – Design, Assessment and Retrofitting of RC Structures – Carpinteri, et al. (eds)
© 2007 Taylor & Francis Group, London, ISBN 978-0-415-44616-7

SIBIE procedure for the identification of ungrouted post-tensioning ducts in concrete

N. Alver & M. Ohtsu

Graduate School of Science and Technology, Kumamoto University, Kumamoto, Japan

ABSTRACT: SIBIE (Stack Imaging of Spectral Amplitudes Based on Impact-Echo) is an imaging procedure applied to the impact-echo data in the frequency domain. The procedure is developed to improve the impact-echo method and to visually identify locations of defects in concrete. SIBIE has been applied to identification of voids within post-tensioning tendon ducts in concrete. In order to investigate the effect of the distance between impact and detection points on SIBIE results and to improve the applicability of the procedure, a concrete slab containing post-tensioning tendon ducts of metal and plastic was tested. Locations of void within both metal and plastic ducts are identified visually by the SIBIE procedure.

1 INTRODUCTION

SIBIE (Stack Imaging of Spectral Amplitudes Based on Impact-Echo) procedure is developed to improve impact-echo method. The impact-echo method is very well-known as a nondestructive testing for concrete structures (Sansalone, 1997a, b). The method has been applied to such types of defects in concrete as thickness measurement of a slab, grouting performance and void detection in a post-tensioning tendon duct, identification of surface-opening crack depth, location of delamination and determination of material properties.

The impact-echo method has been widely applied to identification of void in tendon-ducts (Sansalone, 1997 a, b, Jaeger, 1996). In principle, the location of void is estimated by identifying peak frequencies in the frequency spectrum. However, the frequency spectrum cannot always be interpreted successfully, because many peaks are often observed in the spectrum. Particularly, in the case of a plastic sheath, it becomes more difficult to interpret the frequency spectrum due to the existence of many peaks. This is because a plastic sheath has lower acoustic impedance than concrete or grout. In order to circumvent it, SIBIE procedure is developed (Ohtsu, 2002). SIBIE procedure has been applied to void detection within tendon ducts (Ata et al., 2005, Alver). In this study, the procedure is applied to a concrete specimen containing a metal and a plastic post-tensioning tendon duct. Two-accelerometer system is employed to improve SIBIE results.

The effect of the distance between impact and detection points is also investigated by changing the position of accelerometers at detection points.

2 SIBIE PROCEDURE

Since it is often not easy to identify the particular peaks in the frequency spectrum in the impact-echo method, an imaging procedure is applied to the result of FFT analysis as SIBIE (Stack Imaging of spectral amplitudes based on Impact-Echo). So far, SIBIE is a post-processing technique to impact-echo data. This is an imaging technique for detected waveforms in the frequency domain. In the procedure, first, a cross-section of concrete is divided into square elements as shown in Figure 1. Then, resonance frequencies due to reflections at each element are computed. The travel distance from the input location to the output through the element is calculated as (Ohtsu, 2002),

$$R = r_1 + r_2. \tag{1}$$

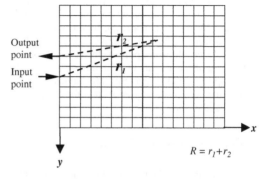

Figure 1. Spectral imaging model.

Resonance frequencies due to reflections at each element are calculated from,

$$f_2' = \frac{C_p}{r_2}, \quad \text{and} \quad f_R = \frac{C_p}{R}. \quad (2)$$

Spectral amplitudes corresponding to these two resonance frequencies in the frequency spectrum are summed up. Thus, reflection intensity at each element is estimated as a stack image. The minimum size of the square mesh Δ for the SIBIE analysis should be approximately equal to $C_p\Delta t/2$, where C_p is the velocity of P-wave and Δt is the sampling time of a recorded wave.

3 EXPERIMENTAL STUDY

The experimental work was carried out in the laboratory. Impact tests were conducted by shooting an aluminum bullet at the surface of a concrete specimen. Dimensions of the specimen are 1000 mm × 400 mm × 260 mm which contains an ungrouted metal sheath and an ungrouted plastic sheath. In order to confirm application of SIBIE to a plastic tendon duct, a specimen containing a plastic sheath was made. The specimen is illustrated in Figure 2. Locations of the tendon ducts are shown in the figure. The aluminum bullet of 8 mm diameter was shot by driving compressed air with 0.05 MPa pressure to generate elastic waves. It is confirmed that the upper bound frequency due to the bullet could cover up to 40 kHz, by using an accelerometer system. Fourier spectra of accelerations were analyzed by FFT (Fast Fourier Transform). Sampling time was 4 μsec and the number of digitized data for each waveform was 2048. The locations of impact and detection are also shown in Figure 2. Two accelerometers were used at the detection points to record surface displacements caused by reflections of the elastic waves. The frequency range of the accelerometer system was from DC to 50 kHz. Locations of accelerometers attached were changed to investigate the effect of the distance between the impact and detection points. These distances are 20 mm, 30 mm, 40 mm, 50 mm and 90 mm. P-wave velocity of the test specimen was obtained as 4025 m/s by the ultrasonic pulse-velocity test. Mixture proportions of concrete are listed in Table 1, along with the slump value and air contents. Mechanical properties of concrete moisture-cured at 20°C for 28 days are summarized in Table 2.

4 RESULTS AND DISCUSSION

Frequency spectra obtained by the impact-test are given in Figure 3. Figure 3a and b are spectra of

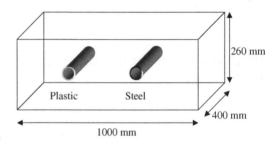

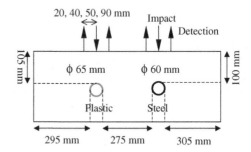

Figure 2. Concrete specimen tested.

Table 1. Mixture proportion and properties of concrete.

Weight per unit volume (kg/m³)				
W/C (%)	Water	Cement	Fine aggregate	Coarse aggregate
55	182	331	743	1159
Admixture (cc)	Slump (cm)	Air (%)	Maximum gravel size (mm)	
132	8	4.5	20	

Table 2. Mechanical properties of concrete at 28-day standard cured.

Compressive strength (MPa)	Young's modulus (GPa)	Poisson's ratio
32.5	29.8	0.28

impact test of plastic sheath, the former is the result obtained by right accelerometer and the latter is the result obtained by left accelerometer. Figure 3c and 3d are results of metal sheath obtained by right and left accelerometers, respectively. Calculated values of the resonance frequencies due to thickness, $f_T = C_p/2T$ and void, $f_{void} = C_p/2d$ and $f'_{void} = C_p/d$ are indicated with lines (Sansalone, 1997, Ohtsu, 2002). It can be seen from the frequency spectra that it is difficult to identify particular peaks since there exist many peaks. Even in the case of a metal sheath, it is still difficult to interpret the frequency spectra.

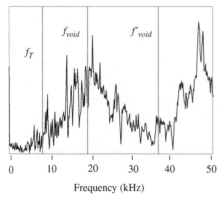

a) Right accelerometer above plastic sheath

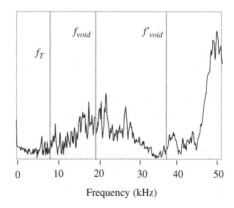

b) Left accelerometer above plastic sheath

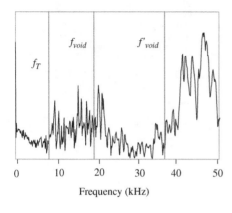

c) Right accelerometer above metal sheath

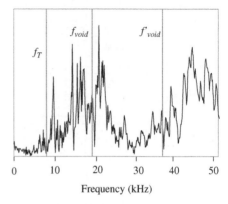

d) Left accelerometer above metal sheath

Figure 3.   Frequency spectra obtained by the impact-test (the distance between impact and detection is 20 mm).

SIBIE analysis was conducted by simply adding two impact-echo results obtained from two accelerometers for each case to visually identify location of void in plastic and metal sheaths. The cross-section of the concrete specimen was divided into square elements to perform the SIBIE analysis. In this study, the size of square mesh for SIBIE analysis was set to 10 mm. SIBIE results for plastic sheath are given in Figure 4, which shows a cross-section of half of the specimen where the plastic sheath is located. In Figure 4a, SIBIE result of impact test for the case of impact-detection distance 20 mm, which is smaller than the plastic sheath diameter, is given. The dark color zones indicate the higher reflection due to the presence of void. It is clearly seen that there is a high reflection in front of the ungrouted plastic sheath. No other high reflections are observed. In Figure 4b, SIBIE result of impact test for the case of impact-detection distance 40 mm, which is slightly larger than

the plastic sheath diameter, is shown. Black color of high reflection is clearly observed in front of the plastic sheath. There is another high reflection observed below the plastic sheath which might be due to a reflection at the bottom of the tendon-duct. The SIBIE result for the case of impact-detection distance 50 mm is shown in Figure 4c. High reflections are clearly observed in front of the plastic sheath. Similar to the former case, there are reflections observed at the bottom of the duct. For the case of impact-detection distance 90 mm, SIBIE result is given in Figure 4d. High reflections are observed in front of the plastic sheath. There are no other high reflections observed at the cross-section. Thus, it is confirmed that SIBIE procedure is available for void detection within plastic sheaths.

SIBIE results for metal sheath are given in Figure 5, which shows a cross-section of half of the specimen where the metal sheath is located. SIBIE result of

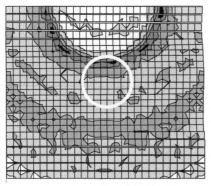

a) Distance between impact and detection is 20 mm.

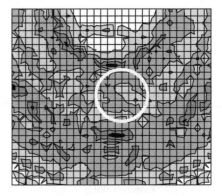

b) Distance between impact and detection is 40 mm.

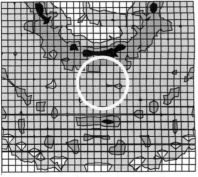

c) Distance between impact and detection is 50 mm.

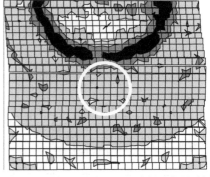

d) Distance between impact and detection is 90 mm.

Figure 4. SIBIE results for plastic sheath (different impact-detection distances).

impact test for the case of impact-detection distance 20 mm, which is smaller than the metal sheath diameter, is given in Figure 5a. High reflections are observed in front of the metal sheath. In case of impact-detection distance 40 mm and 50 mm, SIBIE results are shown in Figure 5b and c, respectively. Reflections are observed in front of the duct and at the bottom of the duct for both cases due to the reflection at the bottom of the metal sheath. For the case of impact-detection distance 90 mm, SIBIE result is given in Figure 5d. High reflections are observed in front of the metal sheath. There are no other high reflections observed at the cross-section. These demonstrate that SIBIE procedure could identify the location of an ungrouted metal duct.

5 CONCLUSION

A concrete specimen containing a metal and a plastic post-tensioning tendon duct was tested by applying the SIBIE procedure. Two-accelerometer system was used to detect elastic waves. Impact was applied at the top of the tendon ducts. Distance of impact and detection locations were changed to investigate the effect of it to SIBIE results. Frequency spectra obtained by the impact test show the difficulty to identify the resonance frequencies of void and thickness only from the spectra due to existence of many peaks. In the case of the impact-detection distance is smaller or relatively larger than the sheath diameter, reflections are observed in front of the duct. In the case of the impact-detection distance is almost equal to the sheath diameter, reflections are observed in front and at the bottom of the duct. The most suitable test configuration is the case of the impact-detection distance is almost equal to the sheath diameter. Reflections from both the top and bottom of the tendon ducts can be observed in this case. In all cases, locations of metal and plastic sheaths were identified by SIBIE. Thus, it is demonstrated that SIBIE procedure is available to detect unfilled ducts in prestressed concrete structures.

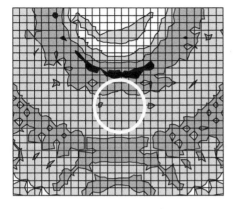

a) Distance between impact and detection is 20 mm.

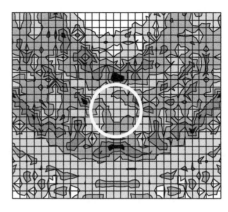

b) Distance between impact and detection is 40 mm.

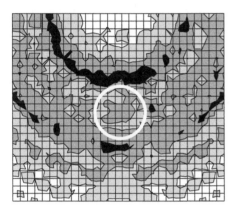

c) Distance between impact and detection is 50 mm.

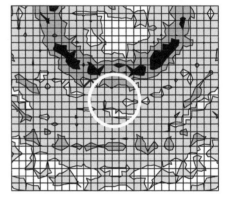

d) Distance between impact and detection is 90 mm.

Figure 5.   SIBIE results for metal sheath (different impact-detection distances).

REFERENCES

Alver, N. & Ohtsu, M. BEM analysis of dynamic behavior of concrete in impact-echo test, *Construction and Building Materials*, Elsevier Science Ltd., (in press).

Ata, N., Mihara, S. & Ohtsu, M. 2005, BEM analysis on dynamic behavior of concrete member due to impact, *Journal of Materials, Concrete Structures and Pavements, JSCE,* Vol. 68, pp. 157–163.

Ata, N., Mihara, S. & Ohtsu, M. Imaging of ungrouted tendon ducts in prestressed concrete by improved SIBIE, *NDT&E International*, Elsevier Science Ltd., (in press).

Jaeger B.J. & Sansalone M.J. 1996. Detecting voids in grouted tendon ducts of post tensioned concrete structures using the impact-echo method, *ACI Structural Journal*, 93(4), 462–472.

Ohtsu M. & Watanabe T. 2002. Stack imaging of spectral amplitudes based on impact-echo for flaw detection, *NDT&E International*, 35, 189–196.

Sansalone M.J. & Streett W.B. 1997. Impact-echo, Ithaca, NY Bullbrier Press.

Sansalone M. 1997. Impact-echo: The complete story, *ACI Structural Journal*, 94(6), 777–786.

Fracture Mechanics of Concrete and Concrete Structures – Design, Assessment and Retrofitting of RC Structures – Carpinteri, et al. (eds)
© 2007 Taylor & Francis Group, London, ISBN 978-0-415-44616-7

Evaluation of the surface crack depth in concrete by impact-echo procedures (SIBIE)

M. Tokai & M. Ohtsu
Kumamoto University, Graduate School of Science and Technology, Kumamoto, Japan

ABSTRACT: The impact-echo method is available for estimating the surface-crack depths. This method is based on monitoring wave motions resulting from a short-duration mechanical impact. Surface displacements are detected by identifying peak frequencies in the frequency spectra. However, the frequency spectra cannot always be interpreted successfully as many peaks are often observed in the spectra. Thus, an imaging procedure to the impact-echo data which is knows as SIBIE procedure is developed. After confirming an applicability of the SIBIE procedure to estimate the depths of open surface cracks, it is clarified that the effect of water filled in the crack is inconsequent. Thus, even though the cracks are filled with water, the depths can be estimated with reasonable accuracy by SIBIE. In addition, the case where the surface cracks are repaired improperly and the void exists around the crack-tip is studied. In all the cases, surface-crack depths could be visually identified by the SIBIE procedure.

1 INTRODUCTION

In ultrasonic testing, time-of-flight technique is used to determine the surface-crack depth. When the distances from the crack to the impact point and from the crack to the sensor are known and the time between the input and the arrival of the diffracted p-wave at the sensor is measured, the depth of the crack can be calculated. However, previous results show the difficulty to measure the time of first-arrived wave. Particularly, in case of water-filled crack, if crack is not fully filled with water, first-arrived wave could give wrong information about the crack depth. In this study, frequency spectrum is analyzed by imaging procedure to visually identify the depth of a surface-crack, which is developed as the SIBIE procedure(Watanabe & Ohtsh, 2000). Thus, surface-crack depths in concrete are visualized by applying SIBIE procedure.

2 IMPACT-ECHO AND RESONANCE FREQUENCY

In the impact-echo method, elastic waves are generated by a short duration mechanical impact. Applying an impact, elastic waves are detected. Then peak frequencies are identified after FFT(Fast Fourier Transform) analysis of detected waves(Sansalone & Lin & Streett, 1998). Theoretically, frequency responses of a concrete member containing defects depend on the size of the member, the location of defects and P-wave velocity. Concerning the frequency responses, following relationships between the resonance frequencies due to the reflections and the depth of a defect are known (Sansalone, & Streett, 1997),

$$f_t = \frac{0.96C_P}{2T} \tag{1}$$

$$f_{crack} = \frac{0.96C_P}{2d} \tag{2}$$

where f_t is the resonance frequency of a plate thickness T, f_{crack} is the resonance frequency of a defect in depth d, C_p is P-wave velocity and 0.96 is a shape factor determined from geometry. The presence of these frequencies is illustrated in Figure 1.

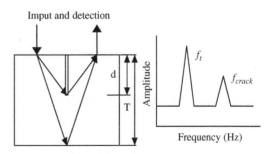

Figure 1. Frequency response of a specimen with crack.

Concerning dynamic motions of concrete members, the dimensional analysis has been carried out (Ohtsu, 1996). Among such parameters as frequency f, characteristic length L and wave velocity V, a following non-dimensional parameter α, as a ratio of characteristic length L to wavelength λ, is obtained.

$$\alpha = \frac{fL}{V} = \frac{L}{\lambda} \qquad (3)$$

In the case that α is larger than 1, it is found that the frequency response is significantly influenced by the characteristic length. Considering the case that α is equal to 1, L is replaced by the thickness T and the depth of a crack d, substituting C_p to V, following relationships are obtained.

$$f'_t = \frac{C_p}{T} \qquad (4)$$

$$f'_{crack} = \frac{C_p}{d} \qquad (5)$$

f'_{crack} and f'_t imply the existence of higher resonance frequencies than those in Equations (4) and (5). Here, the shape factor is not taken into account.

3 SIBIE PROCEDURE

Since it is often not easy to identify the particular peaks in the frequency spectrum in the impact-echo method, an imaging procedure is applied to the result of FFT analysis as SIBIE (Stack Imaging of Spectral Amplitudes Based on Impact-Echo). This is an imaging technique for detected waveforms in the frequency domain. In the procedure, first, a cross-section of concrete is divided into square elements as shown in Figure 2. Then, resonance frequencies due to reflections at each element are computed. The travel distance from the input location to the output through the element is calculated as (Ohtsu & Watanabe, 2002),

$$R = r_1 + r_2 \qquad (6)$$

Resonance frequencies due to reflections at each element are calculated from,

$$f'_{crack} = \frac{C_p}{(R/2)} \qquad (7)$$

$$f_{crack} = \frac{C_p}{R} \qquad (8)$$

Spectral amplitudes corresponding to these two resonance frequencies in the frequency spectrum are summed up. Thus, reflection intensity at each element is estimated as a stack image. The minimum size of the square mesh Δ for the SIBIE analysis should be approximately equal to $C_p \Delta t / 2$, where C_p is the

velocity of P-wave and Δt is the sampling time of a recorded wave.

4 EXPERIMENTS

4.1 Specimen

Experiments were carried out in the laboratory on concrete blocks with surface-cracks of different depths. An artificial surface-crack of 0.5 mm width was located in the middle of the top surface of each specimen by placing a metal plate by casting concrete and by removing it afterwards. A rectangular parallelepiped specimen is of dimensions 400 mm × 250 mm × 300 mm. This had a crack of 100 mm depth. This specimen was tested when the crack was empty (air-filled) and after the crack was filled up to 80% of depth with water. Another specimen is of dimensions 400 mm × 250 mm × 150 mm with a crack of 5 cm depth. The upper half part of the crack was then grouted to test an improperly-repaired crack. The specimens are illustrated in Figure 3. Mixture proportions of concrete are listed in Table 1, along with the slump value and air contents. Mechanical properties of concrete moisture-cured at 20°C for 28 days are summarized in Table 2.

4.2 Impact test

Impact tests were conducted by shooting an aluminum bullet against the concrete surface. An aluminum bullet with a diameter of 8 mm in Figure 4 was shot by driving compressed air with a pressure of 0.05 MPa.

Surface displacements due to the impact were recorded by an accelerometer. The accelerometers

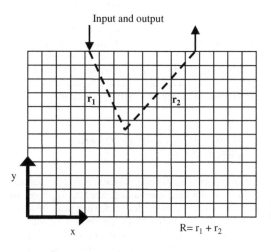

Figure 2. Spectral imaging model.

used at the detection points were of flat type and had sensitivity up to around 50 kHz. The accelerometer and shooting point are arranged as shown in Figure 5. Two cases (Case 1 and Case 2) of 100 mm and 200 mm between impact and detection were tested. Fourier spectra of accelerations were analyzed by FFT (Fast Fourier Transform). Sampling time was 4 μsec and the number of digitized data for each waveform was 2048.

P-wave velocity, C_p, was obtained as 4005 m/s for a rectangular parallelepiped specimen and as 3800 m/s for the specimen with half-grouted crack from the ultrasonic pulse-velocity test.

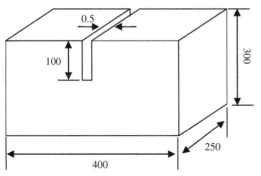

(a) A rectangular parallelepiped specimen

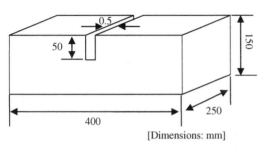

[Dimensions: mm]

(b) A specimen with a partially grouted crack

Figure 3.   Specimens with crack.

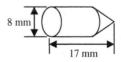

Figure 4.   Configuration of the aluminum bullet.

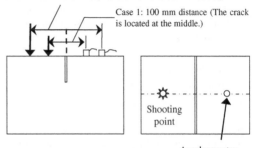

Figure 5.   Location of impact test.

Table 1.   Mixture proportion and properties of fresh concrete.

	Weight per unit volume (kg/m³)				Max. gravel size (mm)	W/C (%)	Air-entrained admixture (cc)	Slump (cm)	Air content (%)
	Water (W)	Cement (C)	Sand (S)	Gravel (G)					
Rectangular parallelepiped specimen	182	331	746	1024	20	55	133	8	6
Specimen with a partially grouted crack	168	305	748	1183	20	55	137	5.2	6.2

Table 2.   Mechanical properties of concrete at 28- day standard cured.

	Compressive strength (MPa)	Poisson's ratio	Young's modulus (GPa)	P-wave velocity (m/sec)
Rectangular parallelepiped specimens	33.0	0.30	30.3	4005
Specimen with a partially grouted crack	29.9	0.22	27.1	3800

5 RESULTS AND DISCUSSION

5.1 *Results of impact tests in a rectangular parallelepiped specimen*

Frequency spectra obtained are shown in Figure 6a–d. The resonance frequencies of the crack depth f_{crack} and f'_{crack} are indicated with arrows. In the Case 1, the resonance frequencies of crack at 100-mm depth, f_{crack} and f'_{crack}, were calculated as 17.9 kHz and 35.8 kHz from Equations (2) and (5). In Figure 6a, although f_{crack} is observed at approximately 17 kHz, f'_{crack} is weaker than f_{crack}. In Figure 6b, f_{crack} is observed at approximately 17 kHz and f'crack is observed at 35 kHz, although f_{crack} is not a strong peak.

In the Case 2, the resonance frequencies of crack at 100-mm depth, f'_{crack} and f'_{crack}, were calculated as 14.2 kHz and 28.3 kHz. In Figure 6c, f'_{crack} is observed at approximately 13 kHz and f'_{crack} is observed at 27 kHz. However, identification is so difficult that there are many peaks in this spectrum.

In Figure 6d, f_{crack} is observed at approximately 13 kHz and f_{crack} is observed at 26 kHz, although f_{crack} is not a strong peak.

5.2 *Results of impact tests in Specimen with a partially grouted crack*

Frequency spectra obtained are shown in Figure 7a–b. The resonance frequencies of the crack depth f'_{crack} and f_{crack} are indicated with arrows. In the Case 1, the resonance frequencies of crack at 50-mm depth, f'_{crack} and f_{crack}, were calculated as 26.9 kHz and 53.7 kHz from Equations (2) and (5). In Figure 7a, although f_{crack} is observed at approximately 27 kHz, f'_{crack} is not observed clearly. In the Case 2, the resonance frequencies of crack at 50-mm depth, f_{crack} and f'_{crack}, were calculated as 17.0 kHz and 34.0 kHz. In Figure 7b, f_{crack} is observed at approximately 15 kHz, but f'_{crack} is not observed clearly.

Because there are many peaks in all frequency spectra, a human error must be included in decision of peak frequency by the impact echo method.

5.3 *Results of SIBIE and discussion*

By using the frequency spectra shown in Figure 6(a)–(d), the SIBIE analysis was performed. Results of SIBIE are shown in Figure 8. The surface crack is indicated as white zone and water part of the crack is indicated as black zone. The mesh elements are arranged at 10-mm pitch evenly. In the figures, dark color regions indicate the high intense regions due to the resonance of diffraction. Arrows show the impact and the detection points.

Results for the specimen with an empty crack are given in Figure 8(a) and (c). Black color indicates

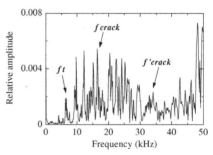

(a) Cross-section with an empty crack (Case 1).

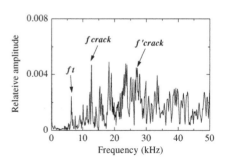

(b) Cross-section with water-filled crack (Case 1)

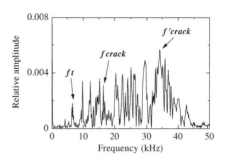

(c) Cross-section with an empty crack (Case 2).

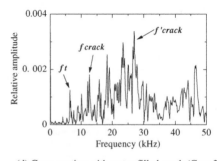

(d) Cross-section with water-filled crack (Case 2)

Figure 6. Frequency spectra of the specimen with a crack of 100-mm depth.

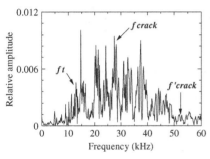

(a) Cross-section with a partially grouted crack (Case 1).

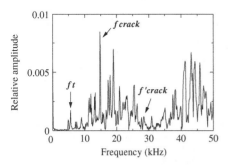

(b) Cross-section with a partially grouted crack (Case 2).

Figure 7. Frequency spectra of eh specimen with a crack of 50-mm depth.

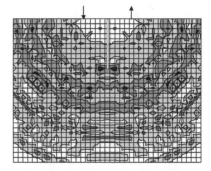

(a) Cross-section with an empty crack (Case 1).

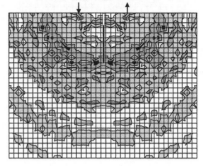

(b) Cross-section with water-filled crack (Case 1).

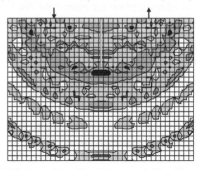

(c) Cross-section with an empty crack (Case 2).

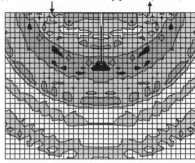

(d) Cross-section with water-filled crack (Case 2).

Figure 8. SIBIE results of the specimens with a crack of 100-mm depth.

the high intense regions which are clearly observed at 100-mm depth in front of the crack tip. Thus, it is demonstrated that the depth of a surface-crack can be visually identified by the SIBIE procedure.

Experiments were repeated on the same concrete specimen after filling 80% of the crack depth with water. By applying frequency spectra obtained from the tests, the SIBIE analysis was performed. Results for specimen with 100-mm crack depth filled with water are given in Figure 8(b) and (d). Dark color regions of diffraction are clearly observed in front of the crack tip in both cases. It is found that water had little effect on the SIBIE results. Thus, high diffractions are observed in front of the crack tip regardless of whether the crack is empty or filled with water.

The SIBIE analysis was performed to determine when the crack is grouted improperly. A result of the SIBIE analysis is shown in Figure 9, where a crack is shown as white zone and grouted part of the crack is indicated with black zone. In Figure 9(a), it is found that the imperfectly grouted crack is visually identified. High diffraction of dark color regions are observed around the ungrouted part of the crack. In Figure 9(b), it is found that diffraction of dark color regions are not clearly observed around the ungrouted

1009

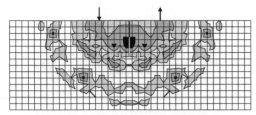

(a) Cross-section with a partially grouted crack (Case 1).

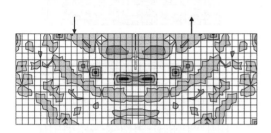

(b) Cross-section with a partially grouted crack (Case 2).

Figure 9.   SIBIE results of the specimens with a partially grouted crack.

part of crack. This might be because that the measurement distance is too far against the crack depth. All SIBIE results of figures 8 and 9 show a completely symmetric picture. This is because the element of identical distance on the travel path is estimated by the same reflection intensity.

6   CONCLUSIONS

Concrete specimens with surface-cracks of varying depths were tested by applying the SIBIE procedure.

The effect of the measurement distance is studied. Conclusions are summarized, as follows;

1. Specimens with a crack, and a water-filled crack were tested. In all the cases, the depths of surface-cracks can be visually identified by the SIBIE procedure.
2. In case of the partially grouted crack, an ungrouted part of the crack is also clearly determined.
   Thus, it is demonstrated that the SIBIE procedure has a great promise to determine the depth of the surface crack.
3. It was found that results of the SIBIE analysis were dependent on the measurement distance. It is necessary to clarify the relation between the measurement distance and the crack depth.
4. In this paper, an artificial surface-crack of 0.5 mm width was estimated. A fine crack is not estimated yet. It should be estimate in a future plan.

REFERENCES

Ohtsu, M. 1996. On high-frequency seismic motions of reinforced concrete structures. *J Mater Concrete Struct Pavements* 544: 277–280

Ohtsu, M & Watanabe, T. 2002. Stack imaging of spectral amplitudes based on impact-echo for flaw detection, *NDT&E International* 35: 189–196

Sansalone, M.J & Streett, W.B. 1997. *Impact-echo, Ithaca, NY:Bullbrier Press*

Sansalone, M.J & Lin, J.M. & Streett, W.B. 1998. Determining the Depth of Surface-Opening Cracks Using Impact-Generated Stress Waves and Time-of-Flight Techniques, *ACI Materials Journal* 95(2): 168–177

Watanabe, T. & Ohtsu M. 2000. Spectral imaging of impact echo technique for grouted duct in post-tensioning prestressed concrete beam, nondestructive testing in civil engineering, *Elsevier*: 453–461

Fracture Mechanics of Concrete and Concrete Structures – Design, Assessment and Retrofitting of RC Structures – Carpinteri, et al. (eds)
© *2007 Taylor & Francis Group, London, ISBN 978-0-415-44616-7*

Determination of surface crack depth and repair effectiveness using Rayleigh waves

T. Shiotani & D.G. Aggelis
Research Institute of Technology, Tobishima Corporation, Noda-shi, Japan

ABSTRACT: Determination of surface crack depth and assessment of repair effectiveness are two non trivial tasks. In the present work, certain correlations between energy related characteristics and crack depth are observed, leading to a multivariate estimation approach. After repair, the same wave parameters are used to characterize the efficiency of repair, since propagation is restored after successful epoxy injection. As for in-situ application, wave inspection is described on a cracked concrete bridge deck, where the cracks were repaired with epoxy agent. It is concluded that the investigation demonstrated the efficiency of the injection, since wave energy and velocity were restored.

1 INTRODUCTION

Surface cracks are the most commonly seen kind of defects in concrete structures. The reason could be overloading, weathering, drying shrinkage, differential settlement, other degradation processes or combination of the above. The major threat they pose concerns the exposure of the metal reinforcement to environmental agents leading to its oxidation (Issa & Debs 2007, Hevin et al. 1998, Ono 1988, Kruger 2005, Shiotani et al. 2005). In order to seal the crack sides and protect the interior, along with structural strength restoration, injection of epoxy or other agent can be used (Aggelis & Shiotani 2006, Binda et al. 1997, Thanoon et al. 2005). However, the effectiveness of the filling as well as the initial determination of crack depth are not trivial tasks. Ultrasonic waves have been used as a non destructive technique in the case of crack characterization. The transit time of longitudinal waves diffracted by the tip of the crack can be used for the estimation of crack depth (Sansalone and Street 1997). However, due to attenuation in many cases the detection of the actual onset of the received waveform could be troublesome, leading to erroneous results. Apart from this, possible bridging points due to reinforcement or other materials could induce further error in the estimation (Liu et al. 2001, Malhotra & Carino 1991). Therefore, the use of Rayleigh waves has been studied extensively (Hevin et al. 1998, Wardany et al. 2004, Zerwer et al. 2005). The advantage of Rayleigh waves is that they carry higher amount of energy than bulk waves as well as their lower geometric spreading, allowing them to propagate at longer distances (Graff

1975). In many of the above studies and others (Doyle & Scala 1978, Achenbach and Cheng 1996, Edwards et al. 2006) amplitude or energy related parameters as well as frequency content has been related to artificial cracks' depth. However, there is no generally accepted and reliable treatment while the issue of non destructive evaluation of repair effectiveness has not been addressed.

In the present paper the correlation of waveform parameters with the depth of artificially machined slots in concrete specimens is studied. Certain trends are observed leading to a multivariate treatment of energy parameters in order to increase the accuracy of depth characterization. Due to the inherent difficulties the accuracy of estimation using one parameter is limited and therefore, combined analysis of different wave features is desirable, as current trends imply (Wu 2006). Additionally, filling of some slots with epoxy was conducted showing that waveform parameters are restored to a great degree while the partial filling of the slot is also examined. As an actual example, the examination of surface opening cracks on a concrete bridge deck before and after repair with epoxy is presented. The ultrasonic measurements reveal the efficiency of repair with the increase in propagation velocity and signal transmission.

2 EXPERIMENTAL SETUP

2.1 Materials and geometry

Two concrete specimens were casted using water to cement ratio of 0.43 and maximum aggregate

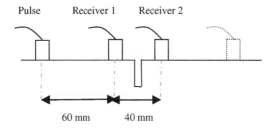

Figure 1. Representation of the experimental setup.

size of 20 mm. The specimens were of prism shape ($150 \times 150 \times 500$ mm³). After the completion of hydration (28 days in water), slots of different depths were machined, from 2 mm up to 23 mm, perpendicular to the longitudinal axis of the specimen. On opposite surfaces of each specimen 2 slots were cut with a sufficient distance of at least 150 mm between them to avoid interactions. The slot width is 4 mm.

2.2 Sensors and excitation

Although different configuration were tested, in this study results from pulse generator C-101-HV of Physical Acoustics Corp, PAC, with a main excitation frequency of around 115 kHz, and the 1910 function synthesizer of NF Electronic Instruments are discussed. Concrete, due to attenuation, limits the propagating frequencies to bands around or below 100 kHz. Therefore, sensitive sensors at this range were used, namely R6, of Physical Acoustics Corp., PAC. However, measurements with broadband sensors, i.e. Fujiceramics FC 1045S, were also conducted to confirm that the resulting trends were not dependent on the sensors frequency response. The representation of the experimental setup can be seen in Figure 1. In order to choose a suitable separation, different distances were tested in sound material (Aggelis & Shiotani, submitted). A separation of 60 mm was chosen to avoid near field effects. The second receiver was placed 40 mm away from the first. It is noted that excitation was conducted from both sides, resulting in similar trends.

3 RESULTS

The aim of this experimental series is twofold; first to search for correlations between slot depth and wave parameters that could lead to characterization and second to check the validity of such a characterization. Therefore, the slots cut on one specimen (2 mm, 9.5 mm, 13 mm, 23 mm) as well as sound material were initially tested to establish some preliminary correlation curves. Afterwards, the slots of the other specimen were tested to examine if the initially observed correlations hold. Additionally, new measurements on the

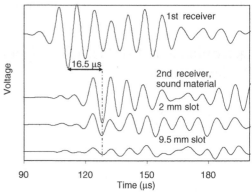

Figure 2. Waveforms collected at wave paths with different slots. The amplitude of the 1st receiver's waveform is reduced to 30% for fitting purposes.

slots of the first specimen were done to check the repeatability of the whole procedure. Therefore, in the following figures, the preliminary curve obtained by the initial data is drawn in dotted line while the correlation curve obtained by the total population of measurements which is therefore considered more reliable is also presented.

In Figure 2 the waveforms of the first and the second receiver are depicted for different slot depths. It is seen that the major part of the energy is due to Rayleigh waves. For the sound material case the Rayleigh burst is clearly observed after the first longitudinal arrivals. The transit time between major peaks corresponds to the Rayleigh wave velocity of 2400 m/s. For the case of 2 mm slot the Rayleigh peaks are also clearly depicted, while for the 9.5 mm their amplitude is certainly lower, although still detectable. For slots of 13.5 mm or higher the Rayleigh burst is not clearly visible. It is noted that the main frequency peak is approximately 115 kHz, corresponding to a wavelength of 20 mm.

As seen in Figure 2, the energy and amplitude of the signal decreases with the slot depth. In Figure 3, the signal amplitude is plotted against the slot depth for all cases examined. This decreasing trend has been stated in previous works (Achenbach and Cheng 1996, Wu et al. 2003) for concrete as well as for metals (Edwards et al. 2006). Despite the experimental scatter, a certain master curve of the form of exponential decrease can be drawn to correlate the slot depth with the amplitude, as seen in Figure 2. It is also seen that up to the slots of approximately 10 mm, there is noticeable difference in amplitude but for deeper slots the amplitude remains approximately constant. The slot of 9.5 mm corresponds to approximately 0.5 of the wavelength as seen on the second horizontal axis of Figure 3, while it is less than 10% of the specimen thickness. The second and third horizontal axes are drawn in order to investigate parameters other than solely the slot

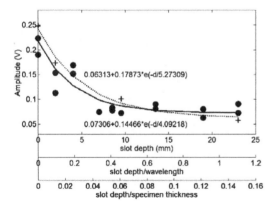

Figure 3. Correlation plot of waveform amplitude vs slot depth for excitation of 115 kHz.

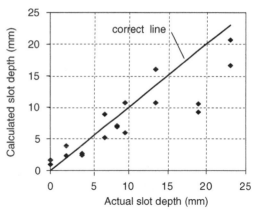

Figure 4. Correlation plot of predicted vs actual values of slot depths.

depth, since the wavelength and the specimen thickness may also influence the Rayleigh amplitude. The slightly decreasing trend of the amplitude for larger slots makes the characterization of larger cracks difficult by using only this parameter. Therefore, in order to characterize sufficiently the slot depth, other features should be used additionally to increase the accuracy.

From the above it is seen that there are correlations between slot depth and certain waveform parameters. However, these correlations are not strong enough to individually lead to accurate characterization. Therefore, a multivariate methodology was applied. For any slot tested, after the waveform was acquired, certain promising parameters were calculated as shown in the above figures. Totally five parameters were included in this analysis. Namely: i) the maximum amplitude, ii) the delay in central time in μs relatively to the central time of the 1st sensor waveform and iii) the inclination of accumulated amplitude, as presented in (Kruger 2005), iv) the absolute central time in μs and v) the time of the first positive peak of the waveform that precedes the Rayleigh peaks which is supposed to be from the refracted longitudinal wave at the tip of the crack.

After obtaining each waveform for a slot, the above mentioned five parameters were extracted. Using the exponential relations, as the example of Figure 3, five values for the depth of each slot were calculated. These were averaged in order to lead to a more reliable result. Applying this methodology for all the tested slots and using at least two individual measurements for each slot, in order to increase the population and check the repeatability, the correlation plot of Figure 4 was produced. There, the calculated depths are plotted vs the actual depths of the slots. It is seen that up to the slot of 13.5 mm, the calculated values do not differ significantly from the actual, having an average error of 1.75 mm. The slot depth of 13.5 mm corresponds to 64% of the major wavelength and 9% of the specimen

thickness. However, for the slots of 19 mm and 23 mm, the predictions are not nearly as accurate.

It is noted that use of only one individual curve for characterization (e.g. the amplitude vs slot depth curve of Figure 3) leads to a typical error of about 4 mm or more. Therefore, the combination and averaging of results of different parameters has certainly positive effect on the accuracy. Additionally, new measurements on different slots would increase the population and thus make the correlation curves more reliable. As seen from Figure 3, the amplitude or any other parameter does not seem sensitive enough for the deepest slots. This is the reason for the reduced accuracy of calculation for slots larger than 13.5 mm.

Considering the frequency content of the waveforms, no stronger correlations were observed. The major peak of 115 kHz does not exhibit any downshift according to the slot depth, while its energy decreases up to the slot of 13.5 mm but is slightly increased for the largest slot of 23 mm. Therefore, this feature is not suitable for characterization.

Since, acceptable characterization is possible up to slot depth equal to about 65% of the wavelength, see Figure 4, slot of 13.5 mm, it was assumed that using longer wavelength could expand the characterization to longer crack depths.

Therefore, the use of wave packets of 50 kHz were applied among others with the function synthesizer 1910 of NF Electronic Instruments. In Figure 5 the amplitude vs slot depth for this excitation is depicted. Although the wavelength is nominally 50 mm, it is seen that the decreasing trend stops again after the slot of 13.5 mm and the larger slots exhibit slightly higher amplitude.

This leads to some considerations concerning the Rayleigh penetration depth. It is generally accepted (Sansalone and Streett, 1997, Wardany et al. 2004, and can be calculated (Aggelis & Shiotani 2006) for

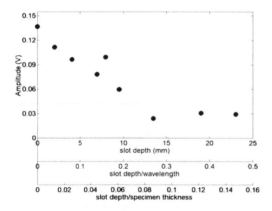

Figure 5. Correlation plot of waveform amplitude vs slot depth for excitation of 50 kHz.

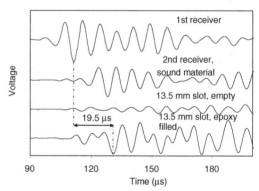

Figure 6. Waveforms after propagation at different wavepaths. The amplitude of the 1st receiver's waveform is reduced to 30% for fitting purposes.

the ideal non attenuative case, that the Rayleigh penetration depth is similar to the wavelength. However, this refers to the amplitude of the wave. The energy carried is proportional to the square of the amplitude. Therefore, taking also into account the attenuation, it is likely that the actual penetration depth is shorter than the one obtained theoretically. Although for the 50 kHz excitation the amplitude decrease is more clear than the 115 kHz case, it seems that the amount of energy penetrating below the shallow layer of 10 mm is not sufficient to distinguish different cracks of 13.5 mm or larger, although the wavelength is nominally 50 mm.

Another argument concerns the dimensions of the specimen. From the above it is seen that using frequency of 115 kHz leads to clear characterization up to slot depth to wavelength ratio 0.5. For the case of 50 kHz, the characterization holds up to depth/wavelength ratio of 0.3, see Figure 5. In both cases however, the decreasing trend stops at around the slot of 13.5 mm. This implies a possible implication of the specimen thickness, since in any case the characterization cannot be performed for slots deeper than 10% of the specimen thickness. As stated in (Zerwer et al. 2005) wavelengths of close to half the specimen thickness do not propagate in the form of Rayleigh waves. For 50 kHz or lower frequency, the wavelength is 50 mm or higher approaching the limit of half the thickness. Therefore, considering that the exponential decay trend concerns Rayleigh waves, it becomes difficult to observe the same trend as the wavelength increases hindering the propagation of Rayleigh waves.

4 REPAIR EFFECT

4.1 Complete filling

A typical treatment of surface cracks is application of epoxy agent that seals the openings and protects

the interior from environmental influence offering also improvement of structural strength (Issa and Debs 2007, Thannon et al. 2005). Two of the cracks were totally filled with two component epoxy adhesive suitable for concrete. The set time is 5 min and sufficient hardening is obtained in 1 hr at 20°C. Measurements were conducted after 1 day when nominally the epoxy exhibits strength of 12 MPa. In Figure 6 the waveform collected at the slot of 13.5 mm after epoxy injection can be compared to the one obtained with the same configuration and setup before the application of epoxy, as well as to the sound material response. It is seen that the injection of epoxy, greatly enhances wave propagation, increasing the transmitted energy to almost the levels of the sound material. It seems that complete filling with the repair agent, restores the propagation since the total energy is of the same level with the sound material. The average increase in energy transmission due to complete filling of the crack with epoxy was 204%. Therefore, surface measurements can be used to characterize the efficiency of repair work. It is mentioned that in the frequency domain the energy was of course increased similarly as in the time domain, without however, any specific peak or band being more characteristic. In other words the filling of the crack enhanced the energy of all frequency bands the same way.

Nevertheless, the Rayleigh peaks are again clearly visible, although delayed compared to the sound material case. They correspond to a velocity of 2050 m/s, about 15% lower than the sound material's Rayleigh velocity. This is expected since concrete elastic properties are higher than epoxy's.

4.2 Partial filling

A case of interest is the partially grouted cracks. In many cases a large part of the crack remains unfilled

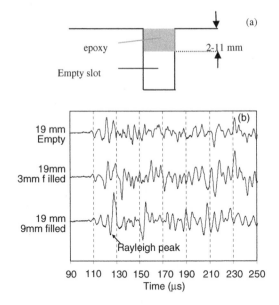

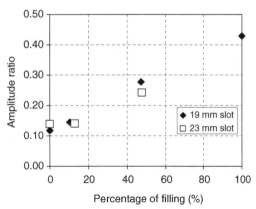

Figure 8. Relation between the waveform amplitude and the degree of epoxy filling of the slot.

Figure 7. (a) Schematic representation of epoxy bridging, (b) waveforms recorded with different slot filling.

especially if the injection is not conducted with an adequate pressure.

However, since the space near the mouth of the crack is filled with grout, this permits the propagation of stress waves resulting in difficulty to obtain information about the extend of grouting by transit time or amplitude data (Sansalone and Streett 1997). To study this case some slots were partially filled with epoxy at different depths. A thick piece of hard paper was wrapped in a Teflon sheet and placed inside the slot leaving an empty space on top. Afterwards, epoxy was placed and was let to set. Then the paper was removed by pulling from one side leaving only the epoxy layer on the top part of the slot as seen in the cross section of Figure 7a.

The cases examined concern filling of 2 and 9 mm to the slot of 19 mm and filling of 3 and 11 mm to the slot of 23 mm. Therefore, the filling ranges from about 10% to 50%. Stress wave measurements were conducted using the configuration with the broadband sensors and the PAC pulser. In Figure 7b one can observe waveforms recorded at the slot of 19 mm with different degree of filling. It is seen that the energy and amplitude of the waveforms increase according to the filling. However, only in the 9 mm filling case a clear Rayleigh wave is visible. The position of this peak corresponds to a velocity of 2040 m/s, being close to the Rayleigh velocity calculated for the fully injected slot of Figure 6. In the case of 3 mm filling, a much weaker peak after the first arrivals is observed while it also arrives earlier than the Rayleigh velocity would

allow. Therefore it should be attributed more likely to contributions of the longitudinal or even shear arrivals passing through the thin layer of epoxy.

In Figure 8, one can observe the increase of amplitude ratio of the second to first receiver according to the degree of filling for the slots of 19 and 23 mm. The value of this ratio after measurement in sound material is 0.43, meaning that this value can be considered maximum. The depths of these slots are quite similar and therefore, measurements results in approximately the same amplitudes. Surface filling of the slot increases slightly the amplitude ratio (1–8%) as seen in Figure 8. Filling up to almost 50% of the initially empty volume of the slot increases further the amplitude ratio. Therefore, approximately half filling of the slot results in raising this parameter to approximately half of its maximum.

Anyway, from the above it can be concluded that surface wave measurements can supply information concerning the degree of filling after repair. From the above experiments it is seen that as the filling percentage gets higher, so do the amplitude and energy of the transmitted wave approaching the energy of the sound case. It is noted that this relationship holds for the specific cases examined and it could be dependent on the initial slot depth, wavelength and member thickness. However, it demonstrates that the characterization of the filling percentage of cracks after repair is possible using wave energy parameters of surface measurements. This correlation should be further examined since there is no other way of obtaining information about the efficiency of repair.

5 IN SITU APPLICATION

As an example of surface measurements conducted for evaluation of the injection efficiency, the following case is presented. Through the thickness cracks

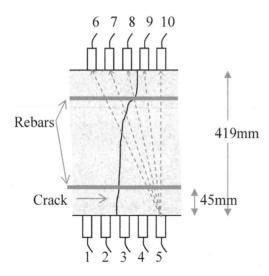

6 7 8 9 10

Rebars

419mm

Crack →

45mm

1 2 3 4 5

Figure 9.   Representation of the sensor arrangement on the concrete deck.

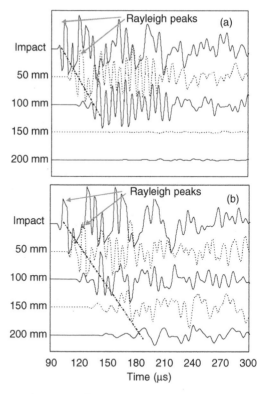

Figure 10.   waveforms collected at different distances from the excitation (a) before and (b) after epoxy injection in material with surface breaking crack between 100 and 150 mm.

were observed in a concrete bridge deck. The engineers at the site decided to treat with epoxy applied from one side by means of syringes through the crack mouth opening of 0.2 mm. Since both the sides of the bridge deck were available, through the thickness measurements were conducted with symmetrically arrayed sensors (Aggelis & Shiotani 2006), see Figure 9. Additionally, surface measurements were also performed before and after the injection using an array of 5 sensors with 50 mm separation distance where the crack was located between the 3rd and 4th.

In Figure 10a the response of the five sensors is depicted before the repair. It is obvious that not enough energy is transmitted through the crack as revealed by the very weak waveforms of sensors 4 and 5 (see 150 and 200 mm in Figure 10a). After the injection, the energy was much enhanced as can be seen in Figure 10b demonstrating that the surface portion through which Rayleigh waves propagate was sufficiently filled. In case the filling was only shallow, as shown in the previous section, no clear Rayleigh wave would be observed. In order to yield more specific information, one should examine the frequency content of the waveforms.

The used excitation provides a wide spectrum of frequencies, with the major peak at 115 kHz. However, propagation in an attenuative and dispersive medium like concrete may induce alterations in frequency content (Jacobs & Owino 2000), mainly causing downshift of the frequencies. Therefore, considering also the time period of the Rayleigh peaks (Aggelis and Shiotani 2006), see the arrows in Figure 10b, it becomes broader with propagation. Thus, it is concluded that the dominant wavelength increases with propagation distance. Specifically, near the impact,

the period of the Rayleigh peak is measured at $8.8\mu s$, (main frequency 113 kHz), while after propagation of 200 mm the period is $23.2\mu s$ (frequency of 43 kHz). Therefore, the wave facing the crack, 125 mm away from the excitation (between 3rd and 4th sensor) exhibits a major wavelength of 35–40 mm considering a Rayleigh velocity of 2600 m/s that was measured in sound material. It is accepted that Rayleigh waves propagate along the surface and sufficient energy penetrates to a depth similar to the wavelength. If this statement holds, the restoration of Rayleigh waves after the epoxy injection is indicative that the surface portion of 35–40 mm of the crack was adequately filled.

The velocity calculated from the peaks is 2414 m/s. Using the same configuration sound material the velocity was measured at 2660 m/s. It is reasonable to expect a drop in velocity of epoxy filled crack compared to the sound material. However, the specific crack's width was 0.2 mm. Considering the propagation velocity of epoxy, sound concrete and the corresponding travel paths at the different materials, the measured Rayleigh velocity for the case of repaired crack should not be reduced by more than some m/s,

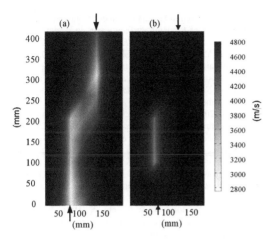

Figure 11. Velocity tomograms of the vicinity of the crack (a) before and (b) after repair. The arrows correspond to the actual position of crack openings.

being within the measurement error. Therefore, the further clear decrease of more than 200 m/s should be attributed to other effects such as the fracture process zone, which expands several cm away from both sides of the crack (Mihashi et al. 1991). In such a case wave propagation takes place not only through the sound concrete and the epoxy filled crack but also through a zone of deteriorated material at both sides of the crack. The thickness of this zone is dependent on the aggregate size and for concrete with 20 mm aggregate, as is the material in the bridge it can be estimated to the order of 4–5 cm (Mihashi et al. 1991). The expansion of this zone to several centimeters justifies the clear decrease in velocity.

In the specific case since both sides were accessible, through the thickness measurements were also conducted to ensure the filling of the interior. Therefore two arrays of sensors were placed on the top and bottom side of the deck. Using pencil lead break excitation at each sensor position, the transit times for the wave to reach each of the opposite side surface sensors were determined and applied to specialized tomography software (Kobayashi et al. 2006).

This way, the velocity structure of the cross section, namely tomogram, is constructed, see Figure 11. The first case (Figure 11a) concerns the cracked situation, while Figure 11b concerns the case after repair. It is seen that the velocity is sufficiently elevated, while only some remaining traces of the crack can be identified after repair. Anyway, due to the different mechanical properties of epoxy compared to those of concrete, it is impossible the velocity to be absolutely restored to the condition of sound material. Also, as mentioned above, except the major crack, the existence of the fracture process zone influences the measurements increasing the overall transit time. Despite these

factors, the calculated velocity after epoxy injection exhibits clear increase, demonstrating that at least a substantial volume of the crack was successfully filled. The use of 5 sensors at each side leads to a number of 50 wave paths examined that provide sufficient information about the interior. However, the density of wave rays near the surface, is not as high as in the interior and therefore the reliability of the results concerning the portion near the surface is not secured. Accordingly, the surface wave examination was essentially conducted in order to obtain information about the surface layer.

6 CONCLUSIONS

In the present paper the non destructive characterization of surface opening cracks is addressed. Surface wave measurements offer a means of non invasive characterization. In such cases, the energy propagates mainly in the form of Rayleigh waves and the amplitude, energy and other features are dependent to a certain degree on the crack depth. Correlation curves between wave parameters and crack depth are established while new measurements on different cracks will increase the reliability of these correlations.

Although larger slot depths are accompanied by lower amplitude and delay in energy transmission, inherent difficulties like the inhomogeneous nature of concrete or ultrasonic coupling variations lead to certain experimental scatter. Therefore, it is reasonable to enhance the characterization by a simple multivariate analysis of 5 different features decreasing the error in depth estimation to less than ±2 mm. The characterization capacity however, is weakened for slot depths approaching the major wavelength. More important however, seem to be the dimensions of the concrete member, since slots exceeding 10% of the thickness result in approximately the same amplitude or energy regardless of the wavelength. For such large slots in many cases the amplitude is unexpectedly increased compared to the smaller cracks.

Further experiments on larger scale would clarify if the observed trends can be generalized for any case sharing the same slot depth to wavelength ratio as is usually suggested (Edwards et al. 2006, Wu 2006). Also search for other wave parameters that are sensitive to the slot depth could increase the accuracy in a more elaborate pattern recognition approach employed. Repair works with injection of epoxy, which is commonly used in practice, can be monitored using the same configuration, since wave propagation takes place also through the new layer of repair material, increasing the energy transmitted to a level similar to that of the sound material. Observation of energy and Rayleigh peaks can yield information concerning the degree of filling. The above findings hold to an actual case of through the thickness cracks in a bridge deck. Surface measurements revealed almost no energy transmitted though the crack in the form of Rayleigh waves initially. Injection with epoxy restored the propagation and Rayleigh waves were clearly observed through the repaired crack, demonstrating that the surface portion was sufficiently filled. If circumstances allow access to both sides of the deck, additional information for the filling of the interior can be supplied, as in the example presented herein. In this case tomography led to a visualization of velocity structure on the cross section revealing the necessary information about the efficiency of repair. Undergoing is the numerical simulation of wave propagation in cases of surface cracks in order to obtain better understanding of the propagation mechanics and increase the characterization accuracy.

REFERENCES

Achenbach, J.D. & Cheng, A. 1996. Depth determination of surface-breaking cracks in concrete slabs using a self-compensating ultrasonic technique. In D.O. Thompson & D.D. Chimenti, (eds), *Review of Progress in Quantitative Nondestructive Evaluation*: 1763–1770, New York and London: Plenum Press.

Aggelis, D.G. & Shiotani, T. 2006. Repair evaluation of concrete cracks using surface and through-transmission wave measurements. (submitted).

Binda, L., Modena, C., Baronio, G. & Abbaneo, S. 1997. Repair and investigation techniques for stone masonry walls. *Construction and Building Materials* 11(3): 133–142.

Doyle P.A. & Scala, C.M. 1978. Crack depth measurement by ultrasonics: a review. *Ultrasonics*, Vol. 16, No. 4, pp. 164–170, 1978.

Edwards, R.S., Dixon, S. & Jian, X. 2006. Depth gauging of defects using low frequency wideband Rayleigh waves. *Ultrasonics* 44: 93–98.

Graff, K.F. 1975. *Wave motion in elastic solids*, New York: Dover Publications.

Hevin, G., Abraham, O., Pedersen, H.A. & Campillo, M. 1998. Characterisation of surface cracks with Rayleigh waves: a numerical model. *NDT&E International* 31(4): 289–297.

Issa, C. A. & Debs, P. 2007. Experimental study of epoxy repairing of cracks in concrete. *Construction and Building Materials* 21: 157–163.

Jacobs, L.J. & Owino, J.O. 2000. Effect of aggregate size on attenuation of Rayleigh surface waves in cement-based materials. *J. Eng. Mech.-ASCE* 26(11): 1124–1130.

Kobayashi, Y., Shiojiri, H. & Shiotani, T. 2006. Damage identification using seismic travel time tomography on the basis of evolutional wave velocity distribution model, *Proc. of Structural Faults and Repair-2006*, Edinburgh, 13–15 June, CD-ROM.

Kruger, M. 2005. Scanning impact-echo techniques for crack depth determination. *Otto-Graf-Journal* 16: 245–257.

Liu, P.L., Lee, K.H., Wu, T.T. & Kuo, M.K. 2001. Scan of surface-opening cracks in reinforced concrete using transient elastic waves. *NDT&E INT* 34: 219–226.

Malhotra, V.M. & Carino, N.J. (eds.). 1991. *CRC Handbook on Nondestructive Testing of Concrete*, Florida: CRC Press.

Mihashi, H., Nomura, N. & Niiseki, S. 1991. Influence of aggregate size on fracture process zone of concrete detected with three dimensional acoustic emission technique. *Cement and Concrete Research* 21: 737–744.

Ono, K. 1998. Damaged concrete structures in Japan due to alkali silica reaction. *The International Journal of Cement Composites and Lightweight Concrete* 10(4):247–257.

Sansalone, M.J. & Streett, W.B. 1997. *Impact-echo nondestructive evaluation of concrete and masonry*. Ithaca, N.Y.: Bullbrier Press.

Shiotani, T., Nakanishi, Y., Iwaki, K., Luo, X. & Haya, H. 2005. Evaluation of reinforcement in damaged railway concrete piers by means of acoustic emission, *Journal of Acoustic Emission* 23: 260–271.

Thanoon, W.A., Jaafar, M.S., Razali, M., Kadir, A. & Noorzaei, J. 2005. Repair and structural performance of initially cracked reinforced concrete slabs. *Construction and Building Materials* 19(8): 595–603.

Wardany, R.A., Rhazi, J., Ballivy, G., Gallias, J.L., & Saleh K. 2004. Use of Rayleigh wave methods to detect near surface concrete damage, *Proc. 16th World Conf. on Non Destructive Testing (WCNDT 2004),Montreal, 30 Aug – 3 September*.

Wu J, Tsutsumi T., Egawa, K. 2003. New NDT method for inspecting depth of crack in concrete using Rayleigh's wave, *Proc. Symp. Japan. Society of Non Destructive Inspection, Tokyo*, 243-252. (in Japanese).

Wu, J. 2006. New NDT method for inspecting depth of crack in concrete using Rayleigh's wave, *Committee report and symposium proceedings on NDT for concrete using elastic wave techniques, Concrete Engineering Series JSCE*, (in press). (in Japanese).

Zerwer, A., Polak, M.A., Santamarina, J.C. 2005. Detection of surface breaking cracks in concrete members using Rayleigh waves, *Journal of Environmental and Engineering Geophysics* 10(3): 295–306.

Fracture Mechanics of Concrete and Concrete Structures – Design, Assessment and Retrofitting of RC Structures – Carpinteri, et al. (eds)
© 2007 Taylor & Francis Group, London, ISBN 978-0-415-44616-7

AE monitoring of a concrete pipe for damage evaluation under cyclic loading

T. Suzuki
Nihon University, Kanagawa, Japan

M. Ohtsu
Kumamoto University, Kumamoto, Japan

ABSTRACT: Deterioration of a pipeline system is normally realized by an accident of water-leakage due to damage accumulation of pipeline materials. In this study, acoustic emission (AE) method was applied to cyclic-loading tests in pipes made of asbestos-cement materials, and quantitative damage evaluation was attempted on the basis of AE parameter analysis. In the cyclic loading test, the damage of the pipe was evaluated from the relationship between *the load ratio* (Ratio of load at the onset of AE activity to previous load) and *the calm ratio* (Ratio of cumulative AE activity under unloading to that of previous maximum loading cycle). It is demonstrated that AE parameter analysis is effective for qualifying the damage levels of pipe materials.

1 INTRODUCTION

Recently, water-leak accidents have been reported in various places for cement-based pipe materials, such as PC (prestressed concrete), RC (reinforced concrete) and AC (asbestos-cement) (Nawa et al 2002). In Most cases damage has accumulated in pipe materials due to aging deterioration, and for quantitatively evaluating pipe damage; a lot of research problems still remain unsolved. In general, material damage is evaluated by its strength, which is the average strength of the pipe materials obtained from external and internal loading tests as stipulated in the codes. However, fracturing behavior of structural materials progresses from micro-cracks in the most deteriorating material. Therefore, the actual pipe strength can not be evaluated accurately only from the average strength of the pipe materials obtained from the strength tests.

In this study, acoustic emission (AE) method is introduced into the external loading tests (compression tests, bending tests) and pipe damage is evaluated using AE monitoring in the fracture process under external pressure. The external loading tests are conducted in two methods: compression tests where uniformly-distributed load is applied to the pipe and bending tests where load is applied at the center of the pipe. The pipe damage is evaluated using AE parameters, *the load ratio* (Ratio of load at the onset of AE activity to previous load) and *the calm ratio* (Ratio of cumulative AE activity under unloading to that of previous maximum loading cycle), which were calculated from the measured AE data.

2 EXPERIMENT METHOD AND MATERIALS

2.1 *Acoustic emission monitoring*

Acoustic emission refers to elastic wave motions observed during the energy release process, such as micro-fracturing within a solid body. In this study, AE sensors detected AE signals generated in the fracture process in pipe materials. Then, based on AE characteristics, the degree of damage is evaluated. For AE measurement, R15 sensors (manufactured by PAC) were installed on the inner surface of the asbestos cement pipes at four locations for the compression tests and eight locations for the bending tests (**Figs. 1 and 2**). DISP-AE system (manufactured by PAC) was used as a measuring device. AE events were amplified with 40 dB gain in a pre-amplifier and 20 dB gain in a main amplifier. From AE measurements, the previous load record was examined based on the Kaiser effect of AE, and the pipe damage was evaluated from the relationship between *the load ratio* and *the calm ratio* (NDIS2421, 2000).

2.2 *Experimental materials*

The test pipe is an asbestos-cement pipe (ACP, Type 4) of 600 mm diameter, which had been used for 32

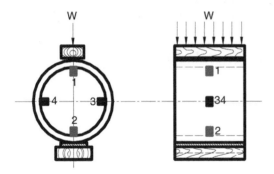

Figure 1. AE sensor arrangements in the compression test.

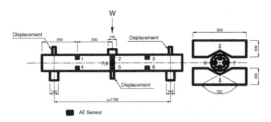

Figure 2. AE sensor arrangements in the bending test.

years. The pipe had been buried underground 1.10 meters below the road and as connected 73 meters long. Each 4.0 meter pipe was cut into two samples as two 1.0 meter samples for a compression test, and one 2.0 meter sample for a bending test.

3 AE GENERATION BEHAVIOR IN FRACTURE PROCESS

AE behavior of a deteriorated pipeline was associated with material damage. AE events were analyzed by using AE parameters. In the compression tests, the strengths of the pipes varied widely from 25.6 kN to 72.2 kN, with the mean value of 55.4 kN. Considering that the number of test specimens was small and the deviation of the strengths was as large as 54% the mean value, it is difficult to evaluate damage of the pipe simply only from the mean value of strengths. In contrast, results of the bending tests show a relatively small variation from 201 kN to 227 kN. Different patterns of AE generating behaviors were observed near the fracture surface and the other regions during the fracture process of the test pipe (**Figs. 3** and **4**). Since these differences in generating patterns were observed regardless of the tests methods (compression test, bending test) and loading cycles (simple loading, cyclic loading), it is considered that the fracture process of the test pipe is so localized that it can be monitored effectively using AE method. In the compression tests, the displacement varied linearly with

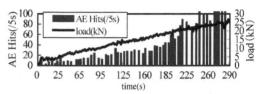

Figure 3. AE generation behaviors in the compression test (fracture position).

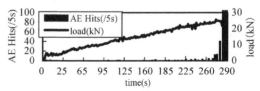

Figure 4. AE generation behaviors in the compression test (sound position).

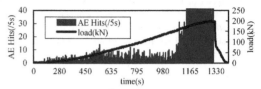

Figure 5. AE generation behaviors in the bending test.

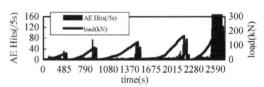

Figure 6. AE generation behaviors in the cyclic bending test.

stress. As the load increased, the number of AE events increased and the events were observed frequently near the ultimate load. The similar process of fracture of the concrete samples was reported in the previous paper (Shigeishi et al 2003, Suzuki et al 2006). In the bending tests, diagonal shear fracture was observed as active AE generation was observed at 44.5 kN in the simple loading tests (**Fig. 5**). This load might be assumed as the most frequent previous load recorded which is discussed below as related to the Kaiser effect. In the cyclic loading tests, frequent AE generation was again observed at about 40 kN, which corresponds to 35.4 N/mm^2 bending stress (**Fig. 6**). From these results, the test pipe is considered to have received 40 to 50 kN external pressure constantly. From AE monitoring in the loading tests, the previous load recorded in the test pipe is assumed to be 40 to 50 kN, which corresponds to 19 to 23% of the maximum load (214 kN mean in the bending tests).

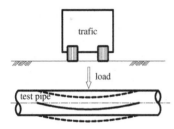

Figure 7. Loaded condition of a tested pipe.

4 DAMAGE EVALUATION BASED ON AE PARAMETER ANALYSIS

The degree of damage to structures is generally evaluated by the mean values of the mechanical properties of the structural materials, such as strength, elastic modulus, etc. However, the mean values of the mechanical properties obtained from the loading tests are not sufficient enough to evaluate the damage of the structures because the actual fracturing behavior of structures progresses from the most deteriorating material, although it is effective to grasp the overall conditions of the structures. From this point of view, the maximum load undergone by the pipes obtained from fracturing behavior and the mechanical properties of the deteriorating structural materials need to be examined in the evaluation.

A maximum previous load recorded in the asbestos-cement pipe was estimated by the Kaiser effect, and then the damage of the pipe was qualified from the relationships between *the load ratio* and *the calm ratio*. The Kaiser effect is defined to no-AE generation behavior until maximum previous load of the test specimen. AE events are detected after maximum previous load by the Kaiser effect.

In the bending tests, the most frequent previous load is found to be 40 to 50 kN. The conditions under which the test pipe had been buried underground are shown in **Fig. 7**. Fracture outline are shown in **Fig. 8**. Because the traffic load applied on the pipe in the transverse direction, the bending tests were conducted. The relationships between the loads in the cyclic loading tests and *the calm ratios* are shown in **Fig. 9**. The high *calm ratio* implies accumulation of damages. When *the calm ratios* do not change due to increased loads, the Kaiser effect is observed. According to the bending tests, cyclic loadings of 0 to 80 kN and 0 to 120 kN do not show any differences in *the calm ratios*. This may suggest that the pipe had the maximum previous load around 100 kN. **Figure 10** shows the relationship between the loads in the cyclic loading tests and *the load ratios*, indicating the decreasing trend. The *load ratios* less than 1.0 mean accumulation of damages. When *the load ratios* do not change due to increased

Figure 8. Result of Fracture in bending test.

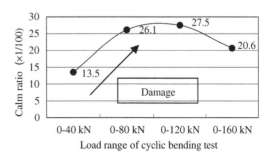

Figure 9. Calm ratios in the cyclic bending test.

loads, the Kaiser effect is observed in the pipe as a stable condition. The figure shows that *the load ratio* decreases sharply at around 100 kN, which reflects the behavior of *the calm ratio*. Based on **Figs. 9** and **10**, it is found that the maximum previous load could be determined based on the Kaiser effect. Then, the test pipe was evaluated from the relationship between *the load ratio* and *the calm ratio*. It is realized that the deterioration was minor at the initial cycle, and it increases from the intermediate level to the heavy (**Fig. 11**).

1021

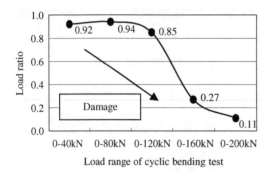

Figure 10.   Load ratios in the cyclic bending.

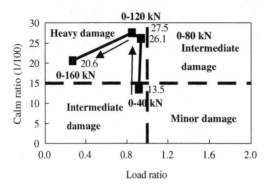

Figure 11.   Relationships between the calm ratio and the load ratio.

5   CONCLUSION

In this study, compression tests and bending tests were carried out on the existing asbestos-cement pipe (Type 4) of 600 mm diameter and the pipe damage was quantitatively evaluated by AE monitoring of the fracture process in the pipe. From the results of the AE measurements, the previous load record and the damage of the pipe were examined. The tested pipe, which had been used for a long time, was confirmed to have been damaged due to aging deterioration (*load ratio*<1.0; damaged). The previous load record showed that the most frequent previous load record was 40 to 50 kN and the load when the Kaiser effect disappeared was about 100 kN at average (50% the maximum previous load record). The relation of *the load ratio* and *the calm ratio* implied the medium to severe level of accumulation of damage. It is difficult to evaluate damage and previous load record of pipes only with conventional external loading tests. However, this study confirmes as the results of the AE measurements, that the damage and the previous load record of the pipe, which had been used for 32 years, could be effectively evaluated based on the Kaiser effect of AE. The pipeline systems are generally buried underground below roads and are assumed to have received repeated loading for a long time. Therefore, AE measurement using the Kaiser effect is effective for evaluating previous load record and damage of existing pipes.

REFERENCES

Nawa, N., Sonoda, K., Iwata H. & Suzuki, T. 2002. Survey and Evaluation Method for Deteriorated Agricultural Pipeline Function, JSIDRE, Vol.70 (12), 31–36.
NDIS2421 2000. Recommended practice for in situ monitoring of concrete structures by acoustic emission.
Shigeishi, M., Nakashima, T. & Ohtsu, M. 2003. Calm Ratio & Load Ratio on A Rainforced Concrete Bridge Main Girder under Cyclic Bending, Proceedings of 2003 National Conference on Acoustic Emission, 165–168.
Suzuki, T., Ohno, K. & Ohtsu, M. 2006. Damage Evaluation of Existing Asbestos-Cement Pipe by Acoustic Emission, The 18th International Acoustic Emission Symposium, 257–262.
Suzuki, T., Ikeda, Y., Komeno, G. & Ohtsu, M. 2004. Damage Evaluation in concrete based on Acoustic Emission Database, Proceedings of the JCI, Vol.26, No.1, 1791–1796.
Suzuki, T., Watanabe, H. & Ohtsu, M. 2002. Damage Evaluation in concrete Using Acoustic Emission Method, The 6th Far-East Conference on Non-Destructive Testing, 111–116.

Fracture Mechanics of Concrete and Concrete Structures – Design, Assessment and
Retrofitting of RC Structures – Carpinteri, et al. (eds)
© 2007 Taylor & Francis Group, London, ISBN 978-0-415-44616-7

Spectral analysis and damage evolution in concrete structures with ultrasonic technique

P. Antonaci, P. Bocca, D. Masera & N. Pugno
Politecnico di Torino – Dipartimento di Ingegneria Strutturale e Geotecnica

M. Scalerandi
Politecnico di Torino – Dipartimento di Fisica

F. Sellone
Politecnico di Torino – Dipartimento di Elettronica

ABSTRACT: The non-linear behavior of both virgin and damaged concrete samples is experimentally investigated by means of ultrasonic tests. Recent theoretical models, indeed, have pointed out that mono-frequency ultrasonic excitations bring to light such phenomena as harmonic generation and sidebands production, which are essentially due to the material non-linearity. The estimation of the harmonic components parameters (amplitudes and phases) is achieved through a signal processing technique based on MUltiple SIgnal Classification (MUSIC) system, which reveals to be optimal for the specific signal model here considered. The experiments described in this paper show that the material non-linear features increase with increasing level of internal deterioration, thus suggesting the possibility to use the ultrasonic signal analysis in the frequency domain as a valuable tool for damage assessment.

1   INTRODUCTION

Standard methods for detecting defects within a concrete structure generally involve the use of drilled core samples. In the case of buildings and monuments of historical interest, it is frequently impossible to extract even small samples. Hence, the need for non-destructive methods in the carrying out of investigations, as for example ultrasonic techniques (UT).

Recently, UT have been widely used in the evaluation of crack growth in concrete to deduce its quality and extent of microcracking. Crack detection is usually carried out by measuring pulse velocity, or resorting to impact echo methods and others (Popovic 1994).

In the present paper a novel approach based on spectral signal-processing and analysis is discussed: it allows the monitoring of the crack growth and damage evolution within concrete structures in term of non-linear response to the application of ultrasonic waves, revealing to be fairly sensitive (Van Den Abeele et al. 2000a-b, Bentahar et al. 2006, Bocca & Rosa 1993, Bocca & Nano 1995, Daponte et al. 1995, Daponte et al. 1990).

An experimental program is planned to evaluate the performances offered by this procedure in damage assessment of granular materials, such as concrete or rocks. Laboratory tests have been conducted on ordinary concrete specimens in order to simulate a possible deterioration process affecting a real structures and consequently investigate its damage level. The final aim of this preliminary study is to develop an effective non-destructive procedure suitable to be used on-site.

The paper is organized as follows: first, the theoretical background concerning the non-linear behavior of concrete is presented; subsequently, the experimental research conducted at the Non-Destructive Testing Laboratory of Politecnico di Torino is described, with detailed reference to the materials tested, the equipment used and the procedure followed; finally, the experimental results are reported and discussed, in terms of repeatability of measurements, signal processing methods and applicability to damage assessment.

2   THEORETICAL FOUNDATIONS

Non-linearity in the elastic behavior of a material is known to be deeply dependent on the level of microcracking and damage affecting the material itself (Pugno & Surace 2000). Indeed, even in such materials as concrete or rocks, that are intrinsically non-linear due to their grain structure, the presence of damage at micro- or meso-scopic level is responsible for a dramatic increase of non-linear features. It follows that the detection of non-linear signatures could be assumed as an indicator of the state of damage.

In particular, it has been observed that material non-linearity causes distortions in the propagation of elastic waves, creating accompanying harmonics, multiplication of waves at different frequencies, and under resonance conditions, changes in resonant frequencies as a function of the driving amplitude.

Hence the increasing importance that Non-linear Elastic Waves Spectroscopy (NEWS) techniques are assuming in diagnostics.

As a first approach, non-linearity can be theoretically modeled by expressing the elastic moduli in a power series of the strain, and considering terms of first or even second order, for highly non-linear materials. Such a power series approach is generally referred to as "classical non-linearity". Though very useful in many cases, however, this theory does not fully explain the behavior of most highly non-linear materials, that exhibit more complicated phenomena in their stress-strain relation, such as hysteresis and discrete memory (Bentahar et al. 2006). Consequently, a theoretical model suitable to describe these materials should contain terms accounting for both classical non-linearity and hysteretic behavior, as well as discrete memory. Accordingly, the one-dimensional constitutive relation between stress and strain to be used in simulation of the dynamic behavior of solids can be expressed as follows:

$$\sigma = \int k(\varepsilon, \varepsilon')d\varepsilon \qquad (1)$$

$k(\varepsilon, \varepsilon')$ being the non linear and hysteretic modulus given by:

$$k(\varepsilon, \varepsilon') = k_0 \{1 - \beta\varepsilon - \delta\varepsilon^2 - \alpha[\Delta\varepsilon + \varepsilon(t)sign(\varepsilon')] + ...\} \qquad (2)$$

In Eq. (2), k_0 turns out to be the linear modulus, $\Delta\varepsilon$ is the local strain amplitude over the previous period, ε' is the strain rate, and $sign(\varepsilon') = 1$ if $\varepsilon' > 0$, while if $sign(\varepsilon') = -1$ if $\varepsilon' < 0$.

The parameters β and δ are the classical non-linear perturbation coefficients, and α is a measure of the material hysteresis.

Due to the various non-linear and hysteretic contributions described above, the harmonic spectrum of a finite amplitude mono-frequency wave propagating through the material may exhibit additional harmonics, whose amplitudes depend on the fundamental strain amplitude in different ways, according to the type of non-linearity. In particular, the classical non-linear theory predicts that the second harmonic is quadratic in the fundamental strain amplitude (slope 2 in a log-log plot), while the third harmonic is cubic (slope 3) and so on. On the contrary, the third harmonic should be quadratic for a purely hysteretic material, thus revealing a different response in comparison with classical non-linear materials.

Table 1. Concrete composition.

Cement type	CEM II A-L 42.5 R
Cement proportions	270 kg/m^3
Sand proportions	1063 kg/m^3
Gravel proportions	799 kg/m^3
Max. aggregate Size	16 mm
Water/cement ratio	0.74

3 EXPERIMENTAL PROGRAM

The theoretical model reported in (Van Den Abeele 2000a-b) was implemented through laboratory experiments. Damage evolution into plain concrete samples was induced by static loading at different stress levels, and subsequently detected using UT.

3.1 Materials and specimens

A concrete slab was produced according to the mix composition given in Table 1.

It was water-cured for 28 days and subsequently air-cured for approximately three years before testing. Core-drilled samples were obtained from this slab. They were consequently cut in order to get cylindrical specimens, approximately 160 mm long and 60 mm in diameter. The mechanical characteristics of the concrete were preliminarily evaluated by means of a uniaxial static compression test, that resulted in a compressive strength of 24 N/mm^2.

3.2 Testing equipment

Static loads were applied by means of a 250 kN servo-controlled material testing machine. The ultrasonic tests for damage characterization were performed by means of the following testing equipment:

– A pair of piezoelectric transducers with a diameter of 35 mm and a work frequency of 55 kHz. One of them was used as the emitting source and the other as the receiving transducer.
– A waveform generator able to produce sinusoidal signals at different frequencies and amplitudes. It was used to drive the emitting source, forcing it to produce a sinusoidal wave at approximately 55 kHz, with varying amplitudes.

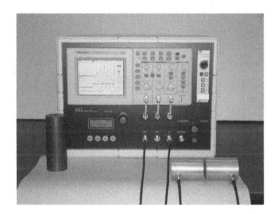

Figure 1. Data acquisition unit and piezoelectric transducers.

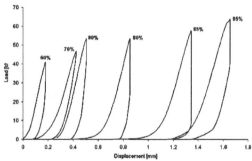

Figure 2. Static load vs. displacement curves for different stress levels.

– A data acquisition unit, equipped with an oscilloscope for real-time data visualization (Fig. 1). A sampling frequency of 1500 kHz was selected, in the respect of Nyquist's theorem.
– A personal computer for post-processing of the acquired data.

3.3 *Testing procedure*

An initially undamaged specimen was first subjected to ultrasonic tests aimed at characterizing its intrinsic non-linearity. Subsequently, it was subjected to a specific load history, simulating a possible damage process affecting a real structure. Such a load history consisted of the following steps:

– Static compressive loading up to 60% of the estimated compressive strength (σ_R) and unloading.
– Static compressive loading up to 70% σ_R and unloading.
– Static compressive loading up to 80% σ_R and unloading.
– Static compressive loading up to 80% σ_R and unloading.
– Static compressive loading up to 85% σ_R and unloading.
– Static compressive loading up to 90% σ_R and unloading.

At the end of each step, the specimen was subjected to ultrasonic tests. The piezoelectric transducers were applied to the transverse surfaces of the specimen, so that the ultrasonic wave traveled in the longitudinal direction. Special care was devoted to ensure that test conditions remained the same throughout the experiments. In particular, the amount of coupling agent (plasticine) to be used was kept constant, as well as the pressure applied to the transducers.

Preliminary experiments have been conducted to verify the repeatability of the measurements. Also, the linearity of the piezoelectric transducers, within the range of voltages used, has been verified by checking the absence of harmonics generation in a linear steel cylinder under the same experimental conditions.

4 RESULTS AND DISCUSSION

In Figure 2, the results of the load tests are reported in terms of load vs. deformation curves. It can be observed that the material stiffness did not significantly vary as a function of the applied stress.

As far as ultrasonic tests are concerned, it shall be recalled that, according to the theoretical model (Van Den Abeele 2000a-b), the input signal is transformed after passing through the material. More specifically, additional harmonic components are generated, thus revealing possible material non-linearity.

In order to get information about such a non-linearity, the harmonic components parameters (i.e. amplitude and frequency) need to be evaluated. In the present study, this task was accomplished using two different estimation techniques: a non-parametric one, the well-known Fast Fourier Transform (FFT) (Manolakis et al. 2005), and a parametric one referred to as MUltiple SIgnal Classification (MUSIC) (Schmidt 1986, Stoica & Nehorai 1989, Stoica & Nehorai 1990, Sellone 2000). When a signal is composed by one harmonic component only buried in white Gaussian noise, the FFT technique corresponds to the maximum likelihood estimation of the amplitude, which is also equal to the Least Square method (LS). Unfortunately, the frequency estimation is limited by the points over which the FFT is evaluated. Moreover, in the experiment considered in this paper, more than one harmonic component are simultaneously present, thus calling for different statistically efficient estimation techniques.

The parameters can be better estimated through a parametric approach such as MUSIC. Indeed, it is an efficient frequency estimator, since it provides asymptotically unbiased estimates of a general set of signal

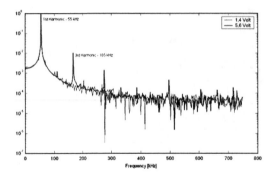

Figure 3. Example of FFT spectrum. The peaks amplitude represents the actual harmonic amplitude.

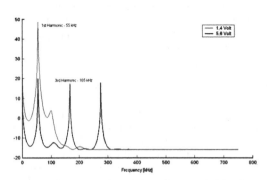

Figure 4. Example of MUSIC pseudospectrum. The peaks amplitude is not related to the actual harmonic amplitude, which is estimated via LS.

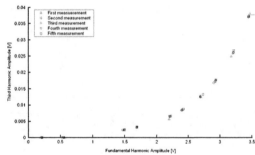

Figure 5. Measurement repeatability.

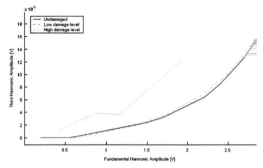

Figure 6. Third vs. fundamental harmonic amplitudes for different damage levels.

parameters, approaching the Cramer Rao bound. By substituting such estimated frequencies back into the original signal model, the problem becomes linear in amplitudes and phases and thus a LS approach can be used to obtain the maximum likelihood estimates. Figures 3 and 4 remark the differences between the two techniques. MUSIC captures more clearly the harmonic components present in the signal.

In both Figures 3 and 4 it is easy to observe harmonics generation, and in particular the presence of the third harmonic which is analyzed here. Odd harmonics are larger than even ones (see Figure 2) indicating dominance of hysteresis or of the second order classical non-linearity (δ term in Eq. (2). The third harmonic increases considerably when increasing the driving amplitude from 4.4 to 8.1 Volts.

It shall be remarked here that in Figure 3 amplitudes are plotted before the LS calculations and therefore they are not related to the actual harmonic amplitude (not even the ratio of amplitudes at different frequencies is meaningful).

Consistently with the above considerations, the fundamental and the third harmonic amplitudes at the end of each load step were evaluated by a software-controlled algorithm implementing MUSIC and LS estimation.

The observations were replicated under constant conditions in order to check measurement repeatability. Accordingly, sets of five observations per driving amplitude have been made on a single specimen, at the end of each load step, taking care that ambient and test conditions were kept constant throughout the testing process. As a result, repeated estimates of the fundamental and third harmonic amplitudes were obtained. An example is shown in Figure 5, where the values are plotted of the five fundamental versus third harmonic amplitudes, resulting from five repeated measurements at the end of a specific load step. A very good repeatability was found to exist, since these values are nearly overlapping.

Once that measurement repeatability was ascertained, the amplitude of the third harmonic (165 kHz) and the amplitude of the fundamental one (55 kHz) were evaluated as a function of the driving strain amplitude, for different damage level.

The results are depicted in Figure 6, where three curves are reported, each one representing the plot of the third harmonic amplitude versus the fundamental one, for varying driving amplitude. The solid line curve corresponds to specimens in the undamaged state, while the dashed-dotted line denotes specimens which had undergone low static loading, that caused a low-damage level, and finally the dotted line corresponds to specimens which had been subjected to a high static stress level, with consequent remarkable damage.

Data at low strain amplitudes indicate that the analyzed concrete is characterized by a non-classical non-linear constitutive equation, with presence of hysteresis. In fact, log-log plots seem to indicate that the third harmonic depends quadratically on the fundamental one, while at larger amplitudes classical non-linearities may be dominant (cubic power law dependence). Further investigation to confirm such observation is in progress.

It can also be observed that for low static stress levels, no apparent damage phenomena occurred. Accordingly, no significant differences may be remarked between 1st–3rd harmonics curves related to specimens in the undamaged state and those related to specimens which had undergone low static loading. On the contrary, more marked differences can be found in visibly damaged specimens, i.e. in specimens subjected to a high static stress level, that caused noticeable macro-cracking occurrences. The increase in non-linearity due to such a damage is revealed by the fact that the third harmonic generation becomes more evident.

5 CONCLUSIONS

The experimental research presented in this paper confirmed the theoretical models concerning the non-linear response of granular materials, such as concrete, when traversed by ultrasonic waves.

The generation of additional harmonic components, and third harmonic in particular, may be assumed as an indicator of the deterioration state, since experimental evidence revealed that it becomes more marked with increasing level of damage.

The use of novel signal processing techniques such as MUSIC makes it possible to obtain more accurate amplitude and phase estimates for sinusoids in white Gaussian noise than classical FFT, thus revealing to be optimal for the specific problem under consideration. It substantially improves the performances of simple data acquisition systems, such as the one used in the course of this experimental study, thus making the proposed damage assessment technique more attractive.

These encouraging findings suggest to continue the research in order to consider additional types of damaging actions (environmental actions, fatigue, creep, etc.) and extend the proposed method to the on-site evaluation of existing structures.

REFERENCES

Van Den Abeele, K. E. A., Johnson, P. A. & Sutin, A. 2000a. Nonlinear Elastic Wave Spectroscopy (NEWS) techniques to discern material damage. Part I: Nonlinear Wave Modulation Spectroscopy (NWMS). *Res Nondestr Eval* vol. 12 (1): 17–30.

Van Den Abeele, K. E. A., Johnson, P. A. & Sutin, A. 2000b. Nonlinear Elastic Wave Spectroscopy (NEWS) techniques to discern material damage. Part II: Single Mode Nonlinear Resonance Acoustic Spectroscopy. *Res Nondestr Eval* vol. 12 (1): 31–42.

Bentahar, M., El Aqra, H., El Guerjouma, R., Griffa, M. & Scalerandi, M. 2006. Hysteretic elasticity in damaged concrete: Quantitative analysis of slow and fast dynamics. *Physical Review B* vol. 73 (14) 014116: 1–10.

Bocca, P. & Rosa, G. 1993. Hysteretic elasticity in damaged concrete: Frequency spectra and ultrasonic pulse attenuation analysis for the assessment of damages in concrete. *Proc. of the International Conference on Non Destructive Testing on Concrete in the Infrastructure, Dearborn, (Michigan, USA), June, 9–11, 1993*: 330–340.

Bocca, P. & Nano, A. 1995. Evoluzione del danneggiamento negli elementi in calcestruzzo: analisi dello spettro di frequenza degli ultrasuoni. *Proc. of the XXIV AIAS National, Parma (Italy), September, 27–30, 1995*: 292–299.

Daponte, P., Maceri, F. & Olivito, R. S. 1995. Ultrasonic signal-processing techniques for the measurement of damage growth in structural materials. *IEEE Trans. on Instrumentation and Measurement* vol. 44 (6): 1003–1008.

Daponte, P., Maceri, F. & Olivito, R. S. 1990. Frequency-domain analysis of ultrasonic pulses for the measure of damage growth in structural materials. *Proc. of the IEEE Ultrasonic Symposium 1990* vol. 2: 1113–1118.

Pugno, N. & Surace, C. 2000. Evaluation of the non-linear dynamic response to harmonic excitation of a beam with several breathing cracks. *Journal of Sound and Vibration* vol. 235 (5): 749–762.

Schmidt, R. 1986. Multiple emitter location and signal parameter estimation. *IEEE Trans. on Antennas and Propagation* vol. 34 (3): 276–280.

Stoica, P. & Nehorai A. 1989. MUSIC, maximum likelihood, and Cramer-Rao bound. *IEEE Trans. on Acoustics, Speech, and Signal Processing* vol. 37 (5): 720–741.

Stoica, P. & Nehorai, A. 1990. MUSIC, maximum likelihood, and Cramer-Rao bound: further results and comparisons. *IEEE Trans. on Acoustics, Speech, and Signal Processing* vol. 38 (12): 2140–2150.

Stoica, P. & Soderstrom, T. 1991. Statistical analysis of MUSIC and subspace rotation estimates of sinusoidal frequencies. *IEEE Trans. on Signal Processing* vol. 39 (8): 1836–1847.

Sellone, F 2006. Robust auto-focusing wideband DOA estimation. *Signal Processing* vol. 86 (1): 17–37.

Manolakis, G., Ingle, V. K. & Kogon, S. M. 2005. *Statistical and Adaptive Signal Processing*, London: Artech House.

Fracture Mechanics of Concrete and Concrete Structures – Design, Assessment and Retrofitting of RC Structures – Carpinteri, et al. (eds)
© 2007 Taylor & Francis Group, London, ISBN 978-0-415-44616-7

Structural health monitoring of strands in PC structures by embedded sensors and ultrasonic guided waves

I. Bartoli
Department of Structural Engineering, University of California, San Diego
DISTART, University of Bologna, Bologna, Italy

P. Rizzo
Department of Civil and Environmental Engineering, University of Pittsburgh

F. Lanza di Scalea
Department of Structural Engineering, University of California, San Diego

A. Marzani, E. Sorrivi & E. Viola
DISTART, University of Bologna, Bologna, Italy

ABSTRACT: The detection of structural defects and the measurement of applied loads in strands is important to ensure the proper performance of civil structures, including post-tensioned concrete structures and cable-stayed or suspension bridges. Ultrasonic guided waves have been used in the past to probe strands. This paper reports on the status of ongoing collaborative studies between the Universities of California, Pittsburgh and Bologna aimed at developing a monitoring system for prestressing strands in post-tensioned structures based on ultrasonic guided waves and built-in sensors.

1 INTRODUCTION

High-strength, multi-wire steel strands are widely used in civil engineering such as in prestressed concrete structures, and cable-stayed or suspension bridges. Material degradation of the strands, usually consisting of indentations, corrosion or even fractured wires, may result in a reduced load-carrying capacity of the structure that can lead to collapse. In a survey involving the study of more than one hundred stay-cable bridges Watson & Stafford (1988) pessimistically reported that most of them were in danger mainly because of cable defects. Strand failures that caused bridge collapses were documented in Wales (Woodward 1988), Palau (Parker 1996), and North Carolina (Chase 2001). Hence the need for developing monitoring systems for strands that can detect, and possibly quantify, structural defects, as well as alert of any prestress loss.

Structural monitoring methods based on Guided Ultrasonic Waves (GUWs) have the potential for both defect detection and stress monitoring. GUWs have been used for the detection of defects in multi-wire strands and reinforcing rods (Kwun and Teller 1994, 1995; Pavlakovic et al. 1999, 2001; Beard et al. 2003; Reis et al. 2005) and for the evaluation of stress levels in post-tensioning rods and multi-wire strands

(Kwun et al. 1998; Chen and Wissawapaisal 2002; Washer et al. 2002). The authors have used GUWs for defect detection and stress monitoring in seven-wire steel and composite strands (Rizzo and Lanza di Scalea 2001, 2004, 2005; Lanza di Scalea et al. 2003).

This paper summarizes representative results obtained to date by the authors on the detection of defects and the monitoring of prestress levels in strands.

2 SAFE METHOD

The Semi-Analytical Finite Element (SAFE) method is an effective tool to model waveguides of arbitrary cross-section (Huang & Dong 1984; Hayashi et al. 2003). The authors have recently extended the method to account for viscoelastic material damping through complex stiffness coefficients (Bartoli et al. 2006). In the SAFE method, at each frequency ω a discrete number of guided modes are obtained. For the given frequency, each mode is characterized by a wavenumber, ξ, and by a displacement distribution over the cross-section. For axisymmetric waveguides, it is convenient to develop the viscoelasticity wave equations by using a cylindrical reference system, with the cross section lying in the r-θ plane and the z-axis being parallel to the waveguide's longitudinal direction (see Figure 1a). The displacement at a point is:

$$\mathbf{u}(r,\theta,z,t) = \tilde{\mathbf{u}}e^{in\theta}e^{i(\xi z - \omega t)} \qquad (1)$$

where t is the time variable, i is the imaginary unit, and n is the circumpherential order of the mode. Subdividing the cross-section into finite elements, the approximate displacement field in the element is:

$$\mathbf{u}^{(e)} = \mathbf{N}(r)\tilde{\mathbf{U}}^{(e)}e^{i(n\theta+\xi z-\omega t)} \qquad (2)$$

where $\mathbf{N}(r)$ is the matrix of the shape functions and $\tilde{\mathbf{U}}^{(e)}$ is the element's nodal displacement vector. Thus the displacement is described by the product of an approximated cross-sectional finite element field with exact time harmonic functions in the propagation direction.

The compatibility and constitutive equations can be written in synthetic matrix forms as:

$$\boldsymbol{\varepsilon} = \mathbf{D}\mathbf{u}, \qquad \boldsymbol{\sigma} = \mathbf{C}^{*}\boldsymbol{\varepsilon} \qquad (3)$$

where $\boldsymbol{\varepsilon}$ and $\boldsymbol{\sigma}$ are the strain and stress vector, respectively, $\mathbf{D}$ is the compatibility tensor and $\mathbf{C}^{*}$ is the complex constitutive linear viscoelastic tensor. More details on the compatibility operator can be found in Hayashi et al. 2003.

The principle of virtual works with the compatibility and constitutive laws leads to the following energy balance equation

$$\int_{\Gamma}\delta\mathbf{u}^{\mathsf{T}}\mathbf{t}d\Gamma = \int_{V}\delta\mathbf{u}^{\mathsf{T}}(\rho\ddot{\mathbf{u}})dV + \int_{V}\delta(\mathbf{u}\mathbf{D})^{\mathsf{T}}\mathbf{C}^{*}\mathbf{D}\mathbf{u}dV \qquad (4)$$

where Γ is the waveguide cross-sectional area, V is the waveguide volume, $\mathbf{t}$ is the external traction vector and the overdot means time derivative. The finite element procedure reduces Equation 4 to the a set of algebraic equations:

$$[\mathbf{A} - \xi\mathbf{B}]_{2M}\,\mathbf{Q} = \mathbf{p} \qquad (5)$$

where the subscript $2M$ indicates the dimension of the problem with M the number of total degrees of freedom of the cross-sectional mesh. Details on the complex matrices $\mathbf{A}$, $\mathbf{B}$ and vector $\mathbf{p}$ can be found in Bartoli et al. (2006). Setting $\mathbf{p}=\mathbf{0}$ in Equation 5, the associated eigenvalue problem can be solved as $\xi(\omega)$. For each frequency ω, $2M$ complex eigenvalues ξ_m and $2M$ complex eigenvectors $\mathbf{Q}_m$ are obtained, corresponding to right-propagating and left-propagating waves. The first M components of $\mathbf{Q}_m$ describe the cross-sectional mode shapes of the m-th mode. Once ξ_m is known, the dispersion curves can be easily computed. The phase velocity can be evaluated by

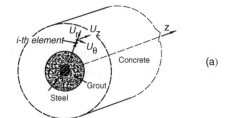

(a)

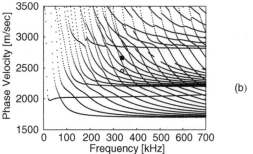

(b)

(c)

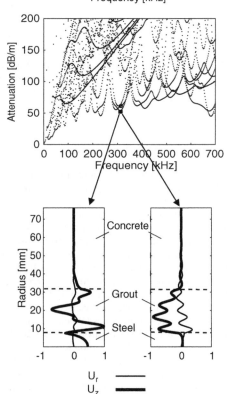

Figure 1. SAFE solutions for a 15.24 mm steel rod embedded in a 63.5 mm grout layer and a 152.4 mm concrete layer (axis-symmetric modes). (a) system modeled; (b) phase velocity curves; (c) attenuation curves; and (d) displacement mode shapes of two modes at low-loss points.

the expression $c_{ph} = \omega/\xi_{real}$, where ξ_{real} is the real part of the wavenumber. The imaginary part of the wavenumber is the attenuation, $att = \xi_{imag}$, in Nepers per meter.

Figure 1 shows the SAFE results for a 15.24 mm (0.6 in)-diameter steel rod embedded in a 63.5 mm (2.5 in)-outer diameter layer of grout and a 152.4 mm (6 in)-outer diameter layer of concrete. By simply discretizing a radius of such multilayer waveguide, the SAFE routine efficiently computed phase velocity (Figure 1b), attenuation (Figure 1c) and cross-sectional mode shapes (Figure 1d). Only axis-symmetric modes are shown. Twenty-five quadratic elements were use to discretize each of the three layers. The particular displacement mode shapes plotted correspond to small attenuation losses of two modes at 310 kHz. It can be seen that no energy is present in the outer concrete layer at both of these low loss points. However, one of the two modes generates substantial displacements within the steel rod. This kind of analysis can help designing a structural monitoring system which concentrates the ultrasonic energy within the strands and provides reasonably large inspection ranges.

3 DEFECT DETECTION

3.1 Experimental setup

Results will be presented for a high-grade steel 270, seven-wire twisted strand with a total diameter of 15.24 mm (0.6 in). This is a typical strand for stay cables and for prestressed concrete structures. The nominal diameter of each of the wires was 5.08 mm (0.2 in). A notch was machined, perpendicular to the strand axis, in one of the six peripheral wires by saw-cutting with depths increasing by 0.5-mm steps to a maximum depth of 3 mm (Figure 2). A final cut resulted in the complete fracture of the helical wire (broken wire, b.w.), which was the largest defect examined. The smallest notch depth of 0.5 mm corresponded to a 0.7% reduction in the strand's cross-sectional area. The largest notch depth of broken wire corresponded to a 15.6 % reduction in the strand's cross-sectional area.

The strand was subjected to a 120 kN tensile load, corresponding to 45% of the material's ultimate tensile strength, that is a typical operating load for stay cables. The load was applied in the laboratory by a hydraulic jack operating in load control.

Magnetostrictive transducers resonant at 320 kHz, were used to excite and detect GUWs in the strand (Figure 2). This frequency was chosen since it is known to propagate with little losses in loaded, free strands (Rizzo & Lanza di Scalea 2004b). The distance between the transmitting and the receiving transducers, d_1 in Figure 2, was fixed at 203 mm (8 in) in all

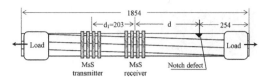

Figure 2. Experimental setup for defect detection in a strand using reflections of guided waves excited and detected by magnetostrictive transducers (dimensions in mm).

tests. By sliding the transmitter/receiver pair along the strand, tests were conducted at the five different notch-receiver distances, d in Figure 2, of 203 mm (8 in), 406 mm (16 in), 812 mm (32 in), 1016 mm (40 in), and 1118 mm (44 in). The latter was the largest distance allowed by the rigid frame of the hydraulic loading.

A National Instruments PXI© unit running under LabVIEW© was employed for signal excitation, detection and acquisition. Five-cycle tonebursts centered at 320 kHz, modulated with a triangular window, were used as generation signals. Signals were acquired at sampling rate equal to 33 MHz and stored after different number of digital averages, namely 500, 50, 10, 5, 2 and 1 (single generation).

3.2 Wavelet feature extraction

Two time windows were selected for the direct signal and the defect reflection measured by the receiver. The gated waveforms were then processed through the Discrete Wavelet Transform (DWT) using the Daubechies of order 40 (db40) mother wavelet. For a 33 MHz sampling frequency, the 320 kHz frequency of interest was contained in the sixth level of DWT decomposition, according to

$$f_j = \Delta \times F / 2^j \qquad (6)$$

relating the reconstructed frequency f_i at level i to the center frequency F of the mother wavelet, the scale 2^j, and the signal sampling frequency Δ. Thus the sixth level was the only one considered in the further analysis (pruning). Representative results of the pruning process are shown in Figure 3, presenting the signals reconstructed from the first six DWT detail decomposition levels (D_1, D_2,..., D_6). The original signal was taken without any averages. The D_6 reconstruction correctly identifies (at around 140 μ sec) the reflection from a 2.5 mm-deep notch in one of the helical wires of the strand. Since levels 1 to 5 will merely reconstruct noise, they will be eliminated in the DWT analysis process.

Subsequently to pruning, the sixth decomposition level was subjected to the thresholding process. The threshold chosen to select the relevant wavelet coefficients is an important variable that affects the sensitivity of the defect sizing. An optimum threshold

1031

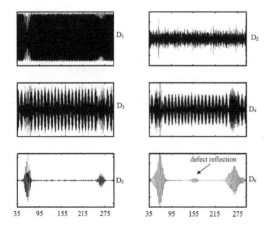

35 95 155 215 275   35 95 155 215 275

Figure 3. Signals reconstructed after pruning the DWT coefficients at the first six decomposition levels.

combination for the direct signal and the defect reflection was searched based on obtaining the largest sensitivity to defect size through a variance-based reflection coefficient.

It was found that the larger sensitivities were obtained when setting more severe thresholds on the defect-reflected signals, with little effect of the thresholds imposed on the direct signal. Based on the findings in Rizzo & Lanza di Scalea (2006), optimum thresholds were fixed at 20% of the maximum wavelet coefficient amplitude for the direct signal, and at 70% of the same quantity for the defect reflection.

A "reflection" Damage Index vector (**D.I.**) was constructed from the ratios between certain features of the reflected signal, $F_{reflection}$, and the same features of the direct signal, F_{direct} :

$$\mathbf{D.I.} = \left[\frac{F_{reflection,i}}{F_{direct,i}} \right] \qquad (7)$$

After parametric studies, the following four features were used to compute a four-dimensional **D.I.**: variance, root mean square, peak amplitude and peak-to-peak amplitude of the thresholded wavelet coefficients at level 6. All **D.I.** components showed a quite linear dependence in a semi-logarithmic scale on the notch depth, and a relatively negligible dependence on the defect position for notches between 1.5 mm and 3 mm in depth. The experimental data for two of these components are shown in Figure 4. The results for very small notches, below 1 mm in depth, were less stable against varying distances due to the poorer SNRs of the defect reflections. The results for the broken wire case (5 mm-deep notch) also showed an increased dependence on the notch-receiver distance, with D.I. components generally increasing for defects located further away from the receiver. This trend is opposite

to what would be expected considering wave attenuation effects, and its origin is probably associated with the interference of multiple propagating modes that is distance dependent. It was also found that the D.I. component based on the variance of the wavelet coefficient vector (Figure 4) had the largest sensitivity to notch depth compared to all other components.

3.3 Statistical defect classification

A multivariate statistical analysis was performed to discriminate the defect indications from random noise which may be present in the measurements. The Mahalanobis Squared Distance (MSD), D_ζ, was used as the discordancy test according to:

$$D_\zeta = \left(\{x_\zeta\} - \{\overline{x}\} \right)^T \cdot [K]^{-1} \cdot \left(\{x_\zeta\} - \{\overline{x}\} \right) \qquad (8)$$

where x_ζ is the potential outlier vector, $\overline{x}$ is the mean vector of the baseline, $[K]$ is the covariance matrix of the baseline and T represents a transpose matrix. In the present study, since the potential outliers were always known a priori, both z_ζ and D_ζ were calculated exclusively without contaminating the statistics of the baseline data.

The baseline distribution was obtained from the ultrasonic signals stored after averaging over ten acquisitions and corrupted by two different levels of white Gaussian noise. The noise signals were created by the MATLAB *randn* function. The random noise increased the sample population and simulated possible variations in SNR of the measurements that can be originated, in practice, by a number of factors including changing sensor/structure ultrasonic transduction efficiency, and changing environmental temperature affecting ultrasonic damping losses. The *randn* function generates arrays of random numbers whose elements are normally distributed with zero mean and standard deviation equal to 1. The function was pre-multiplied by a factor that determines the noise level. Factors equal to 0.01 and 0.1 were considered as "low noise" and "high noise", respectively. For each noise level, 300 baseline samples were created.

The same approach was taken to generate a large number of data for the damaged conditions. Six of the seven total notch sizes discussed in Section 3.1 were considered. The ten-average signals acquired for each of the six defects were corrupted by the low noise level and the high noise level, generating a total of 300 samples for each damage size. These samples represented the testing data of the algorithm. A total 2100 samples data were thus collected for each noise level.

The added noise can be quantified in terms of SNR by the following expression:

$$\mathrm{SNR\ [dB]} = 10\,\mathrm{Log}\left(\sum_{i=1}^{N} s_i^2 / N \middle/ \sum_{i=1}^{N} u_i^2 / N \right) \qquad (9)$$

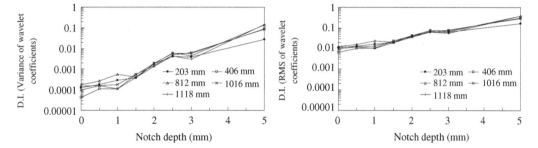

Figure 4.   Components of the Damage Index vector measured from the variance and from the root-mean-square of the thresholdded wavelet coefficients at the sixth decomposition level.

where s_i and u_i are the amplitudes of the ultrasonic signal and of the noise signal, respectively, and N is the number of points. The SNR between the direct signal and the two 0.01 and 0.1 noise levels was about 43 dB and 23 dB, respectively. The SNR between the reflection from the 3 mm-deep notch and the two 0.01 and 0.1 noise levels was about 32 dB and 12 dB, respectively. Clearly, the latter two values decreased with decreasing notch depth.

3.3.1   Defect detection results – "low" noise

The MSD computed from the four-dimensional **D.I.** of all samples, including the baseline data and the damage data, calculated for the low noise level of 0.01 are summarized in Figure 5a. The mean vector and the covariance matrix were determined from the 300 **D.I.** vectors associated with the undamaged condition of the strand. The discordancy values of the damaged conditions were calculated in an exclusive manner. The horizontal line in this figure represents the 99.73% confidence threshold value of 21.579. Eight baseline samples are outliers, thus false positive indications. Clear steps can be seen for increasing levels of damage. All damaged conditions were properly classified as outliers, thus there were no false negative indications. The MSD values showed good discrimination between all defect sizes, including the smallest notch depths, confirming that it is advantageous to combine multiple GUW features to provide a large sensitivity to the defects. Nevertheless, compared to previous multivariate outlier analyses in structural monitoring applications, the dimension of the **D.I.** was still kept at a very low value by selecting only four features of the GUW signals containing the essential information of interest owing to the effectiveness of the DWT decomposition.

3.3.2   Defect detection results – "high" noise

Following the same approach, the MSD results of the **D.I.** corrupted with the high noise level of 0.1 are shown in Figure 5b. The 99.73% confidence threshold was now computed as 18.137. Compared to the low noise results of Figure 5a, it is clear that the heavier

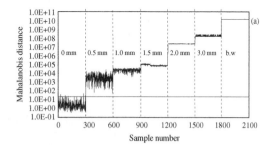

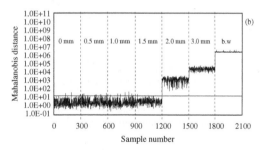

Figure 5.   Mahalanobis squared distance for the baseline (undamaged) and damaged strand data corrupted with the low-level noise (a) and the high-level noise (b).

noise corruption compromises the ability to detect the notch depths below 2.0 mm, corresponding to a 5% reduction in strand's cross-sectional area. The ratios of correctly classified outliers below 5% area reduction were only 12/300, 7/300 and 1/300 for notch depths of 0.5 mm, 1.0 mm and 1.5 mm, respectively. Above the 5% area reduction, the sensitivity to defect detection was also degraded with the increasing noise level; for example, the MSD values for the 2 mm notch depth in Figure 5b are four orders of magnitude smaller than the corresponding values in Figure 5a. The reduced number of false positive indications (three against eight) is the only improvement over the low noise level.

Table 1 summarizes the number of outliers detected in the multivariate analyses for both levels of noise considered; the outliers are false positive indications

Table 1. Results of defect detection for outlier analysis: number of outliers n/300 for the various damage sizes and two levels of noise.

Noise level	Damage Size (notch depth – mm)						
	0	0.5	1.0	1.5	2.0	3.0	5.0 (b.w.)
0.01	8/ 300	300/ 300	300/ 300	300/ 300	300/ 300	300/ 300	300/ 300
0.1	3/ 300	12/ 300	7/ 300	1/ 300	300/ 300	300/ 300	300/ 300

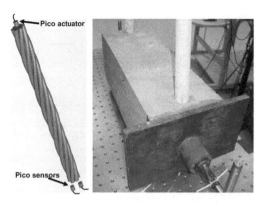

Figure 6. Experimental setup for monitoring prestress levels in a seven-wire strand embedded in a concrete block from through-transmission ultrasonic measurements.

for the baseline data (Damage Size 0) and, instead, correct indications of anomalies for the defect data.

4 PRESTRESS LEVEL MONITORING

Experiments were conducted on three strands, each embedded in a 152 mm (6 in) × 152 mm (6 in) × 1016 mm (40 in) concrete block (Figure 6). A layer of grout was also present in the strand ducts. After the concrete cured, two of the strands were post-tensioned at two different stress levels, namely 70% and 45% of U.T.S. The third strand was left unstressed to provide a total of three different prestressing conditions.

Two novel features of GUWs were considered for this task. The first feature examined was the amount of ultrasonic energy leakage among the individual wires comprising the strand ("interwire" leakage). It was, in fact, anticipated that such leakage increases with increasing level of prestress as a consequence of the increasing interwire stresses. The second feature examined was any shift in the frequency of maximum ultrasonic transmission relative to the excitation frequency. Such shift was expected to occur as a

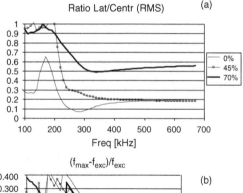

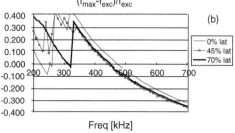

Figure 7. Stress-sensitive features. (a) Energy leakage from center to peripheral wire; and (b) shift between excitation frequency and peak transmission frequency.

result of the changing wave dispersion with increasing interwire contact.

A through-transmission ultrasonic setup, schematized in Figure 6, was adopted for this investigation. A broadband transducer (Pico sensor©, Physical Acoustic Corporation) was used to excite waves in the central wire at one of the strands' ends. The waves were detected by two Pico sensors located on the central wire and on the peripheral wire at the strands' opposite ends. Toneburst signals were excited by sweeping the generation frequency from 100 kHz to 700 kHz.

Figure 7a shows the "energy leakage" feature, here quantified as the ratio between the root-mean-square of the signal detected in the peripheral wire and that of the signal detected in the central wire. This ratio was computed so as to make the measurements robust against changes in the excitation voltage or in the coupling conditions of the excitation transducer. The figure shows that the amount of interwire leakage indeed increases with increasing prestress levels. It is also shown that the sensitivity to prestress level is not uniform throughout the frequency range, with maximum sensitivity to stress being achieved at around 300 kHz.

Figure 7b plots the results of the frequency shift feature, computed as the relative shift between the frequency of the maximum amplitude detected through-transmission (f_{max}) and that of the excitation toneburst (f_{exc}). It can be seen that such shift is sensitive to the presence of prestress in the strands, although it appears less effective than the interwire energy leakage in discriminating the exact level of applied stress. These results, however, suggest that frequency shifts can be very effective for detecting a complete loss of pre-stress. The shifts are uniform in a large frequency range (350 kHz – 700 kHz), which can be useful when considering the use of other sensors (e.g. magnetostrictive) embedded in the grout for monitoring prestressing strands in real structures.

5 CONCLUSIONS

Due to the high probing frequencies and reasonably large propagation distances, Guided Ultrasonic Waves (GUWs) are good candidates for the structural health monitoring of loaded strands.

Modeling the multimode and dispersive character of GUWs in the cylindrical waveguide is important to interpret the measurements and, ultimately, to design an efficient monitoring system. A semi-analytical finite element method was used to model GUW propagation in steel rods embedded in grout and concrete. This technique is being extended to model twisted, embedded waveguides representing seven-wire, prestressing strands in concrete.

A reflection-based Damage Index vector was used to detect notch-like defects in the strands. The Discrete Wavelet Transform (DWT) was employed to compress each GUW measurement to four coefficients. A four-dimensional Outlier Analysis was then performed to discriminate indications of notches from noise simulated in the laboratory. The notches considered were located as far away as 1,100 mm from the sensors. The algorithm was able to properly flag notches as small as 0.5 mm (0.7% strand's area reduction) for SNRs on the order of 32 dB. For higher noise levels, corresponding to SNRs on the order of 12 dB, the properly flagged notches were as small as 2 mm (5% strand's area reduction).

In a parallel study, through-transmission measurements were collected to identify wave features sensitive to prestress levels in strands embedded in post-tensioned concrete blocks. The amount of wave energy leakage between the central wire and the peripheral wires was one stress-sensitive feature identified. The shift between excitation frequency and peak transmission frequency was another stress-sensitive feature identified. Both of these features are being investigated further to assess their applicability to a monitoring system where sensors are embedded along the length of a strand in a real structure.

ACKNOWLEDGMENTS

The strand monitoring project is funded at UCSD by the U.S. National Science Foundation under grant # 0221707 (Dr. S-C. Liu, Program Manager), and by the California Department of Transportation under contract # 59A0538.

(Dr. C. Sikorsky, Program Manager). Funding was also provided by the Italian Ministry for University and Scientific & Technological Research MIUR (40%). The topic of research is part of the research thrusts of the Centre of Study and Research for the Identification of Materials and Structures (CIMEST) at the University of Bologna.

REFERENCES

Bartoli, I., Marzani, A., Lanza di Scalea, F. & Viola, E. 2006. Modeling wave propagation in damped waveguides of arbitrary cross-section, *Journal of Sound and Vibration*, 295: 685–707.

Beard, M.D., Lowe, M.J.S. & Cawley, P. 2003. Ultrasonic guided waves for inspection of grouted tendons and bolts. *ASCE Journal of Materials in Civil Engineering*, 15: 212–218.

Chase, S.B. 2001. Smarter bridges, why and how? *Smart Maerials Bulletin*, 2: 9–13.

Chen, H.-L. & Wissawapaisal, K. 2002. Application of Wigner-Ville transform to evaluate tensile forces in seven-wire prestressing strands. *ASCE Journal of Materials in Civil Engineering*, 128: 1206–1214.

Hayashi, T., Song, W.J. & Rose, J.L. 2003. Guided wave dispersion curves for a bar with an arbitrary cross-section, a rod and rail example. *Ultrasonics*, 41: 175–183.

Huang, K.H. & Dong, S.B. 1984. Propagating waves and edge vibrations in anisotropic composite cylinders. *Journal of Sound and Vibration*, 96: 363–379.

Kwun, H. & Teller, C. M. 1994. Detection of fractured wires in steel cables using magnetostrictive sensors. *Materials Evaluation*, 503–507.

Kwun, H. & Teller, C.M. 1995. *Nondestructive Evaluation of Steel Cables and Ropes Using Magnetostrictively Induced Ultrasonic Waves and Magnetostrictively Detected Acoustic Emissions*. U.S. Patent No. 5,456,113, 1995.

Kwun, H., Bartels, K.A. & Hanley, J.J. 1998. Effect of tensile loading on the properties of elastic-wave in a strand. *Journal of the Acoustical Society of America*, 103: 3370–3375.

Lanza di Scalea, F., Rizzo, P. & Seible, F. 2003. Stress measurement and defect detection in steel strands by guided stress waves. *ASCE Journal of Materials in Civil Engineering*, 15: 219–227.

Parker, D. 1996. Tropical overload. *New Civil Engineer*, 18–21.

Pavlakovic, B.N., Lowe, M.J.S. & Cawley, P. 1999. The inspection of tendons in post-tensioned concrete using guided ultrasonic waves. *Insight-NDT&Condition Monitoring*, 41: 446–452.

Pavlakovic, B.N., Lowe, M.J.S. & Cawley, P. 2001. High-frequency low-loss ultrasonic modes in imbedded bars. *Journal of Applied Mechanics*, 68: 67–75.

Reis, H., Ervin, B.L., Kuchma, D.A. & Bernhard, J.T. 2005. Estimation of corrosion damage in steel reinforced mortar using guided waves. *ASME Journal of Pressure Vessel Technology*, 127: 255–261.

Rizzo, P. & Lanza di Scalea, F. 2001. Acoustic emission monitoring of carbon-fiber-reinforced-polymer bridge stay cables in large-scale testing. *Experimental Mechanics*, 41: 282–290.

Rizzo, P. & Lanza di Scalea, F. 2004. Wave propagation in multi-wire strands by wavelet-based laser ultrasound. *Experimental Mechanics*, 44: 407–415.

Rizzo, P. & Lanza di Scalea, F. 2005. Ultrasonic inspection of multi-wire steel strands with the aid of the wavelet transform. *Smart Materials and Structures*, 14: 685–695.

Rizzo, P. & Lanza di Scalea, F. 2006. Discrete wavelet transform for enhancing defect detection in strands by guided ultrasonic waves. *International Journal of Structural Health Monitoring*, 5: 297–308.

Washer, G., Green, R.E. & Pond, R.B. 2002. Velocity constants for ultrasonic stress measurements in prestressing tendons. *Research in Nondestructive Evaluation*, 14: 81–94.

Watson, S.C. & Stafford, D. 1988. Cables in Trouble. *Civil Engineering*, 58: 38–41.

Woodward, R.J. 1988. Collapse of Ynys-y-Gwas bridge, West Glamorgan. *Proceedings of the Institution of Civil Engineers*, 84: 635–669.

Fracture Mechanics of Concrete and Concrete Structures – Design, Assessment and Retrofitting of RC Structures – Carpinteri, et al. (eds)
© 2007 Taylor & Francis Group, London, ISBN 978-0-415-44616-7

Prediction of repair needs in weathered concrete façades and balconies

I. Weijo, J. Lahdensivu & S. Varjonen
Tampere University of Technology, Institute of Structural Engineering, Tampere, Finland

ABSTRACT: Degradation mechanisms of concrete are fairly well known, but systematic maintenance and repair of concrete structure vary a lot. Consequently the condition of structures may surprise unpleasantly the property owners and the service life of concrete structure can be much shorter than planned. Therefore the building management needs better proceeding to control their real estate. Concerning this problem a research project called Repair Need in Concrete Facades and Balconies has started. The target of the ongoing research is to create data source of detailed information of facades and balconies. Based on this data prediction model will be developed. Using the model property owners can compare their real estates and predict the progress of condition and arrange the stock of buildings into order of repair need. The prediction model gives possibilities to prepare technically and economically to the coming operations beforehand. It enables the building management to plan their maintenance and repair strategies. The data base for this model consists of large number of condition investigations done in Finland.

1 INTRODUCTION

1.1 *The origin of problem*

Since 1960's, a total of over 30 million square metres of concrete-panel facades has been built in Finland as well as over half a million concrete balconies. Especially in 1970's the urbanization caused urgent need for housing production and exponentially growing construction. Concrete and prefabrication became the leading method to construct new centre of population and it became a solution for the settlement of people to the cities. Suburbs were born. Fast and mostly uncontrolled development didn't always lead to good solutions and quality of concrete structures suffered. The quality of concrete itself was poor, details were not properly designed and the deficiency of workmanship affected the final result. Also lack of regulations led to variability in construction. The first Finnish building regulations were created 1976 and have ever since been improved according to the latest and prevailing knowledge of construction. Before that there were more separate directions, like directions concerning the production and use of prefabricated units (1963) and official concrete code, the first one date back to 1954 (Suomen Betoniyhdistys 2002).

Most of the Finnish apartments are privately owned, but there are also big real estate owners, who are possessing relatively large number of apartment houses. During the maintenance and repair of these buildings several problems have occurred. Due to above mentioned issues, the structures have required often unexpected, technically and economically significant repairs sometimes less than in 10 years after their completion. Plenty of new technology and methods for maintenance and repair of concrete structures have been developed in Finland during the last 15 years. These are systematic condition investigation method, different kind of repair alternatives for all kind of damage, high-quality repair products and functional and well-designed structural details among others.

Over one million inhabitants live in concrete buildings in Finland and the renovation of concrete facades affect their cost of living significantly. The target in predicting the repair need of concrete structures is that the condition of Finnish real estates won't make an unpleasant surprise for property owner or people responsible for the building management and that the value of property is preserved.

1.2 *Estimation of residual service life*

Concrete structures deteriorate by several different degradation mechanisms, whose progress depend on many structural, circumstantial and material factors. Hence the service lives of structures vary in practise a lot. The variable structural condition in different houses and the fact that it is often impossible to observe the most significant damage visually before they have progressed too far, makes a thorough condition investigation necessary in most facade repair cases.

Condition investigation means a systematic procedure to find out the condition and performance of a structure or a group of structural parts (e.g. facades and balconies) and their needs for repair, taking into account different degradation mechanisms and using different kinds of research methods. These are field examination, sampling and laboratory tests as well as gathering information from documents, constructions drawings and possible earlier observation data of target. The existence, degree and extent of damage for different degradation mechanisms are determined by condition investigation. Hundreds of condition investigations have been done in Finland during the past 15 years (Suomen Betoniyhdistys 2002).

2 DEGRADATION MECHANISMS

The degradation of concrete structures with age is primarily due to weathering action which deteriorates material properties. Degradation may be unexpectedly quick if used materials or the work performance have been of poor quality or if the structural solutions have been deficient from the beginning. Weathering action may launch several simultaneous deterioration phenomena whereby a facade is degraded by the combined impact of several adverse phenomena. Degradation phenomena proceed slowly initially, but as the damage propagates, the rate of degradation generally increases.

The most common degradation mechanisms causing the need to repair concrete facades, and concrete structures in general, are corrosion of reinforcement due to carbonation or chlorides as well as disintegration of concrete. Disintegration can ensue from freeze-thaw exposure, formation of late ettringite or alkali-aggregate reactions. The most common reason for disintegration in northern Europe is insufficient frost resistance of concrete which can lead to frost damage (Pentti et al., 1998).

These degradation mechanisms may result in, for instance, reduced bearing capacity or fixing reliability of structures. Experience tells that defective performance of structural joints and connection details generally causes localised damage thereby accelerating local propagation of deterioration.

2.1 Reinforcement corrosion

Reinforcing bars within concrete are normally well protected from corrosion due to the high alkalinity of concrete pore water. Corrosion may start when the passivity is destroyed, either by chloride penetration or due to the lowering of the pH in the carbonated concrete.

Carbonation of concrete is a phenomenon that decreases the naturally high alkalinity of concrete, that is, neutralises it. Neutralisation begins at the surface of a structure and propagates as a front at a decelerating rate deeper into the structure. The speed of

Figure 1.   Damage caused by reinforcement corrosion in the window frame.

propagation is influenced foremost by the quality of concrete (proportion of cement and density) as well as rain stress. Heavy rain stress slows down neutralisation by blocking carbon dioxide from penetrating into the structure.

The high alkalinity of concrete protects the reinforcement within from corrosion. When the carbonation front proceeds in concrete to the depth of the reinforcement, the surrounding concrete neutralises and corrosion of re-bars can begin, if there is enough moisture and oxygen available. The rate of corrosion clearly depends on the moisture content of concrete and advances significantly only at over 80% RH. Corrosion decreases the tensile and bond strength of reinforcement while the corrosion products cause pressure to the concrete cover around the reinforcement, inducing cracking.

The passivity of steel may also be destroyed by the presence of chlorine ions derived either from the environment or from the use of contaminated constituents of concrete. It is possible that chlorides have been added to the concrete mix during preparation to accelerate hardening. Chlorides were used mainly in the 1960's in connection with in-situ concreting and in prefabrication plants during the cold season when concrete hardens slowly. Often the amount of salt used as accelerator was multiple compared to steel's corrosion threshold. Chlorides may also penetrate into hardened concrete if the concrete surface is subjected to external chloride aggression, for instance, on bridges where de-icing salts are used. Strong pitting corrosion is typical

Figure 2. In this case water has exposed the structure to freeze-thaw action. Thus the structure has disintegrated.

to chloride corrosion of re-bars, and it may propagate also in relatively dry conditions. Chloride-induced corrosion becomes highly accelerated when carbonation reaches reinforcement depth whereby the extent of visible damage may increase strongly in a short time (Pentti, 1999 and Tuutti, 1982).

2.2 *Frost damage in concrete*

As earlier mentioned, the other main degradation mechanism of concrete is disintegration. Because the most common cause is the freeze-thaw exposure in Finland and other cold countries, it is reviewed here.

Concrete is a porous material whose pore system may, depending on the conditions, hold varying amounts of water. As the water in the pore system freezes, it expands about 9% by volume which creates hydraulic pressure in the system (Pigeon & Pleau, 1995). If the level of water saturation of the system is high, the overpressure cannot escape into air-filled pores and thus it damages the internal structure of the concrete resulting in its degradation. A typical sign for frost damage is widening map cracking. Far advanced frost damage leads to total loss of concrete strength.

The frost resistance of concrete can be ensured by air-entraining which creates a sufficient amount of permanently air-filled so-called protective pores where the pressure from the freezing dilation of water can escape. Finnish guidelines for the air-entraining of facade concrete mixes were issued in 1976 (Suomen Betoniyhdistys 2002).

Moisture behaviour and environmental conditions have a strong impact on stress caused by frost. For instance, the stress that affects balcony structures

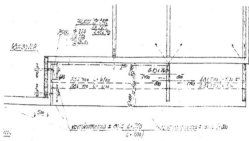

Figure 3. Old construction documents are not always found, often slightly obscure and their reliability is arguable.

depends on the existence of proper waterproofing or protecting balcony glazing.

3 CONDITION INVESTIGATION

The basic aim of the condition investigation is to produce information about the factors affecting the condition and the performance of the structure and consequently about the need and the options of repair for the building management. Condition investigation report gives information about the present state of damage and investigator's estimation about the residual service life of structure and maintenance and possible repair alternatives. A systematic condition investigation consists of fairly simple and clear phases as mentioned later on. The phases are not totally separate and successive, but rather partly overlapping. They also deal more with principles and ideologies than practical actions of condition investigation (Suomen Betoniyhdistys 2002).

3.1 *Determining the structures and materials*

The first phase of the condition investigation is to study what kind of structure or structures are under investigation. This is done by visual inspection and examining the documents and its idea is to get to know what kind of a structural system the object is and what the materials it has been constructed of are. It is self-evident that no successful investigation can be carried out if the investigator does not know the type and the nature of the object. Thus also the person who does the investigation work has to possess wide experience about the behavior of structures under investigation.

The information is gathered from original construction documents and by visual inspection. The information that old documents provide needs to be considered carefully. The structures have not always been constructed exactly according to the original documents. In certain cases the dissimilarity between the structure and the document may be remarkable.

It is also important that different types and parts of structures as well as structures in different exposure conditions are distinguished from each other and that they are also investigated as separate groups of objects. A fairly common error is to examine only some part of the building, for example concrete facade panels, and then generalize the conclusions to apply also to the other parts of the structure, for example to the concrete balconies in the facade. It is important to realize that the results apply only to the structure which they have been obtained from. However, the results between separate groups can be utilized so that if there is no damage in the structures under severe exposure conditions there probably will not be any in the lighter conditions either (Mattila 1998).

3.2 Evaluation of possible deterioration mechanisms

The second phase of condition investigation is to recognize the problems that may exist in the structure. This is received from field and laboratory tests. The information is considered on the basis of the type of structure and materials used in it as well as the exposure conditions of the structure. The possible problems may be caused either by different kinds of deterioration mechanisms or by malfunction of structures, for example problems with moisture. A list of problems and ways of malfunction for example in a concrete panel facade may be as follows:

- reinforcement corrosion due to carbonation of concrete cover or due to chlorides in concrete,
- disintegration of concrete due to freeze-thaw exposure, formation of late ettringite or alkali-aggregate reactions,
- decrease in the bearing capacity of structural members or weakening of fixings or ties in a structure (for example weakening of the ties in sandwich panels due to corrosion),
- malfunction in the moisture behavior of the structure including defects in joints, malfunction of the ventilation inside the structure,
- defects and degradation of paints and coatings,
- defects in facade tiling (ceramic, clay brick or natural stone tiles),
- harmful cracking or deformations in concrete,
- defects due to the use of the structure (for example normal wearing),

All the deterioration mechanisms and types of malfunction have to be considered at least to some extent when aiming at a reliable condition investigation. It is important to notice that the items to be investigated vary widely between different investigation cases, and no fixed set of investigation measures can be used.

The mutual importance and the combined effects of different deterioration mechanisms should also be evaluated carefully. It is self-evident that the factors related to the bearing capacity and safety of fixings is the most important items to be investigated (Mattila 1998).

3.3 Rough evaluation of methods for maintenance or repair

The third phase of a condition investigation is to consider the potential techniques for the maintenance and repair of the structure. This is important to do as early as possible because all kinds of remedial techniques require different amount and type of information of the structure to be repaired.

There is a wide range of options for maintenance and repair of structures ranging from doing nothing to demolition and reconstruction of the structure. The technical and economical comparison between the options is extremely difficult in most cases.

Possible options of repair methods as an example in the case of concrete panel facade are as follows:

1. Doing nothing (nearly always possible at least for a while)
2. Repair by standard painting or by using special protective coatings
3. Careful patch repair by cementitious repair mortars
4. Additional thermal insulation and cladding
5. Realkalisation
6. Cathodic protection
7. Demolition and reconstruction.

The suitable repair method depends on the technical condition of a structure. Usually, the final repair is a combination of some of these options. Some of these options are more used than the rare ones.

These alternatives have to be also re-evaluated from time to time during the investigation process whenever new information about the condition and the need of repair becomes available. It is possible that on the basis of new information about the condition some lighter repair techniques can be used or simultaneously some more investigation measures is needed for evaluation of feasibility of these techniques (Mattila 1998).

3.4 Analysis and report from the gathered information

The fourth phase of the investigation is to gather objective information concerning the deterioration processes and malfunction of the structure and analyze carefully the information.

It is important to notice that the measured values and other observations gathered during the investigation process are not the final results of the investigation, but an analysis to produce the results out of them is always needed.

Practically, the analysis means seeking answers to the following five questions:

1. What kind of problems exists in the structures?
2. What is the extent stage and location of each type of damage and malfunction?
3. What are the reasons for the problems noticed?
4. What kind of effects may the problems have on the structure itself or on the users of the building?
5. How will the damage or malfunction proceed in the future?

The final results can be settled on the basis of the answers to these five questions.

The fifth phase is to prepare a written report in which the results are presented for the client. The report should not consist only of measured values etc. but rather of practical conclusions concerning the alternative practical measures for the client to manage with the structure. There are usually several options for maintenance and repair, and all suitable methods should be evaluated shortly in the report (Mattila 1998).

4 THE PREDICTIVE MAINTENANCE

4.1 *The data base*

Hundreds of condition investigations have been done in Finland. The Institute of Structural Engineering in Tampere University of Technology has done over 200 condition investigation and been in a way a pacemaker developing the procedure. Nowadays there are few large and plenty of small agencies doing only condition investigations for all kinds of facades and balconies. Most of the investigations concern concrete structures as it is the most used construction material in apartment houses.

Out of each investigation a report is written. These reports contain large amount of detailed information concerning concrete structures of different ages and surface types. This information serves at the moment only individual property owner, whereas it also has all the potential to form a massive, constantly growing data base for statistical investigation. Active condition investigation and repair work could produce extensive and reliable data base, which could be used for predicting the repair need of concrete facades and balconies, but which is scattered at the moment and doesn't actually exist, yet.

Gathering up of this database has been started in Tampere University of Technology. In 2006 March started a three-year project, called The Repair Strategies of Concrete Façade and Balconies. Its major aims are to produce above mentioned database and create a sort of predictive model for building management to systematize and add reliability to decision-making concerning maintenance and renovation. To get the

information needed and to get results that really are useful in practise, this work is done together with big property owners and major condition investigation agencies, who are handing over their condition investigation reports and taking part as commentators to this project.

The condition investigation reports include plenty of very detailed information about the present condition of buildings, in a numerical and in a verbal form. To make it comparable and statistically handleable, numerical classification for verbal information has been created. As a result over 60 values per one building and over 20 individual values per one sample taken from the building are registered. The information is assorted according to the surface type of façade and balcony structure as its own file. Because there are about 350–400 reports in the use of this research, the number of values to form statistically reliable damaging and property distributions and comparison is approximately 150 000.

4.2 *Building management*

Building management and the work contribution in building maintenance has essential role on the development of building's condition. With properly timed and right kind of maintenance, it is possible to prolong the service life of structure by postponing or stopping deterioration and protecting structure beforehand. Advanced deterioration happens when there are unidentified problems and numerous buildings to take care of.

By regular maintenance the property owners know better what the state of condition is and are able to budget for bigger operations beforehand. The most important affect is that the service life extends by small-case service operations like coating using protecting products and ensuring the function of flashings and other associated systems. To avoid unexpected renovations, building management needs more preventive maintenance and help to predict coming repair needs.

The substantial factor making systematic building management to success is the identification of damage. A model to identify the damage can be established on the grounds of earlier mentioned data base and it will help to direct the condition investigations' and repair works' timing optimal.

4.3 *A predictive model*

A model to identify damage is actually a large number of distributions of different kind of states of damage and the relation of the damage and other factors. Distributions are classified by surface types and year of construction. One distribution serves as a prediction model of certain structure to which building management can compare their own structure and thus see the odds for what's going on. The property owners use the prediction model as comparative data to their own buildings of similar age and surface types. From the data base the owners find different kind of distributions concerning for example the depths of carbonation front and the average depth of reinforcement.

The distributions help them to estimate what might be going on in buildings of different construction year and surface types. Buildings can be arranged to a rough order of repair urgency. After that they can consider what kind of deterioration might be going on in what kind of structures. They get information, where should they start their inspection and how it is done. Is the expected condition so good that visual evaluation and checking are enough to verify the present condition or is thorough condition investigation needed?

This enables the transition to pro-active maintenance in building management. With that it is possible to estimate the incoming repair needs before damage is visual even for the large number of real estates. Thereby house managers can prepare enough beforehand both technically and financially, by systematization and directing the building management and repair planning. Thus repair work can be optimally timed out taking into consideration the service lives of old structures.

The aim is also to reduce the amount of heavy repair work and at the same time to reduce the need for demolition of deteriorated structure and thereby also reduce the waste that demolition generates. With optimally timed light repair, the heavy repair work can be entirely avoid or at least move to a later point in time. Pro-active building management extends residual service lives of present structures, develops and maintains good, healthy residential environment and prevents the risks of reduced bearing capacity and safety that deterioration causes.

It is important to remember, that using the predictive model does not replace the condition investigation as clarification of the present state of individual structure and building. Condition investigation which is done before engaging to repair operations is the best method to avoid oversized or too risky repair. This predictive model is based on statistical probabilities and function as a help for building management to lead the resources to right targets.

5 CONCLUDING REMARKS

Building management definitely needs technical help organizing their property in the order of repair need. To make sustainable renovation and to ensure continuous safety also between the renovations and service operation, there needs to be reliable information where the decision-making can rely on.

The predictive model is based on over 350 condition reports. Therefore after its completion this model gives reliable data for the use of building management.

REFERENCES

Mattila J. et al. 1998: Condition Investigation Prior to Renovation Process – a Systematic Approach in Saving Buildings. In Central and Eastern Europe IABSE Colloquium Berlin June 4–5, 1998, Pp.68–71.

Pentti M et al. 1998: Repair of Concrete Facades and Balconies. Part 1: Structures, degradation and condition investigation. Tampere. Tampere University of Technology, Structural Engineering. Publication 87. 157 p. (In Finnish)

Pentti, M. 1999: The Accuracy of the Extent-of-Corrosion Estimate Based on the Sampling of Carbonation and Cover Depths of Reinforced Concrete Façade Panels. Tampere, Tampere University of Technology, Structural Engineering, Publication 274. 105 p.

Pigeon M. & Pleau R.1995: Durability of Concrete in Cold Climates. Suffolk. E & FN Spon. 244 p.

Suomen Betoniyhdistys r.y. 2002: Condition Investigation of Concrete Facades by 42. Helsinki, Suomen Betoniyhdistys r.y. 178 p. (In Finnish)

Tuutti K. 1982: Corrosion of Steel in Concrete. Stockholm. Swedish Cement and Concrete Research Institute. CBI Research 4:82. 304 p.

Fracture Mechanics of Concrete and Concrete Structures – Design, Assessment and Retrofitting of RC Structures – Carpinteri, et al. (eds)
© *2007 Taylor & Francis Group, London, ISBN 978-0-415-44616-7*

Safety assessment in fracture analysis of concrete structures

V. Cervenka, J. Cervenka & R. Pukl
Cervenka Consulting, Prague, Czech Republic

ABSTRACT: New safety format suitable for design of reinforced concrete structures using non-linear analysis are required due to the global nature of such approach. Safety formats based on partial factors, global factors and probabilistic analysis are discussed. Their performance is compared on four examples ranging from statically determinate structures with bending mode of failure up to indeterminate structures with shear failure.

1 INTRODUCTION

In recent years, more engineers use non-linear analysis while designing complex buildings, dams, or bridges. This evolution is supported by rapid increase of computational power as well as by new capabilities of the available tools for numerical simulation of structural performance.

The code provisions on the other hand provide very little guidance how to use the results of a non-linear analysis for structural design or assessment. The safety formats and rules that are usually employed in the codes are tailored for classical design procedures based on beam models, hand calculation or linear analysis and local section checks. On the other hand, non-linear analysis is by its nature always a global type of assessment, in which all structural parts, or sections, interact. Until recently the codes did not allow applying the method of partial safety factors for non-linear analysis, and therefore, a new safety format was expected to be formulated. Certain national or international codes have already introduced new safety formats based on over-all/global safety factors to address this issue. Such codes are, for instance, German standard DIN 1045-1 (1998) or Eurocode 2 EN 1992-2, (2005). This paper will try to compare several possible safety formats suitable for non-linear analysis: partial factor method, format based on EN 1992-2, (2005) and fully probabilistic method. A new alternative safety format is also proposed by authors, which is based on a semi-probabilistic estimate of the coefficient of variation of resistance.

Standard design procedure for civil engineering structures based on partial safety factors usually involves the following steps:

(1) Conceptual design with initial dimensioning of structural elements based on estimates and engineering judgment.

(2) Linear elastic analysis of the structure considering all possible load combinations. Results are actions in some critical sections, which could be referred as *design actions* and can be written as

$$E_d = \gamma_{S1} S_{n1} + \gamma_{S2} S_{n2} + \ldots \ldots \gamma_{Si} S_{ni} \qquad (1)$$

They include safety provisions in which the nominal loads S_{ni} are amplified by appropriate partial safety factors for loading γ_{Si}, where index i stands for load type, and their combinations.

(3) *Design resistance* of a section is calculated using design values of material parameters as:

$$R_d = r(f_d, \ldots), f_d = f_k / \gamma_m \qquad (2)$$

The safety provision for resistance is used on the material level. The design value of material property f_d is obtained from the characteristic value f_k by its reduction with an appropriate partial safety factor γ_m.

(4) Safety check of limit state is performed by *design condition*, which requires, that design resistance is greater then design action:

$$E_d < R_d \qquad (3)$$

Note, that in the partial safety factor method the safety of material criteria in local points is ensured. However, the probability of failure, i.e. the probability of violation of the design criteria (3) is not known.

The required reinforcement is designed using steps (2), (3) and (4) in which the resistance function r is changed. At the same time, changes in dimensions may be needed. The whole procedure is repeated until all sections satisfy the design criteria that are usually specified by national or international design codes. The final steps of the design verification process often involve assessment of serviceability conditions, i.e. deflections, crack width, fatigue, etc. In certain cases, these serviceability conditions might be the most important factors affecting the final design.

In the above outlined design procedure, the non-linear analysis should be applied in step (2) to replace the linear analysis. Following the current practice designer will continue to steps (3), (4) and perform the section check using the internal forces calculated by the non-linear analysis. This is a questionable practice due to the following reasons. If design values for material parameters are used in the non-linear analysis, then very unrealistic, i.e. degraded, material is assumed. In statically indeterminate structures, this may result in quite unrealistic redistribution of forces, which may not be necessary on the conservative side. Furthermore, in the non-linear analysis material criteria are always satisfied implicitly by the employed constitutive laws. Therefore, it does not make sense to continue to step (3) and perform section checks. Instead, a global check of safety should be performed on a higher level and not in local sections. This is the motivation for the introduction of new safety formats for non-linear analysis.

Another important factor is that non-linear analysis becomes useful when it is difficult to clearly identify the sections to be checked. This occurs in structures with complicated geometrical forms, with opening, special reinforcement detailing, etc. In such cases, usual models for beams and columns are not appropriate, and non-linear analysis is a powerful alternative.

The above discussion shows that it would be advantageous to check the global structural resistance to prescribed actions rather than checking each individual section. The safety format based on global assessment is more suitable for design approaches based on non-linear analysis. This approach can bring the following advantages:

(a) The nonlinear analysis checks automatically all locations and not just those selected at critical sections.

(b) The global safety format gives information about the structural safety and redundancy. This information is not available in the classical approach of section verification.

(c) The safety assessment on global level can bring, on one hand, more economic solution by exploiting reserves due to more comprehensive design model, on the other hand, the risk of unsafe design is reduced.

However, the above enthusiastic statements should be accepted with caution. There are many aspects of design, which require engineering judgment. Also many empirical criteria must be met as required by codes. Therefore, a global safety assessment based on non-linear analysis should be considered as an additional advanced design tool, which should be used, when standard simple models are not sufficient.

The non-linear analysis offers an additional insight into the structural behavior, and allows engineers to better understand their structures. On the other hand, non-linear analysis is almost always more demanding then a linear analysis, therefore an engineer should be aware of its limits as well as benefits. Other disadvantage is that the force super-position is not valid anymore. The consequence is that a separate non-linear analysis is necessary for each combination of actions.

Finally, a note to terminology will be made. The term for *global* resistance (*global* safety) is used here for assessment of structural response on higher structural level then a cross section. In technical literature, the same meaning is sometimes denoted by the term *overall*. The term *global* is introduced in order to distinguish the newly introduced check of safety on global level, as compared to local safety check in the partial safety factor method. This terminology has its probabilistic consequences as will be shown further in the paper. The proposed global approach makes possible a reliability assessment of resistance, which is based on more rational probabilistic approach as compared to partial safety factors.

2 SAFETY FORMATS FOR NON-LINEAR ANALYSIS

2.1 Design variable of resistance

Our aim is to extend the existing safety format of partial factors and make it compatible with nonlinear analysis. First we introduce a new design variable of resistance $R = r(f, a, \ldots, S)$. Resistance represents a limit state. In a simple case this can be a single variable, such as loading force, or intensity of a distributed load. In general this can represent a set of actions including their loading history. We want to evaluate the reliability of resistance, which is effected by random variation of basic variables f – material parameters, a – dimensions, and possibly others.

The resistance is determined for a certain loading pattern, which is here introduced by the symbol of actions S. It is understood that unlike material parameters and dimensions, which enter the limit state function r as basic variables, the loading is scalable, and includes load type, location, load and includes load type, location, load combination and history. It is the objective of the resistance R to determine the loading magnitude for given loading model.

Random variation of resistance is described by a statistical distribution characterized by following parameters:

R_m mean value of resistance,

R_k characteristic value of resistance, , i.e. 5% kvantile of the resistance

R_d design value of resistance.

The design condition is defined in analogy with partial safety factor method by Eq.(3)

In general, E_d and R_d represent set of actions and the limit state is a point in a multi-dimensional space, respectively. It is therefore useful to define a resistance scaling factor k_R, which describes safety factor with respect to the considered set of design actions. In the simplified form, considering one pair of corresponding components it can be described as:

$$k_R = \frac{R}{E_d} \tag{4}$$

Then, the design condition (3) can be rewritten as:

$$\gamma_R < k_R \tag{5}$$

Where γ_R is required global safety factor for resistance. Factor k_R can be used to calculate the relative safety margin for resistance

$$m_R = k_R - 1 \tag{6}$$

The task now remains to determine the design resistance R_d. The following methods will be investigated and compared:

(a) ECOV method, i.e. estimate of coefficient of variation for resistance.
(b) EN 1992-2 method, i.e estimate of R_d using the overall safety factor from Eurocode 2 EN 1992-2.
(c) PSF method, i.e. estimate of R_d using the partial factors of safety.
(d) Full probabilistic approach. In this case R_d is calculated by a full probabilistic non-linear analysis.

Furthermore, the limit state function r can include some uncertainty in model formulation. However, this effect can be treated separately and shall not be included in the following considerations.

It should be also made clear, that we have separated the uncertainties of loading and resistance (and their random behavior). Our task is reduced to describe the resistance side of design criterion (3).

2.2 ECOV method – estimate of coefficient of variation

This method is newly proposed by the authors. It is based on the idea, that the random distribution of resistance, which is described by the coefficient of variation V_R, can be estimated from mean R_m and characteristic values R_k. The underlying assumption is that random distribution of resistance is according to lognormal distribution, which is typical for structural resistance. In this case, it is possible to express the coefficient of variation as:

$$V_R = \frac{1}{1.65} \ln\left(\frac{R_m}{R_k}\right) \tag{7}$$

Global safety factor of resistance is then estimated as:

$$\gamma_R = \exp(\alpha_R \beta V_R) \tag{8}$$

where α_R is the sensitivity (weight) factor for resistance reliability and β is the reliability index. The above procedure enables to estimate the safety of resistance in a rational way, based on the principles of reliability accepted by the codes. Appropriate code provisions can be used to identify these parameters. According to Eurocode 2 EN 1991-1, typical values are $\beta = 4.7$ (one year) and $\alpha_R = 0.8$. In this case, the global resistance factor is:

$$\gamma_R \cong \exp(-3.76\,V_R) \tag{9}$$

and the design resistance is calculated as:

$$R_d = R_m / \gamma_R \tag{10}$$

The key factor in the proposed method is to determine the mean and characteristic values R_m, R_k. It is proposed to estimate them using two separate nonlinear analyses using mean and characteristic values of input material parameters, respectively.

$$R_m = r(f_m,...), \quad R_k = r(f_k,...) \tag{11}$$

The method is general and reliability level and distribution type can be changed if required. The advantage of this approach is that the sensitivity to individual parameters such as for instance steel or concrete strength can be estimated. The disadvantage is the need for two separate non-linear analyses.

2.3 EN1992-2 method

Design resistance is calculated from

$$R_d = r(\tilde{f}_{ym}, \tilde{f}_{cm}..., S)/\gamma_R \tag{12}$$

Material properties used for resistance function are shown in table above.

The global factor of resistance shall be $\gamma_R = 1,27$ The evaluation of resistance function is done by nonlinear analysis assuming the material parameters according to the above rules.

Table 1. Material parameters used in EN1992-2 method

$\tilde{f}_{ym} = 1.1 f_{yk}$	Steel yield strength
$\tilde{f}_{pm} = 1.1 f_{pk}$	Prestressing steel yield strength
$\tilde{f}_{cm} = 1.1 \frac{\gamma_s}{\gamma_c} f_{ck}$	Concrete compressive strength, where γ_s and γ_c are partial safety factors for steel and concrete respectively. Typically this means that the concrete compressive strength should be calculated as $\tilde{f}_{cm} = 0.843 f_{ck}$

2.4 PSF method – partial safety factor estimate

Design resistance R_d can be estimated using design material values as

$$R_d = r(f_d, ..., S) \tag{13}$$

In this case, the structural analysis is based on extremely low material parameters in all locations. This may cause deviations in structural response, e.g. in failure mode. It may be used as an estimate in absence of a more refined solution.

2.5 Full probabilistic analysis

Probabilistic analysis is a general tool for safety assessment of reinforced concrete structures, and thus it can be applied also in case of non-linear analysis. A limit state function can be evaluated by means of numerical simulation. In this approach the resistance function r (**r**) is represented by non-linear structural analysis and loading function s(**s**) is represented by action model. Safety can be evaluated with the help of reliability index β, or alternatively by failure probability P_f taking into account all uncertainties due to random variation of material properties, dimensions, loading, and other.

Probabilistic analysis based on numerical simulation include following steps:

(1) Numerical model based on non-linear finite element analysis. This model describes the resistance function r (**r**) and can perform deterministic analysis of resistance for a given set of input variables.

(2) Randomization of input variables (material properties, dimensions, boundary conditions, etc.). This can also include some effects of actions, which are not in the action function s (**s**) (for example prestressing, dead load etc.). Random properties are defined by random distribution type and its parameters (mean, standard deviation, etc.). They describe the uncertainties due to statistical variation of resistance properties.

(3) Probabilistic analysis of resistance and action. This can be performed by numerical method of Monte Carlo-type of sampling, such as LHS sampling method. Results of this analysis provide random parameters of resistance and actions, such as mean, standard deviation, etc. and the type of distribution function for resistance.

(4) Evaluation of safety using reliability index β or probability of failure.

Probabilistic analysis can be also used for determination of design value of resistance function r (**r**) expressed as R_d. Such analysis involves the steps (1) to (3) above and R_d is determined for required reliability β or failure probability P_f.

2.6 Nonlinear analysis

Examples in this paper are analysed with program ATENA for non-linear analysis of concrete structures. ATENA is capable of a realistic simulation of concrete beahior in the entire loading range with ductile as well as brittle failure modes as shown in papers by Cervenka (1998), (2002). The numerical analysis is based on finite element method and non-linear material models for concrete, reinforcement and their interaction. Tensile behavior of concrete is described by smeared cracks, crack band and fracture energy, compressive behavior of concrete is described by damage model with hardening and softening. In the presented examples the reinforcement is modelled by truss elements embedded in two-dimensional isoparametric concrete elements. Nonlinear solution is performed incrementally with equilibrium iterations in each load step.

3 EXAMPLES OF APPLICATION

The performance of presented safety formats will be tested on several examples ranging from simple determinate structures with bending failure mode up to statically indeterminate structures with shear failure modes.

Example 1 : simply supported beam in bending

Simply supported beam is loaded by a uniform load. The beam has a span of 6 m, rectangular cross-/section of h = 0.3 m, b = 1 m. It is reinforced with 5 Ø 14 along the bottom surface. The concrete type is C30/37 and reinforcement has a yield strength of 500 Mpa. The failure occurs due to bending with a reinforcement yielding

Example 2 : deep shear beam

Continous deep beam with two spans. It corresponds to one of the beams that were tested at Delft university by Asin (1998). It is a statically indeterminate structure with a brittle shear failure.

Example 3 : bridge pier

This example is chosen in order to verify the behavior of the various safety formats in the case of a problem with second order effect (i.e. geometric nonlinearity). This example is adopted from a practical bridge design in Italy that was published by Bertagnoli et. al. (2004). It is a bridge pier loaded by normal force and moment at the top.

Example 4 : bridge frame structure

The bridge frame structure in Sweeden fails by a combined bending and shear failure. It is an existing bridge

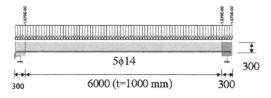

Figure 1. Beam geometry with distributed design load for example 1.

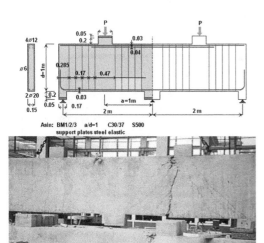

Figure 2. Deam beam geometry for the example 2.

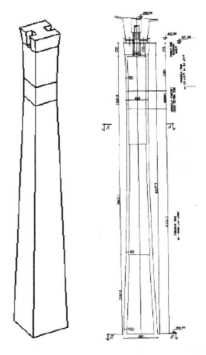

Figure 3. The geometry of the example 2, the bridge pier with second order effecft.

Figure 4. Bridge frame structure, example 4.

that was strengthened by fibre carbon bars, and subjected to a field test up to failure by a single load in the middle of the left span.

The examples are shown in Figure 1 to Figure 4.

In the nonlinear analysis the load is gradually increased up to failure. Typical result from such an analysis is shown in Figure 5 for the case of the example 1. The figure shows the beam response for increasing load using various safety methods presented in Section 2. The straight dashed line represents the load-carrying capacity given by standard design formulas based on beam analysis by hand calculation and critical section check by partial factor method. The other curves corresponds to the analyses with different material properties as specified by the safety format approaches that are presented in Section 2. The curve denoted as PSF, thus corresponds to the partial factor method from Section 2.4, in which the used material parameters are multiplied by the corresponding factors of safety.

The response curve EN1992-2 is obtained from an analysis, where the material parameters are given by Section 2.3. For the ECOV method (Section 2.2), two separate analyses are needed: one using mean material properties, and one with characteristic values. The results from these two analyses are denoted by the

labels "Mean" and "Char." respectively. The ultimate load carrying capacities from each analysis are then used to estimate the design resistance R_d. For all examples the calculated design resistances are shown in Table 2. The design resistances are normalized with respect to the values obtained for PSF method to simplify the comparison.

Typical results from the nonlinear analyses are presented in Figure 6, for the case of the example 2. This figures shows also the comparison of the calculated

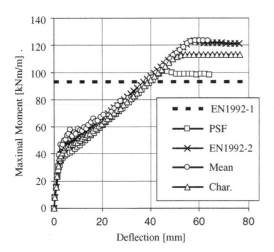

Figure 5. Load-displacement diagrams for bending example 1.

Table 2. Comparison of calculated values for design resistance sing various safety formats.

	PSF	ECOV	EN 1992-2	Probabilistic
Example 1 Bending R_d / R_d^{PSF}	1.0	1.0	0.95	0.96
Example 2 shear beam R_d / R_d^{PSF}	1.0	1.02	0.98	0.98
Example 3 bridge pier R_d / R_d^{PSF}	1.0	1.06	0.98	1.02
Example 4 bridge frame R_d / R_d^{PSF}	1.0	0.97	0.93	1.01

failure mode and the experimental crack pattern. For each example, a full probabilistic analysis was also performed. Each probabilistic analysis consists of several (at least 32 to 64 analyses) nonlinear analysis with different material properties as shown in Figure 7.

4 CONCLUSIONS

The paper presents a comparison of several safety formats for non-linear analysis of concrete structures. The global safety approach is proposed. A new method for verification of ultimate limit state suitable for reinforced concrete design based on non-linear analysis is described. The new method is called ECOV

Figure 6. Shear wall tested in the laboratory, Asin (1999).

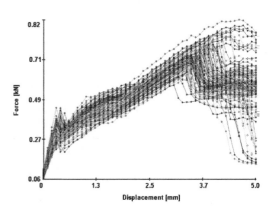

Figure 7. Shear wall tested in the laboratory, Asin (1999).

(Estimate of Coefficient Of Variation). The advantage of the proposed method is that it can capture the resistance sensitivity to the random variation of input variables, and thus it can reflect the effect of failure mode on safety. It requires two nonlinear analyses with mean and characteristic values of input parameters respectively. Other safety formats suitable for non-linear analysis that are based on global resistance are presented. They are: the approach proposed by EN 1992-2, fully probabilistic analysis and a simple approach based on detsign values of input parameters, i.e. characteristic parameters reduced by partial safety factors. The last approach is usually not recommended by design codes, but practicing engineers often overlook this fact, and use this approach if a nonlinear analysis is available in their analysis tools. The consequences are investigated in this paper.

The discussed safety formats are tested on four examples. They include ductile as well as brittle modes of failure and second order effect (of large deformation). For the investigated range of problems, which is quite narrow but still representative, all the methods provide quite reliable and consistent results.

Based on the limited set of examples the following conclusions are drawn:

(e) The proposed EVC method gives consistent results compare to other approaches.
(f) The PSF method, which uses input parameters with partial safety factors appears to be sufficiently reliable and it is a natural extension of the classical approach to the modern design methods based on non-linear analysis.
(g) Fully probabilistic analysis is sensitive to the type of random distribution assumed for input variables. It can provide additional load-carrying capacity if statistical properties of the analyzed system are known or can be accurately estimated.

The methods are currently subjected to further validation by authors for other types of structures and failure modes.

The research presented in this paper was in part resulting from the Grant no. 1ET409870411 of the Czech Academy of Sciences. The financial support is greatly appreciated.

REFERENCES

Asin, M. (1999) The Behaviour of Reinforced Concrete Continuous Deep Beams. Ph.D. Dissertation, Delft Univeristy Press, The Netherlands, 1999, ISBN 90-407-2012-6.

Bertagnoli, G., Giordano L., Mancini, G. (2004), Safety format for the nonlinear analysis of concrete structures. STUDIES AND RESEARCHES –V.25, Politechnico di Milano, Italy.

Cervenka V. (1998): Simulation of shear failure modes of R/C structures. In: Computational Modeling of Concrete Structures (Euro-C 98), eds. R. de Borst, N. Bicanic, H. Mang, G. Meschke, A.A. Balkema, Rotterdam, The Netherlands, 1998, 833–838.

Cervenka V. (2002). Computer simulation of failure of concrete structures for practice. 1st fib Congress 2002 Concrete Structures in 21 Century, Osaka, Japan, Keynote lecture in Session 13, 289–304.

DIN 1045-1 (1998), Tragwerke aus Beton, Stahlbeton und Spannbeton, Teil 1: Bemessung und Konstruktion, German standard for concrete, reinforced concrete and prestressed concrete structures.

EN 1992-2, (2005), Eurocode 2 – Design of concrete structures – Concrete bridges – Design and detailing rules.

Part VIII
Maintenance and retrofitting of RC structures

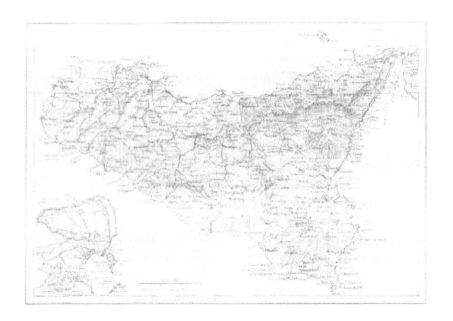

*Fracture Mechanics of Concrete and Concrete Structures – Design, Assessment and
Retrofitting of RC Structures – Carpinteri, et al. (eds)
© 2007 Taylor & Francis Group, London, ISBN 978-0-415-44616-7*

Debonding in FRP strengthened beams: Stress assessment versus fracture mechanics approach

A. Carpinteri, P. Cornetti & N. Pugno

Dipartimento di Ingegneria Strutturale e Geotecnica, Politecnico di Torino, Turin, Italy

ABSTRACT: In the present paper we analyze the debonding failure of a beam strengthened by a fiber reinforced polymer. We propose an analytical approach, whose basic assumptions are that (i) sections inside the beam remain plane and (ii) the adhesive layer acts as a shear lag. A stress concentration is found at the edge of the reinforcement length. Two failure criteria are proposed to study the debonding process. The former is a stress assessment criterion, i.e. failure takes place whenever the maximum tangential stress reaches a limit value (the interfacial bond strength). The latter is an energy, fracture mechanics criterion, i.e. failure takes place as long as the strain energy release rate due to debonding reaches a critical value (the interfacial fracture energy). By means of Castigliano's theorem, the load vs. deflection graph is drawn, showing the possible rising of both snap-back and snap-through instabilities.

1 INTRODUCTION

One of the most promising techniques for strengthening existing structures is the use of strips made of fibre-reinforced polymers (FRP), bonded to the tensile side of the structure. The advantages of this technique are different. FRP strips are easy to install and cause a minimum increase in dimension; furthermore, they have a high strength, a light weight and a long durability.

The behaviour of the strengthened members is substantially different from that of the original structures. Referring to concrete beams, the FRP retrofitting causes a reduced ductility, a different shear response and, more importantly, different failure modes. Among the various failure modes observed, a special interest has been devoted to the debonding of the FRP because of its brittle and catastrophic features, the propagation of the interfacial crack being highly unstable.

In order to predict the critical load at which the debonding phenomenon takes place, several models have been proposed to evaluate the interfacial stresses. They all focus onto the prediction of the stresses in the vicinity of the edge of the FRP strip. These stresses are then used to predict the peak load. A critical review of these models can be found in the paper by Muckopadhyaya & Swamy (2001); the paper concludes that the existing models are too complex for use in practical design.

However, because of the brittleness of the debonding process, an energy approach seems to be more effective, since stress-based failure criteria are more suitable for gradual and ductile failures. An energy-based fracture criterion has recently been proposed by Rabinovitch (2004) and, later, by Colombi (2006), by applying the linear elastic fracture mechanics (LEFM) concept of energy release rate. In other words, the debonding process is assumed to begin when the energy release due to an infinitesimal crack growth is equal or higher than a critical value, i.e. the interfacial fracture energy. The afore-mentioned papers show that simplified models assuming a constant stress field across the adhesive layer thickness can be used to predict the energy release rate. However, its (approximate) evaluation is performed numerically by comparing the energetic state of the whole structure before and after a small interfacial crack growth.

A similar approach, applied to analyse delamination in a different geometry, has been recently proposed by Andrews et al. (2006). Moreover, among recent works based on neighbouring arguments, we cite the papers by Greco et al. (2007) for the evaluation of the strain energy release rate and the paper by Ferracuti et al. (2006) on numerical approaches to FRP debonding.

Aim of the present paper is to introduce a model to analyse FRP strengthened beams. The main simplifying assumption is that the adhesive layer acts as a *shear lag*, i.e. only shear stresses constant over its thickness are considered (for other applications of the shear lag model see, for instance, Stang et al., 1990; Pugno & Carpinteri, 2003). The model is similar to others already available in the literature (Vilnay, 1988; Triantafillou & Deskovic, 1991; Taljsten, 1997; Malek et al., 1998; Smith & Teng, 2001, and references herein). The stress field provided by the model is used to apply the LEFM criterion. With respect to the papers by Rabinovitch (2004) and Colombi (2006), the novelty is that an analytical expression for the energy release rate is provided. It is believed that this can be useful for including debonding failure assessment in practical design codes. Furthermore, the present analytical approach allows one to obtain the complete load vs. displacement diagram, highlighting the possible rising of snap-back and snap-through instabilities (Carpinteri 1984; Carpinteri 1989a, b).

2 EQUIVALENT BEAM MODEL

The easiest analytical model to handle beams strengthened by FRP is the so-called equivalent beam model, based on the assumption of a planar cross section for the whole structure. Let us refer to a beam with a rectangular cross section (Fig. 1), whose width is t. In the following, the quantities with subscript b refer to the beam to be strengthened, the quantities with subscript a refer to the adhesive layer and the ones with subscript r to the reinforcement. Thus E_b, E_r, G_a are the Young moduli of the beam, of the reinforcement and the shear modulus of the adhesive; h_b, h_r, h_a are their respective thicknesses. In order to achieve a dimensionless formulation of the problem, it is convenient to normalise all the quantities with respect to the beam ones, i.e.:

$$n_a = \frac{G_a}{E_b}, \ n_r = \frac{E_r}{E_b}, \ \delta_a = \frac{h_a}{h_b}, \ \delta_r = \frac{h_r}{h_b}, \ \mu = \frac{t}{h_b} \quad (1)$$

For the sake of simplicity, it is convenient to introduce also the mechanical percentage of reinforcement:

$$\rho = \frac{E_r}{E_b} \frac{h_r}{h_b} = n_r \delta_r \quad (2)$$

Wishing to get analytical expressions as simple as possible, we can neglect the heights of the adhesive and of the reinforcement with respect to the beam height as well as the contribution of the adhesive layer to the moment of inertia. Thus, with respect to the bottom of the strengthened beam, the position y_G of the centre of gravity of the whole section is (Fig. 1):

$$y_G = \frac{h_b}{2(1+\rho)} \quad (3)$$

The moment of inertia with respect to the x_G axis reads:

$$I = \frac{1+4\rho}{1+\rho} I_b \quad (4)$$

$I_b = th_b^3/12$ being the moment of inertia of the plain beam section.

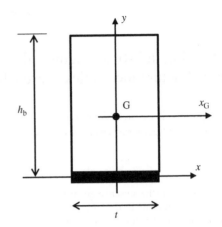

Figure 1.   Geometry of the reinforced cross section.

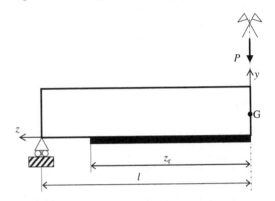

Figure 2.   A FRP reinforced beam under three point bending flexural load. Symmetry is exploited to study only half of the structure.

Let us consider a three point bending geometry (Fig. 2). The beam span is $2l$ and P is the concentrated load. If z is the axial coordinate with origin at the beam mid-span, in the left side of the beam the shear force is $T = P/2$ and the bending moment is $M = -P(l-z)/2$. Therefore, according to the well-known equivalent beam model, the horizontal normal stress σ_r in the reinforcement and the shear stress τ_a in the adhesive layer are, in dimensionless form:

$$\frac{\sigma_r}{E_b} = \frac{3n_r}{1+4\rho} \frac{\Pi\lambda}{\mu} (1-\zeta), \ \ 0 < \zeta < \zeta_r \quad (5)$$

$$\frac{\tau_a}{E_b} = \frac{3\rho}{1+4\rho} \frac{\Pi}{\mu}, \ \ 0 < \zeta < \zeta_r \quad (6)$$

where Π, ζ, λ are respectively the dimensionless load, axial coordinate and length (i.e. the slenderness):

$$\Pi = \frac{P}{E_b h_b^2}, \ \zeta = \frac{z}{l}, \ \lambda = \frac{l}{h_b} \quad (7)$$

The previous expressions are valid for the reinforced zone $(0 < \zeta < \zeta_r = z_r/l)$; outside they vanish and the simple beam theory holds.

3 SHEAR LAG MODEL

With respect to the equivalent beam, a more refined model can be achieved by assuming that the cross sections remain planar after deformation only inside the beam to be strengthened. In fact it is argued that, since the main duty of the adhesive layer is to transfer stresses from the beam to the FRP reinforcement by means of tangential stresses, the shear stress and strain inside the adhesive layer have to be explicitly taken into account to have a more accurate description of the geometry analyzed.

In the following it is assumed that the adhesive layer acts as a shear lag, i.e. no normal stresses are considered within its thickness. Although more complex models can be set by considering also the normal stresses or describing the adhesive layer as a 2D medium, it is argued that the shear lag model could be a reasonable compromise for the description of the FRP debonding. The goal of the present paper is in fact to provide a model that is accurate enough to describe the basic features of the process and, at the same time, simple enough to be handled analytically.

The assumption of planar cross sections reads:

$$w_b(y, z) = w_{b0}(z) + \varphi_b(z) y \tag{8}$$

where w_b is the axial displacement of the beam points, φ is the rotation of the cross section at the distance z from the midspan and w_{b0} is the axial displacement of the points at the bottom of the beam. Denoting by ε_b and ε_r the dilations of the beam points and of the reinforcement and by γ_a the shearing strain of the adhesive, the assumption of a linear elastic behaviour for all the materials composing the structure yields:

$$\sigma_b = E_b \varepsilon_b = E_b \left(\frac{dw_{b0}}{dz} + \frac{d\varphi_b}{dz} y \right) = E_b (\varepsilon_{b0} + \chi_b y) \tag{9}$$

$$\tau_a = G_a \gamma_a = G_a \frac{w_{b0} - w_r}{h_a} \tag{10}$$

$$\sigma_r = E_r \varepsilon_r = E_r \frac{dw_r}{dz} \tag{11}$$

where χ_b is the beam curvature.

The problem can now be solved by imposing the equilibrium. The first two solving equations state the equivalence of the stress distribution with the axial force (which is equal to zero) and with the bending

moment M; the third one represents the differential equilibrium of the reinforcement along z:

$$\int_0^{h_b} \sigma_b \, t \, dy + \sigma_r \, t \, h_r = 0 \tag{12}$$

$$\int_0^{h_b} \sigma_b \, y \, t \, dy = M \tag{13}$$

$$h_r \frac{d\sigma_r}{dz} + \tau_a = 0 \tag{14}$$

By substitution of eqns(9–11) into eqns(12–14) and by a further derivation of the last equation, we get:

$$\begin{cases} \varepsilon_{b0} + \dfrac{h_b}{2} \chi_b + \rho \varepsilon_r = 0 \\[2mm] \dfrac{1}{2} \varepsilon_{b0} + \dfrac{h_b}{3} \chi_b = -\dfrac{P}{2 t h_b^2 E_b}(l - z) \\[2mm] \dfrac{d^2 \varepsilon_r}{dz^2} - \dfrac{G_a}{E_r h_r h_a}(\varepsilon_r - \varepsilon_{b0}) = 0 \end{cases} \tag{15}$$

This is a system of three equations, the first two algebraic and the last one differential. The three unknowns are the deformation functions $\varepsilon_{b0}(z)$, $\chi_b(z)$ and $\varepsilon_r(z)$. By substitution, it is possible to obtain a differential equation, which, in dimensionless form, reads:

$$\frac{d^2 \varepsilon_r}{d\zeta^2} - \beta^2 \varepsilon_r = -\frac{3\beta^2}{1 + 4\rho} \frac{\Pi \lambda}{\mu}(1 - \zeta) \tag{16}$$

where:

$$\beta^2 = \frac{G_a l^2}{E_r h_r h_a}(1 + 4\rho) = \frac{1 + 4\rho}{\rho} \frac{n_a}{\delta_a} \lambda^2 \tag{17}$$

Before solving the differential equation (16), it is interesting to observe that, if the thickness of the adhesive layer tends to zero (i.e. $\beta^2 \to \infty$), the solution tends to the one of the equivalent beam model (i.e. eqn (5)):

$$\varepsilon_r = \frac{3}{1 + 4\rho} \frac{\Pi \lambda}{\mu}(1 - \zeta) \tag{18}$$

except in the neighbourhood of the end of the reinforcement where the second derivative of ε_r is unbounded. Furthermore, it can be easily verified that the equivalent beam solution (18) is a particular integral of the differential equation. As well known, the complete solution is given by the sum of a particular integral and the solution of the associated homogeneous equation. Thus:

$$\varepsilon_r = A\,e^{+\beta\zeta} + B\,e^{-\beta\zeta} + \frac{3}{1+4\rho}\frac{\Pi\lambda}{\mu}(1-\zeta) \qquad (19)$$

In order to determine the constants A and B, two boundary conditions are needed. The first one derives from symmetry considerations, whereas the second one implies a zero normal stress in the FRP at the edge of the reinforced zone:

$$\frac{d\varepsilon_r}{d\zeta} = 0, \quad \text{if } \zeta = 0 \qquad (20a)$$

$$\varepsilon_r = 0, \quad \text{if } \zeta = \zeta_r \qquad (20b)$$

It is interesting to note that both the conditions are violated by the equivalent beam model, where the stress at the edge of the FRP is different from zero and the strain at mid-span is not differentiable. Observe that, at mid-span, one can also set the strain in the reinforcement equal to the value provided by the equivalent beam model. However the effect of the boundary condition at $\zeta=0$ has a negligible effect at the FRP strip edge ($\zeta=\zeta_r$). See Smith & Teng (2001) for a discussion about this point.

By means of the boundary conditions, the final solution is:

$$\varepsilon_r = \frac{3}{1+4\rho}\frac{\Pi\lambda}{\mu}f_\varepsilon \qquad (21a)$$

with:

$$f_\varepsilon = (1-\zeta) - \frac{\beta(1-\zeta_r)\cosh(\beta\zeta) + \sinh[\beta(\zeta_r-\zeta)]}{\beta\cosh(\beta\zeta_r)} \qquad (21b)$$

Function $f_\varepsilon = f_\varepsilon(\zeta, \zeta_r, \beta)$ has been introduced for the sake of simplicity: in fact it is possible to express all the quantities as functions of f_ε and its derivatives and integrals. Through eqns (11) and (14), the horizontal normal stress and the shear stress respectively in the FRP and in the adhesive layer read:

$$\frac{\sigma_r}{E_b} = \frac{3n_r}{1+4\rho}\frac{\Pi\lambda}{\mu}f_\varepsilon \qquad (22)$$

$$\frac{\tau_a}{E_b} = -\frac{3\rho}{1+4\rho}\frac{\Pi}{\mu}\frac{\partial f_\varepsilon}{\partial\zeta} = \frac{3\rho}{(1+4\rho)}\frac{\Pi}{\mu} \times$$
$$\times \left\{1 + \frac{\beta(1-\zeta_r)\sinh(\beta\zeta) - \cosh[\beta(\zeta_r-\zeta)]}{\cosh(\beta\zeta_r)}\right\} \qquad (23)$$

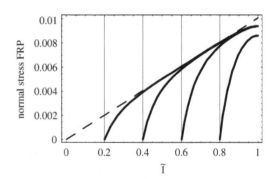

Figure 3. Horizontal normal stress in the FRP versus the axial coordinate (dimensionless quantities). The thick lines refer to a bonded length ζ_r equal to 0.8, 0.6, 0.4, 0.2 from right to left. The dashed line represents the equivalent beam model solution.

while, from the system (15), the beam curvature is:

$$\chi_b h_b = 6\left[\rho\varepsilon_r - \frac{\Pi\lambda}{\mu}(1-\zeta)\right] = \frac{6\Pi\lambda}{\mu}\left[\frac{3\rho}{1+4\rho}f_\varepsilon - (1-\zeta)\right] \qquad (24)$$

In order to have a preliminary check of the capabilities of the model, the main quantities are now plotted with reference to an FRP-strengthened concrete beam. However it should be observed that, when considering a concrete beam to be retrofitted, the results of the present model are essentially indicative since: (i) the contribution of the concrete external cover to the interface compliance cannot be neglected with respect to the contribution of the adhesive layer; (ii) concrete cracking is expected to take place after debonding, whereas in the present model the material is assumed to be linear elastic throughout the debonding process.

Keeping in mind these restrictions, the stress fields (22) and (21) are plotted in Figs. 3 and 4 for different bonded lengths ζ_r and for the following material and geometric properties: $t=100\,\text{mm}$, $l=500\,\text{mm}$; $h_b=120\,\text{mm}$, $h_a=4\,\text{mm}$, $h_r=1.6\,\text{mm}$; $E_b=30\,\text{GPa}$, $G_a=0.72\,\text{GPa}$, $E_r=160\,\text{GPa}$; $P=70\,\text{kN}$. The dashed lines represent the equivalent beam model solution, which is independent of ζ_r. It is evident that, with respect to the simpler beam model, the shear lag model is able to catch the shear stress concentration at the edge of the FRP strip, which is the cause of the FRP debonding. On the other hand, beyond a certain distance from the mid-span and the edge of the reinforcement, the two solutions coincide. More in detail, it can be proved that, for h_a tending to zero, the shear lag solution shows a nonuniform convergence to the equivalent beam solution.

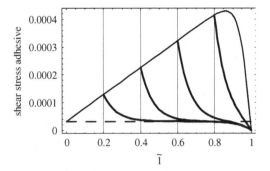

Figure 4. Shear stress in the adhesive layer versus the axial coordinate (dimensionless quantities). The thick lines refer to a bonded length ζ_r equal to 0.8, 0.6, 0.4, 0.2 from right to left. The dashed line represents the equivalent beam model solution. The thin line is the envelope of the maximum values of the shear stress.

4 FAILURE STRESS CRITERION

The maximum value of the shearing stress in the adhesive layer is attained at the end of the reinforced zone, i.e. for $\zeta = \zeta_r$:

$$\frac{(\tau_a)_{max}}{E_b} = \frac{3\rho}{(1+4\rho)} \frac{\Pi}{\mu} \left[1 + \frac{\beta(1-\zeta_r)\sinh(\beta\,\zeta_r) - 1}{\cosh(\beta\,\zeta_r)} \right] \quad (25)$$

Introducing the shearing strength τ_p, the dimensionless load Π_c causing debonding is therefore:

$$\Pi_c = \frac{(1+4\rho)\mu}{3\rho} \frac{\tau_p}{E_b} \frac{\cosh(\beta\,\zeta_r)}{\cosh(\beta\,\zeta_r) + \beta(1-\zeta_r)\sinh(\beta\,\zeta_r) - 1} \quad (26)$$

For the geometry given above, the critical load vs. reinforced zone length is plotted in Figure 5 for a τ_p value equal to 5 MPa. Since the debonded portion of the FRP strip becomes stress free, the plot in Figure 5 can be interpreted as either the graph of the critical loads for different initial lengths of the FRP strip, or the diagram of the load during the debonding process for a given initial FRP strip length. In the latter case, it is interesting to observe that, if the process is load-controlled, the debonding process is unstable until the reinforcement length is much shorter than the beam length (about 20%). Then, an increase in the load is required to have a further debonding; however the strengthening effect of the FRP is rather negligible at that stage.

About the validity of the stress criterion (26), a first drawback is that, for h_a tending to zero, the maximum tangential stress tends to infinity and the critical load vanishes, a result which is physically meaningless. Furthermore, finite element analyses show that, at the edges of the FRP strip, in addition to shear stresses,

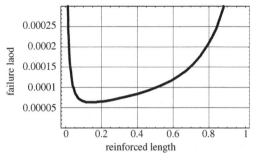

Figure 5. Failure load versus reinforced length ζ_r according to the stress-based failure criterion (dimensionless quantities).

localized vertical normal stresses (along y) are present which are not caught by the present model. Neglecting this local effect could affect the predictive capability of the stress-based failure criterion (26). For these reasons, in the following section a fracture energy criterion, based on LEFM, is put forward. As a matter of fact, it is argued that the energy criterion should be more reliable with respect to the local stress failure criterion since it is based on an overall energy balance. The brittleness of the debonding phenomenon justifies the use of LEFM. Finally, it will be shown that the energy criterion provides a finite failure load for an adhesive layer thickness tending to zero.

5 FRACTURE ENERGY CRITERION AND RELATED SNAP-BACK AND SNAP-THROUGH INSTABILITIES

The computation of the strain energy Φ must be divided in the sum of four terms, i.e. the strain energy within the FRP strip (Φ_r), the adhesive layer (Φ_a), the portion of the beam above the FRP (Φ_{b1}) and the one where there is no reinforcement (Φ_{b2}). The first, the second and the fourth contributions are straightforward:

$$\Phi_r = \frac{h_r}{2E_r} \int_0^{z_r} \sigma_r^2 \, dz \quad (27a)$$

$$\Phi_a = \frac{h_a}{2G_a} \int_0^{z_r} \tau_a^2 \, dz \quad (27b)$$

$$\Phi_{b2} = \frac{1}{2} \int_{z_r}^{l} \frac{M^2}{E_b I_b} \, dz \quad (27c)$$

About the third contribution (Φ_{b1}), it should be noted that, by marking with ε_{bG} the axial dilation of the centre of gravity of the cross section (without the

1057

reinforcement, i.e. $y = h/2$), the classical beam theory yields:

$$\Phi_{b1} = \frac{E_b A_b}{2} \int_0^{z_r} \varepsilon_{bG}^2 \, dz + \frac{E_b I_b}{2} \int_0^{z_r} \chi_b^2 \, dz \qquad (27d)$$

where $A_b = th_b$. Since the axial force is zero, $\varepsilon_{bG} = \rho \varepsilon_r$. Skipping analytical computations, the strain energy of the whole structure can be expressed in dimensionless form ($\tilde{\Phi} = \Phi/E_b h_b^3$) by means of eqns (21–24):

$$\tilde{\Phi} = \frac{\lambda^3 \Pi^2}{2\mu} \left[1 + \frac{9\rho}{1+4\rho} F(\zeta_r, \beta) \right] \qquad (28)$$

where:

$$F(\zeta_r, \beta) = \int_0^{\zeta_r} \left\{ f_\varepsilon \left[f_\varepsilon - 2(1-\zeta) \right] + \frac{1}{\beta^2} \left(\frac{\partial f_\varepsilon}{\partial \zeta} \right)^2 \right\} d\zeta \qquad (29)$$

Function F can be computed analytically; nevertheless, since its expression is rather long, we prefer to omit it. Details will be given elsewhere. Let us observe that the unit term within square brackets in eqn (28) represents the strain energy when no reinforcement is present; hence it is clear that function F is always negative, except for $\zeta_r = 0$, when it is equal to zero.

Once the strain energy is obtained, the energy release rate is provided by deriving the previous expression with respect to the interfacial crack length a. Since $da = -d\zeta_r$, we have:

$$\mathcal{G} = -\frac{1}{t} \frac{d\Phi}{dz_r} \qquad (30)$$

We assume a symmetrical debonding growth with respect to the mid-span. In dimensionless form ($g = \mathcal{G}/E_b h_b$):

$$g = -\frac{1}{\lambda\mu} \frac{d\tilde{\Phi}}{d\zeta_r} = -\frac{9\lambda^2 \Pi^2}{2\mu^2} \frac{\rho}{1+4\rho} F'(\zeta_r, \beta) \qquad (31)$$

where the prime denotes derivative with respect to ζ_r. Observe that, since F' is negative, g is always positive, as we should have expected.

According to LEFM, the energy criterion states that debonding starts whenever the energy release rate reaches its critical value $\mathcal{G}_c$ (called also fracture energy): $\mathcal{G} = \mathcal{G}_c$. The dimensionless critical load Π_c is therefore given by ($g_c = \mathcal{G}_c/E_b h_b$):

$$\Pi_c = \frac{\mu}{3\lambda} \sqrt{-\frac{2(1+4\rho)g_c}{\rho F'(\zeta_r, \beta)}} \qquad (32)$$

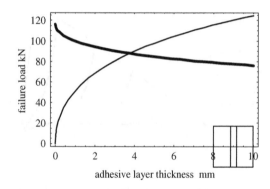

Figure 6. Failure load versus adhesive layer thickness according to the stress criterion (thin line) and to the energy criterion (thick line) for $\zeta_r = 0.8$, $\tau_p = 5$ MPa, $\mathcal{G}_c = 65$ J/m^2; the values of the other parameters are given in the text.

Hence, according to the present LEFM criterion, no FRP debonding is expected if:

$$\Pi < \Pi_c \qquad (33)$$

Qualitatively, the plot of the critical load Π_c vs. ζ_r provided by the energy criterion is similar to the previous one provided by the stress criterion (26): the difference is mainly due to the different values of the parameter governing failure ($\mathcal{G}_c$ instead of τ_p). However, the effect of the adhesive layer thickness is opposite: as shown in Figure 6, according to the energy approach, the failure load usually decreases increasing the thickness, while it increases according to the stress approach. More in detail, while the critical value (26) vanishes for a null adhesive thickness, the critical value (32) tends to a constant (i.e. the same value provided by the equivalent beam model). It is argued that the failure load is the highest among the two predictions, since both energy and stress requirements have to be fulfilled to trigger the debonding process. It is worth noting that, if more failure mechanisms are present (e.g. brittle fracture and plastic collapse) the critical mechanism is the one providing the lowest failure load. However this is not the present case, since the stress criterion (26) is a *local* brittle fracture criterion, i.e. it does *not* coincide with the plastic collapse.

From a practical point of view, in order to be effective, the FRP strip must be glued by a sufficiently thin adhesive layer, i.e. usually the adhesive thickness is such that the energy criterion prevails.

Since the energy release rate provided by the equivalent beam model coincides with the value provided by the shear lag model with a null adhesive thickness, Figure 6 shows also that, if the energy fracture criterion is to be used, the equivalent beam model, for the usual reinforcement lengths, provides an overestimate of the failure load if compared to the more

Table 1. Comparison between analytical approaches and numerical simulations.

	Energy release rate	Maximum shear stress	Mid-span displacement
	J/m^2	MPa	mm
Shear lag model	41.83	3.880	2.832
Equivalent beam model	23.55	0.9689	2.820
Finite element analysis	Not available	3.266	2.818

refined shear lag model; therefore its application may be potentially dangerous.

Because of the above considerations (and the uncertainties related to the stress fields provided by the analytical models highlighted in the previous section), hereafter we make use only of the estimate of the failure load based on the fracture energy criterion (32).

In order to analyse the presence of snap-back and/or snap-through instabilities during the debonding process, the plot of the load P vs. the vertical displacement v at mid-span has to be sought. This can be easily achieved by means of Castigliano's theorem, since we already know the expression of the strain energy of the whole system. Therefore, the strain energy Φ must be derived with respect to the load P:

$$v = 2\frac{d\Phi}{dP} \qquad (34)$$

where the coefficient 2 appears since Φ is the energy contained in half of the beam. In dimensionless form ($\eta = v/h_b$):

$$\eta = 2\frac{d\tilde{\Phi}}{d\Pi} = \frac{2\lambda^3\Pi}{\mu}\left[1 + \frac{9\rho}{1+4\rho}F(\zeta_r,\beta)\right] \qquad (35)$$

In order to have a preliminary check of the analytical results obtained so far, we performed also a finite element analysis of the three point bending geometry described above. In table 1, the numerical results are compared to the ones obtained through the equivalent beam model and the shear lag model. It is evident that the shear lag model shows results which are closer to the numerical simulations and, apart from the mid-span displacement, very different from the values provided by the equivalent beam model.

Substituting eqn (32) into eqn (35), we get the displacement at mid-span at the critical condition:

$$\eta = \frac{2\lambda^2}{3}\sqrt{\frac{2(1+4\rho)g_c}{\rho F'(\zeta_r,\beta)}}\left[1 + \frac{9\rho}{1+4\rho}F(\zeta_r,\beta)\right] \qquad (36)$$

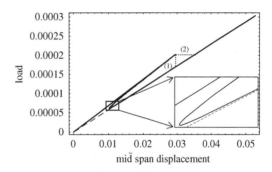

Figure 7. Load versus mid-span deflection (dimensionless quantities) for an initially bonded length ζ_r equal to 0.8. The dashed line represents the behaviour of a beam without reinforcement. The dotted lines represents snap-back (1) and snap-through (2) instabilities.

Eqns (32) and (36) can be seen as the equation of a curve in the plane (Π, η) defined parametrically by means of ζ_r. The curve Π vs. η is plotted in Figure 7 according to the geometrical and material data provided above; furthermore, a fracture energy $G_c = 65$ J/m^2 has been assumed, which corresponds to the concrete fracture energy, since the debonding crack typically runs under the concrete skin. For the sake of clarity, in Figure 7 also the straight lines corresponding to the beam configurations with a completely bonded FRP strip or without reinforcement have been drawn. It is seen that, if the plain beam does not collapse because of other failure mechanisms, the structural behaviour shows both snap-back and snap-through instabilities, i.e. function jumps if the process is displacement or load controlled (Carpinteri 1984; Carpinteri 1989a, b). More in detail, the snap-back instability appears because of the positive slope of the softening branch in Figure 7. The snap-back appears to be rather severe, i.e. the critical load decreases considerably during the debonding process. In other words, the analytical results here presented clearly show that FRP debonding is a highly unstable and brittle phenomenon. Numerical simulations (Carpinteri et al., in press) seem to confirm the presence of snap-back instability detected by the present analytical approach.

6 CONCLUSIONS

In the present paper an analytical approach to study the debonding process of FRP strips from concrete beams has been addressed. The stress field provided by the model has been used to formulate a LEFM failure criterion. With respect to other approaches available in the literature, the present one has the advantage to be analytical: all the main quantities have been expressed as functions of dimensionless (geometrical and mechanical) parameters and of the analytical function f_ε (eqn (21b)). It is believed that these findings can be helpful for standard design codes requirements to avoid FRP debonding. Finally, the instabilities associated with the debonding process have been analyzed.

REFERENCES

Andrews, M.G., Massabò, R. & Cox, B.N. 2006. Elastic interaction of multiple delaminations in plates subject to cylindrical bending. *International Journal of Solids and Structures*. 43:855–886.

Carpinteri, A., Lacidogna, G. & Paggi, M. Acoustic emission monitoring and numerical modeling of FRP delamination in RC beams with non-rectangular cross-section. *RILEM Materials & Structures*, DOI 10.1617/s11527-006-9162-4.

Carpinteri, A. 1985. Interpretation of the Griffith instability as a bifurcation of the global equilibrium. In S.P. Shah (ed.), *Application of Fracture Mechanics to Cementitious Composites*: 287–316; *Proceedings of a NATO Advanced Research Workshop, Evanston, USA, 1984*. Dordrecht: Martinus Nijhoff Publishers.

Carpinteri, A. 1989a. Cusp catastrophe interpretation of fracture instability. *Journal of the Mechanics and Physics of Solids* 37:567–582.

Carpinteri, A. 1989b. Decrease of apparent tensile and bending strength with specimen size: two different explanations based on fracture mechanics. *International Journal of Solids and Structures* 25:407–429.

Colombi, P. 2006. Reinforcement delamination of metallic beams strengthened by FRP strips: Fracture mechanics based approach. *Engineering Fracture Mechanics* 73:1980–1995.

Ferracuti, B., Savoia, M. & Mazzotti, C. 2006. A numerical model for FRP-concrete delamination. *Composites: Part B* 37: 356–364.

Greco, F., Lonetti, P. & Nevone Blasi P. 2007. An analytical investigation of debonding problems in beams strengthened using composite plates. *Engineering Fracture Mechanics* 74:346–372.

Malek, A.M., Saadatmanesh, H. & Ehsani, M.R. 1998. Prediction of failure load of R/C beams strengthened with FRP plate due to stress concentration at the plate end. *ACI Structural Journal* 95:142–52.

Muckopadhyaya, P. & Swamy, N. 2001. Interface shear stress: a new design criterion for plate debonding. *Journal of Composite Construction* 5:35–43.

Pugno, N. & Carpinteri A. 2003. Tubular adhesive joints under axial load. *Journal of Applied Mechanics* 70:832–839.

Rabinovitch, O. 2004. Fracture-mechanics failure criteria for RC beams strengthened with FRP strips – a simplified approach. *Composite Structures* 64:479–492.

Smith, S.T. & Teng, J.G. 2001. Interfacial stresses in plated beams. *Engineering Structures* 23, 857–871, 2001.

Stang, H., Li, Z., Shah, S.P. 1990. Pullout problem: stress versus fracture mechanical approach. *Journal of Engineering Mechanics* 116:2136–2150.

Taljsten, B. 1997. Strengthening of beams by plate bonding. *Journal of Materials in Civil Engineering, ASCE* 9:206–12.

Triantafillou, T.C. & Deskovic, N. 1991. Innovative prestressing with FRP sheets: Mechanics of short-term behavior. *Journal of Engineering Mechanics*. 117:1652–1672.

Vilnay, O. 1988. The analysis of reinforced concrete beams strengthened by epoxy bonded steel plates. *International Journal of Cement Composites and Lightweight Concrete*, 10:73–8.

Fracture Mechanics of Concrete and Concrete Structures – Design, Assessment and Retrofitting of RC Structures – Carpinteri, et al. (eds)
© 2007 Taylor & Francis Group, London, ISBN 978-0-415-44616-7

An experimental study on retrofitted fiber-reinforced concrete beams using acoustic emission

A. Carpinteri, G. Lacidogna & A. Manuello
Department of Structural Engineering and Geotechnics, Politecnico di Torino, Italy

ABSTRACT: Fiber Reinforced Polymer (FRP) sheets are becoming increasingly important as a structural repair method when the load carrying capacity of deteriorated concrete beams is judged to be inadequate. In this field, the repair process must successfully integrate new materials with the old ones, forming a composite system capable of enduring exposure to service loads, environment and time. In this work, a set of Fiber Reinforced Concrete (FRC) beams has been retrofitted with fiber-reinforced polymer (FRP) sheets in order to improve their loading carrying capacity. These beams have been tested up to failure under three-point bending. During the loading tests, the acoustic emission AE technique has been applied to evaluate the damage evolution. By identifying the complete shape of the signals, the three-dimensional locations of damage sources are determined from the AE sensor records. In this connection, the authors have tuned an original procedure using advanced techniques that are usually applied for the analysis of seismic events.

1 INTRODUCTION

A rather recent development in concrete structure technology is represented by steel fiber-reinforced concrete (FRC). By adding steel fibers while mixing the concrete, a so-called homogeneous reinforcement is created. This does not notably increase the mechanical properties before failure, but governs the post-failure behaviour. Thus, plain concrete, which is a quasi-brittle material, is turned to the pseudo-ductile steel fiber-reinforced concrete. In particular, under flexural loading and after matrix crack initiation, the stresses are supported by the bridging fibers and the bending moments are redistributed (RILEM 2002).

As firstly proposed in our experimental research programme, a set of seven FRC beams has been retrofitted with fiber-reinforced polymer (FRP) sheets in order to improve their loading carrying capacity. The use of strengthening techniques may be required in order to reduce the deformations in serviceability state. In this framework, the choice of the proper rehabilitation technique and the assessment of its performance and durability clearly represents outstanding research points. These beams have been tested up to failure under three-point bending (RILEM 1985). The tests programme has permitted to quantify the effect of the FRP reinforcement on the mechanical response of the tested beams. More specifically, an increase in both strength and stiffness of the structural elements has been recognized, although such an increase is realized at the expense of a loss in ductility, or capacity of the structure to deflect inelastically while sustaining a load close to its maximum capacity. Concerning the failure modes of the tested beams, both FRP delamination and concrete shear failure have been observed, although FRP delamination was predominant. Such failure modes give rise to a brittle mechanical response, corresponding to snap-back instabilities (Carpinteri 1989, Carpinteri et al. 2006a).

The experimental method based on the Acoustic Emission (AE) technique (Shigeishi & Ohtsu 2001, Grosse et al. 2003) is proved to be highly effective for the structural monitoring of the tested specimens, especially to check and measure the damage evolution. The AE signals reflecting the energy release taking place during the three point bending tests can be recorded and micro-cracking sources can be localised by measuring time delays by means of spatially distributed AE sensors. In this application, the AE activity is analysed using techniques similar to those employed for detection of earthquake source epicentre (Carpinteri et al. 2006b; Colombo et al. 2003).

2 EXPERIMENTAL SETUP

2.1 Material data

A set of FRC beams measuring $1000 \times 150 \times 150\,\text{mm}^3$ were cast and subsequently retrofitted with FRP sheets at the Fracture Mechanics Laboratory of the Politecnico di Torino. The elastic modulus, E_a, the Poisson's ratio, ν_a, the characteristic compressive stress, as well as the fiber content in the FRC beams are summarized in Table 1.

The reinforcement, externally applied along the lower side of the elements, consists in carbon fiber unidirectional laminate sheets in the form of large flat plates cut into 700 mm long and 100 mm wide elements, having a thickness of 1.4 mm and a Young's modulus of 165 GPa. These sheets have been bonded using an epoxy adhesive on the concrete surface. The mechanical parameters of the FRP sheets are summarized in Table 2.

2.2 Testing machine and AE equipment

The experimental tests were conducted using a servo-controlled machine (MTS) with a closed-loop control (Fig. 1). The three point bending test was realized with a linear actuator (hydraulic jack) with passing stem acting in the middle point of the upper side of the beam element, which was positioned on the base of the frame (see Fig. 1). In particular, two specimens with the same nominal geometric and mechanical properties (TR1 and TR2) are considered and analysed. In both cases we performed the test under displacement control, at the rate of 0.011 mm/min for TR1 and 0.005 mm/min for TR2. In addition, the equipment adopted by the authors consists in six USAM® units used for the AE measurements and six pre-amplified piezoelectric (PZT) AE sensors applied to the external surface of the specimens. Such sensors were calibrated on inclusive frequencies between 50 and 800 kHz. PZT sensors (Fig. 1) were used, thereby exploiting the capacity of certain crystals to produce electric signals whenever they are subjected to a mechanical stress.

The USAM® units were synchronized for multi-channel data processing. The most relevant parameters acquired from the signals were stored in the USAMs memory and then downloaded to a PC for a multi-channel data processing. From this elaboration micro-crack localisation can be performed. The parameters of a signal recorded by a USAM unit include (Fig. 2): the initial instant crossing up-threshold of each P-wave arrival (t_0), the last down-threshold time (t_1) and the number of crossings of the threshold (Fig. 3).

The first arrival time at the sensor, S_i, is the first threshold crossing of the signal and it permits to perform the localisation of an AE source. The duration is the time from the first to the last threshold crossing of the signal. The amplitude is the peak voltage of the signal, while the number of oscillations is the number of threshold crossings by the signal. The threshold is selectable in the range between $100\,\mu V$ and $6.4\,mV$.

Table 2. Mechanical parameters of the FRP sheets.

Young's Modulus (E), ASTM D303	165 GPa
Max. tensile strain at failure	1.8%
Tensile strength, ASTM D3039	2300 MPa
Nominal thickness	1.4 mm

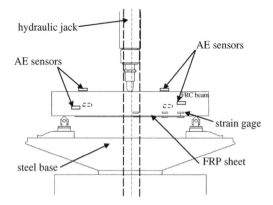

Figure 1. Three point bending experimental setup.

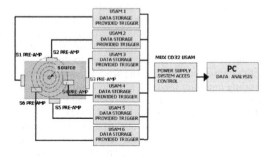

Figure 2. Acoustic emission measurement system.

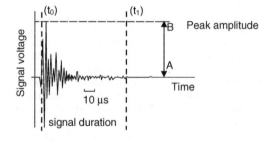

Figure 3. AE signal identified by the transducer.

Table 1. Mechanical parameters of the FRC beams.

Young Modulus (E_a)	35 GPa
Poissons's Ratio (ν_a)	0.18
Characteristic compressive stress	90 MPa
Fiber content	40 Kg/m^3

3 ANALYSIS AND EARLY RESULTS

Considering FRC beams, two typical problems requiring rehabilitation are either too high deformations in service conditions, or an inadequate load carrying capacity.

In retrofitted structures the most frequently observed failure modes are (Arduini et al. 1997, Alaee & Karihaloo 2003, Leung 2004, Taljiesten 1997): (i) FRP debonding or concrete ripping at the ends of the beam, depending on the adhesive properties; (ii) FRP debonding in proximity of flexural cracks; (iii) shear crack propagation in the concrete beam. As a consequence, fracture parameters of the interface and the nonlinear behavior of concrete are expected to exert a key role in the effectiveness of the retrofitting procedure. This is particularly true when ultimate conditions are analyzed. The FRP sheet insertion mainly causes two failure modes: (i) delamination between the FRP sheet and the concrete surface; (ii) a sudden shear failure starting from the cut-off edge of FRP and spreading through the beams.

In the set of FRC beams retrofitted with FRP sheets, the two cases mentioned in the Introduction have been assessed. In particular, specimen TR1 presents a sudden shear rupture localized near one of the two lower supports (Fig. 4.a). The resulting crack growth crosses the whole cross-section, with a peak load approximately equal to 86 kN. For specimen TR2, the failure mode is a delamination with a peak load of about 65 kN. A number of strain gages are placed near the lower side of the beams and along the FRP sheets to record the strain distribution as a function of the applied load. Nevertheless, to evaluate the failure modes only the AE experimental data are considered.

From a qualitative observation of the failed specimens, the cracking pattern for TR1 shows the presence of several cracks along the whole intradox of the specimen. These cracks result into a sudden shear collapse shown in Figure 4a. For specimen TR2 we noticed a smaller number of cracks and a main crack with the mouth located at the end of the delaminated portion of the FRP sheet (Fig. 4b).

4 APPLICATION OF THE AE TECHNIQUE

Acoustic emissions are ultrasonic waves generated by a rapid energy release coming from discontinuities or cracks propagating in the materials subjected to a given stress state. Such waves propagate through the damaged solid in isotropic and homogeneous materials as straight rays moving, at a first approximation, with an isotropic velocity distribution, v. When they reach the outer surface, they are captured by piezoelectric sensors (Carpinteri, et al. 2006a, 2006b). During the loading test, the AE transducers have been positioned in order to investigate on the flexural crack growth and on the progress of the FRP-concrete delamination (Fig. 5).

4.1 AE energy evaluation

The AE energy released during crack growth is an important indicator of the damage process. The cumulative AE energy of the damage process is herein calculated considering that the energy of an AE signal is proportional to the square of the voltage. As a consequence, in our analysis we square and integrate in time the recorded transient voltage of each channel (see also Lin et al 1997):

$$E_c = \frac{1}{C} \sum_i \int_{t_0}^{t_1} A_i(t)^2 dt \,. \tag{1}$$

In Eq.(1) the subscript i denotes the i_{th} event of the recorded transient voltage $A_i(t)$, while t_0 and t_1 represent the starting and ending times of the i_{th} transient voltage recording. C is a constant representing the resistance of the measuring circuit. For the energy calculation the threshold level of the signals recorded by the USAM, is fixed at $100 \, \mu V$. This value is generally adopted by the authors in monitoring AE events in concrete (Carpinteri et al. 2006a). The maximum signal

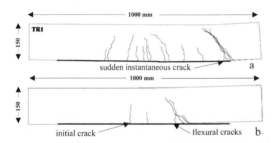

Figure 4. Cracking pattern for specimen TR1 (a) and specimen TR2 (b).

Figure 5. Three point bending tests and AE sensor positions.

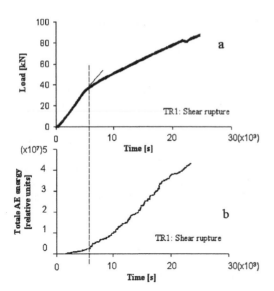

Figure 6. Load versus time diagram for TR1 (a); compared with the total AE energy cumulated during the test (b).

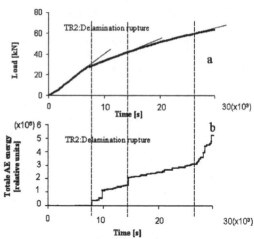

Figure 7. Load versus time diagram for TR2 (a); compared with the total AE energy cumulated during the test (b).

amplitude measured during the tests is about 12.8 mV. In Eq. (1) the cumulated AE energy E_c is obtained adding over all six channels the energy measured for each event.

The total cumulative AE energy computed systematically increases with the different stages of damage, so it could be used to estimate the development of the fracture process. In Figures 6 and 7 the cumulative AE energy is plotted versus time and compared with the load carrying capacity during the three point bending tests for specimens TR1 and TR2.

The cumulative AE released energy exhibits a strong correlation with the decay of the mechanical properties of the specimens. These correspondences are indicated in Figures 6 and 7 with dashed lines. Sudden increases in the released energy level are related to appreciable drops of the Young's modulus during the damage process in specimens TR1 and TR2.

4.2 Localisation of AE sources

The first stage in the localisation method consists in recognising the data needed to identify the AE sources, and is followed by the triangulation procedure (Shah & Li, 1994). During the first stage, the groups of signals, recorded by the sensors, falling into time intervals compatible with the formation of micro-cracks in the analysed volume, are identified. These time intervals, of the order of micro-seconds, are defined on the basis of an *a priori* assured speed of transmission of the *P* waves and of the mutual distance between the applied threshold of 100 μV of the non-amplified signal is appropriate to distinguish between *P*-wave and

S-wave arrival times. In fact, *P*-waves are usually characterized by higher signal amplitudes (Shah & Li 1994, Carpinteri et al. 2006a).

In the second stage, when the formation of microcraks in a three-dimensional space is analysed, the triangulation technique can be applied if the signals recorded by at least five sensors fall into the compatible time intervals. According to this procedure, it is possible to define both the position of the microcracks in the volume and the speed of transmission of the *P*-waves. The localisation procedure can also be performed through numerical techniques using optimisation methods such as the Least Squares Method (LSM) (Carpinteri et al. 2006b).

4.3 Moment tensor analysis

The approach applied in this study relies on the theoretical standpoint on the procedure defined by Shigeishi and Ohtsu (Shigeishi & Ohtsu 2001). This procedure characterizes the AE signal taking into account only the first arrival time of the *P*-waves. The elastic crack displacements $\mathbf{u}(\mathbf{x},t)$, at the points $\mathbf{x}$, which are the sources of the AE signals, are given by:

$$u_i(\mathbf{x},t) = G_{ip,q}(\mathbf{x},\mathbf{y},t)m_{pq} * S(t), \qquad (2)$$

where $G_{ip,q}(\mathbf{x},\mathbf{y},t)$ is the space derivative of the Green's function and the asterisk denotes the convolution operator. The Green's functions describe the elastic displacements $\mathbf{u}(\mathbf{x},t)$ due to a unit displacement applied in $\mathbf{y}$ at time t. In the SiGMA procedure, the magnitude of the elastic displacements, proportional to the amplitudes $A(\mathbf{x})$ of the first *P*-waves reaching the transducers, are given by a modified expression of Eq. (2):

$$A(\mathbf{x}) = \frac{C_s \, REF(\mathbf{t},\mathbf{r})}{R} (r_1 \, r_2 \, r_3) \begin{pmatrix} m_{11} & m_{12} & m_{13} \\ m_{21} & m_{22} & m_{23} \\ m_{31} & m_{32} & m_{33} \end{pmatrix} \begin{pmatrix} r_1 \\ r_2 \\ r_3 \end{pmatrix}, (3)$$

where C_s is a calibration coefficient of the acoustic emission sensors and R is the distance between the AE source at point $\mathbf{y}$ and the sensor located at point $\mathbf{x}$. R represents the distance between the source and the sensor. $REF(\mathbf{t},\mathbf{r})$ is the reflection coefficient of the sensitivity of the sensor depending on the angle between the two direction of the two unit vectors $\mathbf{r}$ and $\mathbf{t}$; these vectors are respectively the unit vector along the R distance and the unit vector along the sensitivity sensor direction.

To represent the moment tensor, it is necessary to determine the six independent unknowns, m_{pq}. The amplitude of the signal $A(\mathbf{x})$ must be received by at least six AE sensors. Through an eigenvalue analysis of the moment tensor, it is possible to determine the type of crack localised. Referring each eigenvalue to the maximum one, it can be written:

$$\frac{\lambda_1}{\lambda_1} = X + Y + Z, \quad \frac{\lambda_2}{\lambda_1} = 0 - \frac{Y}{2} + Z, \quad \frac{\lambda_3}{\lambda_1} = -X - \frac{Y}{2} + Z, (4)$$

where, $\lambda_1, \lambda_2, \lambda_3$ are the maximum, medium and minimum eigenvalues respectively, and X is the component due to shear, Y is the deviatoric tensile component, Z is the isotropic tensile component. Ohtsu classified an AE source with $X > 60\%$ as a shear crack (Mode II), one with $X < 40\%$ and $Y + Z > 60\%$ as a tensile crack (Mode I), and one with $40\% < X < 60\%$ as a Mixed Mode crack (Shigeishi & Ohtsu 2001). Moreover, from an eigenvector analysis it is possible to determine the unit vectors, l and n, which determine the directions of the displacement and the orientations of the crack surface.

4.4 Procedure for 3D AE analysis

Using the USAM® equipment the authors have fine-tuned a computer-based procedure including the AE source location and the moment tensor analysis. This procedure is implemented in an automatic AE data processing program coded in MAPLE 9.5 (FORTRAN77), whose overall flowchart is shown in Figure 8. The whole routine is divided into three parts automatically executed: (i) recognition and identification; (ii) location and moment tensor analysis; (iii) automatic translation of 3D AE pattern. The final output of the code returns a complete description of damage characterization and evolution. The 3D AE source positions obtained are overlapped to the 3D

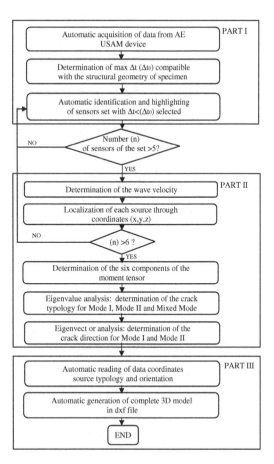

Figure 8. Flowchart of the localization and moment tensor analysis program.

crack patterns deduced from the observation of the cracking map.

In both TR1 and TR2 specimens, damage localization, typology and direction vector of the crack are shown in Figures 9 and 10. The position of each emission source, the typologies for Mode I, Mode II and Mixed Mode and the crack direction for Mode I and Mode II are represented according to the notation listed in Table 3.

4.5 Moment tensor analysis and results

Crack types of AE sources obtained under three point bending tests were analysed as shown in Figures 11 and 12. As regards specimen TR1 (shear rupture), we observe that the percentage of shear cracks (Mode II) becomes progressively comparable with the percentage of tensile cracks (Mode I) during the test (Fig. 11). In fact, beyond a certain distance from the midpoint of the beam, the cracking map shown in Figure 9 denotes

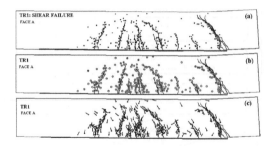

Figure 9. AE sources in TR1 specimen: (a)Localization, (b) crack type for Mode I, Mode II and Mixed Mode, (c) crack direction for Mode I and Mode II.

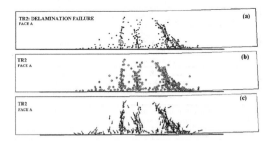

Figure 10. AE sources in TR2 specimen: (a)Localization, (b) crack type for Mode I, Mode II and Mixed Mode, (c) crack direction for Mode I and Mode II.

Table 3. Markers of AE sources and labels identifying crack typology and direction.

Crack Position	•
Crack Typology	
Mode I	⊖
Mode II	⊗
Mixed Mode	⊛
Crack Direction	/

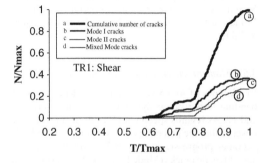

Figure 11. Cumulative number of cracks of different types and total number of cracks for specimen TR1.

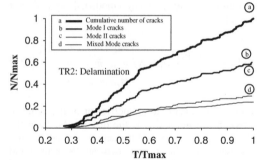

Figure 12. Cumulative number of cracks of different types and total number of cracks for specimen TR2.

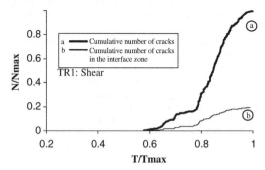

Figure 13. Cumulative number of cracks located in the delamination process zone and total cumulative number of cracks for specimen TR1.

that the micro-cracks propagate along oblique crack planes. On the other hand, for the TR2 specimen failed due to delamination, Mode I cracks are prevalent. At the end of the test, they consist in more than 50% of the total number of the localized events (Fig. 12). This behaviour can be interpreted by noting that, during the progress of damage, the specimen experienced flexural cracks near the mid-span position, that are typically classified as Mode I cracks. The effect of delamination has also been quantified by analysing a set of events located in a portion corresponding to a volume of $25 \times 1000 \times 150 \, \text{mm}^3$ along the intradox of the specimens. This volume has been considered as the damage zone involved in the delamination process between concrete and FRP. As far as the TR1 specimen is concerned, the localized events in this representative zone are less than 20% of the total number of events at the end of the test (see Fig. 13). On the other hand, for the TR2 specimen (see Fig. 14) the percentage of events coming from this zone is significantly higher (50%) than that observed for the TR1 specimen. Considering the total amount of the localized events during the tests, we highlight that the delamination process in the TR2 specimen started earlier than in the TR1 specimen, as confirmed by the comparison between

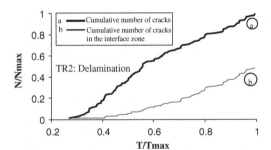

Figure 14. Cumulative number of cracks located in the delamination process zone and total cumulative number of cracks for specimen TR2.

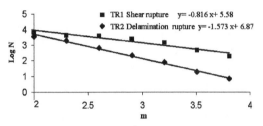

Figure 15. Determination of the "*b*-value" at the end of the tests on the specimens TR1 and TR2.

the percentages of the localized events falling in the delamination process zone. The difference in the failure modes is due to the effectiveness of the bonding between the FRP sheet and the concrete surface for the two specimens.

5 STATISTICAL DISTRIBUTION OF AE EVENTS

By analogy with seismic phenomena, in the AE technique the magnitude may be defined as follows:

$$m = Log_{10}A_{\max} + f(r),\qquad(5)$$

where $A_{\max}$ is the amplitude of the signal expressed in μV, and $f(r)$ is a correction coefficient whereby the signal amplitude is taken to be a decreasing function of the distance r between the source and the AE sensor. The empirical Gutenberg-Richter's law (Richter 1958) provides:

$$Log_{10}N(\geq m) = a - bm \text{ or } N(\geq m) = 10^{\,a-bm},\qquad(6)$$

where N is the cumulative number of earthquakes with magnitude $\geq m$ in a given area and a specific time-range, whilst a and b are positive constants varying from a region to another and from a time interval to another. Eq. (6) has been used successfully in the AE field to study the scaling laws of AE wave amplitude distribution. This approach highlights the similarity between structural damage phenomena and seismic activities in a given region of the earth, extending the applicability of the Gutenberg-Richter's law to Structural Engineering. According to Eq. (6), the "*b*-value" stands for the slope of the regression line in the "log-linear" diagram of AE signal amplitude distribution. This parameter changes systematically at different times in the course of the damage process and therefore can be used to estimate damage evolution modalities. Scale effects on the size of the cracks identified by the AE technique entail, by analogy with

earthquakes (Carpinteri et al. 2006b, Richter 1958), the validity of the following relationship:

$$N(\geq L) = cL^{-2b},\qquad(7)$$

where N is the cumulative number of AE events generated by cracks having a characteristic size $\geq L$, c is the total number of AE events and $D = 2b$ is the non-integer (or fractal) exponent of the distribution. The cumulative distribution (7) is substantially identical to the one proposed by Carpinteri (Carpinteri 1986, 1994), according to which the number of cracks with size $\geq L$ contained in a body is given by:

$$N^*(\geq L) = N_{tot}\, L^{-\gamma}.\qquad(8)$$

In Eq. (8), γ is an exponent reflecting the disorder, i.e., the crack size scatter, and N_{tot} is the total number of cracks contained in the body. By equating distributions (7) and (8), we find that: $2b = \gamma$. At the final collapse, when the size of the largest crack is proportional to the largest size of the body, function (8) is characterised by an exponent $\gamma = 2$, corresponding to $b = 1$. In (Carpinteri 1994) it is also demonstrated that $\gamma = 2$ is a lower limit. This corresponds to the minimum value $b = 1$, observed experimentally when the bearing capacity of a structural member is exhausted. By applying these concepts to the "*b*-value" analysis of specimens TR1 and TR2, it can be seen that the TR1 specimen exhausted its bearing capacity during the loading test, with the formation of cracks of a size comparable to that of the specimen (*b*-value $\cong 1$). On the other hand, the TR2 specimen, characterised by a flexural cracking pattern, was still having a reserve of strength before reaching the final collapse (*b*-value = 1.57). The "*b*-value" for the two specimens is shown in Figure 15.

In Figure 16 we propose a comparison between the two *b*-values of the two specimens considering only the events located in the delamination process zone previously defined. It can be realized that in the delamination process zone the TR1 specimen presents a *b*-value (=1.27) higher than that of the TR2 specimen (=1.01). This fact can be ascribed to the occurrence of delamination failure observed in the TR2 specimen.

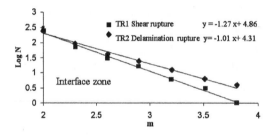

Figure 16. Determination of the "*b*-value" at the end of the tests of the specimens TR1 and TR2, and in relation to the delamination process zone.

6 CONCLUSIONS

A set of FRC beams was cast and subsequently retrofitted with FRP sheets at the Fracture Mechanics Laboratory of the Politecnico di Torino. On the basis of AE monitoring, loading tests have been conducted to analyse the cracking progression in the beams. The experimental results demonstrate the effectiveness of the AE technique for the interpretation and prediction of the failure modes in these composite structures.

ACKNOWLEDGEMENTS

The financial support provided by the Italian Ministry of University and Research (MIUR) is gratefully acknowledged. The authors would like to thank Mr. V. Di Vasto for the technical support and MS eng. R. Gottardo, Technical Manager of the D.L. Building System DEGUSSA Construction Chemicals Italia, for providing the FRP sheets used in the testing programme.

REFERENCES

Alaee, F.J. & Karihaloo, B.L. 2003. Fracture model for flexural failure of beams retrofitted with CARDIFRC. ASCE. *Journal of Engineering Mechanic* 129:1028–1038.

Arduini, M. Di Tommaso, A. & Nanni, A. 1997. Brittle failure in FRP plate and sheet bonded beams. ACI *Structural Journal* 94: 363–370.

Carpinteri, A. 1986. *Mechanical Damage and Crack Growth in Concrete: Plastic Collapse to Brittle Fracture*, Dordrecht: Martinus Nijhoff Publishers.

Carpinteri, A. 1989. Cusp catastrophe interpretation of fracture instability. *Journal of the Mechanics and Physics of Solids* 37: 567–582.

Carpinteri, A. 1994 Scaling laws and renormalization groups for strength and toughness of disordered materials. *International Journal of Solids and Structures* 31: 291–302.

Carpinteri, A. Lacidogna, G. & Paggi, M. 2006a. Acoustic emission monitoring and numerical modelling of FRP delamination in RC beams with non-rectangular cross-section. *Materials and Structures (RILEM)*, in press.

Carpinteri, A. Lacidogna, G. & Niccolini, G. 2006b. Critical behaviour in concrete structures and damage localization by acoustic emission. *Key Engineering Materials* 312: 305–310.

Colombo, S. Main, I.G. & Forde, M.C. 2003. Assessing damage of reinforced concrete beam using b-value analysis of acoustic emission signals. *ASCE Journal of Materials in Civil Engineering* 15: 280–286.

Grosse, C.U. Reinhardt, H.W. & Finck, F 2003. Signal based acoustic emission techniques in civil engineering. ASCE *J Mater Civil Eng.* 15:274–279.

Leung, C.K.Y. 2004. Delamination failure in concrete beams retrofitted with a bonded plate. *Journal of Materials in Civil Engineering* 13:106–113.

Lin, C.K. Berndt, C. Leigh, S. & Murakami, K. 1997. Acoustic Emission Studies of Alumina-1 3% Titania Free-Standing Forms during Four-Point Bend Tests. *Journal of American Ceramic Society* 9: 2382–2394.

RILEM TC TDF-162, 2002. Test and design methods for steel fiber reinforced concrete. Bending test – Final Recommendation,. *Materials and Structures (RILEM)* 35: 579–582.

RILEM FMC1, 1985. Determination of the fracture energy of mortar and concrete by means of three-point bend tests on notched beams. *Materials and Structures (RILEM)*. 18: 285–290.

Richter, C.F. 1958. *Elementary Seismology*. S. Francisco and London: W.H. Freeman and Company.

Shah, S.P. & Li, Z. 1994. Localisation of microcracking in concrete under uniaxial tension. *ACI Materials Journal* 91: 372–381.

Shigeishi, M. & Ohtsu, M. 2001. Acoustic emission moment tensor analysis: development for crack identification in concrete materials. *Construction and Buildings Materials* 15: 311–319.

Taljiesten, B. 1997. Strengthening of beams by plate bonding. *ASCE Journal of Materials in Civil Engineering* 9:206–212.

Fracture Mechanics of Concrete and Concrete Structures – Design, Assessment and Retrofitting of RC Structures – Carpinteri, et al. (eds)
© 2007 Taylor & Francis Group, London, ISBN 978-0-415-44616-7

On the competition between delamination and shear failure in retrofitted concrete beams and related scale effects

A. Carpinteri, G. Lacidogna & M. Paggi
Politecnico di Torino, Department of Structural and Geotechnical Engineering, Torino, Italy

ABSTRACT: In this paper we deal with the most commonly reported failure modes related to interfacial stress concentrations at the FRP cut-off points, i.e. diagonal (shear) crack growth and FRP delamination. Depending on the mechanical properties of the tested beams, their geometry and size, a prevalence of a given failure mode to the other is very often experimentally observed. To analyze this failure mode competition, a combined analytical/numerical model is proposed for the determination of the critical loads required for the onset of delamination or shear failure. In this way, the experimentally detected failure modes observed in RC and FRC beams are reexamined and interpreted in this new framework.

1 INTRODUCTION

Structure rehabilitation is required whenever design mistakes, executive defects or unexpected loading conditions are assessed. In these cases, the use of a strengthening technique may be required in order to either increase the loading carrying capacity of the structure, or to reduce its deformations. The choice of the proper rehabilitation technique and the assessment of its performance and durability clearly represent outstanding research points. Among the different rehabilitation strategies, bonding of steel plates or FRP sheets on the concrete members is becoming increasingly popular (Hollaway & Leeming 1999).

In these situations, the main observed failure modes can be summarized as follows: (a) flexural failure by FRP yielding (Arduini et al. 1997), (b) flexural failure by concrete crushing in compression (Arduini et al. 1997), (c) shear failure (Ahmed et al. 2001), (d) concrete cover separation (David et al. 1993), (e) FRP delamination (Leung 2004a; Leung 2004b), and (f) intermediate crack induced debonding (Alaee & Karihaloo 2003; Wang 2006). Among them, shear failure, concrete cover separation and FRP delamination have been far more commonly revealed in experimental tests. In these cases, damage initiates near the FRP cut-off points due to the presence of a stress concentration or even a stress intensification. As a consequence, either a pure FRP delamination or a diagonal crack growth can occur. Moreover, in the latter case, depending on the amount of steel reinforcement and thickness of concrete cover, the diagonal crack may give rise to either shear failure, or concrete cover separation.

Therefore, since concrete cover separation occurs away from the FRP-concrete interface, this failure mode should be carefully distinguished from the pure FRP delamination. Therefore, it seems to be more appropriate to interpret failure modes (c) and (d) in the same framework.

As regards the mathematical models available in the Literature, most of them focus on the problem of delamination in steel plated and FRP strengthened beams (Smith & Teng 2002; Smith & Teng 2002b). Shearing and peeling stresses in the adhesive layer of a beam with a strengthening plate bonded to its soffit were determined in (Taljsten 1997; Malek et al. 1998; Ascione & Feo 2000; Smith & Teng 2001). The analysis of interface tangential and normal stresses in FRP retrofitted RC beams was also recently reexamined in (Rabinovitch & Frostig 2001; Rabinovitch 2004), along with a fracture mechanics model for the prediction of FRP delamination.

Comparatively, a little attention has been directed toward the analysis of shear crack growth and to the competition between FRP delamination and shear failure. The problem of size-scale effect is also an open issue (Maalej & Leong 2005). To deal with these problems, we propose a combined analytical/numerical model to describe the failure mode competition between FRP delamination and shear failure in reinforced concrete (RC) and fiber-reinforced concrete (FRC) beams. In this way, the experimentally detected failure modes are reexamined and interpreted in this new framework. As regards RC beams, we refer to the data on four-point bending tests reported in (Ahmed et al. 2001), whereas the experimental results

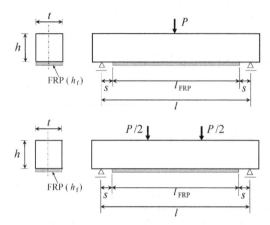

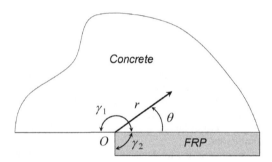

Figure 1. Schemes of three- and four-point bending tests.

Figure 2. Scheme of the bi-material wedge composed of FRP and concrete.

on FRC beams have been determined according to a new testing programme carried out in our laboratory.

2 ANALYTICAL MODEL

In this section, we consider the typical three-point bending and four-point bending tests carried out in the laboratories to assess the mechanical performance of retrofitted beams (see Fig. 1).

2.1 Stress-singularities and generalized stress-intensity factors

According to Linear Elastic Fracture Mechanics, the FRP cut-off point can be a source of stress-singularities due to the mismatch in the elastic properties of concrete and FRP. The geometry of a plane elastostatic problem consisting of two dissimilar isotropic, homogeneous wedges of angles equal to $\gamma_1 = \pi$ and $\gamma_2 = \pi/2$ perfectly bonded along their interface is schematically shown in Figure 2.

In this general case, the singular components of the stress field can be written as follows:

$$\sigma_{ij} = K^* r^{(\text{Re}\lambda - 1)} S_{ij}(\theta), \qquad (1)$$

where K^* is referred to as *generalized stress-intensity factor* (Carpinteri 1987). The parameter λ defines the order of the stress-singularity and can be obtained according to an asymptotic analysis of the stress field, see e.g. (Williams 1952; Bogy 1971; Carpinteri & Paggi 2005; Carpinteri & Paggi 2007) for similar applications.

According to this approach, the parameter λ is determined by solving a non-linear eigenvalue problem resulting from the imposition of the boundary conditions. In the present problem, they consist in the stress-free boundary conditions along the free edges, and in the continuity conditions of stresses and displacements along the bi-material interface. Since we are interested in the analysis of the the singular terms of the stress field, we are concerned only with those values of λ which may lead to singularities. This fact, together with the condition of continuity of the displacement field at the vertex where regions meet, imply that we are seeking for eigenvalues in the range $0 < \text{Re}\lambda < 1$.

Function $S_{ij}(\theta)$ in Eq. (1) is the eigenfunction of the problem and it locally describes the angular variation of the stress field near the singular point, O. It has to be remarked that, since both λ and $S_{ij}(\theta)$ are determined according to the asymptotic analysis, they solely depend on the boundary conditions imposed in proximity of the singular point.

Moreover, from dimensional analysis arguments (Carpinteri 1987), it is possible to consider the following expression for the generalized stress-intensity factor:

$$K^* = \frac{Pl}{th^{1+\text{Re}\lambda}} f\left(\frac{h_f}{h}, \frac{l}{h}, \frac{l_{\text{FRP}}}{h}, \frac{E_f}{E_c}\right), \qquad (2)$$

where function f depends on the boundary conditions far from the singular point and can be determined according to a FE analysis on the actual geometry of the tested specimen. For the sake of generality, this function depends on the relative thickness of the reinforcement compared to the beam depth, h_f/h, on the ratio between the span and the depth of the beam, l/h, on the length of the FRP sheet, l_{FRP}/h, and on the modular ratio between FRP and concrete, E_f/E_c. Parameters P and t denote, respectively, the applied load and the beam thickness.

The critical load corresponding to the onset of delamination can be determined by setting the generalized stress-intensity factor equal to the critical stress-intensity factor for the interface. This approach, well-established for the analysis of bonded joints

(Reedy & Guess 1993; Qian & Akisanya 1999; Carpinteri & Paggi 2006), yields to the following equation:

$$P_C^{del} = K_{C,int}^* \frac{th^{1+Re\lambda}}{l} \frac{1}{f}. \qquad (3)$$

An analogous reasoning can be proposed for the analysis of the onset of shear failure, i.e. before the development of the crack-bridging effect due to steel reinforcement. In this case, we can postulate the existence of a small vertical crack into concrete at the FRP cut-off point simulating an initial defect. This crack may result in a sudden diagonal propagation leading to premature failure of the beam.

The stress field at the crack tip is again singular, but with the order of the singularity typical of a crack inside a homogeneous material (Carpinteri 1987; Bocca et al. 1990):

$$\sigma_{ij} = Kr^{-1/2}F_{ij}(\theta), \qquad (4)$$

where function F locally describes the angular variation of the stress field near the crack tip. From dimensional analysis considerations, it is possible to write the following expression for the Mode I stress-intensity factor:

$$K_I = \frac{Pl}{th^{3/2}}g\left(\frac{a_0}{h}, \frac{h_f}{h}, \frac{l}{h}, \frac{l_{FRP}}{h}, \frac{E_f}{E_c}\right), \qquad (5)$$

where function g depends again on the boundary conditions far from the singular point and can be determined according to a FE analysis on the actual geometry of the tested specimen. The additional parameter a_0 with respect to FRP delamination denotes the initial crack length.

Crack propagation in this case takes place under Mixed Mode, although the Mode I stress-intensity factor is numerically prevailing. Under such assumptions, the critical load corresponding to the onset of shear crack propagation is reached when the Mode I stress-intensity factor equals the critical value of concrete. This condition yields to the following equation:

$$P_C^{shear} = K_{IC}\frac{th^{3/2}}{l}\frac{1}{g}. \qquad (6)$$

2.2 *Size-scale effects and failure modes competition*

For a given tested beam, i.e., for a given beam geometry, the ratio between the critical loads for delamination and shear failure can be written as:

$$\frac{P_C^{del}}{P_C^{shear}} = \left(\frac{K_{C,int}^*}{K_{IC}}\frac{g}{f}\right)h^{(Re\lambda-1/2)}, \qquad (7)$$

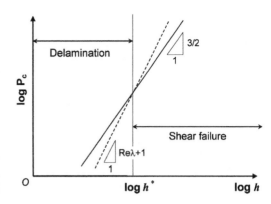

Figure 3. Competition between delamination and shear failure: dashed line corresponds to delamination, solid line corresponds to shear failure.

which is a non-linear function of the beam depth. In addition, it is possible to recast Eqs. (3) and (6) in a logarithmic form:

$$\log P_C^{del} = \log\left(\frac{tK_{C,int}^*}{lf}\right) + (1+Re\lambda)\log h, \qquad (8)$$

$$\log P_C^{shear} = \log\left(\frac{tK_{IC}}{lg}\right) + \frac{3}{2}\log h. \qquad (9)$$

These equations are qualitatively plotted as functions of the beam depth in Figure 3. As expected, the higher the beam depth, for a given ratio l/h, the higher the critical load of failure. The intersection point between the two curves defines the critical beam size corresponding to the transition from pure delamination to shear failure. Moreover, since the real part of the eigenvalue λ is usually higher than 0.5, we expect a prevalence of shear failure in larger beams. In fact, if we consider $E_f = 200$ GPa and $E_c = 30$ MPa as the values representative of the Young's moduli of concrete and FRP, then the asymptotic analysis gives $\lambda = 0.58$.

3 REINFORCED CONCRETE BEAMS

In this section we propose numerical simulations of shear failure and FRP delamination in FRP-retrofitted RC beams. Concerning the tested geometry, we refer to the extensive test programme carried out by Ahmed et al. (2001). More specifically, they tested a set of rectangular beams ($h = 0.225$ m, $l = 1.5$ m, $t = 0.125$ m) under four point bending and considered different FRP lengths ($l_{FRP} = 0.70$ m; 0.65 m; 0.60 m and 0.55 m). Independently of l_{FRP}, all the beams failed due to shear crack propagation originating from the FRP cut-off point. Numerical predictions and experimental results are shown and compared in the sequel.

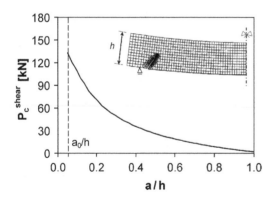

Figure 4. Critical load for shear failure vs. non-dimensional crack length.

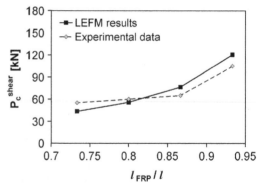

Figure 5. Critical load for shear failure vs. non-dimensional FRP length.

3.1 Shear failure

The simulation of shear failure in RC beams can be performed according to different numerical strategies. For example, Gustraffson (1985) considered a FE formulation taking into account a softening cohesive law for concrete and truss elements connected by springs to the concrete blocks to model the presence of the reinforcement.

To avoid finite element computations, Jenq & Shah (1989) and So & Karihaloo (1993) proposed a semi-analytical model where the concrete contribution to shear strength was evaluated according to LEFM, and the effect of steel-concrete interaction was included using an empirical relationship.

In the present approach, we consider an initial crack length equal to a_0 in correspondence of the FRP cut-off point. Simulation of crack growth is then performed using the FRANC2D finite element code (Wawrzynek & Ingraffea 1987). At each step, the stress-intensity factors are computed using the displacement correlation technique and the direction for crack propagation is determined according to maximum circumferential stress criterion. This approach permits to take into account the contribution of concrete to shear strength, as also done in the approximate model by Jenq & Shah (1989). Since the effect of reinforcement is not considered, the proposed method predicts an unstable crack growth, see e.g. Figure 4 for the case with $l_{FRP} = 0.70$ m.

The computed critical force corresponding to a_0, i.e. to the onset of diagonal failure, is shown in Figure 5 for different FRP lengths. Experimental results by Ahmed et al. (2001) are also reported in the same diagram. As expected, the good agreement between the numerical predictions and the experimental results demonstrate that this approach is suitable for the computation of the value of the critical force corresponding to the onset of crack propagation. Moreover, Figure 5 shows the shorter the FRP sheet, the lower the critical force.

3.2 Delamination

The computation of the critical load required for delamination can be performed according to the analytical model illustrated in Section 2. In this case, the critical stress-intensity factor of the interface can be estimated as $K_{C,int}^* \cong \sqrt{G_C E_a}$, where G_C and E_a are, respectively, the interface fracture energy and the Young's modulus of the adhesive. These parameters can be determined either from experiments (Ferretti & Savoia 2003), or estimated according to the prescriptions reported in design codes and standards (ACI 440R-96 1996; fib Bulletin 2001; JCI 2003).

It is important to observe that the values of P_C^{del} computed for different FRP lengths, say $l'_{FRP} < l_{FRP}$, can also be obtained from the numerical simulation of the delamination process in a reference retrofitted beam with $l'_{FRP} = l_{FRP}$. In fact, considering an interface crack length equal to a, the debonded portion of the FRP sheet is stress-free. Under these conditions, the critical load for interface crack propagation, $P_C^{del}(a)$, corresponds to that for the onset of delamination in a retrofitted beam with a shorter reinforcement length, i.e. $P_C^{del}(l_{FRP} - a)$.

The progress of FRP delamination can be numerically simulated using the finite element method with a cohesive model for the description of the mechanical behavior of the interface. Cohesive models were introduced by Hillerborg et al. (1976) to the analysis of the nonlinear fracture process zones in quasi-brittle materials. Carpinteri firstly applied a cohesive formulation to the study of ductile-brittle transition and snap-back instability in concrete (Carpinteri 1985; Carpinteri et al. 1986; Carpinteri 1989). More recently, we have proposed the use of this approach for the analysis of snap-back instability during FRP delamination (Carpinteri et al. 2006).

Numerical results for the beam with $l_{FRP} = 0.70$ m are reported in Figure 6 in terms of the applied load, P, vs. the mid-span deflection, δ. When the peak

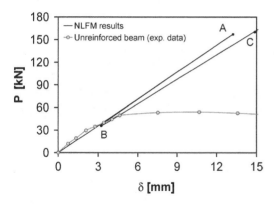

Figure 6. Load vs. non-dimensional deflection during delamination.

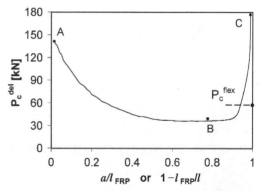

Figure 8. Critical load for delamination vs. non-dimensional crack advancement. P_C^{flex} is the ultimate load of the reference beam without retrofitting.

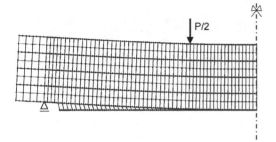

Figure 7. Deformed mesh showing delamination.

load is achieved, point (A), delamination takes place and, using the crack length as a driving parameter, we observe that both the external load and the mid-span deflection of the beam are progressively reduced up to point (B). After that, the progress of delamination requires an increase in the external load, tending asymptotically to the mechanical response of the undamaged RC beam without FRP. From the engineering point of view, this brittle mechanical response is particularly dangerous, since it corresponds to a severe snap-back instability. A deformed mesh showing the progress of delamination is also shown in Figure 7.

The structural response can be represented in terms of critical load vs. crack length or, equivalently, critical load vs. FRP length (see Fig. 8). During crack propagation, i.e. from (A) to (B), we have an unstable crack growth, since the external load is progressively reduced. An increase in the critical load is observed after point (B), i.e. for $a/l_{\text{FRP}} \cong 0.8$. In any case, for shorter FRP lengths, the actual failure load is certainly bounded by flexural or diagonal failures of the concrete beam (see e.g. Figures 6 and 8 where the critical load corresponding to flexural failure of the beam without FRP is reported for comparison).

A comparison between Figures 4 and 8 shows that both the critical loads for shear failure and those for FRP delamination are decreasing functions of l_{FRP} in the usual range of FRP lengths. Moreover, for the case-studies herein analyzed, the critical loads for shear failure are less than those for FRP delamination. This numerical prediction is in agreement with the experimental findings by Ahmed et al. (2001), where the reinforced beams failed due to shear failure. According to Figure 3, a pure FRP delamination is expected for smaller beams.

These results permit to interpret the common experimental observation that shear failure and concrete ripping are the most frequently observed failure modes in RC beams tested in laboratory. FRP delamination is less frequent and should be ascribed to weak bonding properties.

4 FIBER-REINFORCED CONCRETE BEAMS

In this section, we show some of the main results of an experimental programme on fiber-reinforced concrete beams (FRC) retrofitted with FRP. The tests have been conducted in the Laboratory of Materials and Fracture Mechanics of the Department of Structural and Geotechnical Engineering of the Politecnico di Torino.

In this respect, we notice that most of the current research studies on retrofitting techniques deal with standard RC members, whereas FRC beams are analyzed only in a few studies (Yin & Wu 2003). From the engineering point of view, FRC beams can be used for applications requiring high durability and the problem of retrofitting may arise when we are looking at another dimension of such members, namely, their upgrading capability (Shah & Ouyang 1991; Wegian & Abdalla 2005).

In this programme, seven FRC beams have been tested under three-point bending up to failure under

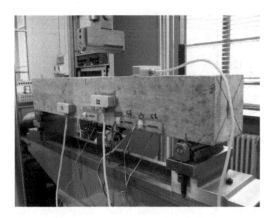

Figure 9. Photo of the testing apparatus.

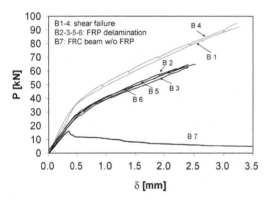

Figure 10. Results of the experimental tests: applied load vs. mid-span deflection.

displacement control (see Fig. 9). The beams are made of high-strength concrete reinforced with standard steel fibers produced by Bekaert. These fibers have a length equal to 50 mm, a diameter equal to 0.50 mm, and a content of 40 kg/m³. The concrete Young's modulus is equal to 35 GPa, with a Poisson's ratio of 0.18. FRP sheets have a Young's modulus equal to 165 GPa, a tensile strength of 2300 MPa and a maximum tensile strain at failure of 1.8%. Concerning the geometrical parameters, we have $l = 100$ cm, $l_{FRP} = 70$ cm, $h = t = 15$ cm, $s = 7.5$ cm and $h_f = 1.4$ mm.

Four retrofitted beams failed due to delamination (B2, B3, B5 and B6), two experienced shear failure (B1 and B4) and the remaining one was tested without FRP (B7) to establish the behavior of the unreinforced beam (see the test results shown in Figure 10). Photos of shear failure and FRP delamination are also shown, respectively, in Figures 11 and 12.

Experimental results indicate that mixing of steel-fibers affects the cracking behavior of concrete, giving

Figure 11. Photo of a failed beam due to shear crack growth.

Figure 12. Photo of a failed beam due to FRP delamination.

rise to distributed crack patterns. The higher frequency of FRP delamination suggests that the failure mode changes from shear failure to pure FRP delamination, as compared to standard RC beams. This result is fully consistent with the analytical model in Section 2. In fact, the use of steel-fibers results into an increased concrete toughness as compared to regular concrete. As a consequence, higher values of K_{IC} correspond to the higher critical loads required for shear crack propagation (see Eq. (6)). Moreover, this confirms that the transition from FRP delamination to shear failure is expected for larger beams.

5 CONCLUSIONS

In this paper we have proposed a combined analytical/numerical approach for the analysis of failure modes in concrete beams. Numerical predictions and experimental results show that shear failure and concrete ripping are more likely to occur in RC beams. In fact, for the analyzed case-studies, the critical load for the onset of shear crack growth is found to be lower than that corresponding to FRP delamination, independently of the reinforcement length. On the other hand, the results of the experimental programme on FRC beams show that in these cases FRP delamination is more frequent than shear failure. This different behavior as compared to RC beams has to be ascribed to the increased value of concrete fracture toughness due to the steel-fiber bridging effect.

Concerning the issue of stability of crack propagation, it has been shown that the process of FRP delamination leads to severe snap-back instabilities, thus resulting into a brittle mechanical response of the retrofitted concrete member. From the numerical point of view, it has been shown that the snap-back instability can be followed either under crack length control, or by computing the critical loads for delamination in correspondence of concrete beams with shorter and shorter FRP lengths. This numerical result gives also the possibility to experimentally follow the snap-back instability by testing concrete beams with different reinforcement lengths.

ACKNOWLEDGEMENTS

The financial support provided by the Italian Ministry of University and Research is gratefully acknowledged. The Authors would like to thank Mr. V. Di Vasto for the technical support provided during the experimental programme and Ing. R. Gottardo, Technical Manager of the D.L. Building Systems DEGUSSA Construction Chemicals Italia, for providing the FRP sheets used in the testing programme.

REFERENCES

Ahmed, O., Van Gemert, D. & Vandewalle, L. 2001. Improved model for plate-end shear of CFRP strengthened RC beams. *Cement & Concrete Composites* 23: 3–19.

Alaee, F. & Karihaloo, B. 2003. Fracture model for flexural failure of beams retrofitted with CARDIFRC. *ASCE Journal of Engineering Mechanics* 129: 1028–1038.

Arduini, M., Di Tommaso, A. & Nanni, A. 1997. Brittle failure in FRP plate and sheet bonded beams. *ACI Structural Journal* 94: 363–370.

Ascione, L. & Feo, L. 2000. Modeling of composite/concrete interface of RC beams strengthened with composite laminates. *Composites Part B: engineering* 31: 535–540.

Bocca, P., Carpinteri, A. & Valente, S. 1990. Size effects in the mixed mode crack propagation: softening and snap-back analysis. *Engineering Fracture Mechanics* 35: 159–170.

Bogy, D. 1971. Two edge-bonded elastic wedges of different materials and wedge angles under surface tractions. *ASME Journal of Applied Mechanics* 38: 377–386.

Carpinteri, A. 1985. Interpretation of the Griffith instability as a bifurcation of the global equilibrium. In S. Shah (Ed.), *Application of Fracture Mechanics to Cementitious Composites (Proc. of a NATO Advanced Research Workshop, Evanston, USA, 1984)*, 287–316. Martinus Nijhoff Publishers, Dordrecht.

Carpinteri, A. 1987. Stress-singularity and generalized fracture toughness at the vertex of re-entrant corners. *Engineering Fracture Mechanics* 26: 143–155.

Carpinteri, A. 1989. Cusp catastrophe interpretation of fracture instability. *Journal of the Mechanics and Physics of Solids* 37: 567–582.

Carpinteri, A., Lacidogna, G. & Paggi, M. 2006. Acoustic emission monitoring and numerical modelling of FRP delamination in RC beams with non-rectangular cross-section. *RILEM Materials & Structures, in press.* doi:10.1617/s11527-006-9162-4.

Carpinteri, A. & Paggi, M. 2005. On the asymptotic stress field in angularly nonhomogeneous materials. *International Journal of Fracture* 135: 267–283.

Carpinteri, A. & Paggi, M. 2006. Influence of the intermediate material on the singular stress field in tri-material junctions. *Materials Science* 42: 95–101.

Carpinteri, A. & Paggi, M. 2007. Analytical study of the singularities arising at multi-material interfaces in 2D linear elastic problems. *Engineering Fracture Mechanics* 74: 59–74.

Carpinteri, A., Di Tommaso, A. & Fanelli, M. 1986. Influence of material parameters and geometry on cohesive crack propagation. In F. Wittmann (Ed.), *Fracture Toughness and Fracture Energy of Concrete (Proc. of Int. Conf. on Fracture Mechanics of Concrete, Lausanne, Switzerland, 1985)*, 117–135. Elsevier, Amsterdam.

David, E., Djelal, C. & Buyle-Bodin, F. 1993. Repair and strengthening of reinforced concrete beams using composite materials. Proceedings of the 2nd international PhD symposium in civil engineering, Budapest.

Ferretti, D. & Savoia, M. 2003. Cracking evolution in R/C members strengthened by FRP-plates. *Engineering Fracture Mechanics* 70: 1069–1087.

Gustafsson, P. 1985. Fracture mechanics studies of non-yielding materials like concrete. *Report TVBM-1007, Div. Build. Mater. Lund. Inst. Tech, Sweden.*

Hillerborg, A., Modeer, M. & Petersson, P. 1976. Analysis of crack formation and crack growth in concrete by means of fracture mechanics and finite elements. *Cement and Concrete Research* 6: 773–782.

Hollaway, L. & Leeming, M. 1999. *Strengthening of reinforced concrete structures*. Cambridge, England: Woodhead Publishing.

Jenq, Y. & Shah, S. 1989. Shear resistance of reinforced concrete beams – a fracture mechanics approach. *Fracture Mechanics: Application to Concrete (Special Report ACI SP-118)*, eds. V.C. Li and Bazant, Z.P., Detroit 25: 327–358.

Leung, C. 2004a. Delamination failure in concrete beams retrofitted with a bonded plate. *ASCE Journal of Materials in Civil Engineering* 13: 106–113.

Leung, C. 2004b. Fracture mechanics of debonding failure in FRP-strengthened concrete beams. In V. Li, C. Leung, K. Willam, and S. Billington (Eds.), *Proceedings of the 5th International Conference on Fracture Mechanics of Concrete and Concrete Structures (FraMCoS-5), Vail, Colorado, USA*, Volume 1, 41–52.

Maalej, M. & Leong, K. 2005. Effect of beam size and FRP thickness on interfacial shear stress concentration and failure mode of FRP-strengthened beams. *Composites Science and Technology* 65: 1148–1158.

Malek, A., Saadatmanesh, H., & Ehsani, M. 1998. Prediction of failure load of R/C beams strengthened with FRP plate due to stress concentration at the plate end. *ACI Structural Journal* 95: 142–152.

ACI 440R-96 1996. State-of-the-art report on fiber reinforced plastic (FRP). Reinforcement for concrete structures.

American Concrete Institute, Committee 440, Michigan, USA.

fib Bulletin 2001. Design and use of externally bonded FRP reinforcement (FRP EBR) for reinforced concrete structures. Bulletin No. 14, prepared by subgroup EBR (Externally Bonded Reinforcement) of fib Task Group 9.3 FRP Reinforcement for Concrete Structures.

JCI 2003. Technical report on retrofitting technology for concrete structures. Technical Committee on Retrofitting Technology for Concrete Structures.

Qian, Z. & Akisanya, A. 1999. An investigation of the stress singularity near the free edge of scarf joints. *European Journal of Mechanics A/Solids* 18: 443–463.

Rabinovitch, O. 2004. Fracture-mechanics failure criteria for RC beams strengthened with FRP strips-a simplified approach. *Composite Structures* 64: 479–492.

Rabinovitch, O. & Frostig, Y. 2001. Delamination failure of RC beams strengthened with FRP strips-A closed form high-order and fracture mechanics approach. *ASCE Journal of Engineering Mechanics* 127: 852–861.

Reedy, J. & Guess, T. 1993. Comparison of butt trensile strength data with interface corner stress intensity factor prediction. *International Journal of Solids and Structures* 30: 2929–2936.

Shah, S. & Ouyang, C. 1991. Mechanical behavior of fiber-reinforced coment-based composites. *Journal of American Ceramics SOciety* 74: 2727–2738.

Smith, S. & Teng, J. 2001. Interfacial stresses in plated beams. *Engineering Structures* 23: 857–871.

Smith, S. & Teng, J. 2002. FRP-strengthened RC beams. I: review of debonding strength models. *Engineering Structures* 24: 385–395.

Smith, S. & Teng, J. 2002b. FRP-strengthened RC beams. II: assessment of debonding strength models. *Engineering Structures* 24: 397–417.

So, K. & Karihaloo, B. 1993. Shear capacity of longitudinally reinforced beams – a fracture mechanics approach. *ACI Journal* 78: 591–600.

Taljsten, B. 1997. Strengthening of beams by plate bonding. *ASCE Journal of Materials in Civil Engineering* 9: 206–212.

Wang, J. 2006. Cohesive zone model of intermediate crack-induced debonding of FRP-plated concrete beam. *International Journal of Solids and Structures* 43: 6630–6648.

Wawrzynek, P. & Ingraffea, A. 1987. Interactive finite element analysis of fracture processes: an integrated approach. *Theoretical and Applied Fracture Mechanics* 8: 137–150.

Wegian, F. & Abdalla, H. 2005. Shear capacity of concrete beams reinforced with fiber reinforced polymers. *Composite Structures* 71: 130–138.

Williams, M. 1952. Stress singularities resulting from various boundary conditions in angular corners of plates in extension. *ASME Journal of Applied Mechanics* 74: 526–528.

Yin, J. & Wu, Z. 2003. Structural performances of short steel-fiber reinforced concrete beams with externally bonded FRP sheets. *Construction and Building Materials* 17: 463–470.

Fracture Mechanics of Concrete and Concrete Structures – Design, Assessment and
Retrofitting of RC Structures – Carpinteri, et al. (eds)
© 2007 Taylor & Francis Group, London, ISBN 978-0-415-44616-7

Size effect on shear strength of concrete beams reinforced with FRP bars

F. Matta
Center for Infrastructure Engineering Studies, University of Missouri-Rolla

A. Nanni
Department of Civil, Architectural and Environmental Engineering, University of Miami

N. Galati
Center for Infrastructure Engineering Studies, University of Missouri-Rolla

F. Mosele
Department of Structural and Transportation Engineering, University of Padova

ABSTRACT: The use of glass fiber reinforced polymer (GFRP) bars as internal reinforcement for portions of massive concrete retaining walls to be penetrated by tunnel boring machines (TBMs), commonly referred to as softeyes, is becoming mainstream. The low shear strength and inherent brittleness of GFRP bars greatly facilitate penetration of the TBM, preventing damage to the disc cutters, and eliminating the risk of costly delays. The safe shear design of softeyes and large members in general must account for the strength decrease due to size effect. To date, this phenomenon has not been documented for FRP reinforced concrete (RC). In this paper, the results of laboratory tests on four large-scale concrete beams reinforced with GFRP bars in flexure and shear are presented and discussed. Preliminary results are reported that indicate a decrease in concrete shear strength attributable to size effect, which is offset by an implicit understrength factor in the current ACI 440 design formula. Further experimental research is ongoing to better characterize the extent of size effect in FRP RC.

1 INTRODUCTION

1.1 Problem statement

The decrease in concrete contribution to the shear capacity at increasing member depth in steel reinforced beams and slabs has been extensively documented (Kani 1967, Shioya et al. 1989, Collins & Kuchma 1999, Frosch 2000, Lubell et al. 2004). Size effect accrues primarily from the larger width of diagonal cracks as the beam effective depth is increased. Contrasting theories on a sound physical modeling of this phenomenon are being debated, primarily the energetic-statistical scaling (Bažant & Kim 1984, Bažant & Yu 2005a, b) and the crack spacing hypothesis incorporated in the Modified Compression Field Theory (Collins et al. 1996).

The issue is also of fundamental and practical relevance in the design of concrete members reinforced with fiber reinforced polymer (FRP) bars, where deeper and wider cracks due to relatively low elastic modulus of the flexural and, when present, shear reinforcement, as well as reduced dowel action contributed by the tension reinforcement, pose safety concerns that must be addressed. The current ACI

"Guide for the Design and Construction of Structural Concrete Reinforced with FRP Bars – ACI 440.1R-06" (ACI 2006) includes a new concrete shear design formula, conceived for application with any reinforcement material. The semi-empirical design equation was rendered in a fringe-type format by calibration on the basis of 370 test results, of which 44 were from FRP reinforced concrete (RC) specimens having a maximum effective depth of 376 mm (Tureyen & Frosch 2003), where size effect is typically negligible. The conservativeness of the design equation for larger FRP RC members remains unproven.

In this paper, the results of the first four large-size glass FRP (GFRP) RC beams tested to date, as part of an extensive research program, are presented and discussed on the basis of the shear strength contribution of the concrete, V_c, and of the transverse reinforcement, V_f. The results indicate the presence of strong size effect on the former. Analysis of the predictions of the ACI formula (ACI 440 2006) compared to experimental results available in the literature and from the present study shows that an implicit understrength factor may offset the strength decrease for effective

Figure 1. Ø32 mm GFRP reinforcing bar.

depths $d \leq 900$ mm, and effective reinforcement ratios ρ_{eff} (i.e. corrected by a factor E_f/E_s to account for the lower FRP stiffness, where E_f = longitudinal elastic modulus of FRP and E_s = elastic modulus of steel) within a range that covers most practical purposes.

1.2 Practical significance

A relevant application of large FRP RC members is in softeye openings in temporary retaining walls for tunneling applications. Softeyes are commonly referred to as the slurry wall sections through which the tunnel boring machine (TBM) penetrates during excavation. The low shear strength and inherent brittleness of glass FRP (GFRP) bars are highly desirable properties for use as softeye reinforcement in lieu of steel. Penetration of the TBMs is greatly facilitated, thereby expediting the field operations, preventing damage to the disc cutters, and eliminating the risk of costly delays. Large-size (Ø32 mm) GFRP bars as shown in Figure 1 are typically required as tensile reinforcement, often in bundles, due to the massive wall dimensions. The technology has been successfully implemented in recent underground projects in North America, Europe, and Asia. Although design principles are fairly well established (Nanni 2003), understanding the implications of size effect is instrumental for the safe design of softeye and other large FRP RC members.

2 DESIGN PROVISIONS

The recently adopted ACI design equation for concrete shear strength is (ACI 440 2006)

$$V_c = \frac{2}{5} k \left(f_c' \right)^{1/2} bd \qquad (1)$$

where $k = [2 \, \rho_f n + (\rho_f n)^2]^{1/2} - \rho_f n$, ρ_f = FRP flexural reinforcement ratio, n = ratio of E_f to the elastic modulus of concrete, f_c' = cylinder compressive strength of concrete in MPa, b = width of rectangular cross section in mm, and d = effective depth of tension reinforcement in mm. The main parameters affecting V_c are recognized as the axial stiffness of the flexural reinforcement, and the concrete tensile strength, herein assumed proportional to $(f_c')^{1/2}$ (ACI 318 2005).

Size effect is not explicitly accounted for in Equation 1. Conversely, specific size effect parameters are incorporated in the following design algorithms. For $d \geq 300$ mm, the Canadian Standard Association (CSA 2004) and ISIS Canada (ISIS 2001) recommend

$$V_c = \left(\frac{130}{1000 + d} \right) \lambda \phi_c \left(f_c' \right)^{1/2} bd$$
$$\geq 0.008 \, \lambda \phi_c \left(f_c' \right)^{1/2} bd \qquad (2)$$

and

$$V_c = \left(\frac{E_f}{E_s} \right)^{1/2} \left(\frac{260}{1000 + d} \right) \lambda \phi_c \left(f_c' \right)^{1/2} bd$$
$$\geq 0.1 \left(\frac{E_f}{E_s} \right)^{1/2} \lambda \phi_c \left(f_c' \right)^{1/2} bd \qquad (3)$$

respectively. The provisions were adopted from the Canadian standard for steel RC (CSA 1994), where λ = modification factor for concrete density = 1 for normal density concrete, and ϕ_c = resistance factor for concrete = 0.83 when no safety factor is applied to predict experimental results for comparison purposes with ACI, to account for different material safety factors for concrete.

The Institution of Structural Engineers (ISE 1999) recommends

$$V_c = 0.79 \left(100 \rho_{eff} \right)^{1/3} \left(\frac{400}{d} \right)^{1/4} \left(\frac{f_{cu}}{25} \right)^{1/3} bd \qquad (4)$$

regardless of beam depth, where f_{cu} = cube compressive strength of concrete. Similarly, the Japanese Society of Civil Engineers (JSCE 1997) proposes

$$V_c = 0.2 \left(100 \rho_{eff} \right)^{1/3} \left(\frac{1000}{d} \right)^{1/4} \left(f_{cd}' \right)^{1/3} bd \qquad (5)$$

where f_{cd} = cylinder compressive strength of concrete, with $(100\rho_{eff})^{1/3} \leq 1.5$, $(1000\,\mathrm{mm}/d)^{1/4} \leq 1.5$ being the size effect parameter based on Weibull statistical theory, and $(f'_c\,\mathrm{MPa}^2)^{1/3} \leq 3.6\,\mathrm{MPa}$.

3 EXPERIMENTAL STUDY

3.1 *Specimen design*

Four large-size beams were designed according to the ACI 440 guide (ACI 440 2006) and constructed. The cross section and reinforcement layout of Specimens I-1, I-2, II-1 and II-2, are illustrated in Figure 2. The overall height of 978 mm and effective depths were selected as replicate of typical full-scale softeyes, thereby providing dimensions where size effect typically becomes of concern in case of steel RC.

Flexural reinforcement consisting of Ø32 mm bars was designed to obtain a nominal GFRP reinforcement ratio $\rho_f = 0.59\%$ for Specimens I-1, I-2 and II-1. The value corresponds to $\rho_{eff} = 0.12\%$, thus below the minimum $\rho_{eff} = 0.15\%$ in studies reported in the literature and used to calibrate the ACI design equation (Tureyen & Frosch 2003), and yet representative of lower-bound real-case scenarios. Bundles of three Ø32 mm bars were used for Specimen II-2, as often encountered in practice, with $\rho_f = 0.89\%$ and $\rho_{eff} = 0.17\%$.

Since at least minimum shear reinforcement is required in most concrete structures, Specimen I-1 was designed to study size effect under such condition, as well as the effectiveness of shear reinforcement in providing postcracking strength. U-shaped Ø16 mm GFRP bars were arranged in the form of closed stirrups spaced at $s \approx s_{min} = 406\,\mathrm{mm}$ on-center, where $s_{min} = A_{fv}\,\min(f_{fv}, f_{fb})/(0.35b) \times \mathrm{mm}^2/\mathrm{N}$, with A_{fv} = area of transverse reinforcement within s_{min}, $f_{fv} = 0.004E_f$ to account for loss of aggregate interlock, and f_{fb} = strength of bent portion of FRP stirrups. The nominal shear strength, V_n, and the shear force associated with the nominal flexural strength, $V(M=M_n)$, were 253.8 kN and 373.1 kN, respectively, assuming $f'_c = 27.6\,\mathrm{MPa}$, and bar strength and axial modulus of 510.2 MPa and 40.7 GPa for the longitudinal reinforcement, and 655 MPa and 40.7 GPa for the shear reinforcement. Shear failure was expected.

Spacing of the shear reinforcement was reduced to 152 mm for Specimens I-2, II-1 and II-2 to further assess the effectiveness of shear reinforcement in providing the required postcracking strength, as well as mitigating the size effect on V_c (Bažant & Sun 1987). Specimen II-1 is replicate of two I-2 sections cast side-by-side and provides a valid counterpart to Specimen I-2, since beam width has negligible effect on V_c (Kani 1967, Sherwood et al. 2006). For Specimens I-2, II-1 and II-2, $V_n = 487.0\,\mathrm{kN}$, 974.0 kN and 1390.0 kN,

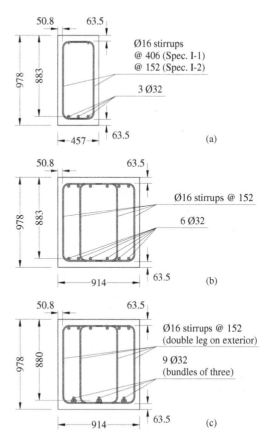

Figure 2. Cross section of GFRP RC Specimens I-1 and I-2 (a), II-1 (b), and II-2 (c). Dimensions in mm.

respectively, thus exceeding $V(M=M_n) = 373.1\,\mathrm{kN}$, 746.2 kN and 1105.3 kN. The expected failure was flexural due to rupture of the longitudinal reinforcement for Specimens I-2 and II-1, and concrete crushing for Specimen II-2.

The total length of each beam was 9.15 m. An anchorage length of 915 mm was provided past the end supports to prevent bar slip.

3.2 *Materials*

E-glass/vinyl ester GFRP bars were used to construct the reinforcement cages for the specimens. Average tensile strength and elastic modulus of eight Ø32 mm bar samples were $f_{fu} = 462.2\,\mathrm{MPa}$ and $E_f = 40.7\,\mathrm{GPa}$ for Specimens I-1 and I-2, and $f_{fu} = 510.2\,\mathrm{MPa}$ and $E_f = 38.0\,\mathrm{GPa}$ for Specimens II-1 and II-2. Average tensile strength and elastic modulus of six Ø16 mm stirrup samples were $f_{fu} = 690.0\,\mathrm{MPa}$ and $E_f = 40.2\,\mathrm{GPa}$, respectively.

Normal weight concrete was used, with average compressive strength $f'_c = 38.8\,\mathrm{MPa}$, 35.4 MPa,

29.0 MPa and 31.5 MPa for Specimens I-1, I-2, II-1 and II-2, respectively, as per cylinder tests performed in accordance with ASTM C 39 at the time of testing.

3.3 Test setup

The beams were tested in four-point bending, with shear span $a = 2743$ mm, thus providing a ratio $a/d = 3.1$ to obtain a lower-bound value for V_c, and constant moment region of 1829 mm. Assemblies including steel cylinders between flat or grooved plates were arranged at the supports in such a manner to simulate a simple support and a hinged support, respectively, and at the loading sections to simulate hinges. Plywood sheets of 6 mm thickness were interposed between the steel plates and the concrete surface at the supports and loading sections. The loads were applied via manually operated hydraulic actuators with capacity of 1780 kN, and measured with an 890 kN load cell placed under each concentrated load.

Direct current voltage transformer (DCVT) sensors and string transducers were used to measure displacements along the length of the beam, at the supports, and at the ends of the flexural reinforcement to capture bar slip. Several strain gauges were used to measure strain in the flexural reinforcement along the beams, in the concrete in compression at midspan, and in the stirrups along the shear spans.

4 RESULTS AND DISCUSSION

4.1 Concrete shear strength

It was found that the difference between the load at which the first inclined ($\geq 45°$) shear crack formed, V_{cr}, and the ultimate strength in six FRP and three steel RC beams without shear reinforcement lay within the 2%–10% range, except in one instance where the difference was 17% for a GFRP RC specimen with $\rho_{eff} = 0.19\%$ and $d = 360$ mm (Tureyen & Frosch 2002). In the analysis of the test results reported herein, the observed V_{cr} was taken as an acceptable indication of a lower bound for V_c (Frosch 2000). In general, relatively larger gaps in the load-deflection curve accompanied by strain increase in the stirrups were observed at the correspondent load levels.

Table 1 compares V_{cr} for each specimen, including the contribution of self-weight computed at a distance d from the center line of the supports, with the predicted V_c per the guidelines reported in Section 2. The average ratio V_{cr}/V_c is 1.11, 0.90, 1.17, 0.88 and 1.02 for ACI 440 (2006), CSA (2004), ISIS (2001), ISE (1999) and JSCE (1997), respectively. It is seen that the predictions per ACI 440 (2006) cannot be seen as unconservative, despite the absence of any size effect parameter in the formulation. The

seeming contradiction is explained by the presence of an implicit understrength factor introduced in Equation 1 (ACI 440 2006). The formula was calibrated by setting K as a constant (2/5) to define a simple and conservative design tool applicable irrespectively of the reinforcement material, being such factor theoretically expressed from equilibrium considerations as

$$K = \left(16 + \frac{4\sigma_m}{3\sqrt{f_c'}} \right)^{1/2} \tag{6}$$

where σ_m = concrete stress in extreme compression fiber of uncracked section (Tureyen & Frosch 2003). For values $\rho_{eff} \leq 0.8\%$, i.e. within a typical design range for under- and over-reinforced FRP RC members, the increase in K resulting from higher flexural stresses σ_m in cracked, lightly reinforced sections, determines significantly higher safety factors with respect to steel RC sections. Such result was desirable due to the relatively small number of test results available to validate the proposed design equation. This is clearly shown in Figure 3, where the ratio between the experimental and the theoretical V_c for Specimens I-1, I-2, II-1 and II-2 and other 52 FRP RC beams found in the literature (Zhao & Maruyama 1995, Deitz et al. 1999, Alkhrdaji et al. 2001, Yost et al. 2001, Tureyen & Frosch 2002, Razaqpur et al. 2004, El-Sayed et al. 2005, 2006) is plotted against ρ_{eff} and d. Since V_{cr} is considered for the results from the present investigation, the corresponding points are plotted including a +17% bar to indicate a reasonable upper bound for V_c.

Formation of the first inclined crack occurred at loads V_{cr} of 1.02, 1.19, 0.98 and 1.25 times the predicted V_c according to ACI 440 (2006) in Specimens I-1, I-2, II-1 and II-2, respectively. Even considering the +17% upper bound for V_c, such values are at least 24% smaller than one would expect at similar levels of ρ_{eff} when size effect is neglected, as illustrated in Figure 3a. This is further substantiated in Figure 3b, as clearly higher experimental versus predicted V_c ratios than that of the present study were reported in the literature for FRP RC specimens having $d \leq 376$ mm, irrespectively of the reinforcement ratio.

The presence of an implicit understrength factor in Equation 1 (ACI 440 2006) for relatively small values of ρ_{eff}, commonly encountered in FRP RC, can be also observed in Figure 4 in the case of large-size cross sections. The range of experimental to predicted V_c for the GFRP RC specimens in the present study tends to lie above that for other 24 large-size steel RC beams in the literature (Kani 1967, Taylor 1972, Kawano & Watanabe 1997, Yoshida 2000, Cao 2001, Angelakos et al. 2001, Lubell et al. 2004), which had d, a/d and ρ in the range 0.9–2.0 m, 2.8–3.0 and 0.50–2.72%, respectively, and were tested in either three- or four-point bending.

Table 1. Shear force at formation of primary shear crack and V_c per existing guidelines.

Specimen	I-1 kN	I-2 kN	II-1 kN	II-2 kN
Test	127.0	144.7	220.0	344.4
ACI (2006) – Eq. 1	124.8	121.8	223.7	275.7
CSA (2004) – Eq. 2	166.9	159.3	288.4	299.8
ISIS (2001) – Eq. 3	130.0	124.1	217.0	225.9
ISE (1999) – Eq. 4*	160.8	155.9	285.1	335.2
JSCE (1997) – Eq. 5	139.0	134.7	246.4	289.7

* Ratio of cylinder to cube compressive strength of 0.75 assumed.

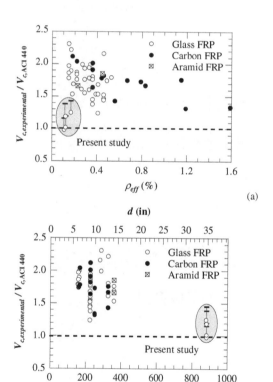

Figure 3. Comparison of experimental and theoretical (ACI 440 2006) V_c in FRP RC members in literature based on effective reinforcement ratio (a) and rectangular section depth (b).

According to the test results from Specimens I-1, I-2, II-1 and II-2, it also appears that the effect of transverse reinforcement on V_c is negligible for any practical purposes, in agreement with a classical assumption in steel and FRP RC design.

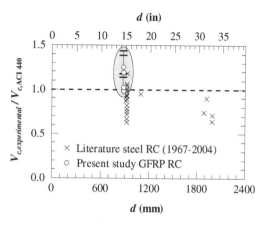

Figure 4. Comparison of experimental and theoretical (ACI 440 2006) V_c in large-size steel RC specimens in literature and GFRP RC specimens in present study.

4.2 Contribution of shear reinforcement

Additional shear strength to that of the concrete is provided by the transverse reinforcement upon its engagement once crossed by a diagonal crack. ACI 440 (2006) follows a common straightforward design approach where such contribution is expressed in the form

$$V_f = A_{fv} \min\left(f_{fv}, f_{fb}\right)\frac{d}{s} \qquad (7)$$

thereby assuming formation of the failure crack at a 45° angle. Since the ratio d/s is not truncated and rendered as an integer number, partial contribution of the stirrups is also admissible, although difficult to justify from a physical standpoint.

Figure 5 shows the load-displacement response of Specimens I-1, I-2 and II-1. Load is measured at the loading section in the beam half where failure occurred. Displacement is measured at the correspondent section of maximum deflection.

Propagation of the primary shear crack deep into the compression zone resulted in failure of Specimen I-1 at a load of 245.5 kN (Figure 6a). The contribution of the GFRP stirrups allowed to attain a total shear strength including self-weight at a distance d from the supports of 273.8 kN $> V_n = 263.1$ kN. The significant size effect on V_c is offset by the implicit understrength factor identified in the design equation, before application of the design strength reduction factor $\phi = 0.75$ to V_n (ACI 440 2006).

Specimens I-2, II-1 and II-2 were designed to fail in flexure, relying upon the additional strength provided by the closely spaced stirrups. In the case of Specimen I-2, the moment capacity was attained at a load of 341.2 kN, above $V(M = M_n) = 305.0$ kN and well

1081

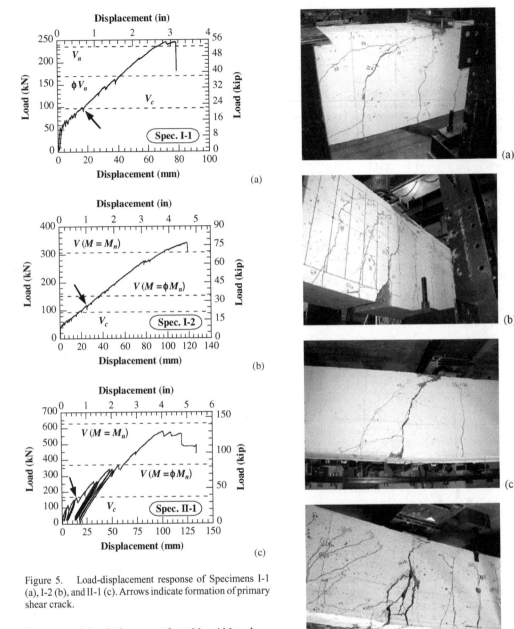

Figure 5.   Load-displacement response of Specimens I-1 (a), I-2 (b), and II-1 (c). Arrows indicate formation of primary shear crack.

Figure 6.   Photos of failure crack in Specimens I-1 (a), I-2 (b), II-1 (c) and II-2 (d).

in excess of the design strength at $M = \phi M_n$, where $\phi = 0.55$ for under-reinforced FRP RC sections (ACI 440 2006). Bar rupture occurred at 305 mm outwards from the nearby loading section (Figures 6b and 7a) due to the combination of tensile and shear stress. However, Specimen II-1 failed in shear compression at a load of 576.5 kN (Figure 5c), fairly close to its nominal strength in flexure. In fact, inspection of the flexural reinforcement upon removal of the surrounding concrete showed some delamination on the surface of the GFRP bars (e.g. in Figure 7b) a clear sign of

impending bar rupture. Ultimately, the stirrups contribution in Specimens I-2 and II-1 allowed to exceed the design strength. Nevertheless, the differences in failure mode between the two parent beams calls for

(a)

(b)

(c)

Figure 7. Photos of longitudinal Ø32 mm GFRP bars at failure section in Specimens I-2 (a), II-1 (b) and II-2 (c).

further investigation on the effectiveness of shear reinforcement in providing the assumed design strength contribution V_f.

Specimen II-2 reached its moment capacity at a load of 861.7 kN, again fairly close to $V(M = M_n) = 960.3$ kN and well above $V(M = \phi M_n) = 505.8$ kN. Failure mode was rupture of the longitudinal bars (Figures 6d and 7c), which is not surprising given the GFRP reinforcement ratio of 0.95 times the value of balanced failure as computed using the material properties determined experimentally (1.05 in original design), and the concrete strain in the equivalent compression stress block at failure typically greater

than the 3000 $\mu\varepsilon$ assumed in design. Further research is needed to characterize the influence, if any, of bundled reinforcement on the structural response of FRP RC members.

5 CONCLUDING REMARKS

Preliminary results have been presented from a pilot investigation aimed at assessing the current ACI 440 (2006) shear design provisions in the case of large-size GFRP RC members, which are increasingly being used worldwide in geotechnical applications, such as softeyes for tunnel excavation, and retaining walls. The following conclusions can be drawn:

1. The concrete shear strength appears to be strongly affected by size effect. With respect to scaled counterparts in the literature, strength reduction of at least 24% has been observed in beams with effective depth of about 880 mm and FRP reinforcement ratio of 0.59% and 0.89%, commonly encountered in practice due to the relatively small axial modulus of GFRP bars.
2. Negligible difference on concrete shear strength has been noted in sections with increased amount of shear reinforcement, in agreement with a classical assumption in steel and FRP RC design.
3. The definition of a simple and conservative design equation for concrete shear strength introduced an implicit understrength factor that offsets size effect. At present, adoption of less conservative approaches should not be considered without explicitly addressing size effect.

The conclusions on size effect based on this study must be further substantiated with results from experiments on large-size FRP RC beams without shear reinforcement, which are ongoing as part of a more extensive research program.

Further research is also needed to evaluate the conservativeness of the design provisions for the stirrups contribution to the shear strength, and for the use of bundled longitudinal FRP reinforcement.

ACKNOWLEDGEMENTS

The financial support of the NSF I/UCRC "Repair of Buildings and Bridges with Composites" (RB^{2}C), and the assistance of the Center's industry member Hughes Brothers, Inc. in supplying the FRP reinforcement are gratefully acknowledged. Special thanks are due to Travis Hernandez, Jason Cox, Preeti Shirgur and the personnel of the UMR Structures Laboratory for their assistance.

REFERENCES

Alkhrdaji, T., Wideman, M., Belarbi, A. & Nanni, A. 2001. Shear strength of GFRP RC beams and slabs. In J. Figueiras, L. Juvandes & R. Furia (eds.), *Composites in Construction – CCC 2001; Proc. intern. conf., Porto, Portugal, 10–12 October 2001*: 409–414.

American Concrete Institute (ACI) Committee 318 2005. *Building code requirements for structural concrete – ACI 318-05*. Farmington Hills, MI: ACI.

American Concrete Institute (ACI) Committee 440 2006. *Guide for the design and construction of structural concrete reinforced with FRP bars – ACI 440.1R-06*. Farmington Hills, MI: ACI.

Angelakos, D., Bentz, E.C. & Collins, M.P. 2001. Effect of concrete strength and minimum stirrups on shear strength of large members. *ACI Structural Journal* 98(3): 290–300.

Bažant, Z.P. & Kim, J.-K. 1984. Size effect in shear failure of longitudinally reinforced beams. *ACI Journal* 81(5): 456–468.

Bažant, Z.P. & Sun, H.-H. 1987. Size effect in diagonal shear failure: influence of aggregate size and stirrups. *ACI Materials Journal* 84(4): 259–272.

Bažant, Z.P. & Yu, Q. 2005a. Designing against size effect on strength of reinforced concrete beams without stirrups: I. Formulation. *Journal of Structural Engineering* 131(12): 1877–1885.

Bažant, Z.P. & Yu, Q. 2005b. Designing against size effect on strength of reinforced concrete beams without stirrups: II. Verification and calibration. *Journal of Structural Engineering* 131(12): 1886–1897.

Canadian Standard Association (CSA) 1994. *Design of concrete structures – CAN/CSA-A23.3-94*. Mississauga, Canada: CSA.

Canadian Standard Association (CSA) 2004. *Design and construction of building components with fibre reinforced polymers – CAN/CSA-S806-02*. Mississauga, Canada: CSA.

Cao, S. 2001. Size effect and the influence of longitudinal reinforcement on the shear response of large reinforced concrete members. *MASc Thesis*. Toronto, Canada: Department of Civil Engineering, University of Toronto.

Collins, M.P. & Kuchma, D. 1999. How safe are our large, lightly reinforced concrete beams, slabs, and footings?. *ACI Structural Journal* 96(4): 482–490.

Collins, M.P., Mitchell, D., Adebar, P. & Vecchio, F.J. 1996. A general shear design method. *ACI Structural Journal* 93(1): 36–45.

Deitz, D.H., Harik, I.E. & Gesund, H. 1999. One-way slabs reinforced with glass fiber reinforced polymer reinforcing bars. In C.W. Dolan et al. (eds.), *Fiber Reinforced Polymer Reinforcement for Reinforced Concrete Structures – FRPRCS-4, ACI SP-188; Proc. int. conf., Baltimore, MD, 31 October–5 November 1999*: 279–286. Farmington Hills, MI: ACI.

El-Sayed, A.K., El-Salakawy, E.F. & Benmokrane, B. 2005. Shear strength of one-way concrete slabs reinforced with FRP composite bars. *Journal of Composites for Construction* 9(2): 147–157.

El-Sayed, A.K., El-Salakawy, E.F. & Benmokrane, B. 2006. Shear strength of FRP-reinforced concrete beams without transverse reinforcement. *ACI Structural Journal* 103(2): 235–243.

Frosch, R.J. 2000. Behavior of large-scale reinforced concrete beams with minimum shear reinforcement. *ACI Structural Journal* 97(6): 814–820.

Institution of Structural Engineers (ISE) 1999. *Interim guidance on the design of reinforced concrete structures using fibre composite reinforcement*. London, UK: ISE.

Intelligent Sensing for Innovative Structures (ISIS) Canada Research Network 2001. *Reinforcing Concrete Structures with Fibre Reinforced Polymers (FRPs) – ISIS Design manual No. 3*. Winnipeg, Canada: ISIS.

Japan Society of Civil Engineers (JSCE) 1997. *Recommendation for Design and Construction of Concrete Structures using Continuous Fiber Reinforcing Materials*. Tokyo, Japan: JSCE.

Kani, G.N.J. 1967. How safe are our large reinforced concrete beams? *ACI Journal* 64(3): 128–141.

Kawano, H. & Watanabe, H. 1997. Shear strength of reinforced concrete columns – Effect of specimen size and load reversal. In *Proc. 2nd Italy-Japan Workshop on Seismic Design and Retrofit of Bridges, Rome, Italy, 27–28 February 1997*: 141–154.

Lubell, A., Sherwood, T., Bentz, E. & Collins, M. 2004. Safe shear design of large, wide beams. *Concrete International* 26(1): 66–78.

Nanni, A. 2003. North American design guidelines for concrete reinforcement and strengthening using FRP: principles, applications and unresolved issues. *Construction and Building Materials* 17(6–7): 439–446.

Razaqpur, A.G., Isgor, B.O., Greenaway, S. & Selley, A. 2004. Concrete contribution to the shear resistance of fiber reinforced polymer reinforced concrete members. *Journal of Composites for Construction* 8(5): 452–460.

Sherwood, E.G., Lubell, A.S., Bentz, E.C. & Collins, M.P. 2006. One-way shear strength of thick slabs and wide beams. *ACI Structural Journal* 103(6): 794–802.

Shioya, T., Iguro, M., Nojiri, Y., Akiyama, H. & Okada, T. 1989. Shear strength of large reinforced concrete beams. *Fracture Mechanics: Application to Concrete – ACI SP 118-12*: 259–279.

Taylor, H.P.J. 1972. Shear strength of large beams. *Journal of the Structural Division* 98(ST11): 2473–2490.

Tureyen, E.J. & Frosch, R.J. 2002. Shear tests of FRP-reinforced concrete beams without stirrups. *ACI Structural Journal* 99(4): 427–433.

Tureyen, E.J. & Frosch, R.J. 2003. Concrete shear strength: another perspective. *ACI Structural Journal* 100(5): 609–615.

Yoshida, Y. 2000. Shear reinforcement for large lightly reinforced concrete members. *MASc Thesis*. Toronto, Canada: Department of Civil Engineering, University of Toronto.

Yost, J.R., Gross, S.P. & Dinehart, D.W. 2001. Shear strength of normal strength concrete beams reinforced with deformed GFRP bars. *Journal of Composites for Construction* 5(4): 268–275.

Zhao, W., Maruyama, K. & Suzuki, H. 1995. Shear behavior of concrete beams reinforced by FRP rods as longitudinal and shear reinforcement. In L. Taerwe (ed.), *Non-Metallic (FRP) Reinforcement for Concrete Structures – FRPRCS-2; Proc. intern. symp., Ghent, Belgium, 23–25 August 1995*: 352–359. London: E&FN Spon.

Fracture Mechanics of Concrete and Concrete Structures – Design, Assessment and
Retrofitting of RC Structures – Carpinteri, et al. (eds)
© 2007 Taylor & Francis Group, London, ISBN 978-0-415-44616-7

An experimental analysis on the time-dependent behaviour of a CFRP retrofitting under sustained loads

F. Ascione
Department of Civil Engineering, University of Rome Tor Vergata, Rome, Italy

G. Mancusi
Department of Civil Engineering, University of Salerno, Fisciano, Italy

ABSTRACT: Over the past few years, the strengthening of concrete structures with fibre reinforced polymer
(FRP) composites has become increasingly user-friendly among engineers worldwide. Within this field one of
the most relevant topics concerns the durability and the reliability of such materials, depending strongly on their
viscous properties. FRPs, in fact, are made up purely of elastic fibres embedded in highly creep sensitive matrices.
Most of the data available in literature about FRP creep tests deal with aeronautical/aerospace or mechanical
applications, while only few concern civil purposes. In this paper the authors present some experimental results
relative to the time-dependent behaviour of FRP pultruded laminates under sustained loads. The experimental
set-up simulates the actual bonding conditions occurring in typical retrofitting interventions. The results present
a part of a two year long experimental programme going on at Salerno University (Italy).

1   INTRODUCTION

Over the past few years, the strengthening of concrete
structures with fibre reinforced polymer (FRP) composites has become increasingly user-friendly among
engineers worldwide.

Nowadays, composite materials are used by civil
engineering for structural rehabilitation together with
traditional materials, like steel.

Within this field there is a great interest in investigating the durability and the reliability of such
materials, mainly with respect to their viscous properties. FRPs, in fact, are made up purely of elastic fibres
embedded in highly creep sensitive matrices.

All current international guidelines for the design of
FRP strengthening interventions assess the importance
of the above problem. In fact several limitations on
FRP stresses at Service Conditions are usually introduced in order to control viscous effects (ACI Committee 440 2000, CEB-FIP 2001, CNR-DT 200/2004,
English translation 2006).

Both theoretical and experimental investigations
on the viscous behaviour of composites used in
the aeronautical and naval fields can be found in
many technical papers or textbooks (Barbero &
Harris 1998, Dutta & Hui 1997, Maksimov & Plume
2001, Pang et al. 1997, Petermann & Schulte 2002,
Scott & Zureick 1998). The common aim is to

introduce constitutive laws accounting for different stress levels as well as different environmental
conditions.

Instead, few studies concerning the time-dependent
behaviour of composite materials used in civil engineering fields are available in literature. In these types
of applications the most relevant phenomenon is the
so-called secondary creep.

Many theoretical investigations designed to study
the effects of secondary creep on the long-term
response of Reinforced Concrete (RC) beams strengthened with FRP laminates have been developed by the
authors. They show the possibility of relevant stresses
migration from the FRP reinforcement to the former
core of the RC beam as time passes. The consequent
increase of the stresses level in the RC beam may compromise the efficacy of the strengthening technique
(Berardi et al. 2003, Ascione et al. 2004).

Therefore it seemed very interesting to further investigate by experimental analyses, the time-
dependent behaviour exhibited by FRP laminates
commonly used in civil applications.

The aim of this paper is to present the experimental results of creep testes carried out on some CFRP
retrofitting systems.

Such results are part of a two year long experimental
programme currently underway at the University of
Salerno.

2 CREEP BEHAVIOUR

When dealing with aged concrete structures, the rate of the viscous effects in the concrete can be considered equal to zero. Consequently, only FRP creep deformations have to be taken into consideration. A similar problem occurs when dealing with steel-concrete composite structures. In fact, in this case, creep phenomena concerns the concrete component, while the steel component does not flow.

The viscous behaviour of FRP materials depends on many factors, like the matrix properties, type of fibre, fibre volume fraction, fibre orientation, load history, environmental conditions (temperature and humidity).

From a theoretical point of view FRP laminates can be classified as orthotropic viscous materials. Consequently, the long term behaviour analysis should require the characterization of all mechanical properties over time.

In particular, with reference to the retrofitting of RC members, the constitutive laws to be investigated is essentially one-dimensional (along the natural direction lying on the longitudinal axis). This allows us to simplify the experimental investigation by the recourse to simple creep tests: a sample is subjected to a constant tensile force and its elongation is monitored over time (Fig. 1). In the first part of the experiment the well known primary creep is dominant. The initial elastic elongation is followed by fast growing deformations. During the next phase the elongation continues on with a constant rate over a longer time period (secondary creep).

A third phase (terziary creep) can also occur, under very high stress or temperature levels: strains grow at an increasing rate until the sample breaks.

The long-term prediction of such effects is usually performed by introducing linear viscoelastic one-dimensional models. Such models are based on either mechanical analogies or experimental data (Barbero & Harris 1998, Dutta & Hui 1997, Maksimov & Plume 2001, Pang et al. 1997, Petermann & Schulte 2002, Scott & Zureick 1998).

Until now, no viscous constitutive law has a widespread validity for composite materials, except in restricted applications of aeronautical, naval or mechanical fields. To characterise the viscous properties of FRPs within the context of civil interventions further experimental investigations are required.

3 EXPERIMENTAL SET-UP

As mentioned above, in order to study the creep behaviour of CFRP laminates, a long-term (two year) experimental programme is being carried out by the authors.

The test involves a system composed of a thin CFRP plate bonded to the top flange of a simply supported titanium beam (Ti-6Al-4V). More details are shown in Fig. 2. By this way the actual mechanical conditions, occurring in retrofitting concrete structures, have been simulated. In fact, the behaviour of the titanium alloy beam is not time-dependent, as is the case of aged concrete structures. Moreover, like FRP, the titanium alloy here considered exhibits a linear elastic response up to very large normal strains. As a consequence, it is possible to evaluate the instantaneous stresses inside the materials by measuring corresponding strains.

The static scheme adopted by the authors is represented in Fig. 2. Its total length is equal to $7300 \, mm = a + 2 \cdot b$, where a is the length of the titanium alloy beam (3500 mm), and b (=1900 mm) denotes the length of the two steel lateral arms. Such arms are connected to the titanium alloy beam by bolted joints (Fig. 3). The bond CFRP/beam has been made by using the adhesive (trade name: Loctite Multibond 330). Finally, $a' = 2900 \, mm$, is the length of the FRP reinforcement: a *CarboDur H514* plate, manufactured by Sika Group. The plate is characterised by a fibre volume fraction higher than 70%, along the longitudinal axis.

Dead loads have been applied at the free ends of the two lateral arms, symmetrically, according to the scheme in Fig. 2.

In particular, $1000 \, mm \times 200 \, mm \times 10 \, mm$ steel sheets, each one about 0,20 kN (Fig. 4), were used.

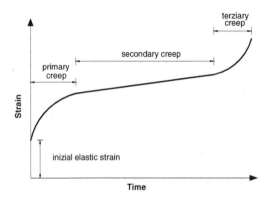

Figure 1.   Creep deformations over time.

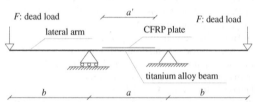

Figure 2.   Experimental scheme.

It is easy to verify that (Fig. 2):

i) a constant bending moment ($M = - F \cdot b$) occurs within the supports;
ii) no shear stresses arise.

The cross-section of the reinforced beam (titanium+CFRP) is shown in Fig. 5 (length unit: mm). The main mechanical properties of the titanium alloy, CFRP and adhesive are described below (Table 1).

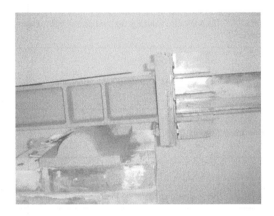

Figure 3. Bolted joints.

Figure 4. Dead loads applied by using steel sheets.

Table 1. Certified mechanical properties (stresses unit: N/mm²).

	Ti-6Al-4V	CFRP	Glue layer
Normal elastic modulus	110000	≥300000	–
Yield normal stress	790 (20°C)	–	–
Ultimate normal stress	895 (20°C)	1450	–
Ultimate shear stress	–	–	≥16,5
Ultimate normal strain	0,1000	≥0,0045	–

The experimental programme has been divided into two phases, each one performed under controlled temperature (20°C ± 1°C) and humidity (50% ± 5%):

– 1st phase: generation of a low stress level in the plate (20% of CFRP tensile strength). Until now this phase, 6-month long, has been completed, and the results obtained by the authors have been recently presented (Ascione & Mancusi, 2006).
– 2nd phase: generation of a higher stress level (70% of CFRP tensile strength). This phase, 18-month long, is still underway, and the results already obtained by the authors, relative to the first six months, are here presented for the first time.

The strain state evolution is monitored by continuous data acquiring hardware/software system equipped with 35 strain-gauges. They are applied both to the titanium beam as well as to the CFRP plate within the cross-section (Fig. 6: positions a, e, f for Ti-6Al-4V; positions b, c, d for CFRP) and along the

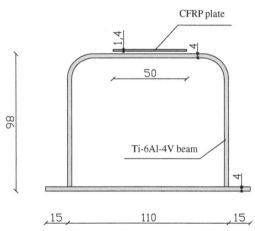

Figure 5. Cross-section (length unit: mm).

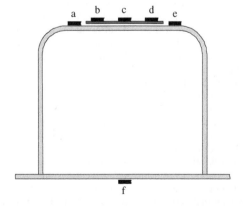

Figure 6. Strain gauges position (mid-span cross-section).

Table 2. Data acquired during the 1st phase.

Elapsed time [days]	Strain gauges position					
	a	b	c	d	e	f
	$[\varepsilon \times 10^6]$					
$t_0 = 0$	848	876	886	872	849	−959
$t_1 = 180$	858	848	867	849	857	−962

Table 3. Stress-strain analysis – 1st phase.

Elapsed time [days]	N_{Ti} [N]	N_{CFRP} [N]	σ_{CFRP} [N/mm^2]	ε_{CFRP} [$\varepsilon \times 10^6$]	E_{CFRP} [N/mm^2]
$t_0 = 0$	−23746	23746	339	878	386105
$t_1 = 180$	−23256	23256	332	855	388304*

*This value results a bit higher than that one at time t_0 only due to experimental error.

Table 4. Data acquired during the 2nd phase (still underway).

Elapsed time [days]	Strain gauges position					
	a	b	c	d	e	f
	$[\varepsilon \times 10^6]$					
$t_2 = 0$	2582	2685	2691	2674	2613	−2933
$t_3 = 15$	3903	1057	1149	1019	3982	−3651
$t_4 = 180$	3939	1011	1077	955	4015	−3638

Table 5. Stress-strain analysis – 2nd phase (still underway).

Elapsed time [days]	N_{Ti} [N]	N_{CFRP} [N]	σ_{CFRP} [N/mm^2]	ε_{CFRP} [$\varepsilon \times 10^6$]	E_{CFRP} [N/mm^2]
$t_2 = 0$	−72406	72406	1034	2683	385389
$t_3 = 15$	−27104	27104	387	1075	360000
$t_4 = 180$	−22693	22693	324	1014	319527

longitudinal axis. Due to the linear elastic response of both materials, stresses can be easily related to strains. Furthermore, an optometric system, OptoNCDT 1401 by $\mu\varepsilon$ Micro Epsilon, is used in order to measure the mid-span deflection over time (in the same position f above mentioned).

Once the normal stresses corresponding to the measured strains (positions a, e and f) have been calculated, it is possible to evaluate the resultant axial force, N_{Ti}, attained within Ti-6Al-4V alloy.

Assuming the supports are frictionless, by equilibrium along the longitudinal axis, the instantaneous axial force within the CFRP plate, N_{CFRP}, is equal to $-N_{Ti}$.

As a consequence of the above hypothesis, the average instantaneous normal stress in the CFRP plate, σ_{CFRP}, can be related to the measured average normal strain $\varepsilon_{CFRP} = (\varepsilon_b + \varepsilon_c + \varepsilon_d)/3$, being ε_i the normal strain at position i ($i = b, c, d$).

4 EXPERIMENTAL RESULTS

As referred above, in this section we discuss some experimental results obtained during the first six months of the 2nd phase, which is still underway at Material and Structures Testing Laboratoty of the Salerno University. Such a phase is characterised by an initial stress level in the CFRP plate equal to about 70% of the CFRP tensile strength. When the 2nd phase was going to start, viscous effects previously provoked in the 1st phase had already affected the plate. Such effects, however, were very negligible as shown below, in Table 2, where some of the data acquired during the 1st phase are summarized (strain unit: $\mu\varepsilon$). They refer to the mid-span cross-section.

In particular, data concerning a, e and f straingauges (Fig. 6 – Table 2) allow us to evaluate the internal axial force, N_{Ti}, within the titanium alloy component. The above assumption of linear-elastic behaviour is utilized.

On the other hand, data concerning b, c and d straingauges allow us to evaluate the average value, ε_{CFRP}, of the instantaneous normal strain in the CFRP plate. The internal axial force acting on the CFRP plate can be assumed equal to $N_{CFRP} = -N_{Ti}$.

Consequently, the average stress value in the CFRP plate can be evaluated as follows:

$$\sigma_{CFRP} = N_{CFRP} / A \qquad (1)$$

where symbol A denotes the FRP cross-section area.

In Table 3 the last column presents the instantaneous values of the following ratio:

$$E_{CFRP} = \sigma_{CFRP} / \varepsilon_{CFRP}. \qquad (2)$$

As can be seen, during the whole 1st phase, the viscous effects were very negligible.

After six months the 1st phase had begun, dead loads were increased, and held constant until today. In Table 4 data acquired during the 2nd phase at three different times are summarized: when 2nd phase started (t_2), 15 days after (t_3), 180 days after (t_4). Likewise in the case of Table 2, they refer to the mid-span cross-section.

Table 5 summarises the correspondent stress-strain analysis. In this case, the viscous effects are more relevant, due to the rheological behaviour of the glue layer

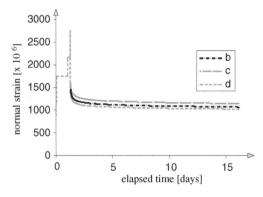

Figure 7a.   Strain analysis within CFRP – range $[t_2, t_3]$.

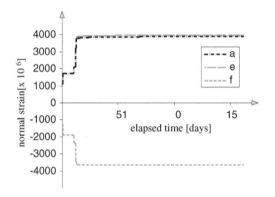

Figure 7b.   Strain analysis within Ti-6A1-4V – range $[t_2, t_3]$.

at FRP to metal interface. In fact, making reference to the creep of the glue layer, it is possible to understand the detected behaviour: after a certain time elapsed, the experiment shows normal strains in the CFRP plate (measurements made at b, c and d positions in Fig. 6) decreased below the values attained in titanium alloy (measurements made at a, e and f positions in Fig. 6).

The following Figs. 7a, b show the response of the system over the time range $[t_2, t_3]$.

Moreover, in the previous Figs. 7a,b close to the time origin, the plots present many discontinuities, corresponding to load steps. In fact, the loading of the retrofitting system was realized by the authors in successive steps, using the above mentioned steel sheets, and took about 24 hours to complete.

Finally, changes in term of the vertical deflection of the system at mid-span cross-section were also measured by the optometric sensor, above mentioned and placed at f position (Fig. 6).

The results of Table 6 confirm the main importance of the first period, $[t_2, t_3]$, on the long-term behaviour, when very high initial internal stresses are present in the CFRP plate.

Table 6.   Changes in term of vertical deflection at mid-span (%).

Time	%
t_3	39,4
t_4	43,0

5   CONCLUSIONS

The experimental results described in the previous section show the investigated CFRP plating exhibits small viscous effects, if the initial stresses are less than 20–25% of the FRP tensile strength.

On the contrary, fast changes in term of system mechanical response (beam + CFRP plate) can be highlighted, if the CFRP internal stresses increase up to 70–75% of the corresponding tensile strength, as it happened during the experiment (2nd phase).

The whole viscous phenomenon, observed in such a phase, occurred within the first 15 days. During this period, a fast decrease of the stresses and strains in the CFRP was observed, while in the subsequent time interval, $[t_3, t_4]$, only negligible changes in term of stresses and strains occurred.

In other words, the viscous effects, detected during the 2nd phase, were relevant and expired within a few days.

The results obtained, suggested to the authors to stop the 2nd phase which was still underway, although it was initially planned over an 18-month period.

Within the next 12 months, similar investigations, under stress levels in the CFRP equal to about 30–50% of the tensile strength will be performed. Such percentage values correspond to the maximum stress levels suggested in the new guide-lines CNR-DT 200/2004, recently edited by Italian National Research Council. An English translation of the guide-lines is available (2006).

The results of the investigation will contribute to assess the safety of the design rules given in such guide-lines.

REFERENCES

ACI Committee 440 2000. *Guide for the design and construction of externally bonded FRP systems for strengthening concrete structures.*

Barbero, E. & Harris, J. S. 1998. Prediction of creep properties from matrix creep data.*Journal of Reinforced Plastics and Composites* 17(4).

Berardi, V. P. & Giordano, A. & Mancusi, G. 2003. Modelli costitutivi per lo studio della viscosità nel placcaggio strutturale con FRP. *XXXII AIAS Conference, Salerno, CD-ROM Proceeding.*

Ascione, L. & Berardi, V. P. & Mancusi, G. 2004. Time-depending behaviour under sustained loads of RC beams externally plated with FRP laminates, IMTCR Conference, Lecce, CD-ROM Proceeding.

Ascione, F. & Mancusi, G. 2006. Long-Term Behaviour of CFRP Laminates: An Experimental Study, CICE Conference, Miami, CD-ROM Proceeding.

CEB-FIP 2001. *Externally bonded FRP reinforcement for RC structures.*

CNR-DT 200/2004 (English translation, 2006). *Guide for the Design and Construction of externally Bonded FRP Systems for Strengthening Existing Structures. Materials, RC and PC structures, Masonry structures.* National Research Council of Italy.

Dutta, K. & Hui, D. 1997. Integrating fire-tolerant design and fabrication of composite ship structures. *Interim report University of New Orleans.*

Maksimov, R. D. & Plume, E. 2001. Long-Term creep of hybrid aramid/glass fiber-reinforced plastics. *Mechanics of Composite Materials* 37(4).

Pang, F. & Wang, C.H. & Baghgate, R.G. 1997. Creep response of woven composites and effect of stitching. *Journal of composites science and technology* 57.

Petermann, J. & Schulte, K. 2002. The effects of creep and fatigue stress ratio on the long-term behavior of angle-ply CFRP. *Composite Structures* 57: 205–210.

Scott, D. & Zureick, A. 1998. Compression creep of a pultruded e-glass/vinylester composite. *Composites Science and Technology* 85: 1361–1369.

Fracture Mechanics of Concrete and Concrete Structures – Design, Assessment and Retrofitting of RC Structures – Carpinteri, et al. (eds)
© *2007 Taylor & Francis Group, London, ISBN 978-0-415-44616-7*

A theoretical model for intermediate debonding of RC beams strengthened in bending by FRP

C. Faella
Dipartimento di Ingegneria Civile, Università di Salerno, Italy

E. Martinelli
Dipartimento di Ingegneria Civile, Università di Salerno, Italy

E. Nigro
Dipartimento di Analisi e Progettazione Strutturale, Università di Napoli "Federico II", Italy

ABSTRACT: In the present paper, a mechanical model accounting for the interface stress-slip relationship is presented with the aim of simulating the behavior of RC beams strengthened by externally bonded FRP systems. Finite element implementation and validation of such a model will be outlined with the aim of pointing out the capability of the proposed procedure of reproducing both end and intermediate debonding phenomena which often affect the overall behavior of FRP-strengthened beams. Behavioral observations will be pointed out for emphasizing the role of the key parameters controlling premature (fracture energy, steel yielding strain, reinforcement amount, etc.) failure due to intermediate debonding.

1 INTRODUCTION

Improving flexural strength in RC members by bonding FRP laminates at their soffit is one of the most common strengthening techniques which can face either a rebar section reduction due to steel corrosion or an increase of actions applied upon the member. Nevertheless the possible premature failure due to debonding between adhesive layer and concrete, which can occur at the beam end (end debonding) or in the cracked zone (intermediate debonding), is one of the failure modes to be prevented.

In the last years huge research efforts have been carried out for understanding the behavior of reinforced concrete beams strengthened by externally bonded FRP. Many of these studies are focused on the formulation of mechanical models, either analytical or numerical in their possible implementation, able to simulate the complex stress and strain distribution throughout the FRP-to-concrete adhesive interface.

Different contributions about this topic have been summarized and compared by Chen & Teng (2001). A simplified model for evaluating interface stresses in FRP (or even steel) strengthened beams has been proposed by Roberts (1988); simplified equations for evaluating shear and normal stresses throughout the FRP-to-concrete interface have been provided by assuming linear elastic behavior of the adhesive interface. Similar relationships, even obtained under simplified hypotheses for interface behavior, have been also provided by Malek et Al. (1996).

The above mentioned papers mainly deal with interface stress distribution in the elastic range, which is an aspect mainly relevant in service conditions. Premature loss of bonding between FRP and concrete can only be simulated by considering a suitable non-linear relationship between interface stresses and strains. Holzenkaempfer (1994) proposed a bi-linear relationship between shear stresses and interface slips; based on such model, Taljsten (1997) determined the expression of the ultimate bearing capacity of FRP-to-concrete joints. Further studies have been devoted to end and intermediate debonding, but nowadays definitive solutions have not yet been reached. Nevertheless, several proposals for quantifying the maximum axial strain developed in FRP at debonding have been derived from simplified mechanical models and calibrated making use of the experimental results available in the scientific literature. Some of the findings of these studies have been also utilized in the following Code of Standards issued in various countries:

– fib bulletin 14 (2001) in Europe;
– ACI 440 (2002) in the United States;
– JSCE Recommendations (2001) in Japan;
– Italian Code CNR DT 200 (2004).

A mechanical model considering non-linear stress-strain relationships for concrete, steel and FRP-to-concrete interface has been already presented by the authors (Faella et Al., 2006), with the aim of simulating the behavior of RC beams strengthened by externally bonded FRP plates. The model, whose hypotheses will be shortly summarized in the following, has been also validated in the same paper by considering almost thirty experimental results obtained by different authors; the experimental-to-numerical comparison pointed out that the model is able to capture both the overall behavior and the failure load. The axial strain developed in FRP at U.L.S. is one of the key parameters to be observed in the analyses; in fact, debonding can occur at the FRP-to-concrete interface reducing the beam bearing capacity with respect to the one obtained for FRP tearing-rupture. The mentioned numerical model is able to assess the FRP effective axial strain at debonding, allowing to point out the key parameters which control debonding failure.

2 THE THEORETICAL MODEL

In the present section an analytical model is presented for simulating the behavior of FRP-strengthened RC beams. Moreover, the formulation of a finite element is presented and a secant procedure for non-linear analysis is also described.

2.1 Analytical formulation within the linear range

A theoretical model can be formulated for simulating the mechanical behavior of RC beams externally strengthened by means of FRP materials. The following assumptions are made:

– the RC beam behaves according to the Bernoulli theory, while FRP plate flexural stiffness is neglected and only axial forces are considered;
– the interaction between the two members is realized through a continuous, linear behaving and thicknessless medium;
– equal transverse displacements, i.e. deflections, occur in the connected members.

The partial interaction between beam and FRP results in an interface slip s which can be expressed as follows if the above hypotheses apply:

$$s = u_{f,sup} - u_{c,inf} =$$
$$= u_f - \varphi \cdot y_{f,sup} - \left(u_c + \varphi \cdot y_{c,inf}\right) = u_f - u_c - \varphi \cdot d \quad (1)$$

with the symbols reported in Figure 1; in particular d is the distance between RC beam and FRP plate centroids. However, assuming the Bernoulli hypothesis for the RC beam, the following equivalence

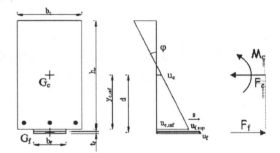

Figure 1. Transverse section of FRP-strengthened RC beam.

between the external bending moment M and the forces represented in Figure 1 can be stated:

$$M = \chi \cdot EI_c + F \cdot d \quad (2)$$

where χ is the curvature and EI_c is the flexural stiffness of unstrengthened RC beam cross section.

The longitudinal shear force per unit length F' depends linearly on the interface slip s:

$$F' = k \cdot s = k_a b_f \cdot s \quad (3)$$

k being the stiffness constant characterizing shear connection; it can be obtained by multiplying the adhesive slip modulus (namely, transverse stiffness) k_a and the width b_f of the adhesive layer.

Using the compatibility equation (1), the equilibrium equation (2) and the interface relationship (3), the following second-order differential equation in terms of curvature may be obtained:

$$\chi'' - \alpha^2 \chi = -\alpha^2 \frac{M}{EI_{full}} - \frac{q}{EI_c} \quad (4)$$

where EI_{full} represents the flexural stiffness of the overall cross-section when no interface slips occur and can be defined as follows if the above mentioned hypotheses apply:

$$EI_{full} = EI_c + \frac{E_f A_f \cdot EA_c}{E_f A_f + EA_c} d^2 = EI_c + EA^* d^2, \quad (5)$$

the term α in equation (4) is defined as follows:

$$\alpha^2 = \frac{k}{EA^*} \cdot \frac{EI_{full}}{EI_c}. \quad (6)$$

The equations briefly outlined above could even be obtained by simplifying the one formulated within the well-known Newmark theory, widely utilized for steel-concrete composite beams (Faella et Al., 2002), and neglecting the flexural stiffness of the bottom element connected to RC beam.

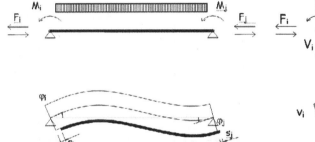

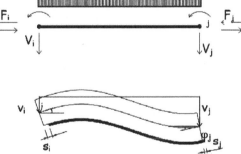

Figure 2. Nodal force and displacement components: simply supported-beam for force based element.

Figure 3. Nodal force and displacement components: unrestrained beam for displacement-based element.

2.2 Outlines of the finite element formulation within the mechanically non-linear range

A finite element can be formulated (among the other possible choices) implementing the exact solution of the structural problem for carrying out linear analyses of RC beams externally strengthened by FRP plates (Faella et Al., 2006). According to the mentioned approach, a formally "force based" finite element can be derived by directly solving equation (4) in order to obtain the various terms of the flexibility matrix **D** and the vector δ_0 of nodal displacements due to distributed loads. The usual relationship of flexibility-based finite elements can be obtained for the simply supported FRP-strengthened beam:

$$\delta = DX + \delta_0 , \qquad (7)$$

X and δ being the vectors of nodal forces and displacements, respectively, whose four components are represented in Figure 2.

The usual displacement-based relationship which relates nodal forces and displacement for the unrestrained FRP-strengthened beam (Figure 3) can be obtained by inverting the flexibility matrix and completing it with nodal shear forces as explained in the mentioned paper (Faella et Al, 2002). The usual relationship between nodal force vectors **Q, Q$_0$**, and nodal displacement vectors **s** (both characterized by the six components represented in Figure 3) can be written by means of the stiffness matrix **K**:

$$Q = Ks + Q_0 . \qquad (8)$$

Closed-form expression for both stiffness matrix and equivalent nodal force vector have been proposed in Faella et Al. (2000).

Non linear behavior of the materials of the FRP strengthened RC beams can be easily introduced within the FE procedure. Several non-linear phenomena have to be considered for simulating the premature failure possibly due to FRP-to-concrete debonding which can occur in an intermediate section or at the

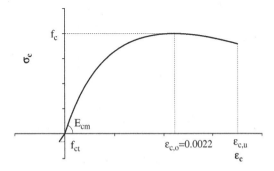

Figure 4. Non-linear stress-strain law for concrete.

FRP cut-off section. The first one deals with the overall behavior of concrete in compression and tension; the rational formula proposed by Saenz (1964) is adopted for concrete in compression while a simple linear relationship up to the tensile strength is considered for concrete in tension (Figure 4).

Moreover, intermediate debonding phenomena in FRP-strengthened beams is hugely controlled by rebar yielding; the typical stress-strain relationship for steel rebars is represented in Figure 5 and will be adopted in the numerical analyses. Strain-hardening in steel is actually neglected because strain values in FRP-strengthened beams are usually not so great for strain hardening to be developed in steel rebars.

The well-established and widely accepted elastic-brittle stress-strain relationship is assumed for FRP plate (Figure 6). Finally, shear behavior of the adhesive interface connecting FRP laminate or fabric to the soffit of the beam can be described by means of the well-known bi-linear elastic-softening curve (Figure 7) introduced by Holzenkaempfer (1994). The linear branch of the interface law is characterized by the shear stiffness $k_a = k/b_f$ which can be related to the slip modulus k introduced in eq. (2).

Fiber discretization (Figure 8) is considered for cross section in order to evaluate the secant stiffnesses

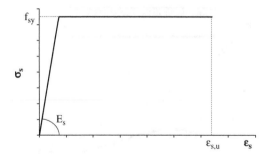

Figure 5. Elastic-perfectly plastic stress-strain law for rebars.

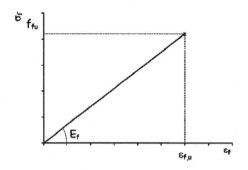

Figure 6. Elastic-brittle stress-strain law for FRP plate.

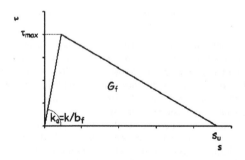

Figure 7. Relationship for FRP-to-Concrete interface.

of reinforced concrete section and the slip modulus of the adhesive interface.

The analysis can be generally pursued until one of the following failure modes is attained:

- concrete crushing, which occurs if the maximum strain value ε_c measured on the concrete fibers at the i-th load step achieves its ultimate value ε_{cu};
- steel rupture, occurring if the steel tensile strain ε_s reaches the corresponding limit value ε_{su};
- FRP plate tearing, if the fiber axial strain ε_f measured in FRP at convergence of the i-th load step achieves the ultimate value ε_{fu};
- FRP debonding, depending on the fact that the maximum interface slip reaches the ultimate value

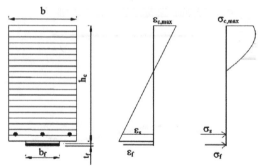

Figure 8. Fiber discretization of the RC beam cross section.

considered in the shear-stress-interface-slip relationship assumed in the analysis.

Further details about the secant procedure implemented for utilizing the proposed Finite Element model along with the relevant results which demonstrate no significant mesh sensitivity of the model can be found in Faella et Al. (2006).

3 DEBONDING FAILURE: BEHAVIORAL OBSERVATIONS

Validation of the proposed procedure can be obtained by comparing numerical and experimental results obtained by one of the various possible campaigns carried out and reported within the scientific literature; the experimental test carried out on the beam A10 in the framework of the campaign reported by Gao et Al. (2004) is considered for validating the numerical procedure. The non-linear stress-strain laws introduced in the previous section are assumed for concrete and steel bars, while elastic-brittle behavior is considered for FRP; the numerical values collected in Table 1 are considered for the corresponding mechanical properties according to data reported by the mentioned authors.

The adhesive interface is modeled by means of the above mentioned bi-linear relationship, whose characteristic parameters are determined according to the fib bulletin 14 – approach 2 proposal reported in the following:

$$\tau_{max} = 0.285 \cdot \sqrt{f_{ck} \cdot f_{ctm}} \quad [\text{MPa}] \qquad (9)$$

$$s_u = 0.185 \quad [\text{mm}]. \qquad (10)$$

Slip modulus k_a has been estimated according to the relationship which accounts for both concrete and adhesive shear stiffness G_c and G_a, as pointed out by Faella et Al. (2003) and adopted in CNR DT200 (2004). The value $k_a = 236.7\,\text{N/mm}^3$ and the value $\tau_{max} = 2.49\,\text{MPa}$ have been assumed for the maximum shear stress.

Table 1. Relevant mechanical properties for validation.

Material	Mechanical property	Numerical value [MPa]
Concrete	f_{cm}	35.7
	E_{cm}	25000
	f_{ctm}	3.57
Rebar steel	f_{sy}	531
	E_s	200000
FRP	f_{fu}	4200
	E_f	235000
Epoxy resin	f_a	30
	E_a	1000

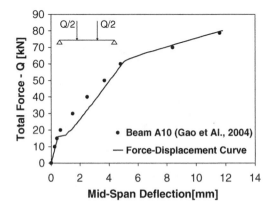

Figure 9. Experimental vs Numerical comparison in terms load-deflection curve (Gao et Al., 2004 – Specimen A10).

Figure 9 shows the curve relating the value of the total load Q and the corresponding midspan deflection of the beam obtained by means of the proposed procedure. The mentioned figure points out that the numerical model can simulate the overall behavior of the considered FRP-strengthened beam.

Initial stiffness, rebar yielding and premature failure due to intermediate debonding are reproduced with a remarkable precision. Only post-cracking stiffness is underestimated because tension-stiffening effect is completely neglected in the analysis as a result of the elastic-brittle branch considered for concrete in tension (Figure 4). Nevertheless, no refinements have been undertaken herein for taking account of tension stiffening because the issue of concern is the possible intermediate debonding failure which usually occurs after yielding and in condition of completely developed cracks.

Finally, comparing numerical results with respect to only one experimental case could be not sufficient for assessing the accuracy of a numerical procedure for simulating the overall behavior of FRP-strengthened beams. Consequently, a complete experimental-to-theoretical comparison can be found in Faella et Al.

(2006) in terms of force, displacement and FRP axial strain at debonding with respect to the results of more than thirty experimental tests taken by the scientific literature.

4 DEBONDING FAILURE: BEHAVIORAL OBSERVATIONS

In the present section, a short discussion is proposed for showing the two possible debonding failure modes induced by bending which has been observed by various authors and can be reproduced by the proposed model. A simply-supported RC beam is considered in the present section; in a first case it will be considered a complete FRP-strengthening throughout all the beam length, while in a second case it will be interrupted at a distance $a = 600$ mm from the theoretical support point. The beam transverse section is rectangular in shape with a depth $h = 600$ mm and the width $b = 300$ mm; reinforcing steel area at the bottom of the section is $A_s = 900$ mm^2 (0.5% of the transverse section area) and the FRP area is 180 mm^2 ($b_f = 75$ mm), which corresponds to 0.1% of A_c. A value $E_f = 205$ GPa is assumed for FRP Young modulus.

For the first case, Figure 10 shows the interface slip evolution under increasing load levels (Figure 10a) and the axial stress in steel rebars and FRP plate at debonding (Figure 10b). They both deal with the beam characterized by FRP strengthening running throughout all the beam length and show how two peak values can be recognized in interface slips. The first and most relevant one is achieved in correspondence of the steel rebar yielding: it increases sharply at the steel yielding section resulting in intermediate debonding. Such local slip growth is due to the fact that load after yielding is only carried out by FRP plate as one can see in Figure 10b, where a sudden increase in FRP strain slope can be observed starting from the above mentioned section. Another peak in slip value occurs near the support due to flexural cracking whose effects can be also observed in terms of FRP and steel axial stresses: slope changes suddenly in the section where cracking begins at about 500 mm from the theoretical support.

Figure 11 shows the relationship between the applied load Q and the maximum strain $\varepsilon_{f,db}$ developed in FRP for the considered beam pointing out the principal states of stress for concrete, steel and the adhesive interface. Concrete cracking induced by bending results in a sudden transition between the uncracked to the cracked state as already pointed out in the previous section devoted to validation. An almost linear branch is followed up to rebar steel yielding which results in a sharp reduction of bending stiffness.

Interface shear stresses abruptly increases after yielding as one can see in terms of interface relative

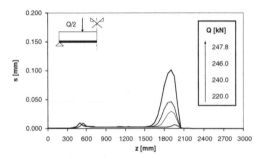

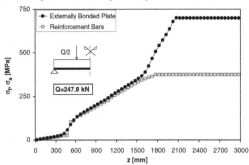

a) interface slips for various load levels up to debonding;

b) axial strains in steel and FRP at debonding ;

Figure 10. Intermediate debonding phenomenon for complete strengthening.

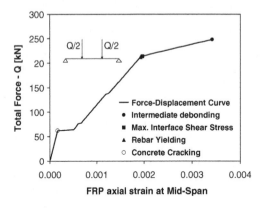

Figure 11. Debonding phenomenon for partial strengthening – Force-FRP strain curve.

displacements in Figure 10a and increments in bending moment can be only carried by FRP. Consequently, after yielding adhesive interface is charged of resisting to a shear stress as great as the total shear force $Q/2$ and the maximum value of interface shear stress τ_{max} is suddenly achieved after yielding, as one can see in and Figure 11 which directly shows the increase in axial strain at midspan after the achievement of the maximum interface shear stress τ_{max}. The great part of numerical models available within the scientific

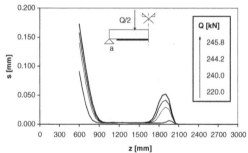

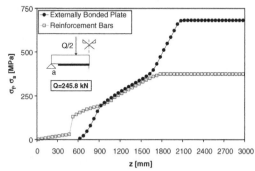

a) interface slips for various load levels up to(end) debonding;

b) axial strains in steel and FRP at (end) debonding;

Figure 12. Debonding phenomenon for partial strengthening.

literature usually neglects the softening branch of the interface stress-strain law, resulting in underestimating the beam response after the achievement of the maximum interface shear stress and, consequently, a quite significant aspect of its mechanical behavior. As shown in Figure 11, a little increase in terms of bearing capacity can be observed after the achievement of τ_{max}; on the contrary, a significant increase in axial deformation developed in FRP can be observed after reaching τ_{max}. This though limited ductility has been experimentally observed by various authors within the scientific literature as one of the key difference between intermediate and end debonding; both these failure modes are premature in nature with respect to the usual crises due to concrete crushing or steel failure, but the first one is at least less fragile than the second one which occurs after steel yielding.

Continuing about the difference between end and intermediate debonding, Figure 12 deals with the case of incomplete FRP-strengthening. A different peak value can be observed in this case either in terms of slip concentration near the FRP cut-off section (Figure 12a) or in terms of FRP stress whose distribution along the beam interface is characterized by two different slope changes (Figure 12b).

In this second case, end-debonding failure occurs before of intermediate debonding because the ultimate

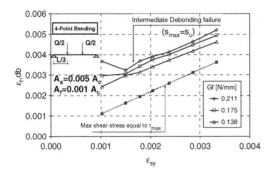

Figure 13. Maximum FRP axial strain at debonding versus rebar yielding strain.

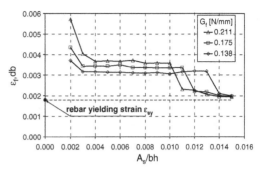

Figure 14. Maximum FRP axial strain at debonding versus rebar area A_s.

slip value is achieved at FRP cut-off section under an ultimate load value slightly lesser than the one obtained for complete strengthening.

Finally, although in the present paper the main emphasis is placed upon intermediate debonding, Figure 10 and Figure 12, among the other ones, show how the proposed procedure can simulate both end and intermediate interface debonding failure.

The above results pointed out that there is a direct relationship between the yielding phenomenon and the occurrence of premature failure due to intermediate debonding. This relationship can be clearly pointed out by considering a parametric variation of the rebar yield stress f_{sy} (and strain ε_{sy}) for the first beam characterized by the complete FRP strengthening. Yielding stress values f_{sy} spanning from 215 to 700 MPa (and ε_{sy} correspondingly between 0.00102 and 0.00333) have been considered along with three possible values for the concrete compressive strength f_{ck} (15 to 25 MPa) which affects the value of fracture energy of the adhesive interface. Figure 13 reports the results of this parametric investigation pointing out the direct relationship between the maximum axial strain $\varepsilon_{f,db}$ developed in FRP at debonding (obtained as the maximum value of the interface slip achieve the ultimate value s_u assumed for the interface relationship) and the value of the axial strain ε_{sy} at yielding. Three series of data have been represented for considering the effect of fracture energy values upon the occurrence of intermediate debonding phenomenon. Although the values attained by $\varepsilon_{f,db}$ in correspondence of the greater values of G_f are always greater than the ones achieved for the lesser ones, the mentioned figure points out that $\varepsilon_{f,db}$ is not as greater as G_f (or its square root, as currently assumed even by the most up-to-date codes of standards), but is hugely affected by the value of ε_{sy}.

Moreover, the same figure also reports the values of axial strains developed2 in FRP when the maximum value of the shear stress $\tau = \tau_{max}$ is achieved at the FRP-to-concrete interface; whatever the fracture

energy be such a value is always strictly related to the occurrence of yielding in steel rebars: consequently, the value of such an axial strain is often close to ε_{sy}. This result points out the need to account for the complete bi-linear law in order to carefully simulate the behavior of the adhesive interface and evaluate a consistent value for $\varepsilon_{f,db}$ (as already shown by Figure 11). The present model (despite the other mentioned ones available in the scientific literature), based on a secant treatment of a closed-form solution for the flexibility and stiffness matrix of the strengthened beam, can easily follow the softening branch of the interface relationship by progressively relaxing the value of the interface stiffning without experiencing problems of localization which possibly affect the performance of finite elements when used for simulating softening behavior.

The distance shown in Figure 13 between the values of $\varepsilon_{f,db}$ and the corresponding axial strains at rebar yielding basically depends upon a series of parameters among which the rebar area; Figure 13 deals with a case of relatively small steel reinforcement area, while Figure 14 shows the possible relationship between the two relevant values achieved by FRP axial strain and the rebar steel area A_s, taken as a percentage of the concrete gross section A_c. The figure shows how the distance between $\varepsilon_{f,db}$ and the corresponding value obtained for $\tau = \tau_{max}$ is as great as the rebar area is small; moreover, for the lower values of A_s, the yielding phenomenon results in a non-relevant increase of interface shear stress (if A_s would be null, no increase occurred at all), and a weaker relationship exists between the two values within this range of A_s.

Load pattern also affects the debonding phenomenon; Figure 15 compared with Figure 13 shows that greater values are developed for FRP strain at debonding $\varepsilon_{f,db}$ especially for the lower values of rebar strain (and stress) at yielding ε_{sy}. Further differences can be observed in the case of distributed load pattern where intermediate debonding phenomenon usually occurs in a section not close to the maximum bending

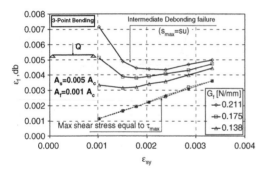

Figure 15. Maximum FRP axial strain at debonding versus rebar yielding strain – Three-Point Bending.

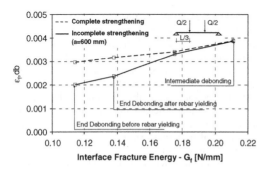

Figure 16. Failure mode versus Fracture energy G_f.

moment, as usually does under point load condition. This aspect, already mentioned in Thomsen et Al. (2004), is deeply examined in Faella et Al. (2007).

Finally, the role of fracture energy G_f as a parameter which affects both end and intermediate debonding is pointed out in Figure 16 which compares the values of maximum FRP strain at debonding $\varepsilon_{f,db}$ developed throughout the adhesive interface for the two cases mentioned at the beginning of the present section.

In the case of complete strengthening, intermediate debonding crisis always occurs while different failure modes can be observed for the case of end debonding as the value of G_f increases. For the lower value of G_f, end debonding prematurely occurs in the RC beam partially strengthened by FRP. Failure of the same beam is less and less premature as G_f increase; increase in G_f (almost) directly result in a corresponding growth of strength against end debonding, while yielding in steel rebars occurs resulting in intermediate failure even for the case of incomplete strengthening for the greater value considered of fracture energy G_f.

In other words, Figure 16 confirms that fracture energy G_f plays a quite different role in controlling end and intermediate debonding; indeed, strength against the former failure mode is directly enhanced as G_f increases, as assumed by both theoretical findings (see

Taljsten, 1997) and the current code of practice. On the contrary, Intermediate debonding, is directly controlled by rebar steel yielding and fracture energy can only partially affect the value of either the load Q achieved at debonding or the maximum strain $\varepsilon_{f,db}$ developed in FRP.

5 CONCLUSIONS

A numerical model has been formulated for simulating the flexural behavior of RC beams strengthened by means of externally bonded FRP systems. Validation of the presented model has been briefly outlined with respect to the results of an experimental test carried out on a beam under four-point bending which failed for intermediate debonding.

Behavioral observations have been drawn with reference to a simply supported beam with the main aim of pointing out the key differences between end and intermediate debonding. Accounting for the softening branch of the shear-stress-interface-slip relationship is of key importance for emphasizing the relative ductility which can be observed in the cases of intermediate debonding failure especially when quantifying FRP maximum strain rather than the maximum force at debonding is of main concern.

Finally, the role of fracture energy G_f has been investigated with respect to both end and intermediate debonding, pointing out its quite diverse influence in controlling these two kinds of premature failure. General remarks should be deduced by these results about the possible enhancement of the simplified formulae available within the scientific literature for evaluating the maximum FRP strain at debonding $\varepsilon_{f,db}$, whose dependence by key parameters like the steel yielding strain ε_{sy} and the amount of rebar are currently neglected. Moreover, the key role played by ε_{sy} and emphasized by the results of the proposed analyses point out the importance of the initial state of stress in the real RC beams which are usually pre-loaded and pre-cracked due to the presence of the self-weights.

ACKNOWLEDGEMENT

The present paper deals with the subject of concern of the Line 8 – Task 2 of the DPC/ReLUIS Research Project that partially granted the research.

REFERENCES

ACI Committee 440.2 R-02 (2002): Guide for the Design and Construction of Externally Bonded FRP Systems for Strengthening Concrete Structures, Revised 28;

Chen J.F., Teng J.G. (2001): Anchorage Strength Models for FRP and Plates Bonded to Concrete, ASCE Journal of Structural Engineering, vol. 127, No. 7, July, 784–791;

CNR DT 200 (2004): Instructions for Design, Execution and Control of Strengthening Interventions by Means of Fibre-Reinforced Composites (in Italian), Italian National Research Council;

Faella C., Martinelli E., Nigro E. (2002): Steel and concrete composite beams with flexible shear connection: "exact" analytical expression of the stiffness matrix and applications, Computer & Structures, Vol. 80/11, pp. 1001–1109;

Faella C., Martinelli E. Nigro E. (2003): Interface Behaviour in FRP Plates Bonded to Concrete: Experimental Tests and Theoretical Analyses, Proceedings of the 2003 ECI Conference on Advanced Materials for Construction of Bridges, Buildings, and Other Structures III, Davos (Switzerland), 7–12 September;

Faella C., Martinelli E., Nigro E. (2006a): Formulation and Validation of a Theoretical Model for Intermediate Debonding in FRP Strengthened RC Beams, Proceedings of the 2nd fib World Conference, Naples (Italy), June 5–8, 2006, Paper 0735;

Faella C., Martinelli E., Nigro E. (2006b): Intermediate Debonding in FRP Strengthened RC Beams: A Parametric Analysis, Proceedings of the 2nd fib World Conference, Naples (Italy), June 5–8, 2006, Paper 0993;

Faella C., Martinelli E., Nigro E. (2007): Intermediate Debonding of RC Beams Strengthened in Bending by FRP: a Theoretical Model and a Simplified Design Approach, Accepted for publication in the Proceedings of the 8th Sym. FRCRCS, Paper 192, Patras (Greece), July 16–18, 2007;

Gao B., Kim J.K., Leung C. K. Y. (2004): Experimental study on RC beams with FRP strips bonded with rubber modified resins, Composites Science and Technology, 64 (2004) 2557–2564;

Holzenkaempfer (1994), Ingenieurmodelle des verbundes geklebter bewehrung fur betonbauteile, Dissertation, TU Braunschweig (in German);

JSCE (2001): Recommendations for upgrading of concrete structures with use of continuous fiber sheets, Concrete Engineering Series 41;

Malek A. M., Saadatmanesh H., Ehsani M. R. (1998), Prediction of failure load of R/C beams strengthened with FRP plate due to stress concentration at the plate end, ACI Structural Journal, Vol. 95, No. 2, 142–152;

Newmark N. M., Siess C.P., Viest I.M. (1951): Tests and Analysis of Composite Beams with Incomplete Interaction, Proc. Soc. Exp. Stress Analysis, 9, 1951, 75–92;

Roberts T. M. (1988): Approximate analysis of shear and normal stress concentrations in the adhesive layer of plated RC beams, The Structural Engineer, Vol. 66, No. 5, 85–94;

Saenz, L.P. (1964): Discussion of Equation for the Stress-Strain Curve of Concrete by Desayi and Krishman, ACI Journal Vol. 61, pp. 1229–1235;

Taljsten B. (1997), Strengthening of Beams by plate bonding, Journal of Materials in Civil Engineering, ASCE, 9 (4), 206–212;

Task Group 9.3 (2001): Externally Bonded FRP Reinforcement for RC Structures, Technical Report Bulletin 14, fib-CEB-FIP;

Teng J.G., Chen J.F., Smith S.T., Lam L. (2002): FRP-strengthened RC Structures, J. Wiley & Sons, GB;

Thomsen H., Spacone E., Limkatanyu S., Camata G. (2004): Failure Mode analysis of Reinforced Concrete Beams Strengthened in Flexure with Externally Bonded Fiber-Reinforced Polymers, ASCE Journal of Composites for Constructions, vol. 8, n. 2, April, pp. 123–131;

Fracture Mechanics of Concrete and Concrete Structures – Design, Assessment and Retrofitting of RC Structures – Carpinteri, et al. (eds)
© 2007 Taylor & Francis Group, London, ISBN 978-0-415-44616-7

Debonding mechanisms in continuous RC beams externally strengthened with FRP

L. Vasseur, S. Matthys & L. Taerwe
Department of Structural Engineering, Ghent University, Magnel Laboratory for Concrete Research

ABSTRACT: One of the biggest challenges in structural engineering nowadays is strengthening, upgrading and retrofitting of existing structures. The use of fibre reinforced polymers (FRPs) bonded to the tension face of the structural member is an attractive technique in this field of application. The strengthening of reinforced concrete (RC) structures by means of externally bonded reinforcement (EBR) is achieved by gluing FRP reinforcement to the concrete substrate. For the efficient utilization of the FRP EBR systems, an effective stress transfer is required between the FRP and the concrete. The paper will discuss the bond behaviour between the FRP and the concrete in the case of flexural strengthening of continuous beams. With respect to this type of beams, few research has been reported. At the Magnel Laboratory currently a test program on flexural strengthening of 2-span continuous beams is ongoing.

1 INTRODUCTION

Structures may need to be strengthened for different reasons, among which a change in function, implementation of additional services or to repair damage. Different strengthening techniques exist. Often applied is externally bonded reinforcement (EBR), based on fibre reinforced polymer (FRP), the so-called FRP EBR.

FRP EBR can be applied for the strengthening of existing structures, enhancing the flexural and shear capacity or to strengthen by means of confinement. This paper discusses flexural strengthening of 2 span reinforced concrete beams. CFRP (Carbon FRP) laminates are glued on the soffit of the spans and/or on the top of the mid-support (Ashour, et al. 2004, El-Refaie, et al. 2003). The efficiency of the FRP EBR strengthening technique is often limited by the capability to transfer stresses in the bond interface. Hereby bond failure between the laminate and the concrete may occur.

For unstrengthened continuous beams a moment redistribution can be observed especially after yielding of one of the critical cross-sections. As a consequence a plastic hinge will be formed. For strengthened continuous beams, after reaching the yield moment, the FRP strengthened cross-section is still able to carry additional load and the formation of a plastic hinge will be restricted.

The aim of this study is to have a better insight in the behaviour of reinforced concrete structures strengthened in flexure in a multi-span situation.

2 CALCULATION MODEL FOR CONTINUOUS BEAMS

2.1 Non-linear moment-curvature diagram

Performing an analysis of a construction according to the linear elasticity theory, a linear relationship between the moment and the curvature is obtained, namely

$$\frac{1}{r} = \frac{M}{EI} \tag{1}$$

with $1/r$ the curvature, M the bending moment and $K = EI$ the bending stiffness.

This stiffness is assumed to be constant and therefore independent of the value of the bending moment. However, for the cross-section of a concrete beam the moment-curvature diagram is non-linear. This non-linear character is caused by the variable bending stiffness, as shown in Figure 1. Two cases are drawn in this graph, a cross-section with externally bonded FRP (strengthened) and a cross-section without FRP (unstrengthened). An important difference between these cases is the bending stiffness (slope of lines K_0, K_1 and K_2). With FRP higher values for K are obtained than without FRP. This different behaviour will influence the moment redistribution of a continuous beam.

If Figure 1 is applied to a continuous beam, we start with the uncracked phase along the whole length of the beam, corresponding to the use of K_0 as bending

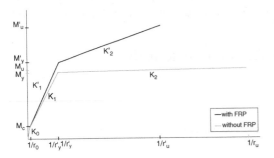

Figure 1. Moment-curvature diagram.

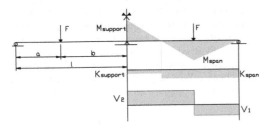

Figure 2. Continuous beam with variable bending stiffness (simplified to 2 stiffness zones).

stiffness. By increasing the load, the beam is characterized by cracked and uncracked zones, each with the related value of bending stiffness. This change of stiffness causes a first redistribution of moments. For the yield load F_y, one or more cross-sections reach the yield moment (M_y). In yield zones without FRP EBR, the bending stiffness K_2 is so small that plastic deformations appear in the critical cross-section and in a restricted area near to it. This is the so-called formation of a plastic hinge. The increasing load is mainly carried by the non plastic zones and during which the bending moment in the plastic hinge stays almost constant ($M_u \approx M_y$) or is slowly increasing. In zones with FRP EBR, the value of the bending stiffness is higher (K'_2). Also plastic deformations appear, but in a more limited way. The yielding zone still carries a significant part of the increasing load and the formation of the plastic hinge is restricted.

2.2 General behaviour of continuous beams

Consider a continuous beam with two identical spans and symmetrical loaded by two point loads (Figure 2). Focused on one span, two zones can be defined, one zone with positive moments (above the mid-support) and another with negative moments (in the spans). It is assumed that in each zone the bending stiffness is constant. So the mid-support zone and the span zone have stiffness $K_{support}$ and K_{span}, respectively.

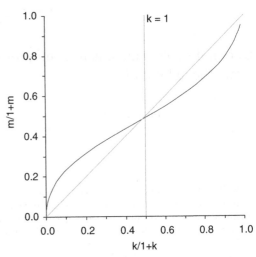

Figure 3. The relation of the moments m in function of the relation of the bending stiffnesses k.

Further, we define:

$$\lambda = \frac{a}{b} \quad m = \frac{M_{sup\,port}}{M_{span}} \quad k = \frac{K_{sup\,port}}{K_{span}} \quad (2)$$

By considering that the angle of rotation above the mid-support equals zero, the following equation can be obtained (Taerwe, et al. 1989):

$$(2 + 3\lambda)m^3 + (3 + 3\lambda - 2\lambda^2)m^2$$
$$- k\lambda(3 + 4\lambda)m - (1 + \lambda)(1 + 2\lambda)k = 0 \quad (3)$$

With Eq. (3) the internal forces in the continuous beam can be calculated. In what follows, calculations are done for $a = 2$ m and $b = 3$ m. Hence with $\lambda = 2/3$ Eq. (3) changes into.

$$36m^3 + (45 - 8k)m^2 - 34km - 35k = 0 \quad (4)$$

This equation is shown in Figure 3. For loads below the cracking moment, the mid-support zone and field span zone are uncracked and the two zones nearly have the same bending stiffness. This condition correspond with $k = 1$. From Eq. (4) we find then $m = 0.9722 = m_{el}$. This value of m corresponds to the moment distribution following the classic theory. Hereby, the relationship between acting load and internal moment is linear, as in the case of isostatic beams. By further increasing the load, the changing bending stiffnesses in different cross-sections modifies k thus the relation between the internal moments m. As a result the moment distribution deviates from the classic theory to the so-called non-linear moment-redistribution.

3 DEBONDING MECHANISMS ON CONTINUOUS BEAMS

3.1 *Overview of different debonding mechanisms*

Bond failure in case of FRP EBR implies the loss of composite action between the concrete and the FRP reinforcement. This type of failure is often very sudden and brittle. According to Matthys (Matthys 2000) different bond failure aspects can be distinguished.

A first type of debonding appears when the externally bonded FRP bridges cracks. This results in elevated shear stresses at the interface and may cause some degree of debonding. In regions with significant shear forces, shear or flexural cracks have a vertical (v) and a horizontal (w) displacement. The vertical displacement of the concrete also causes tensile stress perpendicular to the FRP EBR, which enhances debonding of the laminate (Figure 4).

As second debonding mechanism there is force transfer. Herewith the variation of tensile force in the FRP (ΔN_{fd}) initiates bond shear stresses at the interface due to the composite action between the FRP EBR and the concrete beam. The bond shear stress considered between two sections at a distance Δ_x equals (Figure 5):

$$\tau_b = \frac{\Delta N_{fd}}{w_f \Delta x} \tag{5}$$

with w_f the width of the FRP laminate.

These shear stresses have to be smaller than the bond strength between the concrete and the FRP reinforcement.

A next debonding mechanism is anchorage failure, and relates to curtailment and anchorage length. Theoretically the FRP reinforcement can be curtailed when the axial tensile force can be carried by the internal steel only. The remaining force in the FRP at this point needs to be anchored. The anchorage capacity of the interface is however limited, and hence the FRP may be extended to zones corresponding with low FRP tensile stresses.

At last there is debonding by end shear failure, also known as concrete rip-off. If a shear crack appears at the plate-end, this crack may propagate as a debonding failure at the level of the internal steel reinforcement. In this case the laminate as well as a thick layer of concrete will be ripped off.

3.2 *Avoiding some debonding mechanisms in continuous beams*

To predict the debonding load, the available calculation models (fib 2001) are based on formulas which basically relate to experiments on isostatic reinforced beams and pure shear bond tests.

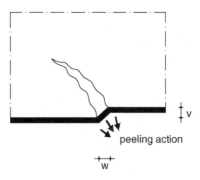

Figure 4. Peeling-off caused at shear cracks.

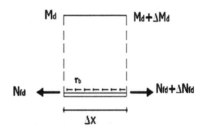

Figure 5. Peeling-off caused by force transfer.

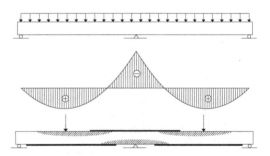

Figure 6. Moments with opposite signs in continuous beams and anchoring laminates into compression zones.

The difference between isostatic beams and continuous beams, which may influence the debonding mechanisms in continuous reinforced concrete beams, is the moment line with opposite signs. Whereas the moment in the span is positive, the moment at the mid-support is negative. As a result, the compression zones in the spans are situated at the top of the beam, at the mid-support the compression zone is situated at the soffit of the beam (shaded zones in Figure 6). This allows in contrast to reinforced isostatic beams, to anchor the CFRP laminates in the compression zones (except for the outer supports) (Figure 6). By extending a laminate into these compression zones, two out of the four different debonding mechanisms will be avoided: debonding by a limited anchorage length and debonding by end shear failure (concrete rip-off).

Debonding by limited anchorage length is prevented by extending the laminate into the compression zone because in this situation the tensile stress in the laminate is gradually reduced to zero, and anchored in a zone with small compressive stresses (no significant risk for buckling).

Debonding by end shear failure occurs when a shear crack appears at the end of the laminate. By extending the laminate into the compression zone, the plate-end reaches a zone where no shear cracks can be formed and neither concrete rip-off will appear.

To anchor the laminate into the compression zone, it is extended beyond the point of contraflexure, which is the location where the internal moment equals zero. For calculating the exact location of this point, it has to be noticed that the point of contraflexure moves with increasing load, due to the non-linear moment redistribution.

3.3 Specific debonding aspects related to continuous beams

In the case of strengthened continuous beams, some particular aspects can be noted, which may also influence the moment of debonding. This is illustrated in the following by means of an analytical study for the beam and strengthening configuration in Figure 7. The applied internal reinforcement is kept constant during the analytical study and is based on the linear theory. In this case almost the same amount of internal reinforcement is used in the spans and in the mid-support (reinforcement ratio's $\rho_{s,span} = 0.68\%$ and $\rho_{s,support} = 0.61\%$). The properties assumed in the analysis are given in Table 1, whereas the amount of FRP strengthening in the spans and mid-support zone is varied (FRP widths of 60 mm, 100 mm, 150 mm and 200 mm are used in this study).

The length of the FRP is chosen in such a way that all four debonding mechanisms can occur. Herewith the laminates are not anchored into the compression zones as described in paragraph 3.2. The length of the laminate at the soffit of the span equals 2000 mm and is applied in such a way that the center of the laminate is just beneath the point load. The laminate at the top of the beam above the mid-support equals 1600 mm (see Figure 7).

The influence of the amount of FRP strengthening on the acting shear forces is illustrated in Figure 8, for a point load F of 100 kN. Herewith, V_1 (solid lines) is the shear force acting between the outer support and the point load. $V_2 = F - V_1$ (dashed lines) is the shear force acting between the point load and the mid-support (Figure 2). As can be noted, the value of V_1 (and hence V_2) is influenced by the FRP reinforcement ratio's of both the span ($\rho_{f,span}$) and the mid-support ($\rho_{f,support}$). By increasing the width of the laminate above the mid-support (increasing $w_{f,support}$

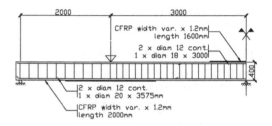

Figure 7. Internal and external reinforcement configuration.

Table 1. Properties of concrete and reinforcement materials.

	Concrete	Steel	CFRP
Compres. strength [N/mm²]	36.0	–	–
Yielding strength [N/mm²]	–	570	–
Yielding strain [%]	–	0.28	–
Tensile strength [N/mm²]	3.3	670	2768
Failure strain [%]	0.35*	12.40	1.46
E-modulus [N/mm²]	32000	210000	189900

* in compression.

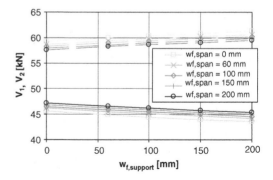

Figure 8. Shear force V_1 in function of width of laminates.

or $\rho_{f,support}$), for $w_{f,span} =$ cte, V_1 decreases and V_2 increases. This is due to the moment redistribution which is dependent on both the external and internal reinforcement ratio used over the length of the beam. Owing to this, also the distribution of the reactive forces in the supports is dependent on the reinforcement ratio's. As a result, the part of the applied load which is carried by the mid-support increases with an increasing amount of FRP at the mid-support. If $w_{f,span}$ (or $\rho_{f,span}$) is increased as well, the decrease of V_1 will be less pronounced and V_1 may even increase (compared to the unstrengthened beam).

As debonding phenomena often relate to the acting shear force, this means that possible debonding of a

FRP laminate not only depends on the FRP configuration at that location, but also on the amount of FRP in the zone with opposite moment sign.

Another significant aspect with respect to the values of $\rho_{f,span}$ and $\rho_{f,support}$ relative to each other, is their influence on the point of contraflexure. Indeed, by increasing $\rho_{f,support}$ (increasing the width of the laminate above the mid-support), $M_{support}$ will increase and M_{span} will decrease. Herewith, the point of contraflexure moves towards the mid-support. On the opposite, by increasing $\rho_{f,span}$ at the soffit of the span, $M_{support}$ will decrease and M_{span} will increase. As a result, the point of contraflexure moves away from the mid-support. Because of this, a change of the distance between the laminate end and the place where the internal moment equals zero (L) can be observed. Indirectly also the anchorage length (l_t) will change. Due to this the debonding mechanisms anchorage failure and concrete rip-off once more will be dependent on the amount of external reinforcement along the beam.

In the following paragraphs, the load at which debonding occurs, for the different debonding mechanisms will be investigated in function of both $\rho_{f,span}$ and $\rho_{f,support}$. Hereby, a differentiation is made between debonding of the top laminate (case A), debonding of the laminate at the soffit of the span between the point load and the mid-support (case B) and debonding of the laminate at the soffit of the span between the point load and the outer support (case C) (Figure 9).

The calculations are performed according to section 2 and fib-bulletin 14 (fib 2001). Results of the debonding load calculations are given as far as they do not exceed the ultimate load of the strengthened beam assuming full composite action. Herewith it is assumed that debonding of the FRP does not occur, and that the construction only can fail by concrete crushing or by exceeding the tensile strength of steel or FRP reinforcement.

In Table 2 a summary is given of the effect of the external reinforcement on the debonding load. In paragraph 3.3.1 till 3.3.4 a more detailed explanation is given about these findings.

3.3.1 Crack bridging

Debonding by crack bridging in case of risk for vertical crack displacement can be modeled according to (fib 2001), based on the following simplified equation:

$$V_{Rpd} = \tau_{Rp} bd \qquad (6)$$

$$\tau_{Rp} = 0,38 + 1,51 \rho_{eq} \qquad (7)$$

with b the width of the beam, d the effective depth of the beam, τ_{Rp} the nominal shear stress corresponding to V_{Rpd}, $\rho_{eq} = (A_s + A_f.E_f/E_s)/bd$ the equivalent reinforcement ratio and A and E the cross-section and modulus of elasticity of the reinforcement (s: steel and f: FRP).

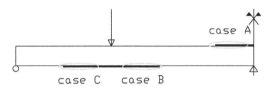

Figure 9. Differentiation between places of debonding.

Table 2. Effect of external reinforcement on debonding load.

		Amount of top laminate above the mid-support ↑	Amount of laminate at the soffit of the span ↑
Crack bridging	A	debonding load ↑	debonding load ↑
	B	debonding load ↓	debonding load ↑
	C	debonding load ↑	debonding load ↑
Force transfer	A	debonding load ↑	debonding load ↑
	B	debonding load ↓	debonding load ↑
	C	debonding load ↑	debonding load ↑
Anchorage failure	A	debonding load ↓	debonding load ↑
	B	debonding load ↑	debonding load ↓
	C	debonding load ↑	debonding load ↓
Concrete rip-off	A	debonding load ↓	debonding load ↑
	B	debonding load ↑	debonding load ↑
	C	debonding load ↑	debonding load ↓

This means that the moment of debonding will be both dependent on the acting shear force (influenced by both $\rho_{f,span}$ and $\rho_{f,support}$) as well as the amount of FRP (A_f) (influenced by $\rho_{f,support}$ in case A and $\rho_{f,span}$ in cases B and C).

The debonding load of the top laminate (case A) is given in Figure 10. By increasing the width of the top laminate, the resistance V_{Rp} increases (increase of A_f in equation 7) and to a lesser extent the acting shear load V_2 also increases (Figure 8). If at the same time the amount of FRP in the spans increases, this further enhances the debonding load due to a reduction of the acting shear force V_2.

The change in debonding load in case B is illustrated in Figure 11. Increasing the widths of the laminate in the span (for $w_{f,support} =$ cte.), increases the debonding resistance (increase of A_f) at one hand and decreases the acting shear load (V_2) at the other hand. This causes a higher debonding load. Increasing also the width of the top laminate will result in a somewhat higher acting shear force V_2 (Figure 8) and has a negative influence on the debonding load.

For the beam considered in this study (Figure 7, Table 1), debonding in case C appeared not governing compared to the ultimate load of the strengthened beam assuming full composite action. Nevertheless, similar to the above argumentation an increased

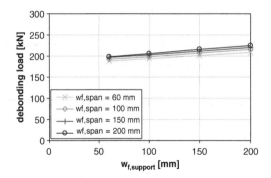

Figure 10. Influence of external reinforcement on the debonding mechanism: Debonding at cracks (case A).

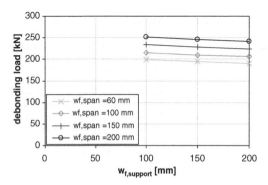

Figure 11. Influence of external reinforcement on the debonding mechanism: Debonding at cracks (case B).

debonding resistance can be expected for a higher width of the laminate in the span. Also increasing the amount of FRP at the mid-support, this will further enhance the debonding load due to a reduction of the acting shear load V_1 (Figure 8).

3.3.2 Debonding due to force transfer

The influence of $\rho_{f,span}$ and $\rho_{f,support}$ on the debonding load of this mechanism is similar as described in the previous section (3.3.1).

Firstly, there is the change of shear force ratio caused by the moment redistribution. By increasing the width of the top laminate above the mid-support, V_2 increase and V_1 decrease (Figure 8). Herewith a lower debonding load is expected in cases A and B and a higher debonding load is expected in case C. By increasing the width of the laminate at the soffit of the span, the opposite effect occurs.

Secondly, the width of the laminate (w_f) is also an important factor in the calculation of the resisting shear force, V_{Rbd} (equation 8) (fib 2001).

$$V_{Rbd} = f_{cbd} \, 0.95 dw_f \tag{8}$$

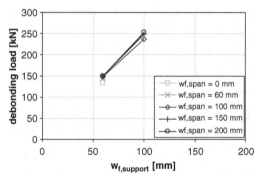

Figure 12. Influence of external reinforcement on the debonding mechanism: Debonding by force transfer (case A).

with $f_{cbd} = 1.8 f_{ctk}/\gamma_c, f_{ctk}$ the characteristic tensile strength of the concrete and γ_c the safety factor ($\gamma_c = 1.50$).

Increasing the width of the laminate has a positive influence on its debonding load. This effect is more pronounced than the acting shear load effect. The combined effect is illustrated in Figures 12 till 14, for cases A, B and C respectively.

Increasing the width of the top laminate results in a large increase of the debonding load for case A (Figure 12). Increasing the width of the laminate in the span also has a favourable influence, yet to a lesser extent.

In Figure 13 the debonding load of case B is illustrated. Here it can be noticed that the increase of the width of the laminate at the soffit of the span has an important positive influence on the debonding load and will delay it, while the increase of the width of the top laminate has a negative influence and results in a lower debonding load.

In Figure 14 the debonding load of case C is illustrated. Here it can be noticed that the increase of the width of both the laminates at the span and at the mid-support increase the debonding load, whereas the laminate at the soffit has the most pronounced effect.

3.3.3 Anchorage failure

Considering debonding by anchorage failure, the variability of the shear forces V_1 and V_2 according to the external reinforcement ratio's is of no importance. What, however, is important, is the redistribution of the internal moments $M_{support}$ and M_{span} and its impact on the location where the internal moment equals M_{curt}. Herewith, M_{curt} is the internal moment for which the axial tensile force can be carried by the internal steel only and consequently the external reinforcement theoretically can be curtailed. M_{curt} can also be seen as the position along the length of the beam where the anchorage length l_t starts. Because of the change in position of M_{curt}, the available anchorage length is dependent

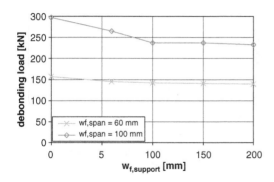

Figure 13. Influence of external reinforcement on the debonding mechanism: Debonding by force transfer (case B).

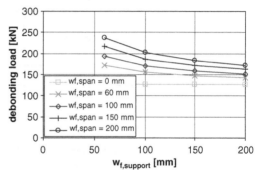

Figure 15. Influence of external reinforcement on the debonding mechanism: Anchorage failure (case A).

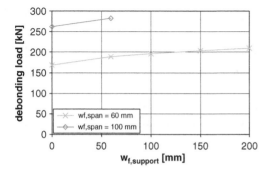

Figure 14. Influence of external reinforcement on the debonding mechanism: Debonding by force transfer (case C).

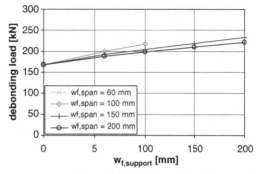

Figure 16. Influence of external reinforcement on the debonding mechanism: Anchorage failure (case B).

on the moment redistribution and consequently on the reinforcement ratio's along the continuous beam.

By increasing the amount of FRP reinforcement above the mid-support, $M_{support}$ increases and M_{span} decreases. This results in a movement of the position where the internal moment equals M_{curt}: (1) towards the laminate end (shorter anchorage length) for the laminate at the top of the beam (case A), and (2) away from the laminate end (larger anchorage length) for the laminates at the soffit of the beam (cases B and C).

By increasing the amount of FRP reinforcement at the soffit of the beam an opposite effect is obtained.

It can be concluded (keeping the total FRP length constant) that a reduced anchorage length is obtained in case A when the amount of FRP above the mid-support is increased and in cases B and C when the amount of FRP at the soffit of the span is increased. Due to this reduction of anchorage length, a lower debonding resistance may be obtained. This influence of the laminate widths (or ρ_f) on debonding at the anchorage zone is illustrated in Figures 15 till 17 for the considered beam and strengthening configuration.

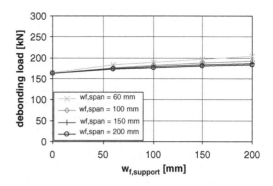

Figure 17. Influence of external reinforcement on the debonding mechanism: Anchorage failure (case C).

3.3.4 Concrete rip-off

To check concrete rip-off, a resistant shear stress, $\tau_{Rd} (= V_{Rd}/bd)$, has to be calculated (equations 9 and 10) (fib 2001). Herewith the changes of the shear

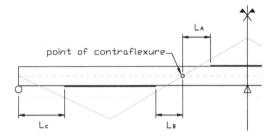

Figure 18. The distance L between the laminate end and the point where the internal moment equals zero.

forces according to the reinforcement ratios are playing an important role. The influence will be similar as earlier discussed in sections 3.3.1 and 3.3.2.

$$\tau_{Rd} = 0,15 \sqrt[3]{3\frac{d}{a_L}\left(1+\sqrt{\frac{200}{d}}\right)} \sqrt[3]{100\rho_s f_{ck}} \qquad (9)$$

$$a_L = \sqrt[4]{\frac{\left(1-\sqrt{\rho_s}\right)^2}{\rho_s}dL^3} \qquad (10)$$

where $\rho_s = A_s/bd$, f_{ctk} is the characteristic compression strength of the concrete and L is the distance between the laminate end and the point where the internal moment equals zero (Figure 18).

By increasing the amount of FRP above the midsupport, a larger part of the applied load is carried by the mid-support. As a result the internal shear force V_1 decreases and V_2 increases. This decrease of V_1 has a positive influence on the debonding load for case C, while the increase of V_2 has a negative influence for cases A and B. By increasing the amount of FRP at the soffit of the beam, the opposite effect is obtained.

In addition to the redistribution of the acting shear force, the debonding load is also governed by the influence of L on the debonding resistance. This distance relates in cases A and B to the point of contraflexure. In case C this distance relates to the outer support (Figure 18). A higher debonding resistance is obtained for decreasing values of L. For the cases A and B, L depends on the location of the point of contraflexure. As a result, the debonding due to concrete rip-off will also depend on the moment redistribution.

Consequently, increasing the amount of FRP above the mid-support moves the point of contraflexure to the left in Figure 18. This causes an increased value L_A, and an equally decreased value L_B (for $L_A + L_B =$ cte.). Hence, a decrease of the resisting debonding force of the FRP laminate at the top of the beam (case A) and an increased value of the debonding resistance of the FRP laminate at the soffit of the beam (case B) are obtained. An opposite effect is obtained when increasing the amount of FRP at the soffit of the beam.

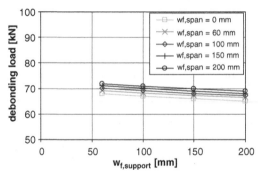

Figure 19. Influence of external reinforcement on the debonding mechanism: Concrete rip-off (case A).

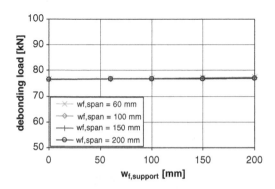

Figure 20. Influence of external reinforcement on the debonding mechanism: Concrete rip-off (case B).

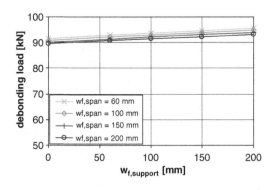

Figure 21. Influence of external reinforcement on the debonding mechanism: Concrete rip-off (case C).

The combined effect of shear and moment redistribution on the concrete rip-off debonding load is illustrated in Figures 19 till 21. For the considered case, and whereby the total length of the FRP is kept constant, the influence of the shear and moment redistribution appears less pronounced than for the other debonding mechanisms. Especially in case B (Figure 20), for which the shear and moment redistribution effect counter act each other, the combined influence is insignificant.

4 CONCLUSION

In continuous beams compression zones are available at which FRP laminates can be anchored. Here two debonding mechanisms (concrete rip-off and anchorage failure) can be avoided.

By means of an analytical study, it has been demonstrated that the debonding loads are also governed by the shear force and moment redistribution. This redistribution is occurring in FRP strengthened continuous beams and depends on the amount of FRP in the spans and at the mid-support, relative to each other. Because of the specific influence on the debonding load, redistribution of internal forces should be considered when verifying the debonding load of strengthened continuous beams. Depending on the situation (amount of FRP, type and location of the debonding phenomenon) both an increased or decreased value of the debonding load may be obtained.

ACKNOWLEDGEMENT

The authors acknowledge the financial support by FWO-Vlaanderen

REFERENCES

Ashour, A. F., El-Refaie, S. A. and Garrity, S. W. 2004 Flexural strengthening of RC continuous beams using CFRP laminates. Cement and Concrete Composites 26(7): p. 765–775.

El-Refaie, S. A., Ashour, A. F. and Garrity, S. W. 2003 Sagging and Hogging Strengthening of Continuous Reinforced Concrete Beams Using Carbon Fiber-Reinforced polymer Sheets. ACI Structural Journal. **Vol. 100**: p. 446–453.

fib 2001 fib bulletin 14, Externally bonded FRP reinforcement for RC structures. International federation for structural concrete, Lausanne.

Matthys, S. 2000 Structural behaviour and design of concrete members strengthened with externally bonded FRP reinforcement. Ghent University, Department of structural engineering.

Taerwe, L. and Espion, B. 1989 Serviceability and the Nonlinear Design of Concrete Structures. IABSE PERIODICA 2/1989.

Fracture Mechanics of Concrete and Concrete Structures – Design, Assessment and Retrofitting of RC Structures – Carpinteri, et al. (eds)
© 2007 Taylor & Francis Group, London, ISBN 978-0-415-44616-7

Repairing and strengthening methods for RC structural members

A. Koçak
Yıldız Technical University, Istanbul, Turkey

M.M. Önal
Gazi University, Ankara, Turkey

K. Sönmez
Civil Engineer, Istanbul, Turkey

ABSTRACT: In this study, the repairing and strengthening methods of reinforced columns and beams which are damaged during earthquakes have been compared. Due to earthquake and/or vertical loads, damaged or insufficient structural components can be repaired and strengthened with various methods. Reinforced jacketing, repairing with steel plates and repairing with FRP components are more used methods. We often come across these three methods in practise. These three methods have advantages and disadvantages. In this study, 21 beams were investigated experimentally. The experimantal research is done on 6 reference beams and 6 cracked beams with its bottom U shaped reinforcements and by adding steel ropes to carrying points. Also 3 beams were retrofitted with steel plates on their tension zone and 6 beams were retrofitted by full jacketing. Strength degradation, energy dissipation, ductility and rigidity of the members using these methods given above are compared in the final section of this experimantal investigation.

1 INTRODUCTION

Nowadays, repairing and strengthening of damaged reinforced-concrete buildings has become an important issue. There are many different kinds of repairing and strengthening techniques in literature. Reinforced jacketing, carbon fiber and reparation with steel plates are the ones that are used the most.

When it comes to reinforced-concrete structures, certain parts or the entire building can be strengthened due to disability, damage or regulation changes. This strengthening process can be done by increasing structure rigidity through adding bearing elements and bearing system or increasing the rigidity and strength of insufficient components in the structure.

In Turkey, which is a seismic country, many structures receive major damage after an earthquake. These structures are demolished if they are highly damaged, but if they receive medium and/or minor damage, they can be repaired or strengthened. Occasionally, bearing members of the structure are repaired individually. Also components of the structure that have less strength and rigidity are retrofitted.

There are many studies in literature which examine not only the strengthening of the damaged structures or those with insufficient rigidity, but also the strengthening of columns and beams that were damaged as a result of an earthquake. There are many different repairments and strengthening methods used in these studies. Also the strengthening methods applied to the beams in this study are examined and tested in various studies. Reinforced-concrete jacketing, fixing with steel plates and carbon fiber procedures are used when beams are repaired individually.

In this study, three different procedures of repairing the damaged beams are examined experimentally [27]. First study includes full jacketing with new longitudinal and wrap reinforcement while the second one uses reinforced-concrete jacketing with tensile reinforcement and U stirrup. The last one is fixing with steel plates.

2 EXPERIMENTAL WORK

2.1 Experimental components

21 beams are examined to find out the efficiency of damaged beams which were repaired. All of these beams have a 2000 mm clear span, and the dimensions of 100*160*2200 mm. The main reinforcement

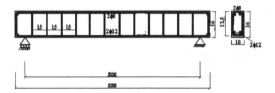

Figure 1. Specimen detail.

Figure 2. Preparing phase of beam specimens.

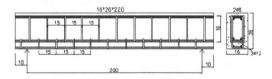

Figure 3. Reparation detail of u-shaped stirrup anchoring.

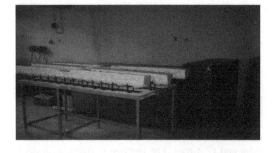

Figure 4. U-shaped stirrup anchoring from underneath.

is 2ϕ12, the mounting reinforcement is 2ϕ8, the stirrups are consisted of 8 mm diameter grade S420 bars and spaced at 150 mm. The concrete grade is C16. (Figure 1, Figure 2).

The beams are produced in five sets. 6 specimens (KM11, KM12, KM13, KM21, KM22, and KM23) that make up the first series are damaged and repaired by U-shaped stirrups. Additional reinforcement is 2ϕ12; Additional stirrups are ϕ8/150 (Figure 3 and Figure 4).

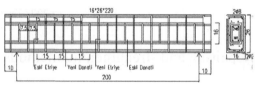

Figure 5. Detail of damaged beam strengthened by jacketing.

Figure 6. Jacketing Process.

Beams making up the second set (KM41, KM42, KM43, KM51, KM52, KM53) are made up of C20 concrete and grade S420 bars with 100*160 mm dimensions and damaged. Grade S420, 2ø12 reinforcement is added up to the tensile zone of damaged beams and the beam is wrapped up by 8 mm diameter stirrups spaced at 150 mm. Also the undersides of the beams are opened up to the cover and additional reinforcements are connected with existing reinforcements. (Figure 5, Figure 6).

Three beams are made as a the third set and repaired by epoxy resin plates. These beams (KM31, KM32, KM33) are repaired by bonding St37, 6*50*1200 mm epoxy resin plates to both sides of the beams' bending zone and tensile zone (Figure 7, 8). The epoxy mortar used to bond the steel plates has two components, namely resin and hardener. The density of epoxy mortar is 1.7 kg/ liter, compressive strength is 65 N/mm², bending strength is 30 N/mm², tensile strength is 20 N/mm², concrete adhesion is 3.5 N/mm² and steel adhesion is 20 N/mm². The epoxy mortar is produced with these characteristics at a 1/3 mixture ratio.

Three specimens of steel plates are prepared for reparation which are grade St37, 400 mm long and have the dimensions of 6*50*1200 mm. The yield strength of the bonding plates are $f_{yk} = 300$ MPa with $f_{su} = 412$ MPa breaking strength and $\varepsilon su = \%12.7$ breaking elongation. The specimen model can be seen in Figure 9.

The reference beams of the fourth set include three beams (RKMk1, RKMk2, RKMk3) are made up of C16 concrete and grade S420 steel. Using the same reinforcement ratio and dimensions (100*160 mm),

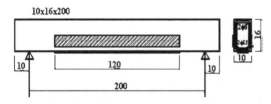

10x16x200

Figure 7. Details of the beam repaired and strengthened by steel plate.

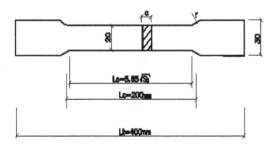

Figure 8. Process of placing the epoxy plates.

Figure 9. Steel specimen used for strengthening.

the fifth set of reference beams (RKMb1, RKMb2, RKMb3) are produced with dimensions that fit after strengthening process (160*260 mm) by using C30 concrete and grade S420 steel (Table 1, Table 2, Table 3). Beams belonging to the first series were loaded to the bearing capacity as calculated in advance; the test was then stopped when a 8 mm displacement was reached at midspan.

The specifications of beams are given below in Table 2. The efficiency of known methods, behaviors of the beams, load-bending relationship after strengthening, rigidity, ductility, load bearing capacity and energy dissipation capacity are examined. The success of the strengthening procedure, efficiency of the method, the values before and after the process and the load against displacement curves are compared.

Table 1. Experiment Components.

Serial Number	Specimen Serial Number	Beam Serial Number	Component Section (mm)	Description
1	1	KM 11	Before,	Jacketing:
2		KM 12	100*160	U-shaped
3		KM 13	after	stirrups to the
4		KM 21	repairing	undersides
5		KM 22	160*260	and 2Φ12
6		KM 23		reinforcement are added to beam.
7	2	KM 41	Before	Jacketing,
8		KM 42	retrofitting,	2ø12 extra
9		KM 43	100*160	reinforcement
10		KM 51	After	and φ8/15
11		KM 52	retrofitting	stirrups
12		KM 53	160*260	added to the undersides of the damaged beams
13	3	KM 31	100*160	Two epoxy
14		KM 32		plates are
15		KM 33		placed to the under and each side of the beams
16	4	RKMk 1	100*160	Reference
17		RKMk 2		beam with
18		RKMk 3		sections 100 × 160 mm
19	5	RKMb 1	160*260	Reference
20		RKMb 2		beam with
21		RKMb 3		sections 160 × 260 mm

2.2 Damaging and strengthening of beam specimens

The beams that were produced as a test specimen had a span length of 2000 mm and were loaded until they broke up with a medium degree of damage. Loads were increased 1962 N every phase. In every load phase, the displacement values of beams having ½ and ¼ points of span length were recorded digitally and by researchers (Figure 10).

The beams that belong to the first series were loaded to a bearing capacity which calculated before, the test was stopped when 8 mm displacement examined at the middle of beam span. Neither a major displacement at the tensile reinforcement nor crushing at the pressure zone was allowed. Theoretic and experimental values of beams that belong to the first series were given in Table 3.

The beams that were made of C20 concrete, grade S420 steel with dimensions of 100*160 mm were damaged and then strengthened. A 2ø12 grade S420

Table 2. Material and geometrical specifications of the beams.

Serial Number	Specimen Number	Dimensions (mm)	Reinforcement	Reinforcement area (mm²)	Reinforcement ratio	Measured d (mm)	f_{ck} (Mpa)	f_{yk} (Mpa)	f_{su} (Mpa)
1	KM11	100 × 160 × 2200	2ø12	226	0.014	130	22.204	529.74	804.42
2	KM12	100 × 160 × 2200	2ø12	226	0.014	135	22.204	529.74	804.42
3	KM13	100 × 160 × 2200	2ø12	226	0.014	137	22.204	529.74	804.42
4	KM21	100 × 160 × 2200	2ø12	226	0.014	135	22.204	529.74	804.42
5	KM22	100 × 160 × 2200	2ø12	226	0.014	135	22.204	529.74	804.42
6	KM23	100 × 160 × 2200	2ø12	226	0.014	135	22.204	529.74	804.42
7	KM 41	100 × 160 × 2200	2ø12	226	0.014	135	22.204	529.74	804.42
8	KM 42	100 × 160 × 2200	2ø12	226	0.014	135	22.204	529.74	804.42
9	KM 43	100 × 160 × 2200	2ø12	226	0.014	136	22.204	529.74	804.42
10	KM51	100 × 160 × 2200	2ø12	226	0.014	135	22.204	529.74	804.42
11	KM52	100 × 160 × 2200	2ø12	226	0.014	130	22.204	529.74	804.42
12	KM53	100 × 160 × 2200	2ø12	226	0.014	130	22.204	529.74	804.42
13	KM31	100 × 160 × 2200	2ø12	226	0.014	135	22.204	529.74	804.42
14	KM32	100 × 160 × 2200	2ø12	226	0.014	130	22.204	529.74	804.42
15	KM33	100 × 160 × 2200	2ø12	226	0.014	130	22.204	529.74	804.42
7	RKMk1	100 × 160 × 2200	2ø12	226	0.014	130	22.204	529.74	804.42
8	RKMk2	100 × 160 × 2200	2ø12	226	0.014	135	22.204	529.74	804.42
9	RKMk3	100 × 160 × 2200	2ø12	226	0.014	135	22.204	529.74	804.42
10	RKMb1	160 × 260 × 2200	4ø12	452	0.01	240	33.27	529.74	804.42
11	RKMb2	160 × 260 × 2200	4ø12	452	0.01	245	33.27	529.74	804.42
12	RKMb3	160 × 260 × 2200	4ø12	452	0.01	245	33.27	529.74	804.42

Table 3. Theoretical and experimental load-bearing capacities of damaged beams.

No	Specimen name	Theoretical Mmax. (Nmm)	Exp Mmax (Nmm)	Theo Pu (kN)	Exp Pu (kN)	Displacement of middle point (mm)
1	KM 11	10630*10³	11000*10³	21,26	22,00	19,00
2	KM 12	10630*10³	12000*10³	21,26	24,00	18,75
3	KM 13	10630*10³	11000*10³	21,26	22,00	22,00
4	KM 21	10630*10³	10000*10³	21,26	20,00	14,00
5	KM 22	10630*10³	10500*10³	21,26	21,00	13,90
6	KM 23	10630*10³	11000*10³	21,26	22,00	10,70
7	KM 41	10630*10³	12000*10³	21.26	24.00	19.40
8	KM 42	10630*10³	11500*10³	21.26	23.00	15.55
9	KM 43	10630*10³	11000*10³	21.26	22.00	14.80
10	KM 51	10630*10³	11500*10³	21.26	23.00	11.20
11	KM 52	10630*10³	11500*10³	21.26	23.00	22.00
12	KM 53	10630*10³	9000*10³	21.26	18.00	15.00
13	KM 31	10630*10³	12000*10³	21.26	24.00	16.00
14	KM 32	10630*10³	10500*10³	21.26	21.00	17.80
15	KM 33	10630*10³	12500*10³	21.26	25.00	19.50
16	RKMk1	10630*10³	9500*10³	21.26	19.00	32.10
17	RKMk2	10630*10³	10500*10³	21.26	21.00	27.60
18	RKMk3	10630*10³	10000*10³	21.26	24.00	31.25

reinforcement was applied to tensile zone and the beam was wrapped up with 8 mm diameter stirrups which were spaced at 150 mm. Also the bottoms of the beams were opened up to the cover and additional reinforcements were connected with existing reinforcements.

The third group of beams that were damaged was repaired by bonding the steel plates, both under and on each side of the beams. The repaired beams were tested and the displacement values were determined every time the load was increased.

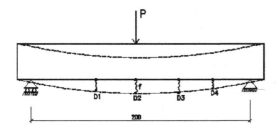

Figure 10. Displacement measurement points and testing system.

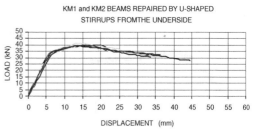

Figure 11. Load-displacement graph of beams which were repaired by jacketing from below.

3 EXPERIMENTAL WORK

The load-displacement curves of test components repaired by jacketing and steel plates are compared with the load-displacement curves of the reference beams (Figures 11, 12, 13, 14) The strength values, rigidity, ductility, load bearing and energy dissipation capacities are examined and the results can be seen at Tables 5, 6 and 7. Also the cracks that emerged during the test process are examined both before and after the strengthening of repaired beams.

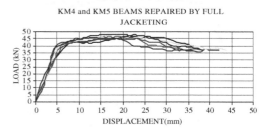

Figure 12. Load-displacement graph of beams which were repaired by reinforced concrete jacketing.

3.1 Load-displacement relationship

The five sets of specimens are exposed to loading and their test values are evaluated and recorded Twelve beams that make up the first two groups are repaired by jacketing; the remaining three specimens are repaired by steel plates and exposed to reloading afterwards. The load-displacement curves that are estimated by the values determined through the loading and breaking of the specimens according to the program, are drawn by both reading from the comparator at points D1, D2 and D3 according to the loading and reading points shown at Table 10 and upon being saved to computer records at point D2 (Figures 11–14). The values used to make up the curves are taken from LVDT that is placed in the middle of the beam span and from comparator. In general, the numbers and widths of cracks at bending zone of the components are similar to each other. The reference beams that make up the second and third group are exposed to loading as well and the displacement values due to loading are recorded. As far as the strengthening through reinforced concrete jacketing goes, the load-displacement curves of the beams that are repaired by damaging ended up similar to the load-displacement curves of the reference beams that are identical to the dimensions after jacketing (Figure 11, 12). The load-displacement curves drawn by repairing the steel plates are similar to the load-displacement curves belonging to the simple beam. However, the results cannot come close to those estimated by reinforced concrete jacketing (Figure 13).

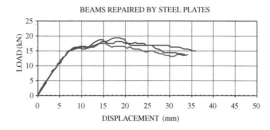

Figure 13. Load-displacement graph of beams which were repaired by steel plates.

3.2 Load bearing of the specimens

During this experimental study, the moment load bearing and horizontal load bearing capacities of the specimens are examined both before and after strengthening. According to the results, when it comes to the beams whose sections, reinforcements and material strengths were the same (Table 2), the cracking load and the yield load values are almost identical (Table 4). In most of the experiments, as far as the specimens go, experimental failure loads are lower than the theoretical failure loads. The breaking loads acquired through jacketing the entire section are close to the load bearing results that are achieved at the reference beams and are higher than the strengthening applied to the U-shaped stirrups. Moreover, the breaking load achieved by strengthening is higher as compared to the initial section and reference beams fitting the initial

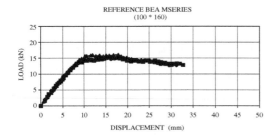

REFERENCE BEAM SERIES
(100 * 160)

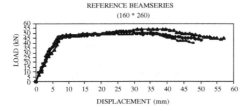

REFERENCE BEAMSERIES
(160 * 260)

Figure 14. Load-displacement graph of reference beams (RKM1, RKM2, RKM3 (100*160) and (160*260)).

Table 4. Theoretical and experimental values of the repaired beams.

Beam serial number	Displacement of yielding moment Δ_y (mm)	Displacement of breaking moment Δ_U (mm)	Ductility $\mu_\Delta = \frac{\Delta_U}{\Delta_y}$
KM11	6,2	31,5	5,1
KM12	6,2	27,8	4,5
KM13	5,6	29,1	5,2
KM21	6,6	26,1	4,0
KM22	6,4	28,5	4,5
KM23	6,4	26,5	4,1
KM41	6.2	35.0	5.6
KM42	6.1	28.4	4.7
KM43	6.0	31.2	5.2
KM51	5.1	35.3	6.9
KM52	5.0	23.8	4.8
KM53	4.6	30.7	6.7
KM31	7.1	30.1	4.2
KM32	7.2	26.4	3.7
KM33	7.0	26.4	3.8
RKM1b	6.9	43.1	6.2
RKM2b	8.0	49.8	6.2
RKM3b	7.0	48.2	6.9
RKM1k	9.1	28.3	3.1
RKM2k	8.1	32.8	4.0
RKM3k	9.0	27.9	3.1

section. The breaking load increase of the specimens subject to experiment by being jacketed on 4 sides as compared to the reference beams corresponding to the initial section (100 × 160 mm) is 200%. When

compared to the reference beams corresponding to the section after strengthening, this increase is around 122%. The increase of the breaking load belonging to the stirrups u-shaped from underneath and one-sided jacketing applied by extra reinforcement is 156% when compared with the initial section and 68% when compared with the reference beams that have repaired beam section. The ratio is 19% when it comes to the repair made by plates.

3.3 The ductility of specimen

For the beams to have a ductile behavior the reinforcement percentage should be within the range specified by the regulations and thus, all the beams are furnished to have the ductility requirement. When exposed to the experiment, all 6 beams provided enough ductility. However there is a decrease in the ductility of some beams because the total ductility ratio of the repaired beams is higher than necessary. In this study, it is estimated that there could be some possible loss of material strength and loss due to loading and bearing conditions, so these elements are taken into consideration. By using the load-displacement curve derived from the loading of the specimens, the shape change of each element due to breaking is divided by the shape change due to creeping, hence, the ductility of each element is calculated (Table 5).

3.4 Bending rigidity of the specimens

The comparisons and comments about the rigidity of the specimens made through the load-displacement curves. The bending rigidities that are calculated are shown in Table 6. There is no big difference when it comes to the rigidity of the reference beams and the beams repaired by jacketing and steel plates. As far as load-displacement relations go, the rigidity is calculated by determining the inclination up to the first crack load. The rigidity loss is calculated by the inclination determined by using failure load on the load-displacement curve and compared with the rigidity during the first crack.

3.5 Energy dissipation capacities of the specimens

By using the load-displacement curve, the energy dissipating capacity is estimated via calculating the area that was under the curve and is shown in Table 7. When we compared the energy dissipation capacity of reference beams, steel plate repairs and beams repaired by jacketing from underside, the best results are achieved by the beam model that is repaired by jacketing with winding the beam up with stirrups. The best energy dissipating increase is achieved by jacketing.

Table 5. Ductility ratio of beams.

No	Specimen name	Theoretical M max. (Nmm)	Exp M max (Nmm)	Theo Pu (kN)	Exp Pu (kN)	Displacement of middle point (mm)
1	KM 11	26000*10³	27300*10³	52,00	34	31
2	KM 12	26000*10³	27850*10³	52,00	33,5	28
3	KM 13	26000*10³	26540*10³	52,00	32,5	29
4	KM 21	26000*10³	25370*10³	52,00	35	26,5
5	KM 22	26000*10³	25500*10³	52,00	34,5	28,4
6	KM 23	26000*10³	25000*10³	52,00	34	27
7	KM 41	22300*10³	23680*10³	44.60	40.0	35.0
8	KM 42	22300*10³	23550*10³	44.60	39.5	28.0
9	KM 43	22300*10³	22700*10³	44.60	38.0	32.0
10	KM 51	22300*10³	22500*10³	44.60	41.0	35.5
11	KM 52	22300*10³	21000*10³	44.60	38.5	24.0
12	KM 53	22300*10³	22800*10³	44.60	42.0	31.3
13	KM 31	11372*10³	12350*10³	22.74	17	30
14	KM 32	11372*10³	12700*10³	22.74	14.5	26.5
15	KM 33	11372*10³	11500*10³	22.74	16.5	26.5
16	RKMk 1	10630*10³	9500*10³	21,26	13,5	28
17	RKMk 2	10630*10³	10500*10³	21,26	13,6	32,5
18	RKMk 3	10630*10³	10000*10³	21,26	13,6	28
19	RKMb 1	26700*10³	27140*10³	53,40	42	43
20	RKMb 2	26700*10³	26970*10³	53,40	46,5	50
21	RKMb 3	26700*10³	27018*10³	53,40	46	43

Table 6. Bending rigidity of specimens.

Beam serial number	Pu, test (kN)	Displacement of middle point (mm)	Energy dissipation capacity (kNmm)
KM11	34	31	114747,74
KM12	33,5	28	109873,69
KM13	32,5	29	113207,57
KM21	35	26,5	101135,62
KM22	34,5	28,4	111074,81
KM23	34	27	101839,96
KM41	40	35	168119.79
KM42	39.5	28.0	133091.99
KM43	38.0	32.0	145486.86
KM51	41	35.5	174353.92
KM52	38.5	24	111402.98
KM53	42	31.3	154353.48
KM31	17	30	97923.54
KM32	14.5	26.5	80696.64
KM33	16.5	26.5	84870.07
RKM1b	42	43	160430.42
RKM2b	46.5	50	195658.12
RKM3b	46	43	164760.39
RKM1k	13.5	28	87021.1
RKM2k	13.6	32.5	102482.6
RKM3k	13.6	28	87251.5

Table 7. Energy dissipation capacities of beams.

Beam serial number	Displacement of mid-point (mm)	Yield rigidity (kN/mm)	Breaking rigidity (kN/mm)	Rigidity decrease (%)
KM11	31	3,27	0,62	81,04
KM12	28	3,27	0,70	78,6
KM13	29	3,07	0,64	79,16
KM21	26,5	2,75	0,75	72,73
KM22	28,4	2,90	0,70	75,86
KM23	27	3,07	0,72	76,55
KM41	35	3.73	0.67	82.01
KM42	28	3.73	0.80	78.56
KM43	32	3.73	0.70	81.24
KM51	35.5	4.33	0.67	84.53
KM52	28	4.70	0.93	80.22
KM53	31.3	4.70	0.75	84.05
KM31	30	2.35	0.64	72.77
KM32	26.5	2.35	0.64	72.77
KM33	26.5	2.35	0.70	70.22
RKM1b	43	3.48	0.57	83.63
RKM2b	50	3.27	0.55	83.18
RKM3b	43	3.73	0.60	83.92
RKM1k	28	1.73	0.53	69.37
RKM2k	32.5	1.80	0.44	75.56
RKM3k	28	1.88	0.55	70.75

4 CONCLUSION AND SUGGESTIONS

In this experimental study, 21 beams with reference beams are damaged, 6 beams of the first set repaired by half jacketing from underneath, the other 6 repaired by full jacketing, 3 are repaired by steel plates and tested again. The values of strengthened beam and reference beam are examined and compared each other. Here are the results:

– Jacketing with additional reinforcement and an additional concrete layer method, as can be deducted from the load-displacement curve, is successful as far as the load bearing values of the repaired beams (Tables 3 and 4), ductility (Table 5), bending rigidities (Table 6) and energy consuming capacities (Table 7) were considered. In other words, the moment load-bearing capacity of the strengthened beams is close to the load bearing capacity of the post-strengthened section (equivalent reference beam). Similarly, repairs with steel plates are also successful.
– Similar to the reference beams, it is observed that capillary cracks that emerged on the beams strengthen with all three methods.
– It is observed that workmanship and the quality of the materials used are two factors that directly effect behavior.
– In the experiment set, the beams that are repaired while still bearing a load on them could not show sufficient strength and could not reach the load bearing values of the reference beams. However, those beams repaired without bearing a load could reach the sufficient load bearing values.
– As far as ductility and energy consuming go, the strengthening-on-four-sides method is more successful than the strengthening-with-a-u-shaped-reinforcement and strengthening-with-steel-plates methods.
– It is observed that increasing the plate thickness is not a successful strength increasing method and causes unwanted failure (brittle).
– Other studies on this subject and the experimental studies disclosed here show that the length of plate effects failure and load bearing capacity of the beam when it comes to the repairs made by steel plates.
– The breaking load increase of the specimens subject to experiment by being jacketed on 4 sides as compared to the reference beams corresponding to the initial section (100 × 160 mm) is 200%. When compared to the reference beams corresponding to the section after strengthening, this increase is around 122%. The increase of the breaking load belonging to the u-shaped stirrups from underneath and one-sided jacketing applied by extra reinforcements is 156% when compared with the initial section and 68% when compared with the reference beams that had been repaired.

– When compared to the reference beams, the ductility increase of the strengthening method of reinforcement with a u-shaped stirrups from underneath is 68% whereas, this increase is 151% with the full jacketing method.

As far as these results are concerned, the following suggestions can be made about strengthening with concrete steel jacketing:

– It should be considered that there could be a decrease of 5–15% in the strength and load bearing capacities of the elements assuming that the application circumstances are not as good as laboratory conditions.
– The number of cracks and the width of the beams that are repaired after having been damaged affect the bending rigidity. Hence, the strengthening of the cracks should be made by injecting epoxy and the rigidity of the elements should be determined without applying jacketing or steel plate bonding.
– During the repair and strengthening of the reinforced concrete load bearing elements, it should not be forgotten that workmanship is of great importance and the details should be applied thoroughly.
– To be able to get the desired results while strengthening the load bearing elements, the element should be free of load.
– The connections of the beam reinforcements at hand and the reinforcements loaded onto the strengthening purpose should be set very well.
– During the experiment including beams that are repaired by adhering strengthening steel, the steel broke off the old concrete. In order to prevent this, anchorage with bolt could be provided.
– Plates used in this study are adhered in a vertical direction and parallel to the beam. Usage of horizontal plate will cause an increase in the experimental results and although they are harder to apply, their usage is also suitable.

5 SYMBOLS

d   : Useful height
f_{ck} : Characteristic compressive strength of concrete
f_{cd} : Concrete strength calculation
f_{yk} : Characteristic compressive strength of reinforcement
f_{yd} : Reinforcement strength calculation
f_{su} : Maximum tensile strength of reinforcement
C   : Concrete class
E   : Elasticity module
M : Moment
S   : Reinforcement class
δ : Lateral displacement
ϕ : Reinforcement diameter
Δy : Yielding moment ductility

Δu : Breaking moment ductility
μΔ : Ductility

REFERENCES

Celep, Z., Boduroğlu, H., 1997, Dinar Öğretmen Evi Kooperatif Binalarõnõn Deprem Davranõşõ ve Yapõlan Güçlendirme Sistemi 4. Ulusal Deprem Mühendisliği Konferansõ, s173–180.

Canbay, E., Sucuoğlu, H., 1998, Seismic Assessment of Damaged/Strengthened Reinforced Concrete Building. Repair and Strengthened of Existing Building, Second Japan-Turkey Workshop on Eartquake Engineering. İstanbul, 132–151.

Atõmtay, E., Tekel, E., 1997, HasarlõBir Yapõnõn Deprem Davranõşõnõn İnclenmesi ve Öğrenilen Dersler, 4. Ulusal Deprem Mühendisliği Konferansõ, İstanbul, 567–576.

Demir, H., 2000, Depremlerden Hasar Görmüş Betonarme Yapõlarõn Onarõm ve Güçlendirilmesi, İstanbul.

Gomes, A.M., Appleton, J., 1998, Repair and Strengthening of Reinforced Concrete Elements under Cyclic Loading, Proceedings of the 11th European Conference on Earthquake Engineering, Rotterdam.

Fukuyama, K., Higashibata, Y., Miyauchi, Y., 2000, Studies on Repair and Strengthening Methods of Damaged Reinforced Concrete Columns, Cement and Concrete Compozites, 22, 1, 81–88.

Ramirez, J.L., 1996, Ten Concrete Colum Repair Methods, Construction and Building Materials, 10,3,195–202.

Frangou, M., Pilakoutas, K., Dritsos, S., 1995, Structural Repair/Strengthening of R/C Columns, Construction and Building Materials, 9, 5, 259–266.

Ersoy, U., Tankut, A.., 1993, Behaviour of Jacketed Columns, Structural Journal ACI, 90, 3.

Aykaç, S., 2000, Onarõlmõş/Güçlendirilmiş Betonarme Kirişlerin Deprem Davranõşõ Doktora Tezi, Gazi Üniversitesi, Fen Bilimleri Enstitüsü, Ankara.

Sharma, A.K., 1986, Shear Strength of Steel Fiber Reinforced Concrete Beams, ACI Proceedings, 83, 624–628.

Can, H., 1994, Deprem Etkisindeki Betonarme Kirişlerin Onarõlmasõ, İMO,Teknik Dergi, 771, Ankara.

Yazar, E., 1997, Hasar Görmüş Kirişlerin Takviyesi, Celal Bayar Üniversitesi, Fen Bilimleri Enstitüsü.

Basunbul, A.I., Gubati, A.A., Al-Sulaimani G.J, Baluch M.H 1990, Repaired Reinforced Concrete Beams, ACI Materõals Journal, s348–354.

Ünsal, Ç.T., 1989, Güçlendirilmiş Betonarme Kirişlerin Davranõş ve Dayanõmlarõ, Yüksek Lisans Tezi, Gazi Üniversitesi, Fen Bilimleri Enstitüsü, Ankara.

Sharif, A., Al-Sulaimani, G.J., Basunbul, I.A., Baluch, M.H., Husain, M., 1995, Strengthening of Shear-Damaged RC Beams by External Bonding of Stell Plates, Magazine of Concrete Research,47, 173, 329–334, London.

Ecemiş, A.,Ş., 2000, Hasar Görmüş Betonarme Kirişlerin Epoksi İla Çelik Levha Yapõştrõrõlarak Güçlendirilmesi Üzerine Deneysel Bir İnceleme, Yüksek Lisans Tezi, Selçuk Üniversitesi Fen Bilimleri Enstitüsü, Konya.

Altõn, S., Anõl, Ö., 2001, "Betonarme Kirişlerin Dõştan Yapõştõrõlan Çelik Plakalarla Kesmeye Karşõ Güçlendirilmesi" Tübitak İnşaat ve Çevre Teknolojileri Araştõrmalarõ Grubu, Yapõ Mkeaniği Laboratuarlarõ-Toplantõsõ,13.

Shahawy, M.A., Beitelman, T., 1996, Flexural Behavior of Reinforced Concrete Beams Strengthened with Advanced Composite Materials, International SAMPE Symposium and Exhibition, Anaheim, USA.

Swamy, R. N., Jones, R., Ang, T.H., 1992, Under and over reinforced concrete beams with glued steel plates, International Journal Cement Composites Ligtweigth Concrete, 4, 19–32.

Swamy, R. N., Jones, R., Bloxham, J. W., 1987, Structural behavior of reinforced concrete beams strengthened by epoxy-bonded steel plates, Structural Engineer, 65, 59–68.

Saafi, M., Toutanji, H. A. And Li, Z., 1999, Behavior of concrete columns confined with fiber reinforced polymer tubes, ACI Materials Journal, Vol. 96, No. 4, pp. 500–509.

Lam, L. And Teng, J. G., 2001, Strength models for FRP-confined concrete, Journal of Structural Engineering, ASCE.

Purba, B.K., Mufti, A.A.,1999, Investigation of the behavior of circular concrete columns reinforced with carbon fiber reinforced polymer (CFRP) jackets, Canadian Journal of Civil Engineering, Vol. 26, pp. 560–596.

Kumbasar, N., İlki, A., 2001, Karbon Lif Takviyeli Polimer Kompozitlerin Yapõ Elemanlarõnõn Onarõm ve Güçlendirilmesinde Kullanõlmasõ, Tübitak İnşaat ve Çevre Teknolojileri Araştõrma Grubu, Yapõ Mekaniği Laboratuvarlarõ Toplantõsõ, 105.

Norris, T., Saadatmanesh, H., Ehsani, M.R., 1997, Shear and flexural strengthening of R/C beams with carbon fiber sheets, Journal of Structural Engineering, 123, 903–911.

Önal, M.,M., 2002, Hasar Görmüş Dikdörtgen Kesitli Kirişlerin Mantolama Yöntemiyle Onarõmõ Üzerine Deneysel Bir Araştõrma, Doktora Tezi, Gazi Üniversitesi.

Fracture Mechanics of Concrete and Concrete Structures – Design, Assessment and Retrofitting of RC Structures – Carpinteri, et al. (eds)
© 2007 Taylor & Francis Group, London, ISBN 978-0-415-44616-7

Seismic performance of RC columns retrofitted with CFRP Strips

H. Araki & R. Sato
Hiroshima University, Japan

K. Kabayama
Shibaura Institute of Technology, Japan

M. Nakata
Kosei Kensetsu Corporation, Japan

ABSTRACT: Reinforced concrete columns of the rail-way structures were severely damaged at 1995 Hyogoken-Nanbu earthquake. Observed main damage patterns in the reinforced concrete structures were shear failures of the columns with poor arrangements of the lateral reinforcement. For those columns, confinement of the column sections is effective to improve the ductility of the columns in avoiding the brittle failure. Authors proposed spacing the carbon fiber reinforced polymer (CFRP) strips for the retrofitting of those columns. This paper presents the seismic performance of reinforced concrete columns retrofitted with CFRP strips comparing with those of the columns with CFRP sheet. In this research program, seismic loading tests were performed using scaled reinforced concrete columns in the laboratory of the Hiroshima University. From test results, The columns retrofitted with CFRP strips showed the very good performance as same as CFRP sheet. Moreover, it is anticipated that there is strong possibility of decreasing the amount of CFRP.

1 INTRODUCTION

The seismic evaluations and the developments of the retrofitting techniques for the existing structures against the major earthquakes had been carried out from 1970 in Japan. In 1995 Hyogoken-Nabu earthquake, reinforced concrete columns of building structures were severely damaged resulting considerable loss of life and wealth. Reinforced concrete columns of the rail-way structure were also damaged at that event, resulting inconvenient of the logistics and the rescue. Observed main damage patterns in the reinforced concrete structures were shear failures of the column due to the poor arrangements of the lateral reinforcement. After that event, numerous retrofitting techniques for the existing reinforced concrete structures and strengthening methods for the damaged columns were proposed and then were speedily used in practice. The most effective method for the brittle failure of the columns was the jacketing column sections with the concrete, the steel plates, and the continuous fiber material. Confinement of the column sections improves the ductility of the column against the seismic loadings in avoiding the brittle failure. While the continuous fiber materials had merits of the high strength and the light weight, researches of the fiber reinforced polymer material for the retrofitting to the existing reinforced concrete structures started at the latter half of the 1980 in Japan, Kastumata. In the first international symposium held in Vancouver on March 1993, many techniques using the fiber reinforced polymer materials were reported. In U.S. and Japan, the guidelines of the design using continuous fiber reinforced polymer were published, ACI Committee 440, JSCE 1996, AIJ Committee 2001.

The carbon or glass fiber sheets as retrofitting materials were mainly used in practice to improve the seismic performance of the existing reinforced concrete structures. Though the CFRP sheet is the superior material to confine the column section from the previous studies, it is difficult to inspect visually the damage level of the columns after the earthquake due to the perfectly wrapped sheet. In point of construction view when using sheet it is needed to cut away or cut off the walls from the columns which has the nonstructural walls in the railway station area.

From the disadvantage of the sheet, authors proposed spacing the CFRP strips of tapes and braiding. Besides the immediate inspections or the construction advantage, while spacing the CFRP strips mitigates the concentration of the stress in the CFRP materials it is expected to avoid the fracture of the retrofitting material, leading to the rapid strength degradations of the columns. In this paper the possibility of decreasing the amount of the CFRP is also discussed because the CFRP is expensive.

This paper presents the seismic performance of reinforced concrete columns retrofitted with CFRP strips of tapes and braiding comparing with that of the column retrofitted with CFRP sheet.

In this research program, seismic loading tests were performed using the seven reinforced concrete columns in the laboratory of the Hiroshima University.

2 EXPERIMENTAL PROCEDURE

2.1 Descriptions of original columns

The configuration and the details of the seven test columns before retrofitting were common and all test columns were designed as the existing reinforced concrete columns of railway bridges and railway stations. Test columns were designed as the shear failure type according to the old Standard Specifications for Concrete Structures(JSCE) and their ratio of the shear strength V_{yd} to the flexural strength V_{rd} was approximately 0.64.

Seven scaled reinforced concrete columns have been assigned for seismic loadings under the constant axial loads. Test columns had clear height of 1,000 mm with a cross section of 300 mm × 300 mm where test columns were 1/3 ~ 1/4 scale of prototype rail way columns. Each column had the heavily reinforced stub, which were 1.2 m × 1.2 m × 0.5 m, at the bottom end of the column to fix rigidly at the reaction floor of the laboratory. Figure 1 shows the details of the original test column. The corners of the column sections were round with radios 32 mm to avoid the rupture of the CFRP. Specified strength of concrete Fc was 27 N/mm². Concrete were cast from the top of the column as the same condition of the actual construction site using the metal form set. All test columns were reinforced with D19 of longitudinal reinforcing bars and $\phi6$ of transverse reinforcing bars. The total longitudinal reinforcement ratios and the transverse reinforcement ratio were 1.47% and 0.094%, respectively. Table 1, Table 2 and Table 3 show the mechanical properties of the used materials, concrete, steel and CFRP, respectively. Young's Modulus E_f of braiding is less than half of that of tape, but the ultimate strain ε_u of braiding is 1.6 times of that of tape.

2.2 Retrofitting procedure

Parameters in the retrofitting procedure for the all test columns are summarized in Table 4. The test columns JRW05-1 was not retrofitted to estimate the effectiveness of the retrofitting with CFRP. The other six

test columns were retrofitted with CFRP sheet and strips(tapes and braiding). Test column JWR05-02 was wrapped with the CFRP sheets which consisted of two kinds of the sheet of 300 g/m² and 400 g/m². Two layers of each sheet were wrapped around the column sections, resulting total amount of the CFRP was 1400 g/m². Test columns JRW05-3 and JRW05-4 were retrofitted with CFRP tapes of which width and unit weight were 30 mm and 30 g/m, respectively. Designed spaces of the tapes were 100 mm and 50 mm and five layers and three layers of tapes were wrapped for the test columns, resulting the total amounts of the

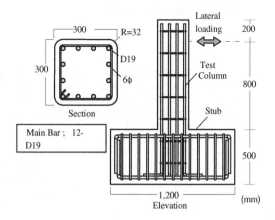

Figure 1. Detail of the test columns.

Table 1. Mechanical properties of concrete.

Test column	Concrete strength σ_B (N/mm²)	Tensile strength σ_t (N/mm²)	Young's modulus E_c (10^4 N/mm²)
JRW05-1	28.36	2.79	3.14
JRW05-2	28.78	2.74	3.14
JRW05-3	29.49	2.76	3.12
JRW05-4	26.19	2.50	3.00
JRW05-5	32.02	2.79	3.26
JRW05-6	27.58	2.97	3.34
JRW05-7	28.54	2.41	2.89

Table 2. Mechanical properties of steel.

Steel bar	Yield strength σ_y (N/mm²)	Ultimate strength σ_u (N/mm²)	Young's modulus E_s (10^5 N/mm²)
Longitudinal steel D19	339	527	1.42
Transverse steel D6	617	713	2.06

CFRP were 1500 g/m² and 1800 g/m². Target ductility of the JRW05-2 ~ 4 were more than ten. Test columns JWR05-5 and JRW05-6 were prepared with the aim of reducing the amount of the CFRP. One layer of the tapes along the total height of the column was wrapped for JRW05-5 of which the target ductility was five. For JRW05-6, five layers of the tapes were wrapped at the critical area of which length was the column depth D from the bottom end of the column and one layer was wrapped the other area.

For JRW05-7, CFRP braiding shown in Figure 2 was newly proposed in the FRP field. Width of the CFRP braiding was 10 mm and its unit weight was the same of the tape.

The amount of CFRP strips were obtained based on the equation (1) proposed for CFRP sheet in Railway Technical Research Institute Guide.

$$\mu = 2.8 + 1.15 \frac{V_{cyd}}{V_{rd}} \quad (1)$$

μ was the target ductility and V_{cyd} was the shear strength of the column after retrofitting, presenting the sum of the shear strength of the concrete column V_{yd} and the strength of the CFRP V_{cf} as equation (2).

$$V_{cyd} = V_{yd} + V_{cf} \quad (2)$$

Table 3. Mechanical properties of CFRP. (including Epoxy).

CFRP	Strength σ_f	Young's modulus E_f	Ultimate strain ε_u
	(N/mm²)	(10⁵ N/mm²)	(%)
Sheet tape	3426	2.39	1.46
Braiding	2248	1.06	2.31

V_{yd} was presented by equation (3) as follows.

$$V_{yd} = V_{cd} + V_{sd} \quad (3)$$

V_{cd} was the strength of the concrete shear resisting mechanisms, and V_{sd} was the strength of arranged transverse reinforcement. V_{cf} was presented by equation (4) as follows.

$$V_{cf} = \frac{0.8 A_{CF} \cdot f_{CFud}}{S_{CF}} \cdot \frac{Z}{\gamma_{bCF}} \quad (4)$$

A_{CF}, S_{CF} and f_{CFud} were the sectional area, the width and the tension strength of the unit of CFRP sheet respectively; $Z = d/1.15$ (d was effective depth of the column); $\gamma_{bCF} = 1.15$.

$$V_{rd} = M_{cf} \Big/ l_a \quad (5)$$

The flexural strength V_{rd} was obtained by dividing the flexural moment capacity M_{cf}, which was calculated

Figure 2. CFRP braiding.

Table 4. List of test columns.

Test column	CFRP	Width (mm)	Space (mm)	Number of layer	Failure mode	V_{cyd}/V_{rd}	Target ductility	Design ductility from Eq.(1)
JRW05-1	–		–	–	Shear	0.64	–	–
JRW05-2	Sheet		–	4*	Flexural	8.02	10	12.02
JRW05-3	Tape	30	100	5	Flexural	8.54	10	12.62
JRW05-4	Tape	30	50	3	Flexural	10.12	10	14.44
JRW05-5	Tape	30	100	1	Flexural	2.22	5	5.35
JRW05-6	Tape	30	100	5&1**	Flexural	8.54	10	12.62
JRW05-7	Braiding	10	100	5	Flexural	4.97	8	8.52

* Two sheets of 400 g/m² and 300 g/m² were layered respectively.
** 5 layers of carbon fiber strips were provided in the area length 1D (column depth) at the bottom end of the column.

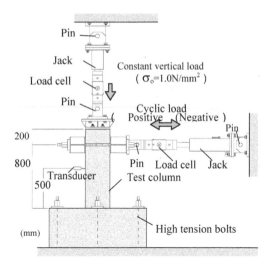

Figure 3. Test setup.

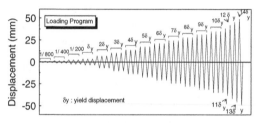

Figure 4. Lateral loading program.

by the fiber model, with shear span $l_a = 800\,\text{mm}$ as equation (5). Target ductility and the design ductility of the test columns obtained from the equation (1) using the specified materials are shown in Table 4. The ductility of the test columns JRW05-7 with CFRP braiding was lower than those of the test columns JRW05-3 in spite of using the same amount of CFRP because the strength for the CFRP braiding of $2248\,\text{N/mm}^2$ was lower than the strength of the CFRP sheet and tapes of $3426\,\text{N/mm}^2$.

2.3 Test setup

The seismic loading tests were conducted in Hiroshima University. The test setup is shown schematically in Figure 3. The stub of the test column was post-tensioned to the reaction floor with the high tension bolts, whose tension force were $300\,\text{kN}$ per a bolt. The lateral and vertical loading systems displaced the actual column in a canti-lever condition. The shear span ratio was $M/QD = 2.67$. All test columns were subjected to reversal lateral loads under the constant axial load which was $90\,\text{kN}(\sigma_o = 1\,\text{N/mm}^2)$. Both ends of the vertical and the lateral loading jacks were supported by the pin joints.

2.4 Measurement system

All test columns were instrumented with number of LVDTs to measure the lateral and the vertical displacement to evaluate the deflection modes and to measure the curvatures of the hinging regions of the columns. The strain gages were mounted on the surfaces of the longitudinal and transverse reinforcing bars to evaluate the yielding and the confinements. Also, the strain gages were mounted on the surface of the CFRP strips

at the all side of the rectangular column sections. Both hydraulic jacks for the lateral and the vertical loads were equipped with load cells at the loading points as shown in Figure 3.

2.5 Loading program

Figure 4 shows the lateral loading program. Before yielding of the longitudinal reinforcing bars lateral loadings were carried out according to the predetermined drift angles, attempting three cycles for the each peak drift angle levels of $R = 0.00125\,\text{rad.}$, $0.0025\,\text{rad.}$, $0.005\,\text{rad.}$ Yielding was monitored by the strain of the longitudinal reinforcing bars at the bottom end of the column. After yielding of the longitudinal reinforcing bars the lateral loadings were carried out under ductility level control, attempting three cycle for each of the peak displacement of ductility levels of $\mu = 1, 2 \sim 10$. After ductility ten only one cycle was applied to the test columns at each displacement level up to the limitation of the loading system.

3 TEST RESULTS

3.1 Yield strength and displacement

Table 5 shows the observed yield strength and yield displacements δy of the test columns at the yielding of the longitudinal reinforcing bars. Yield strengths and displacements of the all teat columns were approximately similar except for JWR05-1. For test column JRW05-1 without retrofitting the yielding were not observed while considerable strength degradations occurred due to the shear failure. Yield strains were observed on the way from drift angle $0.005\,\text{rad.}$ to $0.01\,\text{rad.}$

3.2 Crack pattern and failure modes

The failure patterns for four test columns are illustrated in Figure 5. For JRW05-1 which was not retrofitted diagonal cracks occurred before yielding. The final failure mode was the typical shear failure while scarcely any flexural cracks were observed. For JRW05-2 that was retrofitted with CFRP sheet through the total height of test column, consequently, crack

Table 5. Yield strength and displacement.

Test column	Yield strength (kN)	Yield displacement* (mm)	Drift angle (10^{-3}rad.)
JRW05-1	–	–	–
JRW05-2	190.0	3.23	6.46
JRW05-3	173.5	3.18	6.36
JRW05-4	173.8	3.50	7.00
JRW05-5	185.9	3.49	6.98
JRW05-6	187.8	3.58	7.16
JRW05-7	171.8	3.39	6.78
Ave.	180.5	3.40	6.80

* Displacement measured at the height of the test column 500 mm.

Figure 5. Crack pattern at displacement 10δy.

patterns could not be observed visually. Its failure mode was supposed to be flexural type from the lateral displacement distribution while the horizontal cracks widely opened at the bottom end of the test column. At failure stage the sheet at the bottom portion of the test column expanded to out-plane of the column section with spalling and crashing of concrete cover or buckling of the longitudinal reinforcing bars. Failure patterns for other five test columns retrofitted with CFRP strips were approximately similar. Before lateral displacements 6δy shear cracks were observed as same as JRW05-1 and the flexural cracks in parallel with CFRP strips were progressed at the bottom portions of the test columns.

Width of the shear cracks did not become large due to the confinement of the CFRP strips.

When the displacements became large over 6δy flexural cracks widely opened and crashing of the concrete cover were concentrated at an area that was not confined with the CFRP strips in the lower portions of the test column. The sliding displacement in that area

increased among the total lateral displacement of the test column without the large strength degradations in the shear-displacement curves until the final stages.

For JRW05-3, JRW05-4 and JRW05-7 the lengths of the main damaged area were limited in the bottom area of the columns. For JRW05-05 and JRW05-6 the lengths of the area which the diagonal shear cracks extended were two times of the column depth though the width of the shear cracks did not become so large to lead the strength degradations. Through the lateral loadings the fractures of the CFRP strips did not occur except for JRW05-5 that was retrofitted with only one layer of the CFRP tapes. The partial fracture at the joint of the tapes in JRW05-5 was observed when its ductility extremely exceeded the target ductility.

3.3 Shear force-displacement hysteresis response

The shear force-displacement hysteresis responses of the four test columns are illustrated in Figure 6. Shear force in the Y-axis were corrected for the lateral force caused by the vertical loads. Lateral displacements at the height of 500 mm in the test columns were used in the X-axis. The white circles in the figures indicated 10δy in the plus and minus direction. Seismic performance of JRW05-1 without retrofitting was very poor due to the shear failure before yielding. Rapid degradation of the lateral shear force was observed after the drift angle 0.005rad. and simultaneously the test column lost the vertical load carrying capacity. Hysteresis responses of the other six test columns retrofitted with CFRP were considerably improved by changing of the failure mechanism from the shear failure to flexural failure. Shear force could be transferred smoothly to the basement due to the confinement effects of the CFRP that protected the progress or the expanding of the shear cracks. In the test columns with retrofitting, the significant strength degradations were not observed at the same displacement reversals. Features of hysteresis responses of four test columns JRW05-2, JRW05-3, JRW05-4 and JRW05-7 were spindle types, which showed flexural behavior, up to the large ductility area. JRW05-04 of which the amount of CFRP was the largest in the test columns had the most stable hysteresis loops. For the other two test columns JRW05-5 and JRW05-6 the pinching phenomena were apparently observed near the origin because the damages due to the shear cracks, which occurred allover the column height, were relatively severe.

3.4 Strains in CFRP

The maximum strain distributions of CFRP strips in the perpendicular directions and in the parallel directions of the lateral loading at each displacement level of JRW05-3 retrofitted with CFRP tapes and JRW05-7 retrofitted with CFPR braiding were illustrated in

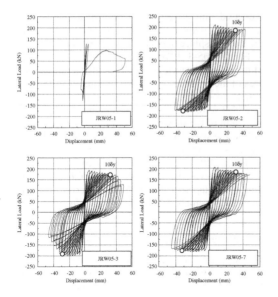

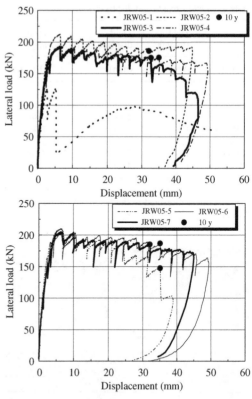

Figure 6. Hysteresis response.

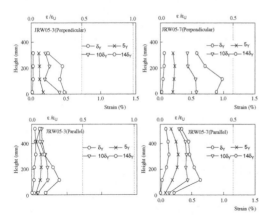

Figure 7. Strain distributions of the CFRP for JRW05-3 and JRW05-7.

Figure 7. Strains in the X-axis were represented as the averages of the maximum strains of the CFRP of both sides of the column sections. Displacement levels shown in the figure were of δy, 5δy, 10δy and 14δy. The X-axis of strain was also represented as the ratio to the ultimate strain ε_u of the CFRP.

Maximum strains gradually increased when the applied displacements increased. For JRW05-3, the maximum strains of the perpendicular direction at 14δy were approximately less than one third of the ultimate strain ε_u of the CFRP. Maximum strain distributions of the parallel direction at 14δy were less than those of the perpendicular direction and were one fourth of the ultimate strain ε_u. Maximum strains of the parallel direction at the bottom end of the columns did

Figure 8. Comparison of the envelope curves.

not become so large because of the restriction from the rigid stub. The strain distributions of the CFRP braiding for JRW05-7 in the Figure 7 were about two times greater than the strain distributions of the CFRP she tapes. It is indicated that the confined effect of the CFRP braiding was approximately the same of the tape or sheet considering Young's Modulus of the both materials.

In comparison with the strains of the both directions, the strains of the perpendicular direction were greater than the strains of the parallel direction. It is indicated that CFRP strips acted more effectively as confinement of the column section rather than shear protection.

4 DISCUSSIONS

4.1 Envelope curves

Figure 8 shows the envelope curves of the hysteresis loops of the all test columns. There was no difference in the initial stiffness in the all test columns.

In the test columns with retrofitting, the peak strength of the applied displacement levels gradually

Table 6. Observed maximum ductility.

Test column	Observed maximum strength (kN)	Maximum strength 80% (kN)	Observed ductility	Design ductility
JRW05-1	127.8	102.2	Less than 1	–
JRW05-2	212.8	170.3	Over 14	12.02
JRW05-3	192.9	154.3	12	12.62
JRW05-4	193.7	155.0	Over 14	14.44
JRW05-5	210.3	168.3	8	5.35
JRW05-6	204.1	163.3	Over 14	12.62
JRW05-7	204.8	163.9	Over 14	8.52

decreased after the maximum strength was recorded. The ductility ten are pointed out in the figure. The strength of JRW05-2 with CFRP sheet was slightly higher than the strength of the other test columns with CFRP strips throughout the lateral loadings. It is indicated that the compressive strength of the concrete and the bond characteristic for the longitudinal reinforcing bars were improved by the CFRP sheet that perfectly confined column sections. In general, large differences between the sheet and the strips were not observed in the envelope curves throughout the lateral loadings. The observed maximum strength and the observed maximum ductility are summarized in Table 6. The maximum ductility was defined as the ductility at the 80% of the maximum strength.

Calculated shear strength V_{yd} and calculated flexural strength V_{rd} using the experimentally obtained material strength were 120 kN and 174 kN, respectively from the equations (3) and (5). The observed maximum strength of JRW05-1 (127.8 kN) was approximately the same of the calculated shear strength V_{yd}. The average of the observed strength (203.1 kN) of the test columns with retrofitting were 1.17 times of the calculated strength V_{rd} because of the confined effects of the CFRP. Observed maximum ductility of test columns with retrofitting exceeded the target ductility shown in the Table 1. Especially the maximum ductility of JRW05-7 with CFRP braiding was 1.75 times of the target ductility. The strength of JRW05-5 continued to sustain the more than 80% of the maximum strength until ductility reached eight over the target ductility five. After ductility ten, the strength rapidly degreased. It is indicated that the design procedure of the retrofitting with the CFRP sheet is applicable to that for CFRP strips.

4.2 Energy dissipation

Figure 9 shows the relationships between the energy dissipation and the applied maximum displacement of the first cycle at each applied displacement levels.

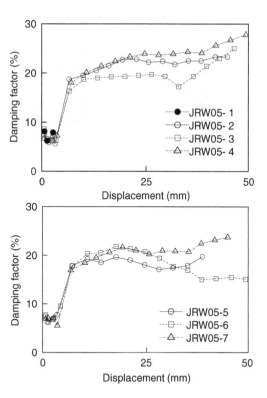

Figure 9. Progress of energy dissipations.

Energy dissipation presented as the equivalent viscous damping factor in this study was one of the most important index to decrease the dynamic responses of the structures during the earthquakes. The equivalent viscous damping factors were obtained by measuring the area of the hysteresis loops.

Initial damping factors of all test columns before yielding were $6 \sim 10\%$ that were the general values of the reinforced concrete structures before cracking. When yielding occurred in the test columns the damping factors increased rapidly up to 20% approximately. After yielding damping factor of the JRW05-2 and JRW05-4 continued to increase gradually up to $25\% \sim 27\%$. The damping factors of other test columns with retrofitting were slightly less than those of the former two columns. This difference came from the pinching or the slip phenomena in the hesteresis loops due to the sliding or the shear behavior at the severely damaged area of the columns.

4.3 Design strength of shear transfer

Before displacement $6\delta y$ the displacement distributions of the test columns with retrofitting were the flexural modes while the rotations at the hinging region were predominant in the displacement distributions.

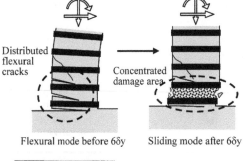

Distributed flexural cracks

Concentrated damage area

Flexural mode before 6δy   Sliding mode after 6δy

Loading direction

Failure pattern at the final stage of the JRW05-3

Figure 10.   Transition of the failure mode.

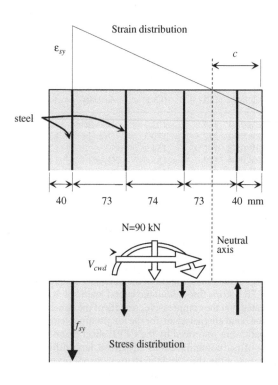

Strain distribution

ε_{sy}

c

steel

40   73   74   73   40 mm

N=90 kN

V_{cwd}

Neutral axis

f_{sy}

Stress distribution

f_{sy}=295 N/mm^2

f_c=27 N/mm^2

ε_{sy}=1.475×10^{-3}

Figure 11.   Stress distribution of the member section.

After displacement 6δy the displacement distributions of the test columns with CFRP strips except for JRW05-4 changed to the sliding mode when the sliding displacement at one concentrated damaged area became large in the total displacement of the test columns as shown in Figure 10.

It is necessary to evaluate the capacity of the shear transfer by the interlocking of the aggregate and the dowel action of the longitudinal reinforcing bars through the severely damaged area.

In this section the design strength of the shear transfer obtained by the equations (6) in JSCE Standard was discussed in comparison with the observed maximum strength in the test columns.

$$V_{cwd} = \frac{(\tau_c + p \cdot \tau_s) A_c}{\gamma_b} \quad (6)$$

τ_c and τ_s are the shear transfer stresses of the concrete and the main reinforcing bars, respectively, obtained by the following equations (7) and (8). γ_b is the member coefficient and A_c is the area of the member section.

$$\tau_c = \mu \cdot f_c^{\ b} (\alpha \cdot p \cdot f_y - \sigma_{nd})^{1-b} \quad (7)$$

$$\tau_s = \frac{0.08 f_y}{\alpha} \quad (8)$$

α in the equation (9) is the reduction factor presenting the degradation of the axial strength for the transverse reinforcing bars and the increase of the

Table 7.   Design strength of shear transfer.

c	P_{sc}	P_{cc}	p
104.4 mm	140.2 kN	387.4 kN	0.0147

α	τ_c	τ_s	V_{cwd}
0.276	8.613	85.51	683.3 kN

shear force. f_c is the concrete strength and f_y is the yield strength of the longitudinal reinforcing bar.

$$\alpha = 0.75 \left\{ 1 - 10 \left(p - 1.7 \frac{\sigma_{nd}}{f_y} \right) \right\} \quad (9)$$

$$0.08 \sqrt{3} \leq \alpha \leq 0.75$$

In the equation, p is the ratio of the tensile reinforcing bars and σ_{nd} is the average compressive stress

1128

perpendicularly acting to the shearing section of the member.

$$\sigma_{nd} = -\frac{1}{2}\frac{\left(P_{sc} + P_{cc}\right)}{A_{cc}} \tag{10}$$

P_{sc} and P_{cc} are the compressive forces at the compressive reinforcing bars and the concrete of the section, respectively. A_{cc} is the compressive area of the member section.

$$b = \frac{2}{3}, \quad \mu = 0.45, \quad \gamma_b = 1.3$$

μ is the average coefficient of friction, b is the coefficient for the cracked section of the member.

The location of the neutral axis of the column section at the yielding of the longitudinal reinforcing bars was calculated from the equilibrium of the axial force of the column section using the design strength of the materials as shown in Figure 11. It is assumed that the Young's Modulus of the steel and the concrete are 2.0×10^5 N/mm^2 and 2.5×10^4 N/mm^2, respectively, and both materials remain in elastic.

The calculated results are summarized in Table 7. Calculated neutral axis located in the assumed area that was 40 mm < c < 113 mm. Calculated design strength of the shear transfer V_{cwd} (=683.3 kN) is over three times of the observed maximum strength which were approximately 200 kN. It is noted that shear force subjected to the columns can be transferred sufficiently through the severely damaged area of the test column to the basement. On the other hand, the average friction coefficient that was assumed to be 0.45 in this case might be evaluated as larger value than the actual value because the concrete of the sliding zone were severely damaged due to the numerous cyclic loadings.

5 CONCLUSIONS

Seismic loading tests of the scaled reinforced columns retrofitted with the CFRP sheet and strips were performed. The following conclusions were made.

(1) Seismic performance of the reinforced concrete columns retrofitted with spaced CFRP strips showed the same performance of the column with CFRP sheet providing the same amount of CFRP. It is indicated that the equations of the evaluation for the CFRP sheet is applicable to that for spaced CFRP strips.

(2) When using spaced CFRP strips the damage level of the column could be easily estimated with visual inspection after earthquake.

(3) Fracture of the CFRP was avoidable while the damages of the column were concentrated to the bare concrete section between spaced CFRP strips.

(4) The displacement distributions changed from the flexural mode to the sliding mode at the concentrated damaged area at the lower portion of the test column retrofitted with spaced CFRP strips.

(5) Calculated design strength of the shear transfer is over three times of the observed maximum strength. The shear force could be sufficiently transferred through the severely damaged area.

(6) It is anticipated that there is strong possibility of decreasing the amount of the CFRP.

ACKNOWLEDGEMENT

The authors would like to thank staffs and graduate students of Structural Earthquake Engineering Laboratory of Hiroshima University.

REFERENCES

ACI Committee 440, "State-of-the Art Report on fiber Reinforced Plastics Reinforcement for Concrete Structure (ACI440R-96)" American Concrete Institute, 1996

AIJ Committee "Design and Construction Guideline of Continuous Fiber Reinforced Concrete" Architectural Institute of Japan, 2001

JSCE "Design and Construction Guideline of Continuous Fiber Reinforced Structure" Japan Society of Civil Engineering, 1996

JSCE "Standard Specifications for Concrete Structures", 2002

Katsumata, H., Kobatake, Y., and Takeda, T. "A Study on Strengthening with Carbon Fiber for Earthquake-Resistant Capacity of Existing Reinforced Concrete Columns "Proceedings of the 9th conference on Earthquake Engineering, Vol.7, Tokyo, Japan, Aug. 1988, pp.517–522

Railway Technical Research Institute : Guide for retrofitting design for the railway column with carbon fiber sheet, JCAA, Tokyo

*Fracture Mechanics of Concrete and Concrete Structures – Design, Assessment and
Retrofitting of RC Structures – Carpinteri, et al. (eds)
© 2007 Taylor & Francis Group, London, ISBN 978-0-415-44616-7*

Effects of concrete composition on the interfacial parameters governing FRP debonding from the concrete substrate

J.L. Pan

*Key Laboratory of Concrete and Prestressed Concrete Structure, Ministry of Education
Southeast University, China*

C.K.Y. Leung

Hong Kong University of Science and Technology, Hong Kong

ABSTRACT: For the concrete beam strengthened with FRP, failure may occur from the bottom of a major flexural/shear crack along the span. As debonding failure occurs within the concrete, interfacial friction resulted from aggregate interlocking between opposing surfaces of the debonded zone plays an important role in the debonding behavior. Interfacial debonding can be analyzed with a three-parameter model. To investigate the effect of concrete composition on debonding behavior, the direct shear test was conducted with ten different compositions of concrete. Interfacial parameters were then derived according to a three-parameter model. The test results show little correlation between interfacial parameters and concrete compressive or splitting tensile strength. However, the debonding initiation strength and residual shear strength correlates well with the surface tensile strength and the maximum crack opening is mainly governed by the aggregate content. According to our results, the composition of the concrete is an important factor that should be considered explicitly in the investigation of FRP debonding from a concrete substrate.

1 INTRODUCTION

The continuous aging of civil infrastructures and the change of load requirements with time give rise to the need of structural strengthening. The bonding of fiber reinforced plastics (FRP) plates on concrete members has been recognized as an effective retrofitting technique. For concrete beams strengthened with FRP, debonding failure may occur from the bottom of a major flexural/shear crack in the span, as shown in Fig. 1. This kind of crack-induced debonding is initiated by the presence of high shear stress concentration at the interface around the bottom of the cracks. As debonding failure occurs within the concrete substrate at a small distance from the

concrete/adhesive interface, interfacial friction resulted from aggregate interlocking between opposing surfaces of the debonded zone plays an important role in the debonding behavior.

Crack-induced debonding failure is often studied with the direct shear test, which involves a FRP plate bonded on a concrete prism. By pulling the FRP plate along the direction of its length, the bond capacity at debonding failure can be obtained. Due to the shear lag phenomenon, the bond capacity approaches a plateau value with increasing bond length. Recently, plenty of experiments have been conducted with single shear tests (e.g. Taljsten 1997, Chajes et al. 1996, Bizindavyi and Neale 1999), double shear tests (e.g. Van Gemert 1980, Neubauer and Rostásy 1997) and modified beam (e.g. Van Gemert 1980, de Lorenzis et al. 2001) to study the bond behavior between FRP and concrete members. These works have led to improved understanding of the failure characteristics of the FRP-to-concrete bond. It is commonly considered that the effectiveness of strengthening depends on the following aspects: (1) surface preparation of concrete; (2) the type of adhesive; (3) geometric factors, such as FRP bond length, thickness of FRP plate, FRP width etc; (4) interfacial fracture energy. The

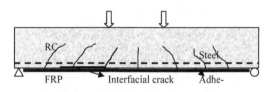

Figure 1.   FRP debonding from the bottom of a major flexural/shear crack.

interfacial fracture energy is often taken to be a function of the compressive or tensile strength of concrete. However, a systematic study on the correlation between the fracture energy and these strength parameters has never been performed. Also, to analyze the debonding process, various investigators have developed models (Holzenkampfer 1994, Yuan et al. 2001, Chen and Teng 2001) based on nonlinear fracture mechanics concepts. The debonded zone is treated as a process zone with residual shear stress, which decreases with interfacial sliding due to effect of abrasion. As aggregate interlocking is related to the size and content of the aggregates in the concrete, the debonding behavior is expected to be affected by the concrete composition, in addition to macroscopic mechanical properties such as compressive and splitting tensile strength.

In the present investigation, the direct shear test is performed on concrete prisms with bonded FRP to study the interfacial debonding behavior. Various strength parameters of the concrete as well as its aggregate content are also measured. The objective of this study is to investigate the effect of (1) concrete properties (such as concrete compressive strength, splitting tensile strength, and concrete surface strength), and (2) the aggregate content, on the debonding behavior. In the testing program, concrete of 10 different compositions are employed. Interfacial debonding is analyzed with a three-parameter model, in which debonding is assumed to initiate once the interfacial shear stress reaches a critical value τ_s. After the initiation of debonding, the interfacial stress drops to a lower value τ_0. On further sliding, the stress will decrease with interfacial sliding at a slope k. Using such a model, the variation of stress and strain along the FRP during the debonding process can be derived. By fitting the theoretical FRP strain distribution to measured values on various specimens, the three model parameters (τ_s, τ_0, k) can be obtained for each concrete composition. The correlations between interfacial parameters and strength parameters as well as aggregate content are then studies, and empirical relations are proposed.

2 EXPERIMENTAL PROGRAM

2.1 Specimen preparation and material properties

To investigate the effect of concrete composition on the debonding behavior, ten batches of concrete with different mixing proportions were employed to prepare the concrete prisms. The mixing proportions for each batch of specimens are listed in Table 1. The size of the concrete prism is 100 mm (width) ×100 mm (depth) ×500 mm (length), as shown in Figure 2. After casting, the specimens were placed in the curing room for 28 days. After curing, FRP sheets can be applied. Before

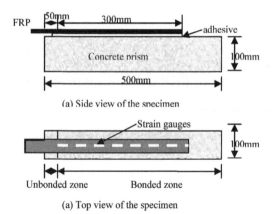

(a) Side view of the specimen

(a) Top view of the specimen

Figure 2. Dimensional information about the specimens.

FRP bonding, the surfaces of concrete prisms were roughened with a needle-gun to expose the aggregates so that a good bond between FRP plate and concrete could be achieved. After the surface preparation, epoxy primer, which is a cohesive liquid at atmosphere temperature, was applied to improve the bond performance. After the primer was hardened, two layers of FRP sheets were bonded to concrete prisms layer by layer using epoxy resin. The bonded part of the FRP plate was 300 mm in extent, starting from a location 50 mm from the edge of concrete prism (Fig. 2). The initial 50 mm is left unbonded to avoid wedge failure of concrete due to shear stress. To ensure full hardening of epoxy, the specimens should be cured for 7 days before testing. In order to record the strain variation along the FRP plate during the loading process, nine strain gauges were placed on the FRP plate with a center-to-center space of 30 mm, as shown in Fig. 2. For the acquisition of strain data, an automatic data logger was employed.

To study the correlations between the concrete material properties and the debonding behavior, the mechanical properties of concrete, including compressive strength, tensile strength and surface tensile strength, were measured together with the ultimate bond capacity. Three small concrete cylinders (100 mm in diameter and 200 mm in height) from each batch of concrete were used to obtain the compressive strength, and another three concrete cylinders with larger size (150 mm in diameter and 300 mm in height) were used to perform splitting tension test. To measure surface tensile strength, a steel plate with dimension of 100 mm × 100 mm was bonded on the concrete prism. The surface tensile strength is then calculated as the ultimate load divided by the total area of the steel plate. In addition, the aggregate content was measured as the area fraction of aggregates (defined as particles with projected dimension larger than 4.75 mm) on the concrete section. The various material properties of the

Table 1. Experimental results of material properties of the concrete.

Concrete composition	Mixing Proportion (C:W:S:A)	Aggregate content	Compr. strength (MPa)	Tensile strength (MPa)	Surface tensile strength (MPa)
M1	1:0.5:1.5:2.6	0.119	43.1	4.18	2.851
M2	1:0.6:1.5:2.6	0.107	35.2	3.17	2.146
M3	1:0.5:1.5:1.5	0.117	57.5	3.69	3.04
M4	1:0.6:1.5:1.5	0.091	38.6	3.38	1.625
M5	1:0.4:1.5:2.6	0.096	61.5	4.48	2.779
M6	1:0.5:1.5:0.5	0.036	47.4	3.76	2.261
M7	1:0.54:2.55:1	0.030	47.13	3.68	2.037
M8	1:0.5:2.25:1.25	0.070	44.73	3.26	2.225
M9	1:0.5:2:1.5	0.103	52.35	3.99	2.730
M10	1:0.5:1.5:2	0.118	57.87	4.49	2.474
Remarks	C:W:S:A is the ratio of cement to water to sand to aggregate				

concrete are summarized in Table 1. The FRP used in the tests is the Reno composite material system, with strength of 4200 MPa in the fiber direction. The Young's modulus of the FRP is about 235 GPa. The FRP thickness is about 0.11 mm per ply according to the production specifications and two plies were employed in the test specimen. In the present experiments, an epoxy-based adhesive was used. According to the manufacturer, the tensile strength of the resin for FRP is about 30 MPa, the tensile modulus is about 3.3 GPa and shear strength is over 10 MPa after curing for 7 days.

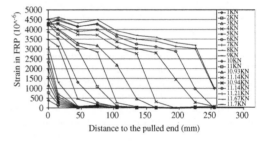

Figure 3. Strain variations at different load values for the specimen M8-2.

2.2 Test setup and test procedure

As for the setup of the direct shear test, a steel frame that can hold the concrete specimen tightly in the vertical direction was designed. The frame was vertically installed in the MTS loading frame that applied a pulling force on the FRP plate. Alignment is important as the force should be along the vertical direction to prevent any horizontal force component that may introduce peeling effect on the FRP plate. During the testing process, a LVDT is used to measure the global displacement of the FRP plate. The test was conducted under displacement control with loading rate of 0.1 mm/min. An automatic data logger was employed to collect the strains along the FRP plate and the displacement from the LVDT, as well as load and stroke data from the MTS machine.

2.3 Strain distributions during the loading process

In the experimental program, a total of 30 specimens (10 for each concrete composition) were tested to investigate the effect of concrete composition on the bond behavior between the FRP plate and concrete. For each group of specimens with the same concrete composition, one specimen was instrumented with nine strain gauges for measuring strain variations during the loading process. Only two strain gauges were put on the other two specimens for checking.

To illustrate the debonding behavior for the specimens under direct shear force, the experimental data of the specimen M7-2 is shown as an example. Fig. 3 shows the strain distributions along the FRP plate at different load values. Each curve corresponds to the strain distribution along the FRP at a particular load. When the load is lower than 8 kN, the strains in the FRP plate decrease quickly with distance from the loaded end. This descending trend is ascribed to the low axial stiffness of the bonded FRP plate with respect to that of the concrete prism. Before initial debonding, increase of the applied load will cause the curve to shift upward, but the shape of the strain distribution doesn't change. However, when the load value goes beyond 8 kN, the interfacial debonding starts to occur from the loaded end, and the shape of the curve starts to change and the slope of the curve near the loaded end tends to decrease. Since the slope of the curve reflects the rate of strain change in the FRP plate (which is proportional to the interfacial shear stress), the decrease of the slope represents shear softening along the debonded interface. As the load increases, the interfacial debonding

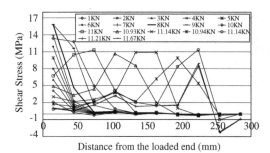

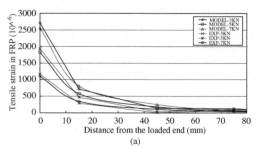

(a)

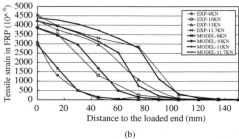

(b)

Figure 4. Shear stress distributions along the concrete/adhesive interface for M7-2.

Figure 5. Comparisons between the fitting results and test results (a) before initial debonding (b) after initial debonding.

tends to propagate to the free end of the FRP plate. This is indicated by the shifting of strain distribution curves towards the free end of the FRP plate. Interestingly, it is found that the maximum strain in the FRP plate stays approximately constant when the debonded zone has extended to a certain distance from the pulled end. This corresponds to the situation with a fully developed debonding zone propagating along the interface.

Fig. 4 shows the approximate shear stress distributions along the concrete/adhesive interface at different load values. The shear stress at each location is calculated as the difference between the tensile stresses at adjacent points divided by the distance between the two points. With this approximate calculation method, the magnitude of the shear stress is not accurate, but the trend of shear stress variation with increased loading can still be revealed. When the applied load is smaller than 8 kN, the shear stress is found to decrease quickly with the distance from the loaded end. When the loading goes beyond 8 kN, location of τ_{max} shifts towards the free end of the plate, indicating the debonding process of the FRP plate from the concrete substrate. As loading continues to increase, the shear stress at points near the pulled end show a decreasing trend, indicating shear softening along the interface in the debonded region.

interfacial shear stress at the concrete/adhesive interface reaches the interfacial shear strength τ_s. After debonding, the residual shear strength (τ_0) at the concrete/adhesive interface is related to the interfacial sliding by:

$$\tau = \tau_0 - ks \qquad (1)$$

where τ_0 and k are initial residual shear strength right after debonding and softening rate in the debonded zone. Based on this interfacial softening relation, the tensile stress or strain distribution along the FRP plate and interfacial shear stress distribution along the concrete/adhesive interface can be derived. Details of the derivation can be found in Leung and Tung (2006).

3 THEORETICAL MODELINNG

3.1 Interfacial shear softening relation

To model interfacial debonding, the shear slip relation along the concrete/adhesive interface is required. With a proper bond shear slip relation, the theoretical stresses or strains along the FRP plate and the interfacial stresses along the concrete/adhesive interface can be obtained. In the present study, the three-parameter model proposed by Leung and Tung (2006) is employed to study debonding behavior. In the model, interfacial debonding is taken to start when the

3.2 Interfacial parameters for various concrete compositions

To illustrate the extraction of interfacial parameters, results for specimen M7-2 is employed. With the three-paramter model, the tensile strain along the FRP plate at different load values are calculated and compared with the test results (Fig. 5). The actual interfacial parameters are the ones that provide good agreement between calculated and measured strain.

Due to the irregular surface of the concrete after roughening, it is very difficult to measure the adhesive thickness accurately. The mean thickness of adhesive

Table 2. Simulation results of the adhesive thickness and interfacial parameters.

	Adhesive thickness (mm)	τ_{max} (MPa)	τ_0 (MPa)	k (MPa/mm)	Max. crack opening(mm)	Exp.P_{ult} (kN)
M1-2	3	16.0	2.5	2.3	1.09	17.46
M2-3	3	13.5	1.6	2.0	0.8	14.75
M3-2	3	17.8	1.8	2.0	0.9	19.43
M4-3	3.5	13.0	1.5	2.3	0.65	15.68
M5	–	–	–	–	–	17.45
M6-3	3.5	14.7	1.5	2.5	0.6	13.31
M7-2	3	11.3	2.0	6.0	0.33	11.71
M8-3	3	13.8	1.3	2.0	0.65	13.07
M9-2	3	16.2	2.0	3.5	0.57	17.48
M10-2	3	15.8	1.8	2.5	0.72	18.07

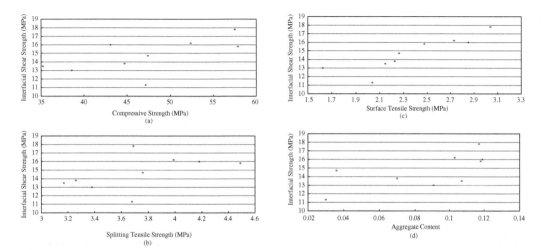

Figure 6. Correlation between the interfacial shear strength and (a) compressive strength, (b) splitting tensile strength, (c) surface tensile strength, (d) aggregate content.

layer is therefore obtained by the fitting of strain distributions along the FRP plate within the elastic stage. Good agreement between calculated and experimental results is obtained when the adhesive thickness is taken to be about 3 mm, as shown in Fig. 5a. The maximum shear stress (τ_s) is determined from the highest applied loading before the elastic stress distribution starts to shift towards the free end of the plate. By fitting the tensile strains along the FRP plate in the nonlinear regime (Fig. 5b), the residual shear strength (τ_0) and the softening rate (k) can be obtained. The maximum sliding is calculated as $\delta = \tau_0/k$, which signifies the sliding distance beyond which the residual stress drops to zero.

The fitting process has been repeated for all specimens of different concrete compositions. The derived adhesive thickness and interfacial parameters are shown in Table 2. For each specimen, the adhesive

thickness is found to be in the range from 3 mm to 3.5 mm, which is consistent with visual inspection.

3.3 Correlations between concrete properties and interfacial shear strength

To study the effect of concrete properties on the debonding behavior, the interfacial shear strength is correlated to the macroscopic properties of concrete. According to the test results, the compressive strength of concrete for each batch ranges from 35.2 MPa to 61.5 MPa. The splitting tensile strength values of the concrete are between 3.17 MPa and 4.49 MPa. The surface tensile strength for each concrete composition varies from 1.625 MPa to 3.04 MPa. Fig. 6 shows the correlations between the interfacial shear strength and concrete material properties. According to the plots, the interfacial shear strength has the best correlation

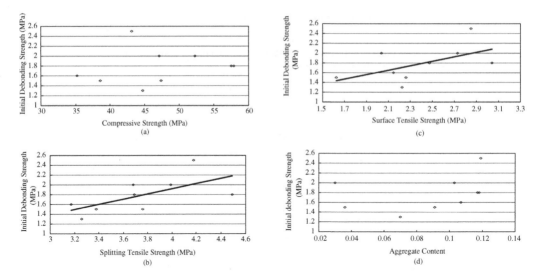

Figure 7. Correlation between the residual debonding strength and (a) compressive strength, (b) splitting tensile strength, (c) surface tensile strength, (d) aggregate content.

with surface tensile strength of concrete. There is also some correlation with the compressive strength, but not as good as that with the surface tensile strength. No clear correlation can be found between the interfacial shear strength and the splitting tensile strength, as well as the aggregate content. According to the test results, it seems that debonding initiation, which is induced by forces acting on the concrete surface only, is not affected by the bulk material properties of the specimen. Also, the aggregate content, which does not govern the surface strength of concrete, has little correlation with the interfacial shear strength.

Based on the above, an empirical equation for the interfacial shear strength can be proposed as:

$$\tau_s = 3.92 f_{ctm} + 5.36 \qquad (2)$$

where 'f_{ctm}' is the concrete surface tensile strength from pull-out test.

3.4 Correlations between residual debonding strength and the concrete properties

As mentioned above, the FRP debonding failure in the direct shear test can be analyzed with the three-parameter model, and the interfacial parameters are obtained by the data fitting with experimental results. The residual shear strength defines the interfacial shear friction between the FRP plate and the concrete right after debonding. It is one of the key parameters determining the debonding behavior. To study the effect of concrete properties on the debonding behavior, the correlations between the residual shear strength and the concrete properties are plotted in Fig. 7. It is found

that the residual shear strength has good correlation with the splitting tensile strength and surface tensile strength of concrete, but has little correlation with concrete compressive strength or aggregate content.

To quantify the effect of concrete properties on the debonding behavior, the residual shear strength is related to the surface tensile strength of concrete f_{ctm} only, and the residual shear strength τ_0 can be given by:

$$\tau_0 = 0.457 f_{ctm} + 0.692 \qquad (3)$$

where f_{ctm} is the concrete surface tensile strength from pull-out test.

3.5 Correlations between the maximum crack opening and the concrete properties

The maximum sliding ($\delta = \tau_0/k$) defines the maximum relative displacement within which interfacial friction still exists between the FRP plate and the concrete. The maximum sliding is an important parameter for analyzing the shear softening in the debonded zone. In Fig. 8, the maximum sliding is plotted against the concrete properties. The results indicate that this parameter correlates better with the surface tensile strength and aggregate content than the compressive or splitting tensile strength of concrete. The strong effect of aggregate content on the maximum crack opening may be resulting from the fact that a higher aggregate content increases the abrasion resistance along the interface and hence delay the shear softening between the FRP and the concrete. The surface tensile strength reflects the quality of concrete surface, and hence it will affect the abrasion behavior during the debonding

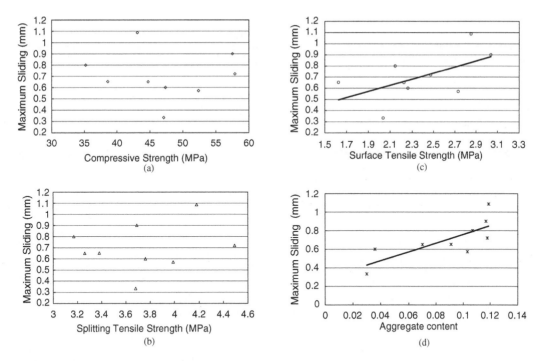

Figure 8.   Correlation between the maximum sliding crack opening and (a) compressive strength, (b) splitting tensile strength, (c) surface tensile strength, (d) aggregate content.

process. To quantify the effect of aggregate content and surface tensile strength, the maximum sliding is modeled with these two parameters. According to linear regression analysis, the empirical model of the maximum crack opening 'δ' in terms of the surface tensile strength 'f_{ctm}' and aggregate content 'a' is given by:

$$\delta = f_{ctm}^{2.6}(-0.583a+0.135) \qquad (4)$$

Note that δ is an important parameter governing the maximum bond force that can be sustained by the FRP. With a high value of δ, a large softening zone can be formed along the interface, resulting in a high load at ultimate debonding failure.

3.6   Determination of ultimate bond capacity with the empirical equations

Using Eqns (2) to (4) derived above, the interfacial parameters for each composition is empirically related to the surface tensile strength and aggregate content. With these parameters, the ultimate bond capacity is derived using the 3-parameter debonding model (with $k = \tau_0/d$). The calculated load capacities are compared with the experimental results in Fig. 9. Most data points locate near the 45-degree line and within the error range of positive and negative 10 percentages. Only

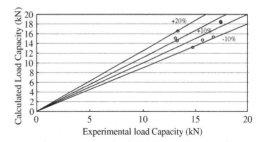

Figure 9.   Comparison of calculated bond capacity with experimental results.

one data point locates at the error line of 20 percentages, which may be resulted from the inconsistency of test data. Generally speaking, the bond capacity is reasonably predicted with parameters derived from the proposed empirical equations.

Since the same data in Fig. 9 has been used to derive the empirical equations, the comparison in Fig. 9 is not an independent validation of the equations. We want to point out, however, that if the aggregate content has not been included in Eqn (4), the resulting fitting will show a much larger coefficient of variation. As a result, the calculated bond capacity will also not be in good agreement with the test results. Our investigation therefore establishes the aggregate content to be an important parameter governing debonding behavior. In other words, the effect of concrete composition should not be neglected in the study of debonding along the FRP/concrete interface.

3.7 Further discussions

The current investigation is certainly far from complete. The proposed empirical equations are meant to illustrate the relations between the interfacial parameters and other concrete parameters. For full verification, more experimental data need to be generated in the future. In the current study, only a single aggregate grading has been employed. In future investigations, the effect of aggregate size distribution should also be considered. Moreover, in this study, the 'aggregate' is arbitrarily defined as particles with projected dimension of 4.75 mm or above on the concrete surface. As only part of an aggregate is exposed at the surface, the actual particle size can be much large than 4.75 mm, which explains the small values of aggregate content given in Table 1. Such a definition for the 'aggregate' should be re-evaluated in future work.

4 CONCLUSIONS

In this paper, experimental investigations have been conducted to study the effect of the concrete composition on FRP debonding behavior. Our results show little correlation between the interfacial parameters and the compressive strength of concrete. However, the interfacial shear strength (τ_s) correlates well with the surface tensile strength and the residual shear strength has good correlation with the surface tensile strength or the splitting tensile strength. The maximum sliding is mostly governed by the surface tensile strength and the aggregate content. Empirical models are hence proposed to relate τ_s and τ_0 to the surface tensile strength and the maximum sliding 'δ' to the aggregate content and the surface tensile strength. While the current study is far from complete, the results indicate that the composition of concrete (specifically, the aggregate content) is an important factor that should be considered explicitly in the investigation of FRP debonding from a concrete substrate.

REFERENCES

Bizindavyi L. and Neale K.W. (1999) "Transfer length and Bond strengths for composites bonded to concrete." *Journal of composites for construction*, Vol. 3, No. 4. pp.153–160.

Chajes, M.J., Finch, W.W. Januszka T.F. and Thomosh, T.A. (1996) "Bond and force transfer of composite material plates bonded to concrete" *ACI Structural Journal*, vol. 93, No. 2, pp. 231–244.

Chen, J.F., and Teng, J.G. (2001) "Anchorage strength models for FRP and steel plates bonded to concrete."*J. Struct.Engrg.*, Vol.127, No.7, pp.784–791.

Holzenkampfer, O. (1994) "Ingenieurmoddelle des Verbundes geklebter Bewehrung fur Betonbauteile, *Dissertation*, TU Braunschweig.

De Lorenzis, L., Miller, B. and Nanni, A. (2001) "Bond of Fiber-reinforced Polymer Laminates to Concrete" *ACI material Journal*, V. 98, No.3, pp. 256–264.

Leung, C.K.Y. Tung, W.K. (2006) "Three-Parameter Model for Debonding of FRP Plate from Concrete Substrate", *ASCE Journal of Engineering Mechanics*, Vol.132, No.5, pp. 509–518 (2006).

Neubauer U. and Rostasy F. S. (1997) "Design aspects of concrete structures strengthened with externally bonded CFRP plates" *Proceeding of the Seventh International conference on structural Faults and Repairs*, Edited by M.C. Forde, Engineering Technics Press, Edinburgh, UK, pp.109–118.

Taljsten, B. (1997) "Strengthening of beams by plate bonding." *J. Mat. in Civ. Engrg.*, ASCE, 9(4), 206–212.

Van Gemert, D. (1980) "Force transfer in epoxy-bonded steel-concrete joints" *International Journal of Adhesion and Adhesive*, No.1, pp. 67–72.

Yuan, H., Wu, Z.S. and Yoshizawa, H. (2001) "Theoretical solutions on interfacial stress transfer of externally bonded steel/composite laminates." *Journal of Structural Mechanics and Earthquake Engineering*, JSCE, No. 675/1–55, pp.27–39.

Fracture Mechanics of Concrete and Concrete Structures – Design, Assessment and Retrofitting of RC Structures – Carpinteri, et al. (eds)
© 2007 Taylor & Francis Group, London, ISBN 978-0-415-44616-7

Modeling and strengthening of RC bridges by means of CFRP

P. Koteš & P. Kotula
University of Žilina, Žilina, Slovakia

ABSTRACT: In the frame of the research work APVT-20-012204, three reinforced concrete girder bridges were diagnosed. The results of the diagnostic will be used for modelling and investigation of the real traffic load effects. The modified reliability levels for the evaluation of existing bridges were used for their evaluation (Koteš 2005, Koteš & Vičan 2006). The different remaining bridge lifetimes for various load models given in the Slovak standard were obtained by calculation.

Here, the paper is focused on just one bridge, which is the most corrupted. This bridge was numerically modeled using software ATENA. The material characteristics needed for modelling were obtained by diagnostic. The bridge should be strengthened due to low load-carrying capacity. The CFRP lamellas for strengthening of the girders subjected to bending will be used in the FEM model. The results of numerical analyses using the FEM model are used for the prepared experiment.

1 INTRODUCTION

The paper presents the objectives and partial results of research work no. APVT-20-012204 "Remaining service life and increase of concrete structure reliability". The optimum methods for increasing of existing concrete structure reliability by the application of the new technologies based on the technical diagnostics method, computer simulations and probabilistic evaluation of the concrete bridges, is the major aim of this project.

The attention of the project is paid to the project concentrates on existing concrete bridge structures, especially to bridge evaluation and the increase of reliability and serviceability for the remaining lifetime (Frangopol & Estes 1997, Nowak 1995, Vičan 1999). There are always applied modern approaches to the structure reliability evaluation in practice in many countries. The approaches are based on probabilistic interpretations and they are supported by computer simulations. The new materials and technologies are used to increase the load carrying capacity and serviceability of concrete bridge structures as well. Many European standards concerning such problems are based on theoretical and experimental test issues.

For successive accepting of the European standards in Slovakia, the research of degradation factors on concrete structures, their accurate evaluation and new technologies are needed to increase the load carrying capacity and serviceability of structures.

2 SELECTION OF CRITERIA AND THE DESCRIPTION OF THE BRIDGE

The criteria for the bridge choice were determined in order to fulfil the aim of the research work. The criteria were focused on the kind of bridge (from the viewpoint of the superstructure type, faults occurrence, accessibility of the bridge), location, importance in the traffic network and adequate traffic action on the bridge.

Three reinforced concrete girder bridges satisfied the determined criteria. The bridges are in the village Kolárovice, part of Škoruby, on the road I / 18 over the Kolárovice river (Fig. 1).

All bridges have similar parameters concerning dimensions, material quality and traffic load, but the reinforcement corrosion and the significant bending and shear cracks were found just on bridge No. 3. Therefore, bridge No. 3 was selected for further investigation (Fig. 2).

The reinforced concrete single span girder bridge has a theoretical span of 10.006 m (the length of the bridge is 10.824 m). The width of the road is 7.51 m and the overall width of the bridge is 9.51 m. The bridge skewness is 45.22°. The superstructure consists of a bridge slab having a thickness of 0.19 m and of six main beams with dimensions of 0.325 / 0.84 m. The end diaphragms have dimensions of 0.58 / 0.84 m and three intermediate diaphragms having dimensions of 0.20 / 0.74 m to ensure the transverse load distribution (Fig. 3).

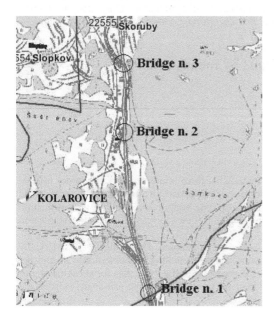

Figure 1. The location of the selected bridges in village Kolárovice.

Figure 2. Bridge superstructure.

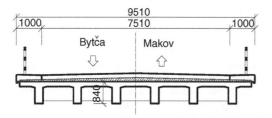

Figure 3. The bridge cross-section.

Table 1. The measured values of geometric and material characteristics.

Variables	The mean value	The standard deviation
Strength of concrete – f_c.[MPa]	49.780	4.8689
Girder height – h [m]	0.8370	0.00356
Girder width – b [m]	0.3222	0.00371
Concrete cover – c [mm]	29.60	1.14
Bar diameter – ϕ [mm]	29.37	0.68
Slab height – h_d [m]	0.1864	0.00230

3 THE RESULTS OF THE DIAGNOSTICS

From the results of the bridge diagnostic follow that the type of the concrete is C 37 and the beams are reinforced by rebar of the type A (10210) in two layers (5ϕA30 in the lower layer and 2ϕA30 in the upper layer). Accordingly, the reinforcement corrosion was indicated. The corrosion caused the diameter loss from the initial value of 30 mm to the actual average value of 29.3 mm (the minimal measured value is 28.7 mm). The concrete cover is 30 mm. The measured values of geometric and material properties concerning all girders and slab are shown in Table 1.

4 THE RELIABILITY-BASED EVALUATION OF THE BRIDGE STRUCTURE

In the frame of the research activities of the Department of structures and bridges at the University of Žilina, the modified reliability levels for the existing bridge evaluation were determined using a theoretical approach taking into account the conditional probability (Koteš 2005). Moreover, the influence of reinforcement corrosion as material degradation was investigated and incorporated into the theoretical approach. The obtained reliability levels depend on the age of the bridge and on the planned remaining lifetime.

The new modified reliability levels given in (Koteš 2005, Koteš & Vičan 2006) were used for determining lower partial safety factors for material and load effects, depending on the bridge age and its remaining lifetime. In the practical design, the reliability levels are transformed to the design values of the material resistance and load effects. In the partial safety factors method, the design values of the material resistance and load effects are determined by means of characteristic values and appropriate partial safety factors.

The selected bridge structure was estimated to a bending load-carrying capacity for recommended planned remaining lifetimes $t_r > 20$ years, $10 \leq t_r \leq 20$ years, $3 \leq t_r < 10$ years a $t_r < 3$ years. The actions given in the Slovak standard were intended as a characteristic variable load. The various lengths of the planned remaining lifetime were taken into account using various partial safety factors of materials (for concrete in compression $\gamma_{c,c}$, concrete in tension $\gamma_{c,t}$ and the reinforcement γ_s) and partial safety factors of loads (permanent loads and variable loads).

From the results of the bridge evaluation it followed that the remaining lifetime for load models given in the Slovak standard was less than 3 years and the load-carrying capacity is equal to $0.82 < 1.0$ (taking into account the reinforcement corrosion). In order to increase the load-carrying capacity and elongation of the remaining lifetime, the urgency of bridge reconstruction or strengthening is evident.

5 STRENGTHENING RC STRUCTURES USING FRP REINFORCEMENT

The rehabilitation of RC structures using FRP (Fibre Reinforced Polymers) materials has become a growing area in the construction industry over the last few years. Many research projects in the world have been carried out to promote this efficient repair technique to extend the service life of existing concrete structures. FRP is a composite material generally consisting of carbon, aramid or glass fibers in a polymeric matrix (e.g. epoxy resin). Among many options, this reinforcement may be in the form of preformed laminates or flexible sheets. The laminates are stiff plates or shells that come pre-cured and are installed by bonding the plate to the concrete surface with epoxy. The sheets are either dry or pre-impregnated with resin (pre-preg) and cured after installation onto the concrete surface. This installation technique is known as wet lay-up (Kotula 2006).

The lightweight and formability of FRP laminates or sheets make these systems easy to install. And since the materials used in these systems are non-corrosive, non-magnetic, and generally resistant to chemicals, they are an excellent option for external reinforcement.

Externally bonded FRP laminates or sheets have shown to be applicable to the strengthening of many types of RC structures such as: columns, beams, slabs, walls, tunnels, chimneys and silos and can be used to improve flexural and shear capacities, and also provide confinement and ductility to compression structural members (Khalifa & Gold & Nanni & Abdel Aziz 1998, Khalifa & Nanni 2002).

Traditional methods such as for example different kinds of reinforced overlays, shotcrete or post tensioned cables are placed on the outside of the structure which normally needs much space. The FRP laminates or sheets do not require much space because they are very thin.

The bending strengthening of the structures is the most common way of the structure strengthening but

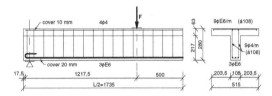

Figure 4. The three times reduced RC girder of RC bridge girder in the village Kolárovice on the road I / 18.

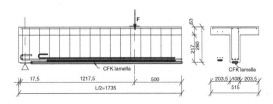

Figure 5. The three times reduced RC girder strengthened with one MBrace®S&P CFK lamella 150/2000.

shear strengthening is also often needed (Täljsten 2001).

6 SIMPLE DESIGN PROCEDURE

In the analysis, two types of beam models were considered. Firstly, the non-strengthened RC girder was calculated. The girder, which is three times reduced compared to the existing RC bridge girder in the village Kolárovice on the road I / 18, is shown in Figure 4. The 1/3 scale beam was considered because of the laboratory possibilities and the beam dimensions.

Next, the model of the strengthened RC girder with the externally bonded one MBrace® S&P CFK lamella 150 / 2000 was calculated (Fig. 5).

Firstly, the resistance bending moment $M_{Rd,0}$ of the un-strengthened girder was calculated. The design moment $M_{Rd,0}$ was determined considering the existing geometry, reinforcement and concrete quality as well as the partial safety factors for material properties. The resistance bending moment $M_{Rd,f}$ of the strengthened cross section was calculated according to the assumption, that the degree of strengthening was $\eta \leq 2$.

The degree of strengthening is defined:

$$\eta = M_{Rd,f}/M_{Rd,0} . \qquad (1)$$

In the following, Figure 6 shows the superposition of the strains and the internal forces acting on a reinforced concrete cross-section.

The required area of lamella and the resistance bending moment of the strengthened girder $M_{Rd,f}$ are derived from the condition of equilibrium $\Sigma F = 0$ and

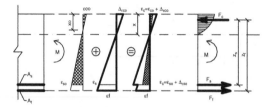

Figure 6. Superposition of the initial strain and additional strain after strengthening.

$\Sigma M = 0$ considering the mechanical behaviour of each material.

Internal forces are defined as:

$$F_s = E_s \cdot A_s \cdot \varepsilon_s \le \frac{f_{yk}}{\gamma_s} \cdot A_s \tag{2}$$

$$F_f = E_f \cdot A_f \cdot \varepsilon_f \tag{3}$$

$$\varepsilon_f \le \varepsilon_{f,lim} \tag{4}$$

$$F_c = \alpha_R \cdot b \cdot x \cdot \frac{\alpha \cdot f_{ck}}{\gamma_c} \tag{5}$$

where F_s = the force in reinforcement; F_f = the force in lamella; F_c = the force in concrete; E_s = the modulus of elasticity of reinforcement; E_f = the modulus of elasticity in fibre direction; A_s = the area of reinforcement; A_f = the area of lamella; ε_s = the strain in reinforcement; ε_f = the strain in lamella; $\varepsilon_{f,lim}$ = the limited strain in lamella; f_{yk} = the characteristic yield strength of reinforcement; f_{ck} = the characteristic compressive cylinder strength of concrete at 28 days; γ_s = the partial safety factor for the properties of reinforcement; γ_c = the partial safety factor for the properties of concrete; x = the neutral axis depth, b = the width of beam (girder); α = the reduction factor for concrete compressive strength; and α_R = the parabolic form parameter.

The presented simple design procedure is implemented in software S&P FRP Lamella (S&P FRP Lamella). The value of the resistant bending moment of the un-strengthened RC girder is equal to $M_{Rd,0} = 3.89$ kNm ($M_{Rk,0} = 4.48$ kNm), to which the maximum force $F_{d,max,0} = 3.33$ kN($F_{k,max,0} = 3.84$ kN) corresponds. Adding one FRP lamella, the value of the resistant bending moment of the strengthened cross section is higher and the value is $M_{Rd,f} = 17.5$ kNm ($F_{d,max} = 14.98$ kN). It means that the degree of strengthening calculated from this program is $\eta = 4.51$. But, the formula (1) limits the degree of strengthening by value $\eta = 2.0$. This means, that the maximum applicable moment for strengthening is equal to value $M_{Rd,f,max} = 7.76$ kNm ($M_{Rk,f,max} = 8.94$ kNm) and the corresponding maximum force is $F_{d,max} = 6.64$ kN ($F_{k,max} = 10.44$ kN).

It should be pointed out that the calculated values of the observed parameters did not include the real adhesion between the concrete surface and the lamella. This means that the adhesion factors between the concrete surface + the epoxy glue + the lamella surface are omitted – only the epoxy glue material is omitted in the software. So, in the next part of the paper we will focus on finding the influence of the adhesion between the concrete surface and the lamella given by the layer of epoxy glue on the real resistance bending moment M_{Rdf} and the maximum loading force F_{max}.

7 NUMERICAL ANALYSIS

7.1 Description of 2D FEM models

The ATENA software (ATENA ver. 3.2.6.) was used for the numerical analyses of RC girders, which are again three times reduced compared to the existing RC bridge girders. For this analysis, two types of half beam FEM models were made. The first type is a FEM model of a half RC girder (Fig. 4). The second type is a FEM model of a half strengthened RC girder with an externally bonded one MBrace® S&P CFK lamella 150 / 2000 (Fig. 5).

The material model of concrete "Concrete-SBETA Material", derived from CEB-FIP MC 90, was applied for concrete. The basic properties of this material model are: tensile strength, fracture energy and the equivalent unaxial law. This material model provides objective results due to the formulations based on energetical principles and its dependency on the finite element grid is negligible. The main reinforcement in the RC girders was modeled as a uni-directional reinforcement element. This material element offers a uni-directional law for a stress–strain diagram in the reinforcement whose form can be: linear, bi–linear and multi–linear. In our case, the bi–linear law was used. The MBrace® S&P CFK lamella 150 / 2000 and the epoxy glue were modeled as the Plane Stress Elastic Isotropic element and alternatively as the 3D Bilinear Steel Von Mises element. The different sizes of the macro elements of the strengthened finite models of the RC girder web were used. The dimensions of these macro elements were: 15, 30 and 50 mm. The dimensions of the macro elements of the girder flange were 30 and 50 mm (Tab. 2.).

The non-strengthened FEM half model of the RC girder is shown in Figure 7.

7.2 Description of 3D FEM models

The same girders (Figs 5–6), which were modeled using 2D FEM models, were modeled by 3D FEM models in ATENA. The reason for the 3D modeling was to compare the results given from the 2D and 3D models and carried out the sensitivity analysis.

Table 2. The review of 2D FEM models.

Model	Descrip. size of FEM grid mm/mm	Type of CFK lamella (epoxy) element
Model 0 (Fig. 4)	F 30/W 15	Non-strengthened
Model 1 (Fig. 5)	F 30/W 15	Plane Stress El. Isotropic
Model 2 (Fig. 5)	F 30/W 15	3D Bilin. Steel Von Mises
Model 3 (Fig. 5)	F 30/W 30	3D Bilin. Steel Von Mises
Model 4 (Fig. 5)	F 50/W 50	3D Bilin. Steel Von Mises

Caption: F – Flange, W – Web of RC girder

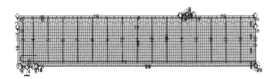

Figure 7. 2D FEM model of RC girder – Model 0.

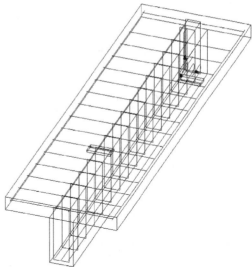

Figure 8. The reinforcement cage model of 3D FEM model.

Table 3. The review of 3D FEM models.

Model	Descrip. size of FEM grid mm/mm	Strengthened/ non-strengthened
Model 5 (Fig. 4)	F 40/W 40	Non-strengthened
Model 6 (Fig. 4)	F 50/W 50	Non-strengthened
Model 7 (Fig. 4)	F 100/W 100	Non-strengthened
Model 8 (Fig. 5)	F 50/W 50	Strengthened
Model 9 (Fig. 5)	F 100/W 100	Strengthened

Caption: F – Flange, W – Web of RC girder

Two types of girders were again modeled – the first type was the non-strengthened FEM 3D model of the half RC girder and the second one was the FEM 3D model of the half strengthened RC girder with an externally bonded one MBrace® S&P CFK lamella 150/2000.

In the case of the 3D modeling, the fracture-plastic material model "3C Nonlinear Cementitious 2" was used because the material "Concrete-SBETA Material" was not available. This type of material is recommended for 3D concrete structures and element modeling (ATENA ver. 3.2.6.).

The reinforcement was modeled in the 3D model as well as in the 2D model. This means, that a uni-directional reinforcement element with a bi–linear law for the stress–strain diagram was used again. The reinforcement cage of the half 3D FEM model is shown in Figure 8.

In the 3D model, it was not possible to take into account the macro elements with a thickness less than 2 mm. So, it was a problem to model the layer of the epoxy glue having a thickness equal to 1 mm and the MBrace® S&P CFK lamella 150 / 2000 having a thickness of 1.2 mm. Therefore, these two macro elements were modeled as one macro element of a shell type having a thickness of 2.2 mm with two layers (layers model). The material characteristic of epoxy glue was assigned to the first layer of the macro element having a thickness of 1.0 mm and the material characteristic of the MBrace® S&P CFK lamella 150 / 2000 (3D Bilinear Steel Von Mises element) was assigned to the second layer of the macro element having a thickness of 1.2 mm.

Once again, the different sizes of the macro elements of non-strengthened and strengthened finite models of the RC girder web and flange were used. The dimensions of these macro elements were: 40, 50 and 100 mm (Tab. 3.). The detail of the epoxy glue and the lamella layers of the 3D model in ATENA is shown in Figure 9.

8 RESULTS OF THE NUMERICAL SENSITIVITY ANALYSIS

8.1 Result from numerical 2D FEM analysis

The analysis was focused on the influence of the externally bonded MBrace® S&P CFK lamella 150 / 2000 on the bending capacity of the RC girder. The value of the additional bending capacity was controlled with the bending crack opening. The value of the deflection in the middle of the RC girder span was the next criterion for the estimation of the strengthening effect. According to the standard (STN P ENV 1992–1–1,

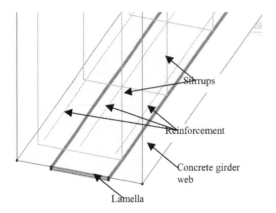

Figure 9. The detail of the strengthened 3D FEM model

Table 4. Results from numerical 2D FEM analysis.

Model	Bending cracks $\omega(i)$ [m]	Deflection in L / 2 $w(i)$ [m]
Model 0	$1.10.10^{-6} < \omega_{\lim}$	$1.356.10^{-3} < w_{\lim}$
Model 1	$1.96.10^{-4} < \omega_{\lim}$	$8.227.10^{-3} < w_{\lim}$
Model 2	$1.97.10^{-4} < \omega_{\lim}$	$8.235.10^{-3} < w_{\lim}$
Model 3	$2.11.10^{-4} > \omega_{\lim}$	$8.208.10^{-3} < w_{\lim}$
Model 4	$2.67.10^{-4} > \omega_{\lim}$	$9.685.10^{-3} < w_{\lim}$

Table 5. Influence of CFK lamella on bending capacity of RC girder – 2D FEM analysis.

Model	Max.load F_{max} [kN]	Model(i)/Model(0) [–]
Model 0	7.685	1.00
Model 1	11.480	1.49
Model 2	11.490	1.49
Model 3	11.570	1.51
Model 4	12.110	1.58

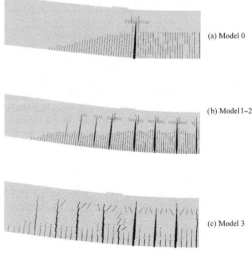

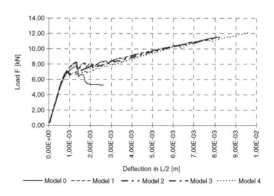

Figure 10. Pattern of bending cracks achieved from 2D FEM models.

Figure 11. Dependency diagram of loading – deflection in L/2 of RC girders – 2D FEM models.

1999), the considered limited deflection is L / 250 which means $w_{\lim} = 13.88$ mm and the limited value of the bending cracks width is $\omega_{\lim} = 0.2$ mm. The results from the numerical 2D FEM analysis are shown in Tables 4–5.

The comparisons between bending crack widths and deflections in L/2 and their limited values are shown in Table 4.

The models of the strengthened RC girders indicate the positive effect of additional reinforcement (MBrace® S&P CFK lamella 150/2000) on the increasing bending capacity of the non-strengthened RC girders. However, the comparison of experiment

measurements and these numerical calculations confirms or refutes the calculated increment of the bending capacity. Figure 11 presents the diagram of the dependency of loading – deflection in L/2 of RC girders.

8.2 *Result from numerical 3D FEM analysis*

In this 3D FEM analysis, the influence of the FEM grid size on the modeling results was detected not only in

1144

Table 6. Results from numerical 3D FEM analysis.

Model	Bending cracks $\omega(i)$ [m]	Deflection in L / 2 w(i) [m]
Model 5	$0.92.10^{-6} < \omega_{lim}$	$0.67.10^{-3} < w_{lim}$
Model 6	$1.76.10^{-5} < \omega_{lim}$	$0.67.10^{-3} < w_{lim}$
Model 7	$0.37.10^{-6} < \omega_{lim}$	$0.56.10^{-3} < w_{lim}$
Model 8	$1.98.10^{-4} < \omega_{lim}$	$5.03.10^{-3} < w_{lim}$
Model 9	$1.97.10^{-4} > \omega_{lim}$	$6.52.10^{-3} < w_{lim}$

Table 7. Influence of CFK lamella to bending capacity of RC girder – 3D FEM analysis.

Model	Max.load F_{max} [kN]	Model(i)/Model(5) [–]
Model 5	5.747	1.00
Model 6	5.288	0.92
Model 7	5.031	0.87
Model 8	8.453	1.47
Model 9	9.043	1.57

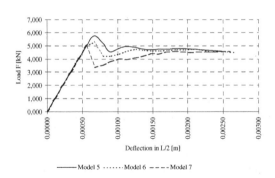

Figure 12. Dependency diagram of load – deflection in L / 2 of RC girders – 3D non-strengthened FEM models.

the case of the strengthened girder, but also in the case of the non-strengthened girder.

The maximum value of the bending crack width $\omega_{lim} = 0.2$ mm and the limited deflection in the middle of the beam $w_{lim} = 13.88$ mm (L/250) were used as the criteria for the girder estimation. The results from the numerical sensitivity 3D FEM analysis are shown in Tables 6–7 and in Figures 12–13.

The models of the strengthened RC girders, like in the 2D FEM models, indicate the positive effect of additional reinforcement (MBrace® S&P CFK lamella 150/2000) on the increasing bending capacity of the non-strengthened RC girders. Figure 12 presents the diagram of the dependency of load – deflection in L/2 of the non-strengthened RC girders (model 5-7) and Figure 13 presents the diagram of the dependency of load – deflection in L/2 of the strengthened RC girders

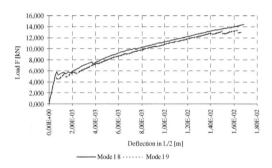

Figure 13. Dependency diagram of loading – deflection in L / 2 of RC girders – 3D strengthened FEM models.

(a) Model 5

(b) Model 6

(c) Model 7

(d) Model 8

(e) Model 9

Figure 14. Pattern of bending cracks achieved from 3D FEM models.

Table 8. Non-strengthened RC girder – comparison of the results.

Model	Max.load F_{max} [kN]	Resistance bending moment $M_{R,0}$ [kNm]
SDC*	3.84	4.48
2D FEM model 0	7.685	8.976
3D FEM model 5	5.747	6.712
3D FEM model 6	5.288	6.176
3D FEM model 7	5.031	6.192

*SDC – simple design procedure, characteristic value

Table 9. Strengthened RC girder – comparison of the results.

Model	Max.load F_{max} [kN]	Resistance bending moment $M_{R,0}$ [kNm]
SDC*	6.64	7.76
2D FEM model 1	11.480	13.409
2D FEM model 2	11.490	13.420
2D FEM model 3	11.570	13.514
2D FEM model 4	12.110	14.144
3D FEM model 8	8.453	9.873
3D FEM model 9	9.043	10.562

*SDC – simple design procedure, characteristic value

(model 8-9). In Figure 14, the influence of the element grid size on the crack pattern (position, number, orientation, …) is shown. The results show that the cracks in the ultimate limit state are not only in the web of the beam, but they reach the flange and they are going almost through the whole flange, because of very small compressive part of the concrete cross-section (3 mm).

9 CONCLUSIONS

The values of the resistance bending moments achieved from the simple design procedure (characteristic value), 2D and 3D FEM models are shown in Table 8.

The value of the resistance bending moment calculated by software S&P FRP Lamella using the code approach is the smallest among all values. Higher values achieved from the ATENA are highly influenced by using the model of concrete material, which better or more realistically describes the real girder behaviour. The characteristic parameters of concrete and reinforcement used in the code approach do not include any properties of real concrete and reinforcement, such as for example tension stiffness, softening law, shear retention factor, crack orientation (fixed or rotated), debonding as a failure mode etc. Next, the influence of the FEM element grid size on the achieved value of the resistance bending moment was demonstrated. This influence was also demonstrated in the case of the crack pattern, this means on position, number and orientation of the cracks. It is valid that the value of the resistance bending moment and the maximum load increase with the decrease of element grid size. The number of cracks and the bending crack width are higher.

Next, the 3D FEM modeling gives the results, which are closer to the code approach than the 2D FEM modeling. The resistance bending moment (maximum force) from the 2D FEM model is 2 times higher than the resistance bending moment achieved from simple design procedure (code approach). In the case of the 3D FEM model, these values are 1.31 – 1.43 times higher.

The results of the strengthened RC girder are shown in Table 9. From the results it can be seen that by using the additional reinforcement (MBrace® S&P CFK lamella 150 / 2000) is increased the bending capacity of the RC girder.

But, the limit degree of the girder strengthening is higher than the value $\eta = 2.0$ neither by using the 2D FEM analysis nor by using the 3D FEM analysis (Tabs 5, 7).

From the results in Table 4 and Table 6 it is seen, that the deflections in L/2 of the non-strengthened RC girders made in the 2D and the 3D FEM model are smaller than the deflections in L/2 of the strengthened RC girders. But, these deflections cannot be compared, because they were achieved at the various maximum loads F_{max} (Tab. 5 and Tab. 7). As a matter of fact, the deflections of the strengthened RC girders are smaller than the deflections of the non-strengthened RC girders if the same maximum loads F_{max} are considered.

Once again, the 3D FEM modeling of the strengthened RC girder gives the results, which are closer to the code approach than the 2D FEM modeling. In this case, the resistance bending moments (maximum force) from the 2D FEM model are 1.73–1.82 times higher than the resistance bending moment achieved from the simple design procedure (code approach). In the case of the 3D FEM model, these values are 1.27–1.36 times higher.

It is necessary to point out that the maximum resistance bending moment M_{Rd} and the maximum load force F_{max} obtained from the ATENA 2D and 3D FEM models are the values corresponding to the limited values. However, the limited values do not correspond to the Ultimate limit state (ULS), but they correspond to the Serviceability limit state (SLS). This means, that the girder does not collapse due to overloading (collapse of the concrete or failure of the lamella or reinforcement), but the girder limit state is defined by exceeding the limited bending crack width.

From the mentioned reasons, the debonding do not appear in the 2D or 3D FEM models because the limited bending crack width was achieved before the debonding in models appears. Also, the debonding is not implemented in software S&P FRP Lamella (S&P FRP Lamella).

The real models of the girders are being prepared at this time and will be tested in the future. So, the achieved results from the ATENA analysis will be compared with the results achieved from the real girder testing. The comparison of the results will be presented as soon as possible. Moreover, the reliability-based solution of the strengthened RC girder is prepared (Omishore & Kala 2006).

ACKNOWLEDGEMENT

This work was supported by the Science and Technology Assistance Agency under the contract No. APVT-20-012204 and by the Slovak Grant Agency, Grant No. 1/3332/06.

REFERENCES

ATENA ver. 3.2.6. *www.cervenka.cz*

Frangopol, D.M. & Estes, A.C. 1997. Lifetime Bridge Maintenance Strategies Based on System Reliability. *Structural Engineering International* 3: 193–198.

Khalifa, A., Gold, W.J. Nanni, A. & Abdel Aziz, M.I. 1998. Contribution of Externally Bonded FRP to Shear Capacity of Flexural Members. *ASCE-Journal of Composites for Construction* 4(2): 195–203.

Khalifa, A. & Nanni, A. 2002. Rehabilitation of Rectangular Simply Supported RC Beams with Shear Deficiencies Using CFRP Composites. *Construction and Building Materials* 3(16): 135–146

Koteš, P. 2005. *Contribution to determining of reliability level of existing bridge structures*. PhD Thesis, Žilina: EDIS (in Slovak)

Koteš, P. & Vičan, J. 2006. Experiences with Reliability-based Evaluation of Existing Concrete Bridges in Slovakia. *The Second International fib Congress 2006, Proceedings of the abstracts, Proceedings of the papers on CD, Napoli, 5–8 June 2006,*

Kotula, P. 2006. *Chosen theoretical and practical problems about strengthening of concrete structures using bonded reinforcement*. PhD Thesis, Žilina: EDIS (in Slovak)

Nowak, A.S. 1995. Calibration of LRFD Bridge Code. *Journal of Structural Engineering*: 1245–1251.

S&P FRP Lamella: *software, www.bow-ingenieure.de*

STN P ENV 1992–1–1, 1999: *Design of concrete structures, Part 1: General rules and rules for buildings*, Bratislava: SÚTN

Täljsten, B. 2001. Strengthening Concrete Beams for Shear with CFRP Sheets. In *Structural Faults and Repair 2001; Proc. of 9th International Conference and Exhibitions, London, 4–6 July 2001, CD–ROM Version*

Vičan, J. at all. 1999. *Guideline – Evaluation Methodology of Existing Concrete Road Bridge. Final Report for Slovak Road Administration*. Žilina: Univerity of Žilina (in Slovak)

Omishore, A. & Kala, Z. 2006. Application Limits of Conventional Models of Systems. *Structures reliability; Proc. VII. Intern. Confer., Prague, 5 April 2006. Prague: CVUT*

*Fracture Mechanics of Concrete and Concrete Structures – Design, Assessment and
Retrofitting of RC Structures – Carpinteri, et al. (eds)
© 2007 Taylor & Francis Group, London, ISBN 978-0-415-44616-7*

Influence of bending and dowel action on the interface bond in RC beams strengthened with FRP sheets: An experimental investigation

J.G. Dai & H. Yokota

LCM Research Center for Coastal Infrastructures, Port and Airport Research Institute, Japan

T. Ueda

Division of Built Environment, Hokkaido University, Japan

B. Wan

Department of Civil and Environmental Engineering, Marquette University, USA

ABSTRACT: This paper presented the test results on a series of RC beams externally bonded with FRP sheets under coupled bending and dowel actions. The main focus was to investigate how the dowel force acting on the FRP sheets influences the shear transfer capacity of the bond between the FRP and concrete. A loading system was developed to simultaneously introduce shear and normal stresses into the FRP-concrete interfaces. Interfacial failure criteria under different loading conditions were discussed. It was found that the dowel force on the FRP sheets affected greatly the initiation of local interface peeling which led to a decrease of the beam's stiffness. However, the ultimate flexural capacity was not significantly influenced by the existence of the dowel action if a sufficient anchorage length was available. The paper also provided a benchmark database that can be used for calibrating the mix-mode bond constitutive laws for FRP-concrete interfaces.

1 INTRODUCTIONS

Extensive tests have indicated that reinforced concrete (RC) beams flexurally strengthened by Fiber Reinforced Polymer (FRP) can hardly reach their full composite capacity due to the premature debonding at the FRP-concrete interface. Up to now, knowledge related to the failure modes, strength and stiffness properties of the FRP strengthened RC beams have been well built up (Saadatmanesh and Mohammad. 1991, Garden et al. 1998, Buyukozturk and Hearing, 1998, Teng et al. 2001). It is noticed that, for the RC beams externally strengthened with FRP sheets, particular attentions will be paid to the failure due to the mid-span debonding of FRP-concrete interface, which is triggered by the stress concentration at the tips of flexural or flexural-shear cracks. The critical crack leading to ultimate debonding failure usually is an inclined one including both crack opening in the direction of parallel to the bond interface and sliding deformation in the direction of perpendicular to the bond interface. The latter component imposes a dowel action on the FRP sheets and eventually causes vertical interface fracture coupled with the slip-induced shear debonding failure. To suppress this kind of mix-mode failure, solutions in the current

design codes are to limit the strain levels in FRP, which are generally derived by considering the shear bond stress-slip law of the FRP sheet-concrete interfaces, flexural crack spacing, anchorage length, etc. (*fib*. 2001, JSCE. 2000, ACI, 2002). In the meantime, it is conceptually suggested that the mix-mode interface bond failure induced by the flexural-shear cracks can be suppressed by additional shear strengthening. However, no direct experimentations have been performed to investigate this interface peeling under the coupled flexural and shearing actions. No any comprehensive mix-mode bond constitutive laws for the FRP-concrete interfaces have been proposed either. Hence the issue of whether or how much the dowel action perpendicular to the FRP sheet-concrete interface influences the interface shear force transfer (flexural strengthening efficiency of the FRP sheets) in the FRP strengthened RC beams remains unclear.

Limited literatures (Karbhari and Engineer 1996, Wan et al. 2004) performed mix-mode loading tests for FRP-concrete interfaces by producing different interface peeling angles between the FRP sheets and concrete in their tests. Their main purposes were to enable the determination of both Mode I and Mode II components of the interfacial fracture energy and to allow a quantitative comparison of interface adhesion

Table 1. Properties of CFRP sheets.

Type	ρ (g/m³)	f_t (MPa)	E_f (GPa)	t_f (mm)	ε_u (%)
FTS-C1-20	200	3550	230	0.11	1.5

Note: ρ = fiber density; f_t = tensile strength; E_f = elastic modulus; t_f = design thickness of FRP sheets; ε_u = ultimate strain of fiber.

mechanisms and energies. Few other researchers (Wu et al. 2005, Dai et al. 2006) conducted other types of mix-mode tests, such as one-dimensional push-off tests or two-dimensional punching shear tests, to investigate the FRP-concrete bond failures under the dowel action. Engineering background of these studies was to bond FRP sheets to concrete structures to prevent the falling of deteriorated concrete blocks. On the whole, there are almost no any direct tests on the FRP/concrete interface subjected to a combined bending and dowel action.

By using a proposed test method, this research aimed to investigate experimentally the interface debonding mechanisms in FRP strengthened RC beams under coupled dowel and bending actions, and also to provide a database for developing more comprehensive fracture criteria for the FRP sheet-concrete bond interface under mix-mode loading conditions.

2 EXPERIMENTAL PROGRAM

2.1 Experimental materials

Properties of Carbon FRP (CFRP) sheets used in this study are shown in Table 1. The applied resin (FR-E3P) and primer (FP-NS) had the elastic modulus of 2.41 GPa and the Poisson ratio of 0.38, respectively. The mixing ratio of resin/hardner by weight was 2:1. High strength concrete with mixing ratio of W:C:S:A = 160:301:742:1160 by weight was used. The concrete had the test compressive strength of 45.0 MPa at the time of beam testing.

2.2 Test setup

A test setup was developed based on a universal test machine namely Autograph system, which had an accurate displacement controlling system. The completed system made it possible to introduce different stress conditions into the FRP sheet-concrete interface through loading the FRP strengthened RC beams in different ways. As shown in Fig. 1, a high strength steel bar was connected to the loading part of the Autograph system. The bar was used to impose the dowel force vertically onto the FRP sheets through a ball hinge and a stiff plate to create a localized mode I stress condition in the bond interface. In addition,

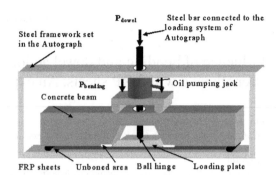

Figure 1. Test setup.

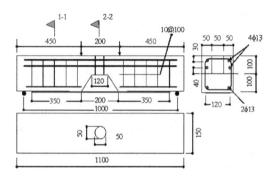

Figure 2. Details of test specimens.

a close-formed steel framework was set inside the Autograph system. This framework provided reactive bending force to the strengthened RC beams. As a result, the pullout force was introduced into the FRP sheets and a mode II interfacial stress condtion was generated. Consequently, the mix-mode loading condition could be achived through exerting the dowel force and bending force couplingly (see Fig. 1).

2.3 Details of specimens

Six RC beams with rectangular section and externally bonded with two layers of CFRP sheets were prepared. CFRP sheets were bonded to RC beams one week after concrete casting and cured for one more week till the tests. Each beam had the span of 1.0 m and the cross-section of 150 × 200 mm. All test beams had two different cross sections (see section 1-1 and 2-2 in Fig. 2). The section 1-1 was larger than section 2-2, at which a 100 high void trapezoid block and a 100 high hollow cylinder were pre-set inside the beams to accommodate all equipments used for imposing the dowel force (See Fig. 1). The trapezoid shape was chosen for the void block to simulate the direction of actual diagonal flexural-shear cracks. Four 13 mm steel bars were arranged in the beams' upper parts to reinforce the concrete around the void cylinder (see Fig. 2). In addition, 10 mm stirrups with the spacing of 100 mm were used to prevent the beams from shear failure before the FRP debonding (see all the details in Fig. 2). A 20 mm long initial crack (unbonded area) was set between the FRP sheets and concrete beams at the outermost of constant moment zone (see Fig. 1). The only test variable was the dowel ratio, which was the ratio of the dowel force imposed onto the interface to its dowel force capacity. Two of the six beams were subjected to only dowel and only bending action, respectively. The remaining four specimens were loaded under a combination of two actions. The way to introduce the coupled bending and dowel actions was to keep the dowel force at a constant level (35%, 50%, 70% and 90% of the interface dowel force capacity respectively as listed in the Table 2) through adjusting the height of the high-strength steel bar when the bending force was added increasingly.

2.4 Test instruments

As shown in Figure 3, transducers were arranged to measure the mid-span deflection of the beams (see 1 and 2 in Fig. 3), the peeing crack opening displacement (hereinafter "the PCOD") that was defined as the interface crack opening displacement between the rigid plate and the concrete beam at the starting point of the un-bonded area (see 10 and 11 in Fig. 3), and the relative displacement (interfacial slip) between the concrete and the FRP sheets at the starting point of bonding area (see 8 and 9 in Fig. 3). Strain gages were mounted on concrete (see 3, 4 and 5 in Fig. 3), steel bars (see 6, 7 in Fig. 3) and the upper and bottom sides of FRP sheets (see 12 in Fig. 3) at the mid-span location. Three gages with long bonding basements were attached on concrete at the location of 15 mm, 45 mm and 75 mm respectively measured from the top of beams. From the starting point of un-bonded area to the beams' supports, strain gages were mounted continuously on the outer surface of FRP with a 10 mm interval from the most inner side of the shear span for the first 120 mm bond length (see 13 in Fig. 3) and with a 20 mm interval for the remaining part (see 14 in Fig. 3).

3 TEST RESULTS

3.1 General description of the failure mode

All the beams failed due to the peeling of FRP sheets with a thin concrete layer from the substrate. Fig. 4 shows the peeled FRP sheets under different loading conditions. It was found that that the volumes of concrete attached to the peeled FRP sheets under dowel, bending and their coupled actions were similar. Bending cracks only occurred within the constant moment zone, where the beam had a small section (see section

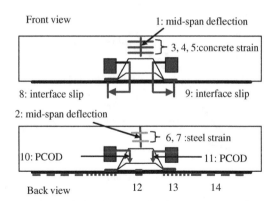

Figure 3. Arrangement of test instruments.

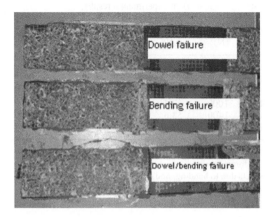

Figure 4. FRP sheets after peeling.

2-2 in Fig. 2). No cracks were observed in the shear span of the RC beams. Also, no bond failure initiated at the end of FRP sheets. So the designed beams succeeded in reproducing an interface debonding failure initiating from the mid-span.

3.2 Bending force versus deformation response

Figure 5 shows the load vs. middle span deflection curves of all tested beams, where the load is expressed as the sum of bending and dowel load. The ultimate bending capacity of each beam and the imposed dowel force are listed in Table 2 as well. It is shown that both stiffness and bending capacity decreased with the increase of dowel forces imposed (see Fig.7 and Table 2). In comparison with the beam (B2) that was only subjected to bending force, the beam with the dowel ratio of 70% (B5) decreased its bending capacity by 13.6%. A more direct way to see how the dowel action influenced the interfacial bond force transfer in the strengthened beams is to compare the maximum tensile strains achieved in the FRP sheets in all

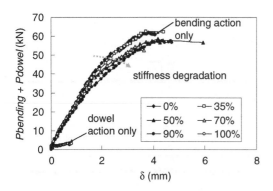

Figure 5.   Load-deflection curves of all tested beams.

Table 2.   Summary of experimental results.

No	P_{dowel} (N)	$P_{bending,max}$ (kN)	$\varepsilon_{frp,max}$	Dowel ratio (%)
B1	2000	0	0	100%
B2	0	62.02	7172	0%
B3	700	61.83	7194	35%
B4	1000	57.4	7021	50%
B5	1400	53.6	6234	70%
B6	1800	55.6	5747	90%

Note: P_{dowel}: dowel force added; $P_{bending,max}$: maximum bending force achieved; $\varepsilon_{frp,max}$: maximum tensile strain of FRP sheets due to bending; and dowel ratio means the ratio of the imposed dowel force to the interface dowel capacity.

test beams since they are direct indices indicating the force transfer capacity in the bond interface, whereas the obtained bending force at the ultimate state also included partially the contribution of reinforcing bars. However, it was difficult to obtain the FRP's tensile strain directly due to the curvature of FRP induced by the dowel action. To solve this difficulty, the maximum tensile strain in the FRP sheets was back-calculated from the ultimate bending load based on plane section assumption. Since the strain distribution profiles of concrete and the steel reinforcement within the constant moment zone were all recorded, the height of neutral axis could be obtained and the only unknown factor for calculating the ultimate bending load was the maximum strain in FRP.

Table 2 lists the calculated maximum bending-induced strains in the FRP sheets in all beams. The back-calculated strains are consequently directly related to flexural strengthening efficiency of FRP sheets. It can be concluded that that coupled bending and dowel actions did not decrease the overall interface shear force transfer significantly. Comparatively, it seems that the dowel action probably had more effects on the member stiffness as shown

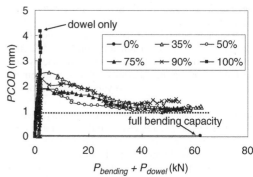

Figure 6.   Load-PCOD relationships under different levels of dowel actions.

in the Fig. 5 because of the dowel induced local debonding.

3.3   Bending force versus peeling crack opening

Extensive reports have shown that the critical peeling of the FRP sheets leading to the failure of whole strengthened system usually initiates from the tip of a flexural-shear crack, and the opening and vertical sliding displacements of the crack tip trigger the horizontal and vertical peeling of FRP sheets from concrete, respectively. Generally, the vertical deformation ability of adhesive layer is very limited. Therefore, a small vertical concrete crack sliding inevitably results in a local peeling of FRP sheets from concrete in vertical direction at the crack location. Then the deformation compatibility to the interface in the vertical direction can be kept from the contribution of FRP sheets' dowel deformation. To prevent this dowel-induced bond deterioration from destroying the overall flexural strengthening system, quantified information is needed to know the effects of tensile stress in the FRP sheet on its dowel deformation ability. By knowing that, engineers can conclude how much the dowel deformation of FRP sheets around cracks can be permitted to achieve expected pullout force capacity in the FRP sheets.

Figure 6 shows the relationships between the added bending load and the dowel deformation of FRP sheets at the crack mouth, namely the interface PCOD. It is clearly seen that the PCOD decreased gradually with the increase of bending load under all dowel ratios. In other words, the higher the pullout force is expected to be achieved in the FRP sheets, the smaller the vertical crack sliding in FRP strengthened RC beams should be allowed. However, it can be seen from Fig. 6 that the increase of the bending load did not influence the PCOD noticeably if the PCOD was under a certain level. The current tests showed that 1.0 mm might be a threshold value for the dowel deformation under

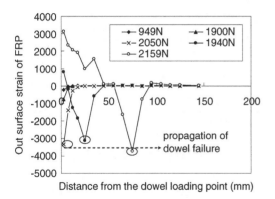

Figure 7. Strain distributions of FRP sheets under dowel actions.

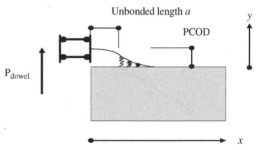

Figure 8. Modeling the initial interface peeling under dowel action.

which the flexural strengthening efficiency would not be reduced in case of using two layers of CFRP sheets. It is considerable that the threshold value relies on the tension/bending stiffness of FRP and the bond length as well. Further experimental testing and numerical simulations are needed to build up their mutual relationships.

3.4 Fracture criteria of the FRP-concrete bond interface

3.4.1 Interface fracture under dowel action

Figure 7 shows the distribution of outer surface strains of FRP sheets under different levels of dowel forces. It was noticed that there was a negative peak value of FRP strain at the initiation of a local dowel failure. The negative strain values showed that the out surface of FRP was subject compressive stresses before dowel failure instead of the tensile stress under pure flexural load. Therefore, it indicated that the local FRP sheet behaved like a beam before the initiation of the dowel failure.

In order to study the initiation of the dowel peeling, the concrete block was assumed to be a rigid block in this study. With this assumption, the FRP sheet can be approximately simulated as a beam on an elastic foundation and the adhesive layer can be simulated as a series of springs as shown in Fig. 8.

The differential equation governing the behaviors of the FRP sheets bonded onto concrete beam can be written as:

$$-EI\frac{d^4y}{dx^4} = q \tag{1}$$

where $q = 0$ at un-bonded area and $q = ky$ at bonded area; EI is the bending stiffness of the FRP sheets; $k = E_a b_f / t_a$, in which E_a and t_a are the elastic modulus and thickness of adhesive layer, respectively; and b_f is the width of FRP sheets.

Through the solution of Eq.1, the relationship between the dowel force and the PCOD (see Fig. 8) can be expressed as following (Dai 2003):

$$PCOD = \frac{P_{dowel}a^3}{6EI} - \frac{(\beta a + 1)(\beta^2 a^2 + 1)P_{dowel}}{4\beta^3 EI} \tag{2}$$

where $\beta^4 = k/4EI$; and a is the un-bonded length.

The interface Mode I fracture energy can be obtained by using the compliance theory:

$$G_{fI} = \frac{P_c^2}{2b_f}\frac{dC}{da} \tag{3}$$

where $C = PCOD/P_{dowel}$, P_c is the peak dowel force.

In this study, values of E_a and t_a were taken as 2.41 GPa and 0.6 mm, respectively, and the length of un-bonded area was 22 mm. The thickness of two layers of FRP including the impregnating resin was taken as 1.8 mm through measuring and EI was 1.64×10^6 N.mm^2 by calculation. The b_f was 120 mm. Therefore, the mode I interfacial fracture energy can be obtained by using Eqs. 2 and 3:

$$G_{fI} = 0.12 N/mm \tag{4}$$

The Mode I fracture energy obtained through the current dowel test is similar to those obtained from three point bending tests (Dai et al. 203, Qiao 2004). This fracture energy can be used to predict the interface dowel force capacity. It should be noted that assuming the FRP sheet as a bending beam is reasonable only when the initial interface crack length (un-bonded length) is short. In such case, the initiation of the dowel failure can be approximately treated as a mode I dominated loading condition. When the dowel-induced peeling continues to propagate in the interface, the FRP sheet-concrete interface is subjected to the mix-mode loading condition. The FRP sheet behaves more like a truss element during the peeling propagation. Detailed debonding mechanisms of FRP

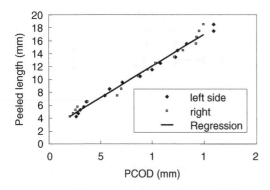

Figure 9. Relationship between peeled length and PCOD.

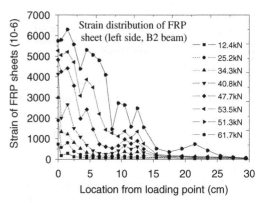

Figure 10. Strain distributions in FRP sheets under different bending load levels.

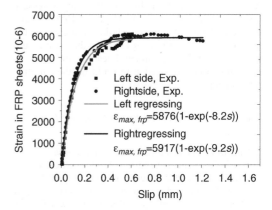

Figure 11. Relationship between the strains of FRP sheets and the slips at the loaded point.

sheet/concrete interfaces and derivation of the Mix-mode interface fracture energy under the dowel action are presented in a paper by Dai et al., 2007.

The movement of the location with the maximum negative strain as shown in Fig. 7 indicates where the initial interface crack had propagated under a certain dowel deformation (PCOD). Therefore, the relationships between the peeled interface length and the PCOD can be drawn in Fig. 9. It is clearly shown that the ratio of the PCOD to the peeled interface length is almost a constant value, indicating there is a critical peeling angle during the propagation of the dowel failure. It is reasonable to extrapolate that the dowel force component will be increased if there is an additional pullout force introduced into the FRP sheets by bending action under the condition of a constant peeling angle. To suppress the dowel-induced interface peeling while maintaining a high pullout force level in the FRP, the solution is to decrease the interface peeling angle by either limiting the dowel deformation or increasing the bond length. On the other hand, in the case of FRP sheet-concrete interfaces with short bond lengths, such as the bond interfaces between two adjacent cracks, neglecting the dowel deformation may lead to over-estimating the bond capacity of resisting debonding failure.

3.4.2 Interface failure under bending action

Bending action on the FRP strengthened RC beam introduced pullout force into FRP sheets and caused a mode II interface failure. Fig. 10 shows the strain distribution in FRP sheets under different bending load levels for B2 beam. The shapes of FRP sheets' strain distribution are almost same as that observed in direct pullout test, indicating that the bending test can generate a shear stress condition similar to that in direct pullout test for the FRP concrete-sheet interface. Generally, the bending test is more convenient to set up in comparison with the direct pullout load test.

Concerning the FRP sheet-concrete interfaces subjected to mode II type loading, Dai et al. (2005)

developed a new approach to define the interfacial fracture energy and the bond stress-slip laws through obtaining the relationship between the pullout force (or the stain ε of the FRP sheets) and the interface slip s at the loaded point of the interface, in other words, at the tip of the initial interface crack. Conceptually, the crack-tip slip can be obtained through either integrating the strains of FRP sheets or reading the displacement transducers (see 8 and 9 in Fig. 3). Fig. 11 presents the experimentally obtained relationship between the strain of FRP and the crack-tip slip, in which the crack-tip slip was calculated by integrating the strains of FRP sheets. According to the methods of Dai et al. (2005), the following formula can be obtained:

$$\varepsilon = \varepsilon_{\max}(1 - \exp(-Bs)) \tag{5}$$

$$G_{fII} = \frac{E_f t_f \varepsilon_{\max}^2}{2} \tag{6}$$

$$\tau = 2G_{fII}B(\exp(-Bs) - \exp(-2Bs)) \tag{7}$$

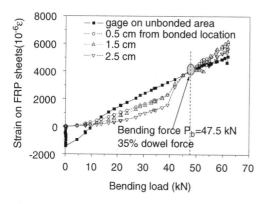

Figure 12. Strains on FRP sheets at different locations in B2.

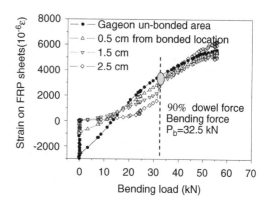

Figure 13. Strains on FRP sheets at different locations in B3.

where G_{fII} is the mode II fracture energy, which was 0.87 N/mm in this research, and B is an empirical factor reflecting the interface bonding stiffness, which was 8.7 mm^{-1} in this research. They can be obtained by regressing the experimental data in Fig. 11. It was noticed that typical values of G_{fII} and B obtained from the pullout test using the same test materials were 1.19 N/mm and 10.4 mm^{-1}, respectively. It seems that the mode II fracture energy obtained in pullout test is higher than that from the current bending test. The change of shear span/height ratio of beam possibly affects the calibration of the Mode II fracture energy.

3.4.3 Interface failure under coupled actions

Discussions on the load-deflection relationships (section 3.2) and the dowel failure characteristics (sections 3.4.1) have indicated that either controlling the dowel deformation or increasing the bond length can suppress the negative effects of dowel action on the shear bond force transfer in the FRP sheet-concrete interfaces. In other words, the sufficiently long bond length may guarantee that interface bond failure is still a shear-dominating one. However, it is different when the local interface bond failure is focused on. As a matter of factor, the combined actions of the dowel and pullout force in FRP sheets influence the initiation of local interface peeling significantly because the interface becomes considerably weaker un der the combination of shear and tensile stress according to the Mohr-Coulomb failure criteria. To know the critical pullout force causing the initial local peeling at different dowel ratios, Fig. 12 and Fig. 13 show the developments of strains on the FRP sheets at different locations with the increase of bending forces. The local peeling of FRP sheets was assumed to occur when two continuous gages on the FRP sheets close to the crack tip suddenly showed almost same strain values. Although the local outer surface strains of the FRP sheets under the coupled dowel and bending actions did not make this change as dramatically as they did

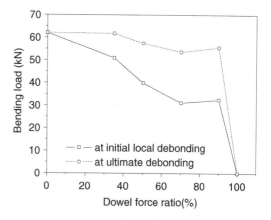

Figure 14. Interface failure envelope curve under the coupled bending and dowel actions.

in the case of dowel action only, the bending force corresponding to the initial local dowel failure was still distinguishable (see circled points in Fig. 12 and Fig. 13). By this way, the relationship between the bending forces correspondent to the initial local peeling and the dowel ratios was obtained and is shown in Fig. 14. The ultimate bending capacity of all tested beams is also given in Fig. 14 for the comparison purpose. It can be seen that the dowel force imposed on FRP sheets affects the local interface debonding significantly although its effect on the ultimate bending capacity is very small. The locally deteriorated interface bond might be the reason why the stiffness of strengthened beams decreases greatly in cases of high dowel force ratios (see Fig. 5).

To look more insightfully at the fracture mechanisms of the FRP sheet/concrete interfaces under the coupled bending and dowel actions, modeling for separating the mode I and mode II fracture energy components and building up fracture energy envelope criteria seems necessary. Based on that a comprehensive interfacial bond constitutive law under the mix-mode loading can be proposed to reproduce and interpret the current test results. These work remains for study in a further step.

4 CONCLUSIONS

1. This research has developed a general test method which can evaluate the bond of FRP/concrete interfaces under different fracture modes, in particular, under the coupled dowel and bending actions.
2. For a local interface bond failure induced by the dowel action only, a model deriving the interface mode I fracture energy was proposed based on the beam on elastic foundation theory and compliance method.
3. The interface mode II fracture energy can be calculated using the maximum tensile strain in the FRP sheets obtained from the bending test. The mode II fracture energy obtained from the bending tests was smaller than that from pullout tests. The beam's shear span/depth may affect the calibration of the mode II fracture energy when a bending test method is applied.
4. If there is a long bond length, existence of dowel action on FRP sheets brings marginal effects to the ultimate interface bonding capacity although it brings an earlier local debonding. However, in the case of short bond length, such as the interface between adjacent flexural-shear cracks, the interface bond force transfer may be significantly affected by the dowel action and the bonded length (crack spacing), both of which affects the interface peeling angle. Therefore, a comprehensive mix-mode interface bond law instead of the pure shear bond stress-slip law should be developed for interface analysis in order to refine the strain limits of FRP for the flexural strengthened RC beams in the design codes.
5. Dowel deformation ability of FRP sheets decreased significantly with the increase of bending force at the beginning stage. However, test results shows there may be a low bound value for the dowel deformation, under which the existence of dowel deformation will not influence the flexural strengthening efficiency of the FRP sheets. This low bound was 1.0 mm in the current study. Obliviously, this value is influenced by the tension stiffness of FRP and the effective bond length.
6. This paper provides a database for calibrating the bond constitutive laws for FRP sheet-concrete interfaces under mix-mode loading. Modeling for separating the mode I and mode II components for the current test method and proposing a universal energy criterion that governs the fracture of

FRP sheet-concrete interfaces under various stress conditions remain for a further-step study.

REFERENCES

ACI committee 440. 2R., 2002, Guide for the Design and Construction of Externally Bonded FRP Systems for Strengthening Concrete Structures. American Concrete Institute, Farmington Hills, MI.

Buyukozturk, O. & Hearing, B. 1998. Fracture Behavior of Precracked Concrete Beams Retrofitting with FRP, ASCE, *Journal of Composites for Construction*, Vol.2, No.3, 138–144.

Dai, J. G., Ueda, T., Muttaqin, H. & Sato, Y. 2003. Mode I Fracture Behaviors of FRP-Concrete Interfaces, *Proceedings of the Japan Concrete Institute*, Vol. 25, 1577–1582.

Dai, J.G. 2003 Interfacial Models for Fiber Reinforced Polymer (FRP) Sheets Externally Bonded to Concrete, Doctoral dissertation submitted to Hokkaido University.

Dai, J.G., Ueda, T. & Sato, Y. 2005. Development of Nonlinear Bond Stress-Slip Model of FRP Sheet-concrete Interfaces with a Simple Method, ASCE, *Journal of Composites for Constructions*, Vol.9, No.1, 52–62.

Dai, J.G., Ueda, T. & Sato, Y. 2007. Bonding Characteristics of Fiber Reinforced Polymer Sheet-Concrete Interfaces under Dowel Load, ASCE, *Journal of Composites for Construction* (in press)

fib bulletin 14., 2001. Externally Bonded FRP Reinforcement for RC Structures.

Garden, H.N., Quantrill, R.J., Hollaway, L.C., Thorne, A.M. & Parke, G.A.R. 1998. An Experimental Study of the Anchorage Length of Carbon Fiber Composite Plates Used to Strengthen Reinforced Concrete Beams, *Construction and Building Materials*, Vol.12, No.4, 203–219.

Japan Society of Civil Engineers, 2000, Recommendations for Upgrading of Concrete Structures with Use of Continuous Fiber Sheets, Concrete Library, Vol.7.

Karbhari, V.M. & Engineer, M., 1996, Investigation of Bond between Concrete and Composites: Use of a Peel Test, *Journal of Reinforced Plastics and Composites*, Vol. 15, No.2, 208–227.

Qiao, P. & Xu, Y. 2004. Evaluation of Fracture Energy of Composite-Concrete Bonded Interfaces Using Three-Point Bend Tests, ASCE, *Journal of Composites for Construction*, Vo. 8, No.4 pp. 352–359.

Saadatmanesh, H. & Ehsani, Mohammad R. 1991, RC Beams Strengthened with GFRP plates, Part I: Experimental Study and Part II: Analysis and Parametric Study, ASCE, *Journal of Structural Engineering*, Vol. 117, No.11, 3417–3455.

Teng, J.G., Chen, J.F., Smith, S.T. & Lam, L. 2001. FRP-strengthened RC Structures, John Wiley & Sons, NY.

Wan, B., Sutton, M. A., Petrou, M. F., Harries, K. A. & Li, N. 2004. Investigation of Bond between Fiber Reinforced Polymer and Concrete Undergoing Global Mixed Mode I/II Loading, ASCE, *Journal of Engineering Mechanics*, Vol. 103, No.12, 1467~1475.

Wu, Z.S., Yuan, H., Asakura, T., Yoshizawa, H., Kobayashi, A., Kojima, Y. & Ahmed, E. 2005, Peeling Behavior and Spalling Resistance of Bonded Bidirectional Fiber Reinforced Polymer Sheets, *Journal of Composites for Construction.*, Vol. 9, No. 3, 214–226.

Fracture Mechanics of Concrete and Concrete Structures – Design, Assessment and Retrofitting of RC Structures – Carpinteri, et al. (eds)
© 2007 Taylor & Francis Group, London, ISBN 978-0-415-44616-7

Repair and retrofitting of structural RC walls by means of post-tensioned tendons

A. Marini
DICATA, University of Brescia, Brescia, Italy

P. Riva
Department of Engineering Design and Technologies, University of Bergamo, Bergamo, Italy

L. Fattori
Structural Engineer, Brescia, Italy

ABSTRACT: The introduction of un-bonded post-tensioned tendons or bars in the critical zone for the structural repair and retrofitting of structural R/C walls is investigated by means of non linear FE analyses. The numerical models are validated through comparison against experimental data on a 1:1 scaled traditional shear walls undergoing cyclic loadings. Comparative pushover and seismic analyses were performed.

The results showed that the introduction of unbonded post tensioned tendons concentrates the damage in a single large crack opening at the wall base section, whereas traditional walls develop an extended crack pattern over the critical zone. Despite similar behaviour is observed in terms of top storey maximum displacement, in case of post-tensioned structural walls the elastic behaviour of the post-tensioned bars, while reducing the energy dissipation capacity of the structure, avoids any residual displacement after a seismic event, thus limiting structural damage even under a design seismic event.

1   INTRODUCTION

Following strong earthquakes, traditional R.C. structural walls may show severe damage characterized by an extended crack pattern along the critical zone and concrete cover spalling close to the base section, followed by the onset of longitudinal reinforcement buckling and the failure of some rebars (Riva et al. 2003). R.C. walls might as well develop a large crack at the base section, causing the shear force to be transferred across the element by dowel effect of the longitudinal rebars only. In this case, sliding shear failure might occur (Riva et al. 2003).

When the residual displacement and the damage are not as severe as to inhibit any further use of the structure, shear wall structural rehabilitation might be considered. To this end it is of outmost importance to identify and test feasible and economic retrofitting techniques.

As for the retrofitting techniques, following the shear wall earthquake damage scenario, reinforcement replacement is a hardly viable solution. Partial replacement of the rebars obtained by means of overlap splicing might suffer problems related to joint efficiency and confinement. Similarly, if

reinforcement is substituted by welding or by clamping part of a new bar, sufficient ductility might not be safely and confidently attained. The results of recent experimental tests on a repaired wall in which the longitudinal reinforcement was partially replaced and connected to the existing one by means of mechanical couplers, showed that these devices lead to an anticipated collapse due to excessive bearing stress close to their end section (Riva et al. 2004).

In this paper, the repair and retrofit of traditional R/C walls by partial or complete substitution of the damaged reinforcement with unbonded post-tensioned reinforcement is studied. Post tensioned reinforcement is introduced in the critical zone only, together with an appropriate shear key device.

This solution might be regarded as an application of the shear wall typology developed for new constructions by Kurama et al. (1999, 2000). Kurama et al. (1999, 2000) introduced prefabricated panels placed one on top of the other and linked by post-tensioned unbonded reinforcement along the entire wall height. These walls exhibit the same initial stiffness and strength of traditional RC walls but maintain a bilinear elastic behaviour throughout the seismic excitation. This way, large displacements can be adsorbed without significant damage, despite having a small energy dissipation capacity. Deformability localizes along the structural horizontal joints between adjacent panels. The structural efficiency of these joints is fundamental for the correct global behaviour of the structures. Transferring shear between the panels or the wall and the foundation has been widely studied (Soudki 1995a, b; Soudki 1996).

A similar approach was adopted for the seismic design of prefabricated frame structures, in which structural components are tightened by unbonded post-tensioned tendons. The small energy dissipation of these structures was proved to be easily overcome by introducing viscous dampers (Pampanin, 2005).

In this paper, the seismic behaviour of structural walls with post-tensioned unbonded high strength bars is investigated and compared to the behaviour of traditional RC walls by means of non linear Finite Element analyses. Post tensioned reinforcements are introduced in the critical zone only in partial or total substitution of the ordinary reinforcement.

In the following, three different solutions are analyzed and compared, namely: (i) shear walls with post-tensioned rebars entirely substituting traditional reinforcements in the critical zone, thus requiring the introduction of special shear resisting devices to allow shear transferring to the wall base, (ii) shear walls with post-tensioned rebars partially substituting traditional reinforcements in the critical zone, thus partially preserving the continuous reinforcement transferring the shear to the wall base (iii) shear walls with traditional reinforcement.

Shear walls were modelled by means of fiber beam elements, implemented within the FE Code MIDAS/Gen (2005). The numerical model was initially validated through comparison with experimental data on the performance of a 1:1 scaled reinforced concrete shear wall subjected to cyclic loadings.

Following analyses were performed on single walls of different typologies, namely with or without unbonded post-tensioned reinforcement, subjected to both pushover tests and to artificial seismic records compatible with Eurocode 8 (2004).

2 DESIGN OF SHEAR WALL RETROFIT

Shear wall retrofitting with unbonded post-tensioned reinforcement is obtained by means of Dywidag® bars placed in the critical zone of the cast-in-place, existing RC element. The unbonded rebars are fixed to the wall foundation and post-tensioned through anchors buried, beyond the critic zone, in the floor concrete slab. Outside the critical zone, the detailing of the traditional RC cast-in-place shear resisting walls is usually left untouched.

Main design unknowns for the retrofitting design are: (i) the initial deformation of the post-tensioned

reinforcement (ε_{sp0}); and (ii) the unbonded reinforcement cross section (A_{sp}). The latter is obtained by enforcing the equality between the resisting and design bending moments at the ultimate limit state. The minimum level of post tension can be determined by imposing a maximum crack opening (i.e. 0.4 mm) at the damage limit state. Furthermore, by imposing the elastic behaviour of the unbonded reinforcement even in the case of a design earthquake excitation ($\varepsilon_{sp} < \varepsilon_{spy}$), no residual deformations are to be expected after the earthquake. Post tension must be adequately increased to account for post-tension losses induced by steel relaxation and creep of concrete. Details on the design of the shear walls can be found in (Oldrati et al., 2004).

The post-tensioned rebars, grease coated and positioned in polymeric extruded sheathings, are placed into confined concrete zones next to the wall edges. Steel spiral hoops are embedded underneath the anchorage plates to ensure adequate concrete confinement.

3 NUMERIC MODELLING AND VALIDATION

Shear walls with either traditional or unbonded post-tensioned reinforcement are modeled by means of force-based fiber elements, based on the Timoshenko beam theory, implemented within the Finite Element Code MIDAS/Gen (2005). Force-based elements are computationally more demanding than displacement-based elements, but they offer the main advantage of being "exact" within the beam theory framework used for the formulation (Spacone et al. 1996). This leads to the use of one element per structural member (beam or column) in a frame analysis, thus requiring a lower number of nodal degrees of freedom. This results in the faster assembly of the numerical mesh, which might in turn be a significant advantage for practitioner engineers.

The single multilayer shear wall (Fig. 1a) is therefore modeled by means of a number of elements equal to the number of floors in the building (Fig. 1b). Each

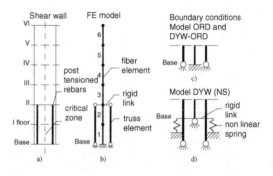

Figure 1. Model A – Force based fiber element mesh.

element has 5 control sections. A simple linear shear force-shear deformation law was used at the section level.

Post-tensioned rebars are modeled by introducing two extra non linear truss elements in the critical zone (Fig. 1b). The trusses are linked to the wall by means of rigid beams. Post-tension is simulated by applying an initial distortion to the truss elements.

As for the boundary conditions, the base node is considered as fixed in case ordinary reinforcement connecting the wall to the foundation is present, i.e. in case of traditional walls and in case of post-tensioned rebars partially substituting the traditional reinforcement (Fig. 1c). On the other hand, in shear walls with post-tensioned rebars only, the rocking mechanism must be allowed, thus the base section must be enabled to rotate. The base node is therefore hinged to the ground (Fig. 1d). To account for resisting actions developed by the compressed concrete at the edges, two nonlinear springs where introduced at the wall base section and rigidly linked to the central base node.

For ordinary concrete, a classical Kent-Park law is adopted. Confined concrete is described by Kent-Park law, followed by a bilinear curve accounting for ductility. For the ordinary steel, Menegotto-Pinto constitutive law is used, whereas unbonded high strength steel is described by a bilinear tension-only law. The material parameters are summarized in Table 1.

A first set of analyses showed that the structural response is affected mainly by the shear capacity, whereas it is basically independent of the shear stiffness. The structural stiffness is governed by bending up to failure.

3.1 Validation of the numerical model

The numerical model (Model A, in the following) is initially validated through comparison against experimental data on the performance of a 1:1 scaled R.C. shear wall, shown in Figure 2, subjected to cyclic loading. The experimental study was carried out at the Laboratory of the University of Brescia. A detailed report of the test setup and results may be found in Riva et al. (2003). The experimental wall is shown in Figure 2a.

Model A results are also compared to the numerical shear wall response obtained by Oldrati et al. (2004), named Model B in the following.

In Model A, only three force-based fiber elements are used: one for the base segment, one for the critical zone, and one element extending to the wall top end (element 1–3, Fig. 2). In the analyses, shear capacity is assumed as equal to the force transmitted by the dowel action of the traditional reinforcement crossing the base section.

Model B refers to the analysis performed by Oldrati et al. (2004) with ABAQUS, using, for the r.c. wall, displacement based user defined fiber elements up to 11.5 m from the restrained end, and linear elastic beam elements beyond that section.

Mechanical properties for the materials, as well as the geometric characteristics of the elements used herein, are those reported in (Riva et al. 2003, Table 1). The same properties were used by Oldrati et al. (2004).

Experimental and numerical responses are shown in Figure 3. The comparison shows a good agreement of numerical and experimental responses in terms of cyclic behavior and resistance.

4 ANALYSIS OF POST-TENSIONED SHEAR WALL

The behaviour of shear walls with post-tensioned, unbonded reinforcement stretching across the critical

Table 1. Material properties.

	σ_{tu} ε_{tu}	σ_{tmax} ε_{tmax}	σ_{tc1} ε_{tc1}	σ_{cy} ε_{cy}	σ_{c1} ε_{c1}	σ_{cu} ε_{cu}
Confined Concrete	−0.145 −0.1%	1.45 −0.027%	23.5 0.107%	47.0 0.7%	47.0 2.46%	4.0 2.71%
Ordinary concrete	−0.145 −0.1%	1.45 −0.27%	15.0 0.047%	30.0 0.2%	30.0 0.35%	4.0 0.7%
	E_s	f_{sy} ε_{sy}	f_{sm} ε_{sm}			
Steel	200000	500 0.25%	580 8%			
	E_{sDYW}	σ_{syDYW}	σ_{suDYW} ε_{suDYW}			
Post tensioned rebars	205000	1080.0	1230.0 8%			

Note: σ [MPa], f [MPa], E[MPa]

Note: σ [MPa], f [MPa], E [MPa]

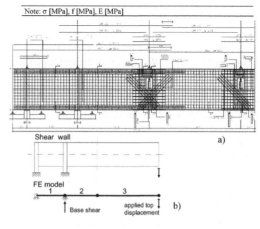

Figure 2. (a) Experimental 1:1 scale R.C. wall (Riva et al. 2000) and (b) numeric model.

1159

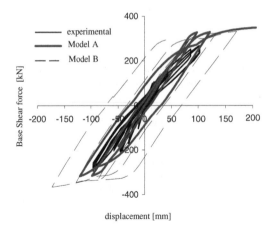

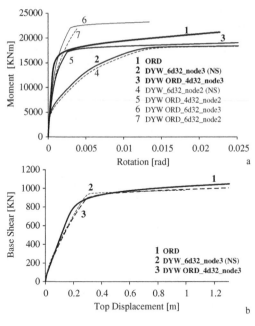

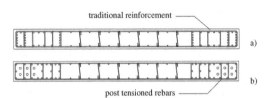

Figure 3. Numerical and experimental response.

Figure 4. Typical critic zone cross section for: a) traditional walls; b) reinforced shear wall with post-tensioned rebars.

Figure 5. Pushover test: a) bending moment versus base section rotation; b) Base shear versus top displacement.

Beside this default setting, the case with four Dywidag bars was also analyzed.

zone is investigated by means of finite element numerical analyses. The study refers to a single shear wall of a six storey building. The wall has a 0.3×3.5 m section, and has a height of 19.2 m. With reference to Figure 1, different models are used to study the three different critical zone reinforcement typology and layout; namely: (i) Model ORD: critical zone with traditional reinforcement (Fig. 4a); (ii) Model DYW-ORD: shear wall critical zone with unbonded rebars integrating the ordinary reinforcement at the base section (Fig. 4b). Ordinary reinforcement is proportioned to resist the shear force according to Eurocode 8 (2004) requirements; (iii) Model DYW: shear wall critical zone with unbonded rebars entirely substituting the longitudinal reinforcement. In this case, additional devices must be designed to allow shear force transfer across the base section. Beyond the critical zone, an ordinary reinforced concrete section is assumed.

Material properties are those described in the previous chapter. Six Dywidag® bars, having a diameter of 32 mm, were introduced at the wall edges in model DYW and four rebars in model DYW-ORD. Each unbonded rebar was post-tensioned to 120 kN. The choice of the post tension level is briefly discussed in section 4.2 and more extensively in Oldrati et al. (2004). The tendons were stretched from the base section to the second floor level (node 3 in Figure 1).

4.1 Non linear "push-over" analysis

A pushover analysis is performed by applying a monotonically increasing lateral displacement to the top storey. This pushover analysis follows a finite element investigation previously performed by Oldrati et al. (2004).

Bending moment versus base section rotation is plotted in Figure 5a for the analyzed shear wall typologies (Curve 1, 2 and 3, for the described reinforcement layout). Ordinary longitudinal reinforcement significantly affects the flexural behaviour of the shear wall. The adoption of unbonded post-tensioned rebars, partially substituting the traditional reinforcement (DYW-ORD), results in a significant increase in the ultimate bending moment when six bars are introduced (curve 6) and it is appreciatively the same when only 4 bars are introduced. The ultimate bending moment of the traditional wall (curve 1) is comparable to the flexural capacity of the wall without ordinary reinforcement (curve 2). Failure of the shear wall with unbonded reinforcement occurs when concrete overcomes its compressive strength, whereas traditional shear walls failure follows reinforcement yielding.

Considering this rebars layout, the yield moment (M_y) is maximum when no shear reinforcement is used (17290 kNm, curve 2), and decreases in case of shear

Table 2. Influence of the base section rotation on the top displacement ($\delta = 0.50$ m).

Model	base rotation ϑ_{nase}[rad]	Rigid displacement induced by base rotation δ_ϑ [m]	δ_ϑ/δ [%]
DYW-ORD	0.88E−02	0.17	34
DYW (NS)	2.00E−02	0.38	76
ORD	0.40E−02	0.07	14

wall with shear reinforcement (16790 kNm, curve 3). The minimum yield moment is reached for traditional shear walls (15240 kNm, curve 1).

Upon yielding of the unbonded bars, an abrupt change in the curve slope is observed. In case of shear walls having both traditional and post-tensioned rebars (DYW-ORD), the change in slope is less pronounced, as the yielding of unbonded reinforcement occurs after that of web reinforcement. On the other hand, traditional walls (ORD) show a smoother transition towards the plastic stage. This is the result of the progressive yielding of the reinforcement occurring along the cross section.

The upper floor maximum displacement is the result of both a rigid translation induced by the base section rotation (rotation times the wall height), and a displacement produced by the flexural deformation of the structure. Pushover analyses were performed by applying an increasing top displacement until concrete reached failure (Fig. 5b). This resulted in unreasonably high values of the total displacement, and the analyses fail to have a physical meaning. Considering a reference value of top displacement of 0.5 m, equivalent to a 5% drift, the base section rotation and the following displacement are listed in Table 2. It is worth noting that when lacking the shear reinforcement (DYW), the damage is localized at the wall base section. In this case, 76% of the total top displacement is induced by the base section rotation. The adoption of both traditional and unbonded rebars (DYW-ORD) limits this effect, despite damage localization is still evident.

Pushover analyses highlighted that, when unbonded rebars integrate the traditional reinforcement, the critical zone might extend well beyond the first floor. The structural response is proved to significantly depend on the length of the wall post-compressed zone. To emphasise this aspect, DYW-ORD models having 6 post-tensioned rebars extending through the first (Fig. 5a, curve 7) or the second floor (Fig. 5a, curve 6) were analyzed. Figure 6 plots the moment versus sectional curvature for the two models (base section A moment values include the unbonded rebars contribution).

When the post compressed zone is limited to the first interstorey height, the base section behaves elastically throughout the analysis because the yielding

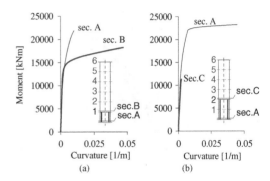

Figure 6. Moment vs curvature for DYW-ORD_6d32 models as a function of the post compressed zone length.

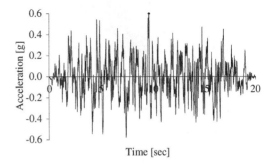

Figure 7. Adopted EC8 compatible earthquake.

moment is first reached outside the post compressed zone, where all damage is accumulated (Fig. 6a). On the other hand, when unbonded rebars are stretched up to the second floor, the element outside the post tensioned zone remain basically elastic and the damage accumulates around the base section (Fig. 6b).

4.2 Dynamic analyses

A EC8 (1997) compatible artificial acceleration time-histories having a peak ground acceleration (PGA) of 0.6 g (5.88 m/s²) was used to perform dynamic analyses of the shear walls (Fig. 7). The large value of PGA is the result of selecting 0.35 g PGA, an Importance Factor of 1.4 and a Type C Soil (Eurocode 8, 2004). Reduced PGA was used to verify the structure under damage limit state seismic excitation.

A pilot parametric study clarified the effect of varying post-tension level (either by varying the tendon post-tensioning, or the tendon number) on the wall structural seismic response. The post-tension level affects the onset of the cracking process. For increasing post tension forces the structure cut the vibratory phenomena more easily, and it exhibits a larger energy dissipation capacity. This is due to the increase in the concrete stress level and in the possible Dywidag® bars plastic deformation. For increasing post tension,

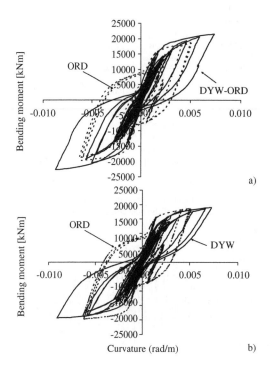

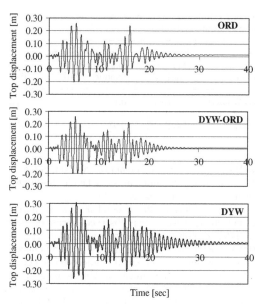

Figure 9. Top displacement versus time for varying critical zone typologies.

Figure 8. Moment versus base rotation curve for: (a) ORD and DYW-ORD; (b) ORD and DYW models.

a delay in the development of the maximum displacements is also observed. Maximum displacements are approximately constant for post tension larger than 120 kN (Oldrati et al, 2004). It is worth noting that an excessive post tension might result in excessive concrete stresses and might jeopardize the capacity of the unbonded bars to allow the elastic recovery of the undeformed shape in ultimate conditions. Based on the parametric study, the optimal structure response was found when adopting 6 and 4 tendons for DYW and DYW-ORD models, respectively. Each Dywidag® rebar post tension level was set to 120 kN, equal to 15% of the yielding stress.

Base bending moment versus base section curvature is shown in Figure 8 for all structural typologies. The curves are characterized by a few large cycles where all energy dissipation is concentrated. It is interesting, however, to note that pinching of the curves is very pronounced in case of unbonded post-tensioned rebars, exhibiting a smaller capacity to dissipate energy but, at the same time, a pronounced self centring behaviour.

Figure 9 shows the top displacement versus time, for the analysed structural typologies. The traditional shear wall (Fig. 9, ORD) shows a greater dissipation capacity with a prompt reduction of the oscillatory phenomena. As a drawback, a larger 12.4 mm residual displacement, caused by the yielding of the reinforcement, is observed.

In case of walls having post-tensioned rebars partially substituting the traditional reinforcement (Fig. 9, DYW-ORD) and tightened to the second floor level, the maximum top displacement is equal to 0.24 m, similar to that of the ORD model, but residual displacements are almost null 10 seconds past the seismic event. Following the seismic event the structure recovers its undeformed shape.

It can be shown that, by limiting the post-compressed zone to the first floor, a residual displacement of 7 mm is recorded after the earthquake. The base section behaves almost elastically throughout the seismic event, whereas large plastic deformation accumulates right beyond the post-compressed zone, proving that the post-tensioned zone must be conveniently extended beyond the first interstorey height.

Walls having unbonded rebars entirely substituting the traditional reinforcement (Fig. 9, DYW), show the largest top displacement (0.3 m), mainly induced by the large rotation at the wall base triggered by the rocking mechanism. The oscillatory phenomena reduce more slowly, showing a smaller energy dissipation capacity, but again no appreciable residual displacements are observed (residual displacement is equal to 20% of the traditional wall residual displacement).

Figures 10 and 11 show base shear force and bending moment versus time histories, respectively, for the analysed structural typologies. Both shear and bending moment are larger for DYW model.

Figure 12 illustrates concrete and reinforcement strain distribution at three different wall heights

1162

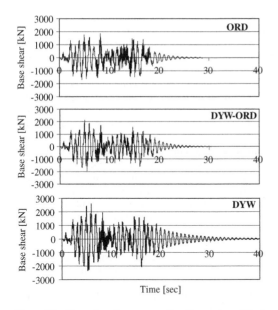

Figure 10. Base shear force versus time for varying critical zone typologies.

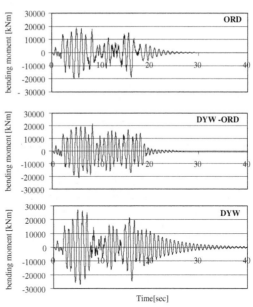

Figure 11. Bending moment versus time for varying critical zone typologies.

(namely: 0.50 m, 1.6 m e 2.7 m from the base section). In case of shear walls with post-tensioned rebars, strain concentration is observed at the base section, compatibly with the large rotation induced by the rocking motion; whereas the traditional wall spreads the strain along the critical zone depth, compatibly with a cantilever resisting mechanism.

In DYW model, damage, in terms of concrete and reinforcement stress level, is confined near the base section because of the prevalence of the base rotation over the flexural deformed shape. Slightly above the base section and beyond, reinforcement behaves elastically and the maximum concrete stresses are never larger than 23 MPa. In DYW-ORD model, damage extends up to 2.5 m above the base section, with traditional reinforcement overcoming the yield stress and maximum concrete stresses equal to 40 MPa. Traditional walls (ORD) exhibit severe damage over the investigated height due to the yielding of the reinforcement, which results in a large residual displacement. In ORD model concrete stress is limited.

Concrete confinement is therefore required anytime unbonded post-tensioned rebars are adopted.

Further analyses allowed verifying that in case of moderate earthquakes (PGA equal to 0.15 g) all models behave elastically, with limited cracking extending over the critical zone. For increasing PGA values (0.25 g), plastic behaviour of the shear reinforcement and larger crack openings can be observed in traditional walls. Walls lacking traditional reinforcement show larger top drifts, mainly induced by the base rotation.

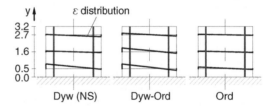

Figure 12. Strain distribution in reinforcement and concrete at different wall heights in case of maximum displacement.

It is worth noting that, in order to reduce the top wall drift, and to cut the oscillatory phenomena more rapidly, additional dissipating devices might be introduced (Kurama 1999).

The behaviour of the three wall typologies at the damage limit state was also investigated. Damage was evaluated by analysing the concrete and steel stress level, as well as the crack opening. In case of shear walls having post-tensioned rebars entirely substituting the traditional reinforcement, the onset of a large crack at the base can be observed as a result of the large rotation required by the rocking mechanism. On the other hand, in case of both post-tensioned rebars partially substituting the traditional reinforcement and in case of ordinary shear walls, diffuse cracking is observed. Crack opening can be calculated by assuming a crack spacing equal to the spacing of the stirrups along the confined zone. The crack opening values obtained by the analyses and the estimate provided by

Table 3. Crack opening at the damage limit state.

Models	Maximum strain	Concentrated crack [mm]	Distributed crack [mm]
DYW-ORD	0.00296	1.5	0.45
DYW	0.01741	8.7	–
ORD	0.00178	0.9	0.27
			(0.23 – EC2)

EC2 (2004) for traditional walls are listed in Table 3. It is worth noting that the 0.4 mm crack opening limit value assumed in the proportioning of the shear wall at the damage limit state, is verified by traditional shear wall only. However, the slightly larger crack opening obtained in case of walls having post-tensioned rebars in partial substitution of the reinforcement is not impairing the structural functionality at the damage limit state.

5 DISCUSSION AND CONCLUDING REMARKS

Following strong earthquakes, traditional R.C. structural walls may show large residual displacements and significant damage extended along the critical zone, with concrete cover spalling, as well as buckling and failure of some rebars. However, when residual displacements and damage are not as severe as to prevent any further usage of the structure, shear wall structural rehabilitation might be considered.

In this paper, the repair and retrofitting of traditional R.C. walls by means of unbonded post-tensioned reinforcements was investigated. Post tensioned rebars were introduced in partial or total substitution of the traditional reinforcement in the critical zone only.

The level of post-tension was evaluated by enforcing crack opening to be smaller than 0.4 mm at the damage limit state, whereas the post-tensioned rebar area was proportioned to ensure the necessary resisting bending moment at the ultimate limit state. Concrete confinement followed the EC8 detailing recommendations.

When subjected to earthquake loadings, unlike traditional R.C. walls, the unbonded, post-tensioned structure are characterized by: i) the elastic behaviour of the unbonded tendons, which results in negligible residual displacement, thus in the recovery of the undeformed shape; ii) the confinement of the damaged zone. The structural response of shear walls with unbonded post tensioned rebars entirely substituting traditional reinforcement is characterized by a concentrated rotation, resulting in a discrete large crack opening at the base section, induced by the rocking motion. In case of partial substitution of the reinforcement, the crack pattern shows both a large crack at the

base and a diffuse crack pattern which extends to a limited zone above the base section.

Given the localization of the damage and the recovery of the undeformed shape after a seismic event, small confined restoration works aimed at restoring the damaged concrete, re-tensioning the unbonded rebars and retrofitting the base shear resistance might be sufficient.

As a drawback, the adoption of this retrofitting structural solution results in a smaller energy dissipation capacity and in larger compression stresses in the concrete. For this solution to be the adopted, special devices ensuring the confinement of the concrete at the base edge, as well as dampers improving the energy dissipating capacity should be introduced.

It is worth noting that the significant upgrade of the base section resistance might cause the critical section to "migrate" to the second floor level. Accordingly, tendons should be stretched beyond the new critical zone in order to avoid anticipated failure of the post tensioned structural wall.

Post-tensioned walls might exhibit a slight increase of the inter-storey drift. This phenomenon, which might adversely affect non structural elements in the building, should be further studied for this solution to be applied.

Future development of the research will focus on the experimental behaviour of retrofitted shear walls, as well as on the design and testing of all structural detailing necessary for the correct structural behaviour.

REFERENCES

Eurocode 2 (2004). Design of concrete structures – Part 1-1: General Rules, and Rules for Buildings, EN 1992-1-1:2004, European Committee for Standardization, December 2004.

Eurocode 8. 2004. Design of structures for earthquake resistance – Part 1: General rules, seismic actions and rules for buildings. En 1998-1:2004, *European Committee for Standardization*.

Kurama, Y., Sause, R., Pessiki, S. and Lu, L. W. 1999. Lateral Load Behavior and Seismic Design of Unbonded Post-Tensioned Precast Concrete Walls, *ACI Structural Journal* 96 (4): 622–632.

Kurama, Y., Sause, R., Pessiki, S. and Lu, L. W. 2000. Seismic Design of Unbonded Post-Tensioned Precast Concrete Walls with Supplemental Viscous Damping, *ACI Structural Journal* 97(4): 648–658.

MIDAS/Gen. 2005. *Analysis Manual for MIDAS/Gen*, Midas information technology Co., Seohyenon-dong.

Oldrati M., Riva P. and Marini A. 2004. Analisi sismica di un edificio con pareti in c.a. con armature post-tese non aderenti. *XI Congresso Nazionale "ANIDIS L'ingegneria Sismica in Italia"*, Genova 25–29 gennaio 2004.

Pampanin, S. 2005. Solutions for high seismic performances of precast/prestressed concrete buildings, *Journal of advanced concrete technology* 3(2): 207–223.

Riva, P., Meda A. and Giuriani, E. 2003. Cyclic behaviour of a full scale R.C. structural wall, *Engineering and Structures* 25: 835–845.

Riva, Meda, A. and Giuriani, E. 2004. Experimental test on a full scale repaired r.c. structural wall, *13th World Conference on Earthquake Engineering*, Vancouver, August 1–6.

Soudki K. A., West, J. S., Rizkalla, S. H. and Blackett, B. 1996 Horizontal connections for Precast Concrete shear wall panels under cyclic shear loading, *PCI Journal*, September–October: 65–76.

Soudki K. A., Rizkalla, S. H. and LeBlanc, B. 1995 a. Horizontal connections for Precast Concrete shear walls subjected to cyclic deformations. Part1: Mild steel connections, *PCI Journal* 40 (3): 78–96.

Soudki K. A., Rizkalla, S.H. and Daikiw, R.W. 1995 b. Horizontal connections for Precast Concrete shear walls subjected to cyclic deformations. Part2: Prestressed connections, *PCI Journal*, September–October: 87–96.

Spacone, E., Filippou, F.C. and Taucer, F.F. 1996. Fibre beam-column model for nonlinear analysis of R/C frames. Part I: formulation, *Earthquake Engineering and Structural Dynamics* 25: 711–725.

*Fracture Mechanics of Concrete and Concrete Structures – Design, Assessment and
Retrofitting of RC Structures – Carpinteri, et al. (eds)
© 2007 Taylor & Francis Group, London, ISBN 978-0-415-44616-7*

Seismic analysis of a RC frame building with FRP-retrofitted infill walls

A. Ilki, C. Goksu & C. Demir
Istanbul Technical University, Structure and Earthquake Engineering Lab., Istanbul, Turkey

N. Kumbasar
Istanbul Technical University, Civil Engineering Faculty, Istanbul, Turkey

ABSTRACT: Although they are not taken into account as load bearing elements, hollow brick infill walls contribute lateral load resistance of existing reinforced concrete buildings, in terms of strength and stiffness. In this study, after a brief summary of an experimental work on reinforced concrete frames with FRP retrofitted infill walls, an existing typical rc frame building characterized by low quality concrete, insufficient confinement of structural members, smooth longitudinal bars, insufficient stiffness and irregular frames was analyzed before and after retrofitting its infill walls using FRP composite sheets. The nonlinear behavior of the building, before and after retrofitting its infill walls with FRP composite sheets, is predicted by push-over analysis. A brief explanation of this retrofitting method as given in the recently published version of Turkish Seismic Design Code is also included as well.

1 INTRODUCTION

Although they are not taken into account as load bearing elements, hollow brick infill walls contribute lateral load resistance of existing reinforced concrete frames, in terms of strength and stiffness. Laboratory tests, as well as on-site observations of structural damages after earthquakes demonstrate the significant contribution of hollow brick infill walls to seismic resistance. For existing structures to benefit from the contribution of infill walls during earthquakes, the walls must be kept in their place and the out-of-plane failure should be prevented. It is clear that any other measure that may enhance the weak tensile properties of the infill walls may further increase the contribution of infill walls to the overall seismic behavior of reinforced concrete frames. For preventing out-of-plane failures and enhancing the tensile characteristics of hollow brick walls, retrofitting the infill walls with fiber reinforced polymer (FRP) composites and connecting infill walls to the reinforced concrete frame using FRP anchorages is a new retrofitting technique. This new retrofitting technique has also been included in the recently published version of the Turkish Seismic Design Code (TSDC 2006).

In this study, after a brief summary of an experimental work on one bay two story reinforced concrete frames with FRP retrofitted infill walls reported by Yuksel et al. 2006 and Erol et al. 2006, an existing typical six story reinforced conrete frame residential building characterized by low quality of concrete, insufficient confinement of structural members, usage of smooth longitudinal bars, insufficient stiffness and irregular frames was analyzed before and after retrofitting its infill walls using FRP composite sheets. The nonlinear behavior of the building, before and after retrofitting the infill walls with FRP composite sheets, is predicted by push-over analysis. A realistic and actually applicable retrofitting scheme is planned during the selection of the infill walls to be retrofitted not to hinder the effective usage of the building. During the analysis, the retrofitted infill walls are represented with diagonal struts, of different characteristics in tension and compression. The stiffness and strength characteristics of these struts used during modeling are taken from TSDC (2006). The push-over behavior of the retrofitted building is then compared with the behavior of original building. It is seen that the investigated retrofit technique, which was proven to be effective experimentally in element basis (Erdem et al. 2006, Yuksel et al. 2006, Binici & Ozcebe et al. 2006, Ozden et al. 2006), is successful in structural basis too. Particularly, by retrofitting the infill walls of reinforced concrete frame structures, it is possible to increase stiffness and lateral strength of the structure, leading to smaller drift and less residual damage during earthquakes. It should be noted that while it depends on the selected retrofitted scheme, the enhancement in lateral strength is more pronounced in the case of current study. Naturally, the increase in stiffness may pose an increase in seismic demand; therefore, similar to other retrofitting techniques, the optimum retrofitting scheme should be sought for adjusting the desired values of strength and stiffness. Besides the significant potential enhancement in stiffness and strength, the easy application of this retrofitting technique is another major advantage. As expected, while stiffness and strength are enhanced, ductility of the structural system is affected negatively due to higher participation of brittle infill walls to the seismic behavior. However since the ductility of the original building is already very poor, the reduction in ductility has marginal effect on the overall behavior. It should be noted that the nonlinear analysis of this building for other different retrofitting schemes were investigated elsewhere, (Ilkõ et al. 2005a and Goksu et al. 2006).

2 PREVIOUS EXPERIMENTAL WORK

The contribution of FRP retrofitted infill walls to the performance of two story, one bay frames were demonstrated experimentally by (Erdem et al. 2006, Yuksel et al. 2006, Binici & Ozcebe 2006 and Ozden et al. 2006). In the study carried out by Yuksel et al. (2006), six reinforced concrete frames including two bare, two infilled frames and two frames with FRP retrofitted infill walls were tested under constant axial load and reversed cyclic lateral loads. The idea was to understand the behavior of FRP retrofitted infilled frames experimentally and collect data to be used in theoretical work. At the end of the tests, it was seen that retrofitting of infill walls with FRP composites in diagonal direction provided significant enhancement in lateral strength and stiffness.

Figures 1 and 2 are the photographs from experimental work carried out in Istanbul Technical University through a joint project with Middle East Technical University and Bogazici University under a NATO Science for Peace Project. As it can be seen in Figure 2, diagonal FRP on both sides of infill were connected to each other by means of anchors made of FRP sheets and FRP diagonals helped the infill wall to be intact even after a considerable damage. Therefore dissipating significant amount of energy, the infill walls may provide an excellent damping effect against the seismic actions. Base shear versus lateral displacement envelopes from the experimental work is given in Figure 3.

As it can be seen in Figure 3, even the frame with infill walls without any retrofit perform better than the bare frame due to significantly increased lateral strength. Introduction of FRP diagonals further increases the lateral strength as well as providing a less steep descending branch in the base shear–displacement relationship.

Figure 1. General appearance of the specimen and the testing setup (Erol et al. 2006).

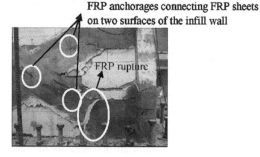

Figure 2. A retrofitted specimen after experiment.

3 OUTLINE OF THE EXISTING BUILDING

The nonlinear behavior of a typical six story reinforced concrete building is investigated by push-over analysis before and after retrofitting. The appearance of the building is given in Figure 4.

The reinforced concrete frame building, which was constructed around 1970s, represents all deficiencies of typical reinforced concrete buildings in Turkey. The building is located in Anatolian part of Istanbul on the highest seismic risk zone and on stiff rock, Figure 5.

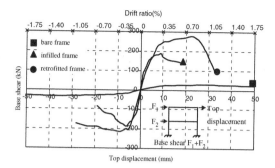

Figure 3. Base shear versus lateral displacement envelopes from experimental work (Yuksel et al. 2006).

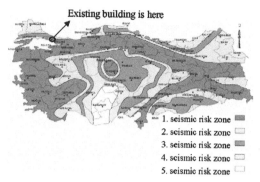

Figure 5. The distribution of seismic risk in Turkey.

Figure 4. Appearance of the existing building.

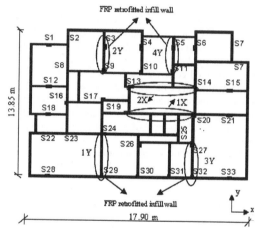

Figure 6. Typical floor plan of the existing building.

The typical floor plan of the building is presented in Figure 6.

All columns are rectangular in cross-section as shown in Figure 6. The structural system is not symmetric in any of the principal directions, many columns are not connected to each other by beams, and the columns and their orientations are not distributed evenly. Cross-sections of beams are 150 mm × 600 mm and cross-sections of columns vary from 240 mm × 240 mm to 240 mm × 600 mm. In addition to these irregularities, the characteristic compressive strength of concrete is as low as 10 MPa, which is a commonly accepted mean value for relatively old existing reinforced concrete structures in

Turkey. Both longitudinal and transverse reinforcement are plain bars with characteristic yield strength of 220 MPa. The transverse reinforcement of the original structure, consisting of 6 mm bars at 300 mm spacing is far from maintaining an adequate confinement required for a ductile behavior.

According to the results of elastic analysis carried out considering the TSDC (1998), the lateral drifts exceed the prescribed limits (relative drifts should be less than 0.0035 and 0.02/R, where R is the seismic load reduction factor based on the ductility and over strength of the structural system) and almost all of the columns are found to be inadequate in terms of flexure in both principal directions. Since lateral stiffness of the structure is quite low due to small cross-sectional areas of columns, poor connectivity of the columns with beams and low concrete quality, the periods of first two modes are found as 1.13 and 1.05 seconds for x and y directions, respectively.

According to TSDC (1998) and TSDC (2006), the design horizontal acceleration is 0.4 g for zones with

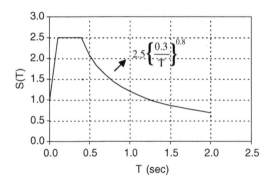

Figure 7. Calculation of spectrum coefficient according to TSDC (1998) and TSDC (2006) for Z_1 type very stiff soil.

Table 1. Dimensions of retrofitted infill walls.

All stories	l mm	h_m mm	t_m mm
1X	5750	2200	200
2X	5650	2200	200
1Y	4100	2200	200
2Y	2400	2200	200
3Y	2760	2200	200
4Y	2450	2200	200

Table 2. Properties of FRP sheets (given by the manufacturer).

Fiber type	t_f mm	T_f N/mm²	E_f N/mm²	ε_f %
Carbon	0.12	4100	231000	1.7

such high seismicity. While determining the equivalent static seismic load, the load reduction factor due to ductility and over strength is taken into account as 4, as mostly done in practice for this type of existing reinforced concrete frame structures. Base shear coefficients can be determined as 0.087 and 0.092 for x and y directions considering the periods of original structure ($T_x = 1.13$ sec, $T_y = 1.05$ sec). It should also be noted that high level of axial stresses on columns reduces the ductility.

The equivalent static seismic base shear force according to TSDC (1998) and TSDC (2006) is calculated by Equation 1:

$$V_t = \frac{WA_o IS(T)}{R_a(T)} \qquad (1)$$

where V_t = base shear force (in this case 1560 kN and 1658 kN for x and y directions, respectively); W = total weight of the structure considering the live load reduction factor (18000 kN for this case with live load reduction factor of 0.3); A_o = effective ground acceleration coefficient (0.4 in this case); I = building importance factor (1 for this case); $S(T)$ = spectrum coefficient (0.87 and 0.92 for x and y directions for this case); $R_a(T)$ = seismic load reduction factor (4 for this case). The spectrum coefficient can be calculated as shown in Figure 7.

4 RETROFITTING SCHEME

Totally 6 infill walls are retrofitted using FRP diagonals, 2 in x direction and 4 in y direction. The retrofitted infill walls which do not have any openings are shown in Figure 6.

The FRP diagonals are assumed to be applied over these infill walls without removing the plasters as also done by Yuksel et al. (2006) in the experimental study. The length, height and thickness of the retrofitted infill walls are shown in Table 1.

where l = length of retrofitted infill wall; h_m = height of retrofitted infill wall; t_m = thickness of retrofitted infill wall.

It is assumed that one ply of FRP sheets are applied over the infill walls in diagonal directions on both faces and FRP sheets are extended over the frame members and sufficiently anchored to them. The width of the FRP sheets is 400–500 mm as a function of the effective width of the compression strut of the infill wall. The properties of FRP sheets are shown in Table 2. where t_f = effective thickness of fabric; T_f = tensile strength of FRP sheet; E_f = tensile elastic modulus of FRP sheet; ε_f = rupture strain of FRP sheet.

5 ANALYSIS

5.1 Outline of recommendations of TSDC (2006)

The most recent version of TSDC (2006) permits retrofitting of reinforced concrete frame buildings using FRP composites through strengthening the infill walls in between the frame members when the ratio of wall length/height is between 0.5 and 2. The connection of retrofitted infill walls to the surrounding reinforced concrete frame is essential since the FRP reinforcement has to prevent the out-of-plane failure of the infill walls. Consequently, the contribution of infill walls to the seismic capacity can be maintained as well as the contribution of FRP reinforcement. In this technique, the retrofitted infill walls are assumed to behave as diagonal compression struts, while FRP sheets are assumed to act as diagonal tension struts. The schematic view and the static model of a retrofitted infill wall are shown in Figures 8 and 10, respectively.

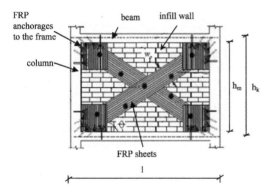

Figure 8. The schematic view of a retrofitted infill wall (Al-Chaar et al. 2002 and TDSC 2006).

Figure 9. FRP anchors connecting the diagonal FRP sheets on two sides of the wall (Yuksel et al. 2006).

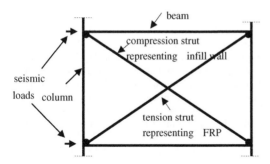

Figure 10. Static model of retrofitted infill walls.

As seen in Figure 8, the FRP composites on both faces of the wall should be connected to each other using the anchors made of FRP sheets. The spacing between these FRP anchors should be less than 600 mm. The connection of the retrofitted infill wall to the surrounding reinforced concrete frame is also to be made by means of FRP anchors as shown in Figure 9.

According to Al-Chaar et al. (2002) and TSDC (2006) the effective width of the compression strut can be calculated by Equation 2:

$$a_m = 0.175(\lambda_m h_k)^{-0.4} r_m \qquad (2)$$

where h_k = the height of column; and r_m = the diagonal length of infill wall (mm).

Table 3. Properties of retrofitted infill walls.

All Stories	a_m mm	w_f mm	t_m mm	E_m N/mm^2	f_m N/mm^2
1X	883	500	200	550	1
2X	741	500	200	550	1
1Y	561	500	200	550	1
2Y	432	400	200	550	1
3Y	401	400	200	550	1
4Y	462	400	200	550	1

The λ_m coefficient can be obtained by using Equation 3:

$$\lambda_m = \left[\frac{E_m t_d \sin 2\theta}{4 E_c I_k h_m} \right]^{0.25} \qquad (3)$$

where E_m = the elasticity modulus of infill wall (MPa); E_c = the elasticity modulus of concrete (MPa); h_m = the height of retrofitted infill wall (mm); I_k = moment of inertia of column (mm^4) and θ = angle of diagonal sheets with respect to the horizontal (degree).

The tensile strength of the tension strut is to be calculated by Equation 4:

$$T_f = 0.003 E_f w_f t_f \qquad (4)$$

where E_f = the elasticity modulus of FRP sheet; w_f = the width of FRP sheet; t_f = the effective thickness of FRP sheet.

Dimensions of the compression and tension struts and mechanical properties of the infill are given in Table 3.

where a_m = the equivalent width of compression strut, w_f = the width of FRP sheet (should be less than a_m according to TSDC (2006)); E_m = the elasticity modulus of infill wall taken as $550 f_m$ (FEMA356 2000); f_m = the compressive strength of the infill wall.

5.2 Assumptions for the analyzed building

Three pushover analyses are carried out for the examined building. The first analysis was carried out for original bare frame structure. In the second analysis, the infill walls, which are to be retrofitted, are included in the structural model without any FRP retrofit. In the third analysis, the FRP retrofit of these infill walls were also taken into account.

Although the TSDC (2006) permits retrofitting of the infill walls, when the length/height ratio is between 0.5 and 2, since the architecture of the analyzed building was not convenient, the length /height ratio of the walls retrofitted in x direction was slightly out

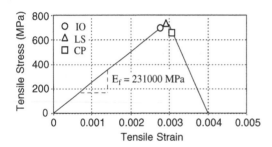

Figure 11. Stress-strain relationship assumed for the FRP diagonals in tension.

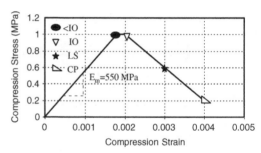

Figure 12. Stress-strain relationship for the infill wall in diagonal compression.

of the permitted ranges (between 2.48 and 2.61). The infill walls, which are retrofitted using FRP sheets in diagonal directions, are shown in Figure 6. During the nonlinear seismic analysis, the behavior of FRP sheets are modeled as tension struts with nonlinear axial hinges at their connections to the reinforced concrete frame. Hinge length has been accepted as half of the length of the tension strut for each hinge. The immediate occupancy (IO), life safety (LS) and the collapse prevention (CP) levels of FRP diagonals are set at tensile strains of 0.00295, 0.0030 and 0.0032, respectively. All three levels are set so close to each other because of very brittle nature of FRP sheets. It should be noted that ultimate tensile strain for FRP sheets given by TDSC (2006) for such applications is 0.003. The stress-strain relationship used for the nonlinear axial hinges representing FRP sheets in tension is shown in Figure 11. The detailed information on the linear elastic stress-strain relationship can be found elsewhere (fib, 2001).

The infill walls are also modeled as diagonal members. However, infill walls are assumed to resist only compressive forces. The behavior of diagonal compression struts formed by infill walls are modeled by axial nonlinear hinges at their connections to reinforced concrete frames. Hinge length has been accepted as half of the length of the compression strut for each hinge. Stress-strain relationships of the infill walls are modeled using the relationship given in Figure 12. The IO, LS and the CP levels of infill walls are set as 0.002, 0.003 and 0.004, respectively.

Application of the equations given in TSDC (2006) for a typical retrofitted infill wall of the building, namely retrofitted wall 1Y is done as follows:

$$\lambda_m = \left[\frac{550 \cdot 200 \cdot \sin 2(28.71)}{4 \cdot 24277 \cdot 54.10^7 \cdot 2200} \right]^{0.25} = 0.0009 \text{ mm}^{-4}$$

$$a_m = 0.175(0.0009 \cdot 2900)^{-0.4} \cdot 4788 = 561 \text{ mm}$$

Plastic hinge characteristics of reinforced concrete columns and beams are obtained through cross-sectional moment-curvature analysis using fiber

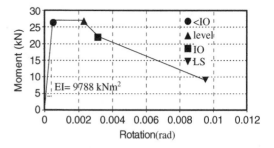

Figure 13. Moment-rotation relationship assigned for a typical column (S28 column-basement).

approach. The details of this approach can be seen elsewhere (Bedirhanoglu & Ilki 2004). While converting the moment-curvature relationship into moment-rotation relationships, plastic hinge lengths of columns and beams are assumed as half of the member depth. The details of moment-curvature analysis of the members of original structure can be found elsewhere (Ilki et al. 2005a, b). Moment-rotation relationship of a typical column is presented in Figure 13. The IO, LS and the CP levels for this column is set to 0.0022, 0.0032 and 0.0096, respectively.

It should be noted that the contribution of FRP reinforcement in compression and the contribution of infill walls in tension are neglected.

5.3 *Analyses results*

The top displacement-base shear relationships obtained by push-over analysis for original bare frame, frame with infill walls and frame with retrofitted infill walls in x and y directions are presented in Figure 14. In this figure, the design base shear forces calculated according to TSDC (1998 and 2006) are also plotted. While calculating the design base shear forces, the seismic load reduction factor is taken into account as four. It should be noted that while analyzing the frame with infill walls, only the infill walls, which are to be retrofitted are included in the model for determining

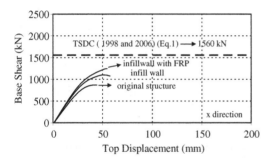

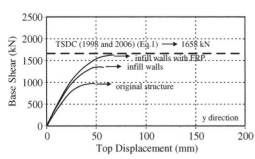

Figure 14. Base shear-top displacement relationships for original and retrofitted structures.

Table 4. Summary of the analysis.

	Bare frame	Infilled frame	Retrofitted frame
Base shear capacity (x direction) (kNm)	872	1101 (26% increase)	1243 (43% increase)
Ultimate disp. (x direction) (mm)	46	58 (26% increase)	55 (20% increase)
Period (x direction) (sec)	1.13	1.07 (4% decrease)	1.05 (6% decrease)
Base shear capacity (y direction) (kNm)	970	1351 (39% increase)	1622 (67% increase)
Ultimate disp. (y direction) (mm)	49	57 (16% increase)	65 (33% increase)
Period (y direction) (sec)	1.05	0.98 (6% decrease)	0.95 (9% decrease)

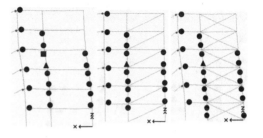

(a) Bare frame (b) Infilled frame (c) Infill walls retrofitted with FRP

δ = 44 mm δ = 44 mm δ = 44 mm
$V_{resisted}$ = 885 kN $V_{resisted}$ = 1064 kN $V_{resisted}$ = 1177 kN

● : < IO level for columns
▲ : >IO and <LS levels for columns
■ : >LS and <CP levels for columns
⬬ : < IO level for infill walls (compression strut)

Figure 15. Damage mechanisms of a retrofitted frame at the ultimate displacement capacity of the original structure (global drift ratio of 0.28%).

the contributions of infill walls and FRP diagonals separately. The other infill walls, which are not retrofitted, are not included in the model assuming that they may prematurely fail due to out of plane effects. Summary of the analyses for all cases is given in Table 4.

Considering the results of the analyses, it is seen that the contribution of infill walls significantly increases the stiffness and lateral load capacity of the structure. Still being below the base shear demand required by TSDC (1998 and 2006), usage of FRP diagonals extends this base shear capacity increase to a further point by increasing the tensile capacity of the infill walls.

As seen in Figure 15, the structure with retrofitted infill walls can resist a higher base shear force with a relatively less damage at the ultimate displacement capacity (global drift ratio of 0.28%) of the original structure. Cross-sections of the columns of external frames and the columns of the third and fourth stories tend to experience larger deformations since the columns of these regions have smaller cross-section areas.

where δ = top displacement; $V_{resisted}$ = resisted base shear force.

Since the stiffness of the retrofitted frames has increased considerably compared to the other frames, the distribution of the plastic hinges is not uniform throughout the structure. Consequently, larger internal forces are exerted to the columns and beams of

the retrofitted frames while almost all critical sections of the remaining frames experience less damage. Damage distribution of an unretrofitted frame axe at the global drift ratio of 0.28% is presented in Figure 16.

Damage mechanisms at ultimate displacement capacities of the structure with infilled frames and the structure with FRP retrofitted infilled frames are also shown in Figure 17. As seen in the given representative frames in Figure 17, the structure with retrofitted infill wall, experience less damage with respect to structure with unretrofitted infill walls.

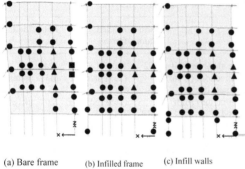

(a) Bare frame   (b) Infilled frame   (c) Infill walls retrofitted with FRP

$\delta = 44$ mm $\delta = 44$ mm $\delta = 44$ mm
$V_{resisted} = 885$ kN $V_{resisted} = 1064$ kN $V_{resisted} = 1177$ kN

Figure 16. Damage mechanism of an unretrofitted frame at the ultimate displacement capacity of the original structure (global drift ratio of 0.28%).

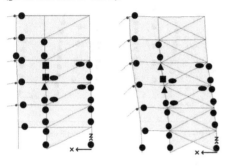

(a) Infilled frame   (b) Frame with FRP retrofitted infill wall

$\delta = 57$ mm $\delta = 54$ mm
$V_{resisted} = 1074$ kN $V_{resisted} = 1243$ kN

Figure 17. Damage mechanism at the ultimate displacement capacities of the analyzed cases with infills and retrofitted infills.

6   CONCLUSIONS

Attempting to analyze the non-linear behavior of a typical existing reinforced concrete structure with various deficiencies, retrofitted with an experimentally verified retrofitting technique, the following conclusions are reached. The investigated retrofitting technique aims to benefit from the existing infill walls against seismic actions, by keeping them in place and preventing out of plane failure by using diagonal FRP sheets applied on both sides of the infill walls as well as introducing a diagonal tension strength to the infill walls through FRP sheets. During the design process of retrofitting the formulations and details mandated by TSDC (2006) were followed.

In order to investigate the effect of infill walls on the overall behavior, two different cases apart from the original reinforced concrete frame structure; namely the frame structure only with infill walls and the frame structure with infill walls retrofitted by FRP diagonal sheets, were analyzed.

Nonlinear push-over analysis showed that the lateral strength and stiffness of the structure with infill walls increased significantly due to the contribution of the infill walls, which are generally neglected due to their tendency to premature strength loss during seismic events. Structural analysis of the structure with FRP retrofitted infill walls exhibited even more increase by means of lateral strength and stiffness, resulting with a better structural performance due to the additional tensile capacity of the FRP sheets.

As a result of the increased stiffness, it is clear that the lateral drifts are limited with respect to the original structure, which in turn limits the residual damage as well. It should be noted that, the alteration of the dynamic characteristics of the structure due to the presence of retrofitted infill walls should be handled carefully since the increase in stiffness may cause an increase in the seismic demand as well.

The investigated retrofitting technique is an easy to apply and occupant friendly technique, which causes less disturbance than many of the available retrofitting techniques. However, it should also be noted that the number of walls suitable for retrofitting is generally limited by architectural reasons such as balconies and door or window openings in the infill walls.

REFERENCES

Al-Chaar, G. K. & Lamb E.G. 2002. Design of fiber-reinforced polymer materials for seismic rehabilitation of infill concrete structures. *Seismic rehabilitation of concrete frames with infill. U.S. army corps of engineers under project 622784AT4.*

Bedirhanoglu, I. & Ilki, A. 2004. Theoretical moment-curvature relationships for reinforced concrete members and comparison with experimental data, *6th international conference on advances in civil engineering, Istanbul, Turkey,* 6–8 October 2004.

Binici B. & Ozcebe, G. 2006. Analysis of infilled reinforced concrete frames strengthened with FRPs. In S.T Wasti & G.Ozcebe (eds), *Advances in earthquake engineering for urban risk reduction, Istanbul, 30 May-1 June 2005.* Netherlands: Springer.

Erdem, I., Akyuz, U., Ersoy U. & Ozcebe, G. 2006. An experimental study on two different strengthening techniques for RC frames. *Engineering structures* 28: 1843–1851.

Erol, G., Demir, C., Ilki, A., Yuksel, E & Karadogan, F. 2006. Effective strengthening rc with and without lap splice problems. 8th *national conference on earthquake engineering, San Francisco, 18–22 April 2006.*

Federal Emergency Management Agency 2000. *FEMA 356 prestandard and commentary for the seismic rehabilitation of buildings.* Washington D.C.

Fib, 2001. Working party of the task group 9.3 FRP reinforcement for concrete structures, Design and use of externally bonded fibre reinforced polymer reinforcement (FRP EBR) for reinforced concrete structures,*Externally bonded FRP reinforcement for rc structures, July 2001*. Switzerland: International federation for structural concrete.

Goksu, C., Demir, C., Darilmaz, K., Ilki, A. & Kumbasar, N. 2006. Static nonlinear analysis of a retrofitted typical reinforced concrete building in Turkey. 8*national conference on earthquake engineering, San Francisco, 18–22 April 2006*.

Ilki, A., Darilmaz, K., Demir, C., Bedirhanoglu, I. & Kumbasar, N. 2005a. Non-linear seismic analysis of a reinforced concrete building with FRP jacketed columns. In G.L. Balazs & A. Borosnyoi (eds.), *Fib symposium keep concrete attractive, 23–25 May 2005*. Budapest: Budapest University.

Ilki, A., Demir,C. & Kumbasar, N. 2005b. Moment-curvature relationships for columns jacketed with FRP composites. In G.L. Balazs & A. Borosnyoi (eds), *Fib symposium keep concrete attractive, Budapest, 23–25 May 2005*. Budapest: Budapest University.

Ozden, S & Akguzel, U. 2006. CFRP overlays in strengthening of frames with column rebar lap splice problem. In S.T Wasti & G.Ozcebe (eds), *Advances in earthquake engineering for urban risk reduction, Istanbul, 30 May-1 June 2005*. Netherlands: Springer.

Turkish Seismic Design Code. 1998. Ministry of Public Works and Settlement, Ankara.

Turkish Seismic Design Code. 2006. Ministry of Public Works and Settlement, Ankara.

Yuksel, E., Ilki, A., Erol, G., Demir, C. & Karadogan, F. 2006. Seismic retrofit of infilled reinforced concrete frames with CFRP composites. In S.T Wasti & G. Ozcebe (eds), *Advances in earthquake engineering for urban risk reduction, Istanbul, 30 May-1 June 2005*. Netherlands: Springer.

Fracture Mechanics of Concrete and Concrete Structures – Design, Assessment and
Retrofitting of RC Structures – Carpinteri, et al. (eds)
© 2007 Taylor & Francis Group, London, ISBN 978-0-415-44616-7

Stress-strain model for compressive fracture of RC columns confined with CFRP jackets

T. Turgay & H.O. Köksal
Yildiz Technical University, Istanbul, Turkey

C. Karakoç
Boğaziçi University, Istanbul, Turkey

Z. Polat
Yildiz Technical University, Istanbul, Turkey

ABSTRACT: Catastrophic earthquakes with quite short periods often occur in Turkey. This fact necessitates and paves the way to the preparations of earthquake scenarios of the existing building stock including structures not properly constructed against earthquakes with loss estimations. There are so many buildings in densely populated areas of Turkey and many other earthquake prone countries needing urgent rehabilitation and strengthening. The use of external FRP jackets can provide confinement for RC columns in a fast and efficient way and improve their performance significantly during an earthquake. The noticeable increase in the compressive strength and ductility of concrete is the main reason for the use of FRP jackets in the seismic retrofitting. For that reason, several researchers have proposed various stress-strain models for describing the behavior of concrete confined with FRP jackets. It is worth mentioning that most of the existing models have been developed to improve or modify the analytical result of Richart's tests on concrete cylinders subjected to fluid pressure. However, in these models, confined compressive strength is predicted without a failure criterion of concrete under multi-axial stress state with the exception of the ones adopting Mander model developed for concrete confined with transverse steel reinforcement. As a preliminary study, the stress-strain relations of some of the experimentally tested RC columns confined with CFRP jackets are successfully predicted adopting Köksal model introducing a new failure criterion of concrete under triaxial compression. This model had been previously developed for RC columns confined with conventional reinforcement of steel.

Keywords: Compressive fracture, confinement, reinforced concrete column, FRP jackets, stress-strain relation, failure criterion.

1 INTRODUCTION

The compressive strength of RC columns increases with increasing confining pressure due to the use of several confinement mechanisms such as stirrups, spirals, FRP composite wraps, steel jackets, etc. The axial behavior of confined concrete was primarily researched by Richart et al. (1928) and the following relation was proposed for the ultimate strength of confined concrete based on the test results:

$$f'_{cc} = f_{co} + k_1 f_l \qquad (1)$$

where f'_{cc} is the compressive strength of confined concrete, f_{co} is the unconfined compressive strength, and

f_l is the effective lateral confining stress. In the recent literature, there are various models for describing the behavior of confined concrete with FRP composite jackets proposed by Saadatmanesh et al. (1994), Karbhari & Gao (1997), Restapol & De Vino (1996), Mirmaran & Shahawy (1998), Samaan et al. (1998), Spoelstra & Monti (1999), Saafi et al. (1999), Toutanji (1999), Xiao and Wu (2000), Lam & Teng (2002), İlki et al. (2004), and Yan & Pantelides (2006). Most of these models adopt the approach of Richardt et al. (1928) and recommend several different expressions for k_1 in Eq (1). Some researchers (Saadatmanesh et al. 1994, Restapol & De Vino 1996, Spoelstra & Monti 1999) have recommended expressions for the prediction of the compressive strength of FRP confined concrete elements, similar to the relation proposed by Mander et al. (1988) adopting William-Warnke failure surface (1975) for tri-axial compression state of circular RC columns with equal effective lateral confining stresses:

$$\frac{f'_{cc}}{f_{co}} = 2.254 \sqrt{1 + 7.94\frac{f_l}{f_{co}}} - 2\frac{f_l}{f_{co}} - 1.254 \qquad (2)$$

Based on tests of FRP confined concrete compression members, Yan & Pantelides (2006) proposed two similar equations separately for hardening and softening behavior, adjusting the failure criterion recommended by William & Warnke (1975).

In this paper, the analytical stress-strain model recommended by Köksal (2006), primarily developed for the prediction of the behavior of RC columns, has been employed for the RC columns confined with FRP composite jackets. The stress-strain curves obtained using the existing models (Saadatmanesh et al. 1994, Samaan et al. 1998, Saafi et al. 1999, Mander et al. 1988) and proposed model (Köksal 2006) are compared with large scale experiments carried out by the first author (Turgay 2007) on four RC columns of square cross-section confined with CFRP jackets.

2 EXPERIMENTAL WORK

The square RC columns in the test program having a cross-section of 200 × 200 mm and 1030 mm nominal overall height, have been tested in the structural laboratory of Yıldız Technical University, as a part of the Ph.D. study of the first author about RC columns confined with CFRP jackets (Turgay 2007). A scheme of the test setup and instrumentation is shown in Figure 1.

Large-scale RC columns have been subjected to monotonic uniaxial compression loading up to the failure. The preparation stages of specimens are consisting of making transverse steel confined square RC columns and hand-apply wrapping the CFRP around these columns. Material properties of CFRP jacket are presented in Table 1.

There is only one type of concrete mix for all test columns. C1, C2, and C3/C5 type columns are tested at the age of 30 days, 60 days and 90 days . The cylindrical compressive strength of the concrete mix of C3/C5 type was 19.36 MPa. All longitudinal bars are 10 mm in diameter, and the total number of bars is illustrated by L4 and L8. The tie spacing is 100 mm, and S8 and S12 denote stirrups with 8 mm and 12 mm in diameter, respectively. The yield strength of reinforcing steel is 422 MPa. In this paper, the test results of four C3 type-columns confined with one layer of CFRP are presented.

For measurement of axial strains, four linear variable displacement transducers (LVDTs) are mounted

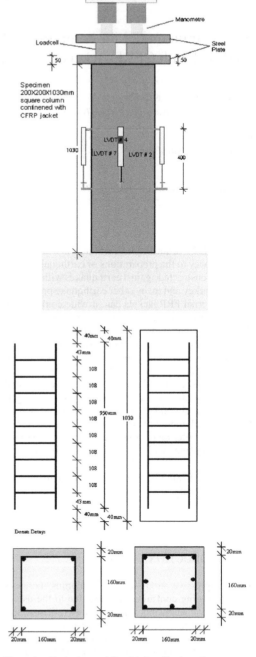

Figure 1. Test setup and the details of test specimens.

on the central 400 mm gage length at each face of a column in a similar way used to eliminate any eccentricity of the applied load as recommended in the study of Shrive et al. (2003). A pre-loading up to the one-fourth of the predicted axial capacity is applied to maintain

Table 1. Material properties of CFRP sheets.

FRP Composite	Tensile Strength (MPa)	Tensile Modulus (MPa)	Maximum Tensile Strain (mm/mm)	Ply Thickness (mm)
CFRP	3430	230000	0.015	0.165

approximately same LVDT readings so that any eccentricity can be eliminated in the linear elastic stage of the overall behavior.

Although the LVDT placing and readings are provided very close to each other, next to the maximum axial load there can be a significant variation between the minimum and maximum values of shortening (Karakoç et al. 2007). Tests are conducted using 2000-kN compression machine. The load is applied to the specimens at an approximately constant rate until the complete failure of columns and is monitored using a load cell. All measured data were collected by a data logger and stored directly in the computer.

3 CONSTITUTIVE MODELS FOR CONFINED CONCRETE

The model proposed by Köksal (2006), is used for the description of the behavior of RC columns confined with CFRP composite jackets. Taking the sign of the compressive stresses as positive because of the dominance of the compressive stresses for the case of RC columns confined with conventional steel reinforcement, and modifying the linear relation of DP criterion, the following failure surface is introduced:

$$f = \sqrt{6}\,\alpha(\xi)\xi + \rho - \sqrt{2}\,k(\sigma_1, \sigma_2, \sigma_3) = 0 \quad (3)$$

where ρ and ξ are deviatoric and hydrostatic lengths, respectively (Köksal 2006). k is considered as a function of the lateral confinement pressure and the cylinder compressive strength of concrete, f'_{cc}:

$$k = k(f_l, f'_c) \quad (4)$$

where f_l is the average of the two principal stresses acting in two orthogonal directions to the crosssection of RC column:

$$f_l = \frac{\sigma_2 + \sigma_3}{2} \qquad if\ \sigma_1 > \sigma_2 > \sigma_3 \quad (5)$$

and the following form is finally proposed for k using the test results of Richardt et al. (1928):

$$k = \left(4.07\frac{f_l}{f'_c} - 0.89\left(\frac{f_l}{f'_c}\right)^2 + 0.807\right)f'_c \quad (6)$$

Material parameter α in Equation 3 is also given in terms of the hydrostatic length:

$$\alpha = \alpha(\xi) = 0.462\,\xi^{-0.2355} \qquad if\ \frac{\xi}{f'_c} \ge 0.58 \quad (7)$$

Since the ultimate hydrostatic pressure is generally between 20 MPa and 60 MPa during the tests of RC columns, $\xi/f'_c \ge 0.58$ draws a reasonable lower limit for the confined concrete strength under compression implying that the mean compression stress $\sigma_m = f'_c/3$. The geometry of the square and rectangular columns does not allow a uniform distribution of lateral pressure. Therefore, k in Equation 6 is simply reduced multiplying by 0.85 in order to reflect this fact.

In this paper, the Saenz's equation (1964) is adopted for describing the monotonic stress-strain relationship of confined concrete:

$$\sigma_1 = \frac{\varepsilon_1 E_0}{1 + \left(\frac{E_0}{E_s} - 2\right)\frac{\varepsilon_1}{\varepsilon_{cc}} + \left(\frac{\varepsilon_1}{\varepsilon_{cc}}\right)^2} \quad (8)$$

where σ_1 and ε_1 are axial compressive stress and strain of concrete, respectively; E_0 is the initial tangent modulus of elasticity in MPa; E_s is the secant modulus at the point of maximum compressive stress f'_{cc} which can be determined using Equations 3, and 5–7. The strain ε_{cc} corresponding to the maximum compressive stress f'_{cc} can be found employing the recommended relations in the pioneering work of Richart et al. (1928):

$$\varepsilon_{cc} = \varepsilon'_c\left(1 + k_2\frac{f_l}{f'_c}\right) \quad (9)$$

where ε'_c is the peak strain at the strength of plain concrete cylinders. Richart et al. (1928) proposed that k_2 could be taken as $5k_1$.

For comparison purposes, the four stress-strain models of concrete under concentric compression and lateral confinement (Saadatmanesh et al. 1994, Samaan et al. 1998, Saafi et al. 1999, Mander et al. 1988) are evaluated throughout the study. First model is the Mander model proposed for concrete under concentric compression and confined with transverse reinforcement consisting of steel stirrups or spirals. The axial stress-strain curve is plotted according to the Popovics equation (1973):

$$\sigma_1 = \frac{f'_{cc}\,xr}{r - 1 + x^r} \quad (10)$$

where

$$x = \frac{\varepsilon_1}{\varepsilon_{cc}} \quad (11)$$

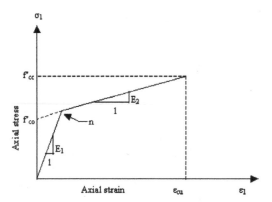

Figure 2. Parameters of bilinear confinement model (Samaan et al. 1998).

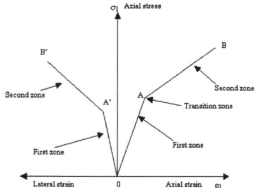

Figure 3. Simplified Stress-strain curves of FRP Grid Confined Concrete (Saafi et al. 1999).

$$r = \frac{E_c}{E_c - E_s} \qquad (12)$$

and ε_{cc} is the strain corresponding to maximum concrete stress f'_{cc}, calculated by the following equation in this model:

$$\varepsilon_{cc} = \varepsilon'_c \left[1 + 5 \left(\frac{f'_{cc}}{f'_{co}} - 1 \right) \right] \qquad (13)$$

For the purpose of determining the confined concrete compressive strength f'_{cc}, the five-parameter multi axial failure surface developed by William and Warnke (1975) is adopted as in Equation 2.

Saadatmanesh et al. (1994) presented analytical models for the analysis of concrete columns under monotonic loading and externally confined with FRP composite straps. Stress-strain model for confined concrete proposed by Mander et al. (1988) is adapted for the analysis of circular and rectangular columns confined with fiber composite straps under a slow strain rate and monotonic loading.

Samaan et al. (1998) proposed an analytical model to predict the complete stress-strain response of FRP-confined concrete columns. For the bilinear stress-strain curves are presented for both axial and lateral directions. The bilinear response of FRP confined concrete; the four-parameter relationship of Richard and Abbott (1975) is adopted and calibrated as follows:

$$\sigma_1 = \frac{(E_1 - E_2)\varepsilon_1}{\left[1 + \left(\frac{(E_1 - E_2)\varepsilon_1}{f_0} \right)^n \right]^{\frac{1}{n}}} + E_2 \varepsilon_1 \qquad (14)$$

where, E_1 and E_2 are first and second slopes respectively; f_o is the reference plastic stress at the intercept of the second slope with stress axis, n is a curve-shaped

parameter that mainly controls the curvature in the transition zone (Fig. 3).

Saafi et al. (1999) suggest that the stress-strain response of FRP-confined specimens is bilinear in nature with a small transition zone between the two linear branches.

Figure 3 shows a simplified stress-strain response of a concrete specimen confined with FRP grid. A confinement effectiveness coefficient k_1 is proposed using a regression analysis of the experimental data:

$$k_1 = 2.2 \left(\frac{f_l}{f'_c} \right)^{-0.16} \qquad (15)$$

By substituting Equation 15 into Eq 1, the compressive strength at every point of the second zone is given by

$$f(\varepsilon_l) = f'_c \left(1 + 2.2 \left(\frac{f_l}{f'_c} \right)^{0.84} \right) \qquad (16)$$

where $f(\varepsilon_l)$ is the axial compressive strength at a radial strain ε_l, and a confining pressure f_l:

$$f_l = \frac{2 t_{com} E_{com} \varepsilon_l}{ds} \qquad (17)$$

where E_{com} is the elastic modulus of the composite grid; t_{com} is composite thickness; d is the diameter of concrete core; and s is the spacing of the circular ribs (Saafi et al. 1999). Substituting Equation 17 into Equation 16, the ultimate compressive strength of confined concrete with FRP grids is given as:

$$f'_{cc} = f'_c \left(1 + 2.22 \left(\frac{2 t f_{com}}{d f'_c} \right)^{0.84} \right) \qquad (18)$$

Figure 4. Failure mode for C3L8S8 and C3L4S12.

According to Saafi et al. (1999) axial strain reaches maximum:

$$\varepsilon_{cc} = \varepsilon_{co}\left(1 + \left(2.6 + 537\varepsilon_{com}\right)\left(\frac{f'_{cc}}{f'_{c}} - 1\right)\right) \qquad (19)$$

where ε_{co} is the axial strain of unconfined concrete, ε_{com} is the ultimate strain of FRP tube.

4 RESULTS

In this paper, results relative to C3 test series of four square RC columns confined with CFRP composite jackets are presented. All columns are confined with one layer of CFRP and have fracture surfaces between the mid-height and top surface except C3L4S8. The fracture region was between the bottom and mid-height of C3L4S8. Actually, fracture of cylinders in compression has even traditionally been a difficult problem because of the way that cracks initiate (Landis et al 2003). Sounds started to be heard during approximately half of the ultimate load and were more frequent and louder very close to the ultimate load at the fracture of CFRP jackets. Coneshaped fracture surface was observed after removing of the CRFP jackets at this region in Figure 4. The failure mode for the confined RC columns was due to the failure of the CFRP jackets and, after this point, the concrete core immediately failed due to the loss of confinement.

As can be seen in Figure 5, the most important observation at the end of the tests is that there is a significant increase up to nearly 100 percent in the deformability of the columns when the diameter of transverse steel changes from 8 mm to 12 mm.

5 CONCLUDING REMARKS

Generally, analytical models developed for concrete confined with FRP composite jackets, are based on the results of small-scale experiments instead of testing

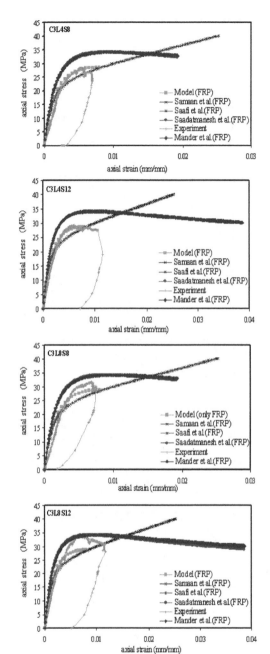

Figure 5. Plots of the analytical models for test columns with the confinement effects of only CFRP.

full-scale columns. For these large-scale specimens the failure mechanisms can be very different compared to the theoretical models (Köksal 2006). Therefore, the validity of the comparison of the proposed models can be best achieved for the experiments of large-scale RC columns.

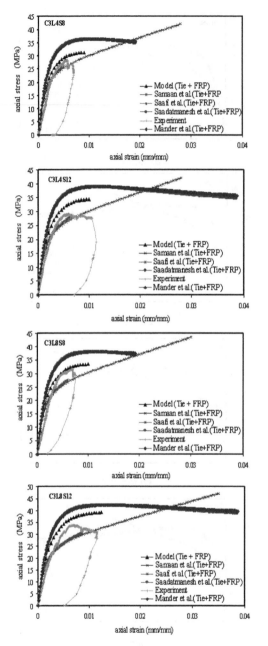

Figure 6. Plots of the analytical models for test columns with the confinement effects of both tie and CFRP.

The models of Mander and Saadatmanesh overestimate the ultimate strength of RC columns especially for the case in which the confinement effects of tie and CFRP are both considered. It can be noted that the predictions of Saadatmanesh and Mander are so close as indicated by the almost same common curve in Figs. 5

and 6. The analytical models of Saafi and Samaan predict the shape of the curve fairly well although Samaan model over-predicts the ultimate strength.

Köksal model predicts not only the ultimate strength of the columns closely but also the trends of the stress-strain plots. Its predictions are somewhat higher for the second case, accounting for the confinement effects of both CFRP jackets and transverse steel, as indicated in Figure 6. The interaction between the confinement mechanisms of transverse steel and FRP composites is a future and important issue to be considered for the load and deformability capacity of RC columns.

ACKNOWLEDGEMENTS

The writers are grateful to YKS Degussa Yapõ Kimyasallarõ and SET BETON for the donation of the FRP materials and ready concrete-mixes.

REFERENCES

Ilki, A., Kumbasar, N., Ozdemir, P., and Fukuta, T. 2004. A trilinear stress-strain model for confined concrete. *Structural Engineering and Mechanics* 18(5): 541–563.

Karakoç, C., Polat, Z., Köksal, H.O., Turgay, T. and Akgün, Ş. 2006. Evaluation of experimental procedures for confined concrete columns using 3D finite element analyses. *Third International Conference on High Performance Structures and Materials, Wessex Institute of Technology* Ostend: Belgium (to be printed in HPSM CMEM-2007)

Karbhari, V.M. and Gao, Y. 1997. Composite jacketed concrete under uniaxial compression-verification of simple design equations. *Journal of Materials in Civil Engineering, ASCE* 9(4): 185–193.

Köksal, H. O. 2006. A failure criterion for RC members under triaxial compression, *Structural Engineering and Mechanics, Techno Press* 24(2): 137–154.

Lam, L. and Teng, J.-G. 2002. Strength models for fiberreinforced-plastic confined concrete. *Journal of Structural Engineering, ASCE* 128 (5): 612–623.

Landis, E.N., Nagy, E.N. and Keane, D.T. 2003. Microstructure and fracture in three dimensions. *Engineering Fracture Mechanics*, 70: 911–925.

Mander, J. B., Priestley, J. N., and Park, R. 1988. Theoretical stress-strain model for confined concrete. *Journal of Structural Engineering, ASCE* 114(8): 1804–1826.

Mirmiran, A., Shahawy, M., Samaan, M. and El-Echary, H. 1998. Effect of column parameters on FRP-confined concrete, *Journal of Composites for Construction, ASCE* 2(4): 175–185.

Popovics, S. 1973. A numerical approach to the complete stress-strain curves for concrete. *Cement and Concrete Research* 3 (5): 583–599.

Restrepol, J.I. and De Vino, B. 1996. Enhancement of the axial load-capacity of reinforced concrete columns by means of fiberglass-epoxy jackets. *Proceedings, Second International Conference on Advanced Composite Materials in Bridges and Structures:* 547–690. Montreal: Canada.

Richard, R M., and Abbott, B. J. 1975. Versatile elastic-plastic stress-strain formula. *J Engr. Mech., ASCE* 101(4): 511–5.

Richart, F.E., Bradtzaeg, A. and Brown, R. L. 1928. *A study of the failure of concrete under combined compressive stresses*, Bulletin No. 185, Engineering experimental station University of Illinois, Urbana, pp. 104.

Saadatmanesh, H., Ehsani, M.R., Li, M.W. 1994. Strength and ductility of concrete columns externally reinforced with fiber composite straps. *ACI Structural Journal* 91(4): 434–47.

Saafi, M., Toutanji, H.A. and Li, Z. 1999. Behavior of concrete columns confined with fiber reinforced polymer tubes, *ACI Materials Journal* 96(4): 500–509.

Saenz, L.P. 1964. Discussion of 'Equation for the stress strain curves of concrete' by Desai and Krishnan. *ACI Journal* 61 (9): 1229–1235.

Samaan, M., Mirmiran, A. and Shahawy, M. 1998. Model of concrete confined by fiber composite. *Journal of the Structural Engineering, ASCE* 124(9): 1025–1031.

Shrive,P.L., Azarnejad, A., Tadros, G., McWhinnie, C. and Shrive, N.G. 2003. Strengthening of concrete columns with carbon fibre reinforced polymer wrap. *Can. J. Civ. Eng*. 30(3): 543–554.

Spoelstra, M.R. and Monti, G. 1999. FRP-confined concrete model. *Journal of Composites for Construction, ASCE* 3(3): 143–150.

Toutanji, H.A. 1999. Stress-strain characteristics of concrete columns externally confined with advanced fiber composite sheets. *ACI Materials Journal* 96(3): 397–404.

Turgay, T 2007. *CFRP Uygulanmöş betonarme elemanlarön performansö* (The performance of FRP strengthened structural members). Ph.D. Thesis, Submitted to Yıldız Technical University.

William, K.J., and Warnke, E.P. 1975. Constitutive model for the triaxial behavior of concrete. *Proc., International Association for Bridge and Struct. Engr.*, 19: 130.

Xiao, Y. and Wu, H. 2000. Compressive behavior of concrete confined by carbon fiber composite jackets. *Journal of Materials in Civil Engineering, ASCE* 12(2): 139–146.

Yan, Z., Pantelides, C.P. and Reaveley, L.D. 2006. Fiberreinforced polymer jacketed and shape-modified compression members: II-Model. *ACI Structural Journal* 103(6): 894–903.

Fracture Mechanics of Concrete and Concrete Structures – Design, Assessment and Retrofitting of RC Structures – Carpinteri, et al. (eds)
© 2007 Taylor & Francis Group, London, ISBN 978-0-415-44616-7

Analysis of the confinement in RC hollow columns wrapped with FRP

G.P. Lignola, A. Prota, G. Manfredi & E. Cosenza
Department of Structural Engineering DIST, University of Naples Federico II, Naples, Italy

ABSTRACT: This paper presents a theoretical model for the analysis of fiber-reinforced polymer (FRP) confined hollow reinforced concrete (RC) sections as found in many tall bridge piers. FRP confinement systems effectiveness have been studied focusing on solid columns while very little has been done about hollow sections. In order to study the behavior of square hollow sections subjected to combined axial load and bending, a total of 7 specimens have been tested, representing, in a scale 1:5, typical square hollow bridge piers. The strengthening scheme consisted of unidirectional Carbon FRP laminates applied in the transverse direction. The proposed model is able to estimate confinement effectiveness in the case of hollow sections. Relevant parameter was the relative wall thickness. Theoretical results, in excellent agreement with authors' sets of experimental data, show that FRP jacketing can enhance the ultimate load and the ductility also in the case of hollow concrete cross sections.

1 INTRODUCTION

1.1 *Hollow cross section confinement*

Compression column elements potentially support a variety of structures such as bridge decks and floor slabs. Columns vary in physical shape depending on their application, although typically they are either circular or rectangular, solid or hollow, for the simplicity of construction.

Various spectacular concrete bridges including hollow piers have been constructed throughout the world particularly in Europe, United States and Japan, where high seismic actions and natural boundaries require high elevation infrastructures. Hollow concrete cross sections are usually found in tall bridge piers. High elevation bridges with very large size columns are constructed to resist high moment and shear demands. In particular, bridge piers designed in accordance with old design codes may suffer severe damage during seismic events, caused by insufficient shear or flexural strength, low ductility and inadequate reinforcement detailing. Many parameters may influence the overall hollow column response such as: the shape of the section, the amount of the longitudinal and transverse reinforcement, the cross section thickness, the axial load ratio and finally the material strength of concrete (core and cover) and steel (reinforcement).

Apart from the possible human victims, severe earthquake damage on bridges results in economic losses in the form of significant repair or replacement costs and disruption of traffic and transportation. For these reasons, important bridges are required to suffer only minor, repairable damage and maintain immediate occupancy after an earthquake to facilitate relief and rescue operations. Most of the existing bridges worldwide were designed before their seismic response had been fully understood and modern codes had been introduced; consequently they represent a source of risk in earthquake-prone regions.

Recent earthquakes in urban areas have repeatedly demonstrated the vulnerability of older structures to seismic actions, also those made by reinforced concrete, with deficient shear strength, low flexural ductility, insufficient lap splice length of the longitudinal bars and, very often, inadequate seismic detailing, as well as, in many cases, insufficient flexural capacity.

The only available answer to the aforementioned problems were either to rebuild the structure or to use standard restoring techniques (i.e. section enlargement, steel jacketing and others) that would have had a high social and economical impact as well as structural consequences such as increase in self weight consequently with a negative contribution to foundations and to the seismic response of the overall structure.

FRP materials as an alternative approached the construction market as a viable, cost and time effective solution for upgrading and retrofitting existing concrete structures. FRP confinement systems effectiveness have been extensively studied in the last years focusing on solid columns while very little research has been done about hollow cross sections.

1.2 *Research impact and objectives*

The objective of the proposed investigation is to evaluate the behavior of square hollow bridge cross sections retrofitted with FRP composites materials used as external jacketing. The influence of external loading conditions, namely pure compression and combined flexure and compression has been studied in order to determine the available ductility of unstrengthened and strengthened rectangular hollow cross sections. This evaluation consists in an experimental phase undertaken in conjunction with analytical studies to predict and to model the results of the former tests. The development of design construction specifications and a refined methodology to design and assess hollow cross section members behavior under combined axial load and bending is the final output of the program.

2 EXPERIMENTAL CAMPAIGN

2.1 Test matrix and test set-up

Full-scale testing of these structures is out of the question, both for logistic and cost reasons, and also because data are needed prior to the design of the structure. Engineers thus have to rely on scale-model tests to predict the behavior of the prototypes and consequently to know how to estimate ultimate strains and stresses in the structures from the results obtained in scaled-down specimens. The probability of finding a serious flaw in a structural member increases with its size and varies with the type of material.

The experimental program has been planned on hollow columns in reduced scale. The scale factor 1:5 has then been chosen and tested specimens, reproducing in scale typical bridge piers, had hollow section external dimensions of $360 \times 360\,\text{mm}^2$ and walls thickness of 60 mm. The internal reinforcement was given by 16-ϕ10 longitudinal bars with 25 mm concrete cover and ϕ4 stirrups at 80 mm on center (Fig. 1).

The selection of the scale factor has been driven by two considerations: the attempt to study specimens whose dimensions were sufficiently large to represent the behavior of real piers and the need to respect laboratory constraints.

The hollow portion of the column had a height of 1.30 m and was made by foam-polystyrene, while the overall height of the specimen is 3 m.

The test matrix was designed in order to assess the FRP wrapping effectiveness in correspondence of three P/M ratios, which are three different neutral axis positions.

Three eccentricities have been selected to study the behavior of the hollow members under P-M combinations carrying the neutral axis at ultimate load external to cross section (e = 50 mm, fully compressed), at mid-height (e = 200 mm), and close to compressed flange (e = 300 mm). Accordingly three specimens are unstrengthened (Series U), while the second three are strengthened (Series S). A compressed unstrengthened specimen was also tested.

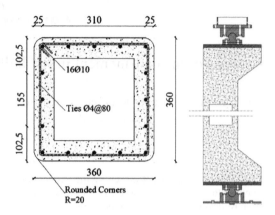

Figure 1. Specimen geometry.

Table 1. Test Matrix.

Specimen Code	Loading Condition	Eccentricity e (mm)
U0 – ...	Pure Compression	0
U1 – S1	Combined Compression	52–80
U2 – S2	and Flexure	200
U3 – S3		300

The matrix of tests done is reported in table 1, where e is the load eccentricity kept constant during each test. Note that a construction issue did not allow testing both specimens U1 and S1 with eccentricity of 50 mm; actual eccentricity was 52 mm and 80 mm.

In the U-series and S-series the two ends have been designed with corbels to allow the application of the axial load with the desired eccentricity as showed in figure 1, whereas the heads had solid sections in order to distribute the load and avoid local failures. In order to apply the axial load with the desired eccentricity, two open hinges have been designed (Fig. 1) to facilitate the centering of the specimen under the testing machine and to apply an eccentric axial load without shear and having free rotations at the two specimen ends.

2.2 Materials characterization

Some tests have been executed on concrete and steel bars: during concrete pouring five specimens of plain concrete have been taken and nine pieces of steel longitudinal bars (ϕ10) picked up. Concrete specimens have been crushed and test results show a mean compressive strength of 32 MPa.

Three series of steel bars have been also tested. The first one have been tested under tension, while the other two under compression with a Length to Diameter L/D ratio of 8 and 16 (to simulate the different free length of the compressed bar). Depending on the stirrup stiffness, the compressed longitudinal bar

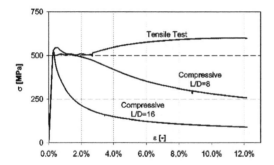

Figure 2. Steel bars characterization.

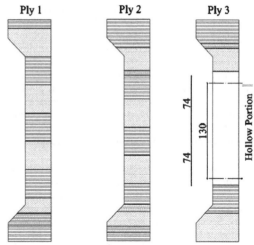

Figure 3. CFRP strengthening scheme.

Table 2. Adopted FRP system materials.

Material	Trade name
Fabric	MapeWrapC Uni-Ax 600/40
Primer	MapeWrap Primer1 A + B
Putty	MapeWrap12 A + B
Resin	MapeWrap31 A + B

between two stirrups has different free length: the distance of the two testing grips have to be one or two times the stirrup spacing to simulate stirrups that are relative stiff or weak, respectively. A completely different behavior can be observed (Fig. 2) in the three cases: the instability in the compression case (buckling) corresponding to L/D = 16 have been accounted for in the following analyses. Test results show a mean tensile strength of 600 MPa and a yield stress of 506 MPa (that in compression roughly correspond to buckling stress in the case of L/D = 16).

2.3 Strengthening scheme

Two plies of CFRP unidirectional fabric with density of 600 gr/m² have been applied in all S specimens for the entire specimen height (Fig. 3). Tensile modulus of elasticity of the CFRP material is 230 GPa and ultimate tensile strength is 3450 MPa. The fabrics have a nominal width of 400 mm and nominal thickness (dry) of 333 μm.

The solid parts only have been further reinforced with a third layer of CFRP (Fig. 3) in order to avoid the occurrence of local failures in zones not of interest for the test but subjected to high stress concentrations.

The CFRP laminates were applied by manual lay-up in the transverse direction. Corners were rounded with a radius of 20 mm as prescribed in many codes and the two plies have been overlapped by 200 mm.

The number of installed plies was considered an upper limit that could be derived from an economical and technical analysis, also accounting for the scale reduction. On this scheme the CFRP reinforcement ratio is four times bigger than the stirrups reinforcement ratio. Nevertheless it has been observed that the influence of the number of layers of FRP on the solid section specimen under eccentric loading is not so pronounced as that of the specimen under concentric loading (Li & Hadi 2003).

The choice of carbon fibers over glass and aramid has been made since the former have not only better mechanical properties with respect to glass and aramid both in terms of ultimate strength and elastic

modulus but also have the best durability performance in exposed environments, guaranteeing the longest life of the intervention.

Trade names of the FRP System materials adopted are reported in table 2.

2.4 Main experimental outcomes

Fundamental results and global outcomes of the experimental research in terms of strength and failure modes of the scaled hollow columns are given in (Lignola et al. 2007a,b). In the following, a brief summary will recall the main results in terms of strength, ductility and failure modes.

The failure of hollow members is strongly affected by the occurrence of premature mechanisms (compressed bars buckling and unrestrained concrete cover spalling). FRP confinement does not change actual failure mode, but it is able to delay bars buckling and to let compressive concrete strains attain higher values, thus resulting in higher load carrying capacity of the column (strength improvement about 7% in the case of larger eccentricity and 19% in the case of smaller eccentricity) and significantly in ductility enhancement.

The ductility increments have been estimated through the comparison of curvature ductility μ_χ. In unstrengthened columns the curvature ductility ranged between 1 (brittle failure of U1 and U2) and 1.54 (specimen U3), while in the case of strengthened columns the curvature ductility increased significantly attaining values ranging between 3.07 and 8.27.

Actually the increment of ductility supplied by confinement corresponds also to a significant gain in terms of area under the curve that is proportional to the specific energy and can be also appreciated comparing the values of these specific energies and remarkable increases of dissipating capabilities for strengthened columns are found.

The maximum energy is recorded for the column S3. For instance the dissipated specific energy of S3 specimen is more than 7 times the dissipated specific energy of the brittle U3 specimen for the same eccentricity. Specific energy increment after peak at 80% of ultimate load is more than 3.71 times the specific energy at peak. This analysis evidenced a remarkable improvement of the seismic response of the wrapped columns: after peak load they kept a good load carrying capability, that is good energy dissipation.

The strength improvement was more relevant in the case of specimens loaded with smaller eccentricity, while the ductility improvement was more relevant in the case of bigger eccentricity. At lower levels of axial load also the brittle effect of reinforcement buckling was less noticeable.

3 THEORETICAL ANALYSIS

3.1 Section analysis: fiber model

Through the use of a fiber model that meshes the concrete cross-sectional geometry into a series of discrete elements/fibers, sections of completely arbitrary cross-sectional shape (including hollow prismatic cross sections) can be modeled. Tension stiffening effect, compressed bars buckling, concrete cover spalling and FRP confinement of concrete are included in the model *(Lignola 2006)*.

The described numerical model uses nonlinear stress-strain relationships for concrete and steel. A reliable stress-strain behavior of concrete is necessary particularly when a member is subjected to combined bending and axial load and confinement effects should be accounted for.

The stress-strain relationship of plain concrete under concentric loading is believed to be representative also of the behavior of concrete under eccentric loading.

3.2 Proposed hollow section confinement model

Two approaches were used to assess the behavior of the FRP confined hollow members. A first approach

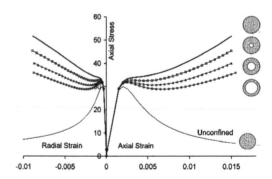

Figure 4. Proposed hollow section confinement model.

that considers the interaction of the four walls forming the hollow member has been proposed. This approach considers the interaction of the four walls forming the hollow member. The walls confinement have been analyzed according to the behavior observed in wall-like columns *(Prota et al. 2006)* similar to the behavior of the walls forming the hollow member. The transverse dilation of the compressed concrete walls stretches the confining device, which along with the other restrained walls applies an inward confining pressure. A detailed description of this approach can be found in *(Lignola 2006)*.

A different approach, instead, that consider the confinement of the whole hollow section has been proposed *(Lignola 2006)*. This confinement model for circular hollow sections has been extended to square hollow ones. The confining pressure is provided by an FRP jacket of the same thickness to an equivalent circular column of diameter D equal to the average side length. The model is able to estimate confinement effectiveness, which is different in the case of solid and hollow sections. The numerical predicted stress strain relationships for a solid section and for a hollow section with different R_i/R_o ratios, but constant relative confinement stiffness $E_f t/(R_o - R_i)$ is depicted in figure 4.

A model based on the assumption that the increment of stress in the concrete is achieved without any out-of-plane strain was proposed *(Braga et al. 2006)*. Plain strain conditions were adopted to simulate the confinement effect.

An elastic model *(Fam & Rizkalla 2001)* based on equilibrium and radial displacement compatibility was presented adopting the equations proposed by *Mander et al. (1988)* through a step-by-step strain increment technique to trace the lateral dilation of concrete.

In the hypothesis of axial symmetry the radial displacement is the only displacement component and stress components (radial and circumferential) can be evaluated according to boundary conditions (i.e. applied external inward pressure and internal outward pressure). The dependence of the lateral strain with the axial strain is explicitly considered through radial equilibrium equations and displacement compatibility. Confining pressure q equation can be explicated in the form $q = q(\varepsilon_c)$, so that at each axial strain ε_c the confining pressure q exerted on concrete by the FRP jacket is associated:

$$q = \frac{v_c}{\dfrac{R_o}{E_f t}(1-v_f) + \dfrac{1+v_c}{E_c}\dfrac{R_o^2}{R_o^2 - R_i^2}\left[(1-2v_c) + \left(\dfrac{R_i}{R_o}\right)^2\right]}\varepsilon_c \quad (1)$$

Previous equation (1) is based on linear elasticity theory for all the involved materials (E_c and v_c are concrete elastic modulus and Poisson's ratio respectively, while E_f and v_f are FRP elastic modulus and Poisson's ratio respectively), R_o and R_i are respectively the outer and inner radius of the hollow circular cross section, t is the thickness of the FRP wrap.

To account for the nonlinear behavior of concrete, a secant approach can be considered. The elastic modulus and the Poisson's ratio are function of the axial strain and of the confinement pressure q.

An iterative procedure is then performed to evaluate, at any given axial strain ε_c, the corresponding stress f_c, pertaining to a Mander's curve at a certain confining pressure $q(\varepsilon_c)$.

To simplify and to avoid an iterative procedure to determine the actual (step i) secant elastic modulus of concrete, this parameter is evaluated as the slope of the line connecting the origin and the previously evaluated (step i-1) stress-strain point.

The secant Poisson's ratio is used to obtain the lateral strain at a given axial strain in the incremental approach. The dilation of confined concrete is reduced by the confinement; therefore, the Poisson's ratio at a given axial strain level is lower in the presence of higher confining pressure. Fitting the results curve of many concrete cylinders tested under different confining hydrostatic pressures q with a second-order polynomial, a simplified linear relationships for ε_c under constant confining pressures is provided, and from regression analysis, it can be considered *(Fam & Rizkalla 2001)*:

$$\frac{v_c}{v_{co}} = 1 + \frac{\varepsilon_c}{\varepsilon_{cc}}\left(0.719 + 1.914\frac{q}{f'_{co}}\right) \quad (2)$$

where v_c is the actual Poisson's ratio at a given axial strain ε_c and actual confining pressure $q(\varepsilon_c)$. The actual peak confined concrete compressive strain (evaluated for the actual confining pressure q) is ε_{cc}. The initial values are the unconfined peak concrete strength (f'_{co}), and the Poisson's ratio (v_{co}) usually ranging between 0.1 and 0.3.

The bigger is the hole, the higher is the deformability of the element and the circumferential stresses

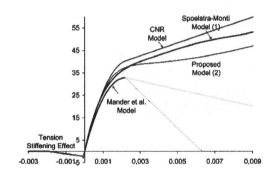

Figure 5. Theoretical confined concrete constitutive laws (hollow section).

compared to the radial component: in the case of solid section the dilation of concrete is restrained by the FRP wraps and this interaction yields a strength improvement, while in the case of thin walls, the larger deformability does not allow to gain such strength improvements, even though a significant ductility development is achieved.

3.3 Adopted confined concrete constitutive laws

To simulate the effect of the FRP confinement the two abovementioned approach has been adopted. *Spoelstra & Monti (1999)* model and the model recommended by the *CNR DT200 (2004)* Italian Instructions have been adapted in the first approach[1].

Compared to these two confinement models[1], adapted to simulate the behavior of hollow square columns by considering the effect of confinement of single walls, the proposed confinement model, denoted with subscript [2], gives a confined concrete strength, corresponding to the peak unconfined strain (0.2%), in-between the previous ones. The main difference is in the subsequent branch, where an almost constant plastic behavior for the hollow section is predicted instead of a hardening branch, common in highly confined solid concrete sections (Fig. 5).

4 THEORETICAL-EXPERIMENTAL COMPARISON

4.1 Global behavior

Only the adapted *Spoelstra & Monti (1999)* model[1] in the following will be considered, because this is an evolution of the *Mander et al. (1988)* model for confined concrete, such as the proposed model[2] is. *Mander et al. (1988)* model has been considered for unconfined concrete coupled with size effect theory after *Hillerborg (1989)* for the concrete post peak softening.

The proposed confinement model predicts quite well the behavior of hollow section confinement, in

Table 3. Experimental-Theoretical failure loads comparison.

Specimen Code		e [mm]	P [kN]	M [kNm]	M Error exper. vs. theor.
U1	exper.	52	2264	117.73	+11.09%
	theor.		2038	106.00	
U2	exper.	200	939	187.73	−0.32%
	theor.		942	188.39	
U3	exper.	300	612	183.64	+7.18%
	theor.		571	171.36	
S1	exper.	80	2138	171.03	+9.87%$_{(1)}$
	theor$_{(1)}$		1946	155.72	+8.36%$_{(2)}$
	theor$_{(2)}$		1973	157.81	
S2	exper.	200	1082	216.48	+3.54%$_{(1)}$
	theor$_{(1)}$		1045	208.91	+2.66%$_{(2)}$
	theor$_{(2)}$		1054	210.82	
S3	exper.	300	697	209.20	+12.06%$_{(1)}$
	theor$_{(1)}$		622	186.52	+13.15%$_{(2)}$
	theor$_{(2)}$		616	184.82	

Table 4. Experimental-Theoretical deformability comparison.

Specimen Code		ε_{max} [‰]	$\varepsilon_{80\%max}$ [‰]	μ_χ [−]	μ_χ Error exper. vs. theor.
U1	exper.	2.2	2.2	1.00	+0.00%
	theor.	2.2	2.4	1.00	
U2	exper.	2.6	2.6	1.00	+0.00%
	theor.	2.4	2.8	1.00	
U3	exper.	2.8	2.8	1.54	−21.13%
	theor.	2.4	3.0	1.95	
S1	exper.	3.1	9.5	4.24	−4.98%$_{(2)}$
	theor$_{(2)}$	3.0	10.0	4.46	
S2	exper.	2.8	5.9	3.02	−40.08%$_{(2)}$
	theor$_{(2)}$	3.2	13.0	5.04	
S3	exper.	3.5	15.4	8.27	−7.59%$_{(2)}$
	theor$_{(2)}$	3.4	15	8.95	

particular the strength increment and the remarkable ductility enhancement.

Experimental and theoretical ultimate axial load P and ultimate flexural capacity M of un-strengthened and strengthened columns corresponding to the eccentricity e related to each specimen are reported in table 3. The proposed model predictions usually underestimate the experimental outcomes with a scatter in the order of about ten percent.

The effect of confinement, on the contrary, is reliably evaluated when theoretical predictions for unstrengthened and corresponding strengthened elements are considered (the same strength increments and curvature ductility increments as experimentally found in performed tests are predicted).

4.2 Deformability

As already mentioned, confinement does not change actual failure mode (steel reinforcement compressive bars buckling and concrete cover spalling), but it is able to delay bars buckling and to let compressive concrete strains attain higher values, thus resulting in higher load carrying capacity of the column and ductility. The increase in confined concrete strength turned into load carrying capacity increase mainly in the columns loaded with small eccentricity (it is clear that close to pure bending load the effect of concrete strength enhancement – i.e. due to confinement – is insignificant because failure swaps to tension side). At lower levels of axial load also the brittle effect of reinforcement buckling is less noticeable. However in small loading eccentricity cases, concrete ductility and stresses are increased significantly thus resulting also in strength and ductility improvements.

The experimental theoretical comparison of concrete strains at peak (ε_{max}) and at 80% of peak load ($\varepsilon_{80\%max}$) on the softening branch is shown in table 4. The model is able to predict quite well this deformability aspect and a clear trend is found: concrete strains both at peak and on the softening branch at 80% of peak load increases when eccentricity increases. This fact is due to the brittle failure mechanisms prevailing when higher level of axial load is applied (this is not confirmed by S2 experimental test and this can be due to the formation of the plastic hinge outside the instrumented portion of the tested element: lower ductility was found experimentally in the instrumented section compared to S1). For example, in S3 column the concrete reached strains up to 15%.

The concrete strains at 80% of peak load are not available in the case of analyses carried out considering adapted *Spoelstra & Monti (1999)* confined concrete model$_{(1)}$ because the post peak predicted behavior was characterized by an inaccurate hardening branch and it was not possible to evaluate the behavior of the hollow section on the softening branch at 80% of peak load.

A comparison between theoretical and experimental strain development is shown in figure 6. The numerical model (dashed line according to the two theoretical approaches) can predict reasonably well the experimental strain evolution (solid line).

One of the major improvements in member behavior due to FRP wrapping is highlighted considering that in unstrengthened columns, when steel reinforcement reaches in compression the buckling stress, as it pushes outward surrounding concrete, the concrete cover spalls out. In the case of members wrapped with FRP, the steel bars, when buckling occurs, push internal concrete unrestrained cover in the inward direction (in the hollow part) only. In the numerical analysis the global response deteriorate when concrete cover starts spalling and compressed steel reinforcement bars buckles. The branch after peak is

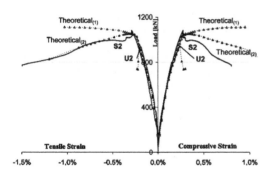

Figure 6. Strain development: U2-S2 Theoretical-experimental comparison.

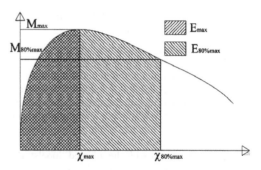

Figure 7. Illustrative Moment vs. Curvature diagram and Specific Energy evaluation.

better predicted adopting stress-strain curves for confined concrete with descending branch after peak (i.e. proposed model$_{(2)}$ prediction).

In the case of flexural elements, the ductility can be evaluated at section level by considering the moment-curvature diagrams (Fig. 7) other than comparing concrete strains at material level.

Rotations in potential plastic hinges are the most common and desirable source of inelastic structural deformations. For elements failing in flexure the curvature ductility μ_χ gives a measure of the ductility of the cross section that gives information about the shape of the descending branch in moment-curvature relationships. The curvature ductility can be defined as the ratio of the curvature on the softening branch at 80% of ultimate load, $\chi_{80\%max}$, and the yielding curvature, χ_y (corresponding to the yielding flexural moment M_y). It is pointed out that the curvature ductility is one for the unstrengthened columns U1 and U2: this is due to the sudden bearing capacity drop that does not allow any curvature increment after-peak (i.e., yielding of steel bars was not attained).

In the strengthened S series, it is clearly pointed out the benefit of confinement. Confinement allows the development of larger curvatures after peak load and curvature ductility is larger than three for all the strengthened columns (Tab. 4).

4.3 Specific Energy

For specimens U3 and S3 it can be observed that the increase of μ_χ from 1.54 to 8.27 corresponds also to a significant gain in terms of area under the curve that is proportional to the specific energy.

The increment of ductility supplied by confinement can be also appreciated comparing the values of the specific energies obtained (Tab. 5) and remarkable increases of dissipating capabilities for strengthened columns are found. Energy values are computed as the area under the Moment vs. curvature diagram at any given χ (refer to illustrative Figure 7).

Table 5. Experimental-Theoretical specific energy comparison

Specimen Code		E_{max} [N]	$E_{80\%max}$ [N]	$\dfrac{E_{80\%max}}{E_{max}}$	E ratio Error exper. vs. theor.
U1	exper.	398	398	1.00	+0.00%
	theor.	393	393	1.00	
U2	exper.	1462	1462	1.00	+0.00%
	theor.	1657	1657	1.00	
U3	exper.	3375	3375	1.00	+0.00%
	theor.	2824	2824	1.00	
S1	exper.	1183	4797	4.05	n.a.$_{(1)}$
	theor.$_{(1)}$	854	n.a.	n.a.	−11.38%$_{(2)}$
	theor.$_{(2)}$	988	4512	4.57	
S2	exper.	2134	7907	3.71	n.a.$_{(1)}$
	theor.$_{(1)}$	2416	n.a.	n.a.	−20.73%$_{(2)}$
	theor.$_{(2)}$	2788	13,042	4.68	
S3	exper.	5615	23,113	4.12	n.a.$_{(1)}$
	theor.$_{(1)}$	4473	n.a.	n.a.	−15.40%$_{(2)}$
	theor.$_{(2)}$	3733	18,176	4.87	

Table 5 shows the comparative study of the theoretical and experimental specific energy, E. The experimental results were noted to be close to the theoretical predictions with a higher scatter in the case of S2 specimen (for the cited effect of plastic hinge formed far from the instrumented portion of the column).

The maximum values of ductility, μ_χ, and specific energy ratio are provided by the column S3. For instance, (Tab. 4–5) when predicted μ_χ is 8.95 (while experimental outcome is 8.27), for the same eccentricity the dissipated specific energy, $E_{80\%max}$, of S3 specimen computed on the softening branch at 80% of ultimate load is more than 7 times the dissipated specific energy at peak, E_{max}, of the brittle U3 specimen. Specific energy increment $E_{80\%max}/E_{max}$, (Tab. 5) in S series' columns, is about 4 emphasizing the meaningful increase of ductility and energy dissipation of the confined structural elements.

4.4 Remarks on the selection of the confined concrete constitutive law

Softening in material constitutive law leads to softening behavior in the section and in the structural element. In a section without rebar buckling, only concrete has a softening behavior: thus moment-curvature relationship softens only in presence of high axial compression force. If steel also has a softening behavior due to buckling, the response of the section becomes softening for any value of axial force. The Mander et al. model compared with the adapted Spoelstra & Monti(1) and the proposed(2) model have been considered in the previous comparisons. Both models are calibrated for solid circular cross sections and have been extended and adapted to the case of those square hollow.

The Spoelstra & Monti model has been adapted also to account for the confinement of a hollow section divided into four connecting walls. The effect of confinement is overestimated in the post-peak branch where in reality the presence of the internal void reduces the efficiency of the confinement exerted by the FRP wraps. This model has been adapted and can be successfully used to predict essentially the strength of the column (corresponding roughly to the occurrence of buckling of the compressed steel reinforcement bars). Since it presents an unrealistic hardening behavior after peak, as a consequence it is not able to simulate reasonably the post peak behavior of hollow core sections.

The proposed confinement model predicts quite well the behavior of hollow section confinement. In contrast to the case of solid section, in the case of thin walls, the larger deformability of the concrete element does not allow to gain significant strength improvements, even though a significant ductility development is achieved.

5 CONCLUSIONS

The proposed confinement model, coupled with the proposed computation algorithm is able to predict the fundamentals of the behavior of hollow members confined with FRP both in terms of strength and ductility giving a clear picture of the mechanisms affecting the response of this kind of element. The same strength increments and curvature ductility increments as experimentally found in performed tests are predicted.

The model is able to trace the occurrence of the brittle mechanisms, namely concrete cover spalling and reinforcement buckling, the evolution of stresses and strains in the confinement wraps and concrete allowing to evaluate at each load step the multiaxial state of stress and the potential failure of the external reinforcement. The main output of the proposed model is also the assessment of the member deformability in terms of both curvature ductility and specific energy. Theoretical results, in satisfactory agreement with authors' experimental data, show that FRP jacketing can enhance the ultimate load and significantly the ductility also in the case of hollow concrete cross sections and under combined compression and flexure loads.

ACKNOWLEDGMENTS

The analysis of test results was developed within the activities of Rete dei Laboratori Universitari di Ingegneria Sismica – ReLUIS for the research program funded by the Dipartimento di Protezione Civile – Progetto Esecutivo 2005–2008. The FRP strengthening of the columns was supported by MAPEI Spa., Milan, Italy.

REFERENCES

Braga, F., Gigliotti, G. & Laterza, M. 2006. Analytical Stress-Strain Relationship for Concrete Confined by Steel Stirrups and/or FRP Jackets. *Journal of Structural Engineering*, 132(9):1402–1416.

CNR-DT 200. 2004. *Guide for the Design and Construction of Externally Bonded FRP Systems for Strengthening Existing Structures*, Published by National Research Council, Roma, Italy.

Cosenza, E. & Prota, A. 2006. Experimental behavior and numerical modeling of smooth steel bars under compression. *Journal of Earthquake Engineering*, 10(3):313–329.

Fam, Amir, Z. & Rizkalla, Sami, H. 2001. Confinement Model for Axially Loaded Concrete Confined by FRP Tubes, *ACI Structural Journal*, 98(4):251–461.

Hillerborg, A. 1989. The compression stress-strain curve for design of reinforced concrete beams. *Fracture Mechanics: Application to Concrete, ACI SP-118*:281–294.

Li, J. & Hadi, M.N.S. 2003. Behavior of externally confined high-strength concrete columns under eccentric loading. *Journal of Composite Structures*; 62:145–153.

Lignola, G.P. 2006. *RC hollow members confined with FRP: Experimental behavior and numerical modeling*. Ph.D. Thesis, University of Naples, Italy.

Lignola, G.P., Prota, A., Manfredi, G. & Cosenza, E. 2007a. Experimental performance of RC hollow columns confined with CFRP. *ASCE Journal of Composites for Construction*, 11(1):42–49

Lignola, G.P., Prota, A., Manfredi, G. & Cosenza, E. 2007b. Deformability of RC hollow columns confined with CFRP. *ACI Structural Journal*, in press.

Mander, J.B., Priestley, M.J.N. & Park, R. 1988. Theoretical stress-strain model for confined concrete. *ASCE Journal of Structural Engineering*, 114(8):1804–1826.

Prota, A., Manfredi, G. & Cosenza, E. 2006. Ultimate behavior of axially loaded RC wall-like columns confined with GFRP. *Composites: pt. B, ELSEVIER*, vol. 37:670–678.

Spoelstra, MR. & Monti, G. 1999. FRP-confined concrete model. *ASCE Journal of Composites for Construction*, 3(3):143–150.

Fracture Mechanics of Concrete and Concrete Structures – Design, Assessment and Retrofitting of RC Structures – Carpinteri, et al. (eds)
© *2007 Taylor & Francis Group, London, ISBN 978-0-415-44616-7*

Testing and analysing innovative design of UHPFRC anchor blocks for post-tensioning tendons

F. Toutlemonde, J.-C. Renaud & L. Lauvin
LCPC, Paris, France

M. Behloul
Lafarge, Paris, France

A. Simon
Eiffage TP, Neuilly-sur-Marne, France

S. Vildaer
VSL, Saint Quentin-en-Yvelines, France

ABSTRACT: Experimental validation of innovative design of anchor blocks for post-tensioning tendons was carried out. The tested specimens were made of UHPFRC, namely Ductal® -FM for 9 blocks and BSI® for 3 other blocks, and did not contain any passive reinforcement normally provided for confining concrete submitted to localized compression, and preventing bursting. They were dedicated to 4T15S, 7T15S and 12T15S prestressing units. Splitting in the middle of block sides, leading to sub-vertical cracks, appears as the dominant critical mechanism. Depending on lateral dimensions of UHPFRC around the reservation, and possible size effects concerning this zone submitted to intense tensile stresses, the maximum load has been obtained from 1,2 to 2,2 times the ultimate force of the tendons F_{prg}, with a typical scatter about ± 10%. While refinements remain possible for optimizing the safety margin depending on project requirements, obtained results validate the innovative (reinforcement-free) design and typical dimensions of these UHPFRC anchor blocks.

1 INTRODUCTION

1.1 Development of UHPFRC post-tensioned structures

Ultra-high performance fiber-reinforced concrete (UHPFRC) represent an important breakthrough for civil engineering, limiting consumption of natural resources and leading to possibly optimal shapes for very durable structures. Important R & D efforts are currently undertaken in order to optimize structural application of these new materials, Bouteille & Resplendino (2005). Valuable structural application of UHPFRC ultra-high compressive strengths together with ductility requirements often leads to pre-stressed or post-tensioned solutions. Moreover, for structures made of pre-cast pre-stressed components, post-tensioning may represent an elegant assembly solution. Therefore, the question of safe design of end blocks for post-tensioning tendons is of high relevance.

In fact, confinement steel is normally provided in such anchor blocks for preventing the concrete bursting due to intense localized compressions, e.g.

according to EN 1992–2: 2005. Specific experimental verifications have recently been carried out for extension of such provisions to very high performance concrete (C80 to C120), Boulay et al. (2004). Due to the high fiber content in UHPFRC, indirect tensile stresses due to the localized compressions may be taken by the fibers, so that conventional transverse confinement steel may not be necessary. Dispensing with conventional steel possibly allows a reduction of transverse dimensions of the end blocks, because the whole UHPFRC part is fiber-reinforced and there is no need for external cover. Optimal dimensions of UHPFRC end blocks can thus be considered as an important step towards rational and valuable development of these materials.

1.2 Research significance

Within the context of French R & D project MIKTI devoted to the development of new steel-concrete composite solutions, an experimental program dedicated to this question was carried out at LCPC. The reason of this comes from the related study of an optimized UHPFRC ribbed deck solution, made of segments assembled by post-tensioning, Toutlemonde et al. (2005).

This program tended to be more generic than previous validations of UHPFRC end blocks related to specific projects: Sherbrooke Footbridge, Ganz & Adeline (1997); Seonyu footbridge, Behloul et al. (2004); Millau toll gate, Hajar et al. (2004). Within the European context, the experiments should also give preliminary indications concerning application of the new standard agreement process, EOTA (2002), to these innovative solutions. Moreover, the results should be used for detailed analysis and possible further blocks optimization, given the companion exhaustive UHPFRC mechanical characterization carried out within the frame of the project.

2 EXPERIMENTAL PROGRAM

2.1 Specimens

The tested anchor blocks were made of UHPFRC, namely Ductal®-FM for 9 blocks and BSI® for 3 other blocks, and did not contain any classical passive reinforcement normally provided for confining concrete submitted to localized compression and preventing bursting of the anchor blocks. They were dedicated to 4T15S, 7T15S and 12T15S post-tensioning units. These units are considered as firstly interesting for possible application in rather thin structural shapes, based on design experience using such materials, Bouteille & Resplendino (2005). It had been chosen to reproduce the tests three times for each configuration, due to unknown scatter of expected results. The blocks had a square cross-section and an aspect ratio equal to 4, for ensuring absence of end effects on the zone which should be submitted to regularized compressive stresses after diffusion of localized compressions.

The innovative design allowed significantly reduced transverse dimensions of the blocks, given in Table 1. Blocks included a real-size sheath and the encased part of the pre-stressing kit in their central part. They were cast horizontally for purpose of being representative and including possible unfavourable orientation of the fibres. The real-size cylindrical plate where strands are clamped to was used for application of the localized compression on the block, with dimension given in Table 1. Due to proprietary units employed, dimensions of the steel end plate were fixed, as given in Table 2.

2.2 Testing procedure

The testing program derived from the standard experimental procedure of ETAG 13 – European Technical Agreement Guide Nr 13 – EOTA (2002). Compressive axial loading was applied on the specimens, located vertically at the thrust center of a 5000 kN-capacity testing machine. During a first phase of the test, at least 10 loading cycles were applied from 0.12 to 0.8 times F_{prg}, until strains stabilization. Servo-control was based on load signal with an imposed rate corresponding to 0.5 MPa/s in the zone of regularized stresses. Then, loading was applied up to failure with a displacement-based control, at a rate of 0.5 mm/mn for the actuator. A safety limit of 10 mm-vertical displacement was authorized.

Centering of the samples was first ensured using lateral stops with respect to the bottom face. Control of lateral dimensions induce an uncertainty of about 1.5% at this step. Axial centering was then verified, and if necessary favored, in case of uneven specimen thickness due to the casting phase (up to 1.5% slope), with respect to the top side corresponding to the post-tensioning end.

Besides load and actuator displacement monitoring, 16 measuring channels were recorded corresponding to identical numerical LVDT sensor equipment for all sides of the prisms. Transverse strains were measured from end to end of each side, at levels distant of 0.7 and 1.5 times the side length from the top of the block, using particular supporting devices already tested in Boulay et al. (2004). At a distance of 0.7 times the side length, tensile stresses are maximal in the reference computation of a homogeneous elastic prism. Vertical strains were measured along the central axis of each side, between levels distant of 0.5 and 1.7 times the side length from the top of the block. Finally, the global vertical settlements were measured along the middle of each block side between the platens of the testing machine. A general view of the testing configuration is given (Fig. 1). One record was taken every 10 kN.

2.3 Material characteristics

Specimens 1 to 9 were made of Ductal®-FM, having a water to cement ratio equal to 0.21 and a volumetric fiber content equal to 2.15%. Thermal treatment was applied at an age of 48 h, consisting in 48 hours exposure at 90°C and 95% RH. Average compressive strength measured on companion cylinders, 70 mm in diameter, was 190 MPa. Average specific gravity determined on the same specimens is 2.53 kg/m³, Young's modulus 55 GPa and Poisson's ratio 0.17. Ductal®-FM characteristics in tension were identified on 7 cm × 7 cm × 28 cm prisms tested under bending. Average limit of linearity identified by 4-point bending of un-notched specimens corresponds to 16.3 MPa which is consistent with a design f_{tj} value equal to 9.6 MPa according to UHPFRC Recommendations, AFGC-SETRA (2002). Modulus of rupture identified by 3-point bending of 6 companion notched specimens reaches 35.4 MPa (average value) and 23.2 MPa (characteristic value).

Table 1. Nominal dimensions of the specimens.

Ref	UHPFRC	Kit type	Unit	F_{prg} kN	Side mm	Heigth mm	Load dia. mm
1	Ductal®-FM	VSL	4T15	1116	160	640	110
2	Ductal®-FM	VSL	4T15	1116	160	640	110
3	Ductal®-FM	VSL	4T15	1116	160	640	110
4	Ductal®-FM	VSL	7T15	1953	185	740	135
5	Ductal®-FM	VSL	7T15	1953	185	740	135
6	Ductal®-FM	VSL	7T15	1953	185	740	135
7	Ductal®-FM	VSL	12T15	3348	255	1020	170
8	Ductal®-FM	VSL	12T15	3348	255	1020	170
9	Ductal®-FM	VSL	12T15	3348	255	1020	170
10	BSI®	Diwydag	7T15	1953	280	1120	130
11	BSI®	Diwydag	7T15	1953	280	1120	130
12	BSI®	Diwydag	7T15	1953	280	1120	130

Table 2. Dimensions of associated post-tensioning devices.

Ref	Kit & unit	End plate diameter mm	End plate thickness mm	Cone-shaped length mm
1 to 3	VSL 4T15	120	10	180
4 to 6	VSL 7T15	145	10	180
7 to 9	VSL 12T15	200	25	360
10 to 12	Diwydag 7T15	170	18	82 + 170

Figure 1. General testing and monitoring setup.

Specimens 10 to 12 were made of BSI®. Average compressive 28 day-strength measured on companion cylinders stored in air, 110 mm in diameter, was 195 MPa (201 MPa when stored at 98% RH). Young's modulus determined on similar cylinders ranges from 64 to 69 GPa. BSI® characteristics in tension were identified on prisms tested under bending. Average limit of linearity identified by 4-point bending of un-notched 7 cm × 7 cm × 28 cm specimens corresponds to 15.9 MPa which is consistent with a design f_{tj} value equal to 9.3 MPa according to UHPFRC Recommendations, AFGC-SETRA (2002). Modulus of rupture identified by 3-point bending of six 10 cm × 10 cm × 40 cm notched specimens reaches 35.6 MPa (average value) and 28.4 MPa (characteristic value).

Blocks were tested at an age of more than one year, so that UHPFRC strength and characteristics can be assumed as stabilized. From the identification hereabove, it can be assumed that a direct tensile stress equal to about 9 MPa should correspond to cracking initiation, and that local hardening behavior with progressive parallel cracking, fibers re-anchoring and apparent ductility can be expected for a certain period

at the structural level due to the significant margin between the maximum equivalent bending stress and the limit of linearity, and due to the significant enough stress gradients at the block scale.

3 GLOBAL RESULTS

3.1 Blocks 1 to 3 for 4T15S units

The global behavior of the blocks is represented in terms of applied load vs. average axial global displacement, including settlement at the ends, for blocks 1 to 3 (Fig. 2), after the first phase of cycles, which explains the non-zero origin of the curves. Maximum loads reached 2439, 2023 and 2085 kN respectively, which represents an average ratio of the ultimate load with respect to F_{prg} equal to 1.955. The minimum to maximum amplitude represents a variation of ±9.5%. In all cases after failure the load could be maintained to

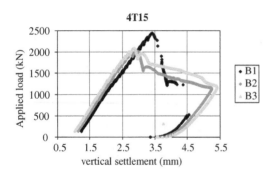

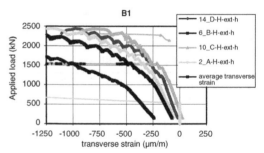

Figure 2.   Load vs. axial global settlement, blocks 1 to 3.

Figure 4.   Block 1. Transverse strains at higher level (from top, 0.7 times the side length).

Figure 3.   Crack pattern evolution. Block 1 side D. (left) at first cracking – (right) after failure.

a value higher than F_{prg} provided displacement is controlled, not force (and even for block 1 some instability was noticed in a load/actuator displacement diagram).

Non-linearity before the peak is hardly significant on this global curve and shall be studied with information from more local survey. Namely, first (fine) cracks have been observed from 0.8 times F_{prg} (blocks 1 and 2) and 1.2 F_{prg} (block 3). As an example for block 1, evolution of this first crack (Fig. 3) in terms of opening and length took place significantly at 1.1 and 1.4 times F_{prg} respectively, and first cracks on another side were visible at 1.6 times F_{prg}. Transverse strains of block 1 at the most critical level (0.7 times the side length from the top side) are represented Fig. 4. Non-linearity is visible for loads ranging from 0.85 to 1.3 times F_{prg}. Moreover, the side B with lowest apparent stiffness did not exhibit visible cracks before the final failure. Whatever the side, transverse displacements from side to side for a load variation equal to F_{prg} were kept below 40 μm, and strain "stabilization" in the sense of ETAG 013 had been obtained even on the less stiff side.

3.2   Blocks 4 to 6 for 7T15S units

The global behavior of the blocks 4 to 6 corresponding to 7T15S units is represented in terms of applied load vs. average axial global displacement, including settlement at the ends (Fig. 5), after the first phase of cycles, which explains the non-zero origin of the curves. Maximum loads reached 3058, 2735 and 2660 kN respectively, which represents an average ratio of the ultimate load with respect to F_{prg} equal to 1.443. The minimum to maximum amplitude represents a variation of ±7.1%. In all cases after failure the load could be maintained to a value close to F_{prg} provided displacement is controlled, not force (yet for blocks 4 and 6 some instability was noticed in a load/actuator displacement diagram).

Non-linearity becomes significant on this global curve somewhat below 2000 kN (close to F_{prg}). Considering information from more local survey, it turns out that first (fine) cracks have been observed from 0.8 times F_{prg} (blocks 5 and 6) and 1.0 F_{prg} (block 6). However even for block 4, non-linear evolution of transverse strains could be observed from about 1550 kN, i.e. 0.8 times F_{prg} (Fig. 6). But whatever the side of the block, transverse strains for a load variation equal to F_{prg} were kept below 250 μm/m, corresponding to transverse displacements from side to side lower than 50 μm, which is close to the conventional limit of one visible crack. Moreover, the first cracks were not observed on the side where non-linearity or non-stabilization of transverse strains first appeared.

3.3   Blocks 7 to 9 for 12T15S units

The global behavior of the blocks 7 to 9 is represented in terms of applied load vs. average axial global displacement, including settlement at the ends, for these blocks corresponding to 12 T15S units (Fig. 7), after the first phase of cycles, which explains the non-zero origin of the curves. Maximum loads reached 4587, 4312 and 4182 kN respectively, which represents an average ratio of the ultimate load with respect to F_{prg}

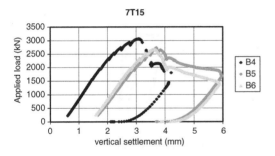

Figure 5. Load vs. axial global settlement, blocks 4 to 6.

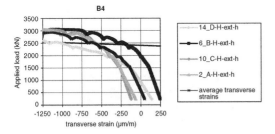

Figure 6. Block 4. Transverse strains at higher level (from top, 0.7 times the side length).

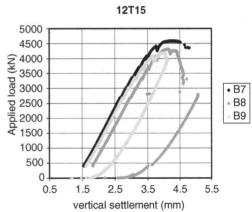

Figure 7. Load vs. axial global settlement, blocks 7 to 9.

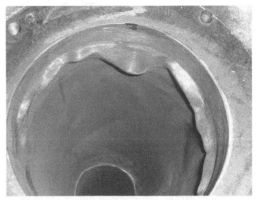

Figure 8. Local buckling of the duct at failure, block 7.

Figure 9. External instability at failure, block 7.

equal to 1.302. The minimum to maximum amplitude represents a variation of ±4.7%. In all cases reaching of the peak load was closely followed by important transverse strains at the top of the block, and concomitant buckling of the conical part of the pre-stressing duct was observed (Fig. 8). Onset of this instability (difficult control of the testing machine actuator, even displacement-controlled, due to transverse possible bursting – Fig. 9 – corresponds to a significantly less efficient confinement effect of the UHPFRC block (relative side dimensions of the block are lower than for Blocks 1 to 6, in correspondence with expected F_{prg}).

Non-linearity of the global curve force vs. axial displacement can be detected at 0.8 times F_{prg}, corresponding to first visible (fine) cracks. More significant non-linearity of transverse strains (Fig. 10) was observed from about 3000 kN (0.9 times F_{prg}). This lower ratio, as compared to 4T15 and 7T15 blocks made of the same material, consistently indicate that thickening of the 12T15 blocks might be desirable for ensuring a comparable pseudo-ductility and margin with respect to the maximum applicable load.

3.4 Blocks 10 to 12 for 7T15S units

The global behavior of the blocks is represented in terms of applied load vs. average axial global displacement, including settlement at the ends, for blocks 10 to 12 corresponding to 7T15S units with the Diwydag system (Fig. 11), after the first phase of cycles, which explains the non-zero origin of the curves. Maximum loads reached 4347, 4042 and 4111 kN respectively, which represents an average ratio of the ultimate load

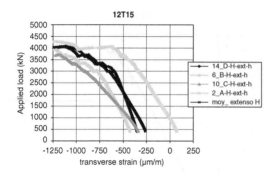

Figure 10. Block 8. Transverse strains at higher level (from top, 0.7 times the side length).

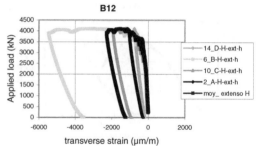

Figure 12. Block 12. Transverse strains at higher level (from top, 0.7 times the side length).

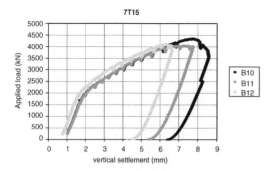

Figure 11. Load vs. axial global settlement, blocks 10 to 12.

Figure 13. Punching of the steel anchor plate. a) Block 10 after failure. b) Block 11 after failure.

with respect to F_{prg} equal to 2.134. The minimum to maximum amplitude represents a variation of $\pm 3.6\%$.

For these blocks non-linearity before the peak is significantly more pronounced, so that the maximum load is reached with corresponding vertical displacements higher than 5 mm. Occurrence of non-linearity takes place between 1550 and 1950 kN (0.8 to 1.0 times F_{prg}). Namely, first (fine) cracks have been observed at 0.8 times F_{prg} (blocks 10 and 11) and 1.0 F_{prg} (block 12). However the load goes on increasing with a still high stiffness and an important safety margin during this phase, and important transverse yielding is observed mainly at the peak (Fig. 12). Due to thick UHPFRC sides around the anchor plate, sudden bursting of the block seems to be prevented and opening of the splitting cracks is rather smoothly controlled for a displacement-controlled phase of the test. Moreover, part of the global axial settlement (about 1 mm) is due to local yielding of the steel anchor plate in correspondence to the cylinder through which load is applied (Fig. 13a, b).

3.5 Crack development – cycles and stabilization

Besides the global behavior of the blocks and the safety margin concerning the maximum load capacity, the

testing procedure of ETAG 13 requires checking the strain stabilization during load cycles from 0.12 to 0.8 times F_{prg}. The criterion is given by (1):

$$\frac{\varepsilon_n - \varepsilon_{n-4}}{\varepsilon_4 - \varepsilon_0} \leq 0.33 \qquad (1)$$

It was not clear whether this verification should be done for each channel corresponding to horizontal or vertical strain ε. Due to the recording process, maximal strain values were determined at the maximum load of the cycle, with a 10 kN precision in terms of load and about 1 μm uncertainty for the distance measurement. In some cases, the strain evolutions from cycle 0 to 4, and possibly for cycle 10 to 14, were so low that each term of the ratio was only some 10^{-6}. It was thus agreed that for a strain evolution from cycle $n - 4$ to n corresponding to less than 10 μm, stabilization could be considered as obtained, whatever the value of the criterion. It was also considered that the 0.01 precision of the ratio could not be expected from the present measurements, therefore when the criterion was equal to 0.35 the stabilization was also deemed as satisfied. Finally, when stabilization was not obtained after 10 cycles, it was decided to apply only 4 more cycles, since the process could not be continued without further indication, and it was decided to go on with the ultimate phase of the test.

Considering the tested blocks, only blocks 4, 5, 8, 9, 11 and 12 had stabilized strains after 10 cycles. But for blocks 1 and 2 only one vertical strain record did not verify the criterion, and for blocks 6 and 10 only one horizontal strain record did not verify it. For block 3, only vertical strains of sides B and D were in excess, which was not related to clear damage. In the case of vertical strains, scatter due to sensor positioning on a rough surface might partially explain the difficulty. As a first experience gained in applying ETAG 13 procedure, it seems that in the present case this verification may have become time-consuming, not fully objective, and not clearly related to critically evolving cracks. In sum, strain stabilization as expected by ETAG 13 seems to hardly make sense for UHPFRC blocks, since possible evolution appears as hardly correlated to presence of the further critical cracks which appear on the upper part of the sides of the blocks due to induced transverse tension. As a clear result of this phase, one would be just able to conclude that for all blocks except 3, 4 and 12, first fine cracks were made visible during this cyclic loading process. Verification of the stabilization criterion hardly depends on the side considered, which may be consistent with the role of casting direction on the fiber orientation and local defects, for these rather thin parts under tension.

3.6 Crack development to failure

Survey of cracks was carried out during the following stage of the tests every 0.2 or 0.1 times F_{prg} depending on their observed progressive development. Crack initiation systematically took place in the middle part of one side. This side corresponded to the side up during casting for two thirds of the blocks (due sometimes to thinner UHPFRC cover of the pre-stressing duct on this side), it could also correspond to the lateral sides during casting with a possibly unfavorable fiber orientation in a straight zone for fresh concrete flow. As expected, cracks were systematically parallel to the loading direction and corresponded to indirect transverse tension within the UHPFRC ligament around the anchoring device. Their progressive extension was noticed especially when the volume of UHPFRC is relatively important (Fig. 14). Local hardening related to fiber anchoring can also have led to multiple parallel cracking (Fig. 15).

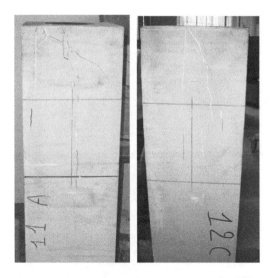

Figure 14. Progressive vertical development of splitting cracks (left) Block 11 side A. (right) Block 12 side C.

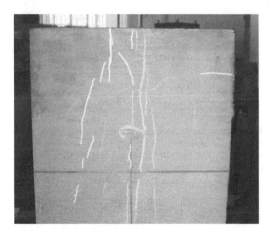

Figure 15. Progressive multiple cracking. Block 9 side C.

Independently of the block scale, final failure, especially in case of instability, produced different superimposed crack patterns, all of them corresponding to UHPFRC lateral or diagonal bursting (edge ejection) associated to punching of the block by the steel anchor plate (Fig. 16).

3.7 Interpretation and sensitivity to test conditions

Due to possible initiation of the cracks by local defects (Fig. 3, Fig. 13) no clear direct correlation was found between the load corresponding to first cracking and the size of the specimen (side, or side minus the duct thickness, or area of the UHPFRC cross-section …). However, for the same material and pre-stressing kits

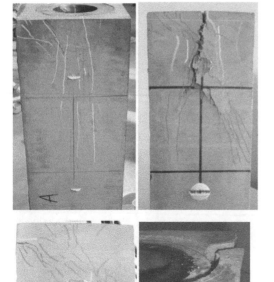

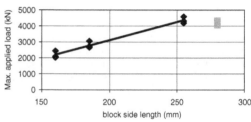

Figure 17. Block capacity as a function of its side length.

Figure 16. First vertical cracks (splitting), multiple parallel branching, and final diagonal cracks due to instable failure mechanism. Top, (left) Block 8. Top, (right) Block 3. Bottom, (left) Block 4. Bottom, (right) Block 7.

of similar shape, the ultimate strengths exhibit linear variations with the side of the block (Fig. 17). Yet for the blocks tested corresponding to 4T15S, 7T15S and 12T15S units, these loads are not in the same proportion as expected design loads F_{prg}. It is thus suggested that, if a similar acceptable safety margin is to be ensured, the side of the blocks should be adapted accordingly, depending on the expected unit and corresponding F_{prg}. Further non-linear F.E. analyses calibrated on present results and accounting for the precise diffusion mechanisms around the pre-stressing anchor, would also be helpful for rationally optimizing the block geometry.

Moreover, as shown Fig. 14, the development of splitting cracks in the middle of the block sides may reach a depth twice as long as the block side. In order to get experimental results not disturbed by end effects, it is recommended that the block aspect ratio is higher than 3, even if due to the conical shape of the prestressing device an aspect ratio of 2 is assumed as sufficient according to ETAG 13.

4 CONCLUSIONS

Experimental validation of innovative design of UHPFRC anchor blocks for post-tensioning tendons was carried out, following provisions of ETAG 13. Some possible improvements of the testing procedure have been identified.

In the tested configurations, fiber capacity within UHPFRC proved to be sufficient for dispensing of classical transverse reinforcement, by ensuring a safety margin ranging from 1.3 up to 2.1 when comparing the ultimate load with design capacity F_{prg}. Especially for the thick enough blocks, scatter of the results is remarkably low.

As expected, the failure mechanism was initiated by indirect tension perpendicularly to the applied loads, cracking was observed in the middle of lateral sides at about 0.7 times the side length from the loaded end. Visible cracks and induced non-linearity due to them were obtained from 0.8 to 1.2 times F_{prg}, yet local hardening UHPFRC behavior proved to be activated for ensuring efficient confinement.

Design, dimension optimization and formal agreement of such blocks, should be pursued, taking benefit of the present results, for delaying crack onset and ensuring a more homogeneous safety margin for the different block sizes corresponding to varied pre-stressing units.

ACKNOWLEDGEMENT

This experimental program has been carried out within the R&D "National Project" MIKTI, funded by the Ministry for Public Works (DRAST/RGCU) and managed by IREX. It has been supervised by a committee chaired by J. Resplendino (CETE de Lyon), also chairman of the *fib* TG 8.6 mirror group. Eiffage Construction, Lafarge and VSL are mentioned for their contribution in specimens preparation. The authors are also pleased to thank A. Mellouk and F.-X. Barin from LCPC Structures Laboratory for their help in the experimental realizations. Advice of R. Chaussin from

ASQPE (French Association for the Quality of Prestressing) for validation of the experimental program is also gratefully acknowledged.

REFERENCES

AFGC-SETRA, 2002. *Ultra High Performance Fibre-Reinforced Concretes. Interim Recommendations.* Bagneux: SETRA.

Behloul, et al. 2004. Seonyu Ductal® footbridge. In *Concrete Structures: the challenge of creativity, Proc. fib Symp. Avignon, 26–28 April 2004*. Paris: AFGC.

Boulay, C. et al. 2004 Safety of VHSC structures under concentrated loading: experimental approach, *Magazine of Concrete Research* 56(9): 523–535.

Bouteille, S. & Resplendino J. 2005. Derniers développements dans l'utilisation des bétons fibrés ultra-performants en France. In *Performance, Durabilité, esthétique, Proc. GC'2005, Paris, 5–6 October 2005*. Paris: AFGC.

EN, 1992–2: 2005. Eurocode 2 – Calcul des structures en béton – Partie 2: Ponts en béton – calcul et dispositions constructives, *CEN.*

EOTA, 2002. Guide d'agrément technique européen sur les kits de mise en tension de structures précontraintes (GATE 013). Bruxelles: EOTA.

Ganz, H. R. & Adeline, R. 1997. Mini-anchorages for Reactive Powder Concrete. In *f.i.p. Int. Conf. On new technologies in structural engineering, Lisbon, July 1997.*

Hajar, Z. et al. 2004. Construction of an ultra-high performance fibre reinforced concrete thin-shell structure over the Millau Viaduct toll gate. In *Concrete Structures: the challenge of creativity, Proc. fib Symp. Avignon, 26–28 April 2004*. Paris: AFGC.

Toutlemonde, F. et al. 2005 Innovative design of ultra-high performance fiber-reinforced concrete ribbed slab : experimental validation and preliminary detailed analyses. In Henry Russel (ed.), *Proc. 7th Int. Symp. On Utilization of High Strength/High Performance Concrete, Washington D.C. (USA), 20–22 June 2005*. ACI-SP 228, 1187–1206.

Fracture Mechanics of Concrete and Concrete Structures – Design, Assessment and
Retrofitting of RC Structures – Carpinteri, et al. (eds)
© 2007 Taylor & Francis Group, London, ISBN 978-0-415-44616-7

Bond behavior between CFRP strips and calcarenite stone

M. Accardi, C. Cucchiara & L. La Mendola
Dipartimento di Ingegneria Strutturale e Geotecnica, Università di Palermo, Italy

ABSTRACT: In this paper a local bi-linear shear stress-slip law with softening is proposed on the basis of
an experimental investigation which reproduces the interface calcarenite-CFRP behavior according to several
codes. The parameters were calibrated on the experimental results relative to double shear pull tests from which
the maximum transferable load in the joint and the strain profiles were obtained. A numerical investigation was
carried out by a finite element software including an interface element modeling interlaminar failure and crack
initiation and propagation, in order to validate the calibration of the parameters defining this interface law and
the accuracy of the recorded measurements.

1 INTRODUCTION

Fiber composite materials are frequently used to
strengthen to flexure and/or shear of reinforced concrete structures. The use of composite material has
recently been extended to the masonry structures of
important historical buildings. The use of composite
materials in reinforcing techniques presents particular advantages which include high strength, lightness,
resistance to corrosion, ease of application and the
possibility of removal.

The effectiveness of this technique is particularly
linked to the bond phenomenon that guarantees the
maximum transferable load by an adequate transfer
length.

Bond behavior between Fiber Reinforced Polymer
(FRP) and masonry can be placed in the wider context of bonding of FRP to quasi-brittle materials like
concrete, mortar and rock. While extensive research
has been conducted and reported in the literature for
reinforced and prestressed concrete structures, much
less has been reported for masonry structures. Many
researchers have studied the characteristics of bonding
between the reinforcement system (FRP or steel plates)
and the basic materials (usually concrete with different
mechanical properties), achieving a good understanding of the related mechanism (Holzenkämpfer (1994);
Chajes et al. (1996); Bizindavyi & Neale (1997, 1999);
Täljsten (1997); Neubauer & Rostasy (1997); Yuan &
Wu (1999); Yuan et al. (2001). The observed modes of
debonding for structural elements strengthened with
FRP can generally be classified into two types: (a)
those associated with high shear and normal stresses
near the bonded plate ends, of which a review of the

experimental and analytical approaches can be found
in Smith & Teng (2001); and (b) those mainly associated with high bond stresses induced by a discontinuity
or pre-existent crack away from the plate ends, of
which a review of the experimental and analytical
approaches can be found in Chen & Teng (2001). These
well-known failure mechanisms for RC structures are
also common for masonry structural elements and they
have been classified according to Italian Recommendations CNR DT 200-2004 as *plate end debonding* and
intermediate crack debonding respectively.

Some codes and recommendations have been introduced in the last few years, in order to regulate the use
of FRP for strengthening existing structures. Among
the experimental tests which have been performed for
bond evaluation (double or single shear pulling test,
double or single shear pushing test, beam test), the
double pull test has been the most usual test method
proposed by the standard codes (JSCE 2001). Two
plates or strips are bonded on opposite sides of a concrete block and equal tensile forces F are applied to
the plates. These forces are balanced by a pulling force
applied in the basic material 2F which may be applied
either through a steel bar embedded in the centre of
the concrete block, or through steel plates bonded to
the sides of the basic material. The double pull test
may not divide the shear force symmetrically on two
sides of the specimen. For this reason it is appropriate
to use the same measurement for each side in order to
verify correct division of the force. Following an extensive literature review it appeared that test specimens
usually failed at few millimeters into basic material
below the adhesive. The most relevant output of a bond
test, beside the bond strength of the specimen, is the

measured relationship between the bond stresses (that is the shear stress along the length of the reinforcement as a response to the external load) and the slip (defined as the relative displacement between the reinforcement and the basic material). Considerations on the local bond stress slip relationship will be presented in this paper. It is worth noting that different local bond-slip curves for FRP-to-concrete bonded joints have been proposed (Nakaba et al. 2001; Savoia et al. 2003; Leung 2004; De Lorenzis et al. 2001).

On the basis of an experimental investigation carried out by the authors in a previous study, an understanding of the interface adherence characteristics is carried out in the present paper, referring to the reinforcing technique of using the wet-lay-up method applied to structural masonry elements consisting of calcarenite stone and bed joint mortar. In particular the focus of this paper is to propose a local stress-slip law calibrated on the experimental results which reproduces the interface behavior between calcarenite and Carbon Fiber Reinforced Polymer (CFRP). The bond stress-slip curves obtained by experimental results can be modelled by a bi-linear law with linear softening, as proposed in several codes. This statement is validated by a numerical analysis based on a finite element method in which an interface element is utilized of which the parameters are maximum bond stress, relative slip and fracture energy.

2 MATERIALS

The study of the bond behavior between FRP and masonry material providing support, requires the physical and mechanical properties of the structural materials involved. The mechanical propriety of calcarenite stone (the resistant element of masonry material) and FRP, determined by standard tests, will be introduced here for this purpose.

2.1 Calcarenite stone

Organogenic calcarenite is a structural stone widely used for historical buildings of architectural interest in the Mediterranean area. It is generally recovered in blocks and is located in a wide and elongated area approximately parallel to the shore line. It has marked heterogeneity and high total open porosity and the colour can change from greyish white to dark red.

The calcarenite stone used in the experimental investigations to which this paper refers, were recovered from a quarry located in western Sicily. The ashlars of this calcarenite stone have average dimensions of $210 \times 160 \times 360$ mm. The cubic and prismatic specimens used to determine the mechanical properties were recovered from the same ashlars to make up the pull test specimens as shown in Figure 1.

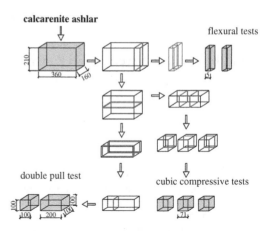

Figure 1. Specimens for mechanical properties and adherence characteristics.

Compressive cubic tests were carried out by a universal testing machine operating in displacement controlled mode with maximum capacity equal to 600 kN. Results are utilized to obtain a correlation between the fracture energy of each pull-out test and the relative cubic compressive strength of the calcarenite ashlar; this correlation is proposed in the following sections.

2.2 CFRP reinforcement

The composite material used here as externally bonded reinforcement consists of two combined materials: high-strength fibers and matrix. The fiber provides the strength of the composite, and the matrix is the product that holds the fibers together and acts as a load transfer medium. The fiber used in the experimental program is a unidirectional carbon fiber fabric, of which the mechanical characteristics, as indicated by the manufacturer, are the following: tensile strength 3450 MPa; elongation break 1.5% and tensile modulus 230 000 MPa. The epoxy resin used in this experimental program is a two-component, 100% solid, non-sag paste and light grey in colour. The mechanical properties of the epoxy resin provided by the manufacturer are the following: tensile strength 30 MPa; elongation break 1.5%; flexural modulus 3800 MPa.

Use of the CFRP system to strengthen structural masonry elements introduces new properties which it is essential to know, starting from the physical and chemical characterization of the interphase CFRP-masonry substrate. The interphase characterization was carried out through microscopic imaging analysis from which it is possible to observe that FRP application on the surface creates a new layer, usually called the interphase, where the matrix used to bond the reinforcement soaks the substrate material (see Fig. 2).

epoxy resin | soaked calcarenite | carbon fibers

Figure 2. Optical microscopy observation of calcarenite-CFRP interface.

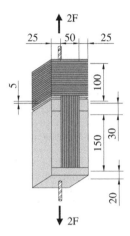

Figure 3. Specimen for pull-out test with bond length 150 mm.

A better understanding of the behavior of this crucial zone, where molecules of the two materials interact creating properties that differ from those of the two separate phases, is vital to improve the way in which materials are combined together.

3 EXPERIMENTAL INVESTIGATION

A synthesis of previous experimental research carried out by the authors is presented in the following paragraphs; details can be found in Accardi & La Mendola 2004.

3.1 Specimen

Double Shear pulling Tests (DST) were carried out with the purpose of determining the pull-out load and the effective transfer length. Specimens consisted of two calcarenite blocks loaded through steel bars previously inserted into them and sealed with epoxy resin (see Fig. 3). In particular the prismatic block was that in which the bond phenomenon was evaluated while the cubic part was the anchorage. The latter was made by wrapping the faces of the cube with the same CFRP strips as used for the bond test. Specimens were prepared with three different bonded length: 50, 100 and 150 mm. The external load yielded two shear forces in the CFRP strips bonded onto each side of the strengthened block.

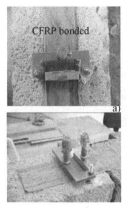

CFRP bonded

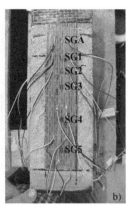

SGA
SG1
SG2
SG3
SG4
SG5
a)
b)

Figure 4. Measurement devices to record (a) slip at the loaded end; (b) strain profiles along the bonded length.

Using this scheme, nine specimens were prepared for each bond length; the same width of CFRP strip was used ($b_f = 50$ mm), as the tests were intended to estimate the effective bond length, i.e. the length beyond which any increase in bond length cannot significantly increase the maximum transferable load.

3.2 Loading and measurement

Each specimen was set in a displacement controlled universal testing machine with maximum capacity equal to 600 kN and subjected to pure tensile force through steel rods causing direct shear on the CFRP strips. Spherical joints were placed at the top and bottom grips to avoid any bending moment caused by eccentricity. For some specimen, in addition to the maximum load-carrying capacity, the total displacement at the section in which the tensile force was applied for each side was measured at each load step by using two LVDTs. Moreover, electrical strain gauges on both of the opposite reinforced sides were fixed along the centre line of the CFRP strip to assess the variation in normal stress along the length of the reinforcement and the bond stress variation at the CFRP-calcarenite interface. Data from load cell, LVDTs and electrical strain gauges were recorded by a data acquisition system. The measurement system of the displacement at the section in which the tensile force was applied and the location of the gauges on one side are shown in Figure 4.

The local measurement system allows the definition of the effective part of the bonded length where the stresses are applied on the reinforcement transfer within the basic material, with a degree of accuracy related to the gauge spacing.

Since the stress gradient in the regions near the strip end is high, the gauge spacing was made smaller to ensure greater accuracy in computing the CFRP axial strains and the transfer length.

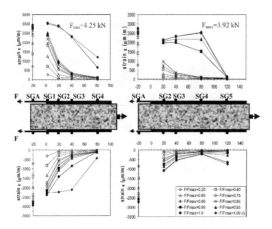

Figure 5. Strain profiles along the length of CFRP strip: (a) L = 100 mm and (b) L = 150 mm.

3.3 Results

All specimens were subjected to tensile force until total bond failure took place. Typical failure started and developed inside the calcarenite block until total debonding of the reinforcement occurred. The thickness torn off the calcarenite block was estimated to be approximately 1.5–2.0 mm, depending on the epoxy resin penetration capability (see Fig. 2). Results of the tests in terms of maximum external load recorded (pull-out load F_u) and corresponding cubic compressive strength are contained in detail in Accardi & La Mendola (2004).

Results in terms of the curves *pull-out load (F) – displacement at the loaded end (U)* of all tests showed similar initial linear elastic behavior with stiffness. $\bar{K} \approx 75000 - 90000 \, N/mm$

Tests with L = 100 and 150 mm bond lengths revealed similar behavior characterized by a constant load increase until the first cracking at the loaded end occurred; this was followed by progressive debonding process up to failure. By contrast, tests with L = 50 mm revealed a lower pull-out load accompanied by very brittle failure, making measurements impossible; these tests did not show any cracking process and the failure mode was quite different than that for specimens with L = 100 and 150 mm.

In Figure 5 strain profiles recorded by strain gauges are reported for two specimens with bond length 100 and 150 mm. Each strain profile is plotted for a given load level. The trend of the strain distribution is in agreement with that recorded by experimental tests carried out on concrete (Bizindavyi & Neale 1999). In the elastic range the strain distribution for both tests follows an exponential decay law from the loaded end to a length between 40 and 80 mm. The elastic limit is about 0.75–0.85 times the maximum load. When

the load increases the curves change shape until their concavity changes.

When total debonding starts, the values of two consecutive strain gauges become almost equal.

The analysis of the recorded data allows the identification of two bond lengths: – in the elastic range the strain profile reaches negligible values between the gauges located at 40 mm and 80 mm; this distance defines the so-called initial transfer length l_b in which the joint is characterized by elastic behavior; – when the load increases the gauges located at 80 mm begin to measure significant strain values; then the effective bond length L_b is identified as the distance between the loaded end and the ultimate point where the strains are not negligible; this measurement was made when the strain value of the first gauge in the bonded length was almost equal to that measured outside.

The effective bond length is evidently longer than the minimum transfer length l_b. After this stage, the load is almost constant but the debonding increases, shifting the effective bond length until the CFRP strip has completely peeled from the matrix. It can be observed that the results presented above lead to the fact that 80 mm $< L_b <$ 120 mm and that the L_b shifting is more evident for the specimen with L = 150 mm. The debonding propagation process is completely absent in the specimens with L = 50 mm; for this reason the strain profiles are not discussed here.

From the strain profiles measured along the CFRP length it is possible to trace an experimental local bond stress-slip curve in the distance between two strain gauges in accordance with Savoia et al. (2003).

By considering an elastic behavior of the CFRP strip, the average value of the bond stress between two subsequent strain gauges $(i, i + 1)$ can be written as a function of the difference of the measured strains as:

$$\tau_{i,av} = \frac{A_f \cdot E_f \left(\varepsilon_{i+1} - \varepsilon_i \right)}{b_f \left(x_{i+1} - x_i \right)} \qquad (1)$$

where A_f, b_f and E_f, are the cross-section area, the width and the elastic modulus of the carbon fibers respectively. Moreover, by assuming that complete compatibility occurs at the first strain gauge position (no slip) and the calcarenite strain is negligible with respect to the CFRP, integration of the strain profiles gives the following expression for the slip at distance $\bar{x}$ from x_i, with $0 \le \bar{x} \le (x_{i+1} - x_i)$:

$$s(\bar{x}) = s(x_i) + \int_0^{\bar{x}} \varepsilon(\bar{x}) d\bar{x} = s(x_i) + \frac{(\varepsilon_{i+1} - \varepsilon_i) \bar{x}^2}{(x_{i+1} - x_i)} \frac{}{2} + \varepsilon_i \bar{x} \qquad (2)$$

Taking $\bar{x}$ as the average value of the strain gauge spacing $\bar{x} = (x_{i+1} - x_i)/2$, Equation (2) yields:

$$s(\bar{x}) = s(x_i) + \frac{1}{2}(x_{i+1} - x_i) \cdot \left[\frac{(\varepsilon_{i+1} - \varepsilon_i)}{4} + \varepsilon_i \right] \qquad (3)$$

1206

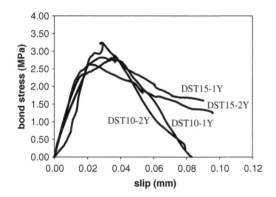

Figure 6. Local bond stress-slip law for specimens with different bonded length.

The average value $s(\bar{x})$ of slip between x_i and x_{i+1} is then computed. For all segments between two consecutive strain gauges locations, couple of values $(\bar{\tau}_{i,av}, s(\bar{x}))$ can be obtained. Results obtained in this way show similar behavior characterized by an elastic ascending branch with slope within $k = 70$–$120\,\text{N/mm}^3$ up to bond strength and afterwards a descending branch. In Figure 6, for two tests with $L = 150\,\text{mm}$ and two tests with $L = 100\,\text{mm}$ the bond stress-slip curves are given for the segment between SG2–SG3.

Tests with bond length $L = 150\,\text{mm}$ exhibit in the softening branch a gradual stress reduction with further sliding. This is probably correlated to the friction present between the two surfaces. Tests with bond length $L = 100\,\text{mm}$ show the same trend as each other but with a different slope in the softening branch if compared with the 150 mm bond length specimens. It is worth noting the absence of the sub-horizontal branch probably because of the shorter length adopted in this test.

These findings were also in the experimental analysis carried out for concrete specimens strengthened with steel plates and reported in Holzenkämpfer (1994).

4 DEBONDING MECHANISM

The available analytical models are basically addressed to evaluating the elastic stiffness (service condition) and bond strength (ultimate condition), using often different approaches for the same problem: *stress approach*, in which the criterion for growth of the debonded FRP-matrix interface is expressed in terms of the interfacial bond stresses; and the *fracture mechanics approach*, in which the criterion for interfacial debonding is expressed in terms of energy

balance (Accardi & La Mendola 2004). But the elastic stiffness, as well as the energy that is available for crack growth and the debonding process, need to be clarified. To gain a clear understanding of load transfer mechanism and crack propagation along the CFRP-calcarenite interface, numerical analyses were carried out to simulate the behavior of the Double Shear pulling Test, based on the LUSAS finite-element program using an appropriate interface element that support a damage model (Qiu et al. 2001).

4.1 Analysis of bonded joint in linear elastic range

The analysis of the bonded joint in the linear elastic range can be carried out by means of a simple shear lag approach.

Täljsten (1996), Wu et al. (2002) suppose uniformly distributed axial stresses in the cross-section of the matrix, neglecting the fact that the bond phenomenon develops around a thin layer near the reinforcement. By contrast, De Lorenzis et al. (2001) suppose rigid block basic material and attribute deformability only to the adhesive and/or primer layer. However, by using F.E.M. in a linear elastic analysis, it is possible to see that the load transfer mechanism is concentrated in a layer near the reinforcement. Usually this layer also involves the basic material. According to the CNR DT 200-2004 Recommendation, it is therefore possible to take a basic material layer into account to define the elastic deformability.

By introducing the values of E_c and v_c for calcarenite stone given in Arces et al. (1998) we obtain:

$$G_c = \frac{E_c}{2(1+v_c)} = \frac{9000}{2(1+0.2)} = 3750\,MPa \qquad (4)$$

Assuming the following expression for the slope of bond stress-slip curve:

$$k = \frac{G_c}{t_c} \qquad (5)$$

and by using the experimental values above obtained it is possible to deduce the thickness $t_c = 31.25$–$53.57\,\text{mm}$ involved in the elastic deformability. The t_c values are in agreement with the elastic FEM analysis, as shown in Figure 7.

4.2 Progressive debonding process

The connections between the reinforcement and the basic material were modelled by the discrete element in the LUSAS program named IPN6 that in a 2D configuration is an interface element between two lines with quadratic interpolation. This type of element inserted at the lines of potential debonding proves to be particularly suitable for modelling inter-laminar failure

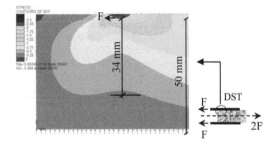

Figure 7. Shear stress contours of DST close to loaded end.

Table 1. Bond properties for DST (numerical parameters)

Specimen code	F_u [N]	s_0 [mm]	τ_u [MPa]	G_f [N/mm]
DST15-1Y	5040	0.053	3.76	0.168
DST15-2Y	3920	0.032	3.82	0.103
DST10-1Y	4930	0.038	4.39	0.163
DST10-2Y	4250	0.050	3.50	0.116

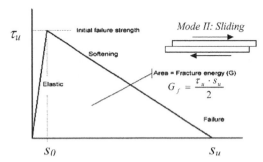

Figure 8. Reinforcement of basic material connection: bi-linear with linear softening interface law and sliding fracture mode.

and crack initiation and propagation. The element has six nodes and no thickness, describing the relationship between the stresses and relative displacements of the nodes connecting the two faces.

These interface elements support a damage model named *Delamination Damage Model* (see Figure 8) characterized by the failure modes, *Opening* and *Sliding*. In the following sections the uncoupled delamination option mode will be chosen in order to avoid the mode I failure. This model reproduces the nonlinear response of a system with potential debonding planes.

The behavior follows a linear law up to the strength threshold value τ_u (initial failure strength), which corresponds to the limit displacement s_0. When this value is exceeded, linear softening behavior occurs up to the point at which the fracture energy G_f is dissipated. Complete separation occurs when the maximum relative displacement (s_u) is reached. Therefore this damage model is capable of describing the mode II failure (sliding) associated with shear interface stresses. Hence the failure mode requires three parameters to describe the interfacial behavior: failure strength (τ_u), corresponding slip (s_0), and fracture energy (G_f).

The τ_u and s_0 parameters can be determined from the experimental values of each bond test and reported

in Table 1 with the maximum load F_u and the fracture energy G_f. The latter is evaluated by using the expression of maximum load F_u obtained from Holzenkämpfer (1994) and Täljsten (1996) by using the fracture mechanics approach considering a Mode II (shear) fracture mechanics of the basic material:

$$F_u = b_f \sqrt{2 E_f t_f G_f} \qquad (6)$$

in which t_f is the thickness of CFRP and the other symbols as defined above

Data for G_f given in Table 1 are obtained from Equation (6), in explicit form:

$$G_f = \frac{F_u^2}{2 b_f^2 E_f t_f} \qquad (7)$$

by introducing the following values: $b_f = 50$ mm, $t_f = 0.13$ mm, $E_f = 230\,000$ N/mm^2.

Using the values identified for each specimen, the comparison in terms of $F - U$ curves is reported in Figures 9 and 10 for the bond tests with bonded length 100 and 150 mm respectively.

The numerical-experimental comparison identifies three different phases: elastic behavior, debonding propagation and snap-back branch; the latter numerically assumed only because it is very difficult to point out experimentally. The numerical model is able to describe the elastic stiffness and to predict the maximum pull-out load because the latter is related to fracture energy only (Yuan et al. 2001).

It is worth noting that the pull-out load and stability of the debonding propagation are influenced by the asperity interlock, the roughness and the shape of the debonded surface.

These aspects determine a different friction between the reinforcement and the debonded surface that could modify the debonding behavior of the specimens as shown in the experimental curves reported in Figures 9 and 10. The friction is more evident for granular material like the calcarenite stone used in the present experimental investigation, whose specimens appear to be mostly made up of small grains joined by calcite with low bond strength.

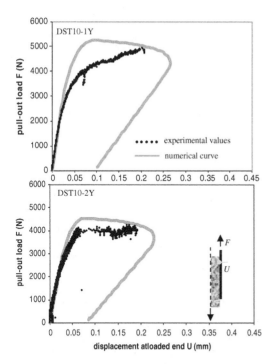

Figure 9. Comparison between pull-out load vs. displacement curves with L = 100 mm bonded length.

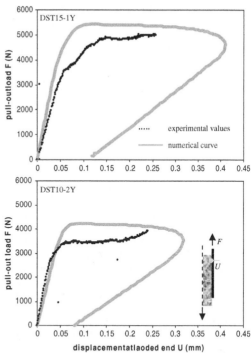

Figure 10. Comparison between pull-out load vs. displacement curves with L = 150 mm bonded length.

4.3 Experimental correlation for debonding limit strain

The results obtained allow the deduction an of important parameter, the fracture energy G_f, which is defined as the energy required to take the area of a bonded surface to complete fracture. It is worth noting that the descending branch of the bond stress-slip curve has a major influence on the fracture energy.

Equation (7) correlates the bond strength of the joint with the fracture energy. However, the failure mode starts and develops within the calcarenite stone, thus the fracture energy depends mainly on the mechanical properties of the base material.

Recently, some studies have been carried out on a procedure to measure the critical mode II strain energy release rate and to identify the fracture energy G_f. In present-day applications identification of the fracture energy is very difficult, especially for existing structural elements. For this reason, the existing design models for concrete (ACI 440-2000; JSCE 2001; fib 2001; CNR-DT 200-2004) and masonry (CNR DT 200-2004) correlate G_f with the mechanical properties of the basic material, as compressive and tensile strengths. Hence, by referring to Equation (7) and by introducing the F_u values obtained by the experimental tests it is possible to find out an experimental correlation between the compressive strength of calcarenite

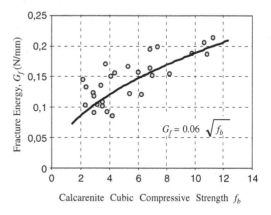

Figure 11. $G_f - f_b$ experimental correlation.

stone used to make up the specimens in the pull tests and the fracture energy. Specifically, the data of the DST with bonded length 150 mm were recorded. The specimens with L = 150 mm have a higher bonded length than the effective transfer length, as pointed out earlier. The values used in Figure 11 are given in Accardi & La Mendola (2004) and Failla et al. (2002).

The choice of cubic compressive strength as a mechanical parameter of the basic material to correlate with the bond strength instead of tensile or shear strength is due to the use of simple and reliable experimental tests, particularly for an anisotropic material like calcarenite stone.

Substituting the correlation indicated in Figure 11 in Equation (6) it therefore yields the following equation:

$$F_u = b_f \sqrt{0.12 E_f t_f \sqrt{f_b}} \qquad (8)$$

which gives the bond strength of the CFRP-calcarenite joint.

Chajes et al. (1996) have already proposed a correlation between the ultimate pull-out load F_u of the FRP-to-concrete bonded joint and the $\sqrt[4]{f_b}$ but only in Chen & Teng (2001) has this been confirmed through an extensive review of available experimental data.

The experimental correlation obtained for Sicilian calcarenite base material shows a possible extension of predictive bond strength models to natural stone treated like concrete with low compressive strength. Nevertheless, the lack of experimental results on various natural stones does not make it possible to generalize the method proposed here. However, it should be noted that the correlation presented here could also be valid for a soft rock like Naples tufa (Cosenza et al. 2000) but is not applicable to harder stone like Leccese stone (Aiello & Sciolti 2004), which instead shows less bond strength than the expected values. The failure mode is also different; for Leccese stone the FRP only involves a very thin layer of stone attached to the composite strip, and this is probably due to the different grain size and porosity of the rocks. For this reason it could be useful to introduce other parameters of the basic material that regard the porosity, the grain size or the structure of the rock.

5 CONCLUSIONS

The experimental and numerical investigations highlighted the behavior of the CFRP-calcarenite bonded joint. The use of external strain gauges placed along the CFRP bond length made it possible to record strain profiles and to deduce a local bond stress-slip law for the CFRP-calcarenite interface.

The numerical model does not show significant differences to predict of the maximum pull-out load and to describe the experimental behavior of the DST specimens. Nevertheless it is worth noting that the experimental measurements recorded for specimens with bond length $L = 100$ mm and $L = 150$ mm showed a different slope in the softening branch that reveal the presence of friction between the two debonded surfaces that could modify the post-peak branch of the local bond-slip law.

It is therefore necessary to consider the role of friction in order to better describe the specimen behavior for different experimental setups and other experimental tests.

REFERENCES

Accardi, M., La Mendola, L. 2004. Stress transfer at the interface of bonded joints between FRP and calcarenite natural stone. In Modena, Laurenço & Roca (eds), *IV Int. Seminar of Struct. Analysis of Hist. Constr.* Padua, 10–13 Nov.: 867–874.

ACI Committee 440. Guidelines for the design and construction of externally bonded FRP system for strengthening concrete structures. American *Concrete Institute*, Detroit, 2000.

Aiello, M.A., Sciolti, M.S. 2004. Analysis of bond performance between CFRP sheets and calcarenite ashlars under service and ultimate condition. In La Tegola & Nanni (eds), *Proc. of the 1st Int. Conf. on Innovative Materials and Techn. for Constr. and Restoration* Vol. 2: 81–96. Lecce, 6–9 June.

Arces, M., Nocilla, N., Aversa, S., Lo Cicero, G. 1998. Geological and geotechnical features of the "Calcarenite di Marsala". In *Proc. of the 2nd Int. Symposium on Hard Soils – soft Rocks:* Naples, Italy 12–14 October.

Bizindavyi, L., Neale, K.W. 1997. Experimental and theoretical investigation of transfer lengths for composite laminates bonded to concrete. In Proc., *Annual Con. of Canadian Society for Civil Engng* Vol. 6: 51–60 Structures-Composites Materials, Structural Systems, Telecommunications Towers, Sherbrooke, Québec, Canada.

Bizindavyi, L., Neale, K.W. 1999. Transfer lengths and bond strengths for composites bonded to concrete. In *J. of Composite for Constr.*, ASCE, 3(4): 153–160.

Chajes, M.J., Finch, W. W., Januszka, T.F., Thonson, T. A. Jr. 1996. Bond and force transfer of composite material plates bonded to concrete. In *ACI Struct. J.*, 93 (2): 208–217.

Chen, J. F., Teng, J. G. 2001. Anchorage strength models for FRP and steel plates bonded to concrete. In *J. of Struct. Engng* ASCE, 127(7): 784–791.

CNR-DT 200/2004. Istruzioni per la progettazione, l'esecuzione ed il controllo di interventi di consolidamento statico mediante l'utilizzo di compositi fibrorinforzati, Roma.

Cosenza, E., Manfredi, G., Occhiuzzi, A., Pecce, M. R. 2000. Toward the investigation of the interface behaviour between tuff masonry and FRP fabrics. In *Mechanics of Masonry Structures Strengthened with FRP Materials: Testing, Design, Control:* 99–108. 7–8 December, Venice.

De Lorenzis, L., Miller, B., Nanni, A. 2001. Bond of FRP Laminates to Concrete. In *ACI Materials J.*, 98 (3): 256–264.

Failla, A., Accardi, M., Rizzo, G., Algozzini, G., Pellitteri, G., Buscaglia, C. (2002). The use of CFRPs in strengthening of historical masonry structures: investigation on durability by accelerated testing. In Benmokrane & El-Salakawy (eds), *2nd Int. Conf. on Durability of fiber reinforced polymer (FRP) composites for construction*, Montreal 29–31 may, 297–303.

fib Bulletin, "Design and use of externally bonded FRP reinforcement for reinforced concrete structures",

Bulletin no. 14, 2001, sub-group EBR of fib Task Group 9.3.

Holzenkämpfer, P. 1994. Ingenieurmodelle des verbundes geklebter bewehrung für betonbauteile. In *PhD dissertation,* TU Braunschweig, (in German).

JSCE Concrete Committee. 2001. Recommendations for upgrading of concrete structures with use of CFRP sheet.

Leung, C. K. Y. 2004. Fracture Mechanics of Debonding Failure. In *FRP-Strengthened Concrete Beams. FraMCoS-5.* April. Vail Cascade Resort, Vail, Colorado: 12–16.

Nakaba, K., Kanakubo, T., Furuta, T. Yoshizawa, H. 2001. Bond Behavior between Fiber-Reinforced Polymer Laminates and Concrete. In *ACI Struct. J.*, 98 (3): 359–367.

Neubauer, U. and Rostasy, F. S. 1997. Design aspects of concrete structures strengthened with externally bonded FRP plates. In *Proc., 7th Int. Conf. on Struct. Faults and Repair*, ECS Publications, Edinburgh, Scotland, 2: 109–118.

Qiu, Y., Crisfield, M. A., Alfano G. 2001. An interface element formulation for the simulation of delamination with buckling. In *Engineering fracture mechanics* 68: 1755–1776.

Savoia, M., Ferracuti, B. and Mazzotti, C. 2003. Non linear bond-slip law for FRP-concrete interface. In *FRPRCS-6*:1–10. Singapore.

Smith, S.T., Teng, J.G. 2002. FRP-strengthened RC beams. I: review of debonding strength models. In *Engng Struct.* 24: 385–395.

Täljsten, B. 1996. Strengthening of concrete prisms using the plate-bonding technique. In *Int. J. of Fract.* 82 (3): 253–266.

Täljsten, B. 1997. Defining anchor lengths of steel and CFRP plates bonded to concrete. In *Int. Journal Adhesion and Adhesives*, 17(4): 319–327.

Wu, Z., Yuan, H., Niu, H. 2002. Stress transfer and fracture propagation in different kinds of adhesive joints. In *Journal of Engineering Mechanics*, ASCE 128 (5): 562–573.

Yuan, H. and Wu, Z. 1999. Interfacial fracture theory in structures strengthened with composite of continuous fiber. In *Proc., Symp. of China and Japan: Sci. and Technol.* of 21st Century, Tokyo, Sept., 142–155.

Yuan, H., Wu, Z., Yoshizawa, H. 2000. Theoretical solutions on interfacial stress transfer of externally bonded steel/composite laminates. In *J. Struct. Mechanics Earthquake Engng.*, 18(1): 27–39. Tokyo.

Fracture Mechanics of Concrete and Concrete Structures – Design, Assessment and Retrofitting of RC Structures – Carpinteri, et al. (eds)
© 2007 Taylor & Francis Group, London, ISBN 978-0-415-44616-7

Influence of concrete thermal degradation on anchorage zones of externally-bonded reinforcement

D. Horak, P. Stepanek & J. Fojtl
Faculty of Civil Engineering, Brno University of Technology, Czech Republic

ABSTRACT: This paper is identification of problem behaviour of anchorage of externally bonded reinforcement on concrete constructions. The tests were realized with FRP strips. The main problem was to determine the influence of thermal degradation of concrete for the load capacity of anchorage zone. This influence was evoked by several different (0, 50, 100 and 200) cycles of freezing of concrete-FRP bond. Tests were performed both as the short-terms and long-terms ones. All tests were made for different anchorage length. The test results were compared with values obtained from analysis of anchorage zones behaviour in nonlinear FEM method. New analytical relationship was also developed to define the behaviour between strip and concrete along the anchorage zone. Finally, control calculations were carried out according two different standards (ACI and Czech standards) and compared with the previous results.

1 INTRODUCTION

Strengthening methods using externally bonded reinforcement are well explored and often used in the Czech Republic and abroad. Within the context of using new technology and new materials it is necessary to adapt the methods of design and application to native conditions in the Czech Republic. Thereby it is possible to restrict damages caused by lack of information about proper design and proper working process.

For that purpose it was necessary to perform series of the short-terms and long-terms tests in local climatic conditions and with using accessible common materials. One range of the research task was focused on anchorage zones of externally bonded reinforcement. Research work was concentrated on following influences:

– influence of anchorage length of externally bonded strip on carrying capacity,
– influence of thermal degradation of bond on deformation (strain) of anchorage zone and its carrying capacity,
– influence of long-term load impact on thermally degraded bond between external reinforcement and concrete.

2 SHORT-TERMS TESTS

The short-terms experiments serve as a base for a design of strengthening. The short-term experiments

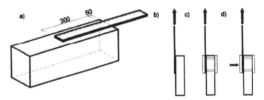

Figure 1. Scheme of the anchorage zone test (3 variations).

involved a series of the tests on a concrete specimen, on which distinct lengths of CFRP glued reinforcement were applied. These specimens demonstrate the behaviour of the reinforcing element under tension and adhesion between individual layers in the reinforcement area.

Anchorage blocks of dimension $150 \times 150 \times 600$ mm were prepared from defined standard concrete quality (C20/25 according to EC2). The CFRP strips of cross-sectional dimension 50×1.2 mm and Young's modulus 155 GPa were bonded to the prepared surfaces of different anchorage lengths – 150, 225 and 300 mm (see Fig. 1a). The aim of the tests was to determine the ultimate axial forces, which could be anchored by different anchorage lengths and by different stress acting on the anchoring area. Tested alternatives are shown in Figure 1b (anchoring without stirrup – i.e. without the cross force acting on the anchorage area), Figure 1c (anchoring with stirrups without perpendicular prestressing on the contact area between the strip and concrete) and Figure 1d (anchoring with prestressed stirrups).

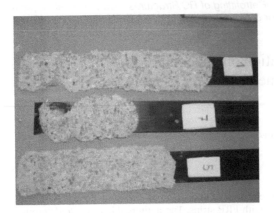

Figure 2. Ripped off strips with the residue thin concrete layer.

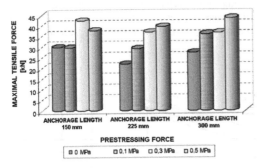

Figure 3. Influence of the prestressing force on carrying capacity of the anchorage zone.

There were three different anchorage lengths (represented with three test specimens with the same anchorage length) in each test alternative. In variation described in Figure 1d there were twice as much specimens and they were tested with different prestressing force.

During the tests it was possible to trace development of stress along the anchorage areas for individual load phases and activation of the anchorage zone. Also the weakest point of this bond system was determined – the failure happened under the bonded strip in surface layer of concrete as can be seen in Figure 2. So the limitation factor of load carrying capacity is the concrete or more precisely the tension capacity of concrete.

The test showed that at the higher level of loading a crack generated in the surface layer of concrete and it shifted active section of anchorage zone. This crack distributed quickly itself and soon after its origin the strip ripped off. Maximal load force could be increased by applying the prestressing force along the anchorage zone. This positive influence is clearly shown in Figure 3.

In terms of theoretical interpretation, for such test specimens it is possible to derive basic differential equations (Brosens K. & Van Gemert D. 1999) for normal and shear stresses in concrete and adhesive, normal stress in a strengthening member (CFRP strip) for several boundary conditions. The elastic behaviour of all materials used (concrete, rebar, adhesive and CFRP strips), full co-operation between bonded strip and concrete and uniformly distributed stresses and strains over the entire width of a cross section of the anchorage area was assumed. The derived equations for normal stress in the CFRP strip $\sigma_p(x)$ and the shear stress $t_p(x)$ in a direction of axis x in the adhesive and the normal stress $\sigma_n(x)$ in the perpendicular direction to the surface of a concrete specimen under glued strip is given by

$$\sigma_p(x) = C_1 e^{Ax} + C_2 e^{-Ax} - \frac{A_c B_2}{A^2}(l - x + a) \tag{1}$$

$$\tau_p(x) = t_p\left[C_1 A e^{Ax} - C_2 A e^{-Ax} - \frac{A_c B_2}{A^2}\right] \tag{2}$$

$$\sigma_n(x) = e^{-\beta x}\left[D_1 \cos(\beta x) + D_2 \sin(\beta x)\right] \tag{3}$$

where

$$A^2 = A_p + A_c B_1 \quad ; \quad A_p = \frac{G_a}{t_a t_p E_p} \quad ; \quad A_c = \frac{G_a}{t_a t_p E_c} \quad ;$$

$$B_1 = -\frac{b_p t_p}{A_{tr}}\left(\frac{z_{ctr}^2}{i_{tr}^2} - 1\right) ; \quad B_2 = F\frac{z_{ctr}^2}{J_{tr}}\frac{1}{l + a + b} \quad ;$$

t_a is the thickness of the adhesive; t_p is the thickness of the CFRP plate; E_p is the Young's modulus of the CFRP strip; E_c is the Young's modulus of concrete; G_a is the shear modulus of the adhesive.

Using the appropriate boundary conditions the constants C_1, C_2, D_1 and D_2 can be identified.

In order to control the comparison of experimentally obtained results and results of the solutions of analytical equations, physically non-linear FEM modelling based on the fracture mechanic model of concrete was used. The behaviour of the other materials (adhesive, CFRP strip) was considered as linear. The calculations were made by software ATENA while deterministic material characteristics for concrete, adhesive and strip were supposed.

The physical-mechanical properties of both strip and glue were assumed according to data provided by producer (they were not tested).

Concrete material parameters used in calculations were taken from the control tests of concrete samples.

The results of experimental, analytical and numerical solution of the strain along the glued length are shown in Figure 4.

Table 1. Material properties for strip and glue used in calculations.

	E [MPa]	μ	Thickness [mm]
carbon strip	155,000	0.3	1.2
glue	8000	0.3	1.4

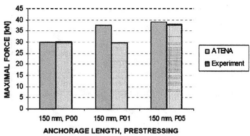

Figure 5. Comparison of the maximal acting force – anchorage length 150 mm, prestressing 0.0 MPa (P00), 0.1 MPa (P01) and 0.5 MPa (P05).

Table 2. Maximal load forces for different anchorage lengths and different level of thermal degradation.

	0 cycles	50 cycles	100 cycles
Anchorage length 225 mm			
Average value	29.428	25.553	27.081
Standard deviation	4.68	1.06	1.03
Anchorage length 300 mm			
Average value	*	36.246	29.748
Standard deviation		4.48	0.51

* Values could not be obtained due to imperfect glue

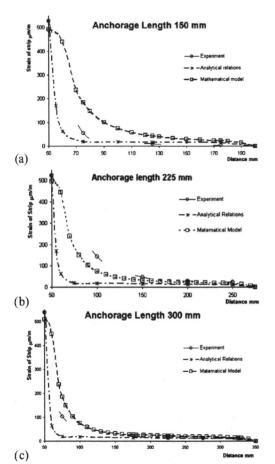

(a)

(b)

(c)

Figure 4. Comparison of the longitudinal strain along the glued length (a) 150 mm (b) 225 mm (c) 300 mm.

The positive effect of perpendicular prestressing acting on the anchorage zone was proved also in numerical solution as the maximal force increased significantly (nearly 30% of initial value). Comparison of results from experiment and numerical model are in Figure 5.

To test the influence of deterioration of concrete caused by the temperature changes, the same specimens with glued lengths 225 and 300 mm were used.

The test specimens with the glued strips were frosted and re-frosted in a water solution. The temperature change moved between −15°C and 15°C, the velocity of the temperature cycle was ±2°C/hour and the maximal and minimal temperature was hold for 4 hours. The specimens were loaded with 0, 50 and 100 temperature cycles, respectively. Each set contained three test specimens.

The results of short-term tests are shown in Table 2. The influence of freezing cycles on the glued connection behaviour is significant. Also by growing number of the freezing cycles the deformation of the strip free end grows (Fig. 6) and the ultimate limit force in the strip decreases.

The failure of the bond system occurred in the same manner as in the short-terms tests, i.e. crack developed in the concrete near surface with bonded strip, quickly extended and caused delamination of the strip. The tensile capacity of the concrete remained the limitation point of this bond system.

The influence of number of freezing cycles to limit load force can be approximately defined by the diagram – Figure 7 – for glued anchorage of

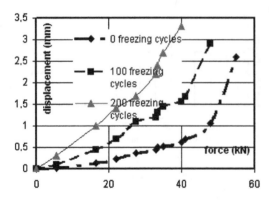

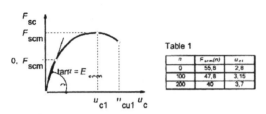

Figure 6. Influence of number of the freezing cycles on dependence of the strip free end displacement and the tension force.

Figure 7. Schematic representation of the expression (4) showing the progress of thermal degradation influence of glued connection.

Table 1

n	$F_{scm}(n)$	u_{c1}
0	55,8	2,8
100	47,8	3,15
200	40	3,7

strip (50/1.2 mm and Young's modulus 155 GPa) on concrete C20/25. The calculation form is

$$\frac{F_{sc}}{F_{scm}} = \frac{k\eta - \eta^2}{1 + (k-2)\eta} \qquad (4)$$

where $F_{sc} = F_{sc}(n)$ is force in strip (depended on n); $\eta = u_c/u_{c1}$; $u_{c1} = u_{c1}(n)$ is the displacement of the free strip end at peak anchoring force $F_{scm} = F_{scm}(n)$; $k = E_{scm} \times |u_{c1}|/F_{scm}$ and values $u_{c1}(n)$, $F_{scm}(n)$ are obtained from experiments – Figure 7.

The expression (4) was verified in terms of set of others 6 test specimens. However the use for the long-terms load acting needs additional tests and verification.

3 LONG-TERMS TESTS

The long-terms tests were performed with specimens loaded with 0 (reference), 100 and 200 freezing cycles described above. All specimens (total lay-out of test specimens see in Table 3) were consequently put into steel construction that allowed to develope permanent

Table 3. Number of test specimens during the long-term test.

Anchoring length	Number of test specimens cycles		
	0 cycles	100 cycles	200 cycles
225 mm	3	3	3
300 mm	3	3	3

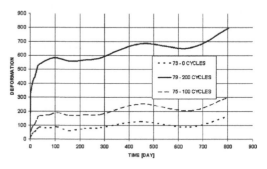

Figure 8. Deformation progress for anchoring length 225 mm.

load to reinforcement. The anchorage zones length were 225 mm and 300 mm for each set of freezing cycles.

All test specimens were made at time t = 0 days. At time t = 28 days the strips were bonded to the concrete blocks. In the same day part of the specimens was stored to the freezing machine and the exposure of the freezing cycles begun. Unexposed specimens were stored separately in standard conditions but they were not loaded. The loading force was applied after all specimens went through prescribed freezing cycles. All 18 specimens were loaded in the same day.

Tension force was put into reinforcement via prestressed spring. Initial force value was 5 kN (i.e. 20% of total load carrying capacity of anchorage zone) and it was slowly increased to present 6.3 kN.

During the experiment strain along the anchorage zone was monitored via strain gauges (same configuration as in the short-terms tests). The acting load force was monitored indirectly by measuring the strain in free end of reinforcement.

Response of seven specimens was measured in 10 minutes intervals by a measuring central. This central recorded also the temperature of surroundings and thus it was possible to eliminate the influences of temperature changes (not-loaded compensation reinforcement elements were measured too). The rest of loaded specimens were measured in the longer time intervals.

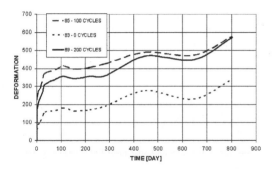

Figure 9. Deformation progress for anchoring length 300 mm.

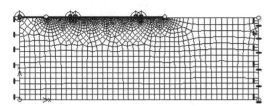

Figure 10. FEM mesh used in calculation.

4 MATHEMATICAL MODELLING

Model of anchorage zone was created in non-linear FEM solver ATENA based on fracture energy of quasi-brittle materials.

Material models for reinforcement and bond material were presumed as elastic with characteristics shown in Table 1 and in addition $\alpha_p = 1.2 \times 10^{-5}$ (for CFRP reinforcement), $\alpha_p = 9 \times 10^{-5}$ (for bond). Some material characteristics for concrete were taken from performed tests (tensile and compression strength) and the rest values were derived through relations in ČSN 73 0038. Concrete material was modelled as non-linear.

Standard Newton-Rhapson method was used for problem solving. FEM mesh was created as quadratic with refinement under problematic zone (Fig. 10). Boundary conditions were the same as in experiment and also positions of monitoring points corresponded with real strain gauges.

In the first step all input data were more precisely identified during comparison of real experiment outputs and numerical solution of the short-terms test of carrying capacity.

Lately two different numerical models were used for creep solution – model based on standard ČSN 73 1201 and model B3 (developed by Bazant and Al Manaseer). The results obtained from mathematical model corresponded well with reality. Thus it is possible to use these creep material models for further solutions

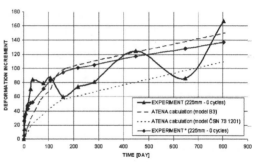

Figure 11. Increment of deformation in anchorage zone (comparison between numerical analysis and two real test specimens).
* Values are re-calculated to eliminate deformations caused by temperature changes.

and construction calculations. This model is also possible to use for verification of developed analytical relations.

Creep rupture of the CFRP strip was not considered in the numerical model. The stress level in bonded strip during the experiment was very low and the ratio of stress level at creep rupture was determined (Yamaguchi 1997) to be 0.91 which value was not reached during test. Also the modulus of elasticity is typically unaffected by environmental conditions (in this case several numbers of freeze-thaw cycles) hence the material properties of CFRP strip were not considered as time-dependent.

The numerical model (using B3 material creep model) shows good correspondence with the real long-term test results obtained on thermally non-degraded specimens.

In the next step it will be necessary to find material model including thermal degradation of concrete. Such simplification (i.e. neglected time-depending changes of material characteristics of CFRP) is possible because of relatively stable material characteristics of CFRP strips unaffected by thermal degradation.

5 CONCLUSIONS

The comparison of the theoretical, measured and numerical non-linear FEM analysis results of stress and strain at the anchorage areas demonstrates good accordance at the linear area of behaviour. The theoretically derived design equations can be used for anchorage of non-prestressed and prestressed CFRP strips.

From the analytical and numerical analysis it is evident, that the anchoring length of CFRP strips on concrete C20/25 is about 365 mm. By increasing the anchoring length the maximum tension force in CFRP strip does not grow.

The failure of anchorage area (of glued contact) starts in concrete – from the beginning of the strip. The main reason for the failure initiation is the normal stress acting perpendicularly to CFRP strip axis.

From the long-terms tests carried out until now it can be stated that there was an evident influence of the number of freezing cycles in the growth of strain in anchorage zone while the permanent constant load. The shorter the anchorage zones the more severe influence. This can be partially restricted by the longer anchorage zone.

While creating numerical model of the long-term loading on thermal degraded structure it is possible to use the simplification and to use only simple material models for externally bond reinforcement (if CFRP materials are used – other materials may show more significant time-depended material degradation). It is enough to consider only concrete degradation as it is the limit factor for carrying capacity of anchorage zone.

ACKNOWLEDGEMENT

This outcome has been achieved with the financial support of the Ministry of Education, Youth and Sports of the Czech Republic, project No. 1M0579, within activities of the CIDEAS research centre. Some results has been achieved within the scientific-research work of the Czech Grant Agency GAČR 103/02/0749, Modern Methods of Strengthening Concrete and Masonry Structures and Optimization of Design" and research work "Progressive reliable and durable structures" (MSM0021630519).

REFERENCES

Stepanek P. & Sustalova I. 2001. Anchoring of nonprestressed and prestressed CFRP Strips at Strengthening of concrete beams. *Proceedings of the International Conference Composites in Material and Structural Engineering*. Klokner Institute, Prague, Czech Republic, June 2001

Stepanek, P. & Svarickova, I. 2003. Some remarks to anchorage of CFRP strips by External strengthening of concrete structures. *Proc. of International Conference Concrete in the 3rd Millennium:* 265–274, Brisbane, Australia

Stepanek P. & Fojtl J. & Dibelka V. & Horak D. 2005. Reinforcing and Additional Strengthening of Concrete Structures with FRP-materials. *1st Central European Congress on Concrete Engineers*: 202–205. Graz, Austria

Brosens K. & Van Gemert D. 1999. Anchoring stresses in the end zones of externally bondedbending reinforcement, *5. Internationales Kolloquium Freiburg*, 1163–1174, Freiburg

ACI 440.1R-03, Guide for the Design and Construction of Externally Bonded FRP Systems for Strengthening Concrete Structures, *American Concrete Institute*, 2003

J. Jang, Y.-F. Wu 2005. Interfacial stresses in FRP Plated Concrete Beams Including Shear Deformation Effect, *Proceeding of the International Symposium on Bond Behaviour of FRP in Structures (BBFS 2005)*, International Institute for FRP in Construction, 169–173

Fracture Mechanics of Concrete and Concrete Structures – Design, Assessment and Retrofitting of RC Structures – Carpinteri, et al. (eds)
© *2007 Taylor & Francis Group, London, ISBN 978-0-415-44616-7*

An experimental study on the long-term behavior of CFRP pultruded laminates suitable to concrete structures

F. Ascione
Department of Civil Engineering, University of Rome "Tor Vergata", Rome, Italy

V. P. Berardi, L. Feo & A. Giordano
Department of Civil Engineering, University of Salerno, Fisciano (SA), Italy

ABSTRACT: The rheological behavior of structural materials has a significant role indeed in Civil Engineering, where concrete and FRP materials undergo creep in normal environmental conditions, while steel exhibits a sizable creep only at high temperature (above 400°C). With reference to RC structures strengthened by means of FRP laminates, FRP creep generally coexists with concrete cracking. The interaction between these phenomena should be taken into account in order to evaluate the structural durability. Here, the first results of a research program on creep in composite pultruded laminates used in Civil Engineering are presented, under various stress levels and in constant environmental conditions (many theoretical and experimental studies on creep have been performed so far in the aerospace and naval fields, but not as many in Civil Engineering). The specimens tested in this project are made of high-modulus carbon fiber-reinforced polymer – CFRP, whose mechanical properties are tailored for Civil Engineering applications. The tests are still in progress in the Materials and Structures Testing Laboratory of the Civil Engineering Department of the University of Salerno (Italy).

1 INTRODUCTION

In the field of Civil Engineering the rheological behavior of materials has an important role. One aspect that is particularly relevant is the creep phenomenon, since the continually increasing strain can compromise the durability of structural elements.

The viscous effects result particularly sizable in the case of fiber reinforced composite materials (FRP – Fiber Reinforced Polymer), due to the polymeric matrix, that is sensitive to viscous phenomena.

Nowadays these innovative materials are mainly utilized in the rehabilitation of damaged Reinforced Concrete (RC) and masonry structures.

Current international guidelines regarding the design of FRP strengthening applications confirm the importance of this problem. In fact, suitable limits have been introduced on the FRP stress state in Serviceability Limit State, in order to limit viscous effects (ACI Committee 440 2000, CEB-FIP 2001, CNR-DT 200/2004 2004).

Both theoretical and experimental studies on the viscous behavior of composites used in the aeronautical and naval fields have been performed (Barbero & Harris 1998, Dutta & Hui 1997, Maksimov & Plume 2001, Pang et al. 1997, Petermann & Schulte 2002, Scott & Zureick 1998).

The main aim of these studies has been to formulate constitutive laws for these materials under different stress states, as well as in different environmental conditions.

The studies developed on composite materials for civil engineering applications, and in particular on their secondary creep are less numerous. As it is well known, the secondary creep mainly occurs during the service life of a structure.

So far, the authors of the present paper have produced several theoretical studies based on the effects that secondary creep has on the long term behavior of RC beams strengthened with FRP laminates. The first results show a sizable influence of the viscous effects on the mechanical behavior of the strengthened structures. This leads to a consistent migration of FRP stresses towards the RC beam. The subsequent increase of the stress state in RC beams may compromise the efficacy of the strengthening technique and produce damage and cracking in the concrete (Ascione & Berardi 2003, Ascione et al. 2004, Berardi et al. 2003).

Therefore it is necessary to develop further theoretical and experimental analyses on this topic, starting from a better characterization of creep properties of FRP laminated composites used in civil applications.

The objective of this work is to present the results of the creep test program on several CFRP pultruded laminates subject to different stress values, in constant environmental conditions. The tests are currently being carried out at the Material and Structures Testing Laboratory of the Civil Engineering Department of the University of Salerno.

2 CREEP BEHAVIOR

The viscous behavior of FRP materials mainly depends on:

– matrix type;
– fiber type;
– fiber volume fraction;
– fiber orientation;
– load history;
– temperature and humidity.

This behavior can be identified by creep tests. A sample is subject to a constant tensile force and its elongation is monitored over time (Fig. 1). In the first part of the test a phenomenon, well known as primary creep, can be investigated. The initial elastic elongation is followed by fast-growing deformations. After this phase, the elongation goes on with an approximately constant rate over a time period longer than the previous one (secondary creep). Finally, if stress or temperature values become very high, the specimen may break (tertiary creep).

From a theoretical point of view FRP laminates are classified as ortotropic viscous materials, whose long term behavior should require the characterization of all the in-time mechanical properties.

In Civil Engineering applications the FRP mechanical behavior is in one-dimension (along one of the natural directions) and the stresses are less than forty per cent of the ultimate values.

These peculiarities simplify FRP long-term analysis, allowing the modeling of these innovative materials through linear viscoelastic one-dimensional models. These models are based on either mechanical analogies or experimental data (Barbero & Harris 1998, Dutta & Hui 1997, Maksimov & Plume 2001, Pang et al. 1997, Petermann & Schulte 2002, Scott & Zureick 1998).

At the moment, there are no constitutive laws of general validity for composite materials.

3 EXPERIMENTAL SET-UP

A creep test program on CFRP unidirectional laminates is being carried out at constant temperature.

The CFRP specimens, characterized by high value of longitudinal Young modulus, are subjected to

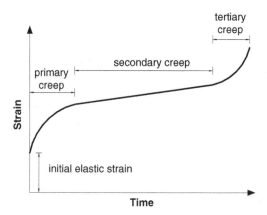

Figure 1. Creep deformations over time.

Table 1. CFRP certified mechanical properties.

E_f [N/mm^2]	f_{fk} [N/mm^2]	ε_{fu} [%]
≥ 300000	1450	0.45

constant different stress values along the longitudinal natural direction.

The experimental equipment includes:

– two identical testing devices able to apply constant loads;
– a data acquisition system.

More specifically, the thickness of all the CFRP laminates is equal to 1.4 mm and their mechanical properties, certified by the producer, are shown in Table 1, where E_f is the longitudinal Young modulus, f_{fk} is the characteristic tensile strength and ε_{fu} is the ultimate strain.

3.1 Testing mechanical device

The dissipation-free testing device allows the application of axial loads to CFRP specimens set through a lever arm.

It is a steel lever arm, whose ends are linked respectively to a dead load and to the CFRP specimens series. Furthermore it includes a fixture composed of a steel structure with roller bearings functioning as the fulcrum (Fig. 2).

The lever arm has been designed to magnify the load applied to the specimens by a factor of 10.

The device is capable of loading three laminates simultaneously at a desired constant load.

The running system requires an efficacious load transfer from the end of the lever arm to the composite specimens. With this aim, the specimen anchorage

Figure 2. Testing device.

Figure 3. Specimens link.

devices consist of two suitable steel plates glued to the laminates and then bolted to each other.

The link between two successive specimens consist of chains and hooks (Fig. 3).

The two end anchorages of each specimen are connected also by means of chains, in order to ensure a

Table 2. Testing stress values.

Testing device			1	2
Specimen	F	[N]	15225	22838
Top	B	[mm]	50	30
	σ	[N/mm^2]	217.5	543.8
	σ/f_{fk}	[%]	15	37.5
Middle	B	[mm]	30	25
	σ	[N/mm^2]	362.5	652.5
	σ/f_{fk}	[%]	25	45
Bottom	B	[mm]	25	15
	σ	[N/mm^2]	435	1087.5
	σ/f_{fk}	[%]	30	75

link in the case of an accidental sliding of the specimen between the anchorage plates.

The axial force applied at the top of CFRP series (F), the CFRP specimen width (B) and the corresponding normal longitudinal stress value (σ), are summarized in Table 2 for each testing device.

3.2 Data acquisition system

The data acquisition system consists of electrical strain gages glued to the external surface of the specimens, thermocouples, scanner and a database management software.

The equipment can automatically record the data corresponding to fixed time steps.

Furthermore the system can correct the strain data by taking into account the temperature effects.

Each specimen is equipped with six electrical strain gages, symmetrically bonded with respect to the middle plane, in order to take care of the measurement errors and of the accidental flexural deformation. In particular, three strain gages are applied on each face, as shown in Figure 4.

This strain gages layout allows both longitudinal and transverse strains to be measured as well as evaluate the Poisson ratio variation over time.

4 EXPERIMENTAL RESULTS

The creep tests on six CFRP specimens, subjected to different tensile stress values, are still being performed through two testing devices (§3.1).

The CFRP strains and the environmental temperature values have been recording since the start time of the test by means of a data acquisition system (§3.2). The test temperature is held constant (20°C) by using an air-condition system.

Starting from the recorded elastic strains values, the average value of Young longitudinal modulus, E_{3m}, and Poisson ratio, ν_{32m}, have been obtained: $E_{3m} = 413383$ N/mm^2, $\nu_{32m} = 0.34$.

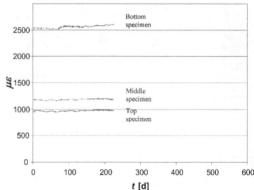

Figure 4. CFRP specimen.

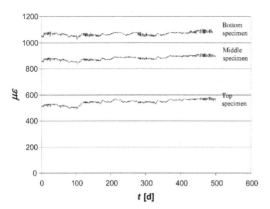

Figure 5. Longitudinal strain over time (testing device n.1).

Figure 6. Longitudinal strain over time (testing device n. 2).

Table 3. Maximum longitudinal strain variation.

	Specimen	σ/f_{fk} [%]	$\Delta\varepsilon/\varepsilon_i$ [%]
Testing device 1	Top	15.0	–
	Middle	25.0	1.35
	Bottom	30.0	1.63
Testing device 2	Top	37.5	1.66
	Middle	45.0	1.71
	Bottom	75.0	1.93

With reference to CFRP specimens tested by the testing device n.2 (Table 2), after about thirty-two hours a sudden rupture of the specimens occurred.

This accidental failure seems to have been caused by a microcrack produced in one of the samples during the cutting phase. A new creep test is being performed on another specimen using the above mentioned testing device.

The average values of specimen longitudinal strains recorded over time, expressed in days, are plotted in Figures 5–6.

5 CONCLUSIONS

In this paper an experimental study on the creep behavior of CFRP pultruded laminates has been presented.

In particular, several creep tests are being performed for different stress values at a constant temperature.

Experimental data relative to the first hours show a limited longitudinal strain variations for the tested specimens, revealing a negligible primary creep phase.

These data confirm the efficacy of carbon fibers in limiting the primary creep strains of CFRP laminates, characterized by high longitudinal Young modulus and high fiber volume fraction, as well known in literature.

The subsequent records for testing device n.1 have highlighted a maximum strain variation after about thirty days (Ascione et al. 2005), starting from the reference time instant $t = 28500$ s (≈ 8 h), which represents the assessment time of the data acquisition system.

The average percent longitudinal strain variations for the middle and the bottom specimens are equal to 1.35% and 1.63%, respectively (Table 3). With reference to the top sample, its strain values, being smaller than the others, are affected by higher measurement errors, and then it is not possible to evaluate with high precision the strain variations. In any way, the test results have highlighted that the creep rate is negligible.

The experimental data recorded until about 500 days have not shown any further relevant increases of the longitudinal strains for all the specimens of testing device n.1.

With reference to the specimens of the testing device n.2, the maximum strain variations with respect to reference time instant, $t = 0$ s, have been recorded after about 76 days. The percent longitudinal strain variations for the top, the middle and the bottom specimens are equal to 1.66%, 1.71% and 1.93%, respectively (Table 3).

The subsequent records for testing machine n.2 have not highlighted any further relevant strain increases.

Starting from the above experimental results, the viscous behavior seems to be depleted, due to the creep rate being negligible.

They would confirm that the CFRP pultruded laminates with high longitudinal Young modulus and high fiber volume fraction highlight limited creep strains.

The analyses above shows the application rules prescribed in the CNR Guidelines about the FRP viscous behavior are confirmed.

They allow one to realize FRP retrofitting by avoiding damage and cracking of the concrete related to the viscous behavior of the composite materials.

The data are still being acquired in order to verify the effective attenuation of creep phenomena.

REFERENCES

ACI Committee 440 2000. *Guide for the design and construction of externally bonded FRP systems for strengthening concrete structures.*

Ascione, L. & Berardi, V.P. 2003. Gli effetti reologici nel placcaggio strutturale di elementi in c.a. con laminati fibrorinforzati. *XVI AIMETA Congress of Theoretical and Applied Mechanics, CD-ROM Proceeding.*

Ascione, L. & Berardi, V.P. & Feo, L. & Giordano, A. 2004. Il fenomeno del creep nel placcaggio strutturale con FRP: progettazione di un dispositivo sperimentale. *XXXIII AIAS Congress, Bari, CD-ROM Proceeding.*

Ascione, F. & Berardi, V.P. & Feo, L. & Giordano, A. 2005. Indagine sperimentale sul comportamento viscoso di laminati pultrusi in fibra di carbonio. *XVII AIMETA Congress of Theoretical and Applied Mechanics, CD-ROM Proceeding.*

Barbero, E. & Harris, J.S. 1998. Prediction of creep properties from matrix creep data. *Journal of Reinforced Plastics and Composites* 17(4).

Berardi, V. P. & Giordano, A. & Mancusi, G. 2003. Modelli costitutivi per lo studio della viscosità nel placcaggio strutturale con FRP. *XXXII AIAS Congress, Salerno, CD-ROM Proceeding.*

CEB-FIP 2001. *Externally bonded FRP reinforcement for RC structures.*

CNR-DT 200/2004 2004 (translated in English 2005). *Istruzioni per la Progettazione, l'Esecuzione ed il Controllo di Interventi di Consolidamento Statico mediante l'utilizzo di Compositi Fibrorinforzati - Materiali, strutture in c.a. e in c.a.p., strutture murarie.*

Dutta, K. & Hui, D. 1997. Integrating fire-tolerant design and fabrication of composite ship structures. *Interim report University of New Orleans.*

Maksimov, R.D. & Plume, E. 2001. Long-Term creep of hybrid aramid/glass fiber-reinforced plastics. *Mechanics of Composite Materials* 37(4).

Pang, F. & Wang, C.H. & Baghgate, R.G. 1997. Creep response of woven composites and effect of stitching. *Journal of composites science and technology* 57.

Petermann, J. & Schulte, K. 2002. The effects of creep and fatigue stress ratio on the long-term behavior of angle-ply CFRP. *Composite Structures* 57: 205–210.

Scott, D. & Zureick, A. 1998. Compression creep of a pultruded e-glass/vinylester composite. *Composites Science and Technology* 85: 1361–1369.

Fracture Mechanics of Concrete and Concrete Structures – Design, Assessment and Retrofitting of RC Structures – Carpinteri, et al. (eds)
© 2007 Taylor & Francis Group, London, ISBN 978-0-415-44616-7

Inverse analysis for calibration of FRP – Concrete interface law

M. Savoia, B. Ferracuti & L. Vincenzi
DISTART- Structural Engineering, University of Bologna, Bologna, Italy

ABSTRACT: Inverse analysis technique is used to derive a non linear mode II interface law for Fiber Reinforced Polymer (FRP) – concrete bonding starting from experimental data. The proposed interface law is based on a fractional formula and includes non linear compliance contributions of adhesive and concrete cover at high shear stresses. It depends on three parameters (maximum shear stress, corresponding slip and an exponent), which are calibrated from experimental results on delamination tests. Values of maximum loads for different bonding lengths and strains profiles along FRP plates are used. Parameter identification is performed by inverse analysis using a Direct Search technique. Considerations on well-posedness of inverse problem adopting different cost functions are given. After parameter identification, numerical results obtained with the proposed interface law are found to be in very good agreement with experimental results.

1 INTRODUCTION

It is well known that failure modes due to FRP delamination may significantly reduce the theoretical bearing capacities of FRP-strengthened R/C beams. A complete knowledge of bond-slip law in terms of fracture energy, peak point and initial stiffness, is fundamental to correctly design FRP retrofitting. First, for sufficiently long bonded lengths, ultimate failure load due to delamination mainly depends on fracture energy of mode II interface law. Moreover, evaluation of effectiveness of strengthening under service loadings requires to calculate stress concentrations close to transverse cracks in concrete and verify it is lower than peak shear stress (Teng et al. 2003, CNR Committee 2006). In the linear range, stress concentration is governed by initial stiffness of the interface, where compliances of both adhesive and external cover of concrete must be taken into account.

In Teng & Smith (2002), some recently proposed interface laws are reviewed. The most common interface law is based on a bilinear shear stress – slip relation, with peak shear stress defined from Mohr-Coulomb criterion for concrete and corresponding slip arbitrarily assumed, and final slip of softening branch defined in order to attain a given value of fracture energy of the law (about 0.15–0.2 mm). Comparison with more sophisticated laws (Ferracuti et al. 2006) showed that this relation is very rough for several reasons: first of all, peak shear stress cannot be obtained from concrete properties only, depending also on surface preparation before bonding and the adopted adhesive; moreover, slope of linear softening

branch has not a physical meaning. As a result, delamination force for short bonded lengths is strongly overpredicted.

A power fractional FRP – concrete interface law has been proposed by Ferracuti et al. (2007), whose parameters (peak shear stress, corresponding slip and an exponent) are calibrated by post-processing experimental data: from equilibrium and compatibility considerations, average shear stress and slip data are computed, and a least-square procedure is adopted to determine the parameters. This study clearly shows that, from the experimental point of view, evaluation of maximum delamination force (global data) or applied force – plate displacement curves only is not sufficient to provide data to define a shear stress – slip interface law. Very accurate tests with measure of strains (local data) along the FRP plate are also needed.

In the present paper, a more complete inverse analysis procedure is used to calibrate the parameters of a non linear constitutive law for FRP – concrete interface. The previously quoted interface law and a bond – slip model recently proposed by the authors (Ferracuti et al. 2006), are adopted.

Inverse analysis technique is applied to a set of experimental results by Mazzotti et al. (2007) on FRP – concrete delamination. Experimental results have been obtained in terms of strains on FRP plate at different values of applied force and delamination forces for different bonding lengths.

Identification method based on genetic algorithms (Storn & Price 1997) is used to perform inverse analysis. The adopted method (Differential Evolution algorithm) is a direct search approach, which is able to avoid convergence on local minima of cost function. In the present problem, the cost function is a weighted sum of errors on prediction of strains along the FRP plate and delamination forces. Considerations on well-posedness of inverse problem are drawn. It is shown that, as already underlined in inverse problems in very different frameworks (see Iacono et al. 2006 and Slowik et al. 2006 for identification of softening law of concrete in tension or Savoia & Vincenzi 2005 for identification of mechanical properties of structures from dynamic tests), well-posedness is achieved only if both global data (e.g. load – deflection curves) and local data (strains along FRP plate) are considered in cost function.

Finally, parameters obtained from inverse analyses have been used to simulate experimental tests and numerical results are found in very good agreement with experimental results.

2 FRP-CONCRETE INTERFACE LAW

The definition of a mode II interface FRP – concrete law is a difficult task for several reasons. From the experimental point of view, tests where force – displacement curves only are determined do not provide for sufficient data to define a local interface law. Measures of strains along FRP plate are also required. On the other hand, transmission length of local stresses at the interface level is very small, less than 80–100 mm from bonding extremity, and several strain gages must be placed within that length.

Moreover, interface law and numerical model are strictly interconnected: kinematic variables the interface law is referred to depend on the features of the adopted model. For instance, in the present model (see Section 4), slip is referred to average displacement of concrete cross-section and, consequently, interface compliance must take shearing deformation of both adhesive and external cover of concrete into account, also at high shear stresses. At load levels typical of service loadings (about 50 per cent of maximum load), maximum slip may already exceed that corresponding to peak shear stress. Softening branch is also very important, since fracture energy of interface law and, consequently, maximum transmissible load strongly depend on it.

Following Ferracuti et al. (2007), the adopted interface law is a power fractional law (see Figure 1):

$$\tau = \bar{\tau} \frac{s_p}{\bar{s}} \frac{n}{(n-1) + (s_p/\bar{s})^n},$$ (1)

where τ, s_p are FRP-concrete shear stress and slip. Moreover, $\bar{\tau}$, $\bar{s}$ indicate maximum shear stress and corresponding slip, whereas $n > 2$ is a parameter mainly governing the softening branch. Equation 1 is

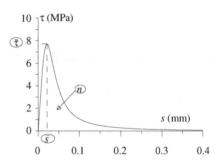

Figure 1. The proposed FRP – concrete interface law (see Equation 1), for $n = 3.008$, $\bar{\tau} = 7.72$ MPa, $\bar{s} = 0.0218$ mm.

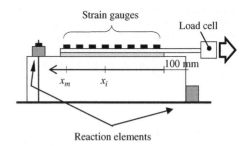

Figure 2. Pull-pull setup for FRP-delamination test (from Mazzotti et al. (2007)).

depicted in Figure 1, adopting values of parameters $(\bar{\tau}, \bar{s}, n)$ obtained in numerical application reported in Section 6.

3 CALIBRATION OF INTERFACE LAW BY INVERSE ANALYSIS

3.1 Experimental test by Mazzotti et al. (2007)

In order to calibrate an interface law, four pull-pull tests from the experimental campaign by Mazzotti et al. (2007) are considered in the present study (Figure 2). Four different bonded lengths BL (50, 100, 200 and 400 mm) were tested, by gluing CFRP plates (plate width $b_p = 80$ mm, thickness $h_p = 1.2$ mm) to concrete blocks ($150 \times 200 \times 600$ mm).

Data obtained from experimental tests are maximum transmissible forces F for different bonding lengths and axial strains ε along the FRP plate for different levels of applied load.

3.2 Global data: maximum force

As well known, increasing the length of the anchorage, transmissible force increases asymptotically up to a maximum value, depending on fracture energy of interface law and mechanical/geometrical properties of the plate. Maximum transmissible force by an anchorage of infinite length is defined as:

$$F_{\max} = b_p \int_0^\infty \tau(x) \cdot dx, \qquad (2)$$

with b_p = plate width, and the following relation holds between F_{max} and fracture energy of interface law G_f (see Ferracuti et al. 2006):

$$F_{\max} = b_p \sqrt{2 E_p h_p G_f} . \qquad (3)$$

where h_p, E_p stand for thickness and elastic modulus of FRP plate. Equation 3 shows that a correct estimate of fracture energy is fundamental for the definition of interface law, because only in this case the value of maximum transmissible load can be accurately predicted.

For the proposed interface law, fracture energy can be expressed in closed form as a function of governing parameters $(\bar{\tau}, \bar{s}, n)$:

$$G_f = \int_0^{+\infty} \tau\,(s_p)\,ds_p = g_f(n)\,\bar{\tau}\,\bar{s} \qquad (4)$$

where g_f is given by:

$$g_f(n) = \pi \left(\frac{1}{n-1}\right)^{1-\frac{2}{n}} \frac{1}{\sin(2\pi/n)} \qquad (5)$$

Equation 5 confirms that Equation 1 requires $n > 2$, otherwise the fracture energy is not a positive and finite quantity.

3.3 Local data: axial strain along FRP plate

As shown in Figure 3b, distribution of strains along the plate for low values of applied load is governed by the initial stiffness of interface law. Initial stiffness of the adopted interface law can be written as a function of three unknown parameters $(\bar{\tau}, \bar{s}, n)$ as:

$$k_p = \lim_{s_p \to 0} \frac{\tau\,(s_p)}{s_p} = \frac{\bar{\tau}}{\bar{s}} \frac{n}{n-1} \qquad (6)$$

On the contrary, for high values of applied load, maximum slope of strain profile along the plate is governed by the maximum shear stress $\bar{\tau}$ (see Figure 3b), as shown by the following equilibrium equation:

$$\bar{\tau} = \frac{A_p}{b_p}\left(\frac{d\sigma_p}{dx}\right)_{\max} = \frac{E_p A_p}{b_p}\left(\frac{d\varepsilon}{dx}\right)_{\max} \qquad (7)$$

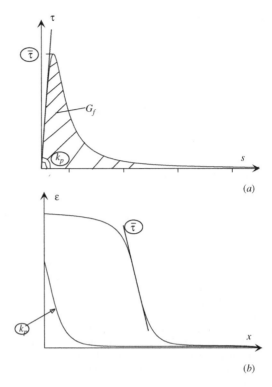

Figure 3. (a) The proposed interface law and (b) a typical distribution of strains along the plate before and after the onset of delamination: k_p = initial stiffness, $\bar{\tau}$ = peak shear stress, G_f = fracture energy of interface law.

where A_p, E_p stand for area and elastic modulus of FRP plate, whereas σ_p, ε are axial stress and strain of FRP.

Therefore, if experimental values of maximum forces at delamination only are adopted for parameter estimation, ill-posedness of identification problem can be expected. In fact, except for the case of very short bonded lengths, maximum transmissible load by an anchorage depends on fracture energy of interface law only, as confirmed by Equation 3. Three unknown parameters cannot be identified using one information only. On the contrary, if values of strains along FRP plate are also used, two additional information are introduced (initial stiffness and peak shear stress), and a well-posed identification problem can be expected.

3.4 Inverse analysis: the cost function

The cost function to be minimized during the identification procedure is the relative error between maximum forces (F_k) and strains (ε_{ijk}) obtained from numerical model (described in Section 4) adopting a

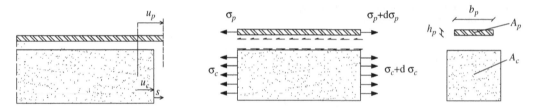

Figure 4. The bond-slip model for FRP – concrete delamination: notation adopted for displacements and stresses.

given set of identification parameters $(n, bar\tau, \bar{s})$, and experimental data $(\bar{F}_k, \bar{\varepsilon}_{ijk})$, i.e.:

$$H = \sum_{k=1}^{4} \left[w_1 \cdot \left(\frac{F_k - \bar{F}_k}{\bar{F}_k} \right)^2 + w_2 \cdot \sum_{i=1}^{m} \sum_{j=1}^{f} \left(\frac{\varepsilon_{ijk} - \bar{\varepsilon}_{ijk}}{\bar{\varepsilon}_{1jk}} \right)^2 \right], \quad (8)$$

where subscripts i, j, k indicate, respectively, i-th position of strain gauge along the plate ($i = 1: m$), j-th level of applied load in a delamination test ($j = 1: f$), and k-th bonded length BL ($k = 1: 4$); moreover, f and m are numbers of selected load levels and strain gauges along the plate, and w_1, w_2 are weight constants for forces and strains. Values are set in a non-dimensional form with respect to experimental force ($\bar{F}_k$) and to strain corresponding to first strain gage ($\bar{\varepsilon}_{1jk}$), respectively.

4 MODEL FOR FRP-CONCRETE DELAMINATION

A non-linear bond – slip model has been recently developed by the authors to study FRP-concrete delamination phenomenon (Ferracuti et al. 2006). Notation adopted for displacements and stresses is reported in Figure 4. Plate and concrete are subject to axial deformation only, i.e., bending of plates is neglected. This assumption is valid in the present case, due to negligible bending stiffness of FRP plate with respect to concrete specimen counterpart.

Axial displacements and forces for concrete and plate are denoted by $u_c, u_p, \sigma_c, \sigma_p$. The governing equations are equilibrium, constitutive and compatibility conditions, which can be written in the form:

$$\frac{du_p}{dx} = \frac{\sigma_p}{E_p}, \quad \frac{d\sigma_p}{dx} = \frac{\tau}{h_p}, \quad (9)$$

$$\frac{du_c}{dx} = \frac{\sigma_c}{E_c}, \quad \frac{d\sigma_c}{dx} = -\frac{b_p}{A_c}\tau, \quad (10)$$

where A and E stand for area and Young modulus, respectively, and subscripts c, p for concrete and FRP plate.

According to notation reported in Figure 4, FRP – concrete interface law is then written in the form:

$$\tau_p = k_p(s_p) \cdot s_p, \quad (11)$$

where $s_p = u_p - u_c$ denotes FRP-concrete slip and $k_p(s_p)$ is the non-linear secant stiffness of interface law.

By substituting Equation 11 in Equations 9, 10, a system of non-linear first order differential equations can be obtained:

$$\frac{d\mathbf{y}(x)}{dx} = \mathbf{A}(y, x)\mathbf{y}(x), \quad 0 \le x \le BL, \quad (12)$$

where BL is the length of the bonded plate, vector $\mathbf{y}$ collects the unknown functions:

$$\mathbf{y}^T = \{N_p, N_c, u_p, u_c\} \quad (13)$$

and the non-linear matrix $\mathbf{A}$ is:

$$\mathbf{A} = \begin{bmatrix} 0 & 0 & b_p\,k_p & -b_p\,k_p \\ 0 & 0 & -b_p\,k_p & b_p\,k_p \\ 1/E_p\,A_p & 0 & 0 & 0 \\ 0 & 1/E_c\,A_c & 0 & 0 \end{bmatrix}. \quad (14)$$

5 INVERSE ANALYSIS BY DIFFERENTIAL EVOLUTION ALGORITHM

Differential Evolution (DE) is a heuristic direct search approach (Storn & Price 1997). A number NP of vectors containing the optimization parameters is adopted:

$$\mathbf{z}_{i,M} = \{n, \bar{\tau}, \bar{s}\}, \quad i = 1, 2, ..., NP$$

Subscript M indicates the M-th generation of parameter vectors, called *population*. The number NP of vectors of the population is kept constant during the minimization process.

In order to minimize the objective function, a direct search method is a strategy that generates variations of parameter vectors. A robust algorithm requires that solution does not converge to a local minimum. Techniques like genetic and evolution algorithms are based on the calculation of several vectors simultaneously. Hence, if some vectors reach local minima, they can be excluded because they are associated with higher values of the cost function.

DE algorithm is briefly described. First of all, initial population is chosen randomly. Then, DE generates a new parameter vector by adding the weighted difference vector between two vectors of the population, so obtaining a third vector (*Mutation* operation). Then, in the *Crossover* operation, a new trial vector is generated by selecting some components of the mutant vector and some of the original vector. If the *trial vector* gives a lower value of objective function than that of the old population, the new generated vector replaces the old vector (*Selection* operation). These operations are described with more details in the following.

5.1 Mutation

For each vector of M-th population $\mathbf{z}_{i,M}$, $i = 1, 2, \ldots,$ NP, a trial vector $\mathbf{v}_{i,M}$ is generated by adding to $\mathbf{z}_{i,M}$ a contribution obtained as the difference between two other vectors of the same population.

According to Storn & Price (1997), the mutant vector is generated according to the expression:

$$\mathbf{v}_{i,M+1} = \mathbf{z}_{best,M} + \gamma \cdot (\mathbf{z}_{r_1,M} - \mathbf{z}_{r_2,M}), \qquad (15)$$

where $r_1, r_2 \in \{1, 2, \ldots, NP\}$ are mutually different integer numbers and $\mathbf{z}_{best,M}$ is the vector giving the minimum value of the object function of the M-th population (see Figure 5) Moreover, $\gamma < 2$ is a positive constant (scale parameter) controlling the amplitude of the mutation.

5.2 Crossover

In order to increase the diversity of the vectors, crossover process is introduced in the DE algorithm.

The *trial vector* $\mathbf{u}_{i,M+1}$ is obtained by randomly exchanging the values of optimization parameters between the original vectors of the population $\mathbf{z}_{i,M}$ and those of mutant population $\mathbf{v}_{i,M+1}$, i.e.:

$$\mathbf{u}_{i,M+1} = \{u_{1i,M+1}, u_{2i,M+1}, u_{3i,M+1}\};$$

where:

$$\left(\mathbf{u}_{i,M+1}\right)_j = \begin{cases} \left(\mathbf{v}_{i,M+1}\right)_j & \text{if } rand(j) \le CR \\ \left(\mathbf{z}_{i,M}\right)_j & \text{if } rand(j) > CR \end{cases} \qquad (16)$$

In Equation 16, $j = 1, 2, 3$ and $(\mathbf{u}_{i,M+1})_j$ is the j-th component of vector $\mathbf{u}_i$. Moreover, $rand(j)$ is the j-th

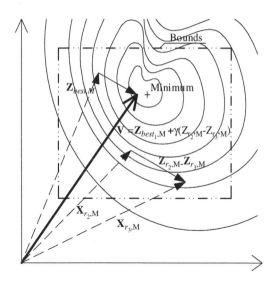

Figure 5. Differential evolution algorithm for parameter identification: mutation operation (according to best combination).

value of a vector of uniformly distributed random numbers, and CR is the crossover constant ($0 < CR < 1$), indicating the percentage of mutations.

5.3 Selection

In order to decide if a vector $\mathbf{u}_i$ may be element of new population of generation $M+1$, each element of the vector $\mathbf{u}_{i,M+1}$ will be compared with the previous vector $\mathbf{z}_{i,M}$. If vector $\mathbf{u}_{i,M+1}$ gives a smaller value of objective function H than $\mathbf{z}_{i,M}$, $\mathbf{u}_{i,M+1}$ is selected as the new vector of population $M+1$.

Otherwise, the old vector $\mathbf{z}_{i,M}$ is retained, i.e.:

$$\mathbf{z}_{i,M+1} = \begin{cases} \mathbf{u}_{i,M+1} & H(\mathbf{u}_{i,M+1}) < H(\mathbf{z}_{i,M}) \\ \mathbf{z}_{i,M} & H(\mathbf{u}_{i,M+1}) \ge H(\mathbf{z}_{i,M}) \end{cases} \qquad (17)$$

with $i = 1, 2, \ldots, NP$.

6 NUMERICAL APPLICATION

Inverse analysis procedure is applied to the present problem, adopting the cost function reported in Equation 8, with weight constants for force and strain contributions equal to $w_1 = 1$ and $w_2 = 1/f$ (f being the number of load levels considered), respectively. For three unknown parameters of interface law (n, $\bar{\tau}$, $\bar{s}$), the values $n = 3.008$, $\bar{\tau} = 7.72$ and $\bar{s} = 0.0218$ have been obtained. In the following, the present solution will be denoted as *Case A*.

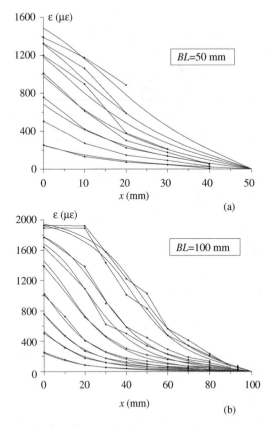

Figure 6. Strains along the bond length at different loading levels: numerical results (—) and experimental data from Mazzotti et al. (2007) (•••): (a) $BL = 50$ mm, (b) $BL = 100$ mm

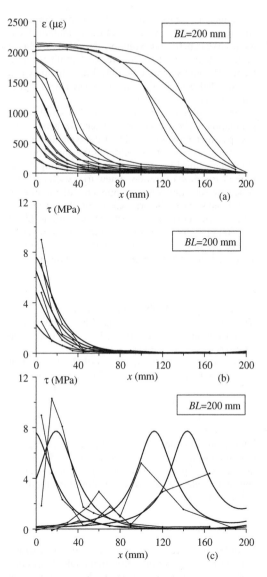

Figure 7. (a) Strains and (b, c) shear stresses at low and high loading levels: numerical results (—) and experimental data from Mazzotti et al. (2007) (•••), for $BL = 200$ mm.

6.1 Numerical against experimental results

Experimental tests have been then numerically simulated with the bond-slip model described in Section 4 and identified values of parameters $(n, \bar{\tau}, \bar{s})$; results have then been compared with experimental data.

Strain distributions in FRP plate along the bonded length are given in Figure 6 for 50 mm and 100 mm bond length, respectively. The highest load level considered is close to failure load obtained experimentally. Numerical results are generally in good agreement with experimental data. For all bonded lengths, the behavior for low loads is very well predicted, so assuring that stiffness of initial (elastic) branch of interface laws is correctly evaluated. Moreover, the bond-slip model is able to follow the growth of delamination at constant load along the bonded length.

FRP strain and shear stress distributions for bonded lengths equal to 200 mm and 400 mm are reported in Figures 7, 8; white dots refer to data obtained during delamination phase. As for shear stresses, numerical and experimental results are in good agreement for low – to – medium loadings. For very high loads, i.e., when plate delamination is in progress, results obtained from experimental data are more irregular. In any case, position of maximum shear stress along the bonded length is well predicted.

Finally, delamination loads obtained numerically as a function of bonded length are compared with experimental results in Figure 9 (see *Case A*). Results confirm that the proposed interface law provides for a good prediction of failure loads. For the smallest

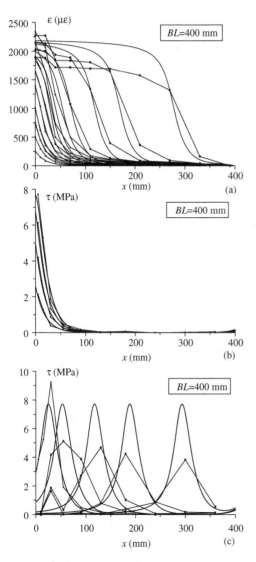

Figure 8.   (a) Strains and (b, c) shear stresses at low and high loading levels: numerical results (—) and experimental data from Mazzotti et al. (2007) (•••), for $BL = 400$ mm.

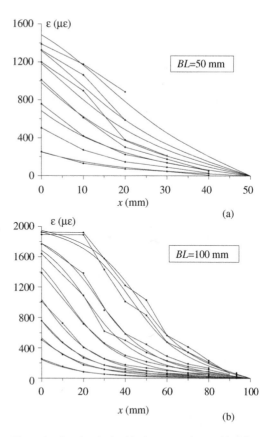

Figure 9.   Results obtained by inverse analyses with different cost functions: maximum force vs bond length compared with experimental results.

bonded length (50 mm), the experimental load is lower than predicted numerically: when the length of the plate is smaller than width, assumption of plane deformation adopted for 1D bond-slip model is not longer valid, and a more complex non linear 3D model should be required.

6.2 Well-posedness of inverse problem

The choice of cost function to be minimized is very important to obtain a well-posed inverse problem for parameter identification. This problem, well known in dynamic identification problems (Savoia & Vincenzi 2005), has been recently treated in Iacono et al. 2006 for parameter identification of non local damage problems.

In the present study, three different cost functions have been considered, with different values of w_1, w_2 weight constants for force and strain contributions in Equation 8.

In the first case (*Case A*), both maximum forces at delamination for different bonding lengths and FRP strains at different force levels have been adopted, with $w_1 = 1$, $w_2 = 1/f$ in Equation 8. This problem has been described in the previous section, and the corresponding solution is considered as the reference solution. In the second case (*Case B*), maximum forces for different bonding lengths only have been considered in the calibration procedure ($w_1 = 1$, $w_2 = 0$). In the third case (*Case C*), strains for different loading levels and bonding lengths only have been considered ($w_1 = 0$, $w_2 = 1/f$).

Table 1. Comparison between results obtained by inverse analysis with different cost function.

Case	n	$\bar{\tau}$	$\bar{s}$	G_f	Min curvature
	–	MPa	mm	MPa mm	–
A	3.01	7.72	0.0218	0.4818	0.0104
B	3.04	6.14	0.0301	0.5173	0.0016
C	2.98	8.11	0.0219	0.5189	0.0098

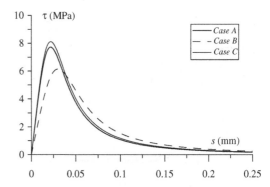

Figure 11. Results obtained by inverse analyses with different cost functions: FRP – concrete interface laws.

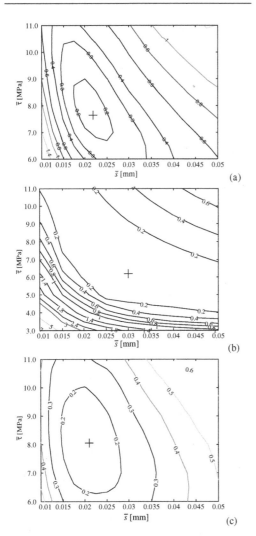

Figure 10. Cost function obtained for (a) *Case A* for $n = 3.00$ (reference solution), (b) *Case B* for $n = 3.04$, (c) *Case C* for $n = 2.98$.

For each cost function, inverse analysis has been performed by DE algorithm. Best values of unknown parameters are reported in Table 1 and interface laws are depicted in Figure 11.

Contour lines of cost function vs. $(\bar{\tau}, \bar{s})$ parameters are shown in Figure 10, setting the value of exponent n equal to the identified value for each individual case. The minimum of the three cost functions is indicated in each figure with a cross.

Figure 10*b* clearly shows that, adopting maximum forces only (*Case B*), a direction exists where the cost function is almost insensitive to variations of identification parameters $\bar{\tau}, \bar{s}$. That direction corresponds to constant values of fracture energy of interface law (see Equation 4). For *Case C* (strain measures only), cost function is similar to reference *Case A*. However, minimum curvature is smaller, so indicating that identification of unknown parameters can be numerically more computationally expensive.

A sensitivity analysis has been also performed to determine minimum curvature of objective functions (see Table 1). Results clearly show that, adopting both forces and strains, minimum curvature of cost function increases by a factor greater than 6 with respect to the case of only forces. Of course, smaller values indicate smaller sensitivity of cost function to parameter variation.

Finally, in Figure 11, debonding forces obtained with interface laws whose parameters have been obtained by adopting different cost functions are compared with experimental results. Smallest error in terms of debonding forces is obtained in *Case B* whose cost function is based on global data (forces) only. Nevertheless, in this case the error on strain profiles is much higher that reference Case A as shown in Figure 12.

7 CONCLUSIONS

Inverse analysis is used to derive a non linear mode II interface law for Fiber Reinforced Polymer (FRP) – concrete bonding starting from experimental data. The proposed law is based on a power fractional

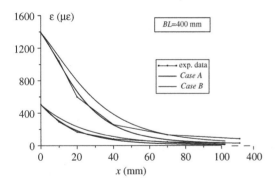

Figure 12.   Results obtained by inverse analyses with different cost functions: distribution of strain along the plate.

law depending on three parameters, maximum shear stress with corresponding slip and an exponent. Pull-pull delamination tests from Mazzotti et al. (2007) are considered: experimental data are both maximum forces and strain profiles along FRP plate. A direct search method (Differential Evolution algorithm) is used to solve the inverse problem. Well-posedness of inverse problems adopting different cost functions is discussed. It is shown that, if only maximum forces are adopted, the problem is ill-posed and cost function is much less sensitive to parameter variation. On the contrary, if both maximum forces and FRP strain are used, sensitivity of cost function to parameter variation is much higher and identification can be performed correctly.

Adopting the so obtained mode II interface law and the bond – slip model described in Section 4, very good agreement is found between numerical and experimental results, in terms of FRP strains, shear stress profiles before and during delamination and values of maximum forces, so assessing the validity of the proposed technique.

ACKNOWLEDGEMENTS

Financial supports of (Italian) Department of Civil Protection (Reluis 2005 Grant – Task 8: Innovative materials for vulnerability mitigation of existing structures) and C.N.R. (PAAS Grant 2001) are gratefully acknowledged.

REFERENCES

Brosens, K. 2001. Anchorage of externally bonded steel plates and CFRP laminates for the strengthening of concrete elements. *PhD. thesis*, Univ. of Leuven, Belgium.

Chen, J.F & Teng, J.G. 2001. Anchorage strength models for FRP and steel plates bonded to concrete. *J. Struct. Eng. ASCE* 127(1): 784–91.

CNR Committee. 2006 *Guide for the design and construction of externally bonded FRP systems for strengthening existing structures,* CNR DT 200/2004 Technical Report.

Ferracuti B., Savoia M. & Mazzotti C. 2006. A numerical model for FRP-concrete delamination. *Composites Part B: Engineering*, 37 (4–5): 356–364.

Ferracuti, B., Savoia, M. & Mazzotti, C. 2007. Interface law for FRP-concrete delamination. *Composite Structures*, in press.

Iacono, C., Sluys L. J., & van Mier J. G. M.. 2006. Estimation of model parameters in nonlocal damage theories by inverse analysis techniques. *Computer Methods in Applied Mechanics and Engineering* 195: 7211–7222.

Lu, X.Z., Teng, J.G., Ye, L.P. & Jiang, J.G. 2005. Bond – slip models for FRP sheets/plates bonded to concrete. *Eng. Struct.* 27(4): 920–937.

Mazzotti, C., Savoia, M. & Ferracuti B. 2007. An Experimental study on delamination of FRP plates bonded to concrete. Submitted.

Neto, P., Alfaiate, J., Almeida, J.R., Pires, E.B. & Vinagre, J. 2004. The influence of the mode II fracture energy on the behaviour of composite plate reinforced concrete. In: Li et al., editors. *Proceedings FraMCoS-5.* Vail, Colorado. USA.

Savoia M. & Vincenzi L. 2005. Differential evolution algorithm in dynamic structural identification, *ICOSSAR – International conference of structural safety and reliability,* Rome, Italy.

Slowik, V., Villmann, B. & Bretschneider, N. 2006. Computational aspects of inverse analyses for determining softening curves of concrete. *Computer Methods in Applied Mechanics and Engineering* 195 (52): 7223–7236.

Storn, R. & Price, K. 1997. Differential Evolution – a simple and efficient heuristic for global optimization over continuous spaces. *Journal of Global Optimization* 11 (4): 341–359.

Teng, J.G. 2001. *FRP composites in civil engineering.* Hong Kong: Elsevier.

Teng, J.G., Smith & S.T. 2002. FRP-strengthened RC beams. I: review of debonding strength models. *Eng. Struct.* 24(4): 385–95.

Teng, J.G., Smith, S.T., Yao, J. & Chen, J.F. 2003. Intermediate crack-induced debonding in RC beams and slabs. *Construction and Building Materials* 17(6–7): 447–62.

Wu, Z., Yuan, H. & Niu, H. 2002. Stress transfer and fracture propagation in different kinds of adhesive joints. *J. Eng. Mech. ASCE* 128(5): 562–73.

Yuan, H., Teng, J.G., Seracino, R., Wu, Z.S. & Yao, J. 2004. Full-range behavior of FRP-to-concrete bonded joints. *Eng. Struct.* 26: 553–65.

Fracture Mechanics of Concrete and Concrete Structures – Design, Assessment and Retrofitting of RC Structures – Carpinteri, et al. (eds)
© 2007 Taylor & Francis Group, London, ISBN 978-0-415-44616-7

Flexural behaviour of RC and PC beams strengthened with external pretensioned FRP laminates

C. Pellegrino & C. Modena
Department of Constructions and Transportation Engineering, University of Padova, Padova, Italy.

ABSTRACT: In this paper some results of an experimental investigation on real-scale RC (Reinforced Concrete) and PRC (Pre-stressed Reinforced Concrete) beams strengthened in flexure with FRP laminates developed at Material Testing Laboratory of the Department of Constructions and Transportation Engineering of the University of Padova are shown. Externally bonded FRP reinforcement is applied with different modalities, using different types of end-anchorage devices and, in some beams, with pre-stress transfer. After the characterisation of the single materials, four points bending tests are executed and failure and cracking modes are studied with particular attention to the behaviour of the anchorages made both with resins and different types of mechanical devices.

1   INTRODUCTION

Externally bonded Fiber Reinforced Polymer (FRP) sheets are currently used to repair and strengthen existing Reinforced Concrete (RC) and Pre-stressed Reinforced Concrete (PRC) structures.

Structural behaviour of FRP strengthened RC elements has been widely studied over the last few years and some studies have resulted in the first design guidelines for strengthened concrete. American ACI 440-02 (ACI Committee 440 2002), European fib –bulletin 14 (fib T.G. 9.3 2001) and Italian Recommendations (CNR-DT 200, 2004) are examples of such guidelines.

Experimental investigations about flexural behaviour of RC beams strengthened with FRP materials have been usually developed on reduced-scale specimens with ordinary FRP laminates/sheets. Few experimental tests (Brena et al. 2003, Chahrour and Soudki 2005, El-Hacha et al. 2003, El-Hacha et al. 2004, Tan et al. 2003, Triantafillou et al. 1992, Wight et al. 2001, Yu et al. 2004) have been developed on real-scale specimens strengthened with pre-tensioned FRP laminates.

A number of experimental programs have been developed in the last years at Material Testing Laboratory of the Department of Constructions and Transportation Engineering of the University of Padova about flexural, shear, axial and bond behaviour of FRP strengthened elements (Pellegrino and Modena 2002, Pellegrino and Modena 2006, Tinazzi et al. 2003, Pellegrino et al. 2004, Pellegrino et al. 2005, Boschetto et al. 2006).

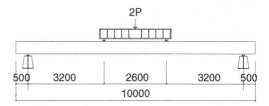

Figure 1.   Load scheme of the beams.

In the present paper, main results of an experimental program about real-scale RC and PRC beams strengthened in flexure with ordinary and pre-tensioned FRP laminates developed at the Material Testing Laboratory of the Department of Constructions and Transportation Engineering of the University of Padova, are shown. Failure and cracking modes are observed and efficiency of different types of mechanical end-anchorages (Pellegrino et al. 2005, Boschetto et al. 2006, El-Mihilmy and Tedesco 2001, Malek et al. 1998, Taljsten 1997) is studied.

2   EXPERIMENTAL PROGRAM

2.1   *Test setup*

Five real-scale beams (four RC beams and one PRC beam with pre-tensioned internal strands) have been tested. Load scheme, dimensions and details of the internal reinforcement of the beams are shown in Figs. 1 and 2.

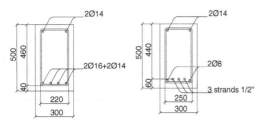

Cross section of RC beams   Cross section of PRC beam

Figure 2.   Cross-sections of the beams.

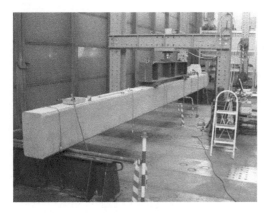

Figure 3.   Typical beam before the execution of the test.

Table 1.   Mechanical properties of concrete.

Mean cubic compressive strength	$R_{cm} = 71\,MPa$
Mean tensile strength	$f_{ctm} = 5.2\,MPa$
Mean elastic modulus	$E_{cm} = 38060\,MPa$

Shear reinforcement consists in stirrups 8 mm diameter with 20 cm spacing (always designed to obtain flexural failure of the specimens). Strands are pre-tensioned with initial stress equal to 1400 MPa.

Beams are instrumented with three strain-gages in the middle cross-section at the upper and bottom face and laterally at the longitudinal reinforcement position and three linear variable differential transformers (LVDT) at midspan and bearings position.

The typical beam before the execution of the test is represented in Fig. 3.

2.2   Materials

Test on the basic materials (concrete, reinforcing and pre-stressing steel) have been developed. The results of the tests are listed in Tabs. 1, 2 and 3.

Unidirectional Carbon Fiber Reinforced Polymer (CFRP) pultruded laminates with $1.2 \times 100\,mm$ and

Table 2.   Mechanical properties of reinforcing steel.

Mean yielding stress	$f_{ym} = 536\,MPa$
Mean ultimate stress	$f_{tm} = 633\,MPa$

Table 3.   Mechanical properties of pre-stressing steel.

Mean yielding stress	$f_{ym} = 1693\,MPa$
Mean ultimate stress	$f_{tm} = 1895\,MPa$

Table 4.   Mechanical properties of CFRP laminate.

Ultimate stress	$f_{fu} = 2780\,MPa$
Elastic modulus	$E_f = 166000\,MPa$
Ultimate strain	$\varepsilon_{fu} = 1.8\,\%$

Figure 4.   Tensile test on CFRP laminate (configurations before and after the test).

$1.2 \times 80\,mm$ areas were respectively used for ordinary and pre-tensioned strengthening.

Tensile tests on CFRP laminate have been also developed. The results of the tests are listed in Tab. 4.

In Fig. 4 the execution of the tensile test on CFRP laminate is shown.

2.3   Characteristics of the specimens

RC-C beam was the control beam without strengthening.

RC-N beam was strengthened with ordinary CFRP laminate. For all strengthened specimens the concrete surface was initially cleaned with an iron brush and then the surface was covered with a layer of primer. Then the CFRP laminates were applied to the concrete prisms with two-component epoxy adhesive

Figure 5. "U"-jacketing at one end of RC-N beam.

Figure 7. Hydraulic jack for pre-stressing CFRP laminate.

Figure 6. Steel bolted plate anchorages at both ends of RC-EA beam.

Figure 8. Steel bolted plate anchorage for RC-PrEA and PRC-PrEA beams.

with a relatively uniform thickness of about 1 mm. Specimens were prepared in laboratory conditions of constant humidity and temperature. One end of the laminate was "U"-jacketed with CFRP sheets for RC-N beam (see Fig. 5).

RC-EA beam was also strengthened with ordinary CFRP laminate at bottom position but mechanical steel bolted plate anchorages were used at both ends (see Fig. 6).

RC-PrEA beam was strengthened with pre-tensioned CFRP laminate. Prestressing was applied with hydraulic jack at one end of the beam (see Fig. 7) while the other end was anchored with a steel bolted plate (see Fig. 8).

When the desired level of pre-stressing was reached the other end of the laminate was also anchored with steel plate. Pre-stressing strain equal to 0.6% was applied to the laminate for RC-PrEA beam.

PRC-PrEA beam (the only PRC beam) was pre-tensioned with the same technique of the previous one applying pre-stressing strain equal to 0.4%.

3 MAIN RESULTS

In Fig. 9 load vs. midspan deflection diagrams are represented for the five beams.

RC-C diagram showed the typical flexural behaviour of RC beams with (I) pre-cracked, (II) cracked and (III) plastic stages.

RC-N diagram showed a brittle behaviour due to sudden delamination of the CFRP laminate starting from the free end and propagating towards the other.

RC-EA diagram showed a similar behaviour with a higher value of the ultimate load due to end anchorage devices. Intermediate delamination of the CFRP occurred in this case with failure of end anchorage (see Fig. 10).

Failure of beams RC-PrEA and PRC-PrEA was still due to delamination of the CFRP but the action of the anchorages delayed the complete failure. Not only a relevant increment of the ultimate load but also an increment of the load at which the first crack appears, occurred for beams with pre-tensioned laminates

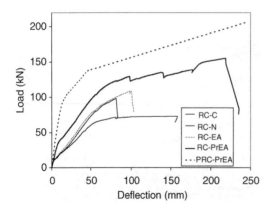

Figure 9.   Load vs. deflection diagrams for the five beams.

Figure 10.   Failure of end anchorage (beam RC-EA).

Figure 11.   Failure of beam PRC-PrEA.

(RC-PrEA and PRC-PrEA) with respect to control beam (RC-C).

In Fig. 11 the delaminated CFRP after failure of PRC-PrEA beam is shown.

In Tab. 5 ultimate values of load and deflection are listed for the five beams.

Table 5.   Ultimate load and maximum deflection for the five beams.

Beam	Ultimate load (kN)	Maximum deflection (mm)	CFRP area (mm²)
RC-C	72.1	158.2	–
RC-N	98.0	81.1	120
RC-EA	107.9	100.4	120
RC-PrEA	155.3	219.2	96
PRC-PrEA	206.1	–	96

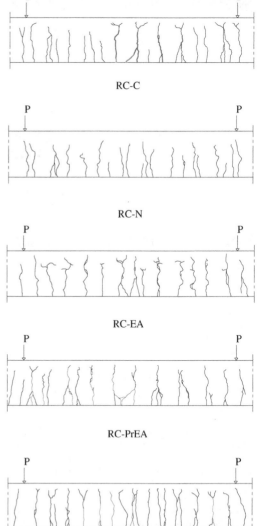

Figure 12.   Cracking patterns at failure.

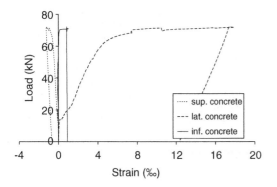

Figure 13.   Load vs. strain diagram for beam RC-C.

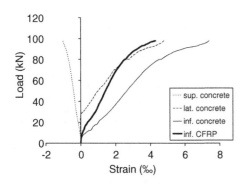

Figure 14.   Load vs. strain diagram for beam RC-N.

In Fig. 12 cracking patterns at failure are shown for the five beams in the zone between the concentrated loads.

More uniform distribution and smaller crack amplitude were detected for strengthened beams with respect to control beam. These effects are more evident for beams with pre-tensioned laminates.

In Fig. 13–17 load vs. midspan strain diagrams are shown for the five beams. In particular strain at superior edge, lateral position at longitudinal reinforcement level and inferior concrete edge are plotted for RC-C beam; strain at superior edge, lateral position at longitudinal reinforcement level, inferior concrete edge and inferior CFRP laminate are plotted for RC-N, RC-EA and RC-PrEA beams; strain of CFRP is plotted for PRC-PrEA beam.

Strain of CFRP laminate (including pre-stressing strain if present) is equal to 0.43% (24% of the ultimate strain) for RC-N beam, 0.58% (32% of the ultimate strain) for RC-EA beam, 1.17% (65% of the ultimate strain) for RC-PrEA beam and 1.35% (75% of the ultimate strain) for PRC-PrEA beam. Therefore pre-tensioning allows a better utilization of the

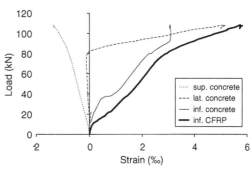

Figure 15.   Load vs. strain diagram for beam RC-EA.

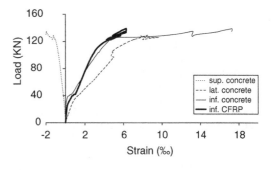

Figure 16.   Load vs. strain diagram for beam RC-PrEA.

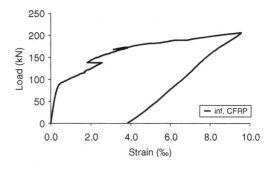

Figure 17.   Load vs. strain diagram for beam PRC-PrEA.

material characteristics with strain values very near to the ultimate.

4   CONCLUSIONS

First results of an experimental investigation on real-scale RC and PRC beams strengthened in flexure with ordinary and pre-tensioned CFRP laminates are shown. The experimental results show that the increments of ultimate capacity vary on the basis of many parameters. In particular, mechanical anchor devices increase ultimate capacity of the structural element delaying delamination.

CFRP pre-tensioning

- increases ultimate capacity of the structural element and load at which first cracking occurs;
- allows reduction of crack amplitudes and more uniform distribution of the cracks;
- allows a better utilization of CFRP material characteristics with strain values very near to the ultimate.

Further investigation is necessary especially about quantification of the increment of capacity given by end anchor devices for which the indications of the principal guidelines (ACI Committee 440 2002, fib T.G. 9.3 2001, CNR-DT 200, 2004) are very scarce or null.

ACKNOWLEDGEMENT

The writers wish to thank Maxfor S.r.l. (Quarto d'Altino, Venice, Italy) for supplying fibers and the adhesion system, and for technical and economical support. They are also writers grateful to E. Bordignon and M. Muner for their experimental work.

REFERENCES

ACI Committee 440, 2002. Guide for the design and construction of externally bonded FRP systems for strengthening concrete structures (ACI 440.2R-02). American Concrete Institute, Farmington Hills, Michigan, USA.

Brena S. F., Bramblett R. M., Wood S. e Kreger M. 2003. Increasing flexural capacity of reinforced concrete beams using carbon fiber-reinforced polymer composites. ACI Structural Journal, 100(1): 36–46.

Boschetto G., Pellegrino C., Tinazzi D., Modena C. 2006. Bond behaviour between FRP sheets and concrete: an experimental study. Proc. of the 2nd fib Congress, Neaples, Italy.

Chahrour A., Soudki K. 2005. Flexural response of reinforced concrete beams strengthened with end-anchored partially bonded carbon fiber-reinforced polymer strips, Journal of Composites for Construction, ASCE, 9(2): 170–177.

Consiglio Nazionale delle Ricerche, Commissione incaricata di formulare pareri in materia di normativa tecnica relativa alle costruzioni 2004. Istruzioni per la progettazione, l'esecuzione ed il controllo di interventi di consolidamento statico mediante l'utilizzo di compositi fibrorinforzati. Materiali, strutture in c.a. e c.a.p., strutture murarie. (CNR-DT 200/2004). Roma, Italy.

El-Hacha R., Wight R. G. e Green M. F. 2003. Innovative system for prestressing fiber-reinforced polymer sheets. ACI Structural Journal, 100(3): 305–313.

El-Hacha R., Wight R. G. e Green M. F. 2004. Prestressed carbon fiber reinforced polymer sheets for strengthening concrete beams at room and low temperatures. Journal of Composites for Construction, ASCE, 8(1): 3–13.

El-Mihilmy M. T., Tedesco J. W. 2001. Prediction of anchorage failure for reinforced concrete beams strengthened with fiber-reinforced polymer plates. ACI Structural Journal, 98(3): 301–314.

fib Task Group 9.3 2001. Externally bonded FRP reinforcement for RC structures. fib bulletin 14, Lausanne, Switzerland.

Malek M., Saadatmanesh H., Ehsani M. 1998. Prediction of failure load of RC beams strengthened with FRP plate due to stress concentration at the plate end. ACI Structural Journal, 95(1): 142–152.

Pellegrino C., Boschetto G., Tinazzi D., Modena C. 2005. Progress on understanding bond behaviour in RC elements strengthened with FRP. International Symposium on Bond Behaviour of FRP in Structures, Hong Kong, China.

Pellegrino C., Modena C. 2002. FRP shear strengthening of RC beams with transverse steel reinforcement. Journal of Composites for Construction, ASCE, 6(2): 104–111.

Pellegrino C., Modena C. 2006. FRP shear strengthening of RC beams: experimental study and analytical modelling. ACI Structural Journal, 103(5): 720–728.

Pellegrino C., Tinazzi D., Modena C. 2004. "Sul confinamento di elementi in c.a. soggetti a compressione", Giornate AICAP 2004, Verona.

Taljsten B. 1997. Strenghtening of beams by plate bonding. Journal of Materials in Civil Engineering, 9(4): 206–212.

Tan K., Tumialan G., Nanni A. 2003. Evaluation of CFRP systems for the strenghtening of RC slabs. University of Missouri-Rolla, CIES 02-38, Final Report.

Tinazzi D., Pellegrino C., Cadelli G., Barbato M., Modena C., Gottardo R. 2003. An experimental study of RC columns confined with FRP sheets, Proc. of Structural Faults & Repair, 10th Int. Conf., London, UK.

Triantafillou T. C., Deskovic N. e Deuring M. 1992. Strengthening of concrete structures with prestressed fiber reinforced plastic sheets. ACI Structural Journal, 89(3): 235–244.

Wight R. G., Green M. F. e Erki A. 2001. Prestressed FRP sheets for poststrengthening reinforced concrete beams. Journal of Composites for Construction, ASCE, 5(4): 214–220.

Yu P., Silva P.F., Nanni A. 2004. Flexural performance of RC beams strengthened with prestressed CFRP sheets. University of Missouri-Rolla.

Fracture Mechanics of Concrete and Concrete Structures – Design, Assessment and Retrofitting of RC Structures – Carpinteri, et al. (eds)
© 2007 Taylor & Francis Group, London, ISBN 978-0-415-44616-7

Long-term properties of bond between concrete and FRP

C. Mazzotti & M. Savoia

DISTART – Structural Engineering, University of Bologna, Bologna, Italy

ABSTRACT: Results of an experimental campaign concerning long-term behavior of bond of FRP plates bonded to concrete are presented. CFRP plates with three different bonded lengths (from 100 mm to 400 mm) have been subject to long term loads by using a mechanical system able to apply a constant traction force. Tests have been carried out in a climatic room with a constant temperature T = 20°C and humidity RH = 60%. Strain profile evolution with time along the bonded length have been recorded by using a number of strain gages placed along the plates. An appreciable stress redistribution with time along the plate has been observed. A simplified visco-elastic analytical model has been finally used to reproduce experimental findings. Good matching between experimental and numerical data has been found.

1 INTRODUCTION

Durability with time of FRP strengthening interventions on reinforced concrete structures strongly depends on interface behaviour between concrete and FRP.

Effectiveness of strengthening can be remarkably reduced due to rheological properties of materials involved (concrete, adhesive, FRP). An important role is also played by the temperature variation. In this framework, only few information can be found in literature concerning the time behaviour of strengthened civil structures (Plevris & Triantafillou 1994; Savoia et al. 2005, Diab & Wu 2006).

In the present paper, results of an experimental campaign concerning long-term behaviour of bond between concrete prisms and FRP plates are presented. Three different bonded lengths (from 100 mm to 400 mm) have been adopted. Seven to eleven strain gauges (depending on bonded lengths) along FRP plates have been used to measure longitudinal strains. A mechanical system able to apply a traction force constant in time to the extremity of plates has been designed. In order to eliminate strain thermal drift, long-term tests have been carried out in climatic room with standard ambient conditions. Strains have been measured during time by using an automatic control system. Strain profile evolutions with time along the bonded length have been recorded. At the moment, time duration of tests is about 650 days.

It is shown that a significant redistribution of shear stresses along the anchorage occurs due to creep deformations at the interface level.

A simplified model has been finally presented, able to describe evolution with time of strain and shear stress distribution along the plate. It is based upon a linear interface model properly modified by introduction of effective longitudinal and shear moduli. Numerical results are in good agreement with experimental data, especially in the initial part of bonded plates, where shear slips are higher.

2 GEOMETRY AND MECHANICAL PROPERTIES OF SPECIMENS

2.1 Specimen preparation

Long-term behaviour of CFRP plates bonded to concrete surfaces has been investigated by considering three different bonded lengths (100, 200 and 400 mm) and two concrete specimens (two anchorages for each specimen).

Concrete specimen dimensions were $150 \times 200 \times 600$ mm. They were fabricated using normal strength concrete. Concrete was poured into wooden forms, externally vibrated. The top was steel – troweled. Five 15 cm – diameter by 30 cm – height standard cylinders were also poured and used to evaluate mechanical properties of concrete, according to Italian standards. Specimens were demoulded after 24 hours and covered with saturated clothes for 28 days; subsequently, they were stored at room temperature and uncontrolled humidity inside the laboratory.

Mean compressive strength $f_{cm} = 52.6$ MPa from compression tests and mean tensile strength $f_{ctm} = 3.81$ MPa from indirect traction tests have been obtained. Mean value of elastic modulus $E_{cm} = 30700$ MPa and Poisson ratio $\nu = 0.227$ have been obtained from preliminary tests. Concrete specimens were sufficiently old before tests (two years) so that shrinkage strains can be neglected.

As for composite plates, CFRP Sika CarboDur S plates, 80 mm wide and 1.2 mm thick, have been used. According to technical data provided by the producer, plates have carbon fiber volumetric content equal to 70 percent and epoxy matrix. Minimum tensile strength is 2200 MPa and mean elastic modulus is $E_p = 165000$ MPa.

Two opposite surfaces of each concrete block have been grinded with a stone wheel to remove the top layer of mortar, until the aggregates were visible (approximately 1 mm). Plates have been bonded to surfaces by using a 1.5 mm thick layer of two – components Sikadur – 30 epoxy adhesive, having mean compressive strength of 95 MPa and mean elastic modulus $E_a = 12800$ MPa, according to producer data. No primer before bonding has been used. Two CFRP plates have been bonded to opposite faces of each specimen, with different bonded lengths (concrete *Block P*1: B.L. = 100, 200 mm; *Block P*2: B.L. = 200, 400 mm), see layout reported in Figure 1a. Curing periods of anchorages of blocks $P1$ and $P2$ were, respectively, 15 and 20 days prior to testing.

2.2 *The bond surface*

Two different positions of bond surface on the concrete specimen were considered in previous experimental tests and FE numerical investigations (Mazzotti et al. 2004), i.e., starting from the loaded end of the specimen or at a given distance from it. These studies showed that if bonding of CFRP plate starts close to the front side of concrete specimen, very high tensile stresses are present in this concrete portion and, typically, an early failure occurs in delamination tests due to concrete cracking of a prism with triangular section. On the contrary, when plate bonded length starts far from the front side, tensile stresses are much smaller and FRP – concrete interface is subject to prevailingly shearing stresses up to delamination failure. Behaviour of the interface is not affected by boundary effects and this test set-up is than more appropriate to obtain data for calibration of interface laws.

In the present experimental investigation, plate bonded length starts 100 mm far from the loaded end of specimen (Fig. 1a).

3 EXPERIMENTAL SETUP AND INSTRUMENTATION

Experimental setup is depicted in Figure 1a. Concrete block was positioned on a rigid frame with a front side reaction element (60 mm height) in order to prevent longitudinal displacements. Free ends of plates were mechanically clamped within a two steel plate systems. A double hinge system allowed for rotations

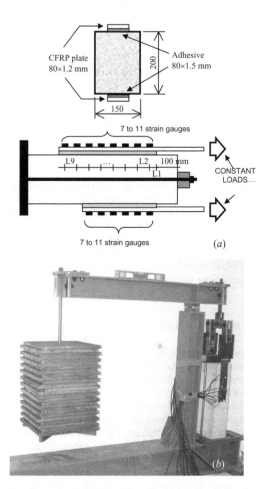

Figure 1. (*a*) Geometry of specimens and (*b*) experimental setup for long – term tests.

around transverse axis. Traction force was applied to steel plates by using a mechanical frame allowing for magnification of applied force at the opposite side of horizontal arm. Force was given as weight of a number of steel plates (Fig. 1b). Amplification factor of mechanical system is about 4. Tests were performed by prescribing constant values of applied force.

Along CFRP plates bonded to concrete, series of seven – to – eleven strain gauges (depending on the plate length) were placed, along the centerline. For each bonded length, spacing between strain gauges is reported in Table 1.

4 RESULTS OF EXPERIMENTAL TESTS

4.1 *Loading program*

Two different loading conditions have been considered: plates of specimen $P1$ have been subject to

Table 1. Distances (mm) between strain gauges along the FRP plate, for different bonded lengths (B.L.).

B.L.	L1	L2	L3	L4	L5	L6	L7	L8	L9	L10	L11
100	10	10	10	15	15	15	15				
200	10	10	10	20	20	20	30	30	40		
400	10	10	20	20	40	40	50	50	50	50	50

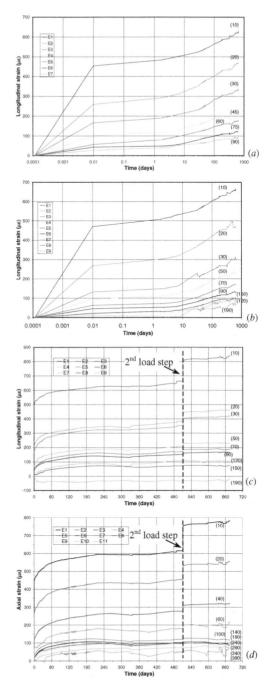

Figure 2. Time evolution of strains in FRP plates at different positions along the anchorage: specimen P1 – (a) B.L. = 100 mm, (b) B.L. = 200 mm; specimen P2 – (c) B.L. = 200 mm, (d) B.L. = 400 mm. Number in parenthesis indicates the distance from the initial section of the anchorage.

a traction force of 12.30 kN (about 50 percent of maximum transmissible load, according to delamination tests on analogous specimens, see Mazzotti et al. 2004, 2005b), followed by 690 days at constant loading. Plates bonded to specimen P2 have been subject to the same sustained load (12.30 kN) for 510 days; after that, a second load step of 3.80 kN has been applied and maintained for further 160 days. Longitudinal strains along the plate at different loading levels and different time intervals have been recorded by an automatic computer system.

Reliability of symmetric double plate system (Fig. 1a), adopted in this experimental campaign, has been verified by comparing strain distribution measured along the plates during loading phase with analogous results obtained from similar specimens, tested according to conventional set-up (Mazzotti & Savoia 2005).

4.2 Long term behaviour

Figures 2a–d show time evolution of longitudinal strains along the plates for 100–200 mm (specimen P1) and 200–400 mm (P2) bonded lengths, respectively. For each strain gauge, distance from the initial section of the ancorage is reported in parenthesis. Time is reported adopting a logarithmic scale for plates bonded to specimen P1: behaviour after small time intervals can then be observed. On the contrary, for plates bonded to specimen P2, linear time scale has been adopted (Figs 2c, d); in this way, behaviour after second load increment (at 510 days) can be observed. It is worth noting that, due to stability of climatic condition and having adopted an automatic data acquisition system during the whole test, smooth curves have been obtained, even for those strain gauges whose strain variation is small.

Reliability of long-term loading system and repeatability of experimental results have been already verified in (Mazzotti & Savoia 2005), where strains evolution with time of two plates with same bonded length of 200 mm (one bonded to specimen P1 and one to specimen P2) have been compared. Very good agreement has been obtained.

From Figures 2a–d, it can be observed that at low stress levels, plates with different bonded length exhibit a similar behaviour: rate of strain starts

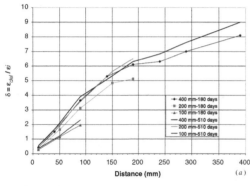

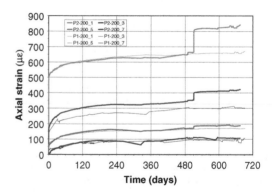

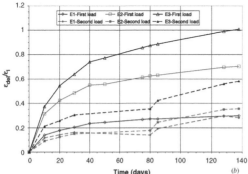

Figure 3. Ratio between delayed and instantaneous strains: (a) along the plate at 180 and 510 days of loading; (b) evolution with time for first three strain gauges (B.L. = 200 mm).

Figure 4. Superposition of strain evolution with time of strain gauges 1,3,5,7 from B.L. = 200 mm of specimens $P1$ and $P2$.

decreasing immediately after load application (as in classical creep tests on concrete specimens). Adopting time log scale, stabilization of delayed strain rate is typically considered when a linear increase of strain is attained. This behaviour has been obtained after about three months of constant loading and can be considered as the beginning of steady-state increase of delayed strain. After 670 days of loading, no remarkable changes in curve slope (with time in log scale) have been observed.

A delayed strain coefficient has been defined as the ratio between delayed strain and instantaneous strain:

$$\delta(t) = \frac{\varepsilon_{del}(t)}{\varepsilon_i}. \tag{1}$$

Delayed strain can be due to both creep compliance and stress variation in the plate, caused by stress redistribution. Of course, if stress distribution was constant in the anchorage during time (no redistribution), and considering linear creep behaviour for both adhesive and concrete cover, delayed strain coefficient at a given time should be equal for all strain gauge positions.

Figure 3a shows, for three bonded lengths, variation of delayed strain coefficient with the distance from

the traction side after 180 and 510 days of loading. Delayed strain coefficient increases significantly far from loaded plate side: delayed deformation is up to 9 times greater than instantaneous deformation. Two types of behaviour can be observed for each bonded length: close to loaded end, delayed strain coefficient increases almost linearly with distance, and slope increases with increasing of bonded length (from 200 to 400 mm). At a distance from loaded end of about 150–200 mm, curves reach a second linear branch with smaller slope up to the end of plates (for 200 mm and 400 mm cases).

Delayed strain coefficients obtained after 510 days of loading are only slightly higher than correspondent values after 180 days, suggesting that both stress redistribution phenomena and creep deformation change only slightly after six months from initial loading.

For specimen $P2$, delayed strain coefficient from Equation 1 has been obtained for two distinct loading steps the specimen was subject to. With reference to bonded plate with B.L. = 200 mm, evolution with time of coefficient $\delta(t)$ has been reported for first three strain gauges in Figure 3b, both after initial (solid lines) and second load step (dashed lines). Time evolution of delayed strain coefficient δ obtained for first strain gauge is very similar between two loading steps; this is due to position of instrument, at the very beginning of bonding length, where influence of viscous properties of plate-concrete interface is very low, creep behaviour of CFRP plate being also very small. On the contrary, for strain gauges far from initial section, a remarkable reduction of coefficient $\delta(t)$ after 2nd load step can be observed. As a confirmation, Figure 4 shows superposition of strain evolution with time of strain gauges belonging to two plates with B.L. = 200 mm; solid lines refer to specimen $P2$ subject to second load step Δp^2 after 510 days from first loading Δp^1 ($\Delta p^2 = 0.35 \Delta p^1$), whereas thin lines refer to specimen $P1$ subject to the initial load step only.

Finally, Figures 5a – d show strain profile along plates at different times, for four different specimens. Strain profiles are quite similar, showing decreasing values of delayed strains with time moving away from the traction side; on the contrary, after a distance of 150–180 mm (realistic value of transfer length), delayed strains increase with time is almost constant.

5 DISCUSSION OF RESULTS

Variation of delayed strain coefficient along the FRP plates is the result of different phenomena.

First of all, CFRP plates exhibit creep deformation when subject to long – term loading. Nevertheless, according to experimental results reported in the literature (Ascione et al. 2005), CFRP plates adopted in the present experimental study is practically insensitive to creep strains, being less than 2 per cent even for medium stress levels.

Secondly, adhesive and concrete cover constituting the interface where load is transferred from plate to concrete are subject to creep strains, increasing compliance of the interface. Increasing with time, plate – concrete slip provides for a shear stress redistribution along the interface: high shear stresses close to loaded end decrease and a longer length of interface is subject to shear stress transfer.

Third, due to very high shear stresses at the adhesive and concrete cover level, close to the loaded end creep phenomenon may be highly non linear. Hence, compliance increase at the beginning of the anchorage is much higher than far from it.

Evolution with time of shear stress redistribution and creep deformation at the interface level (which is variable along the anchorage) may explain the experimental results described in the previous section. In fact, far from the loaded section of the plate, normal stresses in the plate increase and, consequently, the same behaviour is exhibited by axial strains.

Moreover, shear stress redistribution is more evident in the case of long bonded lengths, i.e. longer than transmission length (about 150–180 mm). For this reason, the rate of increase of axial strains in FRP plate is higher for the longest length (400 mm), see Figure 3a. At distance from loaded end greater than 200 mm, shear stresses are small also after long term loading application, and, consequently, the second branch in Figure 3a has smaller slope.

6 SIMPLIFIED CREEP MODEL

In order to describe complex phenomena reported in previous paragraph, visco-elastic constitutive behavior for both interface and CFRP plate must be considered.

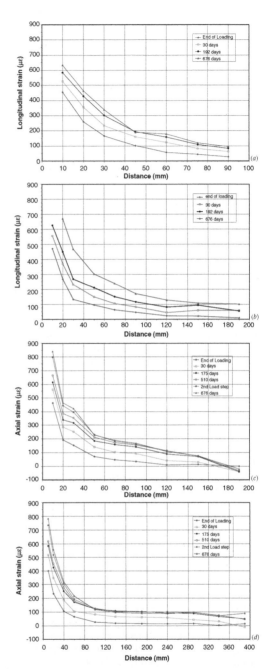

Figure 5. Evolution with time of strains along the bonded length, at different times. Specimen P1: (a) B.L. = 100 mm, (b) B.L. = 200 mm. Specimen P2: (c) B.L. = 200 mm, (d) B.L. = 400 mm.

Due to shear stress redistribution with time, a convolution equation should be solved (Bazant, 1988). The simplest way to solve approximately the convolution integral is the Effective Modulus (EM) method.

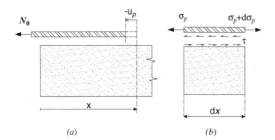

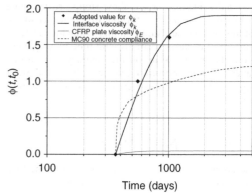

Figure 7. Creep functions adopted for interface (shear) and CFRP plate (axial strain), compared with MC 90 creep function.

(a)   *(b)*

Figure 6.   Bond–slip model for FRP-concrete interface.

According to this theory, long term behavior of a visco-elastic material with a moderate time variation of stress can be described by using pseudo-elastic relations. For the problem at hand, neglecting deformation of concrete specimen, constitutive equations can be written as:

$$\tau - K_p(t)s_p, \quad \sigma_p - E_p(t)\varepsilon_p, \tag{2}$$

where τ and s_p are interface shear stress and slip, respectively, while σ_p and ε_p are plate axial stress and strain. Moreover, $K_p(t)$ and $E_p(t)$ are interface shear effective stiffness and plate longitudinal effective modulus which can be expressed as:

$$K_p(t) = \frac{K_p(t_0)}{1 + \phi_K(t,t_0)} \quad E_p(t) = \frac{E_p(t_0)}{1 + \phi_E(t)}, \tag{3}$$

where $K_p(t_0)$, $E_p(t_0)$ are elastic moduli at the time of loading t_0 and $\phi_K(t,t_0)$, $\phi_E(t)$ are coefficients of viscosity of FRP-concrete interface and CFRP plate, respectively.

The main advantage of EM method is that it is based on linear pseudo-elastic constitutive relations, which can be easily introduced in a bond-slip model.

To this purpose, a linear bond – slip model is adopted. Notation adopted for displacements and stress resultants is reported in Figure 6. Plate is subject to axial deformation only, i.e., bending of FRP reinforcement is neglected. This assumption is valid in the present case, due to negligible bending stiffness of FRP reinforcement with respect to concrete specimen counterpart. Since concrete specimen dimensions are much greater than FRP-reinforcement, strain in concrete can be neglected ($\varepsilon_c = 0$), being very small compared to plate strains. Hence, FRP-concrete slip coincides with FRP axial displacement, i.e. $s_p \cong u_p$.

Governing equations are equilibrium, constitutive and compatibility conditions, which can be written in the form:

$$\frac{du_p}{dx} = \frac{\sigma_p}{E_p(t)}, \quad \frac{d\sigma_p}{dx} = \frac{\tau}{t_p}, \tag{4}$$

where t_p is plate thickness. Derivation of Equation 2b yields:

$$\frac{d^2 s_p}{dx^2} = \frac{d\varepsilon_p}{dx} = \frac{1}{E_p(t)} \frac{d\sigma_p}{dx}. \tag{5}$$

Substituting Equation 2a in 5, governing differential equation and boundary conditions can be written as (see Fig. 6a):

$$\begin{cases} \dfrac{d^2 s_p}{dx^2} - \dfrac{1}{E_p(t)t_f} K_p(t)s(x) = 0 \\ N(0) = N_0 \\ s_p(+\infty) = 0 \end{cases} \tag{6}$$

As for bonding conditions, load application at the initial section ($x = 0$) and null slip at $x \to +\infty$ have been considered. Linear differential Equation 6 can be given a closed form solution, so obtaining plate strain and shear stress, respectively:

$$\varepsilon_p(x,t) = \frac{N_0}{E_p(t)A_p} e^{-\alpha(t)x}$$

$$\tau(x,t) = \alpha(t) \frac{N_0}{b_p} e^{-\alpha(t)x} \tag{7}$$

where

$$\alpha(t) = \left(\frac{1}{E_p(t)t_p} \right)^{1/2} K_p(t)^{1/2}. \tag{8}$$

In order to compare experimental results with numerical predictions, initial tangent modulus has been

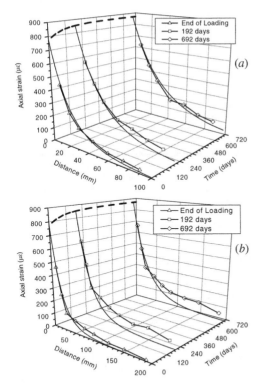

Figure 8. Evolution with time of strains along the bonded length, at different times. Comparison with numerical prediction for specimen $P1$ – (a) B.L. = 100 mm, (b) B.L. = 200 mm.

evaluated from stiffness of interface in the linear range, so obtaining $K_p(t_0) = 450\,\text{N/mm}^3$. As for the interface creep coefficient ϕ_K, different values have been obtained at different time intervals from loading to match with experimental results. An exponential type law has been then calibrated by mean square fitting, so obtaining the expression:

$$\phi_K(t,t_0) = 1.6 \cdot (1 - e^{-\frac{\Delta t}{155}}). \quad (9)$$

For time evolution of CFRP plate creep coefficient $\phi_E(t)$, the following simplified expression has been adopted:

$$\phi_E(t) = 0.05 \cdot (1 - e^{-\frac{\Delta t}{130}}), \quad (10)$$

where asymptotic value 0.05 has been considered from creep tests on CFRP plates (Ascione et al. 2005).

Figure 7 shows both simplified empirical laws for creep coefficients, together with creep coefficient $\phi(t,t_0)$ defined according to Model Code 90 for the

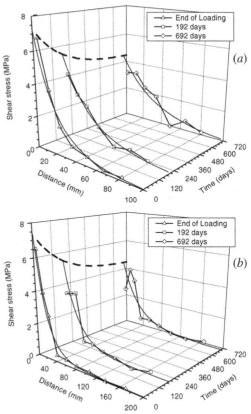

Figure 9. Evolution with time of shear stress along the bonded length, at different times. Comparison with numerical prediction for specimen $P1$ – (a) B.L. = 100 mm, (b) B.L. = 200 mm.

considered concrete. It can be noted that interface compliance is higher than concrete creep, due to additional contribution related with viscosity of adhesive.

Figure 8 shows strain distribution along bonded plate for B.L. = 100, 200 mm of specimen $P1$ at different times from initial loading. In both cases very good correlation between experimental data and numerical predictions have been obtained.

Under constant load, the model properly describes strain distribution along the shortest bonded plate (B.L. = 100 mm), which is smaller than transfer length (100 ÷ 150 mm), and is characterised by a stress redistribution with time throughout all its entire length. In spite of a very small strain variation with time at the beginning of bonded part (dashed line at $x = 0$), due to creep deformation of FRP plate only, an appreciable strain increase can be observed at the opposite end of the plate.

In the second case (B.L. = 200 mm), correlation is good in first part of bonded plate (60 ÷ 90 mm), where stress redistribution occurs. On the contrary, simplified model provides for a poor prediction in the final part of plate, being unable to describe the strain increase recorded by strain gauge with time.

Finally, comparison between experimental shear stress distribution along plates and numerical predictions (Fig. 9) is always very good for both bonded lengths. Experimental shear stresses have been evaluated starting from strain distribution through equilibrium relations (e.g. Mazzotti et al. 2005a) and considering an effective CFRP plate elastic modulus $E_p(t)$, reduced according to expression:

$$E_p(t) = \frac{E_p}{1 + \phi_E}. \tag{11}$$

Stress redistribution with time gives smaller shear stresses close to loaded end and elongation of bonded length subject to shear (effective length).

Effective length is in fact proportional to $\alpha(t)N_0/b_p$, with $\alpha(t)$ defined in Equation 8. This coefficient increases with time because, as shown in Figure 7, creep compliance increase with time of interface K_p is greater than creep compliance of FRP plate itself $(E_p(t))$.

AKNOWLEDGEMENT

The authors would like to thank the Sika Italia S.p.A. for providing CFRP plates and adhesives for experimental tests. The financial supports of (Italian) Department of Civil Protection (RELUIS 2005 Grant – Task 8.2: "Delamination for cyclic actions in r.c. and masonry structures" and C.N.R. (National Council of Research), PAAS Grant 2001, are gratefully acknowledged.

REFERENCES

Ascione, F., Berardi, V.P., Feo, L. & Giordano, A. 2005. Experimental test on long term behaviour of CF pultruded elements (in italian), *Proceedings of XVII AIMETA*, Florence, Italy, September: 1–11 (on CD).

Diab, H.M., & Wu, Z. 2006. Constitutive model for time dependent bonding and debonding along FRP-concrete interface, *Proceedings of Third international Conference on FRP Composites in Civil Engineering (CICE 2006)*, Miami, Florida, USA, December: 25–28.

Mazzotti, C., Ferracuti, B. & Savoia, M. 2004. An experimental study on FRP – concrete delamination, *Proceedings of FraMCoS – 5*, Li et al., eds., Vail, Colorado, U.S.A., V. 2: 795–802.

Mazzotti, C., Savoia, M. & Ferracuti, B. 2005a. A new set-up for FRP-concrete stable delamination test. *Proceedings of FRPRCS7*, Eds. Shield et al., Kansas City, USA, November 2005, 1: 165–180.

Mazzotti, C., Savoia, M. & Ferracuti, B. 2005b. FRP – concrete delamination results adopting different experimental pure shear test setups, *Proceedings of ICF XI*, Turin, Italy, March 2005: 1–6 (on CD).

Mazzotti, C. & Savoia, M. 2005. Long term properties of bond between concrete and FRP. *Proceedings of BBFS05-Bond Behaviour of FRP in Structures*, Chen JF et al. eds., Hong Kong, December 2005: 539–545.

Plevris, N. & Triantafillou, T.C. 1994. Time-dependent behaviour of RC members strengthened with FRP laminates, *J. Struct. Engineering, ASCE*, 120: 1016–1042.

Savoia, M., Ferracuti B. & Mazzotti C. 2005. Long-term creep deformation of FRP-plated r/c tensile members, *J. of Composites for Construction, ASCE*, 9(1): 63–72.

Fracture Mechanics of Concrete and Concrete Structures – Design, Assessment and Retrofitting of RC Structures – Carpinteri, et al. (eds)
© 2007 Taylor & Francis Group, London, ISBN 978-0-415-44616-7

Mode II fracture energy and interface law for FRP – Concrete bonding with different concrete surface preparations

M. Savoia, C. Mazzotti & B. Ferracuti

DISTART – Structural Engineering, University of Bologna, Bologna, Italy

ABSTRACT: Results of an experimental campaign on delamination of CFRP sheets and plates bonded to concrete are presented. In particular, the effect of different surface preparations (grinding or, alternatively, sand blasting) is investigated. Six specimens have been tested by using a particular experimental set-up where the CFRP reinforcement is bonded to concrete and its back side is fixed to an external restraining system. The adopted set-up allows for a stable delamination process, with progressive transition between two limit states (perfect bonding and fully delaminated plate). Both strain gages along the FRP plate and LVDT transducers have been used. Experimental results showed that the technique for surface preparation may have a significant influence on the value of mode II fracture energy. Starting from experimental data, non linear interface laws have been calibrated. It is shown that, when concrete surface has been prepared by sand blasting, high values of fracture energy are attained with a smaller value of peak shear stress and significantly higher slips. Morever, FRP – concrete interface law exhibits a less brittle behavior of the softening branch.

1 INTRODUCTION

When using FRP (Fibre Reinforced Plastic) plates or sheets to strengthen r.c. beams, bonding is very important. Since delamination is a very brittle failure mechanism, it must be avoided in practical applications. Bonding depends on mechanical and physical properties of concrete, composite and adhesive.

Moreover, it is well known that concrete surface preparation and bonding technique are very important for a correct application, in order to avoid premature failures. In recent Italian CNR DT 200/2004 Guidelines (2006) for design of strengthening interventions with FRP, when FRP application is done using certified methods, smaller design coefficients reducing bond strength are allowed. Nevertheless, very few studies can be found on the effect of different preparations of concrete surface before bonding (Toutanji & Ortiz 2001). In Matana et al. (2005), bond strength in peel-off tests has been measured; using different surface treatments, laser profilometer has been used to measure concrete surface roughness (Galecki et al. 2001).

Recent design Guidelines consider mode II fracture energy as the most important material parameter when estimating delamination failure load. Following analogous proposals for mode II fracture energy of concrete, fracture energy of FRP-concrete interface is written as a function of concrete strength through a parameter which must be calibrated experimentally (Ferracuti et al. 2007a). Nevertheless, since that expression

has been calibrated considering all delamination data together, with concrete strength as the only material parameter, coefficient of variation of results is very high and, consequently, design value of fracture energy is much lower than mean value. In order to reduce data scattering, more information of dependence of fracture energy on material parameters are necessary.

In the present paper, a set of experimental results on delamination of FRP reinforcements bonded to concrete is presented. Both CFRP plates and sheets have been tested. Moreover, three different treatments of concrete surface before bonding have been considered: grinding with two types of stone wheel giving different surface roughness or, alternatively, sand blasting.

Tests have been performed by adopting a new experimental set-up, recently developed by the authors, allowing for a stable delamination process. Specimen back-side is fixed to an external retaining system, i.e. concrete and CFRP reinforcement in that section have null displacement. Then, a very stable delamination process occurs at constant applied force, corresponding to delamination force of an anchorage of infinite length. That value is then used to define mode II fracture energy of interface law.

Moreover, a number of closely spaced strain gages is placed along the FRP plate to measure strains. Then, starting from experimental data, average shear stresses between two subsequent strain gages and corresponding shear slips have been computed. These data have been used to calibrate non linear interface laws, according to the procedure described in Savoia et al. (2003) and Ferracuti et al. (2007b).

It is shown that sand blasting of concrete surface gives higher values of mode II fracture energy and correspondingly, of delamination load. Nevertheless, peak shear stress is lower, and higher compliance of FRP-concrete interface is observed. Moreover, softening branch after the attainment of peak shear stress is less brittle. When grinding with stone wheel is used to remove mortar from concrete surface, roughness of the surface and, correspondingly, shear fracture energy are larger than with sand blasting.

Finally, specific fracture energy for sheets bonded to concrete is significantly higher than obtained in the case of CFRP plates.

2 MODE II SHEAR FRACTURE FOR FRP – CONCRETE BONDING

According to theoretical fracture mechanics, fracture propagation direction is governed by the criterion of the maximum release rate (van Mier 1997), as a consequence of basic laws of thermodynamics. Then, usually cracks in concrete specimens are related to tensile failure (Mode I fracture energy being the minimum), and crack propagation direction from the notch tip is taken as normal to the maximum principal stress (van Mier 1997). Also in shear-loaded beams with a start notch in the Mode II direction (Arrea & Ingraffea 1982), if a wide zone of beam is subject to shear, crack typically deviates from Mode II fracture direction to that of maximum principal stress.

Nevertheless, Mode II or shear fracture failures may occur when a narrow region is subject to high shear stresses. For instance, in Bazant & Pfeiffer (1986), shear fracture has been observed in shear-loaded beams with starting notches similar to those tested by Arrea & Ingraffea (1982), but with much smaller distance between applied shear forces: in this case, according to the criterion of the maximum release rate, cracks cannot deviate into a low stress zone of the material, because they would release little energy.

As clearly described in Bazant & Pfeiffer (1986), shear fracture initially forms as a zone of inclined tensile microcracks. Full shearing failure then requires inclined struts between microcracks be finally crushed in compression.

The failure mechanism at the micro-level explain why Mode II fracture energy G_f^{II} is far larger than Mode I (G_f^I), even 25 times according to Bazant & Pfeiffer (1986). For concrete specimens, G_f^I can be considered as a basic material constant whereas G_f^{II} is not, since it can be calculated on the basis of G_f^I, tensile and concrete strength and crack band width.

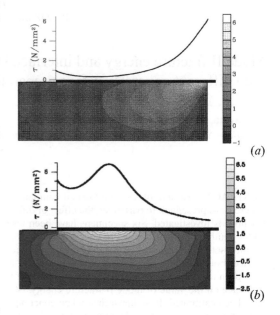

Figure 1. Shear stresses in concrete specimen for FRP plate subject to axial loading calculated at (*a*) 40 percent and (*b*) 100 percent of delamination load (from Freddi et al. 2004).

When a plate is bonded to a concrete specimen and is subject to axial load up to failure, Mode II shear failure occurs (Buyukozturk et al. 2004). In fact, only a small layer of concrete close to the interface is subject to very high shear stresses (see Fig. 1), and criterion of the maximum release rate requires that fracture propagates along it. During delamination, the portion of concrete where shear stresses are transmitted is in fact very small, about 20 centimetres long and 3-5 centimetres depth.

Failure mechanism is similar as described before: according to Mohr-Coulomb criterion, inclined microcracks start locally in mode I condition in the small external layer of concrete because tensile strength is much lower than in adhesive (see Fig. 2). Inclined cracks cannot propagate more than few millimetres inside the concrete specimen, because stresses decrease very rapidly with depth from FRP-concrete interface. Then, a series of inclined struts clamped to concrete substrate are subject to compression and bending. Final failure can be due concrete crushing in compression or transverse cracking on tensile side of concrete struts, depending on dimensions of struts, and a corrugated debonding surface parallel to the interface is typically detected after failure.

As confirmed by several experimental studies (Lu et al. 2005, Ferracuti et al. 2007a), fracture energy due to FRP – concrete debonding is much higher than Mode I fracture energy of concrete. Moreover, Mode II fracture energy may be higher than for plain

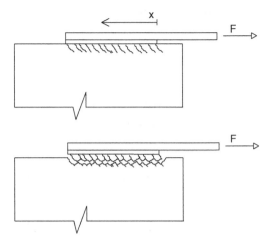

Figure 2. Mode II failure mechanism at the interface level between concrete and FRP reinforcement.

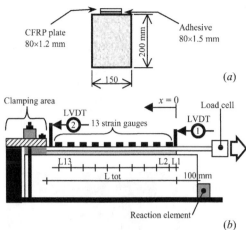

Figure 3. Experimental set-up: (a) Specimen transverse section and (b) side view with instrument positions and CFRP plate clamping system.

concrete, due to penetration of adhesive in the concrete external layer.

Characteristics of debonding failure mechanism explain why mechanical properties of the adhesive, adhesive – concrete compatibility and concrete surface preparation before adhesive application may be very important to increase maximum load against debonding.

Statistical studies on fracture energy of FRP-concrete interface revealed very high coefficient of variation between results obtained from different specimens having same concrete tensile and compression strengths (Ferracuti et al. 2007a), whereas variation is much lower if the same surface preparation and adhesive are adopted.

3 THE EXPERIMENTAL TESTS

3.1 *Mechanical properties of concrete and FRP composites*

Plate – concrete bonding with different surface preparation has been investigated. CFRP sheets and plates have been bonded to concrete blocks and subject to increasing values of axial force up to complete delamination.

Concrete block dimensions were $150 \times 200 \times 600$ mm. They were fabricated using normal strength concrete. Concrete was poured into wooden forms and externally vibrated. The top was steel-troweled.

Mean compressive strength $f_{cm} = 52.7$ MPa from compression tests and mean tensile strength $f_{ctm} = 3.81$ MPa from Brasilian tests have been obtained on cylinders at an age of 20 months. Mean value of elastic modulus was $E_{cm} = 30700$ MPa.

For composites plates, CFRP sheets and plates have been used (Fig. 3a); in Table 1, type and properties of reinforcement have been reported.

A total of ten delamination tests has been performed, two for each type of different reinforcement and surface preparation.

In the present experimental investigation, the plate bonded length starts 100 mm from the front side of specimen and total bonded length is 500 mm (i.e. plate covers the whole concrete block length (Fig. 3b). Previous experimental investigations (Mazzotti et al. 2004) showed that this particular test setup provides for a bond stress-slip behaviour less affected by boundary conditions and more representative of the material behaviour far from cracked sections (i.e. as in the case of plate end debonding).

3.2 *Surface preparation*

Three different techniques for surface preparation have been adopted in order to study the effect of the treatment on delamination force and fracture energy of interface law:

Sand Blasting: Concrete surfaces have been sand blasted in order to remove the whole mortar over the aggregates, so obtaining a very rough (and slightly damaged) concrete surface.

Grinding 1: Top surfaces of concrete blocks have been grinded with a stone wheel to remove the top layer of mortar, just until the aggregate was visible (approximately 1 mm); due to very small dimensions of marble powder glued to the wheel, finished surface was very smooth.

Grinding 2: Top surfaces have been grinded by using a different stone wheel, characterized by a

Table 1. Type and properties of FRP reinforcements considered in experimental tests.

Specimen	Type of Reinforcement	Width b_p (mm)	Thickness h_p (mm)	E_p (MPa)	Surface prep.
P5	Sheet Sika	80	0.13	284000	s. b.
P8	Wrap Hex 230C	80	0.13	291000	gr. 1
P9	Plate Sika	80	1.2	197000	s. b.
P1	Carbodur S	80	1.2	165000	gr. 1
P6		80	1.2	195000	gr. 2

Figure 4. Experimental set-up: view of the clamping system for concrete and CFRP at the opposite sides of the specimen.

Table 2. Spacing between strain gauges (mm) along the CFRP plate.

L1	L2	L3	L4	L5	L6	L7	L8	L9	L10	L11	L12	L13
10	20	20	20	30	30	30	30	30	30	30	30	30

Table 3. Levels of applied force (kN) corresponding to FRP – strain profiles in Figure 3.

Specimen	F1	F2	F3	F4	F5	F6	F7	F8	F9	F10	F11	F12
Sheet (P5A)	2	4	6	8	10	12	14	16	18	20		
Plate (P9A)	4	8	12	16	20	24	28	32	38	42	44	46

coarse iron powder able to provides for a rough finished surface.

3.3 The experimental setup

With classical set-ups (Yao et al. 2005, Ferracuti et al. 2006), complete delamination occurs during a snap-back branch. It is not possible to conduct stable measures because delamination is then a dynamic event, due to instantaneous release of elastic energy of the plate when applied load decreases.

A recently developed experimental setup (Mazzotti et al. 2004) for delamination tests has been adopted. Previous studies showed that this setup allows for a stable and controlled delamination, very important to correctly estimate delamination force as well as to measure strains in FRP during delamination.

The concrete block is positioned on a rigid frame with a front side steel reaction element 60 mm high, to avoid global horizontal translation. Moreover a steel apparatus is clamped to the back side of the specimen in order to prevent, in that section, displacements of both concrete and FRP (Fig. 4). Bonded length L_{tot} from initial to clamped section is 350 mm. In order to apply the load, the opposite side of the reinforcement is mechanically clamped with a two steel plates system free to rotate around the vertical axis. Traction force is then applied to the steel plate system by using a mechanical actuator (Fig. 3b). Tests are then performed under displacement control of the plate free end.

3.4 Instrumentation

A load cell is used to measure the applied traction force during the test. Along the CFRP, a series of thirteen strain gauges is placed on the centerline. In Table 2, spacing between strain gauges is reported, starting from the traction side of bonded part of CFRP plate/sheet. Two LVDTs are also placed at the opposite sides of bonded length in order to measure CFRP elongation and to verify the effectiveness of the clamping system (Fig. 3b).

4 RESULTS OF DELAMINATION TESTS

4.1 The experimental results

Tests have been carried out by performing a first load cycle up to 10 kN of traction force, followed by a monotonic loading at a rate of about 0.2 kN/s. During delamination, plate free end displacement rate was about 50 μm/s, with tests conducted under displacement control.

In Figure 5 results obtained for two specimens with CFRP sheet and plate, respectively, bonded to concrete

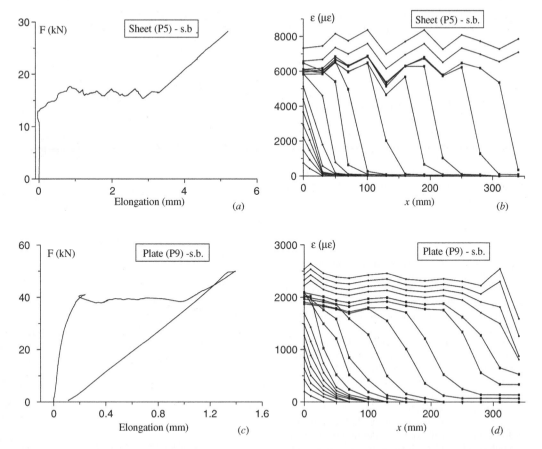

Figure 5.   Results from delamination tests on FRP sheets and plates bonded to concrete: (a,c) force – plate elongation curves, (b, d) strains along FRP reinforcement at different levels of applied force, before and during delamination (s.b. = sand blasting).

surface previously subject to sand blasting treatment are reported (specimens P5A, P9A, respectively). In Figures 5a, c force – CFRP elongation curves of two specimens are reported. Plate elongation is measured as the difference between displacements of initial and end sections of the bonded plate (see Fig. 3). In both cases, three main behaviors can be identified: the first branch is almost linear up to 70–80 percent of maximum transmissible force. Beyond that value, stiffness degradation occurs up to the onset of delamination, when shear strength is attained at the beginning of bonding length. Subsequent delamination occurs at an almost constant value of applied force. Duration of delamination process during test was about 15 seconds. Finally, at the end of complete delamination, the only load-carrying element is the CFRP reinforcement, properly fixed at the extremity, whose behavior is linear elastic.

Some analogies with the tension-stiffening effect can be drawn. Prior to delamination, the specimen is in an uncraked state (usually defined as *State I*,

where both concrete and reinforcement contribute to specimen stiffness). After complete delamination, only CFRP carries the applied load (*State II*); delamination process links these two limit states.

Effectiveness of the clamping system was verified by measurement of absolute displacement of the plate close to restrained section. Maximum displacement recorded by LVDT2 during test was about 0.15 mm.

Longitudinal strain profiles along the CFRP at different loading levels are reported in Figures 5b, d, for sheet and plate bonded to concrete, respectively. Strains at $x = 0$ are obtained from values of external applied force as $\varepsilon_0 = F/E_p A_p$.

Before delamination, FRP strain profile is given by an exponential decay from the loaded extremity up to the end of effective bonding length, where strains are almost zero. When a portion of CFRP is delaminated, no shear stresses are transmitted at the interface level and axial strains are almost constant. In all tests, stable delamination process from the loaded to the clamped end has been clearly observed. Finally, when the whole reinforcement is delaminated, FRP strains are constant along the plate and they grow up linearly together with the applied load.

4.2 Fracture energy of interface law

A further advantage of the proposed set-up is the possibility of obtaining a very stable value of maximum force at delamination F_{max}, corresponding to the asymptotic value of transmissible force by an anchorage of infinite length. Values of delamination forces for ten tests are reported in Table 4. Making use of the following relation (see for instance Brosen 2001, Savoia et al. 2003):

$$F_{max} = b_p \sqrt{2 E_p h_p G_f} ,$$ (1)

where E_p, h_p, b_p are elastic modulus, thickness and width of the reinforcement, respectively, mode II fracture energy G_f of interface law can be obtained from the value of maximum transmissible force F_{max} (see Table 3). Equation (1) is valid for every non linear interface law.

It is worth noting that, if sand blasting is used to remove the top layer of mortar on concrete surface, delamination force is about 15–20 percent greater than grinding the surface with a stone wheel with marble sand (type 1 grinding). In fact, in the first case epoxy resin may penetrate much better in external layer of concrete, even if concrete surface is slightly damaged. On the contrary, with type 1 grinding, concrete surface before adhesive application is very smooth, and interface discontinuity is more pronounced.

This is confirmed by comparison of fracture surfaces after delamination obtained with surface preparation by sand blasting or alternatively stone grinding (see Fig. 6). In the first case (sand blasting), the characteristic inclined cracks on the concrete surface are evident, and a significantly thick and rough layer of concrete is attached to the plate after debonding. In the second case, a thinner layer of concrete is attached to the adhesive, and in some portions there is no concrete at all. Of course, the more irregular the concrete surface after debonding, the higher the mode II fracture energy, as confirmed by results reported in Table 4.

Finally, if grinding is realized with iron powder (type 2), giving a very rough surface with mortar completely removed from aggregate surface, fracture energy is even slightly greater than with sand blasting.

Results reported in Table 4 also show that fracture energy obtained in delamination tests on FRP sheets is 10–25 percent higher than for plates. This is probably due to vanishing in-plane stiffness of FRP sheets with respect to pultruded plates, allowing for a higher redistribution of stresses at the interface level.

Figure 6. Fracture surfaces at the interface level of FRP – concrete specimens after debonding, with different surface preparations before adhesive application: (a) sand blasting and (b) grinding with a stone wheel with marble sand (type 1 grinding).

Table 4. Delamination forces and fracture energy for different concrete surface preparations.

Specimen	Type	F_{max} (kN)	F_{mean} (kN)	G_f (MPa mm)
P5A	Sheet-	16.50	16.95	0.6087
P5B	sand blasted	17.40		
P8A	Sheet-	14.40	14.50	0.4341
P8B	grinded 1	14.60		
P9A	Plate-	37.60	38.35	0.4901
P9B	sand blasted	39.10		
P1A	Plate-	34.50	34.00	0.3903
P1B	grinded 1	33.50		
P6A	Plate-	41.00	39.50	0.5204
P6B	grinded 2	38.00		

5 POST-PROCESSING OF EXPERIMENTAL DATA

Strain data along the FRP plate at different loading levels are used to calculate shear stress – slip data. The origin of the x-axis is taken at the origin of the bonded

Table 5. Parameters of interface laws for different concrete surface preparations.

	$\bar{\tau}$ (MPa)	$\bar{s}$ (mm)	n
Sheet (P5) – sand blasted	5.72	0.069	4.2535
Sheet (P8) – grinded 1	6.22	0.038	3.6862
Plate (P9) – sand blasted	8.00	0.030	3.7063
Plate (P1) – grinded 1	6.43	0.044	4.4370

plate. Considering an elastic behavior for the composite, the average value of shear stress between two subsequent strain gages can be written as a function of the difference of measured strains as:

$$\bar{\tau}_{i+1/2} = \frac{E_p A_p (\varepsilon_{i+1} - \varepsilon_i)}{b_p (x_{i+1} - x_i)}, \tag{2}$$

with A_p, E_p being cross-section and elastic modulus of the composite. Moreover, assuming perfect bonding (no slip) at the end of bonded plate, integration of the strain profile gives the following expression for the slip at x, with $x_i \leq x \leq x_{i+1}$:

$$s(x) = s(x_i) + \frac{(\varepsilon_{i+1} - \varepsilon_i)}{(x_{i+1} - x_i)} \frac{x^2}{2} + \varepsilon_i x, \tag{3}$$

where $s(0) = 0$ is assumed. Average value $\bar{s}_{i+1/2}$ of slip between x_i, x_{i+1} is then computed.

6 CALIBRATION OF A NON LINEAR INTERFACE LAW

According to the procedure described in Savoia et al. (2003), shear stress – slip data are used to calibrate a non linear FRP – concrete interface law. In particular, the interface law recently proposed by the authors (Savoia et al. 2003, Ferracuti et al. 2007b):

$$\tau = \bar{\tau} \frac{s}{\bar{s}} \frac{n}{(n-1) + (s/\bar{s})^n} \tag{4}$$

is adopted, where $\bar{\tau}$ is the peak shear stress, $\bar{s}$ the corresponding slip, and n is a parameter mainly governing the softening branch. Values of $n > 2$ are required in order to obtain positive and finite values of fracture energy.

All shear stress – strain data ($\bar{\tau}_{i+1/2}, \bar{s}_{i+1/2}$) related to both experiments conducted for every specimen are grouped together (Fig. 7). Moreover, fracture energy G_f estimated from the mean value of maximum transmissible force (see Equation 1) is used as a constraint. A least square minimization between theoretical and

experimental shear stress – strain data is then performed to evaluate the three unknown parameters of interface law in Equation 4, i.e., $\bar{\tau}, \bar{s}, n$. Further details on numerical procedure can be found in Savoia et al. (2003) and Ferracuti et al. (2007b).

In Figures 7a–d, shear stress-slip data are reported, together with the corresponding interface law. It is worth noting that the proposed law is in good agreement with experimental data both for slips smaller than $\bar{s}$ and in the softening branch where experimental results are more scattered. Values of interface law parameters obtained from calibration procedure are reported in Table 5. Moreover, in Figure 8, three different interface laws are compared.

With reference to FRP sheet bonded to concrete, it can be verified that sand blasting reduces significantly interface stiffness (about one half) and slightly peak shear stress. Nevertheless, softening branch is less brittle, and corresponding fracture energy is 40 percent greater.

Finally, interface laws obtained from delamination tests on FRP sheet and plates, both with concrete surface sand blasted before resin application, are very different, even is their fracture energies are very close. Plate – concrete interface is much stiffer than sheet – concrete counterpart (more than three times greater), with higher value of peak shear stress but brittle softening branch. This is probably due to the significantly higher in-plane stiffness with respect to FRP sheet.

Differences on stiffness of initial branch of interface laws may have some significant consequences in some applications, because the smaller the initial stiffness, the larger the effective bond length required to transmit the full bonding force.

7 CONCLUSIONS

Results from a set of experimental delamination tests on FRP – concrete specimens with different concrete surface preparation before bonding are presented. Bonding of both CFRP plates and sheets have been tested. A particular set-up has been adopted, allowing for a stable delamination process. Applied force, displacements and strains along FRP plate have been measured. The delamination force is used to estimate the fracture energy of interface law.

Non linear interface shear stress – slip laws have also been calibrated starting from experimental data, adopting the value of fracture energy as a constraint in the minimization procedure between experimental and predicted values.

It is shown that sand blasting of concrete surface before resin application increases delamination force (15–20 percent in the present study), even if it reduces peak shear stress and stiffness of initial branch of interface law.

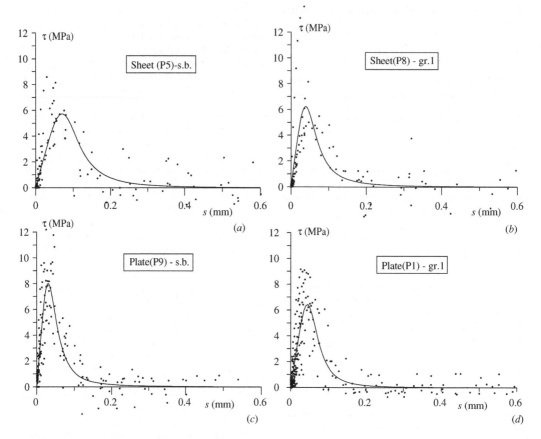

Figure 7. (*a, b, c*) Shear stress-slip data and interpolation curves by post-processing experimental results for different specimens; (*d*) comparison of interface laws (s.b. = sand blasting, gr. 1 = grinding with stone wheel with marble powder).

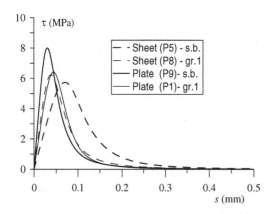

Figure 8. Comparison between interface laws obtained from delamination tests on plate/sheets with different concrete surface preparation (s.b. = sand blasted, gr.1 = type 1 grinding).

ACKNOWLEDGEMENTS

The authors would like to thank the Sika Italia S.p.A. for providing CFRP plates and adhesives for the specimens. Financial support of Department of Civil Protection (Reluis 2005 Grant – Task 8: Innovative materials for vulnerability mitigation of existing structures) and C.N.R., PAAS Grant 2001 are gratefully acknowledged.

REFERENCES

Arrea, M. & Ingraffea, A.R. 1982. Mixed-mode crack propagation in mortar and concrete, Dept. of Structural Engineering, Cornell University, Report n. 81–13.
Bazant, Z.P. & Pfeiffer, P.A. 1986. Shear fracture tests of concrete. *Materials and Structures*, 19: 111–121.
Brosens, K. 2001. Anchorage of externally bonded steel plates and CFRP laminates for the strengthening of

concrete elements. *Doctoral thesis*, University of Leuven, Belgium.

Buyukozturk, O., Gunes, O. & Karaca, E. 2004. Progress on understanding debonding problems in reinforced concrete and steel members strengthened using FRP composites – Short survey, *Construction and Building Materials*, 18(1): 9–19.

CNR Committee 2006. Guide for the Design and Construction of Externally Bonded FRP Systems for Strengthening Existing Structures, *CNR DT 200/2004 Technical Report*.

Ferracuti, B., Savoia, M. & Mazzotti, C. 2006. A numerical model for FRP–concrete delamination. *Composites Part B: Engineering*, 37(4-5): 356–364.

Ferracuti, B., Martinelli, E., Nigro, E. & Savoia, M. 2007*a*. Fracture energy and design rules against FRP-concrete debonding. In T. Triantafillou (ed.), *Proc. of FRPRCS-8 Conference*, Patras, Greece, July 2007.

Ferracuti, B., Savoia, M. & Mazzotti, C. 2007*b*. Interface law for FRP-concrete delamination. *Composite Structures*, in press.

Freddi, F., Salvadori, A. & Savoia, M. 2004. Boundary element analysis of FRP-concrete delamination, In C.A. Brebbia (ed.) *Boundary elements; Proc. int. Symp.*, Southampton, 26: 335–344.

Galecki, G., Maerz, N., Nanni, A. & Myers, J. 2001. Limitations to the use of waterjets in concrete substrate preparation. In *American waterjet conference; Proc. of WJTA*, Minneapolis, Minnesota, USA, 2001: 1–6 (on CD).

Lu, X.Z., Teng, J.G., Ye, L.P. & Jiang, J.J. 2005. Bond-slip model for FRP sheets/plates bonded to concrete, *Engineering Structures*, 27: 920–937.

Matana, M., Galecki, G., Maerz, N. & Nanni, A. 2005. Concrete substrate preparation and characterization prior to adhesion of externally bonded reinforcement. In J.F. Chen & J.G. Teng (eds.), *International symposium on bond behaviour of FRP in structures; Proc. of BBFS,* Hong Kong, December 2005: 1–7 (on CD).

Mazzotti, C., Ferracuti, B. & Savoia, M. 2004. An experimental study on FRP – concrete delamination. In Li et al. (eds.), *Proc. FraMCoS-5*, Vail (CO), 2: 795–802.

Savoia, M., Ferracuti, B. & Mazzotti, C. 2003. Non linear bond- slip law for FRP-concrete interface, In K.H. Tan (ed.), *Proc. of FRPRCS-6 Conference*, Singapore: 1–10.

Toutanji, H., Ortiz, G. 2001. The effect of surface preparation on the bond interface between FRP sheets and concrete members. *Composite Structures*, 53: 457–462.

Van Mier, J.G.M. 1997. *Fracture processes of concrete*, Boca Raton: CRC Press.

Yao, J., Teng, J.G., Chen, J.F. 2005. Experimental study on FRP-to-concrete bonded joints. *Composites Part B: Engineering*, 36(2): 99–113.

Author index